Analysis of Faulted Power Systems

Analysis of Faulted Power Systems

PAUL M. ANDERSON
Power Math Associates , Inc.

IEEE PRESS Power Systems Engineering Series
Paul M. Anderson, Series Editor

IEEE
The Institute of Electrical and Electronics Engineers, Inc., New York

WILEY-INTERSCIENCE
A JOHN WILEY & SONS, INC., PUBLICATION
New York • Chichester • Weinheim • Brisbane • Singapore • Toronto

For ordering and customer service, call 1-800-CALL-WILEY.
Wiley-Interscience-IEEE **ISBN 0-7803-1145-0**

Library of Congress Cataloging-in-Publication Data

Anderson, P. M. (Paul M.)
Analysis of faulted power systems / Paul M. Anderson.
p. cm. — (IEEE Press power system engineering series)
"An IEEE Press classic reissue."
Reprint. Originally published: Iowa State University Press, 1973.
Includes bibliographical references and index.
ISBN 978-0-7803-1145-9
1. Short circuits. 2. Electric circuit analysis. 3. Electric power systems—Mathematical models. I. Title. II. Series.
TK3226.A55 1995
621.319—dc20 95-15246
CIP

Table of Contents

Preface to the IEEE Reissued Edition
Preface

List of Symbols

Chapter 1. General Considerations

1.1	Per Unit Calculations	5
1.2	Change of Base	7
1.3	Base Value Charts	7
1.4	Three-Phase Systems	7
1.5	Converting from Per Unit Values to System Values	10
1.6	Examples of Per Unit Calculations	10
1.7	Phasor Notation	15
1.8	The Phasor a or a-Operator	16
	Problems	17

Chapter 2. Symmetrical Components

2.1	Symmetrical Components of an n-Phase system	19
2.2	Symmetrical Components of a Three-Phase System	23
2.3	Symmetrical Components of Current Phasors	25
2.4	Computing Power of Symmetrical Components	25
2.5	Sequence Components of Unbalanced Network Impedances	27
2.6	Sequence Components of Machine Impedances	30
2.7	Definition of Sequence Networks	31
	Problems	33

Chapter 3. Analysis of Unsymmetrical Faults: Three-Component Method

I.	Shunt Faults	37
3.1	The Single-Line-to-Ground (SLG) Fault	37
3.2	The Line-to-Line (LL) Fault	42
3.3	The Double Line-to-Ground (2LG) Fault	44
3.4	The Three-Phase (3ϕ) Fault	49
3.5	Other Types of Shunt Faults	52
3.6	Comments on Shunt Fault Calculation	53
II.	Series Faults	53
3.7	Sequence Network Equivalents for Series Faults	55
3.8	Unequal Series Impedances	61
3.9	One Line Open (1LO)	63

3.10 Two Lines Open (2LO) 64
3.11 Other Series Faults 66
Problems 66

Chapter 4. Sequence Impedance of Transmission Lines

4.1 Positive and Negative Sequence Impedances of Lines 71
4.2 Mutual Coupling 73
4.3 Self and Mutual Inductances of Parallel Cylindrical Wires 75
4.4 Carson's Line 78
4.5 Three-Phase Line Impedances 81
4.6 Transpositions and Twists of Line Conductors 84
4.7 Completely Transposed Lines 98
4.8 Circuit Unbalance Due to Incomplete Transposition 102
4.9 Sequence Impedance of Lines with Bundled Conductors 106
4.10 Sequence Impedance of Lines with One Ground Wire 112
4.11 Sequence Impedance of Lines with Two Ground Wires 123
4.12 Sequence Impedance of Lines with *n* Ground Wires 128
4.13 Zero Sequence Impedance of Transposed Lines with Ground Wires. 129
4.14 Computations Involving Steel Conductors 133
4.15 Parallel Transposed and Untransposed Multicircuit Lines 137
4.16 Optimizing a Parallel Circuit for Minimum Unbalance 143
Problems 145

Chapter 5. Sequence Capacitance of Transmission Lines

5.1 Positive and Negative Sequence Capacitance of Transposed Lines .. 152
5.2 Zero Sequence Capacitance of Transposed Lines 156
5.3 Mutual Capacitance of Transmission Lines 158
5.4 Mutual Capacitance of Three-Phase Lines without ground Wires ... 163
5.5 Sequence Capacitance of a Transposed Line without Ground Wires. 166
5.6 Mutual Capacitance of Three-Phase Lines with Ground Wires 168
5.7 Capacitance of Double Circuit Lines 172
5.8 Electrostatic Unbalance of Untransposed Lines 177
Problems 181

Chapter 6. Sequence Impedance of Machines

I. Synchronous Machine Impedances 183

6.1 General Considerations 183
6.2 Positive Sequence Impedance 189
6.3 Negative Sequence Impedance 199
6.4 Zero Sequence Impedance 201
6.5 Time Constants 202
6.6 Synchronous Generator Equivalent Circuits 205
6.7 Phasor Diagram of a Synchronous Generator 207
6.8 Subtransient Phasor Diagram and Equivalent Circuit 214
6.9 Armature Current Envelope 219
6.10 Momentary Currents 221

II. Induction Motor Impedances 222

6.11 General Considerations 222
6.12 Induction Motor Equivalent Circuit 222

6.13 Induction Motor Subtransient Fault Contribution 226
6.14 Operation with One Phase Open 227
Problems 228

Chapter 7. Sequence Impedance of Transformers

I. Single-Phase Transformers 231

7.1 Single-Phase Transformer Equivalents 231
7.2 Transformer Impedances 233
7.3 Transformer Polarity and Terminal Markings 234
7.4 Three-Winding Transformers 236
7.5 Autotransformer Equivalents 239
7.6 Three-Phase Banks of Single-Phase Units 243

II. Three-Phase Transformers 247

7.7 Three-Phase Transformer Terminal Markings 247
7.8 Phase Shift in Y-Δ Transformers 248
7.9 Zero Sequence Impedance of Three-Phase Transformers 251
7.10 Grounding Transformers 255
7.11 The Zigzag-Δ Power Transformer 257

III. Transformers in System Studies 260

7.12 Off-Nominal Turns Ratios 260
7.13 Three-Winding Off-Nominal Transformers 265
Problems 265

Chapter 8. Changes in Symmetry

8.1 Creating Symmetry by Labeling 273
8.2 Generalized Fault Diagrams for Shunt (Transverse) Faults 273
8.3 Generalized Fault Diagrams for Series (Longitudinal) Faults .. 278
8.4 Computation of Fault Currents and Voltages 280
8.5 A Fundamental Result: The Invariance of Power 284
8.6 Constraint Matrix K 286
8.7 Kron's Primitive Network 288
8.8 Other Useful Transformations 291
8.9 Shunt Fault Transformations 294
8.10 Transformations for Shunt Faults with Impedance 298
8.11 Series Fault Transformations 304
8.12 Summary 304
Problems 305

Chapter 9. Simultaneous Faults

I. Simultaneous Faults by Two-Port Network Theory 308

9.1 Two-Port Networks 308
9.2 Interconnection of Two-Port Networks 319
9.3 Simultaneous Fault Connection of Sequence Networks 323
9.4 Series-Series Connection (Z-Type Faults) 325
9.5 Parallel-Parallel Connection (Y-Type Faults) 330
9.6 Series-Parallel Connection (H-Type Faults) 334

II. Simultaneous Faults by Matrix Transformations 336

9.7 Constraint Matrix for Z-Type Faults 337
9.8 Constraint Matrix for Y-Type and H-Type Faults 339
Problems 341

Chapter 10. Analytical Simplifications
10.1 Two-Component Method 345
I. Shunt Faults 347
10.2 Single-Line-to-Ground Fault 347
10.3 Line-to-Line Fault 349
10.4 Double Line-to-Ground Fault 350
10.5 Three-Phase Fault 352
II. Series Faults 353
10.6 Two Lines Open (2LO) 353
10.7 One Line Open (1LO) 354
III. Changes in Symmetry with Two-Component Calculations 355
10.8 Phase Shifting Transformer Relations 356
10.9 SLG Faults with Arbitrary Symmetry 357
10.10 2LG Faults with Arbitrary Symmetry 358
10.11 Series Faults with Arbitrary Symmetry 360
IV. Solution of the Generalized Fault Diagrams 362
10.12 Series Network Connection—SLG and 2LO Faults 362
10.13 Parallel Network Connection—2LG and 1LO Faults 363
Problems 363

Chapter 11. Computer Solution Methods Using the Admittance Matrix
11.1 Primitive Matrix 366
11.2 Node Incidence Matrix 369
11.3 Node Admittance and Impedance Matrices 373
11.4 Indefinite Admittance Matrix 375
11.5 Definite Admittance Matrix 386
Problems 389

Chapter 12. Computer Solution Methods Using the Impedance Matrix
12.1 Impedance Matrix in Shunt Fault Computations 393
12.2 An Impedance Matrix Algorithm 401
12.3 Adding a Radial Impedance to the Reference Node 401
12.4 Adding a Radial Branch to a New Node 402
12.5 Closing a Loop to the Reference 403
12.6 Closing a Loop Not Involving the Reference 404
12.7 Adding a Mutually Coupled Radial Element 408
12.8 Adding a Group of Mutually Coupled Lines 415
12.9 Comparison of Admittance and Impedance Matrix Techniques 418
Problems 419

Appendix A. Matrix Algebra 421

Appendix B. Line Impedance Tables 436

Appendix C. Trigonometric Identities for Three-Phase Systems 467

Appendix D. Self Inductance of a Straight Finite Cylindrical Wire 470
Appendix E. Solved Examples.................................. 474
Appendix F. Δ-Y Transformations 502
Bibliography.. 503
Index 507

Preface to the IEEE Reissued Edition

Many textbooks have been written to describe a power system operating under balanced or normal conditions. A few texts also deal, at least in an elementary manner, with the unbalanced system conditions. The primary object of this book is to provide a text to be used by graduate students and to challenge the student to draw upon a background of knowledge from earlier studies. Since this is intended for advanced studies, the student reader should be able to use circuit concepts at the advanced undergraduate or beginning graduate level and should have a working knowledge of matrix notation. Particular stress here is placed upon a clear, concise notation, since it is the author's belief that this facilitates learning.

Although the thrust of the book is toward the solution of advanced problems, a thorough background is laid in the early chapters by solving elementary configurations of unbalanced systems. This serves to establish the algebraic style and notation of the book as well as to provide the reader with a growing knowledge and facility with symmetrical components. It also introduces the elementary concepts for those not familiar with this discipline. In either case, this background information of the early chapters should be studied since it establishes certain conventions used throughout the book.

The text contains sufficient material for an entire year of study on the subject. The first three chapters are introductory and may be covered rapidly in a graduate level class where the students have already been introduced to symmetrical components. This material should be reviewed, at least in a quick reading, as the matrix notation used throughout the book is introduced in these chapters.

The middle portion of the book, Chapters 4 through 7, treats the subject of *power system parameters*. Here, the sequence impedances of transmission lines, machines, and transformers are developed in detail. This is important material and is often omitted in the education of power engineers. Methods are used here that permit exact solutions of very general physical problems, such as finding the impedance of untransposed or partially transposed lines. Matrix methods are used to clarify computations and adapt them to computer usage.

The final portion of the book presents the application of symmetrical components to a variety of problems and provides an introduction to computer solution methods for large networks. Here, one learns to appreciate the use of matrix algebra in the solution of complex problems. This section also reinforces the engineer's appreciation of symmetrical components as a problem-solving technique.

In preparing the manuscript of a book of this type, one stands on the shoulders of giants. Several excellent books introduced these concepts in the first half of the twentieth century, in particular those by C. F. Wagner and R. D. Evans, G. O. Calabrese, and the books by Edith Clarke, are classics on symmetrical components and are still used by many of us. Their basic ideas are enlarged upon here, and the presentation simplified by the use of matrix methods, thereby aiding understanding and making computer solution much easier. The author's colleagues at Iowa State University played an important role in the development of the book. They used the book in its early stages, found many problems, and offered helpful suggestions for areas needing improvement. The electrical engineering department at Iowa State University was also helpful and understanding of the need for this effort and provided support in many ways.

This new printing of the book by IEEE Press provides software that was especially developed at Iowa State University for the solution of exactly the kind of problems introduced in this book. The program PWRMAT was developed by the author, fellow faculty members, and graduate students for the purpose of providing the engineer with a convenient method of solving the many problems associated with matrices of complex numbers. The program has been improved over the years by Iowa State University and has been used and enjoyed by many students for at least twenty years. Iowa State University has graciously agreed that the program could be distributed with this printing of the book to make problem solving easier. This is important, since the drudgery of solving these complex problems by calculator distracts the engineering student from the objective of learning the concepts. The software and text file versions of the user's manual are attached to the book on diskettes. The software is easy to master and to use. It permits the user to write small programs in a simple language that can read a user-created data file to solve the problem at hand and print the computed results in an orderly matrix notation. Operations such as matrix inversion or reduction are accomplished with ease. It is hoped that the addition of this software will help many new readers in their desire to become more proficient with the important subjects covered in this book.

P. M. Anderson
San Diego, California

Preface

Many textbooks have been written dealing with the power system that operates under balanced or normal conditions. A few, particularly the excellent recent text by W. D. Stevenson [9], deal in an elementary way with both normal and faulted systems. Most of the books treating unbalanced and faulted systems, however, have been in print for years and are inadequate for several reasons. Nevertheless, in spite of their date of copyright, the serious student should become familiar with the famous works of C. F. Wagner and R. D. Evans [10], the outstanding volumes by Edith Clarke [11] and the more recent work of Calabrese [24].

The goal here is to produce a text to be used by graduate students, one that can draw upon a background of knowledge from previous courses. Since this is an advanced text, the student should be able to employ circuit concepts not usually taught to undergraduates and should recognize the beauty and simplicity of matrix notation. Particular stress is placed upon a clear, concise notation since it is the author's belief that this facilitates learning.

Although the thrust of the book is toward the solution of advanced problems, a thorough background is laid by solving elementary configurations of unbalanced systems. This serves to establish the algebraic style and notation of the book. It also serves to introduce the elementary concepts to the uninitiated. Thus for some it will be an organized review and for others an introduction to the solution of faulted networks. In either case this background should be studied since it establishes certain conventions used later.

The text contains sufficient material for a two-semester or three-quarter treatment of the subject. The first three chapters are introductory and may be covered rapidly in a graduate class where the students have already been introduced to symmetrical components. This material should be reviewed, at least in a quick reading, as the matrix notation used throughout the book is introduced in these chapters.

The middle portion of the book, Chapters 4 through 7, treat the subject of *power system parameters*. Here the sequence impedances of lines, machines, and transformers are developed in detail. This is important material and is often omitted in the education of power engineers. Methods are used here which permit exact solutions of very general physical problems such as finding the impedance of untransposed or partially transposed lines. Matrix methods are used to clarify these computations and adapt them to computer usage.

The final portion of the book presents the application of symmetrical components to a variety of problems and provides an introduction to computer solution of large networks. Here one learns to appreciate the use of matrix algebra in the solution of complex problems. This section also reinforces the engineer's appreciation of symmetrical components as a problem-solving technique.

At Iowa State University we have found it convenient to cover most of the first 10 chapters in a two-quarter sequence, leaving computer applications as a separate course. This means that some of the sections in Chapters 1-10 must be omitted, but the student is encouraged to pursue these on his own. In this two-quarter presentation Chapters 1-3 are skimmed quickly since the course carries an undergraduate prerequisite which introduces symmetrical components. Then the balance of the first quarter is spent on power system parameters, leaving the applications for the second quarter.

This book would not have been possible without the unique contribution of many individuals to whom the author is greatly indebted. Several Iowa State University colleagues, particularly W. B. Boast, J. W. Nilsson, and J. E. Lagerstrom (now of the University of Nebraska), are largely responsible for the author's interest in the subject. These three were also responsible for the organization and teaching of a short course in symmetrical components, taught in connection with the Iowa State University A-C Network Analyzer for 10 years or so. The author's interest in this course, first as a student and later as a teacher, helped him gain competence in the subject. Indeed, many ideas expressed here are taken directly or indirectly from the short course notes. The influence of the late W. L. Cassell must also be mentioned, for his insistence on a clear notation has contributed to the education and understanding of many students, the author included. The author is particularly indebted to David D. Robb who used much of the book in a graduate class and made countless valuable suggestions for improvements. Portions of the computer solutions presented are the work of J. R. Pavlat and G. N. Johnson, and these contributions are gratefully acknowledged.

Finally, I wish to express my thanks to the Electrical Engineering Department of Iowa State University and to W. B. Boast, head of the department, for giving me the opportunity to prepare this material. Special thanks are due to my wife, Ginny, who provided both moral support and expert proofreading, and to my editor, Nancy Bohlen, who is a marvel with both mathematical notation and eccentric authors.

List of Symbols

1. CAPITALS

A .... ampere, unit symbol abbreviation for current
A .... complex transformation matrix; transmission parameter matrix; node incidence matrix
A .... magnetic vector potential
$\mathcal{A}$ .... inverse transmission parameter matrix
B .... $= \mathcal{I}m\, Y$, susceptance
B .... complex transformation matrix; shunt susceptance matrix
C .... capacitance
C .... coulomb, unit symbol abbreviation for charge
C .... complex transformation matrix; Maxwell's coefficients; capacitance coefficients
D .... distance or separation
E .... source emf; voltage
E .... primitive source voltage vector
F .... farad, unit symbol abbreviation for capacitance
F, F' .... fault point designation
F-D-Q .... rotor circuits of a synchronous machine
G .... $= \mathcal{R}e\, Y$, conductance
G .... inverse hybrid parameter matrix
GMD, GMR .... mutual geometric mean distance, geometric mean radius
H .... henry, unit symbol abbreviation for inductance
Hz .... hertz, unit symbol abbreviation for frequency
H .... hybrid parameter matrix
I .... rms phasor current
$\mathbf{I}_{abc}$ .... $= [I_a\ I_b\ I_c]^t$, line current vector
$\mathbf{I}_{012}$ .... $= [I_{a0}\ I_{a1}\ I_{a2}]^t$, sequence current vector
I_B .... base line current, A
J .... joule, unit symbol abbreviation for energy
J .... primitive current source vector
K .... dielectric constant
K .... Kron's transformation or connection matrix
L .... inductance
LL .... line-to-line
LN .... line-to-neutral
L .... inductance matrix
M .... $= 10^6$, mega, a prefix

M .... mutual inductance
M_{ij} .... minor of a matrix
$\mathbf{M}$ .... two-port network vector
N .... newton, unit symbol abbreviation for force
N .... zero potential bus designation
$\mathbf{N}$ .... two-port network vector
$\mathcal{P}$ .... phasor operator
P .... average power; transformer circuit designation
$\mathbf{P}$ .... Vandermonde matrix; potential coefficient matrix; Park's transformation matrix
Q .... average reactive power; transformer circuit designation; total charge; phasor charge density
R .... $= \mathcal{R}e\, Z$, resistance; transformer circuit designation
$\mathbf{R}$ .... resistance matrix
S .... $= P + jQ$, complex apparent power
S_B............... base apparent power, VA
SLG .... single-line-to-ground
T .... time; time constant; torque; equivalent circuit configuration
T_B............... base time, s
$\mathbf{T}_\phi$............... twist matrix
$\mathbf{U}$ .... unit matrix
V .... rms phasor voltage
V .... volt, unit symbol abbreviation for voltage
VA .... voltampere, unit symbol abbreviation for apparent power
$\mathbf{V}_{abc}$ .... $= [V_a\ V_b\ V_c]^t$, phase voltage vector
$\mathbf{V}_{012}$ .... $= [V_{a0}\ V_{a1}\ V_{a2}]^t$, sequence voltage vector
V_B .... base voltage, V
W .... watt, unit symbol abbreviation for power
Wb .... weber, unit symbol abbreviation for magnetic flux
X .... $= \mathcal{I}m\, Z$, reactance
$\mathcal{Y}$ .... primitive admittance matrix
Y .... $= G + jB$, complex admittance
Y_B .... base admittance, mho
$\mathbf{Y}$ .... admittance matrix
$\mathcal{Z}$ .... primitive impedance matrix
Z .... $= R + jX$, complex impedance
Z_B .... base impedance, Ω
$\mathbf{Z}$ .... impedance matrix

2. LOWERCASE

a .... $= e^{j2\pi/3}$, 120° operator
ac .... alternating current
a-b-c .... stator circuits of a synchronous machine; phase designation
adj .... adjoint (of a matrix)
b .... $= \omega c$, line susceptance per unit length
ber, bei .... real, imaginary Bessel functions
c .... capacitance per unit length
dc .... direct current

d_0, d_2 . . . . zero, negative sequence electrostatic unbalance factors
d, q . . . . stator circuits, referred to the rotor
det . . . . determinant (of a matrix)
e . . . . base for natural logarithms
f . . . . frequency
f_k . . . . kth fraction of total line length
g . . . . ground terminal
h . . . . two-port hybrid parameter designation
h . . . . a constant (1 or $\sqrt{3}$)
i . . . . instantaneous current
i . . . . instantaneous current vector
j . . . . $= \sqrt{-1}$, 90° operator
k, k′ . . . . constants used in computing L, C
k . . . . $= 10^3$, kilo, a prefix
k . . . . $= \sqrt{3/2}$, a constant used in synchronous machine theory
ℓ . . . . inductance per unit length; leakage inductance
ln, log . . . . natural (base e), base 10 logarithms
m . . . . $= 10^{-3}$, milli, a prefix; a constant used in computing skin effect
m . . . . mutual inductance per unit length
m_0, m_2 . . . . zero, negative sequence electromagnetic unbalance factors
m . . . . complex transformation ratio
n . . . . number of phases; number of nodes; $= 10^{-9}$, nano, a prefix
n . . . . neutral terminal; neutral voltage; turns ratio; number of turns
pu . . . . per unit
p . . . . instantaneous power
q . . . . linear charge density of a wire
q . . . . vector of linear charges on a group of wires
r . . . . radius; internal (source) resistance; resistance per unit length
s . . . . line length, length of section k, slip of an induction motor
s . . . . speed voltage vector
t . . . . time
u . . . . unit step function
v . . . . instantaneous voltage
v . . . . instantaneous voltage vector
x . . . . line reactance per unit length; internal (source) reactance
y . . . . two-port admittance parameter designation
z . . . . two-port impedance parameter designation; internal (source) impedance; impedance per unit length
$\bar{z}$ . . . . transmission line primitive impedance

3. UPPERCASE GREEK

Δ . . . . delta connection; determinant (of a matrix)
Σ . . . . summation symbol
Ω . . . . ohm, unit symbol abbreviation for impedance

4. LOWERCASE GREEK

α .... phase angle
α_R, α_L .... ac/dc skin effect ratios
δ .... torque angle of a synchronous machine
δ_{ij} .... Kronecker delta
ϵ .... $= \epsilon_0 \kappa$, permittivity
θ .... phase angle; rotor angle of a synchronous machine
κ .... dielectric constant
λ .... element of Vandermonde matrix; flux linkage
μ .... $= \mu_0 \mu_r$, permeability (μ_0, free space; μ_r, relative)
μ .... $= 10^{-6}$, micro, a prefix
π .... pi, 3.14159265...
ρ .... resistivity
τ .... time constant
ϕ .... magnetic flux; phase angle
ω .... radian frequency; synchronous machine speed

5. SUBSCRIPTS

a .... phase a; armature
A .... phase a
b .... phase b
B .... phase b
B .... base quantity
c .... phase c; core loss quantity
C .... phase c; transformer circuit designation
d .... direct axis; direct axis circuit quantity
D .... direct axis damper winding quantity
e .... excitation quantity, of a transformer
eq .... equivalent circuit quantity; equivalent spacing
env .... envelope of an ac wave
F .... referring to the fault point; field winding
f .... referring to the fault point
g .... referring to the fault point
H .... transformer winding designation
LN .... line-to-neutral
LL .... line-to-line
m .... magnetizing quantity (in a transformer); motor quantity; mutual (coupling or GMD)
max .... maximum
min .... minimum
M .... mutual (frequently $M0, M1, M2$)
n .... neutral
q .... quadrature axis; quadrature axis circuit quantity
Q .... quadrature axis damper winding quantity
r .... rotor quantity
R .... rotor quantity
s .... source quantity; stator quantity; self (GMD)

S .... stator quantity; transformer circuit designation; self (frequently $S0, S1, S2$)
sym .... symmetrical
T .... transformer circuit designation
u .... per unit
X .... transformer winding designation
Y .... transformer winding designation
1ϕ, 3ϕ .... single-phase; three-phase
0, 1, 2 .... zero, positive, negative sequence quantity
$0, \Sigma, \Delta$ .... zero, sum, difference sequence quantity
$0, \Delta$ .... initial condition; change condition

6. SUPERSCRIPTS

$(\ )^t$ .... transpose (of a matrix)
$(\ )^{-1}$ .... inverse (of a matrix)
$(\tilde{\ })$ .... (tilde), distinguishing mark for various quantities
$(\hat{\ })$ .... (circumflex), distinguishing mark for various quantities
$(\dot{\ })$ .... $= d/dt$, derivative with respect to time
$(\ )^*$.............. conjugate, of a phasor or a matrix
$(\ )'$.............. (prime), transient
$(\ )''$.............. (double prime), subtransient

7. NUMERAL SYMBOLS

1ϕ............... single-phase
2LG .... double-line-to-ground
3ϕ............... three-phase
1LO .... one line open
2LO .... two lines open

Analysis of Faulted Power Systems

chapter 1

General Considerations

The analysis of power systems usually implies the computation of network voltages and currents under a given set of conditions. In many cases the computation is organized to give a particular kind of information for a special purpose. For example, it may be desirable to determine the current flowing through the ground in a particular situation to facilitate the setting of a ground relay. Figure 1.1 presents an organization for power system computation. The class of problems to the left in the figure are called "steady state" because they are to be solved by algebraic equations. This is not to imply that the system is static or unchanging at the moment in time for which a solution is sought. On the contrary, the system may be undergoing very rapid changes, for example, in a faulted condition. The point is that algebraic equations are much easier to solve than differential equations. We therefore have learned to make good use of the steady state solutions in system planning and for determining system protection. This is like taking a set of photographs of the system behavior under certain specified conditions. From these photographs we can design system additions and protective schemes and can learn a great deal about the system strengths and weaknesses.

The dynamic problems shown to the right in Figure 1.1 are what we usually call "stability problems" in power system work. Here we solve a set of differential equations to determine the behavior of voltages, currents, and other variables as functions of time.

Both the steady state and dynamic problems are usually of large dimension for power system work. Networks of a few hundred nodes are quite common and many machines may be represented in dynamic problems. Thus in both kinds of problems we must be oriented toward computer solutions of some kind.

The emphasis in this text is on the faulted network, both balanced and unbalanced. As will be shown later, the techniques used for unbalanced faulted networks will work equally well for unbalanced normal networks. In most cases we will consider the system to be linear and will use algebraic equations exclusively to take "photographs" of voltage-current relationships under certain given conditions. There is much we can do with these photographs, and these ideas will be explored in some depth.

Unbalanced system operation is one we would usually prefer to ignore. Often when dealing with normal or near-normal operation, this is precisely what we do—ignore it. That is, we assume balanced operation, solve the network on a per phase basis, and extrapolate to obtain information for the remaining two

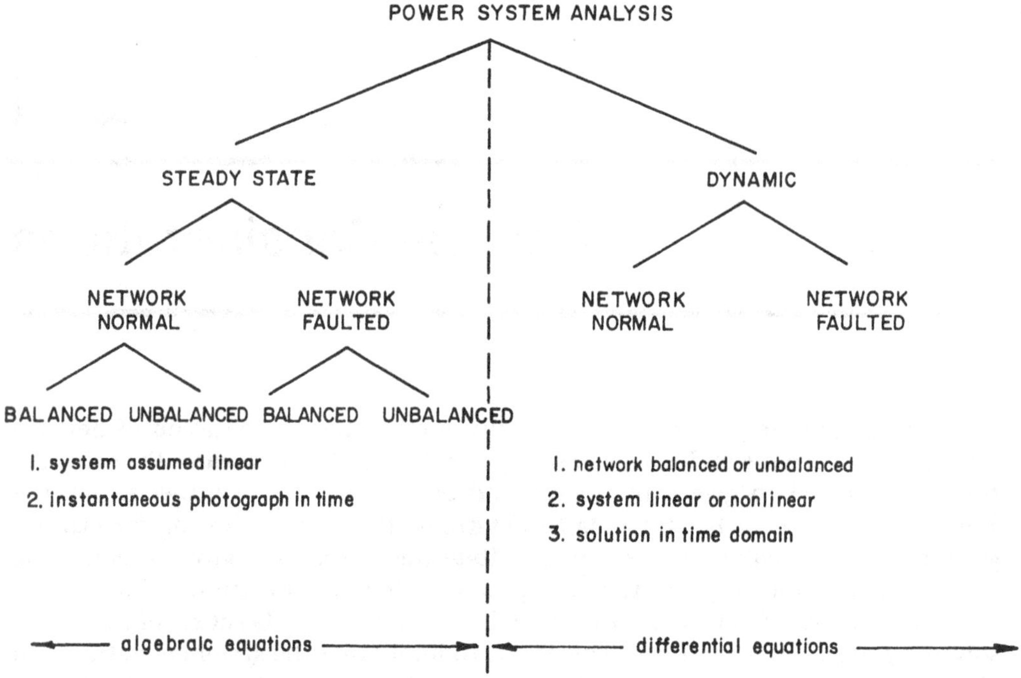

Fig. 1.1. An organization of power system analysis problems.

phases.[1] This results in a great saving in time and effort and usually gives results of reasonable accuracy if the system is nearly normal, i.e., nearly balanced. When the system is obviously unbalanced, other methods must be used. The method generally favored is that of "symmetrical components" as proposed by Fortescue in 1918 [1]. This method permits us to extend the per phase analysis to systems with unbalanced loads or with unbalanced termination of some kind, such as a short circuit or fault. The method is of limited value when the system itself is unbalanced. Our analysis will make extensive use of symmetrical components and will do so with the aid of matrix notation whenever possible. In doing so, we will assume a balanced system ($Z_a = Z_b = Z_c$) with unbalanced loading unless special note is made of a system or network unbalance. The only reason for doing this is to extend the simplicity of the per phase representation to unbalanced system terminations. It may be possible in the future by the use of a large computer to completely represent all three phases of a system and to handle unbalances in either the system or the load in a more direct manner.

The method of symmetrical components has led to the development of useful apparatus to measure or use symmetrical component variables. One example is the negative sequence relay used to protect generators from overheating in the event of unbalanced loading. The positive sequence segregating network is sometimes used to supply the sensing voltage to generator voltage regulators. Certain connections of instrument current and potential transformers develop zero sequence quantities that are used in protective ground relaying schemes.

Finally, a comment concerning the use of digital computers is necessary.

[1]A three-phase system is assumed here since they are by far the most common. When unspecified, a three-phase system will always be implied.

Since power systems are really very large electric networks, it is usually impossible to perform a complete solution by hand computation because of the size of the problem. Thus any solution or technique which we devise must be applicable to the digital computer. The power system engineer should be capable of bridging this gap between theory and computer utilization, and this subject will be emphasized throughout the text.

1.1 Per Unit Calculations

The calculation of system performance conveniently uses a per unit representation of voltage, current, impedance, power, reactive power, and apparent power (voltampere).[2] The numerical per unit value of any quantity is its ratio to the chosen base quantity of the same dimensions. Thus a per unit quantity is a "normalized" quantity with respect to a chosen base value.

Any quantity can be converted to a per unit quantity by dividing the numerical value by a selected base quantity of the same dimensions. In an electrical network five quantities are usually involved in the calculation of networks. These are shown in Table 1.1 together with their dimensions. Our use of the term "dimension" here is admittedly loose since I and V can be specified dimensionally in terms of force, length, time, and charge, for example. The meaning is clear, however. In calculations of steady state phenomena the time dimension is suppressed in phasor notation. Of the five remaining quantities one is dimensionless and the other four are completely described by only two dimensions. Thus an *arbitrary* choice of two quantities as base quantities will automatically fix the other two. In power system calculations the nominal voltage of lines and equipment is almost always known, so the voltage is a convenient base value to choose. A second base is usually chosen to be the apparent power (voltampere). In equipment this quantity is usually known and makes a convenient base. In a system study the voltampere base can be selected to be any convenient value such as 100 MVA, 200 MVA, etc.

Table 1.1. Electrical Quantities and Their Dimensions

Quantity	*Symbol*	*Dimension*
Current, A	I	$[I]$
Voltage, V	V	$[V]$
Voltamperes, S	$S = P + jQ$	$[VI]$
Impedance, Ω	$Z = R + jX$	$[V/I]$
Phase angle	ϕ, θ, etc.	dimensionless
Time, sec	t	$[T]$

The same voltampere base is used in all parts of the system. One base voltage is selected arbitrarily. All other base voltages must be related to the arbitrarily selected one by the turns ratios of the connecting transformers. Special treatment must be given to load tap changing transformers and to network loops wherein the net product of turns ratios around the loop is not equal to unity. This matter is treated in Chapter 7. In three-phase networks the turns ratios used to relate the several base voltages are those of Y-Y or equivalent Y-Y transformers.

[2]Portions of this section are derived from [2], prepared by several authors, particularly W. B. Boast, J. E. Lagerstrom, and J. W. Nilsson.

If we designate a base quantity by the subscript B, we have on a per phase basis

$$\text{base voltamperes} = S_B \quad \text{VA} \tag{1.1}$$

and

$$\text{base voltage} = V_B \quad \text{V} \tag{1.2}$$

Then the base current and base impedance are computed as

$$\text{Base } I = I_B = \frac{S_B}{V_B} \quad \text{A} \tag{1.3}$$

$$\text{Base } Z = Z_B = \frac{V_B}{I_B} = \frac{V_B^2}{S_B} \quad \Omega \tag{1.4}$$

Similarly, we define a base admittance as

$$\text{Base } Y = Y_B = \frac{S_B}{V_B^2} \quad \text{mho} \tag{1.5}$$

Having defined the base quantities, we can normalize any system quantity by dividing by the base quantity of the same dimension. Thus the per unit impedance Z is defined as

$$Z = \frac{Z \text{ ohm}}{Z_B} \quad \text{pu} \tag{1.6}$$

Note that the dimensions (ohm) cancel, and the result is a dimensionless quantity whose units are specified as per unit or pu. At this point one might question the advisability of using the same symbol, Z, for both the per unit impedance and the ohmic impedance. This is no problem, however, since the problem solver always knows the system of units he is using and it is convenient to use a familiar notation for system quantities. Usually we will remind ourselves that a solution is in per unit by affixing the "unit" pu as in equation (1.6). Furthermore, where there may be a question of units, we will always identify a *system* quantity by adding the appropriate units, such as the notation (Z ohm) of equation (1.6).

Since we may write $Z = R + jX$ in ohms, we may divide both sides of this equation by Z_B with the result

$$Z = R + jX \quad \text{pu} \tag{1.7}$$

where

$$R = \frac{R \text{ ohm}}{Z_B} \quad \text{pu} \tag{1.8}$$

and

$$X = \frac{X \text{ ohm}}{Z_B} \quad \text{pu} \tag{1.9}$$

In a similar manner we write $S = P + jQ$ voltampere, and dividing through by S_B, we have

$$S = P + jQ \quad \text{pu} \tag{1.10}$$

where

$$P = \frac{P \text{ watt}}{S_B} \quad \text{pu} \tag{1.11}$$

and

$$Q = \frac{Q \text{ var}}{S_B} \quad \text{pu} \tag{1.12}$$

1.2 Change of Base

The question sometimes arises that given a pu impedance referred to a given base, what would be its pu value if referred to a new base? To answer this question, we first substitute equation (1.4) into (1.6) with the result

$$Z = (Z \text{ ohm}) \frac{S_B}{V_B^2} \quad \text{pu} \tag{1.13}$$

Two such pu impedances, referred to their respective base quantities, are now written, using the subscript o for old and n for new.

$$Z_o = (Z \text{ ohm}) \frac{S_{Bo}}{V_{Bo}^2}$$

$$Z_n = (Z \text{ ohm}) \frac{S_{Bn}}{V_{Bn}^2} \tag{1.14}$$

But the system or ohmic value must be the same no matter what the base. Equating the (Z ohm) quantities in (1.14), we have

$$Z_n = Z_o \left(\frac{V_{Bo}}{V_{Bn}}\right)^2 \left(\frac{S_{Bn}}{S_{Bo}}\right) \quad \text{pu} \tag{1.15}$$

Equation (1.15) is very important since it permits us to change base without knowledge of the ohmic value (Z ohm). Note that the pu impedance varies directly as the chosen (new) voltampere base and inversely with the square of the voltage base.

1.3 Base Value Charts

In most power system problems the nominal transmission line voltages are known. If these nominal voltages are chosen as the base voltages, an arbitrary choice for the S_B will fix the base current, base impedance, and base admittance. These values are tabulated in Table 1.2 for several values of system voltage and for several convenient MVA levels. A more extensive list of base values is given in Appendix B.

1.4 Three-Phase Systems

The equation derived for pu impedance (1.13), or its reciprocal for pu admittance, is correct only for a single-phase system. In three-phase systems, however, we often prefer to work with three-phase voltamperes and line-to-line voltages. We investigate this problem by rewriting (1.13) using the subscript LN for "line-to-neutral" and 1ϕ for "per phase," with the result

Table 1.2. Base Current, Base Impedance, and Base Admittance for Common Transmission Voltage Levels and for Selected MVA Levels

	Base Kilovolts	*Base Megavolt-Amperes* 5.0	10.0	20.0	25.0	50.0	100.0	200.0	250.0
Base current in amperes	34.5	83.67	167.35	334.70	418.37	836.74	1673.48	3346.96	4183.70
	69.0	41.84	83.67	167.35	209.19	418.37	836.74	1673.48	2091.85
	115.0	25.10	50.20	100.41	125.51	251.02	502.04	1004.09	1255.11
	138.0	20.92	41.84	83.67	104.59	209.18	418.37	836.74	1045.92
	161.0	17.93	35.86	71.72	89.65	179.30	358.60	717.21	896.51
	230.0	12.55	25.10	50.20	62.76	125.51	251.02	502.04	627.55
	345.0	8.37	16.74	33.47	41.84	83.67	167.35	334.70	418.37
	500.0	5.77	11.55	23.09	28.87	57.74	115.47	230.94	288.68
Base impedance in ohms	34.5	238.05	119.03	59.51	47.61	23.81	11.90	5.95	4.76
	69.0	952.20	476.10	238.05	190.44	95.22	47.61	23.81	19.04
	115.0	2645.00	1322.50	661.25	529.00	264.50	132.25	66.13	52.90
	138.0	3808.80	1904.40	952.20	761.76	380.88	190.44	95.22	76.18
	161.0	5184.20	2592.10	1296.05	1036.84	518.42	259.21	129.61	103.68
	230.0	10580.00	5290.00	2645.00	2116.00	1058.00	529.00	264.50	211.60
	345.0	23805.00	11902.50	5951.25	4761.00	2380.50	1190.25	595.13	476.10
	500.0	50000.00	25000.00	12500.00	10000.00	5000.00	2500.00	1250.00	1000.00
Base admittance in micromhos	34.5	4200.80	8401.60	16803.19	21003.99	42007.98	84015.96	168031.93	210039.91
	69.0	1050.20	2100.40	4200.80	5251.00	10502.00	21003.99	42007.98	52509.98
	115.0	378.07	756.14	1512.29	1890.36	3780.72	7561.44	15122.87	18903.59
	138.0	262.55	525.10	1050.20	1312.75	2625.50	5251.00	10502.00	13127.49
	161.0	192.89	385.79	771.58	964.47	1928.94	3857.88	7715.75	9644.69
	230.0	94.52	189.04	378.07	472.59	945.18	1890.36	3780.72	4725.90
	345.0	42.01	84.02	168.03	210.04	420.08	840.16	1680.32	2100.40
	500.0	20.00	40.00	80.00	100.00	200.00	400.00	800.00	1000.00

$$Z = \frac{S_{B\text{-}1\phi}}{V_{B\text{-}LN}^2} (Z \text{ ohm}) \text{ pu} \tag{1.16}$$

and

$$Y = \frac{V_{B\text{-}LN}^2}{S_{B\text{-}1\phi}} (Y \text{ mho}) \text{ pu} \tag{1.17}$$

But using LL to indicate "line-to-line" and 3ϕ for "three-phase," we write for a balanced system

$$V_{B\text{-}LN} = \frac{V_{B\text{-}LL}}{\sqrt{3}} \text{ V} \tag{1.18}$$

and

$$S_{B\text{-}1\phi} = \frac{S_{B\text{-}3\phi}}{3} \text{ VA} \tag{1.19}$$

Making the appropriate substitutions, we compute

$$Z = \frac{S_{B\text{-}3\phi}}{V_{B\text{-}LL}^2} (Z \text{ ohm}) \text{ pu} \tag{1.20}$$

and

$$Y = \frac{V_{B\text{-}LL}^2}{S_{B\text{-}3\phi}} (Y \text{ mho}) \text{ pu} \tag{1.21}$$

These equations are seen to be identical with the corresponding equations derived from per phase voltampere and line-to-neutral voltages. This is fortunate since it makes the formulas easy to recall from memory.

A still more convenient form for (1.20) and (1.21) would be to write voltages in kV and voltamperes in MVA.

Thus we compute

$$Z = \frac{\text{Base MVA}_{3\phi}}{(\text{Base kV}_{LL})^2} (Z \text{ ohm}) \text{ pu} \tag{1.22}$$

The admittance formula may be expressed two ways, the choice depending upon whether admittances are given in micromhos or as reciprocal admittances in megohms. From (1.21) we compute

$$Y = \frac{(\text{Base kV}_{LL})^2 \ (Y\ \mu\text{mho})}{(\text{Base MVA}_{3\phi})\ (10^6)} \text{ pu} \tag{1.23}$$

or

$$Y = \frac{(\text{Base kV}_{LL})^2 \ (10^{-6})}{(\text{Base MVA}_{3\phi})\ (Z\ \text{M}\Omega)} \text{ pu} \tag{1.24}$$

Equations (1.23) and (1.24) are both useful in transmission line calculations where the shunt susceptance is sometimes given in micromhos per mile and sometimes in megohm-miles.

Usually the subscripts LL and 3ϕ can be omitted without ambiguity, but it is wise to use caution in these simple calculations. Many a system study has been made useless by a simple error in preparing data, and the cost of such errors can easily run into thousands of dollars.

For transmission lines it is possible to further simplify equations (1.22) to (1.24). In this case the quantities usually known from a knowledge of wire size and spacing are

1. the resistance R in ohm/mile at a given temperature
2. the inductive reactance X_L in ohm/mile at 60 Hz
3. the shunt capacitive reactance X_C in megohm-miles at 60 Hz

We make the following arbitrary assumptions:

$$\text{Base MVA}_{3\phi} = 100 \text{ MVA}$$

$$\text{line length} = 1.0 \text{ mi} \tag{1.25}$$

Values thus computed are on a per mile basis but can easily be multiplied by the line length. Likewise, for any base MVA other than 100 the change of base formula (1.15) may be used to correct the value computed by the method given here. Thus for one mile of line

$$Z = \frac{(Z\ \Omega/\text{mi})\ (\text{Base MVA}_{3\phi})}{(\text{Base kV}_{\text{LL}})^2} = (Z\ \Omega/\text{mi})K_Z \quad \text{pu} \tag{1.26}$$

$$K_Z = \frac{\text{Base MVA}_{3\phi}}{(\text{Base kV}_{\text{LL}})^2} = \frac{100}{(\text{Base kV}_{\text{LL}})^2} \tag{1.27}$$

Similarly, we compute

$$B = \frac{(\text{Base kV}_{\text{LL}})^2\ (10^{-6})}{(\text{Base MVA}_{3\phi})\ (X_C\ \text{M}\Omega\text{-mi})} = \frac{K_B}{X_C} \quad \text{pu} \tag{1.28}$$

where

$$K_B = \frac{(\text{Base kV}_{\text{LL}})^2\ (10^{-6})}{100} = 10^{-8}\,(\text{Base kV}_{\text{LL}})^2 \tag{1.29}$$

The constants K_Z and K_B may now be tabulated for commonly used voltages. Several values are given in Table 1.3.

1.5 Converting from Per Unit Values to System Values

Once a particular computation in pu is completed, it is often desirable to convert some quantities back to system values. This conversion is quite simple and is performed as follows:

$$(\text{pu } I)\ (\text{Base I}) = I \quad \text{A}$$

$$(\text{pu } V)\ (\text{Base V}) = V \quad \text{V}$$

$$(\text{pu } P)\ (\text{Base VA}) = P \quad \text{W}$$

$$(\text{pu } Q)\ (\text{Base VA}) = Q \quad \text{var} \tag{1.30}$$

Usually there is no need to convert an impedance back to ohms, but the procedure is exactly the same.

1.6 Examples of Per Unit Calculations

To clarify the foregoing procedures some simple examples are presented.

Table 1.3. Values of K_Z and K_B for Selected Voltages

Base kV	K_Z	K_B
2.30	18.903592	0.0529×10^{-6}
2.40	17.361111	0.0576
4.00	6.250000	0.1600
4.16	5.778476	0.1731
4.40	5.165289	0.1936
4.80	4.340278	0.2304
6.60	2.295684	0.4356
6.90	2.100399	0.4761
7.20	1.929012	0.5184
11.00	0.826446	1.2100
11.45	0.762762	1.3110
12.00	0.694444	1.4400
12.47	0.643083	1.5550
13.20	0.573921	1.7424
13.80	0.525100	1.9044
14.40	0.482253	2.0736
22.00	0.206612	4.8400
24.94	0.160771	6.2200
33.00	0.091827	10.8900
34.50	0.084016	11.9025
44.00	0.051653	19.3600
55.00	0.033058	30.2500
60.00	0.027778	36.0000
66.00	0.022957	43.5600
69.00	0.021004	47.6100
88.00	0.012913	77.4400
100.00	0.010000	100.0000
110.00	0.008264	121.0000
115.00	0.007561	132.2500
132.00	0.005739	174.2400
138.00	0.005251	190.4400
154.00	0.004217	237.1600
161.00	0.003858	259.2100
220.00	0.002066	484.0000
230.00	0.001890	529.0000
275.00	0.001322	756.2500
330.00	0.000918	1089.0000
345.00	0.000840	1190.2500
360.00	0.000772	1296.0000
362.00	0.000763	1310.4400
420.00	0.000567	1764.0000
500.00	0.000400	2500.0000
525.00	0.000363	2756.2500
550.00	0.000331	3025.0000
700.00	0.000204	4900.0000
735.00	0.000185	5402.2500
750.00	0.000178	5625.0000
765.00	0.000171	5852.2500
1000.00	0.000100	10000.0000
1100.00	0.000083	12100.0000
1200.00	0.000069	14400.0000
1300.00	0.000059	16900.0000
1400.00	0.000051	19600.0000
1500.00	0.000044	$22500.0000 \times 10^{-6}$

Example 1.1

Power system loads are usually specified in terms of the absorbed power and reactive power. In circuit analysis it is sometimes convenient to represent such a load as a constant impedance. Two such representations, parallel and series, are possible as shown in Figure 1.2. Determine the per unit R and X values for both the parallel and series connections.

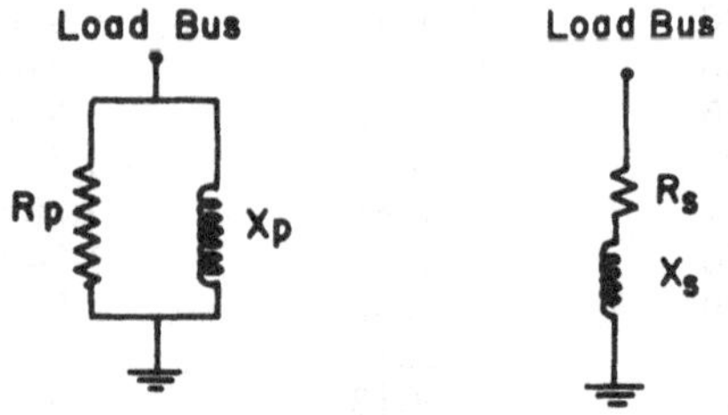

Fig. 1.2. Constant impedance load representation: left, parallel representation; right, series representation.

Solution

Let

$$P = \text{load power in W}$$
$$Q = \text{load reactive power in var}$$
$$R_p \text{ or } R_s = \text{load resistance in } \Omega$$
$$X_p \text{ or } X_s = \text{load reactance in } \Omega$$
$$V = \text{load voltage in V}$$

Parallel Connection. From the parallel connection we observe that the power absorbed depends only upon the applied voltage, i.e.,

$$P = V^2/R_p \tag{1.31}$$

From equation (1.13) we have

$$R_u = \frac{R_p\ (S_B)}{(V_B)^2} \text{ pu} \tag{1.32}$$

where the value subscripted u is a pu value. Substituting R_p from (1.31), we compute

$$R_u = (V/V_B)^2\ (S_B/P) = V_u^2/P_u \text{ pu} \tag{1.33}$$

and we note that (1.33) is the same as (1.31) except that all values are pu. Similarly, we find the expression for pu X to be

$$X_u = (V/V_B)^2\ (S_B/Q) = V_u^2/Q_u \text{ pu} \tag{1.34}$$

Series Connection. If R and X are connected in series as in Figure 1.2 b, the problem is more difficult since the current in X now affects the absorbed power P. In terms of system quantities, $I = V/(R_s + jX_s)$. Thus

$$P + jQ = VI^* = \frac{VV^*}{R_s - jX_s} = \frac{|V|^2}{R_s - jX_s} \tag{1.35}$$

Multiplying (1.35) by its conjugate, we have

$$P^2 + Q^2 = \frac{|V|^4}{R_s^2 + X_s^2} \tag{1.36}$$

Also, from (1.35)

$$P + jQ = \frac{|V|^2 (R_s + jX_s)}{R_s^2 + X_s^2} \tag{1.37}$$

Substituting (1.36) into (1.37), we compute

$$P + jQ = \frac{(R_s + jX_s)(P^2 + Q^2)}{|V|^2}$$

Rearranging,

$$R_s + jX_s = \frac{|V|^2}{P^2 + Q^2}(P + jQ)\ \Omega \tag{1.38}$$

Equation (1.38) is the desired result, but it is not in pu. Substituting into (1.13), we have

$$R_u + jX_u = \frac{(R_s + jX_s) S_B}{V_B^2}$$

Then we compute from (1.38)

$$R_u = \frac{V_u^2\ S_B\ (P \text{ watt})}{P^2 + Q^2} \text{ pu} \tag{1.39}$$

$$X_u = \frac{V_u^2\ S_B\ (Q \text{ var})}{P^2 + Q^2} \text{ pu} \tag{1.40}$$

Example 1.2

Given the two-machine system of Figure 1.3, we select, quite arbitrarily, a base voltage of 161 kV for the transmission line and a base voltampere of

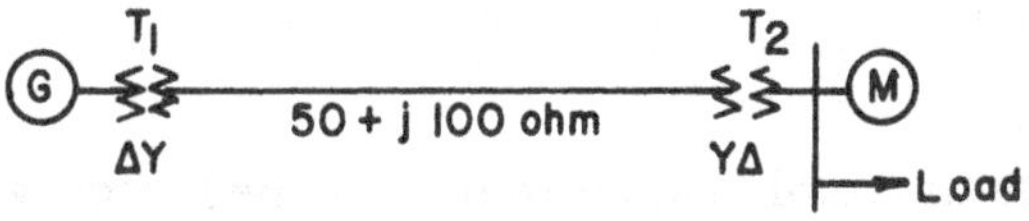

Fig. 1.3. A two-machine system.

20 MVA. Find the pu impedances of all components referred to these bases. The apparatus has ratings as follows:

Generator: 15 MVA, 13.8 kV, $x = 0.15$ pu
Motor: 10 MVA, 13.2 kV, $x = 0.15$ pu
T1: 25 MVA, 13.2–161 kV, $x = 0.10$ pu
T2: 15 MVA, 13.8–161 kV, $x = 0.10$ pu
Load: 4 MVA at 0.8 *pf lag*

Solution

Using equation (1.15), we proceed directly with a change in base for the apparatus.

$$\text{Generator: } x = (0.15)\left(\frac{20}{15}\right)\left(\frac{13.8}{13.2}\right)^2 = 0.2185 \text{ pu}$$

$$\text{Motor: } x = (0.15)\left(\frac{20}{10}\right)\left(\frac{13.2}{13.8}\right)^2 = 0.2745 \text{ pu}$$

$$\text{T1: } x = (0.10)\left(\frac{20}{25}\right)\left(\frac{161}{161}\right)^2 = 0.08 \text{ pu}$$

$$\text{T2: } x = (0.10)\left(\frac{20}{15}\right)\left(\frac{161}{161}\right)^2 = 0.1333 \text{ pu} \tag{1.41}$$

For the transmission line we must convert from ohmic values to pu values. We do this either by dividing by the base impedance or by application of equation (1.22). Using the latter method,

$$Z = \frac{(50 + \text{j}100 \text{ ohm})(20)}{(161)^2} = 0.0386 + \text{j}0.0771 \text{ pu} \tag{1.42}$$

For the load a parallel R-X representation may be computed from equations (1.32) and (1.34)

$$\begin{aligned} S = P + \text{j}Q &= |S|(\cos\theta + \text{j}\sin\theta) \\ &= 4(0.8 + \text{j}0.6) \\ &= 3.2 + \text{j}2.4 \text{ MVA} \end{aligned}$$

Then

$$R_u = \frac{V_u^2 S_B}{P} = \frac{V_u^2(20)}{3.2} = 6.25\, V_u^2 \quad \text{pu} \tag{1.43}$$

Similarly, $X_u = 8.33\, V_u^2$ pu.

Example 1.3

Suppose in Example 1.2 that the motor is a synchronous machine drawing 10 MVA at 0.9 *pf lead* and the terminal voltage is 1.1 pu. What is the voltage at the generator terminals?

Solution

First we compute the total load current. For the motor, with its voltage taken as the reference, i.e., $V = 1.1 + \text{j}0$, we have

$$I_M = \frac{P - \text{j}Q}{V^*} = \frac{9 - \text{j}(-10 \sin 25.9^\circ)}{20(1.1)} = 0.409 + \text{j}0.1985 \text{ pu}$$

For the static load

$$I_L = \frac{3.2 - \text{j}2.4}{20(1.1)} = 0.1455 - \text{j}0.109 \text{ pu}$$

Then the total current is $I_M + I_L$ or

$$I = 0.5545 + \text{j}0.0895 \text{ pu} \tag{1.44}$$

From Example 1.2 we easily find the total pu impedance between the buses to be the total of T1, T2, and Z (line); $Z = 0 + j0.213$ pu. Note that the transmission line impedance is negligible because the base is small and the line voltage high for the small power in this problem. Thus the generator bus voltage is

$$\begin{aligned} V_g &= 1.1 + j0 + (0 + j0.213)(0.5545 + j0.0895) \\ &= 1.1 - 0.0191 + j0.118 = 1.08 + j0.118 \text{ pu} \\ &= 1.087 \underline{/6.24^\circ} \text{ pu on 13.2 kV base} \\ &= 14.32 \text{ kV} \end{aligned}$$

1.7 Phasor Notation

In this book we will deal with voltages and currents which are rms *phasor* quantities. This implies that all signals are pure sine waves of voltage or current but with the time variable suppressed. Thus only the magnitude (amplitude) and relative phase angle of the sine waves are preserved. To do this we use a special definition.

Definition: A phasor A is a complex number which is related to the time domain sinusoidal quantity by the expression[3]

$$a(t) = \mathcal{R}e(\sqrt{2}A\,e^{j\omega t}) \tag{1.45}$$

If we express A in terms of its magnitude $|A|$ and angle α, we have $A = |A|e^{j\alpha}$ and

$$a(t) = \mathcal{R}e(\sqrt{2}\,|A|e^{j(\omega t+\alpha)}) = \sqrt{2}\,|A|\cos(\omega t + \alpha) \tag{1.46}$$

Thus equations (1.45) or (1.46) convert the rms phasor (complex) quantity to the actual time domain variable. We seldom have occasion to do this since it is generally preferable to work exclusively with complex rms phasor quantities.

Note the simplicity of the definition. The phasor is *not* a "rotating vector." It is simply a complex number having the same dimensions (ampere, volt) as the time domain quantity. Note that a single constant frequency (constant ω) is implied by the definition. Since we are dealing with steady state solutions of networks, ω is constant and impedance is a constant ratio of V/I and is itself a phasor (complex) quantity. There is, however, no occasion to convert impedance to the time domain since this is meaningless.

From our phasor definition we may write

$$\begin{aligned} \sqrt{2}\,|A|e^{j(\omega t+\alpha)} &= \sqrt{2}\,|A|\cos(\omega t + \alpha) + j\sqrt{2}\,|A|\sin(\omega t + \alpha) \\ &= a(t) + jb(t) \end{aligned} \tag{1.47}$$

If we differentiate (1.46) with respect to time, we have

$$\dot{a}(t) = -\omega\sqrt{2}\,|A|\sin(\omega t + \alpha)$$

or from (1.47),

$$\dot{a}(t) = -\omega b(t) \tag{1.48}$$

Substituting $b(t)$ from (1.48) into (1.47), we have the differential equation

$$\dot{a}(t) + j\omega a(t) = j\omega\sqrt{2}A e^{j\omega t}$$

[3]The definition could just as well use the imaginary part of $(\sqrt{2}Ae^{j\omega t})$, but either definition should be used consistently.

which has the solution

$$a(t) = \left[a(0) - \frac{A}{\sqrt{2}}\right]e^{-j\omega t} + \frac{A}{\sqrt{2}}e^{j\omega t}$$

This can be rearranged to the form

$$A = \frac{a(t) - a(0)\,e^{-j\omega t}}{j\sqrt{2}\sin\omega t} \tag{1.49}$$

which is a formula for computing a phasor A, based on definition (1.45) and given the time domain sinusoid $a(t)$.

It is convenient to think of (1.49) as a "phasor transformation" which we designate by the script letter $\mathcal{P}$ according to the relation[4]

$$\mathcal{P}[a(t)] = \mathcal{P}[\sqrt{2}|A|\cos(\omega t + \alpha)] = |A|e^{j\alpha} = A \tag{1.50}$$

Then the inverse transformation is

$$\mathcal{P}^{-1}[A] = \mathcal{R}e(\sqrt{2}A\,e^{j\omega t}) = a(t) \tag{1.51}$$

Either the symbolic notation of (1.50) or the exact formula (1.49) may be used. Note that the expression (1.50) is a linear transformation [4].

1.8 The Phasor a or a-Operator

In working with three-phase quantities, it is convenient to have a simple phasor operator which will add 120° to the phase angle of a phasor and leave its magnitude unchanged. A common notation for this is to define

$$a = 1\underline{/120^\circ} = e^{j2\pi/3} \tag{1.52}$$

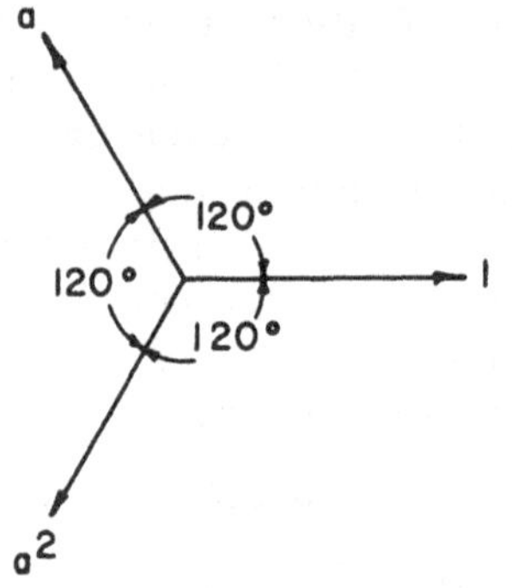

Fig. 1.4. Phasor diagram of the a-operator.

It is convenient to think of the a-operator as one which rotates a phasor by +120° and the a^2-operator as one which rotates a phasor by $-120°$. It is also clear that (see Fig. 1.4)

$$a^2 = 1\underline{/-120^\circ},\quad a^3 = 1\underline{/0^\circ},\quad a^4 = a$$
$$a^5 = a^2,\quad a^6 = 1,\quad \text{etc.} \tag{1.53}$$

At times it is convenient to know various linear combinations of a-operators. Some familiar combinations are given in Table 1.4.

[4]This notation is due to Nilsson [3].

Table 1.4. a-Operator Functions

Function	*Polar*	*Rectangular*
a	$1\angle 120^\circ$	-0.5 + j0.866
a^2	$1\angle 240^\circ$	-0.5 - j0.866
a^3	$1\angle 0^\circ$	1.0 + j0
a^4	$1\angle 120^\circ$	-0.5 + j0.866
$1 + a = -a^2$	$1\angle 60^\circ$	0.5 + j0.866
$1 + a^2 = -a$	$1\angle -60^\circ$	0.5 - j0.866
$1 - a$	$\sqrt{3}\angle -30^\circ$	1.5 - j0.866
$1 - a^2$	$\sqrt{3}\angle 30^\circ$	1.5 + j0.866
$a - 1$	$\sqrt{3}\angle 150^\circ$	-1.5 + j0.866
$a^2 - 1$	$\sqrt{3}\angle -150^\circ$	-1.5 - j0.866
$a - a^2$	$\sqrt{3}\angle 90^\circ$	0.0 + j1.732
$a^2 - a$	$\sqrt{3}\angle -90^\circ$	0.0 - j1.732
$a + a^2$	$1\angle 180^\circ$	-1.0 + j0
$1 + a + a^2$	0	0

Problems

1.1. Convert all values to pu on a 10 MVA base with 100 kV base voltage on the line.

Fig. P1.1.

Generator: 15 MVA, 13.8 kV, $X = 0.15$ pu
Motor: 10 MVA, 12 kV, $X = 0.07$ pu
T1: 20 MVA, 14-132 kV, $X = 0.10$ pu
T2: 15 MVA, 13-115 kV, $X = 0.10$ pu
Line: 200 + j500 Ω

1.2. Prepare a per phase schematic of the system shown and give all impedances in pu on a 100 MVA, 154 kV transmission base.

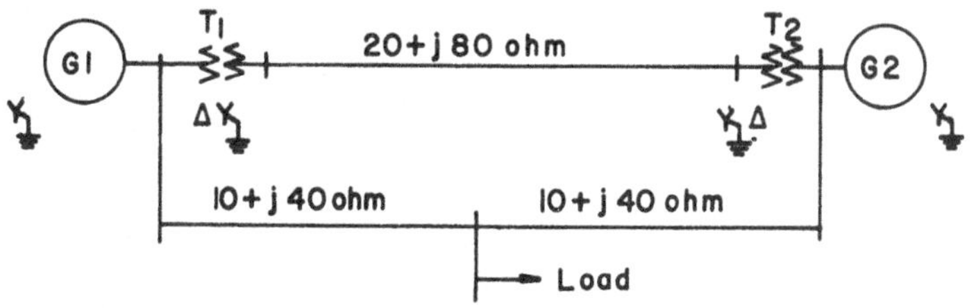

Fig. P1.2.

G1: 50 MVA, 13.8 kV, $X = 15\%$
G2: 20 MVA, 14.4 kV, $X = 15\%$
T1: 60 MVA, 13.2-161 kV, $X = 10\%$
T2: 25 MVA, 13.2-161 kV, $X = 10\%$
Load: 15 MVA, 80% *pf lag*

1.3. Draw a per phase impedance diagram for the system shown. Assume that the load impedance is entirely reactive and equal to j1.0 pu. Find the Thevenin equivalent, looking

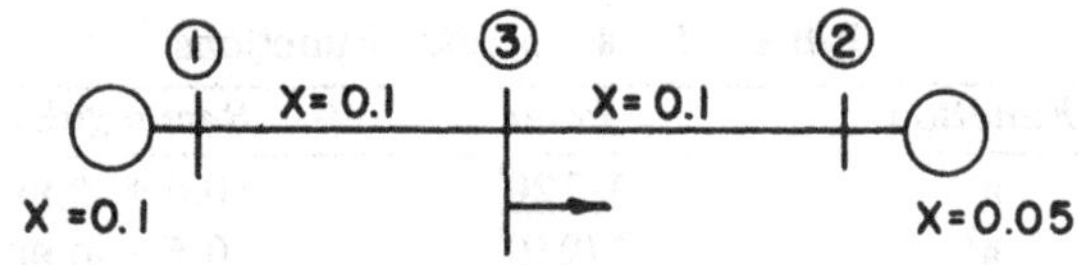

Fig. P1.3.

into this system from an external connection at bus 3 if

(a) Generated voltages V_1 and V_2 are equal.

(b) Generated voltages V_1 and V_2 are not equal.

1.4. The following table of values has been prepared for the various line sections in a small electric system. Find the total pu impedance and shunt susceptance of each line on a 10 MVA base, using the line nominal voltage as a voltage base.

Nominal Voltage (kV)	*Line Length (mi)*	*Wire Size*	R (Ω/mi)	X (Ω/mi)	X_C ($M\Omega$-mi)
13.8	5.0	4/0 cu	0.278	0.690	0.160
13.8	2.0	4 cu	1.374	0.816	0.193
13.8	3.9	4/0 A	0.445	0.711	0.157
13.8	6.2	336.4 A	0.278	0.730	0.172
13.8	7.3	556.5 A	0.088	0.330	0.142
69.0	10.0	4/0 A	0.445	0.711	0.157
69.0	25.0	336.4 A	0.278	0.730	0.172

1.5. Verify equation (1.49) and compute A for the following time domain sinusoids:

(a) $A_m \cos \omega t$

(b) $A_m \cos(\omega t + \alpha)$

(c) $B_m \sin \omega t$

(d) $B_m \sin(\omega t + \beta)$

1.6. Prove that the transformation $\mathcal{P}[a(t)]$ is a linear transformation.

1.7. Given that $a(t)$ is a sinusoid given by equation (1.46), find the phasor transform of $b(t)$ where $b(t)$ is the derivative of $a(t)$ with respect to time.

1.8. Compare the phasor transform $\mathcal{P}$ to the Laplace transform $\mathcal{L}$.

chapter 2

Symmetrical Components

It was pointed out in Chapter 1 that a per phase representation of power systems is preferred because of its simplicity. In solving problems of balanced three-phase networks this is easily accomplished by changing all delta connections to equivalent wye connections and solving one leg (one phase) of the resulting network. Since the system is balanced, the results for the remaining two phases differ from the first by ±120° in phase and the problem is solved. Such a solution avoids the complexity of detailing each terminal and makes use of inherent system symmetry.

The method of symmetrical components provides a means of extending per phase analysis to systems with unbalanced loads. This is possible because of a property of unbalanced phasors discovered by Fortescue [1]. He observed that a system of three unbalanced phasors can be broken down into two sets of balanced phasors plus an additional set of single-phase phasors. If the voltages and currents of the problem are represented in this way, a per phase representation is adequate for each component and the desired simplification has been achieved. Such a system will be analyzed in detail later. First, as an introduction to the subject consider the more general problem of an n-phase system.

2.1 Symmetrical Components of an n-Phase System

In a brilliant paper published in 1918 [1], C. L. Fortescue proposed a system whereby an unbalanced set of n phasors may be resolved into $n - 1$ balanced n-phase systems of different phase sequence and one zero-phase sequence system. By his definition a zero-phase sequence system is one in which all phasors are of equal magnitude and angle or they are all identical. We denote Fortescue's system mathematically as follows.

Consider the n-dimensional system of phasors defined by the following equations:

$$
\begin{aligned}
V_a &= V_{a1} + V_{a2} + V_{a3} + \cdots + V_{an} \\
V_b &= V_{b1} + V_{b2} + V_{b3} + \cdots + V_{bn} \\
&\cdots\cdots\cdots\cdots\cdots\cdots\cdots\cdots \\
V_n &= V_{n1} + V_{n2} + V_{n3} + \cdots + V_{nn}
\end{aligned}
\tag{2.1}
$$

where

$V_a, V_b, \ldots, V_n$ = an *unbalanced* set of phasors
$V_{a1}, V_{b1}, \ldots, V_{n1}$ = the first set of n balanced phasors with an angle $2\pi/n$ between components $a, b, \ldots, n$

$V_{a2}, V_{b2}, \ldots, V_{n2}$ = the second set of n balanced phasors with an angle $4\pi/n$ between components $a, b, \ldots, n$

. . . .

$V_{a(n-1)}, V_{b(n-1)}, \ldots, V_{n(n-1)}$ = the $(n-1)$st set of n balanced phasors with an angle $2\pi\ (n-1)/n$ between components $a, b, \ldots, n$

$V_{an}, V_{bn}, \ldots, V_{nn}$ = the final or nth set of n balanced phasors with an angle $n(2\pi/n) = 2\pi$ between components $a, b, \ldots, n$; or the components of this last set are all identical

Equation (2.1) may be simplified by adopting a notation similar to the a-operator of section 1.8. Suppose we generalize the definition of a to be

$$\mathrm{a} = e^{j2\pi/n} \tag{2.2}$$

where n = the system dimension or number of phases. Then

$$\begin{aligned}
\mathrm{a}^2 &= e^{j4\pi/n} \text{ rotates a phasor by } 4\pi/n \text{ radians}\\
\mathrm{a}^3 &= e^{j6\pi/n} \text{ rotates a phasor by } 6\pi/n \text{ radians}\\
&\cdots\cdots\cdots\cdots\cdots\cdots\\
\mathrm{a}^n &= e^{j2\pi} \quad \text{rotates a phasor by } 2\pi \text{ radians.}
\end{aligned}$$

Negative powers of a are also easily defined and amount to negative (clockwise) rotation of phase angles.

Using the definition (2.2), we operate on (2.1) by multiplying the equations by $1, \mathrm{a}, \mathrm{a}^2, \ldots, \mathrm{a}^n$ with the result

$$\begin{aligned}
V_a &= V_{a1} + V_{a2} + \cdots + V_{an}\\
\mathrm{a}V_b &= \mathrm{a}V_{b1} + \mathrm{a}V_{b2} + \cdots + \mathrm{a}V_{bn}\\
\mathrm{a}^2V_c &= \mathrm{a}^2V_{c1} + \mathrm{a}^2V_{c2} + \cdots + \mathrm{a}^2V_{cn}\\
&\cdots\cdots\cdots\cdots\cdots\cdots\\
\mathrm{a}^{n-1}V_n &= \mathrm{a}^{n-1}V_{n1} + \mathrm{a}^{n-1}V_{n2} + \cdots + \mathrm{a}^{n-1}V_{nn}
\end{aligned} \tag{2.3}$$

By our definition of the phase sequence systems $1, 2, \ldots, n$ we immediately recognize that

$$V_{a1} = \mathrm{a}V_{b1} = \mathrm{a}^2V_{c1} = \cdots = \mathrm{a}^{n-1}V_{n1} \tag{2.4}$$

This is because we defined this sequence to have exactly an angle $2\pi/n$ between components and, since the components are balanced (equal magnitudes and phase displacements), rotating V_{b1} by $2\pi/n$ radians makes it coincide with V_{a1}, etc. Note also that the sum of all the components of equation (2.4) is nV_{a1}.

Now add the equations (2.3) together and consider this sum. For the right-hand side we have nV_{a1} for the first (subscript 1) group as previously noted. All other groups add to zero. This can be shown by adding vertically the subscript 2 group as an example.

$$\begin{aligned}
V_{a2} &= & &= V_{a2}\\
\mathrm{a}V_{b2} &= \mathrm{a}\mathrm{a}^{-2}V_{a2} & &= \mathrm{a}^{-1}V_{a2}\\
\mathrm{a}^2V_{c2} &= \mathrm{a}^2\mathrm{a}^{-4}V_{a2} & &= \mathrm{a}^{-2}V_{a2}
\end{aligned}$$

$$\frac{a^{n-1} V_{n2} = a^{n-1} a^{-2(n-1)} V_{a2} = a^{-(n-1)} V_{a2}}{V_{a2} + aV_{b2} + a^2 V_{c2} + \cdots + a^{n-1} V_{n2} = 0} \tag{2.5}$$

The sum is zero because the right-hand side of (2.5) is a balanced set of n voltages whose sum is zero. In a similar way we show that all groups on the right side of (2.3) except the first add to zero since they are all balanced sets of voltages. Adding equations (2.3) then gives the result

$$V_a + aV_b + a^2 V_c + \cdots + a^{n-1} V_n = nV_{a1} \tag{2.6}$$

or

$$V_{a1} = \frac{1}{n}(V_a + aV_b + a^2 V_c + \cdots + a^{n-1} V_n) \tag{2.7}$$

In a similar way we show that

$$V_{a2} = \frac{1}{n}(V_a + a^2 V_b + a^4 V_c + \cdots + a^{2(n-1)} V_n)$$

$$V_{a3} = \frac{1}{n}(V_a + a^3 V_b + a^6 V_c + \cdots + a^{3(n-1)} V_n)$$

$$\cdots\cdots\cdots\cdots\cdots\cdots\cdots\cdots\cdots\cdots\cdots$$

$$V_{an} = \frac{1}{n}(V_a + V_b + V_c + \cdots + V_n) \tag{2.8}$$

Matrix notation provides a convenient way of writing equations (2.7) and (2.8). We list the equation for V_{an} first since it is the simplest, and we call it V_{a0}. Then

$$\begin{bmatrix} V_{a0} \\ V_{a1} \\ V_{a2} \\ \cdots \\ V_{a(n-1)} \end{bmatrix} = \frac{1}{n} \begin{bmatrix} 1 & 1 & 1 & \cdots & 1 \\ 1 & a & a^2 & \cdots & a^{n-1} \\ 1 & a^2 & a^4 & \cdots & a^{2(n-1)} \\ \cdots & \cdots & \cdots & \cdots & \cdots \\ 1 & a^{n-1} & a^{2(n-1)} & \cdots & a^{(n-1)(n-1)} \end{bmatrix} \begin{bmatrix} V_a \\ V_b \\ V_c \\ \cdots \\ V_n \end{bmatrix} \tag{2.9}$$

which we may write in matrix notation as

$$\not{\mathbf{V}} = \mathbf{C}\,\mathbf{V} \tag{2.10}$$

Here we use the notation V-slash ($\not{\mathbf{V}}$) to indicate the voltage **V** split into symmetrical components of phase a and where **V** is an array of the original n unbalanced phasors. The matrix **C** is a matrix of operators, or it is a transformation whereby a set of phasors **V** may be resolved into a new set of phasors which define the symmetrical components $\not{\mathbf{V}}$. Actually, **C** determines only the symmetrical components of phase a, namely $V_{a0}, V_{a1}, \ldots, V_{a(n-1)}$, but from these the components of the other phases are easily found by symmetry.

Since (2.10) defines a transformation, the uniqueness of this operation is of immediate interest. The transformation is unique if and only if **C** is nonsingular [5, 6]. If **C** is nonsingular, its inverse

$$\mathbf{C}^{-1} = \mathbf{A} \tag{2.11}$$

exists and we may write from (2.10)

$$\mathbf{V} = \mathbf{A}\,\mathbf{\mathbb{V}} \tag{2.12}$$

Equation (2.12) indicates that the phase voltages may be synthesized from the symmetrical components of phase a, a most important result. Our problem, then, is to show that **C** is nonsingular.

It is helpful to recognize that **C** is a special case of the Vandermonde matrix [4, 7, 8] **P** where

$$\mathbf{P} = \begin{bmatrix} 1 & 1 & 1 & \cdots & 1 \\ \lambda_1 & \lambda_2 & \lambda_3 & \cdots & \lambda_n \\ \lambda_1^2 & \lambda_2^2 & \lambda_3^2 & \cdots & \lambda_n^2 \\ \cdots & \cdots & \cdots & \cdots & \cdots \\ \lambda_1^{n-1} & \lambda_2^{n-1} & \lambda_3^{n-1} & \cdots & \lambda_n^{n-1} \end{bmatrix} \tag{2.13}$$

In the case of matrix **C** we have exactly the form **P** with $\lambda_1 = 1$. The determinant of **P** is given by

$$\det \mathbf{P} = \prod_{i>j} (\lambda_i - \lambda_j) = (\lambda_2 - \lambda_1)\cdots(\lambda_n - \lambda_1)(\lambda_3 - \lambda_2)\cdots(\lambda_n - \lambda_{n-1}) \tag{2.14}$$

and we see that **P** is nonsingular as long as the λ_i are distinct. In the case of **C**

$$\lambda_i = e^{j2\pi(i-1)/n} \tag{2.15}$$

are all distinct, being complex quantities of magnitude 1 but differing in argument by $2\pi/n$ radians, as shown in Figure 2.1. Thus the transformation (2.10) is indeed unique, and its inverse transformation (2.12) exists.

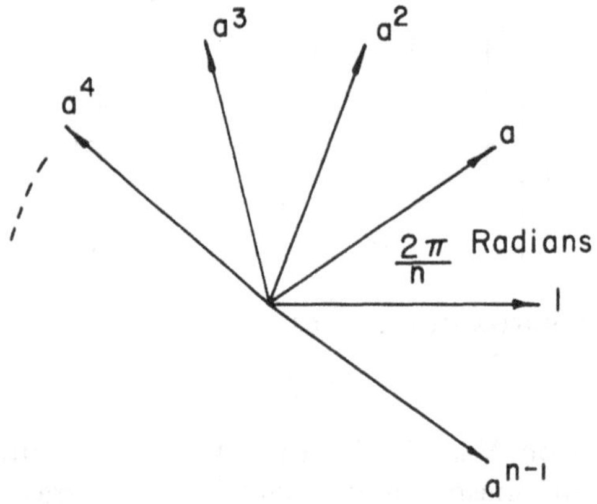

Fig. 2.1. Complex operators from row 2 of C.

To find the inverse transformation, we examine equation (2.1) where $V_{a0} = V_{an}$. Since all symmetrical components may be written in terms of the phase a components, we rewrite (2.1) entirely as a function of phase a. Thus

$$\begin{aligned} V_a &= V_{a0} + V_{a1} + V_{a2} + \cdots + V_{a(n-1)} \\ V_b &= V_{b0} + V_{b1} + V_{b2} + \cdots + V_{b(n-1)} \\ V_c &= V_{c0} + V_{c1} + V_{c2} + \cdots + V_{c(n-1)} \\ &\cdots\cdots\cdots\cdots\cdots\cdots \\ V_n &= V_{n0} + V_{n1} + V_{n2} + \cdots + V_{n(n-1)} \end{aligned} \tag{2.16}$$

First, we note that $V_{a0} = V_{b0} = V_{c0} = \cdots = V_{n0}$. Also, from (2.4)

$$V_{a1} = aV_{b1} = a^2 V_{c1} = \cdots = a^{n-1} V_{n1}$$

Similarly, $V_{a2} = a^2 V_{b2} = a^4 V_{c2} = \cdots = a^{2(n-1)} V_{n2}$ with similar expressions applying for the remaining sequences. These relationships are established by definition of sequence quantities. Substituting into (2.16), we have

$$\begin{aligned}
V_a &= V_{a0} + V_{a1} + V_{a2} + \cdots + V_{a(n-1)} \\
V_b &= V_{a0} + \mathrm{a}^{-1} V_{a1} + \mathrm{a}^{-2} V_{a2} + \cdots + \mathrm{a}^{-(n-1)} V_{a(n-1)} \\
V_c &= V_{a0} + \mathrm{a}^{-2} V_{a1} + \mathrm{a}^{-4} V_{a2} + \cdots + \mathrm{a}^{-2(n-1)} V_{a(n-1)} \\
&\cdots\cdots\cdots\cdots\cdots\cdots\cdots\cdots \\
V_n &= V_{a0} + \mathrm{a}^{-(n-1)} V_{a1} + \mathrm{a}^{-2(n-1)} V_{a2} + \cdots + \mathrm{a}^{-(n-1)^2} V_{a(n-1)}
\end{aligned} \tag{2.17}$$

This equation may be simplified by changing the negative exponents of "a" to positive exponents by the relation $\mathrm{a}^{-k} = \mathrm{a}^{in-k}$, $0 < k < n$, $i = 1, 2, \ldots$ which amounts to adding 360° to a negative argument to make it positive. If this is done, (2.17) becomes

$$\begin{aligned}
V_a &= V_{a0} + V_{a1} + V_{a2} + \cdots + V_{a(n-1)} \\
V_b &= V_{a0} + \mathrm{a}^{n-1} V_{a1} + \mathrm{a}^{n-2} V_{a2} + \cdots + \mathrm{a} V_{a(n-1)} \\
V_c &= V_{a0} + \mathrm{a}^{n-2} V_{a1} + \mathrm{a}^{n-4} V_{a2} + \cdots + \mathrm{a}^2 V_{a(n-1)} \\
&\cdots\cdots\cdots\cdots\cdots\cdots\cdots\cdots \\
V_n &= V_{a0} + \mathrm{a} V_{a1} + \mathrm{a}^2 V_{a2} + \cdots + \mathrm{a}^{n-1} V_{a(n-1)}
\end{aligned} \tag{2.18}$$

or in matrix form

$$\begin{bmatrix} V_a \\ V_b \\ V_c \\ \cdots \\ V_n \end{bmatrix} = \begin{bmatrix} 1 & 1 & 1 & \cdots & 1 \\ 1 & \mathrm{a}^{n-1} & \mathrm{a}^{n-2} & \cdots & \mathrm{a} \\ 1 & \mathrm{a}^{n-2} & \mathrm{a}^{n-4} & \cdots & \mathrm{a}^2 \\ \cdots & \cdots & \cdots & \cdots & \cdots \\ 1 & \mathrm{a} & \mathrm{a}^2 & \cdots & \mathrm{a}^{n-1} \end{bmatrix} \begin{bmatrix} V_{a0} \\ V_{a1} \\ V_{a2} \\ \cdots \\ V_{a(n-1)} \end{bmatrix} \tag{2.19}$$

This is identical with (2.12) $\mathbf{V} = \mathbf{A} \mathbf{\mathcal{V}}$ from which we conclude that

$$\mathrm{A} = \begin{bmatrix} 1 & 1 & 1 & \cdots & 1 \\ 1 & \mathrm{a}^{n-1} & \mathrm{a}^{n-2} & \cdots & \mathrm{a} \\ 1 & \mathrm{a}^{n-2} & \mathrm{a}^{n-4} & \cdots & \mathrm{a}^2 \\ \cdots & \cdots & \cdots & \cdots & \cdots \\ 1 & \mathrm{a} & \mathrm{a}^2 & \cdots & \mathrm{a}^{n-1} \end{bmatrix} \tag{2.20}$$

2.2 Symmetrical Components of a Three-Phase System

The n-phase system described in the preceding paragraph is of academic interest only. We therefore move directly to a consideration of the more practical three-phase system. Other useful systems, such as the two-phase system, may be described as well but are omitted here in order to concentrate on the more common three-phase problem.

If $n = 3$, the phasor a-operator rotates any phasor quantity by 120° and the identities of section 1.8 apply. We are directly concerned with the "analysis and synthesis equations," (2.10) and (2.12) respectively in the three-phase system. From section 2.1, with $n = 3$, we have for the *analysis* equation

$$\begin{bmatrix} V_{a0} \\ V_{a1} \\ V_{a2} \end{bmatrix} = \frac{1}{3} \begin{bmatrix} 1 & 1 & 1 \\ 1 & a & a^2 \\ 1 & a^2 & a \end{bmatrix} \begin{bmatrix} V_a \\ V_b \\ V_c \end{bmatrix} \tag{2.21}$$

or

$$\mathbf{V}_{012} = \mathbf{C}\,\mathbf{V}_{abc} \tag{2.22}$$

where we have written $\mathbf{\breve{V}}$ as $\mathbf{V}_{012}$ and $\mathbf{V}$ as $\mathbf{V}_{abc}$ since the latter notation is more descriptive and is not unduly awkward when $n = 3$. We will use either notation, however, depending upon which seems best in any given situation. Also note that the subscript 012 establishes a notation used by many authors [9, 10, 11], for example, where

0 refers to the "zero sequence"
1 refers to the #1 or "positive sequence"
2 refers to the #2 or "negative sequence"

The names zero, positive, and negative refer to the sequence (of rotation) of the phase quantities so identified. This is made clear if typical sequence quantities are sketched as shown in Figure 2.2. There it is quite plain that the positive sequence

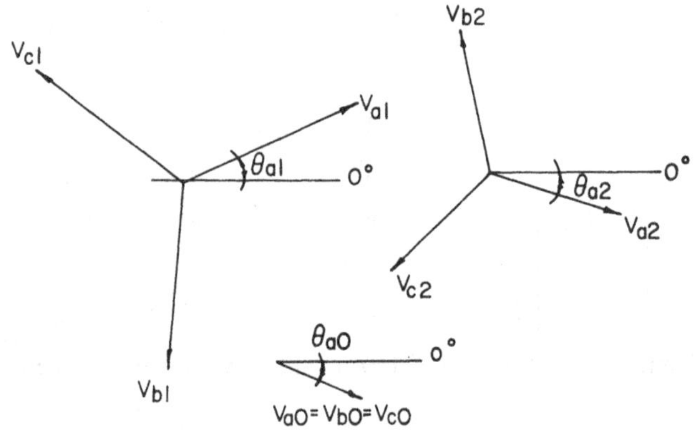

Fig. 2.2. A typical set of positive, negative, and zero sequence voltages.

set (V_{a1}, V_{b1}, V_{c1}) is the same as the voltages produced by a generator which has phase sequence *a-b-c*, which we denote here as *positive*. The negative sequence set (V_{a2}, V_{b2}, V_{c2}) is seen to have phase sequence *a-c-b* which we denote as *negative*. The zero sequence phasors (V_{a0}, V_{b0}, V_{c0}) have zero-phase displacement and thus are identical.

The synthesis equation for a three-phase system, corresponding to equation (2.19) is

$$\begin{bmatrix} V_a \\ V_b \\ V_c \end{bmatrix} = \begin{bmatrix} 1 & 1 & 1 \\ 1 & a^2 & a \\ 1 & a & a^2 \end{bmatrix} \begin{bmatrix} V_{a0} \\ V_{a1} \\ V_{a2} \end{bmatrix} \tag{2.23}$$

or

$$\mathbf{V}_{abc} = \mathbf{A}\,\mathbf{V}_{012} \tag{2.24}$$

By direct application of (2.23) we "synthesize" the phase quantities as shown graphically in Figure 2.3. Note that we could also apply equation (2.21) to the unbalanced phasors of Figure 2.3 by either analytical or graphical means to obtain the symmetrical component quantities.

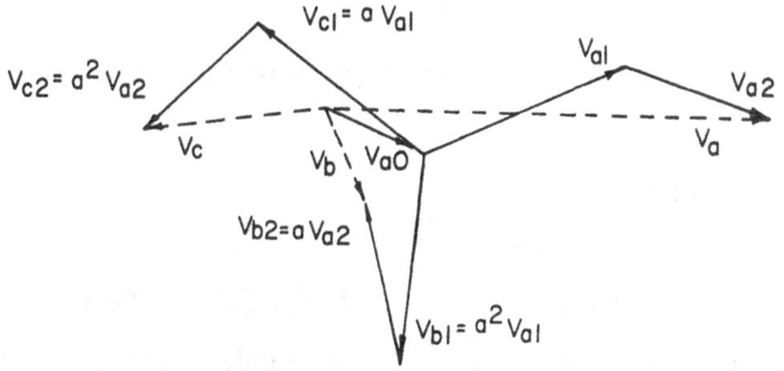

Fig. 2.3. Synthesis of sequence quantities to give phase quantities.

2.3 Symmetrical Components of Current Phasors

Up to this point we have considered the symmetrical components of line-to-neutral voltage phasors only. Symmetrical components of line-to-line voltages may also be computed and used. These same techniques, however, apply to any unbalanced set of three-phase (or n-phase) quantities. Thus for currents we have relations identical to (2.22) and (2.24), viz.,

$$\mathbf{I}_{012} = \mathbf{C}\mathbf{I}_{abc} \tag{2.25}$$

and

$$\mathbf{I}_{abc} = \mathbf{A}\mathbf{I}_{012} \tag{2.26}$$

2.4 Computing Power from Symmetrical Components

For any three-phase system the total power at any point is the sum of the powers computed in the individual phases, i.e., with $P_{3\phi}$ indicating the average value of the three-phase power

$$P_{3\phi} = \Re e\, \mathbf{V}_{abc}^{t}\mathbf{I}_{abc}^{*} \tag{2.27}$$

where

$$\mathbf{V}_{abc}^{t} = \text{transpose of } \mathbf{V}_{abc} = [V_a\, V_b\, V_c], \text{ a row vector}$$

$$\mathbf{I}_{abc}^{*} = \begin{bmatrix} I_a^* \\ I_b^* \\ I_c^* \end{bmatrix} = \text{the conjugate of } \mathbf{I}_{abc}$$

But from (2.24) and (2.26) we may write the power in terms of symmetrical components as

$$P_{3\phi} = \Re e\, \mathbf{V}_{012}^{t}\,\mathbf{A}^{t}\mathbf{A}^{*}\,\mathbf{I}_{012}^{*} \tag{2.28}$$

Since $\mathbf{A}$ is a symmetric matrix, $\mathbf{A}^t = \mathbf{A}$. We examine the matrix product $\mathbf{AA}^*$

where we note that

$$a^* = a^2, \qquad a^{2*} = a \tag{2.29}$$

Performing the operation indicated by the matrix product, we conclude that

$$AA^* = 3U \tag{2.30}$$

where

$$U = \begin{bmatrix} 1 & 0 & 0 \\ 0 & 1 & 0 \\ 0 & 0 & 1 \end{bmatrix} = \text{the unit matrix}$$

Substituting this result into (2.28), we have

$$\begin{aligned} P_{3\phi} &= 3\,\Re e\, V_{012}^t\, U I_{012}^* \\ &= 3\,\Re e\,(V_{a0} I_{a0}^* + V_{a1} I_{a1}^* + V_{a2} I_{a2}^*) = 3\,\Re e\, V_{012}^t\, I_{012}^* \end{aligned} \tag{2.31}$$

This is an important result and shows that the total three-phase power is a function of the symmetrical components of voltage and current of the same phase sequence. Thus there is no "coupling" of power from the negative sequence current reacting with the positive or zero sequence voltage, for example.

Equation (2.31) shows clearly that a change in the power relationship has occurred by virtue of the transformation from the *a-b-c* frame of reference of the 0-1-2 frame of reference. At times it would be convenient to work with a *power-invariant* transformation which we may denote as $\hat{A}$. For power invariance we require that

$$\hat{A}\hat{A}^* = U \tag{2.32}$$

such that $P_{3\phi} = \Re e(\hat{V}_{a0}\hat{I}_{a0}^* + \hat{V}_{a1}\hat{I}_{a1}^* + \hat{V}_{a2}\hat{I}_{a2}^*)$. We assume that $\hat{A}$ differs from A only by a constant. If we let $\hat{A} = A/h$ then we easily show that $\hat{A}^{-1} = hA^{-1}$. We also compute

$$\hat{A}\hat{A}^* = \frac{1}{h^2} AA^* = \frac{1}{h^2} 3U \tag{2.33}$$

But from (2.32) this quantity should be equal to the identity matrix. From this we conclude that $h = \sqrt{3}$ for power invariance. This means that we have the transformation

$$\hat{A} = \frac{1}{\sqrt{3}} \begin{bmatrix} 1 & 1 & 1 \\ 1 & a^2 & a \\ 1 & a & a^2 \end{bmatrix} \tag{2.34}$$

and its inverse

$$\hat{A}^{-1} = \frac{1}{\sqrt{3}} \begin{bmatrix} 1 & 1 & 1 \\ 1 & a & a^2 \\ 1 & a^2 & a \end{bmatrix} \tag{2.35}$$

Transformation (2.34) is often found in recent power literature.

This result suggests that a new definition of the A matrix may be desirable. Suppose we define

$$A = \frac{1}{h}\begin{bmatrix} 1 & 1 & 1 \\ 1 & a^2 & a \\ 1 & a & a^2 \end{bmatrix} \tag{2.36}$$

such that

$$A^{-1} = \frac{h}{3}\begin{bmatrix} 1 & 1 & 1 \\ 1 & a & a^2 \\ 1 & a^2 & a \end{bmatrix} \tag{2.37}$$

Then we may set

$$\begin{aligned} h &= 1 \text{ for the Fortescue transformation} \\ &= \sqrt{3} \text{ for the power invariant transformation} \end{aligned} \tag{2.38}$$

In this book definition (2.36) will be used. The reader may substitute his own value of "h" from (2.38). In this way we let the same transformation serve both those who prefer the older definition and those who insist upon power invariance. Using (2.36) then, we compute

$$P_{3\phi} = \frac{3}{h^2}\,\mathcal{R}e\,(V_{a0}I_{a0}^* + V_{a1}I_{a1}^* + V_{a2}I_{a2}^*) \tag{2.39}$$

2.5 Sequence Components for Unbalanced Network Impedances

Consider the three-phase system shown in Figure 2.4, where each current encounters an impedance in its phase conductor and where, in general, the self and mutual impedances are unequal, i.e.,

$$Z_{aa} \neq Z_{bb} \neq Z_{cc}, \qquad Z_{ab} \neq Z_{bc} \neq Z_{ca} \tag{2.40}$$

Thus both the self and mutual impedances constitute sets of unbalanced or unequal complex impedances. Even balanced currents will therefore produce unequal voltage drops between *m* and *n*.

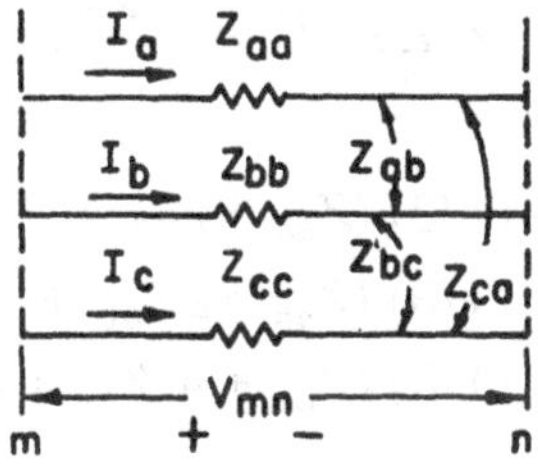

Fig. 2.4. A three-phase system with series Z.

Using equation (2.22), we were able to find symmetrical components for a set of three unbalanced (complex) phasor voltages. We apply a similar technique to investigate the possibility of finding symmetrical components *of* the unbalanced impedances.[1]

[1] We emphasize the word *of* because later we will define and use sequence impedances Z_0, Z_1, and Z_2 which will express impedance *to* the flow of I_{a0}, I_{a1}, and I_{a2} respectively. These are not the impedances under investigation here.

We begin by writing the voltage *drop* equation from m to n of Figure 2.4, which we will call $\mathbf{V}_{mn}$.

$$\mathbf{V}_{mn} = \begin{bmatrix} V_{mn\text{-}a} \\ V_{mn\text{-}b} \\ V_{mn\text{-}c} \end{bmatrix} = \begin{bmatrix} Z_{aa} & Z_{ab} & Z_{ac} \\ Z_{ba} & Z_{bb} & Z_{bc} \\ Z_{ca} & Z_{cb} & Z_{cc} \end{bmatrix} \begin{bmatrix} I_a \\ I_b \\ I_c \end{bmatrix} \tag{2.41}$$

where $\mathbf{V}_{mn}$ is actually $\mathbf{V}_{mn\text{-}abc}$. Applying the synthesis transform (2.24) to both sides of (2.41), we have

$$\mathbf{A}\,\mathbf{V}_{mn\text{-}012} = \mathbf{Z}\,\mathbf{A}\,\mathbf{I}_{012} \tag{2.42}$$

or the symmetrical components of the voltage drop are given by

$$\mathbf{V}_{mn\text{-}012} = \mathbf{A}^{-1}\,\mathbf{Z}\,\mathbf{A}\,\mathbf{I}_{012} \tag{2.43}$$

Suppose we define

$$\mathbf{V}_{mn\text{-}012} = \mathbf{Z}_{mn\text{-}012}\,\mathbf{I}_{012} \tag{2.44}$$

Then $\mathbf{Z}_{mn\text{-}012}$ is defined by the similarity transformation [7]

$$\mathbf{Z}_{mn\text{-}012} = \mathbf{A}^{-1}\,\mathbf{Z}\,\mathbf{A} \tag{2.45}$$

The concept of the similarity transformation gives us mathematical insight into the operations being performed. We interpret A as a linear operator which transforms currents or voltages from one coordinate system (0-1-2) to another (*a-b-c*). Z is another transform which takes a current vector $\mathbf{I}_{abc}$ into a voltage drop vector $\mathbf{V}_{mn}$, both being in the *a-b-c* coordinate system. The operator A, since it is nonsingular, may be used to find the transformation of similarity to take currents into voltage drops in the 0-1-2 coordinate system. Since $\mathbf{A}^t \neq \mathbf{A}^{-1}$, the transformation is not orthogonal [7].

The new matrix $Z_{mn\text{-}012}$ may be found directly as indicated by equation (2.45) with the result

$$\mathbf{Z}_{mn\text{-}012} = \begin{bmatrix} (Z_{S0} + 2Z_{M0}) & (Z_{S2} - Z_{M2}) & (Z_{S1} - Z_{M1}) \\ (Z_{S1} - Z_{M1}) & (Z_{S0} - Z_{M0}) & (Z_{S2} + 2Z_{M2}) \\ (Z_{S2} - Z_{M2}) & (Z_{S1} + 2Z_{M1}) & (Z_{S0} - Z_{M0}) \end{bmatrix} \tag{2.46}$$

where we have defined

$$\begin{aligned} Z_{S0} &= (1/3)\,(Z_{aa} + Z_{bb} + Z_{cc}) \\ Z_{S1} &= (1/3)\,(Z_{aa} + \mathrm{a}Z_{bb} + \mathrm{a}^2 Z_{cc}) \\ Z_{S2} &= (1/3)\,(Z_{aa} + \mathrm{a}^2 Z_{bb} + \mathrm{a}Z_{cc}) \end{aligned} \tag{2.47}$$

and

$$\begin{aligned} Z_{M0} &= (1/3)\,(Z_{bc} + Z_{ca} + Z_{ab}) \\ Z_{M1} &= (1/3)\,(Z_{bc} + \mathrm{a}Z_{ca} + \mathrm{a}^2 Z_{ab}) \\ Z_{M2} &= (1/3)\,(Z_{bc} + \mathrm{a}^2 Z_{ca} + \mathrm{a}Z_{ab}) \end{aligned} \tag{2.48}$$

In these definitions we have taken advantage of the fact that mutual impedances of passive networks are reciprocal [12], i.e., $Z_{ab} = Z_{ba}$, etc. Note that we have followed the form of equation (2.21) in grouping terms for definitions (2.47) and (2.48). We find a curious symmetry in these definitions. Since equation (2.44) relates symmetrical components of phase a, similarly (2.47) and (2.48) pivot on phase a quantities as a point of symmetry. Note that these results are identical for either h from (2.38). (Why?)

Substituting the matrix (2.46) into equation (2.44), we see immediately a problem which we should like to avoid. Examine, for example, the equation for the positive sequence component of the voltage drop $\mathbf{V}_{mn\text{-}012}$:

$$V_{mn\text{-}1} = (Z_{S1} - Z_{M1}) I_{a0} + (Z_{S0} - Z_{M0}) I_{a1} + (Z_{S2} + 2Z_{M2}) I_{a2} \tag{2.49}$$

Obviously, the positive sequence voltage drop depends upon not only I_{a1} but I_{a0} and I_{a2} as well. This means there is mutual coupling *between sequences*—a rather disturbing result. Furthermore, $\mathbf{Z}_{mn\text{-}012}$ is *not* symmetric, therefore this mutual-effect is not reciprocal.

There are several special cases in which the matrix $\mathbf{Z}_{mn\text{-}012}$ is simplified. Specifically, there are three special cases of both self and mutual impedances which cause significant changes in this matrix.

Special case 1—zero impedance. The absence of any impedance in either the self or mutual case is obviously a simplifying assumption. In many cases the mutual impedances are neglected since they are often small compared to self impedances. Note however that the matrix $\mathbf{Z}_{mn\text{-}012}$ is nonsymmetric with respect to *both* Z_S and Z_M terms and is not made symmetric by eliminating either the self or mutual terms. Since elimination of self impedance terms must generally be rejected, this is not a satisfactory simplification to make.

Special case 2—equal impedance. In many problems the self or mutual impedances may be equal in all three phases. In such cases equations (2.47) and (2.48) become

$$\begin{aligned} Z_{S0} &= Z_{aa} \\ Z_{S1} &= Z_{S2} = 0 \end{aligned} \tag{2.50}$$

and

$$\begin{aligned} Z_{M0} &= Z_{bc} \\ Z_{M1} &= Z_{M2} = 0 \end{aligned} \tag{2.51}$$

Referring to (2.46), we note that the simultaneous application of (2.50) and (2.51) eliminates the off-diagonal terms of $\mathbf{Z}_{mn\text{-}012}$. This means that $\mathbf{Z}_{mn\text{-}012}$ is not only reciprocal but that there is zero coupling between sequences.

Note that the application of *either* (2.50) *or* (2.51) alone will still result in a nonreciprocal $\mathbf{Z}_{mn\text{-}012}$.

Special case 3—symmetric impedance. A less restrictive case than those above is one in which the self or mutual impedances are symmetric with respect to phase a, i.e.,

$$\begin{aligned} Z_{bb} &= Z_{cc} \\ Z_{ab} &= Z_{ca} \end{aligned} \tag{2.52}$$

With these restrictions the self impedances become

$$\begin{aligned} Z_{S0} &= (1/3)(Z_{aa} + 2Z_{bb}) \\ Z_{S1} &= Z_{S2} = (1/3)(Z_{aa} - Z_{bb}) \end{aligned} \tag{2.53}$$

and the mutual impedances are

$$Z_{M0} = (1/3)(Z_{bc} + 2Z_{ab})$$
$$Z_{M1} = Z_{M2} = (1/3)(Z_{bc} - Z_{ab}) \tag{2.54}$$

This restriction, applied with respect to self or mutual impedances, makes the $\mathbf{Z}_{mn\text{-}012}$ matrix symmetric with respect to that kind of impedance. However, the application of one without the other still leaves the matrix nonsymmetric.

Summary. Since we have four possible patterns for two kinds of impedances there are 16 different combinations of interest. These are shown in Table 2.1 where we use the notation:

zero = all matrix elements are zero
diag = the matrix is diagonal
sym = the matrix is symmetric
non = the matrix is nonsymmetric

Table 2.1. Summary of Matrix Conditions Based on Impedances

Mutual Impedances	*Self Impedances*			
	Zero	*Equal*	*Symmetric*	*Arbitrary*
Zero	zero	diag	sym	non
Equal	diag	diag	sym	non
Symmetric	sym	sym	sym	non
Arbitrary	non	non	non	non

If the impedance matrix is *zero*, the voltage drops vanish and the problem is a degenerate one. A *diagonal matrix* means that the sequences are uncoupled, i.e., currents from one sequence produce voltage drops only in that sequence—a very desirable characteristic. A *symmetric matrix* means that there is mutual coupling between sequences but that it is reciprocal, e.g., the coupling from positive to negative sequences is exactly the same as the coupling from negative to positive. This situation can be simulated by a passive network. A *nonsymmetric matrix* means that the mutual coupling is not the same between two sequences. This situation generally requires controlled sources for laboratory simulation, but its mathematical representation is no more difficult than the symmetric case. It does require the computation of all matrix elements, whereas the symmetric case requires the computation of only the upper (or lower) triangular matrix because of symmetry. In most of our computations we will consider the self impedances to be *equal*. Thus, except for the case of arbitrary (nonsymmetric) mutuals, the problem is one of a diagonal or symmetric matrix representation. If the self impedances are only symmetric or are arbitrary, we often conclude that the method of symmetrical components is too complicated to be of value. This was particularly true in the days of network analyzer simulation. With digital computer solutions the representation of nonreciprocal networks should be a much less formidable problem, and we may soon be dealing with arbitrary impedances as a commonplace occurrence.

2.6 Sequence Components of Machine Impedances

In the case of synchronous or induction machines we have a very special case of section 2.5. Lewis and Pryce [13] have asserted that in this case the impedance matrix corresponding to the **Z** matrix of (2.42) is the "circulant matrix"

$$\mathbf{Z} = \begin{bmatrix} Z_k & Z_m & Z_n \\ Z_n & Z_k & Z_m \\ Z_m & Z_n & Z_k \end{bmatrix} \tag{2.55}$$

This matrix form is the result of the self impedances being equal and the mutual impedances being circular, i.e., Z_m is clockwise (*a-b-c*) and Z_n is counterclockwise (*a-c-b*).

It is easy to show that a similarity transform $\mathbf{Z}_{012} = \mathbf{A}^{-1}\mathbf{Z}\mathbf{A}$ diagonalizes **Z**, i.e.,

$$\mathbf{Z}_{012} = \mathbf{A}^{-1}\mathbf{Z}\mathbf{A} = \begin{bmatrix} Z_0 & 0 & 0 \\ 0 & Z_1 & 0 \\ 0 & 0 & Z_2 \end{bmatrix} \tag{2.56}$$

This means that transforming from *a-b-c* coordinates to 0-1-2 coordinates completely decouples the sequences, a fact that makes symmetrical components extremely useful in dealing with machines. Performing the operation indicated by (2.56), we easily show that

$$\begin{aligned} Z_0 &= Z_k + Z_m + Z_n \\ Z_1 &= Z_k + a^2 Z_m + a Z_n \\ Z_2 &= Z_k + a Z_m + a^2 Z_n \end{aligned} \tag{2.57}$$

It is also shown in [13] that Z_0, Z_1, and Z_2 are the eigenvalues of **Z**. Thus we note that the eigenvalues are distinct and that the columns of **A** are the eigenvectors of **Z**.

2.7 Definition of Sequence Networks

In many problems the unbalanced portion of the physical system can be isolated for study, with the rest of the system considered as balanced. This is the case for an unbalanced load or fault supplied by a system of balanced or equal-phase impedances. In such problems we attempt to find the symmetrical components of voltage and current at the point of unbalance and synthesize these sequence quantities to determine system (*a-b-c*) quantities. To find the sequence quantities is the major objective in many problems, and it is helpful to introduce the concept of sequence networks to do this.

Before defining the sequence network, we define the fault point of a network. The *fault point* of a system is that point to which the unbalanced connection is attached in an otherwise balanced system. Thus, a one-line-to-ground fault at bus K defines bus K as the fault point. Similarly, an unbalanced three-phase load at bus M defines bus M as the fault point. We take a liberal definition of what constitutes a "fault," it being any connection or situation which causes an unbalance among the three phases.

A *sequence network* is a copy of the original balanced system to which the fault point is connected and contains the same per phase impedances as the physical balanced system, arranged in the same way, the only difference being that the value of each impedance is a value unique to each sequence. This really is not as complicated as it sounds. Since the positive and negative sequence currents are both balanced three-phase current sets, they see the same impedance in a passive three-phase network. The zero sequence currents generally see a different impedance from positive and negative sequence currents and may sometimes see an infinite impedance. Machine impedances are usually different for all three sequences. The topic of sequence impedances will be greatly expanded upon later.

Each sequence network, once the fault point is determined, can be analyzed by Thevenin's theorem by considering it to be a two-terminal (one-port) network where one terminal is the fault point, designated F, and the other terminal is the "zero-potential bus," designated N. This is done in the usual way, looking back from terminals F and N to find the open circuit voltage and the short circuit current or equivalent impedance (for that sequence). The impedance seen is usually designated Z_0, Z_1, and Z_2 for the zero, positive, and negative sequence networks and represents the Thevenin equivalent impedance *to* the flow of I_{a0}, I_{a1}, and I_{a2} respectively.[2] The Thevenin equivalent voltage in the positive sequence network is the open circuit voltage at the fault point and is a phasor quantity. Thevenin equivalent or open circuit voltages in the negative and zero sequence networks are zero because *by definition the only voltages generated in the three-phase system are positive sequence* (sequence *a-b-c*) *voltages.*

Sequence networks are usually designated schematically by boxes in which the fault point F, the zero-potential bus N, and the Thevenin voltage and impedance are shown. Such a set of boxes is shown in Figure 2.5 for the zero, positive, and negative sequences.

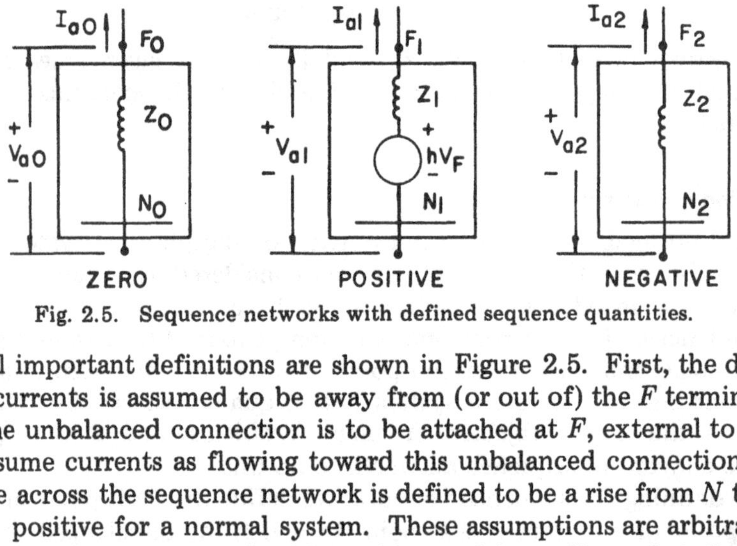

Fig. 2.5. Sequence networks with defined sequence quantities.

Several important definitions are shown in Figure 2.5. First, the direction of sequence currents is assumed to be away from (or out of) the F terminal. This is because the unbalanced connection is to be attached at F, external to the boxes, and we assume currents as flowing toward this unbalanced connection. Second, the voltage across the sequence network is defined to be a rise from N to F as this makes V_{a1} positive for a normal system. These assumptions are arbitrary but are necessary in order to proceed with problem solving. Note that the Thevenin equivalent voltage is shown as V_F which is the prefault (F open) voltage of phase a at F. (This voltage is sometimes designated E_{a1} or simply as E_g.) From Figure 2.5 we write the important equations for the voltage drop from F to N as

$$\begin{bmatrix} V_{a0} \\ V_{a1} \\ V_{a2} \end{bmatrix} = \begin{bmatrix} 0 \\ \mathrm{h}V_F \\ 0 \end{bmatrix} - \begin{bmatrix} Z_0 & 0 & 0 \\ 0 & Z_1 & 0 \\ 0 & 0 & Z_2 \end{bmatrix} \begin{bmatrix} I_{a0} \\ I_{a1} \\ I_{a2} \end{bmatrix} \tag{2.58}$$

[2]Note that these are *not* the same impedances as Z_{S0}, Z_{S1}, and Z_{S2} or Z_{M0}, Z_{M1}, and Z_{M2} of section 2.5 and are not the machine impedances alone of section 2.6.

or

$$\mathbf{V}_{012} = \mathbf{E} - \mathbf{Z}_{012}\mathbf{I}_{012} \tag{2.59}$$

Example 2.1

Suppose that a certain unsymmetrical condition gives the following data at the fault point.

$$V_F = 1.0\underline{/0^\circ}, \qquad I_{a1}Z_1 = (0.2\underline{/0^\circ})\text{h}$$
$$I_{a2}Z_2 = (0.2\underline{/0^\circ})\text{h}, \qquad I_{a0}Z_0 = (0.6\underline{/0^\circ})\text{h}$$

Find the phase and line-to-line voltages at F.

Solution

From equation (2.58)

$$\mathbf{V}_{012} = \begin{bmatrix} V_{a0} \\ V_{a1} \\ V_{a2} \end{bmatrix} = \text{h}\begin{bmatrix} 0 \\ 1.0\underline{/0^\circ} \\ 0 \end{bmatrix} - \text{h}\begin{bmatrix} 0.6\underline{/0^\circ} \\ 0.2\underline{/0^\circ} \\ 0.2\underline{/0^\circ} \end{bmatrix} = \text{h}\begin{bmatrix} -0.6\underline{/0^\circ} \\ 0.8\underline{/0^\circ} \\ -0.2\underline{/0^\circ} \end{bmatrix}$$

Then, from (2.24) $\mathbf{V}_{abc} = \mathbf{A}\,\mathbf{V}_{012}$,

$$\begin{bmatrix} V_a \\ V_b \\ V_c \end{bmatrix} = \frac{1}{\text{h}}\begin{bmatrix} 1 & 1 & 1 \\ 1 & \text{a}^2 & \text{a} \\ 1 & \text{a} & \text{a}^2 \end{bmatrix} \text{h} \begin{bmatrix} -0.6\underline{/0^\circ} \\ 0.8\underline{/0^\circ} \\ -0.2/0^\circ \end{bmatrix} = \begin{bmatrix} 0 \\ 1.25\underline{/-136.2^\circ} \\ 1.25\underline{/136.2^\circ} \end{bmatrix} \tag{2.60}$$

We see that $V_a = 0$, which indicates that phase a is shorted to ground at the fault point F. Note that both V_b and V_c are 125% of normal. This condition results in line-to-ground faults whenever Z_0 is greater than Z_1.

To compute the line-to-line voltages, use the results of (2.60) and write

$$V_{ab} = V_a - V_b = 0 - 1.25\underline{/-136.2^\circ} = 1.25\underline{/43.8^\circ} \text{ pu}$$
$$V_{bc} = V_b - V_c = (-0.9 - \text{j}0.866) - (-0.9 + \text{j}0.866) = -\text{j}1.732 = 1.732\underline{/-90^\circ} \text{ pu}$$
$$V_{ca} = V_c - V_a = V_c = 1.25\underline{/136.2^\circ} \text{ pu}$$

where these values are all in pu on a phase voltage base. On a line-to-line voltage base, which may be more meaningful, we divide by $\sqrt{3}$ to get

$$V_{ab} = 0.722\underline{/43.8^\circ} \text{ pu}$$
$$V_{bc} = 1.0\underline{/-90^\circ} \text{ pu}$$
$$V_{ca} = 0.722\underline{/136.2^\circ} \text{ pu} \tag{2.61}$$

Problems

2.1. Sketch a set of symmetrical component phasors for a four-phase system, i.e., $n = 4$. Indicate clearly the angle between components V_{ax}, V_{bx}, V_{cx}, and V_{dx} in each set. Compute the a-operator angle and show that equations (2.4) and (2.5) are satisfied.

2.2. Repeat problem 1 but with $n = 5$.

2.3. Show that C of equation (2.10) is nonsingular by performing elementary operations [7] on its rows and columns.

2.4. Evaluate a, C, det C, and $\mathbf{C}^{-1} = \mathbf{A}$ for $n = 2, 3$.

2.5. Write out matrix C for $n = 2, 3, 4, 5, 6, \ldots$ and reduce each element to its simplest form;

for example, when $n = 5$, set $a^6 = a$, $a^8 = a^3$, $a^9 = a^4$, etc. Can you see a pattern emerging as to the construction of C insofar as a-exponents are concerned?

2.6. Show that the A matrix defined by equation (2.20) has rank n.

2.7. Apply the analysis equation (2.21) to the unbalanced phasors of Figure 2.3 to obtain the symmetrical component phasor quantities.

2.8. Given that $V_a = 100 + j0$, $V_b = j80$, and $V_c = -60$, determine the symmetrical components V_{a1}, V_{a2}, and V_{a0}.

2.9. Given that $V_{a0} = 100 + j0$, $V_{a1} = 200 - j100$, and $V_{a2} = -100 + j0$, find the phase voltages V_a, V_b, and V_c. Let h = 1.

2.10. If the load is unbalanced, neutral current will exist. Find the relation between neutral current I_n and phase a zero sequence current I_{a0}.

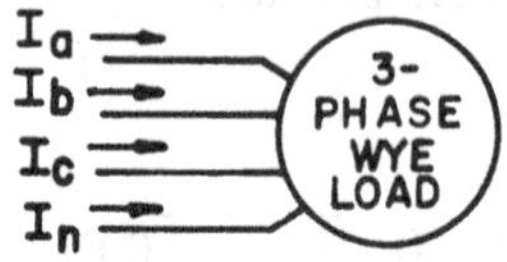

Fig. P2.10.

2.11. If the load is unbalanced, its neutral n' will not be at the same potential as the source neutral n. Find the relation between the neutral shift $V_{n'n}$ and the zero, i.e., $V_{an'0}$.

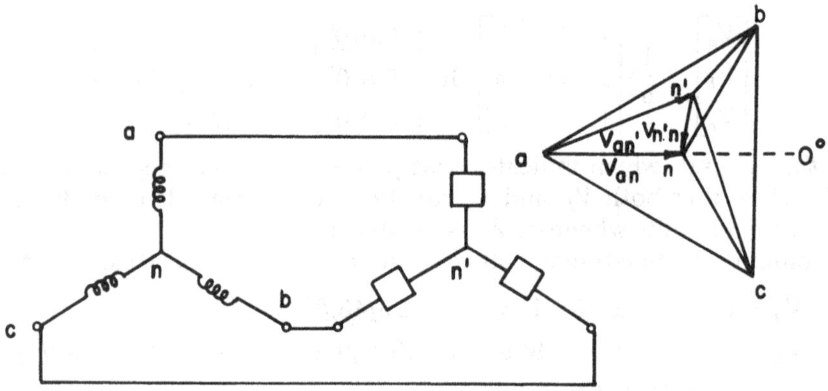

Fig. P2.11.

2.12. Each current transformer has 400/5 A ratio. The currents in the ammeters are as shown on the phasor diagram. Find all five remaining currents. Relate I_4, I_n, and I_{a0}.

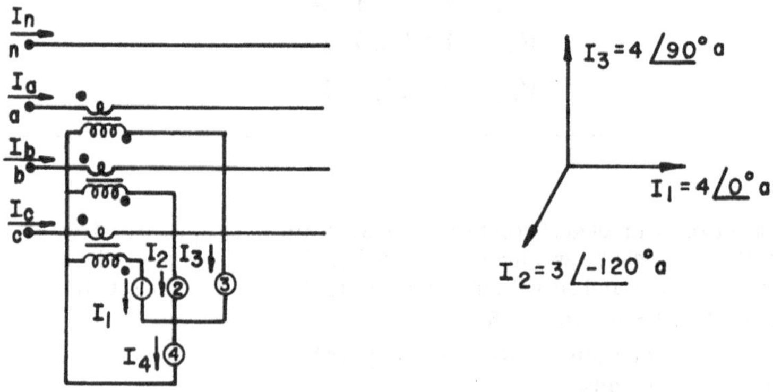

Fig. P2.12.

2.13. This isolated neutral system has a ground fault on phase a. When ungrounded, all line-to-neutral (also line-to-ground) voltages are balanced. If helpful, the balanced line capacitances to ground may be used to aid in visualizing the line-to-neutral voltages of this delta circuit.

Find the symmetrical components of the phase a line-to-neutral voltages when the ground fault is applied, i.e., V_{an0}, V_{an1}, V_{an2}.

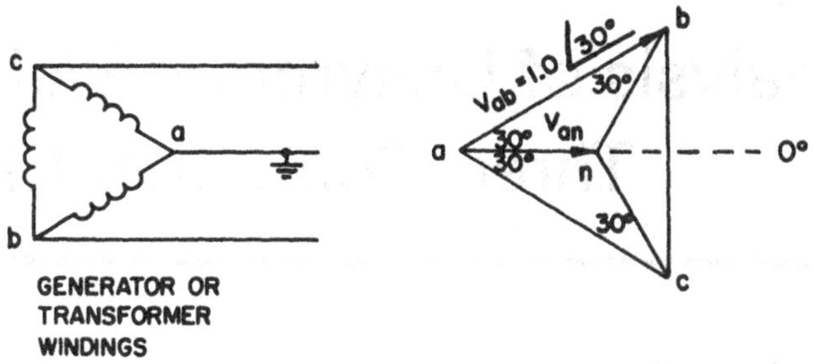

Fig. P2.13.

2.14. Prove that power is invariant under the transformation to the 0-1-2 frame of reference if the value of "h" is taken to be $\sqrt{3}$.

2.15. Consider an impedance matrix which is symmetric with all off-diagonal terms equal. Find a transformation which will diagonalize this impedance matrix and show that such a transformation is real. *Hint*: Find the eigenvalues and eigenvectors of the Z matrix and examine the unitary matrix of eigenvectors. (See [4, 5, 6].)

chapter 3

Analysis of Unsymmetrical Faults: Three-Component Method

Having introduced the symmetrical component notation and defined the sequence networks, we are ready to evaluate the way in which unsymmetrical conditions may be represented. In doing so, we will proceed in a very orderly manner, evaluating the conditions at the fault point and then deriving the exact circuit representation in the 0-1-2 coordinate system. We call this type of analysis the *three-component method* to distinguish it from other methods to be introduced later.

In what follows we will refer to the unbalanced condition at the fault point as a "fault." It should be understood that this term is intended to mean any unbalanced situation and may be an unbalanced load or other unsymmetrical condition.

It is also convenient to distinguish between shunt and series unbalances or faults. A *shunt fault* is an unbalance between phases or between phase and neutral. A *series fault* is an unbalance in the line impedances and does not involve the neutral or ground, nor does it involve any interconnection between phases. We will consider these items separately.

Our objective here is to determine exactly how the sequence networks are related or how they are interconnected for various kinds of fault situations. Since we must do this for several different situations we establish the following procedure.

1. Sketch a *circuit diagram* of the fault point showing all phase connections to the fault. Label all currents, voltages, and impedances, carefully noting assumed positive directions and polarities. Such a sketch is shown in Figure 3.1. It is assumed that a "normal" system consisting of only balanced impedances is connected to the left and right of the fault point and that the Thevenin equivalent looking in at this point is known. Note that phase voltages are defined as drops from line to ground at this point and that currents are defined as flowing from the system toward the fault.
2. Write the *boundary conditions* relating known currents and voltages for the type of fault under consideration.
3. *Transform* the currents and/or voltages of 2 from the *a-b-c* to the 0-1-2 coordinate system by use of the transformation $\mathbf{A}$ or $\mathbf{A}^{-1}$.
4. Examine the *sequence currents* to determine the proper connection of the F or N terminals of the sequence networks, satisfying 3.

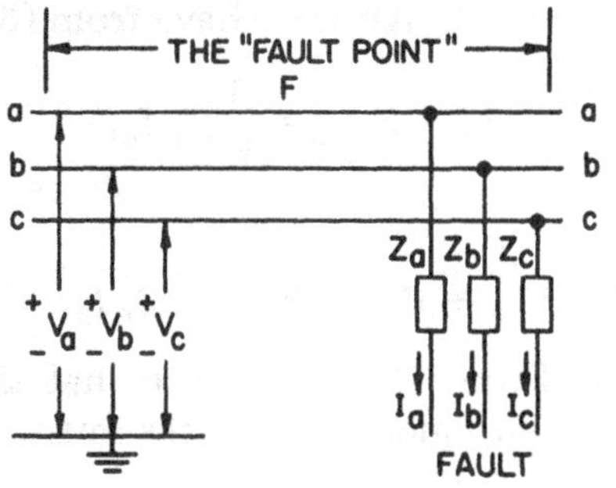

Fig. 3.1. Circuit diagram of the fault point.

5. Examine the *sequence voltages* to determine the connection of the remaining terminals of the sequence networks, adding impedances as required to satisfy 3 and 4.

These five steps will be followed rigorously for each type of fault except where noted to the contrary.

I. SHUNT FAULTS

Shunt faults are an important class of faults and include various kinds of "short circuits" as well as unbalanced loads.

3.1 The Single Line-to-Ground (SLG) Fault

1. *Circuit diagram*: See Figure 3.2.
2. *Boundary conditions*: By inspection of Figure 3.2,

$$I_b = I_c = 0 \tag{3.1}$$

and

$$V_a = Z_f I_a \tag{3.2}$$

3. *Transformation*: From equation (2.25) we write $\mathbf{I}_{012} = \mathbf{A}^{-1}\,\mathbf{I}_{abc}$, or from (3.1), with A defined from (2.36)

$$\mathbf{I}_{012} = \frac{\mathrm{h}}{3}\begin{bmatrix} 1 & 1 & 1 \\ 1 & \mathrm{a} & \mathrm{a}^2 \\ 1 & \mathrm{a}^2 & \mathrm{a} \end{bmatrix}\begin{bmatrix} I_a \\ 0 \\ 0 \end{bmatrix} = \frac{\mathrm{h}}{3} I_a \begin{bmatrix} 1 \\ 1 \\ 1 \end{bmatrix} \tag{3.3}$$

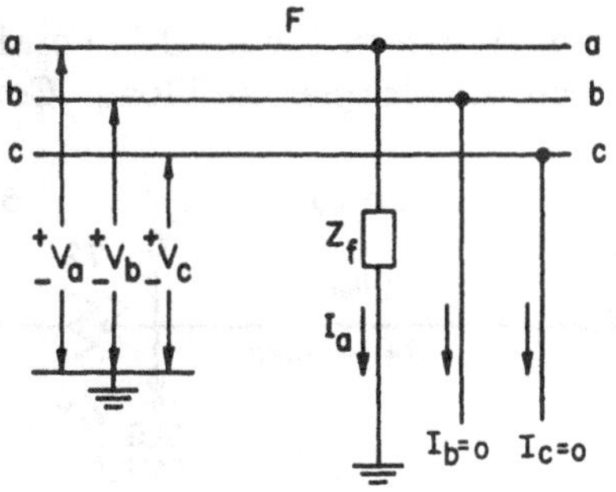

Fig. 3.2. Diagram of a SLG fault at *F*.

or all sequence currents are equal. Also, we have from (3.2) and (3.3),

$$V_a = Z_f I_a = \frac{3}{h} Z_f I_{a1}$$

which we may write as

$$V_{a0} + V_{a1} + V_{a2} = 3Z_f I_{a1} \tag{3.4}$$

4. *Sequence currents*: From (3.3) we note that the sequence currents are equal. This implies that the sequence networks must be connected in series, as shown in Figure 3.3.

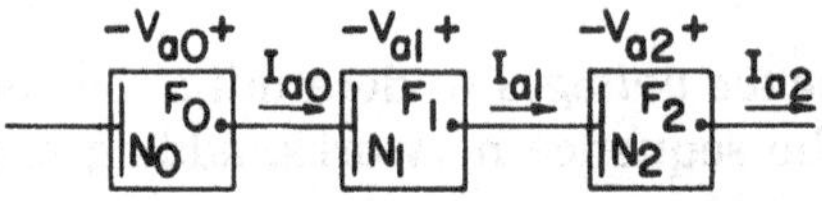

Fig. 3.3. Sequence network partial connection specified by the current equation.

5. *Sequence voltages*: From (3.4) we see that the sequence voltages add to $3Z_f I_{a1}$. This requires the addition of an external impedance as noted from Figure

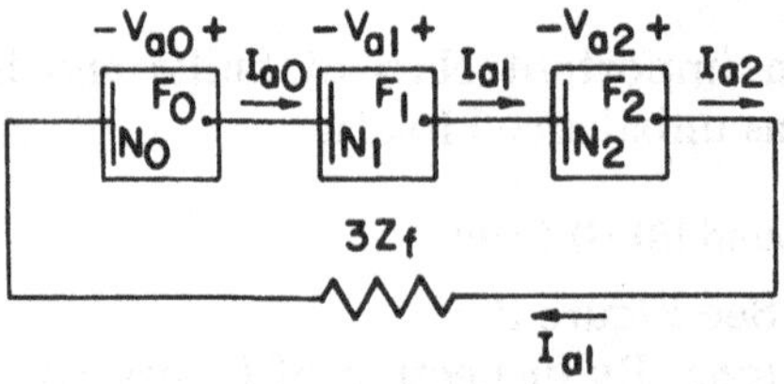

Fig. 3.4. Sequence network connection for a SLG fault.

3.3. The final connection is shown in Figure 3.4. With this connection we compute

$$I_{a0} = I_{a1} = I_{a2} = \frac{hV_F}{Z_0 + Z_1 + Z_2 + 3Z_f} \tag{3.5}$$

and knowing the sequence currents we easily find the sequence voltages from equation (2.58).

Example 3.1

The simple power system shown in Figure 3.5, consists of a generator, transformer, transmission line, load transformer, and load. Consider a SLG fault at bus

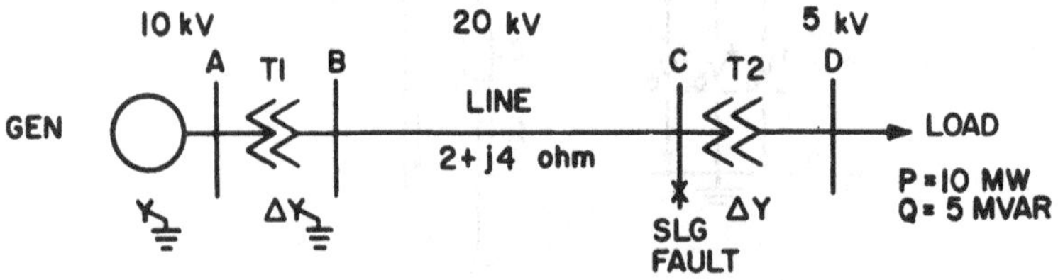

Fig. 3.5. Power system for Example 3.1.

C with a fault resistance of 4 ohms. The following data concerning the system is known.

Generator: 25 MVA, 10 kV, x = 0.125 pu, connected Y-grounded
T1: 30 MVA, 10–20 kV, x = 0.105, connected Δ-Y-grounded
Line: $Z = 2 + j4$ Ω
T2: 20 MVA, 5–20 kV, x = 0.05 pu, connected Y-Δ
Load: static (constant z) load of 10 + j5 MVA at 5 kV

Solution

Select S_B = 20 MVA, a load voltage of 5 kV, and compute all system impedances. Let h = 1.

Generator: $x = (0.125)(20/25) = 0.10$ pu
T1: $x = (0.105)(20/30) = 0.07$ pu
Line: $z = [(2 + j4)(20)]/(20)^2 = 0.1 + j0.2$ pu
T2: $x = 0.05$ pu

Load (as series impedance):

$$R = \frac{(V_u)^2 (S_B) P}{P^2 + Q^2}$$

$$= \frac{(1.0)^2 (20 \times 10^6)(10 \times 10^6)}{(10 \times 10^6)^2 + (5 \times 10^6)^2} = \frac{200}{125} = 1.6 \text{ pu}$$

Similarly, X = 100/125 = 0.8 pu. Then the positive sequence network for a SLG fault at bus C is represented as shown in Figure 3.6.

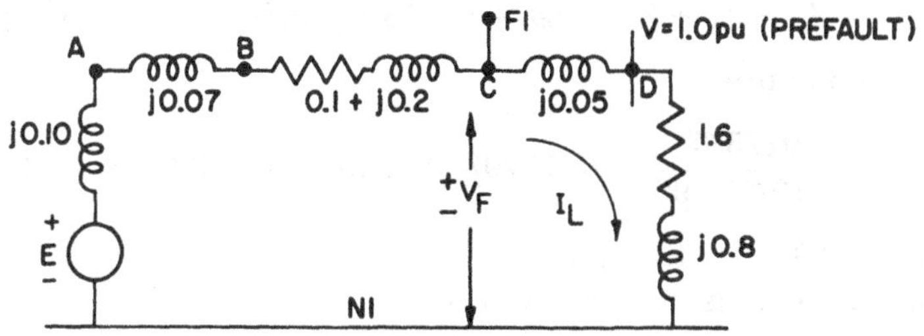

Fig. 3.6. One-line diagram of the positive sequence network.

The load current I_L is (with V as the reference phasor)

$$I_L = \frac{P - jQ}{V^*} = \frac{10 - j5}{20} = 0.5 - j0.25 \text{ pu}$$

The Thevenin voltage at bus C is

$$V_F = 1.0 + j0 + (0.5 - j0.25)(j0.05)$$
$$= 1.0125 + j0.025 = 1.0125\underline{/1.27^\circ} \text{ pu}$$

We set $V_F = 1.0125\underline{/0^\circ}$ and it becomes the reference phasor in the fault calculations. The impedance seen looking in at $F1$ with E shorted is 0.1 + j0.37 on the left in parallel with 1.6 + j0.85 on the right, or

$$Z_1 = \frac{(0.1+j0.37)(1.6+j0.85)}{1.7+j1.22} = 0.3319\underline{/67.1905} = 0.1287+j0.3059 \quad \text{pu}$$

This is the impedance to the flow of both positive and negative sequence currents.

The zero sequence current, for reasons to be investigated later, sees an open circuit to the left of bus A, with A grounded (because of the Y-grounded connection) and an open circuit to the right of bus C. Thus Z_0 is the sum of the line and $T1$ impedances, or $Z_0 = 0.1 + j0.27$ pu.

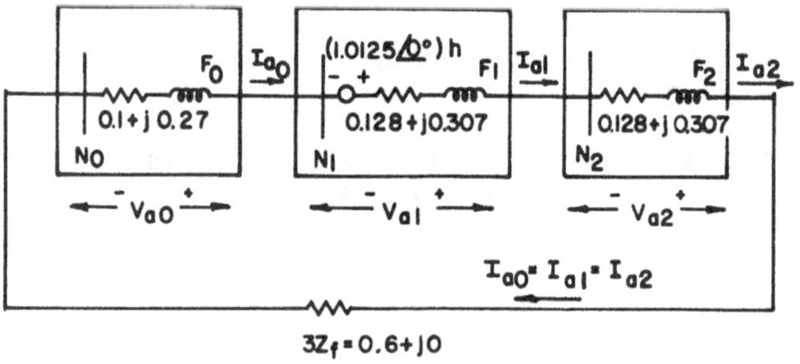

Fig. 3.7. Sequence networks for SLG fault at bus C.

The complete connection of sequence networks is shown in Figure 3.7, where Z_f is shown to be

$$Z_f = \frac{4+j0}{Z_B} = \frac{4}{20} = 0.2 \quad \text{pu}$$

The total circuit impedance is

$$Z_t = Z_0 + Z_1 + Z_2 + 3Z_f = 0.956 + j0.884 = 1.3\underline{/42.8^\circ} \quad \text{pu}$$

From (3.5) we compute

$$I_{a1} = \frac{1.0125\underline{/0^\circ}}{0.3319\underline{/67.1905}} = 0.7779\underline{/-42.6496} = 0.5722 - j0.5271 \quad \text{pu}$$

Then $I_a = 3I_{a1} = 2.3338\underline{/-42.6496}$ pu

Since the fault occurs on a 20 kV bus,

$$I_B = S_B/V_B = \frac{20 \times 10^6}{\sqrt{3}(20 \times 10^3)} = 577.3503 \quad \text{A}$$

and $I_a = 1347.4$ A. We may also synthesize the phase voltages by first computing the sequence voltages.

$$\begin{aligned} V_{a0} &= -Z_0 I_{a0} = -(0.2879\underline{/69.677})(0.7779\underline{/-42.6496}) \\ &= 0.2240\underline{/-152.9727} = -0.1995 - j0.1018 \\ V_{a1} &= V_F - Z_1 I_{a1} = (1.0125+j0) - (0.3319\underline{/67.1905})(0.7779\underline{/-42.6496}) \\ &= (1.0125+j0) - (0.236+j0.1075) = 0.7765 - j0.1075 \\ V_{a2} &= -Z_2 I_{a2} = -(0.3319\underline{/67.1905})(0.7779\underline{/-42.6496}) \\ &= -0.2348 - j0.1072 = 0.2582\underline{/-155.4591} \end{aligned}$$

Thus

$$V_a = V_{a0} + V_{a1} + V_{a2} = 0.3433 - j0.3162 = 0.4668\underline{/-42.6496}$$

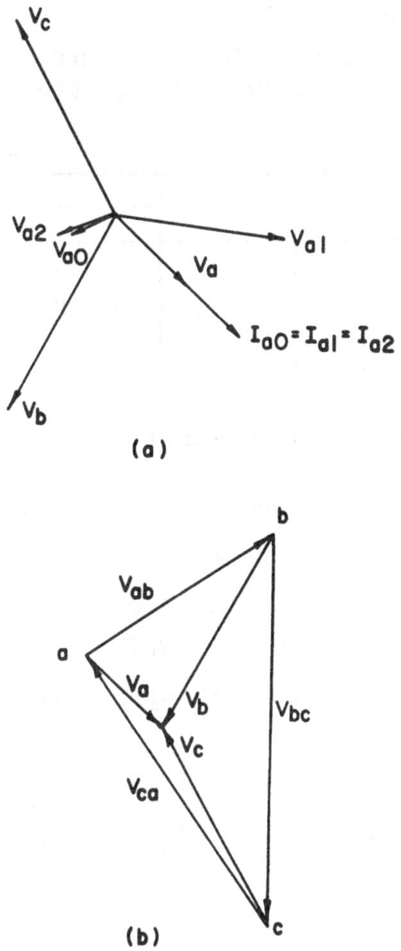

Fig. 3.8. Currents and voltages for the SLG fault: (a) sequence quantities (b) postfault LL voltages at the fault.

which should check with

$$V_a = Z_f I_{a1} = (0.6)(0.7779\underline{/-42.6496}) = 0.4667\underline{/-42.6469}$$

This is a good check considering the algebra involved. From (2.23)

$$\begin{aligned} V_b &= V_{a0} + \alpha^2 V_{a1} + a V_{a2} \\ &= -0.1995 - j0.1018 - 0.4817 - j0.6199 + 0.2102 - j0.1497 \\ &= -0.4709 - j0.8714 = 0.9905\underline{/-118.3879} \\ V_c &= V_{a0} + a V_{a1} + a^2 V_{a2} \\ &= -0.1995 - j0.1018 - 0.2960 + j0.7271 + 0.0246 + j0.2569 \\ &= -0.4709 + j0.8823 = 1.0001\underline{/+118.0913} \end{aligned}$$

These voltages are shown in Figure 3.8 together with the line-to-line voltages at bus C, where

$$V_{ab} = V_a - V_b, \qquad V_{bc} = V_b - V_c, \qquad V_{ca} = V_c - V_a$$

3.2 The Line-to-Line (LL) Fault

1. *Circuit diagram:* See Figure 3.9 and note that the LL fault is placed between lines *b* and *c* to retain symmetry with respect to phase *a*.

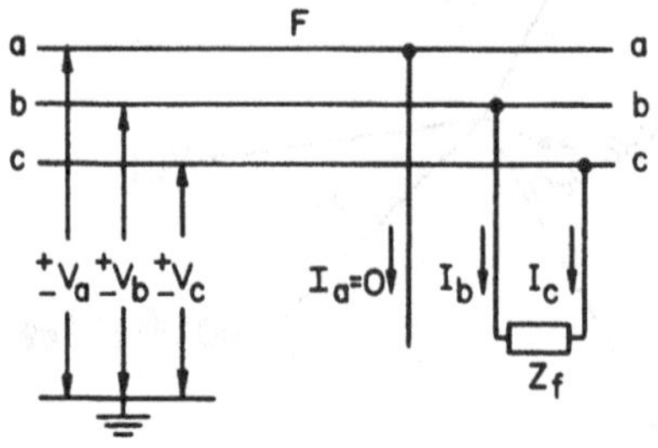

Fig. 3.9. Diagram of a LL fault at *F*.

2. *Boundary conditions:* By inspection of Figure 3.9,

$$I_a = 0 \tag{3.6}$$

$$I_b = -I_c \tag{3.7}$$

$$V_b - V_c = I_b Z_f \tag{3.8}$$

3. *Transformation:* From the analysis equation $\mathbf{I}_{012} = \mathbf{A}^{-1}\mathbf{I}_{abc}$ and incorporating (3.6) and (3.7),

$$\mathbf{I}_{012} = \frac{h}{3}\begin{bmatrix} 1 & 1 & 1 \\ 1 & a & a^2 \\ 1 & a^2 & a \end{bmatrix}\begin{bmatrix} 0 \\ I_b \\ -I_b \end{bmatrix} = \frac{jhI_b}{\sqrt{3}}\begin{bmatrix} 0 \\ 1 \\ -1 \end{bmatrix} \tag{3.9}$$

Also from (3.8) we write

$$Z_f I_b = V_b - V_c = \frac{1}{h}(V_{a0} + a^2 V_{a1} + aV_{a2}) - \frac{1}{h}(V_{a0} + aV_{a1} + a^2 V_{a2})$$

$$\frac{1}{h} Z_f (I_{a0} + a^2 I_{a1} + aI_{a2}) = \frac{(a^2 - a)}{h} V_{a1} + \frac{(a - a^2)}{h} V_{a2}$$

From (3.9) $I_{a0} = 0$ and $I_{a1} = -I_{a2}$. Substituting this information into the above, we have for any h,

$$Z_f(a^2 - a) I_{a1} = (a^2 - a) V_{a1} - (a^2 - a) V_{a2}$$

or

$$Z_f I_{a1} = V_{a1} - V_{a2} \tag{3.10}$$

4. *Sequence currents:* From (3.9), $I_{a0} = 0$, so the zero sequence network is open. Also from (3.9) $I_{a1} = -I_{a2}$, which requires the connection shown in Figure 3.10.

5. *Sequence voltages:* From equation (3.10) and Figure 3.10 we see that the remaining connection must be as shown in Figure 3.11. Then

$$I_{a1} = \frac{hV_F}{Z_1 + Z_2 + Z_f} \tag{3.11}$$

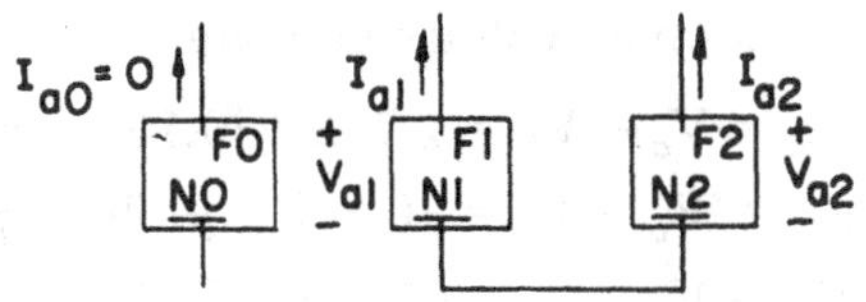

Fig. 3.10. Sequence network partial connection specified by the current equation.

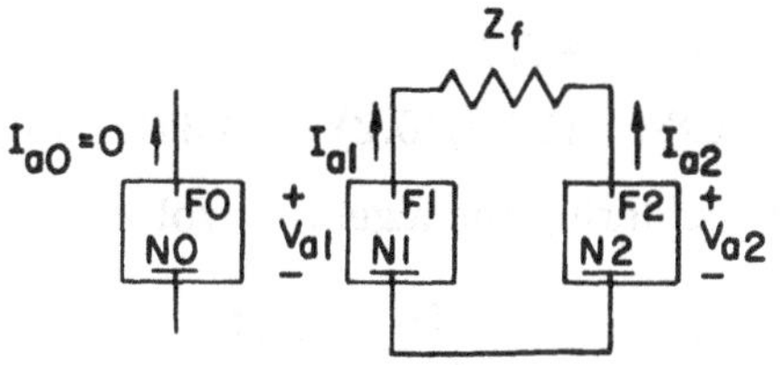

Fig. 3.11. Sequence network connection for a LL fault.

Example 3.2

Compute the phase voltages and currents for a LL fault at bus *C* of Figure 3.5, where a fault impedance of 4 ohms is assumed between phases *b* and *c*. Let h = 1.

Solution

The sequence networks are exactly as shown in Figure 3.7, but their interconnection is that of Figure 3.11. With the new connection the total impedance is

$$Z_t = Z_1 + Z_2 + Z_f = 0.456 + j0.614 = 0.765\underline{/53.4^\circ} \quad \text{pu}$$

Then

$$I_{a1} = -I_{a2} = hV_F/Z_t = \frac{1.0125\underline{/0^\circ}}{0.765\underline{/53.4^\circ}}$$

$$= 1.325\underline{/-53.4^\circ} = 0.788 - j1.065 \quad \text{pu}$$

From equation (3.9) with h = 1

$$I_b = -I_c = -j\sqrt{3}I_{a1} = -1.86 - j1.38 = -2.32\underline{/36.6^\circ} \quad \text{pu} = 1320 \quad \text{A}$$

This system voltages may also be synthesized from a knowledge of the sequence currents and sequence network connections. Thus with h = 1,

$$V_{a1} = hV_F - Z_1 I_{a1} = 1.0125 - (0.332\underline{/67.3^\circ})(1.325\underline{/-53.4^\circ})$$

$$= 1.0125 - 0.427 - j0.1055 = 0.5855 - j0.1055 \quad \text{pu}$$

$$V_{a2} = -Z_2 I_{a2} = -(0.332\underline{/67.3^\circ})(-1.325\underline{/-53.4^\circ})$$

$$= 0.44\underline{/13.9^\circ} = 0.427 + j0.1055 \quad \text{pu}$$

and we compute

$$V_a = V_{a1} + V_{a2} = 1.0125 + j0$$

$$V_b = a^2 V_{a1} + aV_{a2} = -0.389 - j0.448 - 0.310 + j0.317$$

$$= -0.699 - j0.131 = -\ 0.705\underline{/10.6^\circ} \quad \text{pu}$$

$$V_c = aV_{a1} + a^2 V_{a2} = -0.196 + j0.554 - 0.118 - j0.423$$

$$= -0.314 + j0.131 = -0.342\underline{/-22.6^\circ} \quad \text{pu}$$

The line-to-line voltages on a phase voltage base are

$$V_{ab} = V_a - V_b = 1.712 + j0.131 \quad \text{pu}$$
$$V_{bc} = V_b - V_c = -0.385 - j0.262 \quad \text{pu}$$
$$V_{ca} = V_c - V_a = -1.326 + j0.131 \quad \text{pu}$$

We make a check on V_{bc} from equation (3.8)

$$V_{bc} = V_b - V_c = I_b Z_f$$
$$= (-1.86 - j1.38)(0.2) = -0.372 - j0.276 \quad \text{pu}$$

which is a good check, considering the algebra involved.

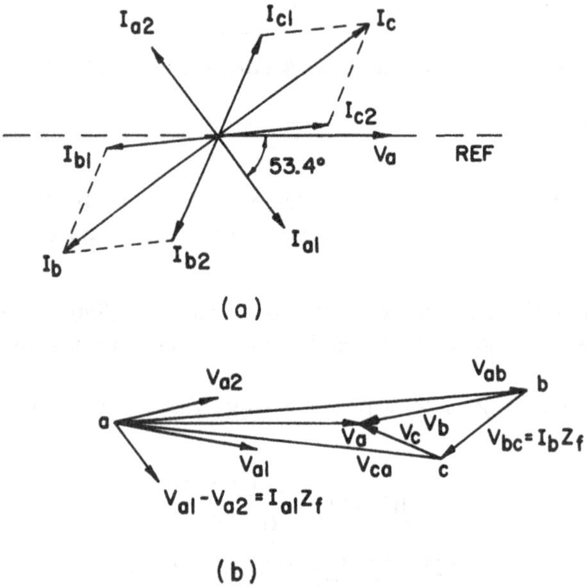

Fig. 3.12. Phasor diagrams for a LL fault: (a) currents (b) voltages.

Phasor diagrams of currents and voltages are shown in Figure 3.12. Note that if Z_f were zero the voltage "triangle" would collapse to zero on the *b-c* side and the $V_{a1} - V_{a2}$ phasor would go to zero.

3.3 The Double Line-to-Ground (2LG) Fault

1. *Circuit diagram:* The fault connection is shown in Figure 3.13. Note that symmetry with respect to phase *a* is obtained by faulting phases *b* and *c*.

2. *Boundary conditions:* By inspection of Figure 3.13 we have

$$I_a = 0 \tag{3.12}$$

$$V_b = (Z_f + Z_g) I_b + Z_g I_c \tag{3.13}$$

$$V_c = (Z_f + Z_g) I_c + Z_g I_b \tag{3.14}$$

3. *Transformation:* From (3.12) we write

$$I_a = 0 = \frac{1}{h}(I_{a0} + I_{a1} + I_{a2}) \tag{3.15}$$

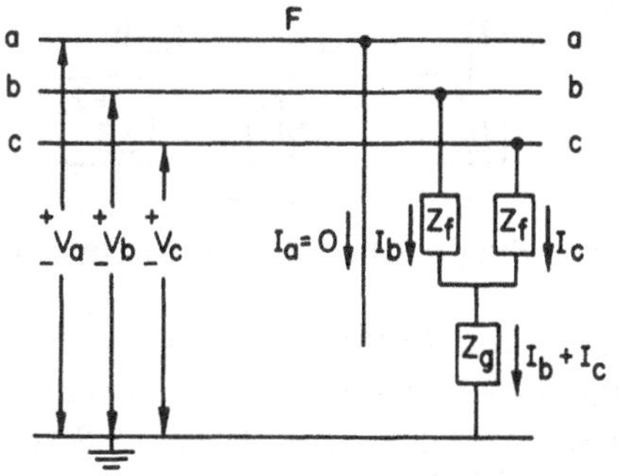

Fig. 3.13. Diagram of a 2LG fault at F.

From (2.24) we write

$$V_b = \frac{1}{h}(V_{a0} + a^2 V_{a1} + aV_{a2}) \tag{3.16}$$

$$V_c = \frac{1}{h}(V_{a0} + aV_{a1} + a^2 V_{a2}) \tag{3.17}$$

and find the difference

$$V_b - V_c = -\frac{j\sqrt{3}}{h}(V_{a1} - V_{a2}) \tag{3.18}$$

But from (3.13) and (3.14) we also find that

$$V_b - V_c = Z_f(I_b - I_c) \tag{3.19}$$

Substituting (3.18) and a similar relation for $I_b - I_c$ into (3.19), we have for any h, $V_{a1} - V_{a2} = Z_f(I_{a1} - I_{a2})$, or

$$V_{a1} - Z_f I_{a1} = V_{a2} - Z_f I_{a2} \tag{3.20}$$

Also, adding equations (3.16) and (3.17), we find the sum

$$V_b + V_c = \frac{1}{h}[2V_{a0} - (V_{a1} + V_{a2})] \tag{3.21}$$

which we equate to the sum of (3.13) and (3.14)

$$V_b + V_c = \frac{Z_f}{h}[2I_{a0} - (I_{a1} + I_{a2})] + \frac{Z_g}{h}[4I_{a0} - 2(I_{a1} + I_{a2})] \tag{3.22}$$

Since (3.21) and (3.22) are equal, we collect terms to write for any h,

$$2V_{a0} - 2Z_f I_{a0} - 4Z_g I_{a0} = V_{a1} + V_{a2} - Z_f(I_{a1} + I_{a2}) - 2Z_g(I_{a1} + I_{a2})$$

This may be simplified by noting that $I_{a1} + I_{a2} = -I_{a0}$ and using the result (3.20). Rearranging then, we have

$$V_{a0} - Z_f I_{a0} - 3Z_g I_{a0} = V_{a1} - Z_f I_{a1} \tag{3.23}$$

4. *Sequence currents:* From (3.15) we see immediately that the N terminals of the sequence networks must be connected to a common node as shown in Figure 3.14.

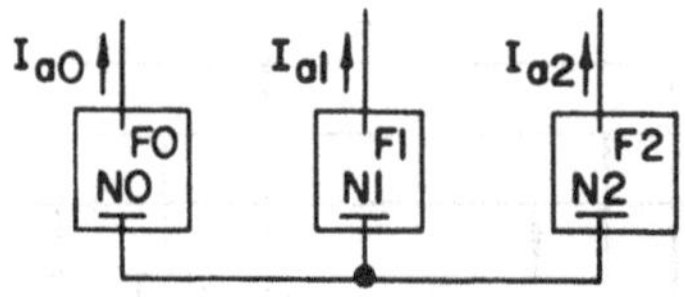

Fig. 3.14. **Sequence network partial connection specified by the current equation.**

5. *Sequence voltages:* From (3.20) we observe that the voltages across the positive and negative sequence networks are equal if an external impedance of Z_f is added in series with each network. Similarly, (3.23) requires that an external impedance of $Z_f + 3Z_g$ be added to the zero sequence network. The final connection is shown in Figure 3.15. Then

$$I_{a1} = \frac{hV_F}{Z_1 + Z_f + \dfrac{(Z_2 + Z_f)(Z_0 + Z_f + 3Z_g)}{Z_0 + Z_2 + 2Z_f + 3Z_g}} \tag{3.24}$$

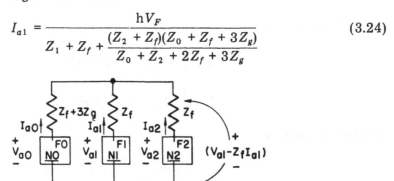

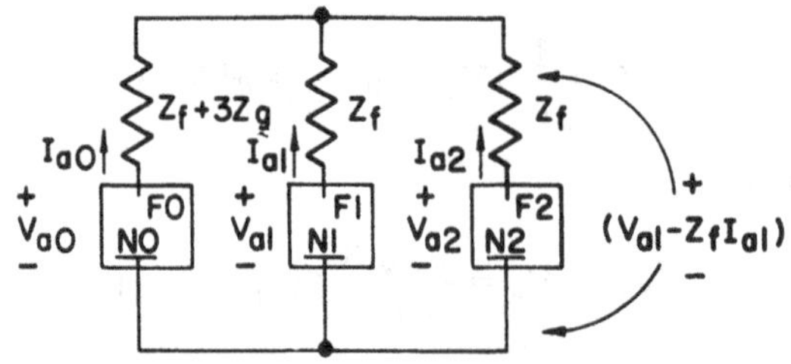

Fig. 3.15. **Sequence network connection for a 2LG fault.**

Example 3.3

Refer again to the system of Example 3.1, this time with a 2LG fault and with fault impedances of Z_f = 4 ohms and Z_g = 8 ohms. The fault is at bus C, so the sequence networks are internally the same as those of Figure 3.7. Let h = 1.

Solution

The pu impedances are

$$Z_0 = 0.1 + j0.27, \quad Z_1 = Z_2 = 0.128 + j0.307$$
$$Z_f = 0.2 + j0, \quad Z_g = 0.4 + j0$$

As before, hV_F = 1.0125 + j0 pu. The network to be solved appears schematically as shown in Figure 3.16.

The total impedance seen by current I_{a1} is

$$Z_t = 0.328 + j0.307 + \frac{(0.328 + j0.307)(1.5 + j0.27)}{1.828 + j0.577}$$
$$= 0.617 + j0.515 = 0.805\underline{/39.8^\circ} \text{ pu}$$

Then I_{a1} is

$$I_{a1} = \frac{1.0125\underline{/0^\circ}}{0.805\underline{/39.8^\circ}} = 1.259\underline{/-39.8^\circ} = 0.965 - j0.805 \text{ pu}$$

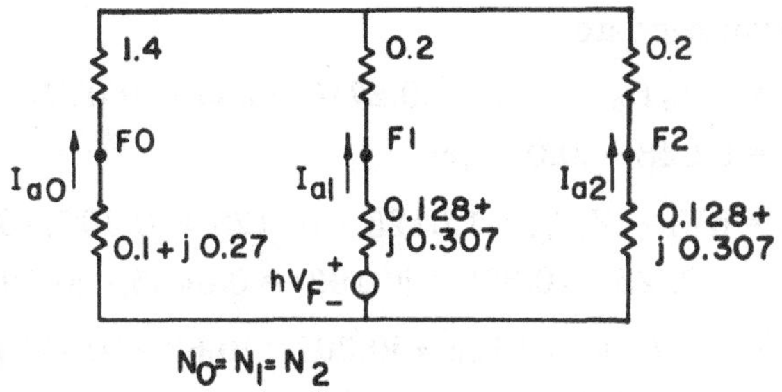

Fig. 3.16. A 2LG fault network.

By inspection of Figure 3.16 we compute

$$I_{a0} = -\frac{Z_2 + Z_f}{Z_2 + Z_0 + 2Z_f + 3Z_g} I_{a1}$$

$$= -\frac{0.449\underline{/43.1^\circ}}{1.92\underline{/17.5^\circ}} (1.259\underline{/-39.8^\circ}) = -0.294\underline{/-14.2^\circ} = -(0.285 - j0.072) \text{ pu}$$

Similarly,

$$I_{a2} = -\frac{Z_0 + Z_f + 3Z_g}{Z_2 + Z_0 + 2Z_f + 3Z_g} I_{a1} = -\frac{1.525\underline{/10.2^\circ}}{1.92\underline{/17.5^\circ}} (1.259\underline{/-39.8^\circ})$$

$$= -1.00\underline{/-47.1^\circ} = -(0.68 - j0.732) \text{ pu}$$

Checking these results, we compute $I_a = I_{a0} + I_{a1} + I_{a2} = 0 - j0.001 \cong 0$. Synthesizing the other phase currents, we have

$$I_b = I_{a0} + a^2 I_{a1} + aI_{a2}$$

$$= -0.285 + j0.072 - 1.181 - j0.432 - 0.293 - j0.955$$

$$= -1.759 - j1.315 = -2.195\underline{/36.8^\circ} \quad \text{pu}$$

and

$$I_c = I_{a0} + aI_{a1} + a^2 I_{a2}$$

$$= -0.285 + j0.072 + 0.215 + j1.24 + 0.977 + j0.210$$

$$= 0.907 + j1.522 = 1.774\underline{/59.2^\circ} \quad \text{pu}$$

Thus $I_b + I_c = -0.852 + j0.207$ pu. But we may also show from (2.26) and (3.15) that

$$I_b + I_c = 3I_{a0} = -0.855 + j0.216 \tag{3.25}$$

which is a good check.

The sequence voltages are

$$V_{a0} = -\ Z_0 I_{a0} = -(0.1 + j0.27)(-0.285 + j0.072)$$
$$= 0.048 + j0.07 \quad \text{pu}$$

$$V_{a1} = hV_F - Z_1 I_{a1} = 1.0125 - (0.128 + j0.307)\ (0.965 - j0.805)$$
$$= 1.0125 - (0.370 + j0.193) = 0.6425 - j0.193 \quad \text{pu}$$

$$V_{a2} = -Z_2 I_{a2} = -(0.128 + j0.307)(-0.68 - j0.732)$$
$$= 0.311 + j0.115 \quad \text{pu}$$

and the drop across parallel networks is

$$V_{a1} - Z_f I_{a1} = 0.6425 - j0.193 - (0.2)(0.965 - j0.805)$$
$$= 0.4495 - j0.032 \quad \text{pu}$$

The phase voltages are synthesized from equation (2.24).

$$V_a = V_{a0} + V_{a1} + V_{a2} = 1.001 - j0.008 \text{ pu}$$

which is close to the desired value of 1.0125 + j0.

$$V_b = V_{a0} + a^2 V_{a1} + aV_{a2}$$
$$= 0.0479 + j0.07 - 0.488 - j0.4595 - 0.2552 + j0.212$$
$$= -0.6953 - j0.1775 = -\ 0.717\underline{/14.3^\circ} \quad \text{pu}$$

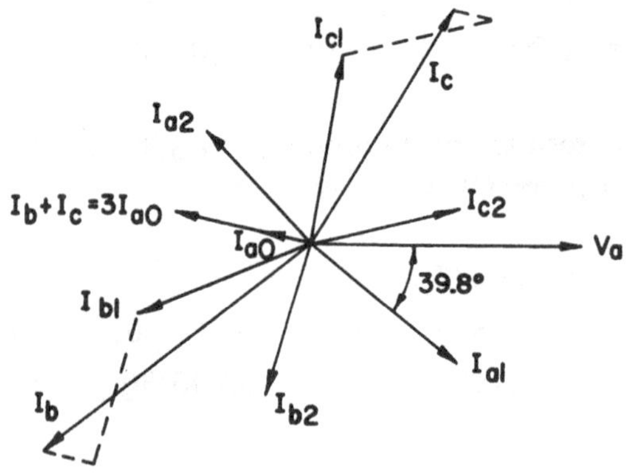

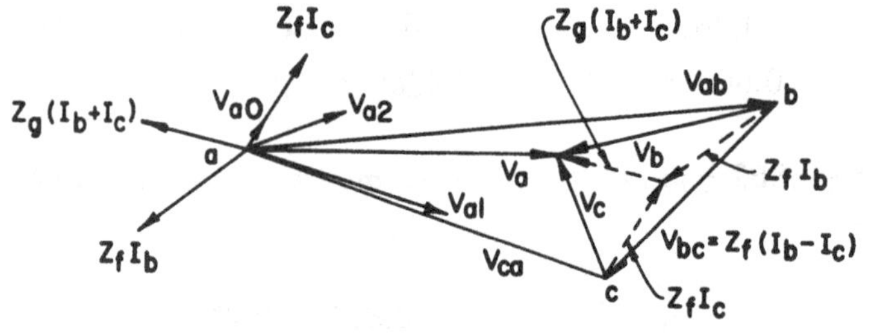

Fig. 3.17. Current and voltage phasors for a 2LG fault: (a) currents (b) voltages.

$$
\begin{aligned}
V_c &= V_{a0} + aV_{a1} + a^2 V_{a2} \\
&= 0.0497 + j0.07 - 0.154 + j0.6525 - 0.0558 - j0.3270 \\
&= -0.1619 + j0.3955 = -0.427 \,\underline{/-67.7^\circ} \quad \text{pu}
\end{aligned}
$$

As a check we compute from (3.13)

$$
\begin{aligned}
V_b &= (Z_f + Z_g) I_b + Z_g I_c = Z_f I_b + Z_g (I_b + I_c) \\
&= (0.2)(-1.759 - j1.315) + (0.4)(-0.852 + j0.207) \\
&= (-0.352 - j0.263) + (-0.341 + j0.083) = -0.693 - j0.180 \quad \text{pu}
\end{aligned}
$$

a close check. The current and voltage phasors are plotted in Figure 3.17.

3.4 The Three-Phase (3ϕ) Fault

The three-phase fault, although not an unbalanced one, is analyzed here to complete the "family" of shunt unbalances of greatest common interest. This fault is important for several reasons. First, it is often the most severe type[1] and hence must be checked to verify that circuit breakers have adequate interrupting rating. Second, it is the simplest fault to determine analytically and is therefore the only one calculated in some cases when complete system information is lacking. Finally, it is often assumed that other types of faults, if not cleared promptly, will develop into 3ϕ faults. It is therefore essential that this type of fault be computed in addition to the other types.

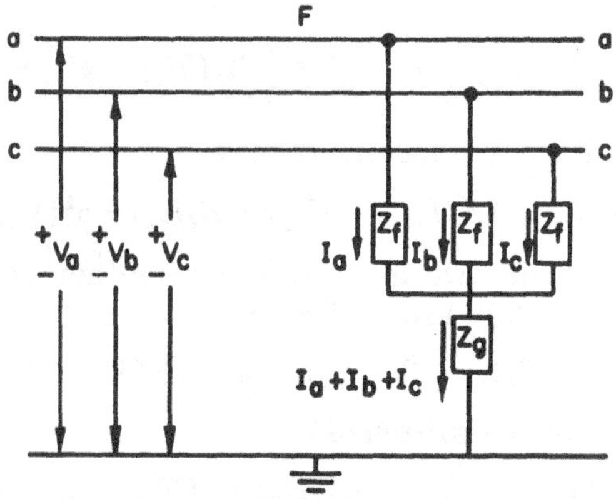

Fig. 3.18. Diagram of a 3ϕ fault at F.

1. *Circuit diagram:* The 3ϕ fault connection is shown in Figure 3.18. The fault is shown as a 3ϕ-to-ground fault with individual phase impedances Z_f and neutral-to-ground impedance Z_g.

2. *Boundary conditions:*

$$V_a = Z_f I_a + Z_g (I_a + I_b + I_c) \tag{3.26}$$

$$V_b = Z_f I_b + Z_g (I_a + I_b + I_c) \tag{3.27}$$

[1] Exceptions: The SLG fault is more severe than the 3ϕ fault in cases where (1) the generators have solidly grounded neutrals or low-impedance neutral impedances and (2) on the Y-grounded side of Δ-Y-grounded transformer banks.

$$V_c = Z_f I_c + Z_g(I_a + I_b + I_c) \tag{3.28}$$

3. *Transformation:* First, we write (3.26)-(3.28) in terms of the symmetrical components of phase a, recognizing that $I_a + I_b + I_c = (3/\mathrm{h})I_{a0}$, by definition. Thus

$$V_a = \frac{1}{\mathrm{h}}(V_{a0} + V_{a1} + V_{a2}) = \frac{1}{\mathrm{h}} Z_f(I_{a0} + I_{a1} + I_{a2}) + \frac{3}{\mathrm{h}} Z_g I_{a0} \tag{3.29}$$

$$V_b = \frac{1}{\mathrm{h}}(V_{a0} + \mathrm{a}^2 V_{a1} + \mathrm{a}V_{a2}) = \frac{1}{\mathrm{h}} Z_f(I_{a0} + \mathrm{a}^2 I_{a1} + \mathrm{a}I_{a2}) + \frac{3}{\mathrm{h}} Z_g I_{a0} \tag{3.30}$$

$$V_c = \frac{1}{\mathrm{h}}(V_{a0} + \mathrm{a}V_{a1} + \mathrm{a}^2 V_{a2}) = \frac{1}{\mathrm{h}} Z_f(I_{a0} + \mathrm{a}I_{a1} + \mathrm{a}^2 I_{a2}) + \frac{3}{\mathrm{h}} Z_g I_{a0} \tag{3.31}$$

Subtracting (3.31) from (3.30), we find V_{bc} for any h to be

$$V_{bc} = V_b - V_c = (\mathrm{a}^2 - \mathrm{a})V_{a1} + (\mathrm{a} - \mathrm{a}^2)V_{a2} = Z_f[(\mathrm{a}^2 - \mathrm{a})I_{a1} + (\mathrm{a} - \mathrm{a}^2)I_{a2}] \tag{3.32}$$

Since $(\mathrm{a}^2 - \mathrm{a}) = -\mathrm{j}\sqrt{3}$, (3.32) may be simplified as follows:

$$V_{bc} = -\mathrm{j}\sqrt{3}\, V_{a1} + \mathrm{j}\sqrt{3}\, V_{a2} = Z_f(-\mathrm{j}\sqrt{3}\, I_{a1} + \mathrm{j}\sqrt{3}\, I_{a2})$$

or $V_{a1} - V_{a2} = Z_f(I_{a1} - I_{a2})$. This may be written more conveniently as

$$V_{a1} - Z_f I_{a1} = V_{a2} - Z_f I_{a2} \tag{3.33}$$

Now add equations (3.29) and (3.30) and recognize that $1 + \mathrm{a}^2 = -\mathrm{a}$ and $1 + \mathrm{a} = -\mathrm{a}^2$. Then

$$V_a + V_b = \frac{1}{\mathrm{h}}(2V_{a0} - \mathrm{a}V_{a1} - \mathrm{a}^2 V_{a2}) = \frac{1}{\mathrm{h}} Z_f(2I_{a0} - \mathrm{a}I_{a1} - \mathrm{a}^2 I_{a2}) + \frac{6}{\mathrm{h}} Z_g I_{a0}$$

Rearranging and canceling the h factor, we have

$$2(V_{a0} - Z_f I_{a0} - 3Z_g I_{a0}) = \mathrm{a}(V_{a1} - Z_f I_{a1}) + \mathrm{a}^2(V_{a2} - Z_f I_{a2}) \tag{3.34}$$

From (3.33) we see that the two quantities in parentheses on the right-hand side of (3.34) are equal. Thus (3.34) simplifies to

$$2(V_{a0} - Z_f I_{a0} - 3Z_g I_{a0}) = (\mathrm{a} + \mathrm{a}^2)(V_{a1} - Z_f I_{a1})$$

Since $\mathrm{a} + \mathrm{a}^2 = -1$, this may be written as

$$2(V_{a0} - Z_f I_{a0} - 3Z_g I_{a0}) = -(V_{a1} - Z_f I_{a1}) \tag{3.35}$$

Now add equations (3.30) and (3.31) to find $V_b + V_c$ and recall that $\mathrm{a} + \mathrm{a}^2 = -1$. Then

$$V_b + V_c = 2V_{a0} - V_{a1} - V_{a2} = Z_f(2I_{a0} - I_{a1} - I_{a2}) + 6\, Z_g I_{a0}$$

which we rearrange to write

$$2(V_{a0} - Z_f I_{a0} - 3Z_g I_{a0}) = (V_{a1} - Z_f I_{a1}) + (V_{a2} - Z_f I_{a2})$$

and utilizing (3.33) again,

$$V_{a0} - Z_f I_{a0} - 3Z_g I_{a0} = V_{a1} - Z_f I_{a1} \tag{3.36}$$

4. *Sequence currents*: There are no sequence current equations for this fault.
5. *Sequence voltages*: From equations (3.33) and (3.36) we deduce that each

sequence network with series impedances of $Z_f + 3Z_g$, Z_f, and Z_f for the zero, positive, and negative sequences are in parallel, exactly as in Figure 3.15. However, upon further examination of equations (3.35) and (3.36),

$$V_{a1} - Z_f I_{a1} = -2(V_{a0} - Z_f I_{a0} - 3Z_g I_{a0}) \tag{3.37}$$

$$V_{a1} - Z_f I_{a1} = V_{a0} - Z_f I_{a0} - 3Z_g I_{a0} \tag{3.38}$$

we see a contradiction. Obviously, these equations can be satisfied simultaneously only if $I_{a0} = 0$, in which case $V_{a0} = 0$ also. But this requires that (3.33) also be zero, i.e.,

$$V_{a1} - Z_f I_{a1} = V_{a2} - Z_f I_{a2} = 0 \tag{3.39}$$

If we short out the negative sequence network, $I_{a2} = 0$ and $V_{a2} = 0$ as well. Thus our sequence network connections are those shown in Figure 3.19.

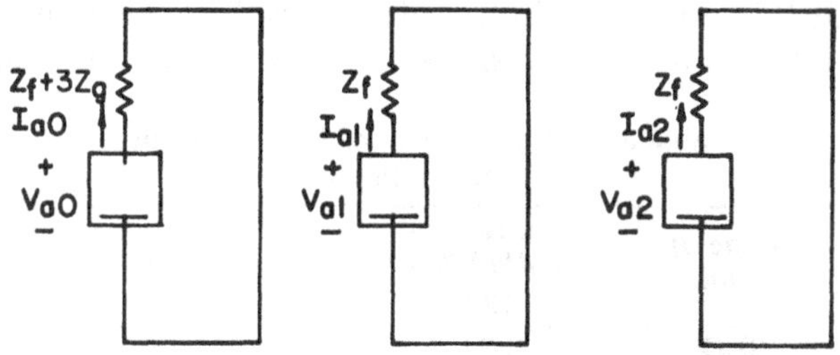

Fig. 3.19. Sequence network connections for a 3ϕ fault.

Note that the result in Figure 3.19 could have been obtained by inspection of Figure 3.18. Since the "load" is Z_f in each phase and the applied voltages are balanced, the currents are obviously balanced. Then

$$(3/\text{h})I_{a0} = I_a + I_b + I_c = 0 \tag{3.40}$$

Also, since the currents are balanced,

$$\mathbf{I}_{012} = \frac{\text{h}}{3}\begin{bmatrix} 1 & 1 & 1 \\ 1 & \text{a} & \text{a}^2 \\ 1 & \text{a}^2 & \text{a} \end{bmatrix}\begin{bmatrix} I_a \\ I_b \\ I_c \end{bmatrix} = \begin{bmatrix} 0 \\ I_{a1} \\ 0 \end{bmatrix} \tag{3.41}$$

or $I_{a1} = \text{h}I_a$, $I_{a0} = I_{a2} = 0$.

The impedance Z_f is really the load impedance of a balanced three-phase load. If this impedance is small, however, we would consider this situation to be a short circuit. In either event the computation is the same.

Example 3.4

Consider a 3ϕ fault with fault impedance of 4 ohms ($Z_f = 4$ ohm) at bus C of Figure 3.5. Let h = 1.

Solution

The pu fault impedance is $Z_f = 4/Z_B = 4/20 = 0.2$ pu. Thus

$$I_{a1} = I_a = \frac{\text{h}V_F}{Z_1 + Z_f} = \frac{1.0125 + \text{j}0}{0.328 + \text{j}0.307}$$

or

$$I_{a1} = \frac{1.0125/\underline{0^\circ}}{0.449/\underline{43.1^\circ}} = 2.255/\underline{-43.1^\circ} = 1.647 - j1.54 \quad \text{pu}$$

The voltage at the fault is

$$V_a = V_{a1} = Z_f I_{a1} = (0.2)(1.647 - j1.54) = 0.329 - j0.308 = 0.451/\underline{-43.1^\circ} \quad \text{pu}$$

Currents and voltages in phases *b* and *c* are found by applying phase rotations of -120° and +120° respectively to the above results.

It is interesting to compare the results of Examples 3.1–3.4 to get some idea as to the relative severity of the various faults. This comparison is somewhat arbitrary because of the choice of the fault impedances. For example, in the 2LG fault the impedance Z_g was chosen to be 0.4 pu, whereas Z_f was chosen as 0.2 pu in all cases. With this understanding that the results are subject to choice of impedance, we have the comparisons shown in Table 3.1. It is clear from this table

Table 3.1. Comparison of Fault Currents and Voltages

Type of Fault	*Current in Faulted Phase* (pu)	*Lowest Voltage at Fault* (pu)
SLG*	2.34	0.468
LL*	2.32	0.342
2LG†	2.195	0.303
3φ*	2.255	0.451

*Z_f = 0.2 pu, †Z_f = 0.2 pu and Z_g = 0.4 pu.

that the SLG fault is the most severe from the standpoint of current magnitudes. This is not to imply that this would always be the case by any means. The circuit parameters of the examples used here are not chosen to be typical of any physical system but merely to illustrate the procedure. In a physical system each fault at each fault location must be computed on the basis of actual circuit conditions. When this is done, it is often the case that the SLG fault is the most severe, with the 3φ, 2LG, and LL following in that order. This was not true in the examples because of the large value of Z_g chosen.

3.5 Other Types of Shunt Faults

The four types of faults discussed above are the types of most general interest. The SLG fault is usually assumed to be by far the most prevalent, making up perhaps 70% of all transmission line faults.[2] Occasionally, a fault configuration other than these four is of interest, e.g., in special cases where it is considered necessary to analyze a given unusual occurrence. In such cases a straightforward application of steps 1-4 will permit the evaluation of any situation. Reference [14] gives the sequence network connections for a number of these special cases, and some of them are recommended for study in the problems at the end of the chapter.

[2] The "Westinghouse Transmission and Distribution Reference Book" [14, p. 358] gives a typical frequency of occurrence for 3φ, 2LG, LL, and SLG faults as 5%, 10%, 15%, and 70% respectively.

3.6 Comments on Shunt Fault Calculation

In all the above the computation of fault currents and voltages are based upon finding the open circuit (Thevenin) voltage hV_F at the fault point F. This voltage and the Thevenin impedance are then connected to external faults, i.e., to fault connections external to the positive sequence network. This equivalent circuit gives correct results for the current flowing external to F, namely I_{a1}.

If the distribution of I_{a0}, I_{a1}, and I_{a2} *inside* the sequence networks is desired, this must be found by Kirchhoff's laws applied to the actual branches within the three sequence networks. Then the currents in any branch may be synthesized by $\mathbf{I}_{abc} = \mathbf{A}\,\mathbf{I}_{012}$ to find the branch phase currents at that location which contribute to the fault. Note that this branch current does *not* include the balanced load current which was flowing prior to the fault. However, since the network is linear, the fault and load currents may be added by linear superposition to obtain the total current flowing after the fault is applied.

Example 3.5

Compute the total current flowing from bus B to bus C in the system of Figure 3.5 for the SLG fault condition of Example 3.1.

Solution

In Example 3.1 we computed the load current to be $I_L = 0.5 - j0.25$ pu. This is based on a reference voltage at bus D of $1.0 + j0$ pu. The SLG fault current is $I_a = 1.713 - j1.965$ pu, based on a reference voltage $hV_F = 1.0125 + j0$ at bus C. Actually there is only 1.27° difference in the reference voltages and this will be ignored. If it were significant, the fault current could easily be rotated to agree with the load reference. The load and fault currents are superimposed as shown in Table 3.2.

Table 3.2. Total Current Leaving Bus B for a SLG Fault at Bus C

Phase	*Load Current*	*Fault Current*	*Total Current*
a	0.5 - j0.25	1.434 - j1.554	1.934 - j1.804
b	-0.467 - j0.308	0.142 - j0.008	-0.324 - j0.316
c	-0.033 + j0.558	0.142 - j0.008	0.109 + j0.550

II. SERIES FAULTS

Next we consider a group of unbalanced conditions which do not involve any connection between lines or between line and neutral at F. These are referred to as series faults because there is generally an unbalanced series impedance condition.

Since the unbalance for this type of problem is in series with the line, there is no "fault point" in the sense described in Part I. Rather, there are two "fault points," one on either side of the unbalance. Thus the sequence network is still that of a completely symmetric system, and the unbalanced portion is isolated *outside* the sequence network.

Our notation for this situation will be to consider the system connections to the point of unbalance as shown in Figure 3.20, where the two sides of the fault point are F and F'. Current direction is assumed to be from F to F', and a voltage *drop* is shown in the assumed direction of current.

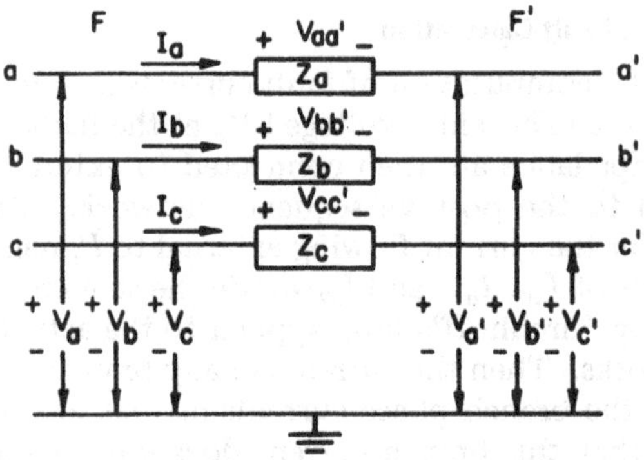

Fig. 3.20. Voltages and currents at the fault point F-F'.

The sequence networks contain the symmetrical portions of the system, looking back to the left of F and to the right of F'. These symmetrical portions may or may not be interconnected. We shall represent the sequence networks schematically as shown in Figure 3.21. Note that the voltage polarities and current directions are consistent with those defined in Figure 3.20.

As shown in Chapter 2 the general case of series unbalance (i.e., with $Z_a \neq Z_b \neq Z_c$) generates a condition where coupling exists between sequence networks. Such a condition would require a special connection from, say, I_{a1} to the network of I_{a2}, with possibly an isolation and phase shifting transformer. We will generally avoid such situations since they introduce as many problems as they solve. The problems under study here are the simple cases of series unbalance, but they include some very important special cases. One class of special problems which are of particular interest is that of open lines. This is easily solved by our usual technique, as shown below.

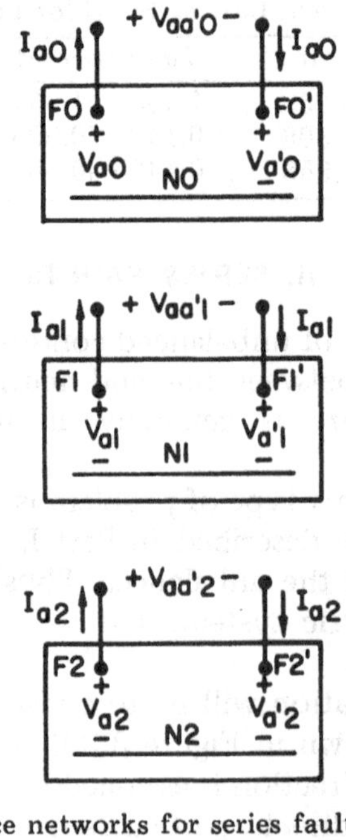

Fig. 3.21. Sequence networks for series faults between F and F'.

3.7 Sequence Network Equivalents for Series Faults

Before proceeding with the analysis of series faults, we examine briefly the nature of the sequence networks of Figure 3.21. For a shunt fault the situation is simpler since only one point in the balanced network, namely F, is of interest, and a straightforward application of Thevenin's theorem derives a simple equivalent circuit. With series faults the situation is changed. Now there are two points in the balanced (unfaulted) system which are of direct interest, F and F'. Hence we need a two-port Thevenin's equivalent. Since Thevenin's theorem is not commonly used for more than one port, we digress briefly to establish a method of obtaining such an equivalent.[3]

The zero and negative sequence networks—being entirely linear, bilateral, and passive—are no problem insofar as their two-port description is concerned. Many authors, [5] for example, examine two-port parameters of such networks in great detail, and six possible sets of parameters are derived—Y, Z, H, G, $ABCD$, and $\mathcal{ABCD}$. For our work we will be most interested in the Y and Z parameters described by the two equations

$$\mathbf{V} = \mathbf{Z}\,\mathbf{I} \tag{3.42}$$

and

$$\mathbf{I} = \mathbf{Y}\,\mathbf{V} \tag{3.43}$$

We establish the nature of these parameters by referring to the general two-port network shown in Figure 3.22, where we define the currents to be entering

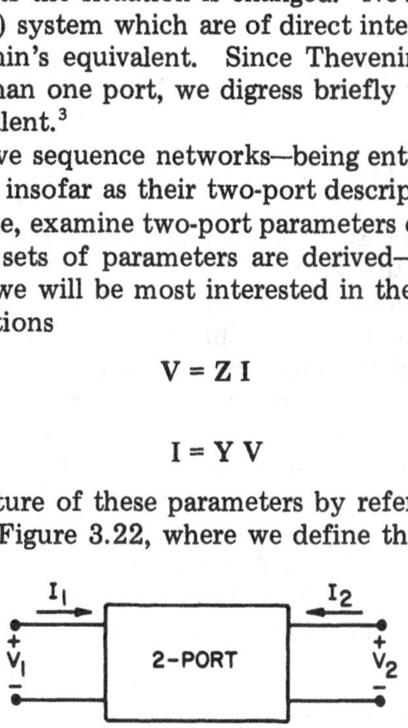

Fig. 3.22. General two-port network.

the network as shown and the voltages to be drops in the direction of the current through the network. Then we write,

$$\begin{bmatrix} V_1 \\ V_2 \end{bmatrix} = \begin{bmatrix} Z_{11} & Z_{12} \\ Z_{21} & Z_{22} \end{bmatrix} \begin{bmatrix} I_1 \\ I_2 \end{bmatrix} \tag{3.44}$$

or

$$\begin{bmatrix} I_1 \\ I_2 \end{bmatrix} = \begin{bmatrix} Y_{11} & Y_{12} \\ Y_{21} & Y_{22} \end{bmatrix} \begin{bmatrix} V_1 \\ V_2 \end{bmatrix} \tag{3.45}$$

The parameters of (3.44) are called the "open circuit impedance parameters" since they are found by leaving the ports "open." Thus

$$\begin{aligned} Z_{11} &= \left.\frac{V_1}{I_1}\right]_{I_2=0} & Z_{12} &= \left.\frac{V_1}{I_2}\right]_{I_1=0} \\ Z_{21} &= \left.\frac{V_2}{I_1}\right]_{I_2=0} & Z_{22} &= \left.\frac{V_2}{I_2}\right]_{I_1=0} \end{aligned} \tag{3.46}$$

[3] This subject is explored in greater depth in [15] where Ward applies Thevenin's theorem to n ports and derives quite general results. Our analysis is restricted to two ports.

and these equations show exactly how the parameters may be found. For example, to find Z_{11}, apply a current, say 1 ampere, at port 1 and leave port 2 open. Then measure or compute V_1 to find the ratio V_1/I_1. Note that a current source has infinite impedance, so that both ports are open; hence the name "open circuit Z parameters."

Similarly, the Y parameters of (3.45) are found by shorting terminals. Thus we have

$$Y_{11} = \left.\frac{I_1}{V_1}\right]_{V_2=0} \qquad Y_{12} = \left.\frac{I_1}{V_2}\right]_{V_1=0}$$
$$Y_{21} = \left.\frac{I_2}{V_1}\right]_{V_2=0} \qquad Y_{22} = \left.\frac{I_2}{V_2}\right]_{V_1=0} \tag{3.47}$$

and we note that in this case we apply a voltage source of zero impedance to one port while shorting the other port.

Note that the networks with which we deal are linear, bilateral, passive networks. In such cases we are assured that $Z_{12} = Z_{21}$ and $Y_{12} = Y_{21}$. We also note that the lower (negative) terminals are common to ports 1 and 2 in power networks since this connection represents the system neutral. A port description often ignores impedances in this "neutral" in any event [5]. Also, it is easy to show that $\mathbf{Y} = \mathbf{Z}^{-1}$. Hence the name "short circuit Y parameters."

The Y parameters are really more useful for finding an equivalent circuit since the elements of the $\mathbf{Y}$ matrix give the admittance connected between nodes of the network. This is because each Y is found with the opposite port grounded so that any impedance at that port is shorted out. Thus we may redraw Figure 3.22 in a general way, using admittances computed in (3.47) to find the equivalent circuit of Figure 3.23. We see that Y_{11} is the sum of admittances connected to the +1 terminal, Y_{22} is the sum of admittances connected to the +2 terminal, and Y_{12} is the negative of the admittance between +1 and +2.

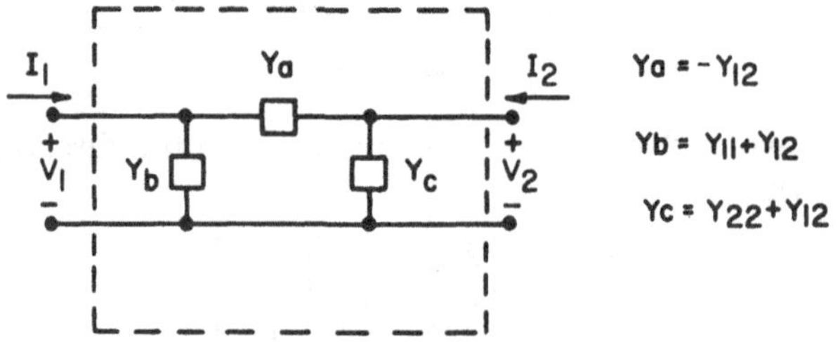

Fig. 3.23. Two-port equivalent circuit.

Returning to the sequence networks of Figure 3.21, we see that for the zero sequence network,

$$I_1 = -I_{a0}, \qquad I_2 = I_{a0}, \qquad V_1 = V_{a0}, \qquad V_2 = V_{a'0} \tag{3.48}$$

and similarly for the negative sequence network. Thus, for example, we write

$$\begin{bmatrix} -I_{a0} \\ I_{a0} \end{bmatrix} = \begin{bmatrix} Y_{11-0} & Y_{12-0} \\ Y_{21-0} & Y_{22-0} \end{bmatrix} \begin{bmatrix} V_{a0} \\ V_{a'0} \end{bmatrix} \tag{3.49}$$

where we continue to associate the left (number 1) terminal with F and the right (number 2) terminal with F'.

The positive sequence network description is the same as that given above except that the positive sequence network has internal sources. If we consider these sources as independent,[4] the network equations become [5]

$$\mathbf{V} = \mathbf{Z}\,\mathbf{I} + h\mathbf{V}_s \tag{3.50}$$

and

$$\mathbf{I} = \mathbf{Y}\,\mathbf{V} + h\mathbf{I}_s \tag{3.51}$$

where $\mathbf{V}_s$ and $\mathbf{I}_s$ are due exclusively to the internal sources, and $\mathbf{Z}$ and $\mathbf{Y}$ are the same as before. More specifically $\mathbf{V}_s$ in equation (3.50) consists of the open circuit voltages measured at ports 1 and 2 (F and F'), such voltages being due to internal (generator) sources. Likewise $\mathbf{I}_s$ in (3.51) consists of the currents which would flow *into* ports 1 and 2 (F and F') if these ports were shorted. Since

$$\mathbf{Y} = \mathbf{Z}^{-1} \tag{3.52}$$

we easily show that

$$\mathbf{I}_s = -\mathbf{Y}\,\mathbf{V}_s \tag{3.53}$$

It is reasonable that these two source descriptions should be related. Note that the scalar multiplier h is required on the source terms due to the way the symmetrical component transformation $\mathbf{A}$ was defined in (2.36).

To construct an equivalent circuit for the positive sequence network, it is easiest to work with the admittance description. Thus from equation (3.51) we write

$$\begin{bmatrix} -I_{a1} \\ I_{a1} \end{bmatrix} = \begin{bmatrix} I_1 \\ I_2 \end{bmatrix} = \begin{bmatrix} Y_{11\text{-}1} & Y_{12\text{-}1} \\ Y_{21\text{-}1} & Y_{22\text{-}1} \end{bmatrix} \begin{bmatrix} V_{a1} \\ V_{a'1} \end{bmatrix} + h \begin{bmatrix} I_{S1} \\ I_{S2} \end{bmatrix} \tag{3.54}$$

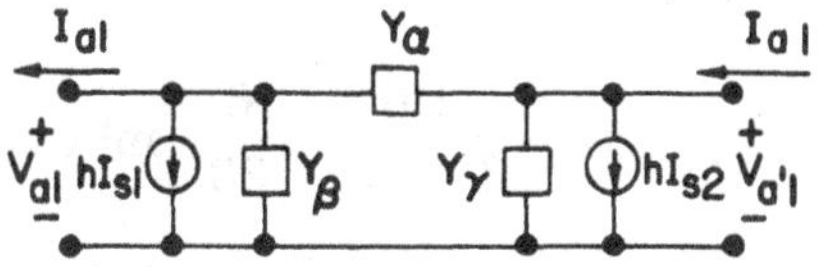

Fig. 3.24. Equivalent circuit for positive sequence network.

The equivalent circuit corresponding to equation (3.54) is shown in Figure 3.24, where we use the simplification

$$Y_\alpha = -Y_{12\text{-}1}, \qquad Y_\beta = Y_{11\text{-}1} + Y_{12\text{-}1}, \qquad Y_\gamma = Y_{22\text{-}1} + Y_{12\text{-}1} \tag{3.55}$$

where Y_α, Y_β, and Y_γ are elements of the equivalent circuit but the quantities on the right of (3.55) are Y matrix elements.

Example 3.6

The system shown in Figure 3.25 is to be studied for a series fault between buses F and F'. Find the equivalent circuits for the positive and negative sequence networks.

[4]It could be argued that the terminal voltage of a synchronous generator depends both upon its current and perhaps upon some remote voltage. For simplicity we assume constant voltage sources. Otherwise the Y and Z matrices become complicated and are no longer reciprocal [5].

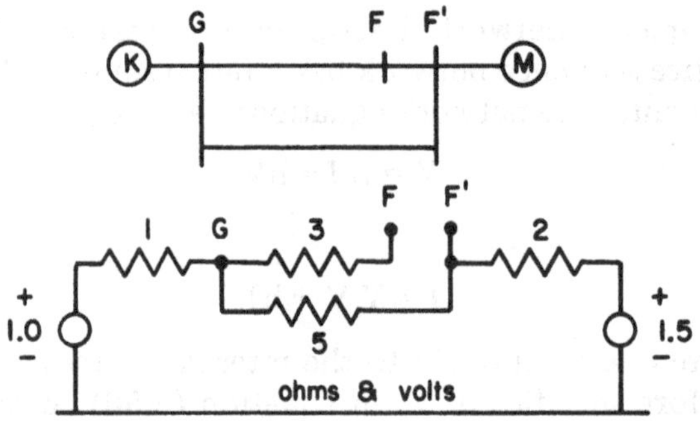

Fig. 3.25. System diagram for Example 3.6.

Solution

Following the notation of Figure 3.22, we sketch the system as shown in Figure 3.26.

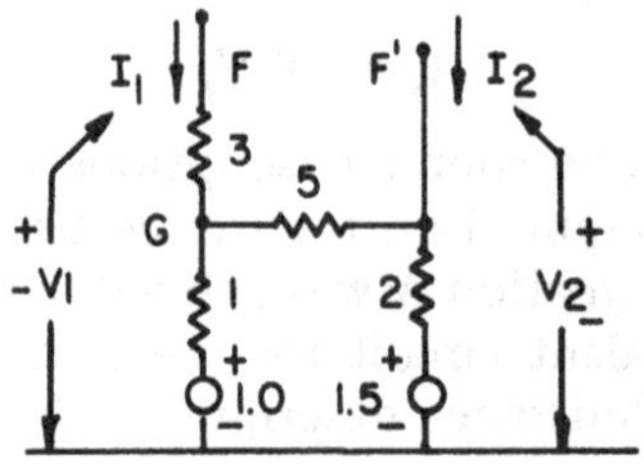

Fig. 3.26. Currents and voltages of the system under study.

To find the elements of the Y matrix, we first remove the internal sources. Since these are voltage sources, they are shorted. Then, with $V_2 = 0$, apply $V_1 = 1.0$ and compute

$$Y_{11} = I_1 = \frac{1.0}{3 + 5/6} = \frac{6}{23} \text{ mho}$$

and

$$Y_{21} = I_2 = -(1/6)\, I_1 = -1/23 \text{ mho}$$

Now, with $V_1 = 0$ and $V_2 = 1.0$, compute

$$Y_{22} = I_2 = \frac{2 + 23/4}{2\,(23/4)} = \frac{31}{46} \text{ mho}$$

and

$$Y_{12} = I_1 = -\frac{1}{4}\left(\frac{2}{2 + 23/4}\right)\left(\frac{31}{46}\right) = -\frac{1}{23} \text{ mho (check)}$$

Thus

$$\mathrm{Y} = \begin{bmatrix} 6/23 & -1/23 \\ -1/23 & 31/46 \end{bmatrix} \text{ mho}$$

Now, with both F and F' shorted, we compute the source currents as follows for h = 1 (using superposition):

$$I_{S1} = I_{S1(1.0)} + I_{S1(1.5)}$$

$$= -\frac{5}{8}\left(\frac{1.0}{1 + 15/8}\right) - \frac{1}{4}(0) = -\frac{5}{23} \text{ A}$$

$$I_{S2} = I_{S2(1.0)} + I_{S2(1.5)}$$

$$= -\frac{3}{8}\left(\frac{1.0}{1 + 15/8}\right) - \frac{1.5}{2} = -\frac{3}{23} - \frac{3}{4} = -\frac{81}{92} \text{ A}$$

Thus the positive sequence equivalent circuit is shown in Figure 3.27. The negative sequence network equivalent is the same as Figure 3.27, assuming that the negative sequence impedances are the same as positive sequence impedances, except the current sources are missing.

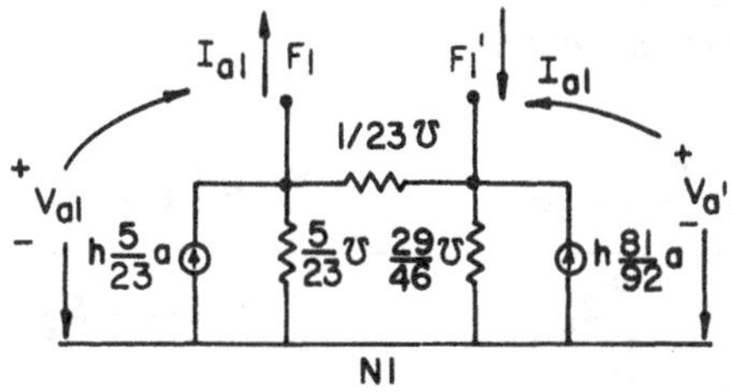

Fig. 3.27. Positive sequence network equivalent.

One additional important result may be established for the two-port Thevenin equivalent. Our two-port network is a special one since if any connection is to be made at all between F and F', it is made in such a way that $I_a = I_{a'}$. Then $\mathbf{I}_{012} = \mathbf{I}_{012'}$ or $\mathbf{I}_{012}$ at $F = \mathbf{I}_{012}$ at F', and the currents entering the ports of the sequence networks have a special constraint. Thus, for example, I_{a1} leaves the network at F and enters at F'. Note that this constraint is not due to any internal network condition but is due entirely to the anticipated external connection. In other words, we assume that the phases do not become crossed in establishing the series unbalance. (See [16] for a case where the phases do become crossed.) This equality of sequence currents is recognized in equation (3.54) and may also be noted in connection with the Z description

$$\begin{bmatrix} V_{a1} \\ V_{a'1} \end{bmatrix} = \begin{bmatrix} Z_{11\text{-}1} & Z_{12\text{-}1} \\ Z_{21\text{-}1} & Z_{22\text{-}1} \end{bmatrix} \begin{bmatrix} -I_{a1} \\ I_{a1} \end{bmatrix} + \text{h} \begin{bmatrix} V_{S1} \\ V_{S2} \end{bmatrix} \tag{3.56}$$

but this equation simplifies immediately to

$$\begin{bmatrix} V_{a1} \\ V_{a'1} \end{bmatrix} = \text{h} \begin{bmatrix} V_{S1} \\ V_{S2} \end{bmatrix} - \begin{bmatrix} (Z_{11\text{-}1} - Z_{12\text{-}1}) I_{a1} \\ -(Z_{22\text{-}1} - Z_{12\text{-}1}) I_{a1} \end{bmatrix} \tag{3.57}$$

Thus the voltage V_{a1} may be written in terms of the equivalent voltage V_{S1} and the current I_{a1}, exactly as in the case of shunt faults where we write

$$V_{a1} = \text{h} V_F - Z_1 I_{a1}$$

This means that the ports of the network are completely uncoupled and may be represented by the circuit of Figure 3.28.

Using the Y parameters, which are preferred when finding the equivalent circuit, we solve (3.54) for V_{a1} and $V_{a'1}$ by premultiplying by $\mathbf{Y}^{-1}$. The result is given by

$$\begin{bmatrix} V_{a1} \\ V_{a'1} \end{bmatrix} = \mathbf{Y}^{-1}\left\{\begin{bmatrix} -I_{a1} \\ I_{a1} \end{bmatrix} - \mathrm{h}\begin{bmatrix} I_{S1} \\ I_{S2} \end{bmatrix}\right\}$$

$$= \frac{1}{\det \mathbf{Y}}\begin{bmatrix} \mathrm{h}(Y_{12\text{-}1} I_{S2} - Y_{22\text{-}1} I_{S1}) - (Y_{12\text{-}1} + Y_{22\text{-}1}) I_{a1} \\ \mathrm{h}(Y_{12\text{-}1} I_{S1} - Y_{11\text{-}1} I_{S2}) + (Y_{12\text{-}1} + Y_{11\text{-}1}) I_{a1} \end{bmatrix} \quad (3.58)$$

where $\det \mathbf{Y} = Y_{11\text{-}1} Y_{22\text{-}1} - Y_{12\text{-}1}^2$. Thus for the uncoupled representation of Figure 3.28 the $\mathbf{Z}$ parameters are preferred.

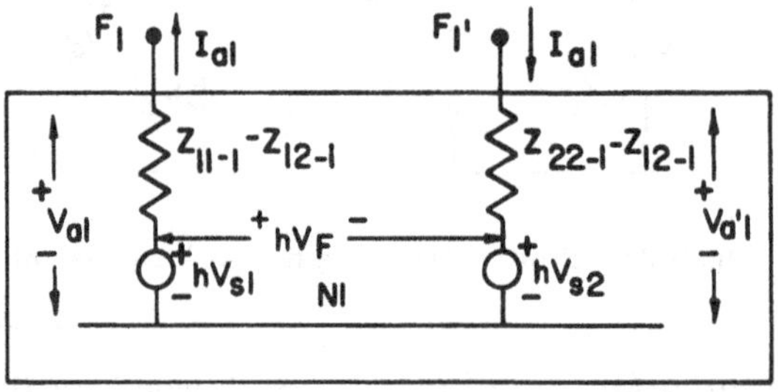

Fig. 3.28. Uncoupled positive sequence network.

Example 3.7

Find the uncoupled representation similar to Figure 3.28 for the system of Example 3.6.

Solution

The determinant of Y is

$$\det \mathbf{Y} = (6/23)(31/46) - (1/23)^2 = 4/23$$

Thus from (3.58)

$$\begin{bmatrix} V_{a1} \\ V_{a'1} \end{bmatrix} = \frac{23}{4}\begin{bmatrix} \mathrm{h}[(-1/23)(-81/92) - (31/46)(-5/23)] - [(-1/23) + (31/46)]\, I_{a1} \\ \mathrm{h}[(-1/23)(-5/23) - (6/23)(-81/92)] + [(-1/23) + (6/23)]\, I_{a1} \end{bmatrix}$$

$$= \begin{bmatrix} 1.063\mathrm{h} - 3.625\, I_{a1} \\ 1.375\mathrm{h} + 1.250\, I_{a1} \end{bmatrix}$$

3.8 Unequal Series Impedances

One case of unequal series impedance where phase a is symmetric is that shown in Figure 3.29.

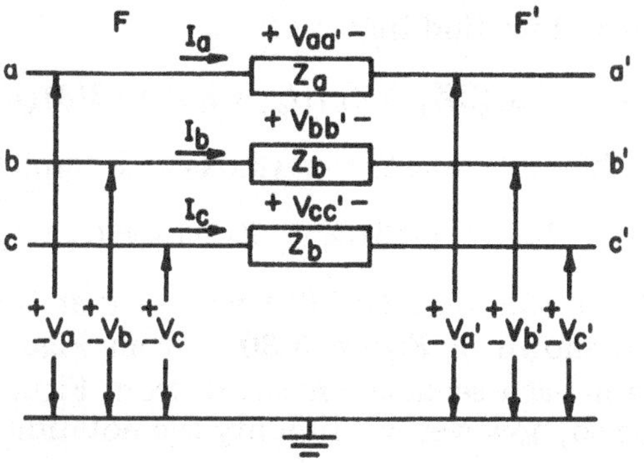

Fig. 3.29. Circuit diagram for unequal series impedances at F-F'.

1. *Circuit diagram:* See Figure 3.29 and note that $Z_a \neq Z_b$.
2. *Boundary conditions:* By inspection of Figure 3.29

$$\begin{bmatrix} V_{aa'} \\ V_{bb'} \\ V_{cc'} \end{bmatrix} = \begin{bmatrix} V_a \\ V_b \\ V_c \end{bmatrix} - \begin{bmatrix} V_{a'} \\ V_{b'} \\ V_{c'} \end{bmatrix} = \begin{bmatrix} Z_a & 0 & 0 \\ 0 & Z_b & 0 \\ 0 & 0 & Z_b \end{bmatrix} \begin{bmatrix} I_a \\ I_b \\ I_c \end{bmatrix} \tag{3.59}$$

or

$$\mathbf{V}_{abc} - \mathbf{V}_{a'b'c'} = \mathbf{Z}_{abc}\mathbf{I}_{abc} \tag{3.60}$$

3. *Transformation:* We transform (3.60) from the a-b-c coordinate system to the 0-1-2 coordinate system by a similarity transformation. Thus

$$\mathbf{V}_{aa'\text{-}012} = \mathbf{V}_{012} - \mathbf{V}'_{012} = \mathbf{Z}_{012}\,\mathbf{I}_{012} \tag{3.61}$$

where

$$\mathbf{Z}_{012} = \mathbf{A}^{-1}\mathbf{Z}_{abc}\mathbf{A} \tag{3.62}$$

Performing the indicated matrix multiplication, we find

$$\mathbf{Z}_{012} = \frac{1}{3}\begin{bmatrix} Z_a + 2Z_b & Z_a - Z_b & Z_a - Z_b \\ Z_a - Z_b & Z_a + 2Z_b & Z_a - Z_b \\ Z_a - Z_b & Z_a - Z_b & Z_a + 2Z_b \end{bmatrix} \tag{3.63}$$

4. *Sequence currents:* There are no sequence current equations for this case.
5. *Sequence voltages:* Since there are no sequence current equations, we must completely determine the sequence network connections by considering only the voltage equation (3.61).

From equation (3.61) we compute row 1 − row 2 to find

$$V_{aa'0} - V_{aa'1} = Z_b(I_{a0} - I_{a1})$$

or

$$V_{aa'0} - Z_bI_{a0} = V_{aa'1} - Z_bI_{a1} \tag{3.64}$$

Similarly, from row 2 - row 3 we see that $V_{aa'1} - Z_b I_{a1} = V_{aa'2} - Z_b I_{a2}$, which combines with (3.64) for the important equation

$$V_{aa'0} - Z_b I_{a0} = V_{aa'1} - Z_b I_{a1} = V_{aa'2} - Z_b I_{a2} \tag{3.65}$$

Also, taking row 1 + row 2 we find that

$$V_{aa'0} + V_{aa'1} = (1/3)(2Z_a + Z_b)(I_{a0} + I_{a1}) + (2/3)(Z_a - Z_b)I_{a2}$$

Substituting for $V_{aa'0}$ the value taken from (3.64) and simplifying, we find that

$$V_{aa'1} - Z_b I_{a1} = (1/3)(Z_a - Z_b)(I_{a0} + I_{a1} + I_{a2}) \tag{3.66}$$

From the two equations (3.65) and (3.66) we see that the sequence networks must be connected as shown in Figure 3.30. From Figure 3.30 (knowing the equivalent impedance in each sequence network from Figure 3.28) we may compute I_{a1}. Before doing so, however, we simplify the notation.

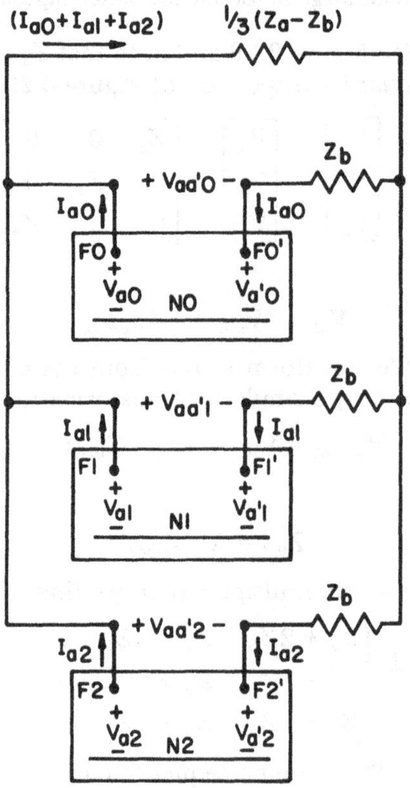

Fig. 3.30. Sequence network connection for unequal series impedances at F–F'.

In working with shunt faults, we defined Z_1 to be the impedance to the flow of I_{a1} in the positive sequence network. Similar definitions applied for Z_2 and Z_0. If we follow this same rule for series faults,

$$\begin{aligned} Z_0 &= (Z_{11\text{-}0} - Z_{12\text{-}0}) + (Z_{22\text{-}0} - Z_{12\text{-}0}) \\ &= Z_{11\text{-}0} + Z_{22\text{-}0} - 2Z_{12\text{-}0} \end{aligned} \tag{3.67}$$

Similarly,

$$Z_1 = Z_{11\text{-}1} + Z_{22\text{-}1} - 2Z_{12\text{-}1} \tag{3.68}$$

and

$$Z_2 = Z_{11\text{-}2} + Z_{22\text{-}2} - 2Z_{12\text{-}2} \tag{3.69}$$

We also defined $\mathrm{h}V_F$ as a voltage rise in the direction of I_{a1} flow for shunt faults, or the open circuit voltage drop from F to neutral. In the case of series faults we have

$$\mathrm{h}V_F = V_{S1} - V_{S2} \tag{3.70}$$

or $\mathrm{h}V_F$ is the open circuit voltage drop from F to F', or

$$\mathrm{h}V_F = V_{aa'1}]_{I_{a1}=0} \tag{3.71}$$

With this notation established, we compute

$$I_{a1} = \frac{\mathrm{h}V_F}{Z_t} = \frac{\mathrm{h}(\text{open circuit } V_{FF'})}{Z_t} \tag{3.72}$$

where

$$Z_t = Z_b + Z_1 + Z \tag{3.73}$$

and where

$$Z = \frac{(Z_a - Z_b)(Z_b + Z_0)(Z_b + Z_2)}{(Z_b + Z_2)(Z_a - Z_b) + (Z_b + Z_0)(Z_a - Z_b) + 3(Z_b + Z_0)(Z_b + Z_2)} \tag{3.74}$$

By inspection of Figure 3.30 we see that

$$I_{a2} = -I_{a1}Z/(Z_b + Z_2) \tag{3.75}$$

and

$$I_{a0} = -I_{a1}Z/(Z_b + Z_0) \tag{3.76}$$

where Z is defined as in (3.74).

3.9 One Line Open (1LO)

One open line conductor in phase a is a special case of the previous case in which

$$Z_a = \infty, \qquad Z_b \text{ is finite} \tag{3.77}$$

No additional computation is necessary to show that the sequence network connections are as shown in Figure 3.31. In this case the branch $(1/3)(Z_a - Z_b)$ which shunts all three sequence networks is missing. The sequence network connection of Figure 3.31 is similar to the 2LG shunt fault connection except the parallel connections are made between F and F' rather than between F and N.

By inspection of Figure 3.31 we compute

$$I_{a1} = \mathrm{h}V_F/Z_t \tag{3.78}$$

where

$$Z_t = Z_b + Z_1 + Z \tag{3.79}$$

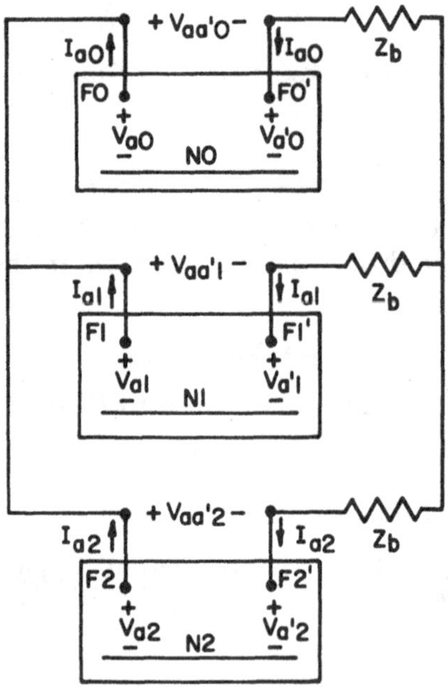

Fig. 3.31. Sequence network connection for line a open at F-F'.

and

$$Z = \frac{(Z_b + Z_0)(Z_b + Z_2)}{2Z_b + Z_2 + Z_0} \tag{3.80}$$

Then,

$$I_{a2} = -I_{a1} Z/(Z_b + Z_2) \tag{3.81}$$

and

$$I_{a0} = -I_{a1} Z/(Z_b + Z_0) \tag{3.82}$$

3.10 Two Lines Open (2LO)

If two lines are open, $Z_b = \infty$. Note that the connection of Figure 3.30 will not suffice for this situation. We adopt the usual analytical technique.

1. *Circuit diagram:* See Figure 3.32 and note that $Z_b = Z_c = \infty$.

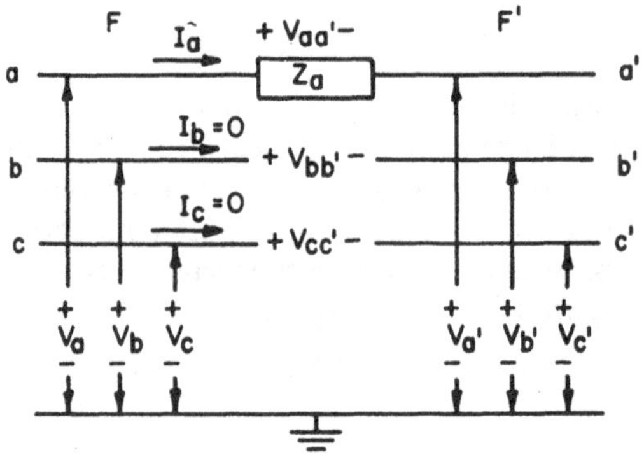

Fig. 3.32. Circuit diagram for two lines open.

2. *Boundary conditions:* By inspection of Figure 3.32 we write

$$I_b = I_c = 0, \qquad V_{aa'} = Z_a I_a \tag{3.83}$$

3. *Transformation:* The sequence currents are $\mathbf{I}_{012} = \mathbf{A}^{-1}\mathbf{I}_{abc}$. But from (3.83) $\mathbf{I}_{abc} = [I_a \quad 0 \quad 0]^t$ so that $\mathbf{I}_{012}$ reduces to

$$\mathbf{I}_{012} = \frac{h}{3} I_a \begin{bmatrix} 1 \\ 1 \\ 1 \end{bmatrix} \tag{3.84}$$

and the sequence currents are all equal. This is somewhat analogous to the SLG fault, where we found this same condition to be true.

Also from (3.83) $V_{aa'} = Z_a I_a$ or

$$\frac{1}{h}(V_{aa'0} + V_{aa'1} + V_{aa'2}) = \frac{Z_a}{h}(I_{a0} + I_{a1} + I_{a2})$$

Rearranging, we have for any h

$$(V_{aa'0} - Z_a I_{a0}) + (V_{aa'1} - Z_a I_{a1}) + (V_{aa'2} - Z_a I_{a2}) = 0 \tag{3.85}$$

4. *Sequence currents:* Equation (3.84) requires that the three sequence networks be in series.

5. *Sequence voltages:* Equation (3.85) requires that an impedance with total value $3Z_a$ be inserted in series with the sequence networks. The equation also suggests that an impedance Z_a could be associated with each network as shown in Figure 3.33.

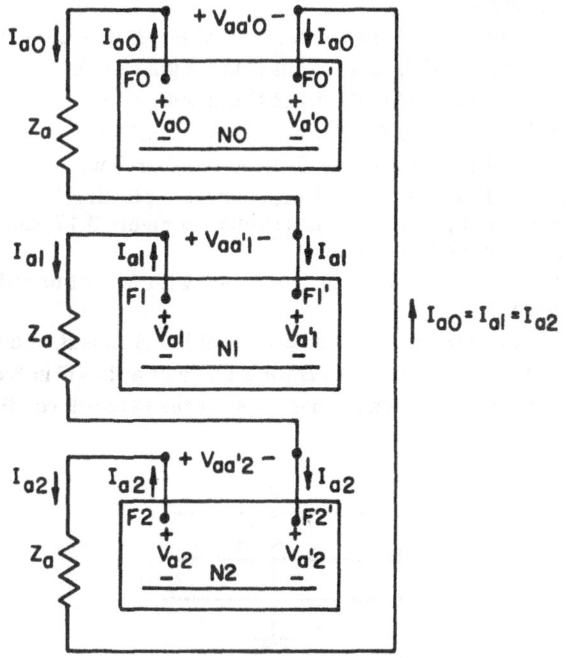

Fig. 3.33. Sequence network connections for two lines open.

To compute the sequence currents, we write from Figure 3.33

$$I_{a0} = I_{a1} = I_{a2} = \frac{hV_F}{Z_0 + Z_1 + Z_2 + 3Z_a} \tag{3.86}$$

3.11 Other Series Faults

Nearly all cases of series faults of general interest are special cases of the three types considered in the preceding paragraphs. The one notable exception is the case where a neutral impedance exists in the zero sequence network. If it is desirable to account for the effect of such neutral impedance separately from Z_0, a separate impedance $3Z_g$ may be placed in series with the zero sequence network. The multiplier "3" is required because a neutral current, if it exists, is always $3I_{a0}$ and since we define the zero sequence network to have only a current I_{a0}, we may associate the factor 3 with the impedance. Then the voltage drop across the zero sequence network and Z_g is

$$V_{a0} + 3Z_g I_{a0} = -(Z_0 + 3Z_g) I_{a0} \tag{3.87}$$

Ordinarily, the impedance $3Z_g$ will be a part of Z_0, in which case we write simply $V_{aa'0} = -Z_0 I_{a0}$.

Problems

3.1. Repeat Example 3.1 for a SLG fault at bus C but with zero fault impedance.
3.2. Repeat Example 3.1 for a SLG fault at bus C but with a fault impedance of 8 ohms. Plot I_a versus fault impedance.
3.3. Analyze the system of Figure 3.5 for a SLG fault at bus B with zero fault impedance.
3.4. Repeat Example 3.2 for a LL fault at bus C but with zero fault impedance.
3.5. Repeat Example 3.2 for a LL fault at bus C but with a fault impedance of 8 ohms. Plot I_b versus fault impedance.
3.6. Analyze the system of Figure 3.5 for a LL fault at bus B with zero fault impedance.
3.7. Repeat Example 3.3 for a 2LG fault at bus C but with $Z_g = 0$.
3.8. Repeat Example 3.3 for a 2LG fault at bus C but with $Z_f = 0$.
3.9. Repeat Example 3.3 for a 2LG fault at bus C but with $Z_f = Z_g = 0$.
3.10. Analyze the system of Figure 3.5 for a 2LG fault at bus B with $Z_f = Z_g = 0$.
3.11. Graph the results of Example 3.3 and problems 7-9 plotting I_b versus Z_f, then I_b versus Z_g. Also plot a series of phasor diagrams similar to Figure 3.17, showing how the various phasors are changed by fault impedance.
3.12. Repeat Example 3.4 for a 3ϕ fault at bus C but with a fault impedance of 0 and then 8 ohms. Plot I_a versus Z_f.
3.13. Analyze the system of Figure 3.5 for a 3ϕ fault at bus A, bus B, and bus D.
3.14. Compute the total current leaving bus B for a LL fault at bus C in Example 3.2.
3.15. Determine the sequence network connections for the network condition shown in Figure P3.15.

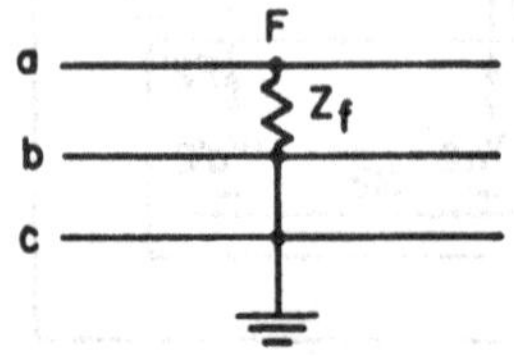

Fig. P3.15.

3.16. Determine the sequence network connections for the network condition shown in Figure P3.16.

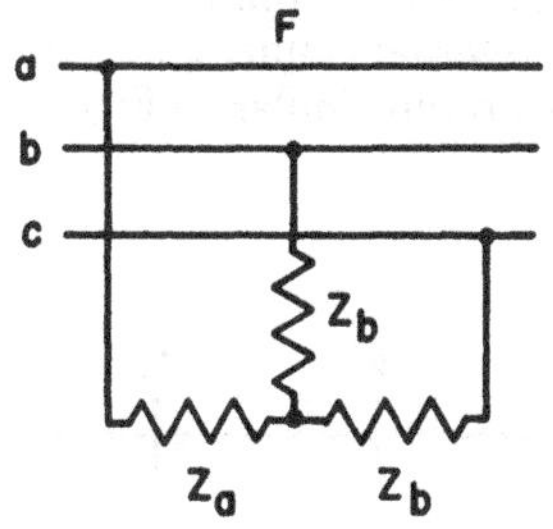

Fig. P3.16.

3.17. Consider an unbalanced Δ load at point *F*. Under what conditions can such a load be represented by symmetrical components? Explain fully.

3.18. Determine the sequence network connections for the network condition shown in Figure P3.18.

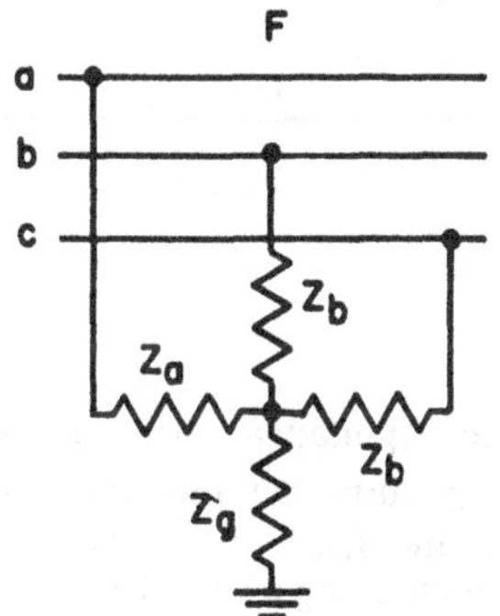

Fig. P3.18.

3.19. Consider the problem of Figure P3.18 but where the phase impedances are Z_a, Z_b, and Z_c, with $Z_b \neq Z_c$ as in problem 3.18. How can such a problem be solved by symmetrical components?

3.20. Find the two-port Z parameters for the circuits shown in Figure P3.20. All impedances are in ohms.

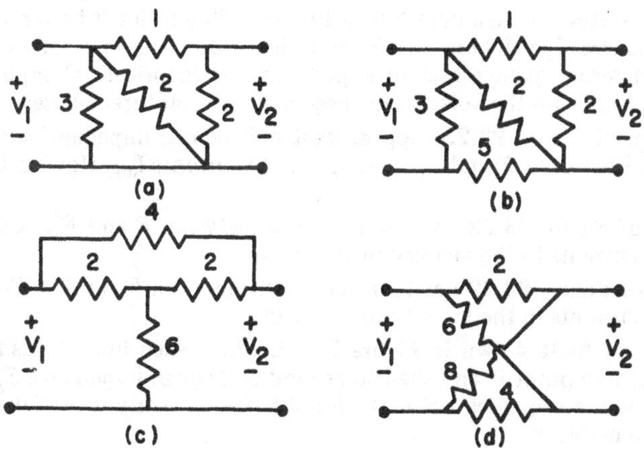

Fig. P3.20.

3.21. Find the Y parameters for the circuits shown in Figure P3.20. Show that $\mathbf{Y} = \mathbf{Z}^{-1}$ in every case.

3.22. Determine the two-port equivalent for the circuit of Figure P3.22 with terminals F and F' brought out as the only external connections. Find both the positive and negative sequence equivalents, with your result similar to Figure 3.24.

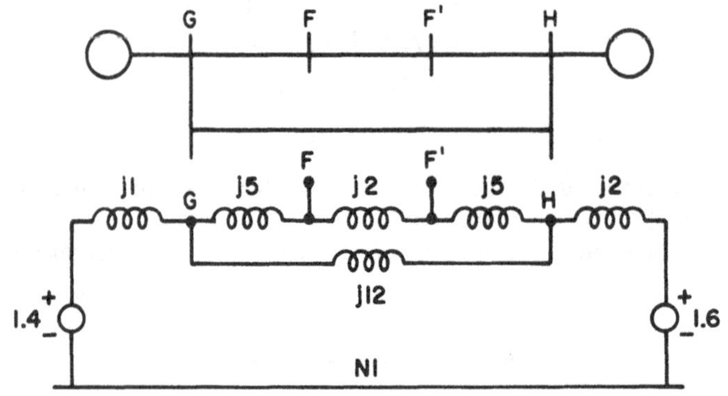

Fig. P3.22.

3.23. Repeat problem 3.22, with the impedance between F and F' removed.

3.24. In the case of series faults, we propose a connection be made from F to F' external to the normal system in such a way that $\mathbf{I}_{abc} = \mathbf{I}'_{abc}$. That is, if $\mathbf{I}_{abc}$ leaving F has any value, it will have that same value upon entering at F'. Use this to establish that the sequence currents are also equal, i.e., $\mathbf{I}_{012} = \mathbf{I}'_{012}$.

3.25. Suppose that (contrary to the usual assumption with a series fault that $\mathbf{I}_{abc} = \mathbf{I}'_{abc}$) we let $I_a = I'_a$ and permit phases b and c to become crossed, i.e., $I_b = I'_c$ and $I_c = I'_b$. Derive the sequence network connection to represent such a condition. See [16].

3.26. Find the uncoupled two-port equivalent, similar to Figure 3.28, for the system of Figure P3.22.

3.27. Find the uncoupled two-port equivalent, similar to Figure 3.28, for the system described in problem 3.23.

3.28. Consider the system of Example 3.6 in Figure 3.25 and let it be terminated with unbalanced impedances $Z_a = 1$ ohm and $Z_b = 3$ ohm $= Z_c$. Suppose further that the positive, negative, and zero sequence system impedances are identical. Compute the system currents from F to F' and the voltages at these nodes for all three phases.

3.29. In the system of Figure P3.22 suppose that unbalanced impedances $Z_a = j1$ and $Z_b = j5$ are connected between F and F'. Compute the currents $\mathbf{I}_{abc}$ flowing toward the fault at F.

3.30. The system of Figure P3.22 has one line open between F and F'. Compute the system voltages and currents in the vicinity of the fault.

3.31. The system of Figure P3.22 has two lines open between F and F'. Compute the system voltages and currents in the vicinity of the fault.

3.32. Consider the 3ϕ fault shown in Figure P3.32, where each line at F is faulted through an impedance Z_f to a point W and then to ground G through impedance Z_g.
   (a) Determine the sequence networks for this fault, clearly identifying points F, W, and G in each network.
   (b) Compute the voltage $V_{FW} = V_{AF} - V_{WG}$.
   (c) Distinguish between the ground G and the neutral W in each sequence network.

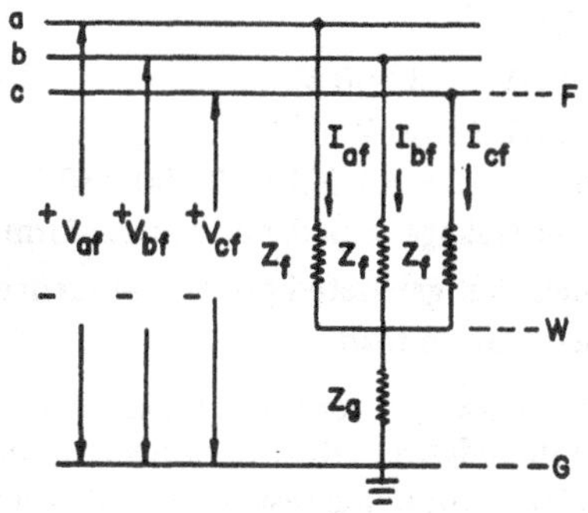

Fig. P3.32.

3.33. Generator G supplies its load through a power transformer bank T. In order to limit the SLG fault current for faults on the generator leads, the generator neutral is "high-resistance" grounded through the grounding transformer TG and resistor R. The balanced bank of capacitors C represents surge protection capacitors plus the lumped capacitances to ground of the generator leads, generator windings, and transformer T windings. See Figure P3.33.

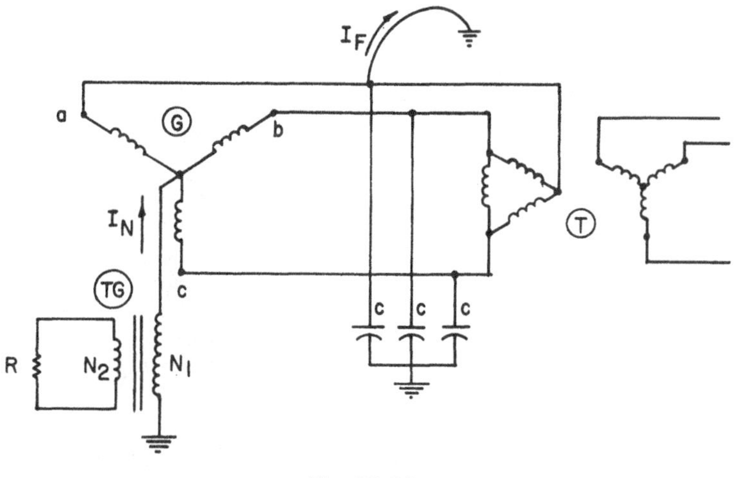

Fig. P3.33.

In order to minimize the danger of damaging transient overvoltages, the following criterion for sizing R is sometimes advocated: Select R so that the power dissipated in it during a SLG fault on the generator leads is equal to ΣQ_c where ΣQ_c is equal to total reactive power (magnetizing vars) generated in all capacitances during the SLG fault.

Capacitance: $C = 0.40$ μF/phase

Generator:

13,800 V (LL)
25,000 kW (3ϕ)
31,250 kVA (3ϕ)
60 Hz
$X_1 = 1.20$ pu based on generator ratings
$X_2 = 0.10$ pu based on generator ratings
$X_0 = 0.05$ pu based on generator ratings

Grounding transformer TG:

Rated HV = 13,800 V

Rated LV = 480 V

Turns ratio $a = N1/N2 = 13{,}800/480$

Neglect leakage impedance of transformer

Generator excitation is such that generated positive sequence voltage is $E_{a1} = 1.00$ pu. The SLG is assumed to occur at no load.

(a) Draw the sequence networks properly interconnected to represent the SLG fault on phase a. Evaluate all impedances either in ohms or in pu.
(b) Calculate the value of the grounding resistor R that will satisfy the given criterion.
(c) Calculate the fault current I_F and the generator neutral current I_N, that correspond to the value of R found in (b).
(d) Find the minimum kVA rating of the distribution transformer TG.

chapter 4

Sequence Impedance of Transmission Lines

In the previous chapter the various common shunt and series fault connections are analyzed and sequence network connections are derived for each situation. The application of these ideas to physical systems requires detailed knowledge of the actual impedances which make up each of the sequence networks. This means that each component of the power system—the lines, machines, transformers, and loads must be analyzed to determine the impedance of each to the flow of positive, negative, and zero sequence currents.

The approach taken here to the understanding and computation of sequence impedances is a practical and at times heuristic one. A more rigorous and exhaustive treatment of the subject is to be found in many references, particularly those by Fortescue [1], Wagner and Evans [10], Clarke [11], Kimbark [17, 18, and 19], the Westinghouse Corporation [14] and the General Electric Company [20]. Most of these resources provide additional references. Stevenson [21] also gives an excellent brief review of the subject. Since the subject is so well documented, the emphasis here will be on providing insight and understanding and on establishing correct, expedient methods for determining sequence impedances.

The subject is conveniently divided as follows. Chapters 4 and 5 consider the sequence impedance and admittance of transmission lines. Machines are discussed in Chapter 6 and transformers in Chapter 7.

This analysis of transmission line impedance is restricted to the case of three-phase lines unless otherwise noted. Even with this restriction there are several important areas to investigate. The most obvious is the self impedance of the line to the flow of balanced three-phase currents and to uniphase or zero sequence currents. The effect of line transpositions is important. Since it is becoming more common to *not* transpose lines, what effect does this have on the phase impedances? The computation of mutual impedances is discussed, and the effect of currents in nearby circuits or in the ground below an overhead line is also examined.

4.1 Positive and Negative Sequence Impedances of Lines

A transmission line is a passive device, and if transposed it presents identical impedances to the flow of currents in each of the phase conductors. Furthermore, the phase sequence of the applied voltage makes no difference since the voltage drops are the same for an *a-b-c* sequence as for *a-c-b*. Therefore the positive and negative sequence impedances are identical, or

$$Z_1 = Z_2 = R_1 + jX_1 \quad \Omega/\text{phase} \tag{4.1}$$

Usually we compute the impedance on a per unit length basis and multiply by the line length[1]

$$Z_1 = (r_1 + jx_1)s \quad \Omega/\text{phase} \tag{4.2}$$

where

r_1 = line resistance to positive sequence currents of each phase, Ω/unit length
x_1 = line reactance to positive sequence currents of each phase, Ω/unit length
s = line length

The resistance r_1 is simply the resistance of one phase conductor or bundle per unit length. It is usually tabulated by wire size as a function of temperature and for various frequencies of interest.

We assume that this resistance is the same in all three phases. The reactance x_1 is often thought of as having two components: one due to all flux linked internally and externally out to a radius of one unit of length (one foot, for example) and the other due to flux external to one unit of length and out to the (geometric mean) center of the conductor carrying the return current. Thus, from [9], using English units,[2]

$$\ell_1 = 0.3219 \ln \frac{D_m}{D_s} \quad \text{mH/mi}$$

or

$$x_1 = 4.657 \times 10^{-3} f \log_{10} \frac{D_m}{D_s}$$

$$= 2.020 \times 10^{-3} f \ln \frac{D_m}{D_s} \quad \Omega/\text{mi} \tag{4.3}$$

where

D_m = mutual geometric mean distance (GMD)
D_s = self geometric mean distance, or geometric mean radius (GMR)

These terms need further amplification (see Stevenson [9] and Woodruff [22]). In general terms when a phase conductor consists of several cylindrical, nonmagnetic strands, we compute D_m and D_s between two conductor groups x (consisting of m strands) and y (consisting of n strands) as follows:

$$D_m = \begin{pmatrix}\text{product of distances from all } m \text{ strands} \\ \text{of } x \text{ to all } n \text{ strands of } y\end{pmatrix}^{1/mn}$$

$$D_s = \begin{pmatrix}\text{product of distances from all } m \text{ strands} \\ \text{of } x \text{ to itself and to all other strands of } x\end{pmatrix}^{1/m^2}$$

where the distance from a strand "to itself" is sometimes called the self GMD and equals $0.7788r$ (r = radius) for a cylindrical wire. To be more specific, for three

[1] We use the common notation here of capital letters R, X, and Z for total line ohms and lower case r, x, and z for ohms/unit length.
[2] We use the notation "ln" for "log to the base e."

phase conductors separated by distances D_{ab}, D_{bc}, and D_{ca} center to center, we compute

$$D_m = (D_{ab}D_{bc}D_{ca})^{1/3} \triangleq D_{eq} \tag{4.4}$$

which is sometimes called the "equivalent spacing." Also, for a transposed line the self GMD is

$$D_s = (D_{s1}D_{s2}D_{s3})^{1/3} \tag{4.5}$$

where D_{si} = self GMD of phase a in position i of a transposition and where the units of D_s and D_m must be the same. For example, if the frequency is 60 Hz and length is in miles, equation (4.3) may be written in the more useful forms

$$x_1 = 0.2794 \log_{10} \frac{D_m}{D_s} \quad \Omega/\text{mi} \tag{4.6}$$

or

$$x_1 = 0.1213 \ln \frac{D_m}{D_s} \quad \Omega/\text{mi} \tag{4.7}$$

Then we define at 60 Hz

$$x_a = x_1 \text{ (1 ft spacing)} = 0.1213 \ln \frac{1}{D_s} \quad \Omega/\text{mi} \tag{4.8}$$

$$x_d = x_1 \text{ (spacing factor)} = 0.1213 \ln D_m \quad \Omega/\text{mi} \tag{4.9}$$

Obviously equation (4.8) is a function only of the conductor and conductor arrangement if bundled. For single-wire conductors x_a is tabulated in wire tables as shown in Tables B.4 to B.15 of Appendix B. Equation (4.9), on the other hand, does not depend on the conductor type at all and is a function only of D_m. Values for x_d are tabulated in Table B.24 and are referred to as "inductive reactance spacing factors." These values need only be added to find x_1, i.e.,

$$x_1 = x_a + x_d \quad \Omega/\text{mi} \tag{4.10}$$

4.2 Mutual Coupling

One of the problems inherent in representing transmission lines is the situation wherein any wire of a group of parallel wires carries a nonzero current. In such cases any conductor which parallels the given current-carrying wire will experience an induced voltage for each unit of length of the parallel because the flux linkages of that nearby circuit are not zero. In terms of field concepts we write the flux linkages λ_{21}, or the flux linking circuit 2 due to a current in circuit 1, as [12]

$$\lambda_{21} = \oint_{c_2} A \cdot ds_2 > 0 \quad \text{Wb turn} \tag{4.11}$$

where

$$A = \text{magnetic vector potential} = \frac{\mu_0 I_1}{4\pi} \int_{c_1} \frac{1}{r}\, ds_1 \quad \text{Wb/m}$$

ds_2 = element of length along circuit 2

ds_1 = element of length along circuit 1 (4.12)

Obviously, a mutual coupling exists any time the magnetic vector potential A is greater than zero. When circuit 1 (the inducing circuit) is a balanced three-phase circuit and I_1 is considered the superposition of the three phase currents, A is zero and no mutual induction takes place. In the case of zero sequence currents, however, $I_1 = 3I_{a0}$ and the mutually induced voltage may be large. We represent this mutual coupling as a mutual inductance M where

$$M = \lambda_{21}/I_1 \quad \text{H} \tag{4.13}$$

and treat this problem circuitwise in much the same way that we analyze a transformer.

Suppose that two parallel circuits are designated a and b, as shown in Figure 4.1, and have self impedances Z_{aa} and Z_{bb} respectively and mutual impedance Z_{ab}. Let currents I_a and I_b enter the unprimed ends of each circuit. We assume that the circuit representation of "ground" implies a perfectly conducting plane. Also, as noted in Figure 4.1, we designate voltage drops to ground at each end of

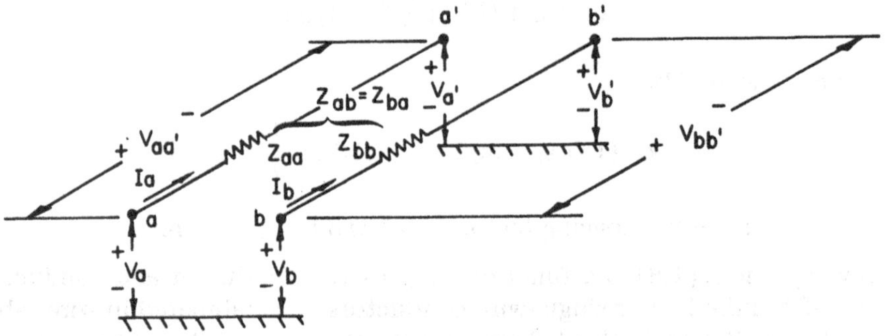

Fig. 4.1. Two circuits a and b with mutual coupling.

the two circuits and also identify a voltage drop in the direction of the assumed current. The equation for these voltage drops along the wires is

$$\begin{bmatrix} V_{aa'} \\ V_{bb'} \end{bmatrix} = \begin{bmatrix} V_a \\ V_b \end{bmatrix} - \begin{bmatrix} V_{a'} \\ V_{b'} \end{bmatrix} = \begin{bmatrix} Z_{aa} & Z_{ab} \\ Z_{ab} & Z_{bb} \end{bmatrix} \begin{bmatrix} I_a \\ I_b \end{bmatrix} \tag{4.14}$$

where we recognize that $Z_{ab} = Z_{ba}$ in a linear, passive, bilateral network. Equation (4.14) can be justified and further insight gained by viewing this situation as shown in the one-turn transformer equivalent of Figure 4.2. Here the wires a-a' and b-b' are viewed as turns of a one-turn (air core) transformer the "core" of which is shown. If we apply a voltage $V_{aa'}$ with polarity shown and this causes a current I_a to increase in the direction shown, a flux ϕ_{ba} (flux linking coil b due to I_a) will be increasing in the direction shown. Then, by Lenz's law a flux ϕ_{ab} will be established to oppose ϕ_{ba}. This requires that the b terminal be positive with respect to b', or an induced current I_b' will flow if the circuit b-b' is closed through a load. This is indeed a voltage drop in the direction b-b' and adds to the self impedance drop $I_b Z_{bb}$ of that circuit as in the V_b equation (4.14). A similar argument establishes the validity of the V_a equation (4.14). Following the usual dot convention, we can dot the a and b ends of the two lines to indicate the polarity of the induced voltage. The dot convention is particularly convenient in situations where the polarity is not obvious. Figure 4.1 or 4.2 and equation (4.14) are all

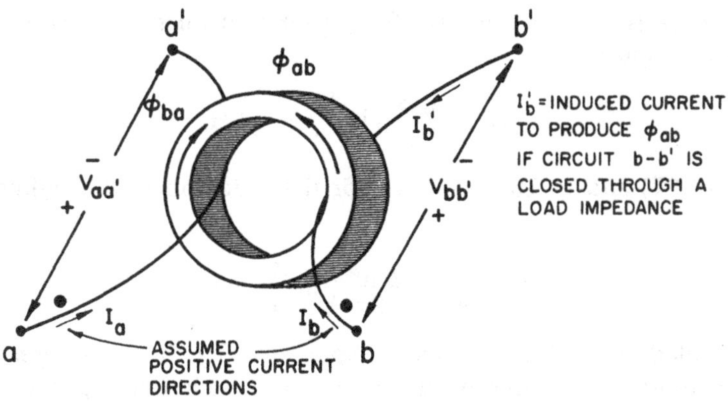

Fig. 4.2. A one-turn transformer equivalent.

that we need to establish the correct voltage and current relationships in mutually coupled circuits.

4.3 Self and Mutual Inductances of Parallel Cylindrical Wires

Inductance is usually defined by dividing the flux linkages by the current. For self inductance the flux linkages of a given circuit are divided by the current in that circuit, i.e.,

$$L = \lambda_{11}/I_1 \quad \text{H} \tag{4.15}$$

where λ_{11} is the flux linkage in weber turns of circuit 1 due to current I_1 in amperes. Mutual inductance is similarly defined, as given by equation (4.13).

The computation of the self inductance of a straight finite cylinder is shown in several references ([23], for example) and is usually divided into two components

$$L = L_i + L_e \quad \text{H} \tag{4.16}$$

where L_i is the partial self inductance of the wire due to internal flux linkages and L_e is the partial self inductance due to flux linkages outside the wire. Then it can be shown that for uniform current density

$$L_i = \mu_w s/8\pi \quad \text{H} \tag{4.17}$$

where

μ_w = permeability of the wire H/m
= $4\pi \times 10^{-7}$ H/m for nonferrous materials
s = length of wire in meters

The external partial flux linkages are [23]

$$\lambda_e = \frac{\mu_m I_1}{2\pi}\left(s \ln \frac{s + \sqrt{s^2 + r^2}}{r} - \sqrt{s^2 + r^2} + r\right) \quad \text{Wb turn} \tag{4.18}$$

where

μ_m = permeability of the medium surrounding the wire
= $4\pi \times 10^{-7}$ H/m for air
r = radius of the wire in meters

If $r << s$, as is always the case for power transmission lines, (4.18) may be simplified to compute

$$L_e = \frac{\mu_m s}{2\pi}\left(\ln\frac{2s}{r} - 1\right) \text{ H} \qquad (4.19)$$

Combining (4.17) and (4.19), we have for the inductance of a cylindrical wire s meters long

$$L = \frac{\mu_w s}{8\pi} + \frac{\mu_m s}{2\pi}\left(\ln\frac{2s}{r} - 1\right) \text{ H} \qquad (4.20)$$

(A more detailed derivation is given in Appendix D.) In most cases we are concerned with nonferrous wires in an air medium such that μ_w and μ_m are both equal to $4\pi \times 10^{-7}$ henry/meter. This assumes, for composite conductors like ACSR, that the ferrous material carries negligible current. With this approximation we write the inductance as

$$\begin{aligned} L &= \frac{1}{2} \times 10^{-7} s + 2 \times 10^{-7} s \left(\ln\frac{2s}{r} - 1\right) \\ &= 2 \times 10^{-7} s \left[\frac{1}{4} + \left(\ln\frac{2s}{r} - 1\right)\right] \text{ H} \qquad (4.21) \end{aligned}$$

But

$$\frac{1}{4} = \ln\frac{1}{e^{-1/4}} = \ln\frac{1}{0.779} \qquad (4.22)$$

and we recognize for cylindrical wires that $0.779r = D_s$. Using this relationship, we further simplify the inductance formula to write

$$\ell = 2 \times 10^{-7}\left(\ln\frac{2s}{D_s} - 1\right) \text{ H/m} \qquad (4.23)$$

where both D_s and s are in meters or any other consistent units of length.

Some engineers prefer the inductance to be specified in henry/mile, while some prefer henry/kilometer. Also, many prefer using base 10 logarithms instead of the natural (base e) logarithms. Both of these choices affect the constant in equation (4.23). Suppose we replace the quantity 2×10^{-7} by a constant "k" to write

$$\ell = k\left(\ln\frac{2s}{D_s} - 1\right) \text{ H/unit length} \qquad (4.24)$$

then the constant k can be chosen according to the user's preference for English or metric units and for base e or base 10 logarithms. Several choices of the constant k are given in Table 4.1. Obviously, s and D_s must always be in the same units. It seems odd that this inductance should be a function of s, the line length. We will show later that the numerator $2s$ cancels out in every practical line configuration where a return current path is present.

Following a similar logic, we define mutual inductance to be [23, 24]

$$m = k\left(\ln\frac{2s}{D_m} - 1\right) \text{ H/unit length} \qquad (4.25)$$

Table 4.1. Inductance Multiplying Constants

Constant	*Unit of Length*	*Natural Logarithm* (ln)	*Base 10 Logarithm* ($\log_{10}$)
k	km	0.2000×10^{-3}	0.4605×10^{-3}
	mi	0.3219×10^{-3}	0.7411×10^{-3}
2πk	km	1.257×10^{-3}	2.893×10^{-3}
	mi	2.022×10^{-3}	4.656×10^{-3}
f = 50 Hz			
fk	km	0.01000	0.02302
	mi	0.01609	0.03705
ωk	km	0.06283	0.1446
	mi	0.10111	0.2328
f = 60 Hz			
fk	km	0.01200	0.02763
	mi	0.01931	0.04446
ωk	km	0.07539	0.1736
	mi	0.12134	0.2794

Note: 1.6093 km = 1.0 mi; f = 50 Hz, ω = 314.159 rad/sec; f = 60 Hz, ω = 376.991 rad/sec.

where D_m is the geometric mean distance between the conductors. This definition comes directly from equation (4.13) where the flux linkage term is similar to (4.18) except r (radius) is replaced by D (the distance between wires). Boast [12] shows that this distance is actually D_m, the geometric mean distance or the distance between the centers of cylindrical wires.

All of the foregoing assumes that the current has uniform density, an assumption that is often accepted. This assumption introduces a slight error for large conductors, even at power frequencies. This complication, due to skin effect and proximity effect, is discussed in many references such as Rosa and Grover [25], Stevenson [9], Calabrese [24], Attwood [23] and Lewis [26]. Briefly stated, skin effect causes the current distribution to become nonuniform, with a larger current density appearing on the conductor surface than at the center. This reduces the internal flux linkages and lowers the internal inductance as compared to the uniform current density (dc) case given by (4.17). It also increases the resistance. These effects are summarized in Figure 4.3 for solid round wires. Stevenson [9] gives convenient formulas for both the resistance and internal inductance ac to dc ratios as (subscript 0 indicates dc value)

$$\alpha_R = \frac{R}{R_0} = \frac{mr}{2}\left[\frac{\text{ber}\, mr\, \text{bei}'\, mr - \text{bei}\, mr\, \text{ber}'\, mr}{(\text{ber}'\, mr)^2 + (\text{bei}'\, mr)^2}\right] \tag{4.26}$$

$$\alpha_L = \frac{L_i}{L_{i_0}} = \frac{4}{mr}\left[\frac{\text{ber}\, mr\, \text{ber}'\, mr + \text{bei}\, mr\, \text{bei}'\, mr}{(\text{ber}'\, mr)^2 + (\text{bei}'\, mr)^2}\right] \tag{4.27}$$

where r is the conductor radius and m, in units of $(\text{length})^{-1}$, is defined as

$$m = \sqrt{\omega\mu\sigma} \tag{4.28}$$

where $\mu = \mu_r\mu_o$. Stevenson [9] writes

$$mr = 0.0636\sqrt{\mu_r f/R_0} \tag{4.29}$$

where f is the frequency in Hz, μ_r is the relative permeability, and R_0 is the dc resistance in ohm/mile. The Bessel functions used in (4.26) and (4.27) were origi-

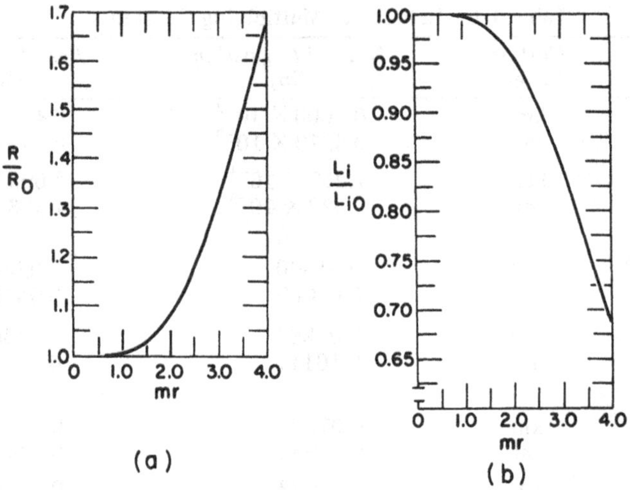

Fig. 4.3. Ratios of (a) ac to dc resistance and (b) internal inductance as a function of the parameter *mr*. (From *Elements of Power System Analysis* by W. D. Stevenson. Copyright McGraw-Hill, 1962. Used with permission of McGraw-Hill Book Co.)

nally derived by Lord Kelvin [25] and are defined as follows [9]:

$$\text{ber}\ mr = 1 - \frac{(mr)^4}{2^2 \cdot 4^2} + \frac{(mr)^8}{2^2 \cdot 4^2 \cdot 6^2 \cdot 8^2} - \cdots$$

$$\text{bei}\ mr = \frac{(mr)^2}{2^2} - \frac{(mr)^6}{2^2 \cdot 4^2 \cdot 6^2} + \cdots \tag{4.30}$$

and

$$\text{ber}'\ mr = \frac{d}{d(mr)}\ \text{ber}\ mr = \frac{1}{m}\frac{d}{dr}\ \text{ber}\ mr$$

$$\text{bei}'\ mr = \frac{d}{d(mr)}\ \text{bei}\ mr = \frac{1}{m}\frac{d}{dr}\ \text{bei}\ mr \tag{4.31}$$

Using the above refinements for internal inductance we may rewrite (4.17) as

$$\ell_i = \alpha_L \mu_w / 8\pi = \mathrm{k}\alpha_L/4 \quad \text{H/unit length}$$

This changes the inductance formula (4.24) to

$$\ell = \mathrm{k}\left(\ln \frac{2se^{\alpha_L}}{D_s} - 1\right) \text{ H/unit length} \tag{4.32}$$

For many computations sufficient accuracy may be obtained by ignoring this effect.

4.4 Carson's Line

A monumental paper describing the impedance of an overhead conductor with earth return was written in 1923 by Carson [27]. This paper, with certain modifications, has since served as the basis for transmission line impedance calculations in cases where current flows through the earth.

Carson considered a single conductor *a* one unit long and parallel to the ground as shown in Figure 4.4. The conductor carries a current I_a with a return

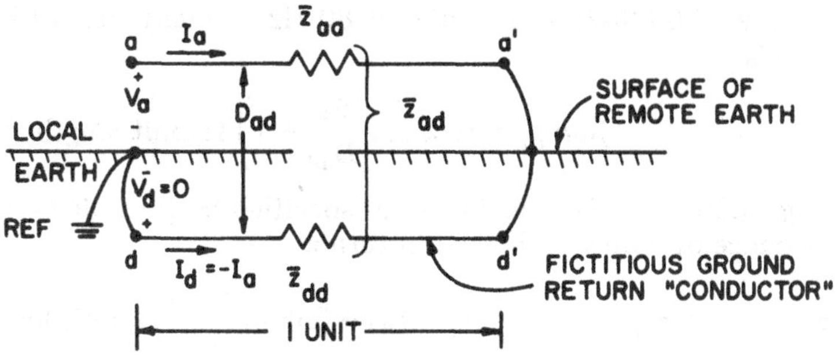

Fig. 4.4. Carson's line with ground return.

through circuit d-d' beneath the surface of the earth. The earth is considered to have a uniform resistivity and to be of infinite extent. The current I_a in the ground spreads out over a large area, seeking the lowest resistance return path and satisfying Kirchhoff's law to guarantee an equal voltage drop in all paths. Wagner and Evans [10] show that Carson's line can be thought of as a single return conductor with a self GMD of 1 foot (or 1 meter), located at a distance D_{ad} feet (or meters) below the overhead line, where D_{ad} is a function of the earth resistivity ρ. The parameter D_{ad} is adjusted so that the inductance calculated with this configuration is equal to that measured by test.

From equation (4.14) we write for Carson's line,

$$\begin{bmatrix} V_{aa'} \\ V_{dd'} \end{bmatrix} = \begin{bmatrix} V_a - V_{a'} \\ V_d - V_{d'} \end{bmatrix} = \begin{bmatrix} \bar{z}_{aa} & \bar{z}_{ad} \\ \bar{z}_{ad} & \bar{z}_{dd} \end{bmatrix} \begin{bmatrix} I_a \\ -I_a \end{bmatrix} \text{ V/unit length} \tag{4.33}$$

where V_a, $V_{a'}$, V_d and $V_{d'}$ are all measured with respect to the same reference. Since $V_d = 0$ and $V_{a'} - V_{d'} = 0$, we solve for V_a by subtracting the two equations to find

$$V_a = (\bar{z}_{aa} + \bar{z}_{dd} - 2\bar{z}_{ad}) I_a = z_{aa} I_a \tag{4.34}$$

by definition, where we clearly distinguish between $\bar{z}$ and z. Thus

$$z_{aa} \triangleq \bar{z}_{aa} + \bar{z}_{dd} - 2\bar{z}_{ad} \quad \Omega\text{/unit length} \tag{4.35}$$

Using equations (4.24) and (4.25) and ignoring skin effect, we may write the self and mutual impedances of equation (4.33) as follows. The self impedance of line a is

$$\bar{z}_{aa} = r_a + j\omega \ell_a = r_a + j\omega k \left(\ln \frac{2s}{D_{sa}} - 1 \right) \Omega\text{/unit length} \tag{4.36}$$

Similarly,

$$\bar{z}_{dd} = r_d + j\omega k \left(\ln \frac{2s}{D_{sd}} - 1 \right) \Omega\text{/unit length} \tag{4.37}$$

where we arbitrarily set $D_{sd} = 1$ unit length. Carson found that the earth resistance r_d is a function of frequency and he derived the empirical formula

$$\begin{aligned} r_d &= 1.588 \times 10^{-3} f \quad \Omega\text{/mi} \\ &= 9.869 \times 10^{-4} f \quad \Omega\text{/km} \end{aligned} \tag{4.38}$$

which has a value of 0.09528 ohms/mile at 60 Hz. Finally, from (4.25) the mutual impedance is

$$\bar{z}_{ad} = j\omega m_{ad} = j\omega k \left(\ln \frac{2s}{D_{ad}} - 1\right) \ \Omega\text{/unit length} \tag{4.39}$$

Combining (4.36), (4.37), and (4.39) as specified by formula (4.35), we compute the impedance of wire a with earth return as

$$z_{aa} = \bar{z}_{aa} + \bar{z}_{dd} - 2\bar{z}_{ad} = (r_a + r_d) + j\omega k \ln \frac{D_{ad}^2}{D_{sa}} \ \Omega\text{/unit length} \tag{4.40}$$

As pointed out in (4.37), it is common to let $D_{sd} = 1$ unit of length (in the same units as D_{ad} and D_{sa}) such that the logarithm term is written as

$$\ln \frac{D_{ad}^2}{D_{sa}D_{sd}} = \ln \frac{D_{ad}^2}{(D_{sa})(1)}$$

The argument of this logarithm has the dimension (length^2/length^2) or it is dimensionless. However it *appears* to have the dimension of length. For this reason it has become common practice to define a quantity D_e as

$$D_e = D_{ad}^2/D_{sd} \quad \text{(unit length)}^2\text{/unit length} \tag{4.41}$$

Then we write (4.40) as

$$z_{aa} = (r_a + r_d) + j\omega k \ln \frac{D_e}{D_{sa}} \ \Omega\text{/unit length} \tag{4.42}$$

This expression, or one similar to it, is commonly found in the literature. It could be argued that D_e is not a distance since it is numerically equal to D_{ad}^2, which has dimensions (unit length)^2. We will take equation (4.41) as the definition of D_e.

The self impedance of a circuit with earth return depends upon the impedance of the earth which in turn fixes the value of D_e. Wagner and Evans [10] discuss this problem in some detail and offer a physical explanation of Carson's original work. Table 4.2 gives a summary of their description for various earth conditions.

Table 4.2. D_e **for Various Resistivities at 60 Hz**

Return Earth Condition	*Resistivity* (Ωm)	D_e *(ft)*	D_{ad} *(ft)*
Sea water	0.01–1.0	27.9–279	5.28–16.7
Swampy ground	10–100	882–2790	29.7–52.8
Average damp earth	*100*	*2790*	*52.8*
Dry earth	1000	8820	93.9
Pure slate	10^7	882,000	939
Sandstone	10^9	8,820,000	2970

The quantity D_e is a function of both the earth resistivity ρ and the frequency f and is defined by the relation

$$D_e = 2160\sqrt{\rho/f} \ \text{ft} \tag{4.43}$$

If no actual earth resistivity data is available, it is not uncommon to assume ρ to be 100 ohm-meter, in which case the italic quantities in Table 4.2 apply. Wagner and Evans [10] provide data for ρ at various locations in the United States.

4.5 Three-Phase Line Impedances

To find the impedance of a three-phase line we proceed in exactly the same way as for the single line in the previous section. The configuration of the circuits is shown in Figure 4.5 where the impedances, voltages, and currents are identified.

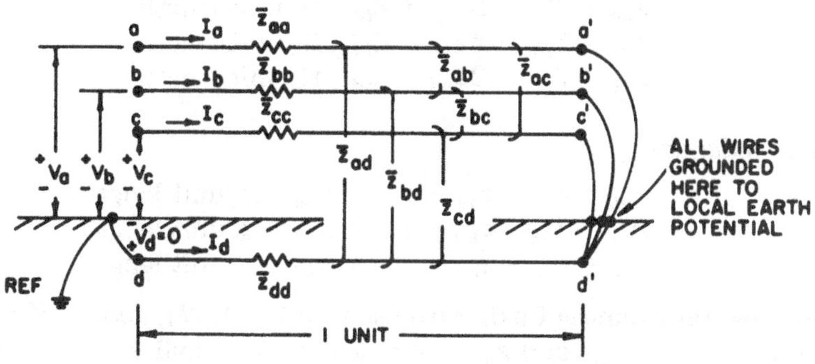

Fig. 4.5. Three-phase line with earth return.

Since all wires are grounded at the remote point a'-b'-c', we recognize that

$$I_d = -(I_a + I_b + I_c) \tag{4.44}$$

Then, proceeding as before, we write the voltage drop equations in the direction of current flow as follows:

$$\begin{bmatrix} V_{aa'} \\ V_{bb'} \\ V_{cc'} \\ V_{dd'} \end{bmatrix} = \begin{bmatrix} V_a - V_{a'} \\ V_b - V_{b'} \\ V_c - V_{c'} \\ V_d - V_{d'} \end{bmatrix} = \begin{bmatrix} \bar{z}_{aa} & \bar{z}_{ab} & \bar{z}_{ac} & \bar{z}_{ad} \\ \bar{z}_{ab} & \bar{z}_{bb} & \bar{z}_{bc} & \bar{z}_{bd} \\ \bar{z}_{ac} & \bar{z}_{bc} & \bar{z}_{cc} & \bar{z}_{cd} \\ \bar{z}_{ad} & \bar{z}_{bd} & \bar{z}_{cd} & \bar{z}_{dd} \end{bmatrix} \begin{bmatrix} I_a \\ I_b \\ I_c \\ I_d \end{bmatrix} \text{ V/unit length} \tag{4.45}$$

We call these equations the "primitive voltage equations." The impedance of the line is usually thought of as the ratio of the voltage to the current seen "looking in" the line at one end. We select a voltage reference at the left end of the line and solve (4.45) for the voltages V_a, V_b, and V_c. We can do this since current I_d is known and because we may write

$$V_{a'} - V_{d'} = 0, \quad V_{b'} - V_{d'} = 0, \quad V_{c'} - V_{d'} = 0 \tag{4.46}$$

for the condition of the connection at the receiving end of the line. Since $V_d = 0$, we subtract the fourth equation of (4.45) from the first with the result:

$$V_a - (V_{a'} - V_{d'}) = (\bar{z}_{aa} - 2\bar{z}_{ad} + \bar{z}_{dd}) I_a + (\bar{z}_{ab} - \bar{z}_{ad} - \bar{z}_{bd} + \bar{z}_{dd}) I_b$$
$$+ (\bar{z}_{ac} - \bar{z}_{ad} - \bar{z}_{cd} + \bar{z}_{dd}) I_c$$

For convenience we write this result as $V_a = z_{aa}I_a + z_{ab}I_b + z_{ac}I_c$, where we have defined new impedances z_{aa}, z_{ab}, and z_{ac}. Note that when $I_b = I_c = 0$, z_{aa} is exactly the impedance for the single line with earth return (4.40). If we repeat the above operation for the b and c phases, we have the result

$$\begin{bmatrix} V_a \\ V_b \\ V_c \end{bmatrix} = \begin{bmatrix} z_{aa} & z_{ab} & z_{ac} \\ z_{ab} & z_{bb} & z_{bc} \\ z_{ac} & z_{bc} & z_{cc} \end{bmatrix} \begin{bmatrix} I_a \\ I_b \\ I_c \end{bmatrix} \text{ V/unit length} \tag{4.47}$$

where we recognize the reciprocity of mutual inductances in a linear, passive, bilateral network ($z_{ab} = z_{ba}$, etc.). The impedance elements of (4.47) are readily found to be

Self impedances

$$\begin{aligned} z_{aa} &= \bar{z}_{aa} - 2\bar{z}_{ad} + \bar{z}_{dd} \quad \Omega/\text{unit length} \\ z_{bb} &= \bar{z}_{bb} - 2\bar{z}_{bd} + \bar{z}_{dd} \quad \Omega/\text{unit length} \\ z_{cc} &= \bar{z}_{cc} - 2\bar{z}_{cd} + \bar{z}_{dd} \quad \Omega/\text{unit length} \end{aligned} \tag{4.48}$$

Mutual impedances

$$\begin{aligned} z_{ab} &= \bar{z}_{ab} - \bar{z}_{ad} - \bar{z}_{bd} + \bar{z}_{dd} \quad \Omega/\text{unit length} \\ z_{bc} &= \bar{z}_{bc} - \bar{z}_{bd} - \bar{z}_{cd} + \bar{z}_{dd} \quad \Omega/\text{unit length} \\ z_{ac} &= \bar{z}_{ac} - \bar{z}_{ad} - \bar{z}_{cd} + \bar{z}_{dd} \quad \Omega/\text{unit length} \end{aligned} \tag{4.49}$$

To examine these impedances further we use (4.36), (4.37), and (4.39) to identify elements similar to $\bar{z}_{aa}$, $\bar{z}_{dd}$ and $\bar{z}_{ad}$ respectively. We shall call these impedances the "primitive impedances." All are listed below in terms of the physical distances involved.

Primitive self impedances

$$\begin{aligned} \bar{z}_{aa} &= r_a + j\omega k\left(\ln\frac{2s}{D_{sa}} - 1\right) \Omega/\text{unit length} \\ \bar{z}_{bb} &= r_b + j\omega k\left(\ln\frac{2s}{D_{sb}} - 1\right) \Omega/\text{unit length} \\ \bar{z}_{cc} &= r_c + j\omega k\left(\ln\frac{2s}{D_{sc}} - 1\right) \Omega/\text{unit length} \\ \bar{z}_{dd} &= r_d + j\omega k\left(\ln\frac{2s}{D_{sd}} - 1\right) \Omega/\text{unit length} \end{aligned} \tag{4.50}$$

Primitive line-to-line mutual impedances

$$\begin{aligned} \bar{z}_{ab} &= j\omega k\left(\ln\frac{2s}{D_{ab}} - 1\right) \Omega/\text{unit length} \\ \bar{z}_{bc} &= j\omega k\left(\ln\frac{2s}{D_{bc}} - 1\right) \Omega/\text{unit length} \\ \bar{z}_{ca} &= j\omega k\left(\ln\frac{2s}{D_{cd}} - 1\right) \Omega/\text{unit length} \end{aligned} \tag{4.51}$$

Primitive line-to-earth mutual impedances

$$\begin{aligned} \bar{z}_{ad} &= j\omega k\left(\ln\frac{2s}{D_{ad}} - 1\right) \Omega/\text{unit length} \\ \bar{z}_{bd} &= j\omega k\left(\ln\frac{2s}{D_{bd}} - 1\right) \Omega/\text{unit length} \\ \bar{z}_{cd} &= j\omega k\left(\ln\frac{2s}{D_{cd}} - 1\right) \Omega/\text{unit length} \end{aligned} \tag{4.52}$$

From these primitive self and mutual impedances we compute the circuit self and mutual impedances using equations (4.48) and (4.49). For simplicity we use the approximations

$$\sqrt{D_e} = D_{ad} = D_{bd} = D_{cd}, \qquad D_s = D_{sa} = D_{sb} = D_{sc} \tag{4.53}$$

and use the definition $D_{sd} = 1$. Then we compute

$$z_{aa} = (r_a + r_d) + j\omega k \ln\frac{D_e}{D_s} \quad \Omega/\text{unit length}$$

$$z_{bb} = (r_b + r_d) + j\omega k \ln\frac{D_e}{D_s} \quad \Omega/\text{unit length}$$

$$z_{cc} = (r_c + r_d) + j\omega k \ln\frac{D_e}{D_s} \quad \Omega/\text{unit length} \tag{4.54}$$

$$z_{ab} = r_d + j\omega k \ln\frac{D_e}{D_{ab}} \quad \Omega/\text{unit length}$$

$$z_{bc} = r_d + j\omega k \ln\frac{D_e}{D_{bc}} \quad \Omega/\text{unit length}$$

$$z_{ca} = r_d + j\omega k \ln\frac{D_e}{D_{ca}} \quad \Omega/\text{unit length} \tag{4.55}$$

This result is interesting since the mutual impedance terms have a resistance component. This is due to the common earth return.

Example 4.1

Compute the phase self and mutual impedances of the 69 kV line shown in Figure 4.6. Assume that the frequency is 60 Hz ($\omega = 377$) and the phase wires are 19-strand 4/0, hard-drawn copper conductors which operate at 25 C. Ignore the ground wire entirely and assume that the phase wires have the configuration shown for the entire length of the line. Assume that the earth resistivity ρ is 100 ohm-meter and that the line is 40 miles long.

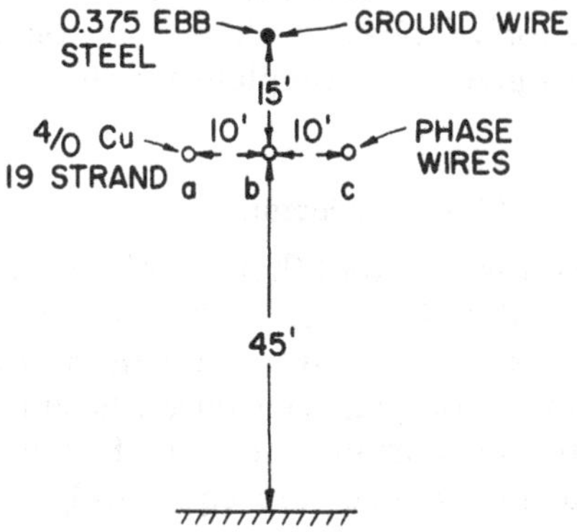

Fig. 4.6. Line configuration of a 69 kV circuit.

Solution

From Table B.4 we find the conductor values

$$r_a = r_b = r_c = 0.278 \quad \Omega/\text{mi}$$

$$D_{sa} = D_{sb} = D_{sc} = 0.01668 \text{ ft}$$

From Table 4.2 we find $D_e = 2790$ ft. At 60 Hz, $r_d = 0.09528\ \Omega$/mile and the constant ωk is, from Table 4.1, $\omega k = 0.12134$. Then, from (4.54) the self impedance terms are

$$z_{aa} = z_{bb} = z_{cc} = r_a + r_d + j\omega k \ln \frac{D_e}{D_s}$$

$$= (0.278 + 0.09528) + j(0.12134) \ln \frac{2790}{0.01668}$$

$$= 0.3733 + j1.4594 \quad \Omega/\text{mi}$$

The mutual impedances are computed from (4.55) as follows.

$$z_{ab} = r_d + j\omega k \ln \frac{D_e}{D_{ab}}$$

$$= 0.09528 + j(0.12134) \ln \frac{2790}{10} = 0.0953 + j0.6833 \quad \Omega/\text{mi}$$

$$z_{ac} = 0.09528 + j(0.12134) \ln \frac{2790}{20} = 0.0953 + j0.5992 \quad \Omega/\text{mi}$$

$$z_{bc} = z_{ab} = 0.0953 + j0.6833 \quad \Omega/\text{mi}$$

For 40 miles of line we multiply the above values by 40 to write, in matrix notation

$$\mathbf{Z}_{abc} = \begin{bmatrix} (14.932 + j58.376) & (3.812 + j27.332) & (3.812 + j23.968) \\ (3.812 + j27.332) & (14.932 + j58.376) & (3.812 + j27.332) \\ (3.812 + j23.968) & (3.812 + j27.332) & (14.932 + j58.376) \end{bmatrix} \Omega$$

It is now appropriate to ask the question, What is the impedance of phase *a*? From the matrix equation (4.47) it is apparent that the voltage drop along wire *a* depends upon all three currents. Thus the ratio of V_a to I_a does not give the entire picture of the behavior of phase *a*. The question then is best answered by equation (4.47) since this gives the complete picture of the line behavior for all conditions.

4.6 Transpositions and Twists of Line Conductors

From equation (4.47) it is apparent that the phase conductors of a three-phase circuit are mutually coupled and that currents in any one conductor will produce voltage drops in the adjacent conductors. Furthermore, these induced voltage drops may be unequal even for balanced currents since the mutual impedances depend entirely on the physical arrangement of the wires. From equation (4.55) we note that the mutual impedances are equal only when $D_{ab} = D_{bc} = D_{ac}$, an equilateral triangular spacing. In practice such a conductor arrangement is seldom used.

One means of equalizing the mutual inductances is to construct transpositions or rotations of the overhead line wires. A transposition is a physical rotation of the conductors, arranged so that each conductor is moved to occupy the next physical position in a regular sequence such as *a-b-c*, *b-c-a*, *c-a-b*, etc. Such a transposition arrangement is shown in Figure 4.7, where the conductors begin at lower

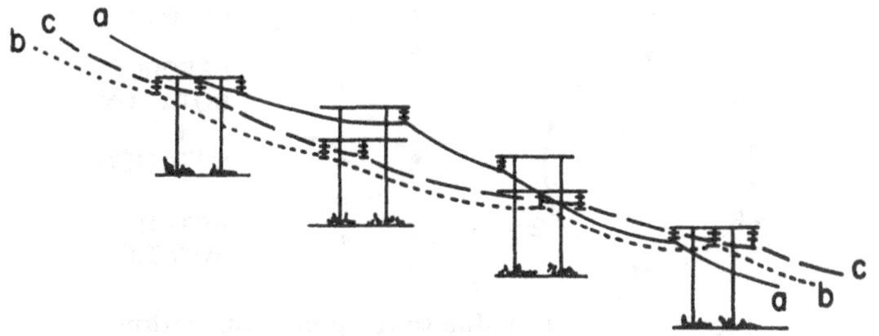

Fig. 4.7. An overhead line transposition or rotation.

right in an *a-b-c* horizontal arrangement and emerge at upper left in a *b-c-a* horizontal arrangement. If a section of line is divided into three segments of equal length separated by rotations, we say that the line is "completely transposed." Under this arrangement the current in conductor *a* would see the impedances in the first column of the impedance matrix of (4.47) for one-third the total length but would then see the impedances of the second column and finally of the third column, all in equal amounts.

The usual terminology for this rotation of conductors is to refer to each rotation such as that in Figure 4.7 as a "transposition" and to a series connection of three sections, separated by two successive rotations, as a "transposition cycle" or a "completely transposed line." We will find it convenient to refer to the rolling of three conductors as in Figure 4.7 by the name "rotation" to distinguish it from a situation where only two wires are transposed. This latter arrangement will be called a "twist." Both operations are identified in the literature as "transpositions" but we must distinguish more clearly between the two if we are to accurately compute the exact phase impedances.

4.6.1 Rotation of line conductors using $\mathbf{R}_\phi$

Mathematically we may introduce a rotation by means of a simple matrix operation. We define for this purpose a "forward (or clockwise) rotation matrix" $\mathbf{R}_\phi$ where

$$\mathbf{R}_\phi = \begin{matrix} & \begin{matrix} 2 & 3 & 1 \end{matrix} \\ \begin{matrix} 1 \\ 2 \\ 3 \end{matrix} & \begin{bmatrix} 0 & 0 & 1 \\ 1 & 0 & 0 \\ 0 & 1 & 0 \end{bmatrix} \end{matrix} \tag{4.56}$$

Premultiplying an impedance matrix by this rotation matrix has the effect of "rotating" the elements as shown in Figure 4.8; i.e., the self and mutual impedances for the conductor in position 1 are moved in the matrix to the positions initially occupied by the self and mutual impedances of conductor 2, etc. In applying the rotation matrix, we shall consider only the case where all phase wires are identical. Then

$$r_a = r_b = r_c = r, \qquad D_{sa} = D_{sb} = D_{sc} = D_s \tag{4.57}$$

Under these conditions the impedances depend, not on the phase designation, *a-b-c*, but only on the *position* the wire occupies on the tower. We designate these positions 1-2-3 as in Figure 4.8.

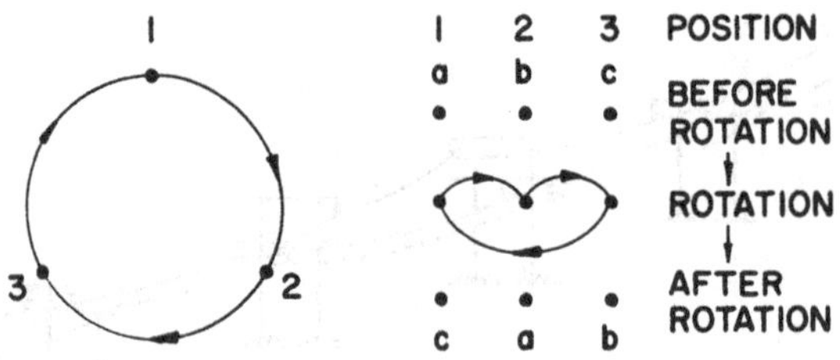

Fig. 4.8. Rotation due to $\mathbf{R}_\phi$ premultiplication.

It is helpful to recognize that the inverse of the matrix $\mathbf{R}_\phi$ exists and is equal to

$$\mathbf{R}_\phi^{-1} = \begin{array}{c} \\ 1 \\ 2 \\ 3 \end{array}\!\!\begin{array}{c} \begin{array}{ccc} 3 & 1 & 2 \end{array} \\ \begin{bmatrix} 0 & 1 & 0 \\ 0 & 0 & 1 \\ 1 & 0 & 0 \end{bmatrix} \end{array} = \mathbf{R}_\phi^t \tag{4.58}$$

From Figure 4.8 we recognize this to be a reverse or counterclockwise rotation since it rearranges the matrix elements in exactly the reverse of the reordering produced by $\mathbf{R}_\phi$. Also, we may compute the quantities

$$\mathbf{R}_\phi^2 = \mathbf{R}_\phi^{-1}, \qquad (\mathbf{R}_\phi^{-1})^2 = \mathbf{R}_\phi \tag{4.59}$$

Thus two rotations produce the same effect as one rotation in the opposite direction. This fact may be verified intuitively by an examination of Figure 4.8.

Mathematically, a rotation is the result of a linear transformation. If equation (4.47) is rewritten with numerical subscripts to reference the quantities to a physical position rather than a terminal (*a-b-c*) connection, we have

$$\begin{bmatrix} V_1 \\ V_2 \\ V_3 \end{bmatrix} = \begin{bmatrix} z_{11} & z_{12} & z_{13} \\ z_{21} & z_{22} & z_{23} \\ z_{31} & z_{32} & z_{33} \end{bmatrix} \begin{bmatrix} I_1 \\ I_2 \\ I_3 \end{bmatrix} \tag{4.60}$$

or in matrix form,

$$\mathbf{V}_{123} = \mathbf{z}_{123}\mathbf{I}_{123} \tag{4.61}$$

This equation may be transformed by the linear transformation (4.56) to compute

$$\mathbf{R}_\phi \mathbf{V}_{123} = \mathbf{R}_\phi \mathbf{z}_{123}\mathbf{I}_{123} \tag{4.62}$$

or

$$\begin{bmatrix} V_3 \\ V_1 \\ V_2 \end{bmatrix} = \begin{bmatrix} z_{31} & z_{32} & z_{33} \\ z_{11} & z_{12} & z_{13} \\ z_{21} & z_{22} & z_{23} \end{bmatrix} \begin{bmatrix} I_1 \\ I_2 \\ I_3 \end{bmatrix}$$

The result of this operation is to place the bottom row of $\mathbf{z}_{123}$ on top and move the other two rows down. Note however from (4.62) that we may compute

$$\mathbf{R}_\phi \mathbf{V}_{123} = (\mathbf{R}_\phi \mathbf{z}_{123} \mathbf{R}_\phi^{-1})\mathbf{R}_\phi \mathbf{I}_{123} \tag{4.63}$$

Carrying out the indicated operation, we have

$$\begin{bmatrix} V_3 \\ V_1 \\ V_2 \end{bmatrix} = \begin{bmatrix} z_{33} & z_{31} & z_{32} \\ z_{13} & z_{11} & z_{12} \\ z_{23} & z_{21} & z_{22} \end{bmatrix} \begin{bmatrix} I_3 \\ I_1 \\ I_2 \end{bmatrix} \tag{4.64}$$

or

$$\mathbf{V}_{312} = \mathbf{z}_{312}\mathbf{I}_{312} \tag{4.65}$$

The result of postmultiplying a matrix by $\mathbf{R}_\phi^{-1}$ is to move the third column to the first position. These operations are noted compactly by (4.65).

In a similar way we may show that the transformation $\mathbf{R}_\phi^{-1}$ produces the following result.

$$\mathbf{R}_\phi^{-1}\mathbf{V}_{123} = (\mathbf{R}_\phi^{-1}\mathbf{z}_{123}\mathbf{R}_\phi)\mathbf{R}_\phi^{-1}\mathbf{I}_{123}$$

$$\begin{bmatrix} V_2 \\ V_3 \\ V_1 \end{bmatrix} = \begin{bmatrix} z_{22} & z_{23} & z_{21} \\ z_{32} & z_{33} & z_{31} \\ z_{12} & z_{13} & z_{11} \end{bmatrix} \begin{bmatrix} I_2 \\ I_3 \\ I_1 \end{bmatrix} \tag{4.66}$$

or

$$\mathbf{V}_{231} = \mathbf{z}_{231}\mathbf{I}_{231} \tag{4.67}$$

The effect of these various matrix manipulations are shown in Table 4.3.

Table 4.3. The Effect of Transposition Matrix Operations

Premultiply by $\mathbf{R}_\phi$	Move 3rd row to position 1	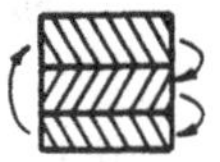
Postmultiply by $\mathbf{R}_\phi^{-1}$	Move 3rd column to position 1	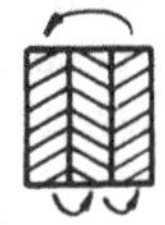
Premultiply by $\mathbf{R}_\phi^{-1} = \mathbf{R}_\phi^2$	Move 1st row to position 3	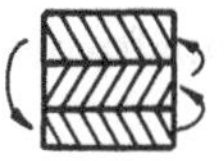
Postmultiply by $\mathbf{R}_\phi$	Move 1st column to position 3	

We now apply the rotation matrix to the problem of computing the impedance of a line which includes transpositions or rotations. Consider the line shown in Figure 4.9 where each current sees the impedance of each wire position for a portion of the line length..

In *section 1* we observe that positions 1-2-3 correspond to phases *a-b-c* such that the voltage equations (4.68) apply for the section.

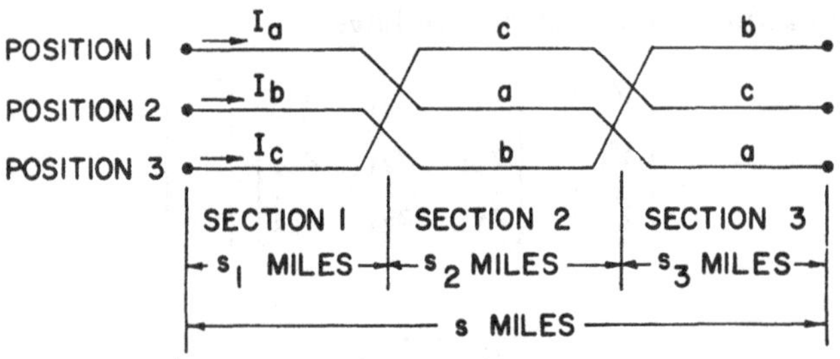

Fig. 4.9. A complete transposition or rotation cycle.

$$\begin{bmatrix} V_a \\ V_b \\ V_c \end{bmatrix} = \begin{bmatrix} V_1 \\ V_2 \\ V_3 \end{bmatrix} = \begin{bmatrix} Z_{11\text{-}1} & Z_{12\text{-}1} & Z_{13\text{-}1} \\ Z_{21\text{-}1} & Z_{22\text{-}1} & Z_{23\text{-}1} \\ Z_{31\text{-}1} & Z_{32\text{-}1} & Z_{33\text{-}1} \end{bmatrix} \begin{bmatrix} I_1 = I_a \\ I_2 = I_b \\ I_3 = I_c \end{bmatrix} \text{ V} \tag{4.68}$$

or

$$\mathbf{V}_{abc} = \mathbf{V}_{123} = \mathbf{Z}_{123}\mathbf{I}_{123} = \mathbf{Z}_{123}\mathbf{I}_{abc} \tag{4.69}$$

where the impedances here are the total impedances for line section 1. We use the notation

$$Z_{ij\text{-}k} = f_k Z_{ij} \tag{4.70}$$

where

$f_k = s_k/s$
$i, j = 1, 2, 3$ = position indicator
$k = 1, 2$, or 3 = line section identifier
s_k = length of line section k
s = total length of line

Here Z_{ij} is the total line impedance corresponding to positions i, j and f_k is the fraction of the total line length in section k.

For *section 2* phases *c-a-b* correspond to positions 1-2-3 such that the voltage equation may be written as

$$\mathbf{V}_{cab} = \mathbf{V}_{123} = \mathbf{Z}_{123}\mathbf{I}_{123} = \mathbf{Z}_{123}\mathbf{I}_{cab} \text{ V} \tag{4.71}$$

Applying the transformation $\mathbf{R}_\phi^{-1}$, this equation becomes

$$\mathbf{V}_{abc} = \mathbf{Z}_{231}\mathbf{I}_{abc} \tag{4.72}$$

$$\begin{bmatrix} V_a \\ V_b \\ V_c \end{bmatrix} = \begin{bmatrix} Z_{22\text{-}2} & Z_{23\text{-}2} & Z_{21\text{-}2} \\ Z_{32\text{-}2} & Z_{33\text{-}2} & Z_{31\text{-}2} \\ Z_{12\text{-}2} & Z_{13\text{-}2} & Z_{11\text{-}2} \end{bmatrix} \begin{bmatrix} I_a \\ I_b \\ I_c \end{bmatrix} \text{ V} \tag{4.73}$$

This equation may be verified by an inspection of Figure 4.9.

Section 3 has the correspondence *b-c-a* to 1-2-3 or

$$\mathbf{V}_{bca} = \mathbf{V}_{123} = \mathbf{Z}_{123}\mathbf{I}_{123} = \mathbf{Z}_{123}\mathbf{I}_{bca} \text{ V} \tag{4.74}$$

Transforming by $\mathbf{R}_\phi = (\mathbf{R}_\phi^{-1})^2$, we compute

$$\mathbf{V}_{abc} = \mathbf{Z}_{312}\mathbf{I}_{abc} \text{ V} \tag{4.75}$$

or

$$\begin{bmatrix} V_a \\ V_b \\ V_c \end{bmatrix} = \begin{bmatrix} Z_{33\text{-}3} & Z_{31\text{-}3} & Z_{32\text{-}3} \\ Z_{13\text{-}3} & Z_{11\text{-}3} & Z_{12\text{-}3} \\ Z_{23\text{-}3} & Z_{21\text{-}3} & Z_{22\text{-}3} \end{bmatrix} \begin{bmatrix} I_a \\ I_b \\ I_c \end{bmatrix} \text{ V} \tag{4.76}$$

To compute the total voltage drop along the line, we add the drop in each section given by equations (4.68), (4.73), and (4.76), with the result

$$\begin{bmatrix} \Sigma V_a \\ \Sigma V_b \\ \Sigma V_c \end{bmatrix} = \begin{bmatrix} (Z_{11\text{-}1} + Z_{22\text{-}2} + Z_{33\text{-}3}) & (Z_{12\text{-}1} + Z_{23\text{-}2} + Z_{31\text{-}3}) & (Z_{13\text{-}1} + Z_{21\text{-}2} + Z_{32\text{-}3}) \\ (Z_{21\text{-}1} + Z_{32\text{-}2} + Z_{13\text{-}3}) & (Z_{22\text{-}1} + Z_{33\text{-}2} + Z_{11\text{-}3}) & (Z_{23\text{-}1} + Z_{31\text{-}2} + Z_{12\text{-}3}) \\ (Z_{31\text{-}1} + Z_{12\text{-}2} + Z_{23\text{-}3}) & (Z_{32\text{-}1} + Z_{13\text{-}2} + Z_{21\text{-}3}) & (Z_{33\text{-}1} + Z_{11\text{-}2} + Z_{22\text{-}3}) \end{bmatrix} \begin{bmatrix} I_a \\ I_b \\ I_c \end{bmatrix} \text{ V} \tag{4.77}$$

It may be convenient to write (4.77) in terms of the total line impedance in each position multiplied by the appropriate fraction f_k defined in equation (4.70). The result is

$$\begin{bmatrix} \Sigma V_a \\ \Sigma V_b \\ \Sigma V_c \end{bmatrix} = \begin{bmatrix} (f_1 Z_{11} + f_2 Z_{22} + f_3 Z_{33}) & (f_1 Z_{12} + f_2 Z_{23} + f_3 Z_{31}) & (f_1 Z_{13} + f_2 Z_{21} + f_3 Z_{32}) \\ (f_1 Z_{21} + f_2 Z_{32} + f_3 Z_{13}) & (f_1 Z_{22} + f_2 Z_{33} + f_3 Z_{11}) & (f_1 Z_{23} + f_2 Z_{31} + f_3 Z_{12}) \\ (f_1 Z_{31} + f_2 Z_{12} + f_3 Z_{23}) & (f_1 Z_{32} + f_2 Z_{13} + f_3 Z_{21}) & (f_1 Z_{33} + f_2 Z_{11} + f_3 Z_{22}) \end{bmatrix} \begin{bmatrix} I_a \\ I_b \\ I_c \end{bmatrix} \text{ V} \tag{4.78}$$

From (4.54) and (4.55)

$$\begin{aligned} Z_{ij} &= (r_a + r_d)\, s + j\omega ks \ln \frac{D_e}{D_s} \quad \Omega,\ i = j \\ &= r_d\, s + j\omega ks \ln \frac{D_e}{D_{ij}} \quad \Omega,\ i \neq j \end{aligned} \tag{4.79}$$

It is apparent that for identical conductors

$$Z_{11} = Z_{22} = Z_{33} \triangleq Z_s \quad \Omega \tag{4.80}$$

Also, in a linear, passive, bilateral network $Z_{ij} = Z_{ji}$. Using this fact we may define three kinds of mutual impedance terms as follows:

$$\begin{aligned} Z_{k1} &= f_1 Z_{12} + f_2 Z_{23} + f_3 Z_{13} \quad \Omega \\ Z_{k2} &= f_1 Z_{13} + f_2 Z_{12} + f_3 Z_{23} \quad \Omega \\ Z_{k3} &= f_1 Z_{23} + f_2 Z_{13} + f_3 Z_{12} \quad \Omega \end{aligned} \tag{4.81}$$

Then (4.78) becomes

$$\begin{bmatrix} \Sigma V_a \\ \Sigma V_b \\ \Sigma V_c \end{bmatrix} = \begin{bmatrix} Z_s & Z_{k1} & Z_{k2} \\ Z_{k1} & Z_s & Z_{k3} \\ Z_{k2} & Z_{k3} & Z_s \end{bmatrix} \begin{bmatrix} I_a \\ I_b \\ I_c \end{bmatrix} \text{ V} \tag{4.82}$$

In terms of the geometry of the line, impedances (4.81) may be written as

$$Z_{k1} = r_d s + j\omega ks \left(f_1 \ln \frac{D_e}{D_{12}} + f_2 \ln \frac{D_e}{D_{23}} + f_3 \ln \frac{D_e}{D_{13}} \right) \Omega$$

$$Z_{k2} = r_d s + j\omega k s\left(f_1 \ln \frac{D_e}{D_{13}} + f_2 \ln \frac{D_e}{D_{12}} + f_3 \ln \frac{D_e}{D_{23}}\right) \ \Omega$$

$$Z_{k3} = r_d s + j\omega k s\left(f_1 \ln \frac{D_e}{D_{23}} + f_2 \ln \frac{D_e}{D_{13}} + f_3 \ln \frac{D_e}{D_{12}}\right) \ \Omega \tag{4.83}$$

Example 4.2

Compute the total impedance of the line in Example 4.1 if the line is transposed by two rotations such that

$$\begin{aligned} s_1 &= 8 \text{ mi} & f_1 &= s_1/s = 0.2 \\ s_2 &= 12 \text{ mi} & f_2 &= s_2/s = 0.3 \\ s_3 &= 20 \text{ mi} & f_3 &= s_3/s = 0.5 \end{aligned}$$

Solution

The impedance computed in Example 4.1 for the configuration given in Figure 4.6 will be taken as the impedance per mile of the first section $f_1 s = 8$ miles in length. Then from (4.68) we premultiply by the length, 8 miles, to compute

$$\text{for } s_1 : \mathbf{Z}_{123} = \begin{bmatrix} (2.986 + j11.675) & (0.762 + j5.466) & (0.762 + j4.794) \\ (0.762 + j5.466) & (2.986 + j11.675) & (0.762 + j5.466) \\ (0.762 + j4.794) & (0.762 + j5.466) & (2.986 + j11.675) \end{bmatrix} \Omega$$

The second section is 12 miles long and is transposed to a 2-3-1 configuration as given by (4.73). Thus, for this section premultiplication by 12 $\mathbf{R}_\phi^{-1}$ gives

$$\text{for } s_2 : \mathbf{Z}_{231} = \begin{bmatrix} (4.479 + j17.513) & (1.143 + j8.199) & (1.143 + j8.199) \\ (1.143 + j8.199) & (4.479 + j17.513) & (1.143 + j7.190) \\ (1.143 + j8.199) & (1.143 + j7.190) & (4.479 + j17.513) \end{bmatrix} \Omega$$

The third section is 20 miles long and is transposed to a 3-1-2 configuration as in (4.76). Then premultiplication by 20 $(\mathbf{R}_\phi^{-1})^2$ gives

$$\text{for } s_3 : \mathbf{Z}_{312} = \begin{bmatrix} (7.466 + j29.188) & (1.906 + j11.983) & (1.906 + j13.665) \\ (1.906 + j11.983) & (7.466 + j29.188) & (1.906 + j13.665) \\ (1.906 + j13.665) & (1.906 + j13.665) & (7.466 + j29.188) \end{bmatrix} \Omega$$

Then the total impedance is the sum of the impedances of the three sections, viz.,

$$\mathbf{Z}_{abc} = \begin{bmatrix} (14.932 + j58.376) & (3.812 + j25.650) & (3.812 + j26.659) \\ (3.812 + j25.650) & (14.932 + j58.376) & (3.812 + j26.323) \\ (3.812 + j26.659) & (3.812 + j26.323) & (14.932 + j58.376) \end{bmatrix} \Omega$$

Note that the diagonal terms are the same as for the untransposed line. Similarly, the off-diagonal *resistances* are the same as in Example 4.1. The off-diagonal *reactances*, however, are more nearly equal or balanced than before. In Example 4.1 the difference between the largest and smallest mutual reactance is

$$\text{diff} = 27.332 - 23.968 = 3.364 \ \Omega$$

In the transposed line this largest difference is diff = 26.659 - 25.650 = 1.009 Ω. Thus, even very unequal transpositions help a great deal in equalizing the mutual impedances.

Example 4.3

Consider another case of transposing the line of Example 4.1 but using only two transposition sections rather than three. Let the sections be defined by

$$f_1 = 0.4 = 16 \text{ mi}, \qquad f_2 = 0.6 = 24 \text{ mi}, \qquad f_3 = 0$$

Solution

This example is solved in the same way as the previous one except that only two impedance matrices need to be computed. For the first section we have

$$\text{for } s_1 : \mathbf{Z}_{abc} = \begin{bmatrix} (5.973 + j23.350) & (1.525 + j10.933) & (1.525 + j9.587) \\ (1.525 + j10.933) & (5.973 + j23.350) & (1.525 + j10.933) \\ (1.525 + j9.587) & (1.525 + j10.933) & (5.973 + j23.350) \end{bmatrix} \Omega$$

And for the second section, again assuming a rotation $\mathbf{R}_\phi^{-1}$, we have

$$\text{for } s_2 : \mathbf{Z}_{abc} = \begin{bmatrix} (8.959 + j35.025) & (2.287 + j16.398) & (2.287 + j16.398) \\ (2.287 + j16.398) & (8.959 + j35.025) & (2.287 + j14.380) \\ (2.287 + j16.398) & (2.287 + j14.380) & (8.959 + j35.025) \end{bmatrix} \Omega$$

Adding the two sections, we have

$$\mathbf{Z}_{abc} = \begin{bmatrix} (14.932 + j58.376) & (3.812 + j27.332) & (3.812 + j25.986) \\ (3.812 + j27.332) & (14.932 + j58.376) & (3.812 + j25.314) \\ (3.812 + j25.986) & (3.812 + j25.314) & (14.932 + j58.376) \end{bmatrix} \Omega$$

The largest reactance difference now is diff = 27.332 - 25.314 = 2.018 Ω, which is still a substantial improvement over the case with no transpositions.

Suppose that instead of a rotation $\mathbf{R}_\phi^{-1}$ we rotated in the opposite direction, i.e., $\mathbf{R}_\phi$. Then the impedance of the second section would be, from (4.64)

$$\text{for } s_2 : \mathbf{Z}_{abc} = \begin{bmatrix} (8.959 + j35.025) & (2.287 + j14.380) & (2.287 + j16.398) \\ (2.287 + j14.380) & (8.959 + j35.025) & (2.287 + j16.398) \\ (2.287 + j16.398) & (2.287 + j16.398) & (8.959 + j35.025) \end{bmatrix} \Omega$$

Then

$$\mathbf{Z}_{abc} = \begin{bmatrix} (14.932 + j58.376) & (3.812 + j25.314) & (3.812 + j25.986) \\ (3.812 + j25.314) & (14.932 + j58.376) & (3.812 + j27.332) \\ (3.812 + j25.986) & (3.812 + j27.332) & (14.932 + j58.376) \end{bmatrix} \Omega$$

This result is the same as the previous one except that the matrix elements have been moved.

Equation (4.82) is the desired equation for a line which has undergone a complete transposition or rotation cycle. Note that the line impedance includes the effect of any return current through the earth. In the case of balanced currents, as in the positive and negative sequence cases, this earth impedance will disappear since no current flows through the earth. When zero sequence currents flow, however, all three currents return through the earth and the earth impedance is very important.

To compute the sequence impedances, we turn to the definitions of equations (2.46)-(2.48) to compute

$$Z_{S0} = Z_s = (r_a + r_d)s + j\omega ks \ln \frac{D_e}{D_s} \quad \Omega$$

$$Z_{S1} = Z_{S2} = 0 \text{ (phase wires identical)}$$

$$Z_{M0} = r_d\, s + j\omega ks \ln \frac{D_e}{D_{eq}} \quad \Omega$$

$$Z_{M1} = j\omega ks \left[\frac{(f_3 + af_2 + a^2 f_1)}{3} \ln \frac{D_e}{D_{12}} + \frac{(f_1 + af_3 + a^2 f_2)}{3} \ln \frac{D_e}{D_{23}} + \frac{(f_2 + af_1 + a^2 f_3)}{3} \ln \frac{D_e}{D_{13}} \right] \Omega$$

$$Z_{M2} = j\omega ks \left[\frac{(f_3 + a^2 f_2 + af_1)}{3} \ln \frac{D_e}{D_{12}} + \frac{(f_1 + a^2 f_3 + af_2)}{3} \ln \frac{D_e}{D_{23}} + \frac{(f_2 + a^2 f_1 + af_3)}{3} \ln \frac{D_e}{D_{13}} \right] \Omega \qquad (4.84)$$

where in Z_{M0} we use the quantity D_{eq}, the "equivalent spacing," of three conductors defined as the GMD of the three distances or

$$D_{eq} = (D_{12} D_{23} D_{13})^{1/3} \qquad (4.85)$$

and where all distances are in the same units throughout.

Applying equation (2.46) we write the sequence impedances as

$$\begin{bmatrix} V_{a0} \\ V_{a1} \\ V_{a2} \end{bmatrix} = \begin{bmatrix} (Z_s + 2Z_{M0}) & -Z_{M2} & -Z_{M1} \\ -Z_{M1} & (Z_s - Z_{M0}) & 2Z_{M2} \\ -Z_{M2} & 2Z_{M1} & (Z_s - Z_{M0}) \end{bmatrix} \begin{bmatrix} I_{a0} \\ I_{a1} \\ I_{a2} \end{bmatrix} \text{V} \qquad (4.86)$$

Since Z_{M1} and Z_{M2} are nonzero, there is coupling between the sequence networks in the general case of a transposed line. The self impedance to the flow of I_{a0} and I_{a1} are equal respectively to

$$Z_0 = Z_s + 2Z_{M0} = s \left[(r_a + 3r_d) + j\omega k \ln \frac{D_e^3}{D_s D_{eq}^2} \right] \Omega$$

$$Z_1 = Z_2 = Z_s - Z_{M0} = s \left(r_a + j\omega k \ln \frac{D_{eq}}{D_s} \right) \Omega \qquad (4.87)$$

where we note that the earth resistance has vanished in Z_1 as predicted. Later we will show that (4.86) is simplified in the special case where $f_1 = f_2 = f_3$.

4.6.2 Computation of sequence impedances by $\mathbf{R}_{012}$

The method used above to compute the sequence impedance of a transposed line was to first identify the phase impedances of each line section, add these section impedances, and finally transform the sum to the 0-1-2 frame of reference. Stated mathematically, this operation consisted of the following:

$$\Sigma \mathbf{V}_{012} = \mathbf{A}^{-1} \Sigma \mathbf{V}_{abc} = s\mathbf{A}^{-1} (f_1 \mathbf{z}_{123} + f_2 \mathbf{z}_{231} + f_3 \mathbf{z}_{312}) \mathbf{I}_{abc} \quad \text{V} \qquad (4.88)$$

Expanding this equation, we write

$$\begin{aligned} \Sigma \mathbf{V}_{012} &= s\, \mathbf{A}^{-1} (f_1 \mathbf{z}_{123} + f_2 \mathbf{R}_\phi^{-1} \mathbf{z}_{123} \mathbf{R}_\phi + f_3 \mathbf{R}_\phi \mathbf{z}_{123} \mathbf{R}_\phi^{-1}) \mathbf{A}\mathbf{A}^{-1} \mathbf{I}_{abc} \\ &= s[f_1 \mathbf{A}^{-1} \mathbf{z}_{123} \mathbf{A} + f_2 (\mathbf{A}^{-1} \mathbf{R}_\phi^{-1} \mathbf{A})(\mathbf{A}^{-1} \mathbf{z}_{123} \mathbf{A})(\mathbf{A}^{-1} \mathbf{R}_\phi \mathbf{A}) \\ &\quad + f_3 (\mathbf{A}^{-1} \mathbf{R}_\phi \mathbf{A})(\mathbf{A}^{-1} \mathbf{z}_{123} \mathbf{A})(\mathbf{A}^{-1} \mathbf{R}_\phi^{-1} \mathbf{A})] \, \mathbf{I}_{012} \end{aligned} \qquad (4.89)$$

Suppose we define the quantity

$$\mathbf{R}_{012} = \mathbf{A}^{-1}\mathbf{R}_{\phi}\mathbf{A} = \begin{bmatrix} 1 & 0 & 0 \\ 0 & a & 0 \\ 0 & 0 & a^2 \end{bmatrix} \tag{4.90}$$

Then we readily verify that

$$\mathbf{R}_{012}^{-1} = \mathbf{A}^{-1}\mathbf{R}_{\phi}^{-1}\mathbf{A} = \begin{bmatrix} 1 & 0 & 0 \\ 0 & a^2 & 0 \\ 0 & 0 & a \end{bmatrix} \tag{4.91}$$

We also recognize that for the first line section, using (2.45)

$$\mathbf{z}_{012} = \mathbf{A}^{-1}\mathbf{z}_{123}\mathbf{A} \quad \Omega/\text{mi} \tag{4.92}$$

Using (4.90)-(4.92), equation (4.89) becomes

$$\Sigma\, \mathbf{V}_{012} = s(f_1\mathbf{z}_{012} + f_2\mathbf{R}_{012}^{-1}\mathbf{z}_{012}\mathbf{R}_{012} + f_3\mathbf{R}_{012}\mathbf{z}_{012}\mathbf{R}_{012}^{-1})\mathbf{I}_{012} \tag{4.93}$$

This result is better understood when it is recognized that

$$\mathbf{R}_{012}^{2} = \mathbf{R}_{012}^{-1}, \quad (\mathbf{R}_{012}^{-1})^2 = \mathbf{R}_{012} \tag{4.94}$$

so that the third term may be thought of as two rotations of the initial configuration.

One may easily verify that

$$\mathbf{R}_{012}\mathbf{z}_{012}\mathbf{R}_{012}^{-1} = \begin{bmatrix} z_{00} & a^2 z_{01} & a z_{02} \\ a z_{10} & z_{11} & a^2 z_{12} \\ a^2 z_{20} & a z_{21} & z_{22} \end{bmatrix} \tag{4.95}$$

$$\mathbf{R}_{012}^{-1}\mathbf{z}_{012}\mathbf{R}_{012} = \begin{bmatrix} z_{00} & a z_{01} & a^2 z_{02} \\ a^2 z_{10} & z_{11} & a z_{12} \\ a z_{20} & a^2 z_{21} & z_{22} \end{bmatrix} \tag{4.96}$$

where from (2.46) we compute

$$z_{S0} = (r_a + r_d) + j\omega k \ln \frac{D_e}{D_s} \quad \Omega/\text{unit length}$$

$$z_{S1} = z_{S2} = 0 \quad \Omega/\text{unit length}$$

$$z_{M0} = r_d + j\omega k \ln \frac{D_e}{D_{eq}} \quad \Omega/\text{unit length}$$

$$z_{M1} = j\frac{\omega k}{3}\left(\ln\frac{D_e}{D_{23}} + a\ln\frac{D_e}{D_{13}} + a^2\ln\frac{D_e}{D_{12}}\right) \quad \Omega/\text{unit length}$$

$$z_{M2} = j\frac{\omega k}{3}\left(\ln\frac{D_e}{D_{23}} + a^2\ln\frac{D_e}{D_{13}} + a\ln\frac{D_e}{D_{12}}\right) \quad \Omega/\text{unit length} \tag{4.97}$$

such that the elements of $\mathbf{z}_{012}$ are defined to be

$$z_{00} = (r_a + 3r_d) + j\omega k \ln \frac{D_e^3}{D_s D_{eq}^2} \quad \Omega/\text{unit length}$$

$$z_{01} = -z_{M2} = z_{20}, \qquad z_{02} = -z_{M1} = z_{10} \quad \Omega\text{/unit length}$$

$$z_{11} = z_{22} = r_a + j\omega k \ln \frac{D_{eq}}{D_s} \quad \Omega\text{/unit length}$$

$$z_{12} = +2z_{M2}, \qquad z_{21} = +2z_{M1} \quad \Omega\text{/unit length} \tag{4.98}$$

Combining as specified in (4.93) gives the same results for the entire line as (4.84)-(4.86) of the previous section.

Example 4.4

Compute the sequence impedance matrix $\mathbf{Z}_{012}$ for the untransposed transmission line of Example 4.1 and also for the transposed lines of Examples 4.2 and 4.3.

Solution

Using the computer programs of Appendix A, we compute the sequence impedances by the similarity transformation (4.92).

1. For the line of Example 4.1 with no transpositions,

$$f_1 = 1.0, \qquad f_2 = 0, \qquad f_3 = 0$$

we compute

$$\mathbf{Z}_{012} = \begin{bmatrix} (22.556 + j110.80) & (0.971 - j0.561) & (-0.971 - j0.561) \\ (-0.971 - j0.561) & (11.120 + j32.165) & (-1.942 + j1.121) \\ (0.971 - j0.561) & (1.942 + j1.121) & (11.120 + j31.165) \end{bmatrix} \Omega$$

2. For the line of Example 4.2 with two rotations

$$f_1 = 0.2, \qquad f_2 = 0.3, \qquad f_3 = 0.5$$

we compute

$$\mathbf{Z}_{012} = \begin{bmatrix} (22.556 + j110.80) & (-0.291 - j0.056) & (0.291 - j0.056) \\ (0.291 - 0.056) & (11.120 + j32.165) & (0.583 + j0.112) \\ (-0.291 - j0.056) & (-0.583 + j0.112) & (11.120 + j32.165) \end{bmatrix} \Omega$$

3. For the line of Example 4.3 with transposition

$$f_1 = 0.4, \qquad f_2 = 0.6, \qquad f_3 = 0$$

we compute

$$\mathbf{Z}_{012} = \begin{bmatrix} (22.556 + j110.80) & (0.388 + j0.449) & (-0.388 + j0.449) \\ (-0.388 + j0.449) & (11.120 + j32.165) & (-0.777 - j0.897) \\ (0.388 + j0.449) & (0.777 - j0.897) & (11.120 + j32.165) \end{bmatrix} \Omega$$

The transpositions are effective in reducing the coupling between sequence networks. Even one rotation produces a substantial reduction in the magnitude of the off-diagonal terms of $\mathbf{Z}_{012}$ and two rotations practically eliminate this coupling. In a later example we show that a completely transposed line has zero coupling between sequences.

4.6.3 Twisting of line conductor pairs

Occasionally only two of the three line conductors may be transposed as shown in Figure 4.10. This may be thought of as a single-phase transposition and results in a reversal of phase sequence from *a-b-c* to *a-c-b*. Mathematically this rearrangement may be accomplished by means of a twist matrix T_ϕ, where we define

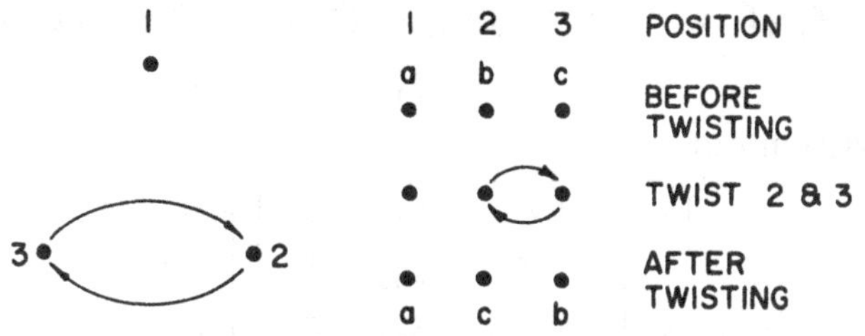

Fig. 4.10. Twisting of conductors in positions 2 and 3.

$$T_{\phi 23} = \begin{array}{c} \\ 1 \\ 2 \\ 3 \end{array} \begin{array}{c} \begin{array}{ccc} 1 & 3 & 2 \end{array} \\ \begin{bmatrix} 1 & 0 & 0 \\ 0 & 0 & 1 \\ 0 & 1 & 0 \end{bmatrix} \end{array} \tag{4.99}$$

such that

$$T_{\phi 23} I_{acb} = I_{abc} \tag{4.100}$$

Actually there are three twist matrices, the other two being defined as

$$T_{\phi 12} = \begin{array}{c} \\ 1 \\ 2 \\ 3 \end{array} \begin{array}{c} \begin{array}{ccc} 2 & 1 & 3 \end{array} \\ \begin{bmatrix} 0 & 1 & 0 \\ 1 & 0 & 0 \\ 0 & 0 & 1 \end{bmatrix} \end{array} \tag{4.101}$$

and

$$T_{\phi 13} = \begin{array}{c} \\ 1 \\ 2 \\ 3 \end{array} \begin{array}{c} \begin{array}{ccc} 3 & 2 & 1 \end{array} \\ \begin{bmatrix} 0 & 0 & 1 \\ 0 & 1 & 0 \\ 1 & 0 & 0 \end{bmatrix} \end{array} \tag{4.102}$$

Obviously, twisting any pair twice restores the configuration to its original state, i.e.,

$$T_{\phi 12}^2 = T_{\phi 23}^2 = T_{\phi 13}^2 = \begin{bmatrix} 1 & 0 & 0 \\ 0 & 1 & 0 \\ 0 & 0 & 1 \end{bmatrix} = U_3 \tag{4.103}$$

We also readily verify that each twist matrix is equal to its own inverse, i.e.,

$$T_{\phi 23} = T_{\phi 23}^{-1}, \qquad T_{\phi 12} = T_{\phi 12}^{-1}, \qquad T_{\phi 13} = T_{\phi 13}^{-1} \tag{4.104}$$

This is logical since it makes no difference which direction the two wires are twisted, the results being the same in either case.

If a matrix equation is transformed by the twist matrix, a change in the equation occurs which depends upon which of the three twist matrices is used. For example, suppose the transformation $\mathbf{T}_{\phi 23}$ is used to premultiply both sides of the equation, $\mathbf{V}_{123} = \mathbf{Z}_{123}\mathbf{I}_{123}$. Then $\mathbf{T}_{\phi 23}\mathbf{V}_{123} = (\mathbf{T}_{\phi 23}\mathbf{Z}_{123}\mathbf{T}_{\phi 23}^{-1})(\mathbf{T}_{\phi 23}\mathbf{I}_{123})$ or

$$\mathbf{V}_{132} = \mathbf{Z}_{132}\mathbf{I}_{132} \tag{4.105}$$

which may be written out as

$$\begin{bmatrix} V_1 \\ V_3 \\ V_2 \end{bmatrix} = \begin{bmatrix} Z_{11} & Z_{13} & Z_{12} \\ Z_{31} & Z_{33} & Z_{32} \\ Z_{21} & Z_{23} & Z_{22} \end{bmatrix} \begin{bmatrix} I_1 \\ I_3 \\ I_2 \end{bmatrix} \tag{4.106}$$

In general, we may easily verify the following *twist rules*: (1) premultiplication by $\mathbf{T}_{\phi ij}$ interchanges rows i and j (2) postmultiplication by $\mathbf{T}_{\phi ij}$ interchanges columns i and j. Note that (2) is the same as postmultiplication by $\mathbf{T}_{\phi ij}^{-1}$.

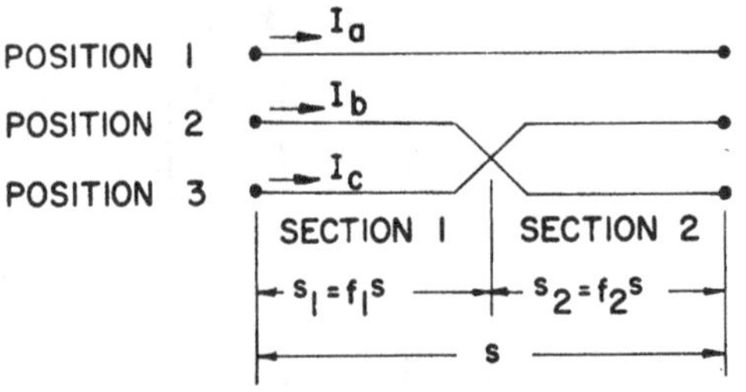

Fig. 4.11. A line with wires 2 and 3 twisted.

As an example of twisting consider the line shown in Figure 4.11 where wires 2 and 3 are twisted. In the *section 1* we write

$$\begin{bmatrix} V_a \\ V_b \\ V_c \end{bmatrix} = \begin{bmatrix} V_1 \\ V_2 \\ V_3 \end{bmatrix} = f_1 \begin{bmatrix} Z_{11} & Z_{12} & Z_{13} \\ Z_{21} & Z_{22} & Z_{23} \\ Z_{31} & Z_{32} & Z_{33} \end{bmatrix} \begin{bmatrix} I_1 = I_a \\ I_2 = I_b \\ I_3 = I_c \end{bmatrix} \tag{4.107}$$

Where Z_{ij} is the total impedance of s miles, based on the configuration of section 1.

For *section 2* we begin by writing

$$\mathbf{V}_{acb} = \mathbf{V}_{123} = f_2\mathbf{Z}_{123}\mathbf{I}_{acb} \tag{4.108}$$

which we transform by $\mathbf{T}_{\phi 23}$ to compute

$$\begin{bmatrix} V_a \\ V_b \\ V_c \end{bmatrix} = f_2 \begin{bmatrix} Z_{11} & Z_{13} & Z_{12} \\ Z_{31} & Z_{33} & Z_{32} \\ Z_{21} & Z_{23} & Z_{22} \end{bmatrix} \begin{bmatrix} I_a \\ I_b \\ I_c \end{bmatrix} \tag{4.109}$$

Adding (4.107) and (4.109) we find the total impedance to be

$$\begin{bmatrix} \Sigma V_a \\ \Sigma V_b \\ \Sigma V_c \end{bmatrix} = \begin{bmatrix} Z_{11} & (f_1 Z_{12} + f_2 Z_{13}) & (f_1 Z_{13} + f_2 Z_{12}) \\ (f_1 Z_{12} + f_2 Z_{13}) & (f_1 Z_{22} + f_2 Z_{33}) & Z_{23} \\ (f_1 Z_{13} + f_2 Z_{12}) & Z_{23} & (f_1 Z_{33} + f_2 Z_{22}) \end{bmatrix} \begin{bmatrix} I_a \\ I_b \\ I_c \end{bmatrix} \tag{4.110}$$

where the reciprocity of mutuals has been recognized to write $Z_{ji} = Z_{ij}$ in all cases. Obviously, any twisted pair could be analyzed in this same way. Combining the rotation and the twist, we may compute the line impedance for any physical situation. Note that this method allows an exact accounting of every wire configuration and the length of line with a particular arrangement. It therefore comprises an exact method of finding not only phase impedances but sequence impedances as well, taking into account the exact amount of mutual coupling between phases or between sequence networks.

Example 4.5

Repeat Example 4.3 with line sectioning

$$f_1 = 0.4, \qquad f_2 = 0.6, \qquad f_3 = 0$$

except twist the wires by applying $\mathbf{T}_{\phi 23}$ instead of transposing by $\mathbf{R}_\phi^{-1}$.

Solution.

The impedance for the first section is exactly the same as that computed in Example 4.3, viz., for

$$s_1: \mathbf{Z}_{123} = \begin{bmatrix} (5.973 + j23.350) & (1.525 + j10.933) & (1.525 + j9.587) \\ (1.525 + j10.933) & (5.973 + j23.350) & (1.525 + j10.933) \\ (1.525 + j9.587) & (1.525 + j10.933) & (5.973 + j23.350) \end{bmatrix} \Omega$$

According to (4.109) the second section has the impedance matrix for

$$s_2: \mathbf{Z}_{132} = \begin{bmatrix} (8.959 + j35.026) & (2.287 + j14.381) & (2.287 + j16.399) \\ (2.287 + j14.381) & (8.954 + j35.026) & (2.287 + j16.399) \\ (2.287 + j16.349) & (2.287 + j16.399) & (8.959 + j35.026) \end{bmatrix} \Omega$$

Then

$$\mathbf{Z}_{abc} = \begin{bmatrix} (14.932 + j58.376) & (3.812 + j25.314) & (3.812 + j25.986) \\ (3.812 + j25.314) & (14.932 + j58.376) & (3.812 + j27.332) \\ (3.812 + j25.986) & (3.812 + j27.332) & (14.932 + j58.376) \end{bmatrix} \Omega$$

This result is the same numerically as a rotation $\mathbf{R}_\phi$, but the phase position changes to *a-c-b* on the pole.

In the foregoing analysis we have considered three different twist matrices $\mathbf{T}_{\phi 23}$, $\mathbf{T}_{\phi 12}$, and $\mathbf{T}_{\phi 13}$. It can be easily shown that only one is necessary since the other two may be defined in terms of a rotation-and-twist or twist-and-rotation combination. Suppose we define the only twist matrix to be $\mathbf{T}_\phi$ where

$$\mathbf{T}_\phi = \mathbf{T}_{\phi 23} \tag{4.111}$$

Then it is easy to show that

$$\mathbf{T}_{\phi 12} = \mathbf{R}_\phi \mathbf{T}_\phi = \mathbf{T}_\phi \mathbf{R}_\phi^{-1}, \qquad \mathbf{T}_{\phi 13} = \mathbf{R}_\phi^{-1} \mathbf{T}_\phi = \mathbf{T}_\phi \mathbf{R}_\phi \tag{4.112}$$

Furthermore, if it is desirable to compute the 0-1-2 impedances for section 2 directly, we may write

$$\mathbf{V}_{012} = f_2 (\mathbf{T}_{012} \mathbf{Z}_{012} \mathbf{T}_{012}^{-1}) \mathbf{I}_{012} \tag{4.113}$$

where we define

$$\mathbf{T}_{012} = \mathbf{A}^{-1} \mathbf{T}_\phi \mathbf{A} \tag{4.114}$$

It is easy to show that

$$\mathbf{T}_{012} = \mathbf{T}_{\phi} = \mathbf{T}_{\phi}^{-1} = \mathbf{T}_{012}^{-1} \tag{4.115}$$

Had we defined $\mathbf{T}_{\phi}$ to be a choice other than (4.111), then (4.114) would be complex.

Using identities (4.112) we may compute the effect of twisting any pair of wires. Thus we shall hereafter refer to only one twist matrix (4.111), although in practice the use of all three may be preferred by some.

4.7 Completely Transposed Lines

In some problems the lines may be completely transposed or may be assumed so to simplify computations. In such problems the impedance calculation is a special case of equation (4.78) where

$$f_1 = f_2 = f_3 = 1/3 \tag{4.116}$$

Then, from (4.81) $Z_{k1} = Z_{k2} = Z_{k3} = (1/3)(Z_{12} + Z_{23} + Z_{13}) = Z_k$ by definition, where we compute, from (4.83),

$$Z_k = r_d s + j\omega k s \ln \frac{D_e}{D_{eq}} \tag{4.117}$$

and write

$$\begin{bmatrix} V_a \\ V_b \\ V_c \end{bmatrix} = \begin{bmatrix} Z_s & Z_k & Z_k \\ Z_k & Z_s & Z_k \\ Z_k & Z_k & Z_s \end{bmatrix} \begin{bmatrix} I_a \\ I_b \\ I_c \end{bmatrix} \tag{4.118}$$

This simplifies the sequence impedances since from (2.48) $z_{M1} = z_{M2} = 0$, and we write

$$\begin{bmatrix} V_{a0} \\ V_{a1} \\ V_{a2} \end{bmatrix} = \begin{bmatrix} Z_0 & 0 & 0 \\ 0 & Z_1 & 0 \\ 0 & 0 & Z_1 \end{bmatrix} \begin{bmatrix} I_{a0} \\ I_{a1} \\ I_{a2} \end{bmatrix} \tag{4.119}$$

where from (4.87) for one unit length line

$$z_0 = (r_a + 3r_d) + j\omega k \ln \frac{D_e^3}{D_s D_{eq}^2}$$

$$z_1 = r_a + j\omega k \ln \frac{D_{eq}}{D_s} = z_2 \qquad \Omega/\text{unit length} \tag{4.120}$$

Observe that when the line is completely transposed, the impedance matrix (4.119) is diagonal and there is no coupling between sequence networks. Note also that the positive and negative sequence impedances are equal, a fact that is intuitively obvious.

The zero sequence impedance is sometimes seen in the literature in a form somewhat different from that of (4.120). Suppose we write the zero sequence impedance as

$$z_0 = (r_a + 3r_d) + j3\omega k \ln \frac{D_e}{D_s^{1/3} D_{eq}^{2/3}} \qquad \Omega/\text{unit length} \tag{4.121}$$

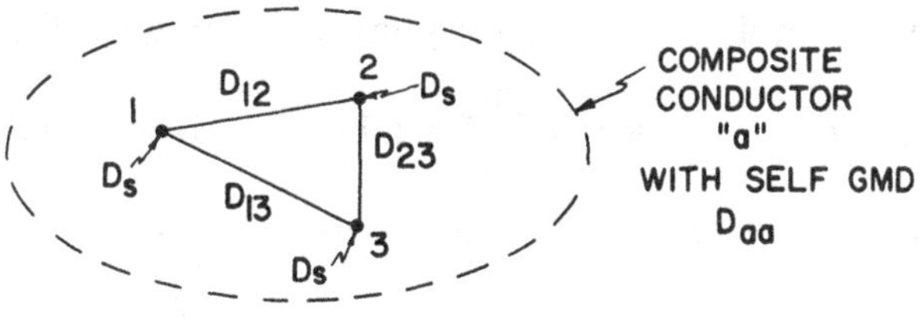

Fig. 4.12. A fictitious composite wire a.

We can write the denominator of the logarithm in the form

$$D_s^{1/3} D_{eq}^{2/3} = D_s^{1/3} (D_{12} D_{23} D_{13})^{2/9} = [D_s^3 (D_{12} D_{23} D_{13})^2]^{1/9} \tag{4.122}$$

This quantity is the self GMD of a composite conductor, which we shall call "a," consisting of three single wires of GMR equal to D_s and separated by distances D_{12}, D_{23}, and D_{13} as shown in Figure 4.12. Then the quantity (4.122) is clearly the self GMD of this composite conductor or

$$D_{aa} = D_s^{1/3} D_{eq}^{2/3} \tag{4.123}$$

The zero sequence impedance may then be written as

$$z_0 = (r_a + 3r_d) + j3\omega k \ln \frac{D_e}{D_{aa}} \quad \Omega/\text{unit length} \tag{4.124}$$

This view of a transposed line behaving as a composite conductor is a useful concept. In fact, the zero sequence may be computed directly from this idea. Suppose that zero sequence currents flow in the three phase conductors and return in the earth as shown in Figure 4.13. To compute the inductance of the composite line conductor a under these conditions, we recognize two components of inductance—that due to I_a and that due to I_e as in Figure 4.13. The inductance due to I_a is

$$L_a = k \ln \frac{D_m}{D_s} = k \ln \frac{D_{ae}}{D_{aa}} \quad \text{H/unit length} \tag{4.125}$$

Since $I_e = I_a$ and $D_{se} = 1.0$, the inductance due to I_e is,

$$L_e = k \ln \frac{D_{ae}}{1.0} \quad \text{H/unit length} \tag{4.126}$$

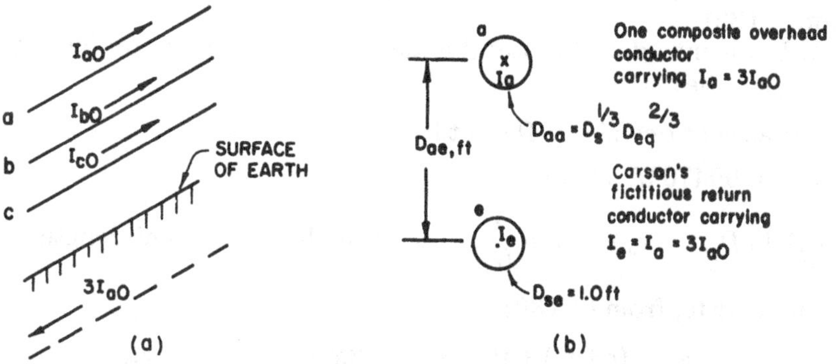

Fig. 4.13. Circuit for which zero sequence impedance is desired: (a) pictorial view (b) equivalent arrangement for computing inductance.

Adding, we have the total inductance for phase a which we call L_{aa}.

$$L_{aa} = L_a + L_e = \text{k} \ln \frac{D_{ae}^2}{D_{aa}} = \text{k} \ln \frac{D_e}{D_{aa}} \quad \text{H/unit length} \tag{4.127}$$

Then the total self impedance of line a and earth return is

$$z_{aa} = r_{aa} + \text{j}\, x_{aa} = (r + r_d) + \text{j}\omega\text{k} \ln \frac{D_e}{D_{aa}} \quad \Omega/\text{mi} \tag{4.128}$$

If $f = 60$ Hz and the length is in miles,

$$z_{aa} = (r + 0.09528) + \text{j}0.1213 \ln \frac{D_e}{D_{aa}} \quad \Omega/\text{mi}$$

This is the zero sequence impedance of *one line with earth return.*

In this problem the composite line a consists of three conductors, each carrying a current I_{a0}. If we consider the three conductors to be arranged as shown in Figure 4.12, then D_{aa} for equation (4.127) is given by equation (4.123). If we let r_a be the resistance in ohms per mile of any one conductor a, b, or c of composite conductor a, then $r = r_a/3$ and the impedance seen by the current $3I_{a0}$ is

$$z_{aa} = \left(\frac{r_a}{3} + r_d\right) + \text{j}\omega\text{k} \ln \frac{D_e}{D_{aa}} \quad \Omega/\text{unit length} \tag{4.129}$$

On a *per phase* basis, since each phase is in parallel with the other two, the impedance seen by I_{a0} is

$$z_0 = 3z_{aa} = (r_a + 3r_d) + \text{j}3\omega\text{k} \ln \frac{D_e}{D_{aa}} \quad \Omega/\text{unit length} \tag{4.130}$$

For example with $f = 60$ Hz and the length in miles,

$$z_0 = (r_a + 0.286) + \text{j}0.364 \ln \frac{D_e}{D_{aa}} \quad \Omega/\text{mi/phase}$$

Equation (4.130) is exactly the same as (4.124) which was derived by a different method. The GMD method is much simpler and is usually recommended for completely transposed lines. The positive (or negative) sequence impedance is also computed using the GMD method for transposed lines, and the subject is discussed thoroughly in many references, for example, Stevenson [9], Westinghouse [14] or Woodruff [22].

W. A. Lewis introduced a useful notation [10, p. 419] as follows. Define the following quantities:

r = resistance of one phase, Ω/unit length

$r_e = 3r_d = 0.004764f \quad \Omega/\text{mi}$

$$x_e = 3\omega\text{k} \ln D_e, \qquad x_a = \omega\text{k} \ln \frac{1}{D_s}, \qquad x_d = \omega\text{k} \ln D_{eq} \quad \Omega/\text{unit length} \tag{4.131}$$

Then we may write, from (4.130)

$$z_0 = (r + r_e) + \text{j}(x_e + x_a - 2x_d) \quad \Omega/\text{unit length} \tag{4.132}$$

Example 4.6

Compute the sequence impedance of the 40-mile transmission line of Example 4.1 assuming that the line is completely transposed, i.e., $f_1 = f_2 = f_3 = 1/3$.

Solution

The self impedance term Z_s in (4.118) is exactly the same as that computed in Example 4.1.

$$Z_s = 14.932 + j58.376 \ \Omega$$

The off-diagonal terms are all equal to Z_k of (4.117). To find this value, we first compute $D_{eq} = [(10)(10)(20)]^{1/3} = 12.6$ ft. Then

$$\begin{aligned} Z_k &= r_d s + j\omega ks \ln \frac{D_e}{D_{eq}} \\ &= (0.09528)(40) + j(0.12134)(40) \ln \frac{2790}{12.6} \\ &= 3.812 + j26.211 \ \Omega \end{aligned}$$

In matrix notation

$$\mathbf{Z}_{abc} = \begin{bmatrix} (14.932 + j58.376) & (3.812 + j26.211) & (3.812 + j26.211) \\ (3.812 + j26.211) & (14.932 + j58.376) & (3.812 + j26.211) \\ (3.812 + j26.211) & (3.812 + j26.211) & (14.932 + j58.376) \end{bmatrix} \Omega$$

It is interesting to compare this result with those computed previously for the partially transposed line. Only the off-diagonal reactance changes, and these changes are shown in Table 4.4. The sequence impedances are computed from (4.119) or by a similarity transformation of $\mathbf{Z}_{abc}$ with the result

$$\mathbf{Z}_{012} = \begin{bmatrix} (22.554 + j110.79) & (0.000 + j0.000) & (0.000 + j0.000) \\ (0.000 + j0.000) & (11.120 + j32.166) & (0.000 + j0.000) \\ (0.000 + j0.000) & (0.000 + j0.000) & (11.120 + j32.166) \end{bmatrix} \Omega$$

Table 4.4. Comparison of Results—Examples 4.1 to 4.6

Example Number	*Transposition Type*	*Section Factors* f_1	f_2	f_3	X_{ab} *(ohms)*	X_{bc} *(ohms)*	X_{ca} *(ohms)*	*Max* ΔX *(ohms)*
4.1	none	1.0	0	0	27.332	27.332	23.968	3.364
4.2	$R_{\bar{\phi}}^{1}$	0.2	0.3	0.5	25.650	26.323	26.659	1.009
4.3	$R_{\bar{\phi}}^{1}$	0.4	0.6	0	27.332	25.314	25.986	2.018
4.3a	R_{ϕ}	0.4	0.6	0	25.314	27.332	25.986	2.018
4.5	$T_{\phi 23}$	0.4	0.6	0	25.314	27.332	25.986	2.018
4.6	$R_{\bar{\phi}}^{1}$	0.33	0.33	0.33	26.211	26.211	26.211	0

Table 4.5. Comparison of Sequence Coupling Impedances

Transposition Type	*Section Factors* f_1	f_2	f_3	Z_{01}, Z_{10}, Z_{02}, *or* Z_{20}	Z_{12} *or* Z_{21}
none	1.0	0	0	±0.971 − j0.561	±1.942 + j1.121
$R_{\bar{\phi}}^{1}$	0.2	0.3	0.5	±0.291 − j0.056	±0.583 + j0.112
$R_{\bar{\phi}}^{1}$	0.4	0.6	0	±0.388 + j0.449	±0.777 − j0.897
$R_{\bar{\phi}}^{1}$	0.33	0.33	0.33	0.000 + j0.000	0.000 + j0.000

Note that for this completely transposed line the off-diagonal terms are all zero. Comparison of the off-diagonal sequence impedances computed in Examples 4.4 and 4.6 is shown in Table 4.5.

4.8 Circuit Unbalance Due to Incomplete Transposition

Section 4.7 notes that a complete transposition of the phase wires, with $f_1 = f_2 = f_3$, eliminates coupling between sequence networks because the sums of all phase-to-phase mutuals are equal. Suppose that the transposition is only partially complete ($f_1 \neq f_2 \neq f_3$) or omitted entirely. What effect does this have on the sequence coupling and on the system operation?

To answer this question it is useful to examine what we mean by the equation for the voltage drop along a line. Equation (4.82) is the voltage at the "sending end" of a line which is transposed but *not* such that $f_1 = f_2 = f_3$. In fact these factors may take on any value between 0 and 1 as long as their sum is unity. Equation (4.82) is therefore a perfectly general equation for a transmission line, providing the line has equal self impedance in each phase. If we omit the factor s, the line length, we may write the voltage drop per unit length as

$$\begin{bmatrix} \Sigma V_a \\ \Sigma V_b \\ \Sigma V_c \end{bmatrix} = \begin{bmatrix} z_s & z_{k1} & z_{k2} \\ z_{k1} & z_s & z_{k3} \\ z_{k2} & z_{k3} & z_s \end{bmatrix} \begin{bmatrix} I_a \\ I_b \\ I_c \end{bmatrix} \text{ V/unit length} \tag{4.133}$$

where

$$z_s = (r_a + r_d) + j\omega k \ln \frac{D_e}{D_s} \quad \Omega\text{/unit length} \tag{4.134}$$

and

$$\begin{bmatrix} z_{k1} \\ z_{k2} \\ z_{k3} \end{bmatrix} = \begin{bmatrix} r_d \\ r_d \\ r_d \end{bmatrix} + j\omega k \begin{bmatrix} f_1 & f_2 & f_3 \\ f_2 & f_3 & f_1 \\ f_3 & f_1 & f_2 \end{bmatrix} \begin{bmatrix} \ln \frac{D_e}{D_{12}} \\ \ln \frac{D_e}{D_{23}} \\ \ln \frac{D_e}{D_{13}} \end{bmatrix} \quad \Omega\text{/unit length} \tag{4.135}$$

Obviously the voltage drop in each phase involves the impedance of the earth, even though the voltage of the earth "conductor" at the point of voltage measurement is zero.

It is convenient to think of impedances (4.134) and (4.135) as the effective phase self and mutual impedances of a physical system in which the ground is a perfectly conducting plane, as shown in Figure 4.14. Here the earth impedances are included as a part of the phase impedances. Then the impedance of the line may be computed by placing a three-phase short circuit on the remote end of the line and observing the voltages and currents at the sending end. Equation (4.133) expresses this relationship, and from this equation we may extract the impedance data we seek.

To determine the amount of unbalance in a given line, it is convenient to solve (4.133) for the currents since we often measure unbalance by specifying the per

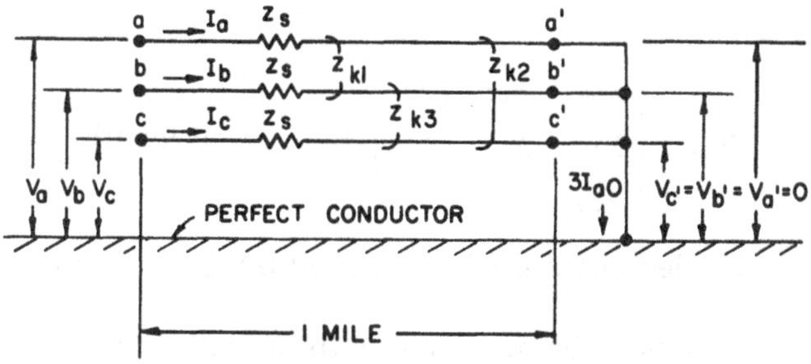

Fig. 4.14. Equivalent circuit for equation (4.133).

unit negative or zero sequence currents when the applied voltage is strictly positive sequence. Transforming (4.133), we write

$$\mathbf{V}_{012} = \mathbf{z}_{012}\mathbf{I}_{012} \text{ V/unit length} \tag{4.136}$$

where

$$\mathbf{z}_{012} = \begin{bmatrix} z_{00} & z_{01} & z_{02} \\ z_{10} & z_{11} & z_{12} \\ z_{20} & z_{21} & z_{22} \end{bmatrix} \Omega\text{/unit length} \tag{4.137}$$

The elements of $\mathbf{z}_{012}$ are completely defined by (2.46)-(2.48). Applying these formulas we compute

$$\begin{aligned} z_{S0} &= z_s \quad \Omega\text{/unit length} \\ z_{S1} &= z_{S2} = 0 \quad \Omega\text{/unit length} \\ z_{M0} &= (1/3)(z_{k1} + z_{k2} + z_{k3}) \quad \Omega\text{/unit length} \\ z_{M1} &= (1/3)(\mathrm{a}^2 z_{k1} + \mathrm{a} z_{k2} + z_{k3}) \quad \Omega\text{/unit length} \\ z_{M2} &= (1/3)(\mathrm{a} z_{k1} + \mathrm{a}^2 z_{k2} + z_{k3}) \quad \Omega\text{/unit length} \end{aligned} \tag{4.138}$$

From these quantities we apply (2.46) to compute

$$\begin{aligned} z_{00} &= z_s + (2/3)(z_{k1} + z_{k2} + z_{k3}) \quad \Omega\text{/unit length} \\ z_{11} &= z_{22} = z_s - (1/3)(z_{k1} + z_{k2} + z_{k3}) \quad \Omega\text{/unit length} \\ z_{01} &= z_{20} = -(1/3)(\mathrm{a} z_{k1} + \mathrm{a}^2 z_{k2} + z_{k3}) \quad \Omega\text{/unit length} \\ z_{02} &= z_{10} = -(1/3)(\mathrm{a}^2 z_{k1} + \mathrm{a} z_{k2} + z_{k3}) \quad \Omega\text{/unit length} \\ z_{12} &= -2z_{01} \quad \Omega\text{/unit length} \\ z_{21} &= -2z_{02} \quad \Omega\text{/unit length} \end{aligned} \tag{4.139}$$

Since the amount of unbalance is determined from the unbalanced current flowing when balanced voltages are applied, it is convenient to invert the voltage equation (4.136) to find

$$\mathbf{I}_{012} = \mathbf{y}_{012}\mathbf{V}_{012} \text{ A} \tag{4.140}$$

where

$$\mathbf{y}_{012} = \mathbf{z}_{012}^{-1} = \frac{1}{\det \mathbf{z}_{012}} \begin{bmatrix} (z_{11}z_{22} - z_{21}z_{12}) & (z_{02}z_{21} - z_{01}z_{22}) & (z_{01}z_{12} - z_{11}z_{02}) \\ (z_{20}z_{12} - z_{10}z_{22}) & (z_{00}z_{22} - z_{20}z_{02}) & (z_{10}z_{02} - z_{00}z_{12}) \\ (z_{10}z_{21} - z_{20}z_{11}) & (z_{20}z_{01} - z_{00}z_{21}) & (z_{00}z_{11} - z_{10}z_{01}) \end{bmatrix} \text{mho} \cdot \text{unit length} \tag{4.141}$$

If we define

$$\mathbf{y}_{012} = \begin{bmatrix} y_{00} & y_{01} & y_{02} \\ y_{10} & y_{11} & y_{12} \\ y_{20} & y_{21} & y_{22} \end{bmatrix} \text{ mho} \cdot \text{unit length} \tag{4.142}$$

then every admittance element is easily identified by the corresponding impedance quantities of (4.141).

Since the unbalance is to be measured with only positive sequence voltage applied, we write (4.140) as

$$\begin{bmatrix} I_{a0} \\ I_{a1} \\ I_{a2} \end{bmatrix} = \begin{bmatrix} y_{00} & y_{01} & y_{02} \\ y_{10} & y_{11} & y_{12} \\ y_{20} & y_{21} & y_{22} \end{bmatrix} \begin{bmatrix} 0 \\ V_{a1} \\ 0 \end{bmatrix} = \begin{bmatrix} y_{01} \\ y_{11} \\ y_{21} \end{bmatrix} V_{a1} \tag{4.143}$$

We then define the per unit unbalances for zero sequence and negative sequence currents as [28]

$$m_0 = \frac{I_{a0}}{I_{a1}} = \frac{y_{01}}{y_{11}}, \qquad m_2 = \frac{I_{a2}}{I_{a1}} = \frac{y_{21}}{y_{11}} \text{ pu} \tag{4.144}$$

In terms of the $\mathbf{z}_{012}$ matrix elements these factors are written as

$$m_0 = \frac{z_{02} z_{21} - z_{01} z_{22}}{z_{00} z_{22} - z_{20} z_{02}}, \qquad m_2 = \frac{z_{20} z_{01} - z_{00} z_{21}}{z_{00} z_{22} - z_{20} z_{02}} \text{ pu} \tag{4.145}$$

Gross and Hesse [28] note that in physical systems

$$z_{22} >> z_{02}, z_{21}, \qquad z_{00} >> z_{20}, z_{01}$$

If we divide the m_0 equation by z_{22} and the m_2 equation by z_{00}, we have the approximate formulas

$$m_0 \cong -\frac{z_{01}}{z_{00}}, \qquad m_2 \cong -\frac{z_{21}}{z_{22}} \text{ pu} \tag{4.146}$$

and substituting from (4.139) we write

$$m_0 \cong -\frac{(\mathrm{a} z_{k1} + \mathrm{a}^2 z_{k2} + z_{k3})}{3z_s + 2(z_{k1} + z_{k2} + z_{k3})} \text{ pu}$$

$$m_2 \cong \frac{2(\mathrm{a}^2 z_{k1} + \mathrm{a} z_{k2} + z_{k3})}{3z_s - (z_{k1} + z_{k2} + z_{k3})} \text{ pu} \tag{4.147}$$

Gross and Hesse [28] give values of m_0 and m_2 for various configurations and also examine the effect of ground wires on these unbalance factors. These computations show that for most commonly used configurations, m_0 is nearly constant at about 1% but is increased a great deal by the addition of ground wires, the effect of which should not be neglected. The factor m_2, on the other hand, varies over a wider range, say from about 3 to 20%, but is only slightly affected by ground wires. Gross and Nelson [29] give quick estimating curves for these factors at a variety of spacings and configurations. Gross et al. [30] extend this investigation to include double circuit lines.

The unbalance factors computed in equations (4.144)–(4.147) are due to the line only. This is much greater than the total unbalance of the system, assuming that only this one line is unbalanced. To compute the system unbalance, we should include the effect of the Thevenin equivalent impedance, looking into the network at each end of the line. If we designate these Thevenin impedances and voltages by the primed and unprimed notation of Figure 4.14, we have for the sequence voltage equation

$$\begin{bmatrix} 0 \\ E_{t1} - E'_{t1} \\ 0 \end{bmatrix} = \begin{bmatrix} Z_{t0} + Z_{00} + Z'_{t0} & Z_{01} & Z_{02} \\ Z_{10} & Z_{t1} + Z_{11} + Z'_{t1} & Z_{12} \\ Z_{20} & Z_{21} & Z_{t2} + Z_{22} + Z'_{t2} \end{bmatrix} \begin{bmatrix} I_{a0} \\ I_{a1} \\ I_{a2} \end{bmatrix}$$

where the Thevenin voltages E_{t1} and E'_{t1} are strictly positive sequence and the Thevenin equivalent impedances Z_{t0}, Z_{t1}, Z_{t2} and Z'_{t0}, Z'_{t1}, Z'_{t2} are assumed to be balanced and therefore uncoupled. Since the unbalance factors have the **Z** matrix diagonal terms in the denominator, the system unbalance is smaller than the unbalance due to this one line alone. However, since the Thevenin impedances are functions of the system condition, it may be preferable to compute the unbalance for the line only, as in (4.144) or (4.145), with the understanding that the values computed will be pessimistic. If done in this way, the unbalance factor may be considered as a parameter of the line itself. It may be desirable, however, to perform computations of system unbalance in a particular situation.

Example 4.7

Compute the unbalance factors of previous examples using both the exact and the approximate formulas.

Solution

From Examples 4.4 and 4.6 we find $\mathbf{Z}_{012}$ matrices which we invert as follows:

1. From Example 4.1 with no transpositions and $f_1 = 1.0$, $f_2 = 0$, $f_3 = 0$ we compute

$$\mathbf{Y}_{012} = \begin{bmatrix} (1.768 - j8.671) & (0.311 + j0.010) & (-0.146 - j0.274) \\ (-0.146 - j0.274) & (9.717 - j27.846) & (-1.933 - j0.284) \\ (0.311 + j0.010) & (0.720 + j1.816) & (9.717 - j27.846) \end{bmatrix} \text{mmho}$$

2. From Example 4.2 with two rotations and $f_1 = 0.2$, $f_2 = 0.3$, $f_3 = 0.5$ we compute

$$\mathbf{Y}_{012} = \begin{bmatrix} (1.764 - j8.667) & (-0.057 - j0.052) & (0.074 + j0.026) \\ (0.074 + j0.026) & (9.609 - j27.776) & (0.336 + j0.388) \\ (-0.057 - j0.052) & (-0.456 - j0.234) & (9.608 - j27.776) \end{bmatrix} \text{mmho}$$

3. From Example 4.3 with one rotation and $f_1 = 0.4$, $f_2 = 0.6$, $f_3 = 0$ we compute

$$\mathbf{Y}_{012} = \begin{bmatrix} (1.765 - j8.668) & (0.024 + j0.148) & (-0.144 + j0.054) \\ (-0.143 + j0.053) & (9.633 - j27.791) & (-0.046 - j1.023) \\ (0.023 + j0.148) & (1.006 - j0.197) & (9.633 - j27.791) \end{bmatrix} \text{mmho}$$

Table 4.6. Comparison of Sequence Unbalance Computations

Example Number	Section Factors			Exact Method		Approximate Method	
	f_1	f_2	f_3	% m_0	% m_2	% m_0	% m_2
4.1	1.0	0.0	0.0	$1.054\angle 72.68^\circ$	$6.624\angle 139.14^\circ$	$0.992\angle 71.51^\circ$	$6.590\angle 139.07^\circ$
4.2	0.2	0.3	0.5	$0.265\angle -67.00^\circ$	$1.746\angle -81.86^\circ$	$0.262\angle -67.60^\circ$	$1.743\angle -81.86^\circ$
4.3	0.4	0.6	0.0	$0.510\angle 151.89^\circ$	$3.484\angle 59.83^\circ$	$0.525\angle 150.61^\circ$	$3.487\angle 59.96^\circ$
4.3a	0.4	0.6	0.0	$0.521\angle -9.55^\circ$	$3.484\angle -141.71^\circ$	$0.525\angle -7.60^\circ$	$3.487\angle -141.82^\circ$
4.5	0.4	0.6	0.0	$0.521\angle -9.55^\circ$	$3.434\angle -141.71^\circ$	$0.525\angle -7.60^\circ$	$3.487\angle -141.82^\circ$

4. From Example 4.6 with complete transposition and $f_1 = f_2 = f_3 = 1/3$ we compute

$$\mathrm{Y}_{012} = \begin{bmatrix} (1.764 - j8.667) & (0.000 + j0.000) & (0.000 + j0.000) \\ (0.000 + j0.000) & (9.608 - j27.775) & (0.000 + j0.000) \\ (0.000 + j0.000) & (0.000 + j0.000) & (9.608 - j27.775) \end{bmatrix} \text{ mmho}$$

The unbalance factors are now easily computed from equations (4.144) and (4.146). Results are shown in Table 4.6 and show that the approximate formulas (4.146) for computing the unbalance factors are quite accurate and will suffice for most computations.

4.9 Sequence Impedance of Lines with Bundled Conductors

Consider the transmission line shown in Figure 4.15 consisting of four overhead wires with common earth return. This system is similar to that of the three-phase line of Figure 4.5, and the various primitive impedances are computed from relations similar to (4.50)-(4.52) which may be written by inspection. Also, as in the previous case, we let

$$I_a + I_b + I_c + I_x = -I_d \text{ A} \tag{4.148}$$

Then we may write the primitive equation

$$\begin{bmatrix} V_{aa'} \\ V_{bb'} \\ V_{cc'} \\ V_{xx'} \\ V_{dd'} \end{bmatrix} = \begin{bmatrix} V_a - V_{a'} \\ V_b - V_{b'} \\ V_c - V_{c'} \\ V_x - V_{x'} \\ 0 - V_{d'} \end{bmatrix} = \begin{bmatrix} \bar{z}_{aa} & \bar{z}_{ab} & \bar{z}_{ac} & \bar{z}_{ax} & \bar{z}_{ad} \\ \bar{z}_{ba} & \bar{z}_{bb} & \bar{z}_{bc} & \bar{z}_{bx} & \bar{z}_{bd} \\ \bar{z}_{ca} & \bar{z}_{cb} & \bar{z}_{cc} & \bar{z}_{cx} & \bar{z}_{cd} \\ \bar{z}_{xa} & \bar{z}_{xb} & \bar{z}_{xc} & \bar{z}_{xx} & \bar{z}_{xd} \\ \bar{z}_{da} & \bar{z}_{db} & \bar{z}_{dc} & \bar{z}_{dx} & \bar{z}_{dd} \end{bmatrix} \begin{bmatrix} I_a \\ I_b \\ I_c \\ I_x \\ I_d \end{bmatrix} \text{ V/unit length} \tag{4.149}$$

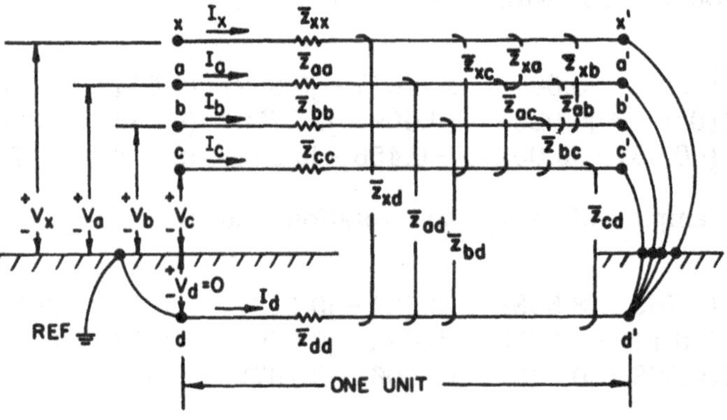

Fig. 4.15. Four conductors with earth return.

where the elements of the primitive impedance matrix are all defined by

$$\bar{z}_{pp} = r_p + j\omega k \left(\ln \frac{2s}{D_{sp}} - 1 \right) \ \Omega/\text{unit length}$$
$$p = a, b, c, x \tag{4.150}$$

and

$$\bar{z}_{pq} = j\omega k \left(\ln \frac{2s}{D_{pq}} - 1 \right) \ \Omega/\text{unit length}$$
$$p, q = a, b, c, x, d$$
$$p \neq q \tag{4.151}$$

Since the line currents sum to the negative of I_d, the last equation may be subtracted from the other four to write the line equations

$$\begin{bmatrix} V_a \\ V_b \\ V_c \\ V_x \end{bmatrix} = \begin{bmatrix} z_{aa} & z_{ab} & z_{ac} & z_{ax} \\ z_{ba} & z_{bb} & z_{bc} & z_{bx} \\ z_{ca} & z_{cb} & z_{cc} & z_{cx} \\ z_{xa} & z_{xb} & z_{xc} & z_{xx} \end{bmatrix} \begin{bmatrix} I_a \\ I_b \\ I_c \\ I_x \end{bmatrix} \ \text{V/unit length} \tag{4.152}$$

where the matrix elements are defined in terms of the primitive impedances, viz.,

$$z_{pq} = \bar{z}_{pq} - \bar{z}_{pd} - \bar{z}_{dq} + \bar{z}_{dd} \ \ \Omega/\text{unit length}$$
$$p, q = a, b, c, x \tag{4.153}$$

Now suppose that wire x is connected in parallel with wire a such that their voltage drops are identical or $V_{xx'} = V_{aa'}$ from which we find that

$$V_x - V_a = 0 \tag{4.154}$$

To make use of this property, we replace the V_x equation in (4.152) by a new equation computed from (4.154). Then (4.152) becomes

$$\begin{bmatrix} V_a \\ V_b \\ V_c \\ 0 \end{bmatrix} = \begin{bmatrix} z_{aa} & z_{ab} & z_{ac} & z_{ax} \\ z_{ba} & z_{bb} & z_{bc} & z_{bx} \\ z_{ca} & z_{cb} & z_{cc} & z_{cx} \\ (z_{xa} - z_{aa}) & (z_{xb} - z_{ab}) & (z_{xc} - z_{ac}) & (z_{xx} - z_{ax}) \end{bmatrix} \begin{bmatrix} I_a \\ I_b \\ I_c \\ I_x \end{bmatrix} \tag{4.155}$$

Since wires x and a are in parallel, they form a new phase a composite or "bundled" conductor as indicated in Figure 4.16 where we define the new phase a

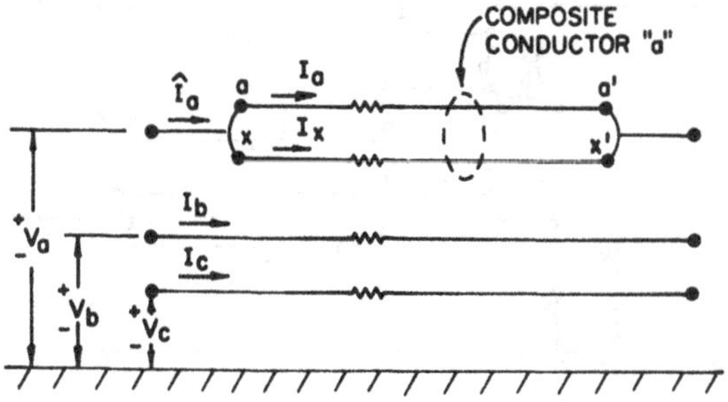

Fig. 4.16. Bundling of a and x to form the new phase a composite conductor.

current to be

$$\hat{I}_a = I_a + I_x \quad \text{A} \tag{4.156}$$

Now a zI_x product in each of the *a-b-c* equations may be added and subtracted, leaving the equations unchanged. This amounts to replacing I_a in (4.155) by (4.156) and replacing the fourth column of the impedance matrix by the difference between column 4 and column 1. The result is

$$\begin{bmatrix} V_a \\ V_b \\ V_c \\ \hline 0 \end{bmatrix} = \left[\begin{array}{ccc|c} z_{aa} & z_{ab} & z_{ac} & (z_{ax} - z_{aa}) \\ z_{ba} & z_{bb} & z_{bc} & (z_{bx} - z_{ba}) \\ z_{ca} & z_{cb} & z_{cc} & (z_{cx} - z_{ca}) \\ \hline (z_{xa} - z_{aa}) & (z_{xb} - z_{ab}) & (z_{xc} - z_{ac}) & \hat{z}_{xx} \end{array}\right] \begin{bmatrix} I_a + I_x \\ I_b \\ I_c \\ \hline I_x \end{bmatrix} \tag{4.157}$$

where $\hat{z}_{xx} = z_{xx} - z_{ax} - z_{xa} + z_{aa}$.

Writing (4.157) in partitioned matrix form we have

$$\begin{bmatrix} \mathbf{V}_{abc} \\ 0 \end{bmatrix} = \begin{bmatrix} \mathbf{z}_1 & \mathbf{z}_2 \\ \mathbf{z}_3 & \mathbf{z}_4 \end{bmatrix} \begin{bmatrix} \hat{\mathbf{I}}_{abc} \\ \mathbf{I}_x \end{bmatrix} \tag{4.158}$$

We then compute by matrix reduction,

$$\mathbf{V}_{abc} = (\mathbf{z}_1 - \mathbf{z}_2 \mathbf{z}_4^{-1} \mathbf{z}_3) \hat{\mathbf{I}}_{abc} \tag{4.159}$$

The effect of adding the one wire x to phase a is to greatly increase the GMR of phase a. This reduces the impedance of phase a but also reduces all other self and mutual impedances. The amount of the reduction is given by the matrix $\mathbf{z}_2 \mathbf{z}_4^{-1} \mathbf{z}_3$ each term of which, for this simple case, may be computed from the formula

$$(\mathbf{z}_2 \mathbf{z}_4^{-1} \mathbf{z}_3)_{pq} = \frac{(z_{px} - z_{pa})(z_{xq} - z_{aq})}{z_{xx} - z_{ax} - z_{xa} + z_{aa}}$$

$$p, q = a, b, c \tag{4.160}$$

This same idea can be extended to any number of added wires which may then be paralleled with any phase. These additions may be made either one at a time or simultaneously. Of particular interest is the case where three wires are added to the *a-b-c* configuration with one wire to be added to each phase. The circuit is shown in Figure 4.17 where wires r, s, and t are bundled with wires a, b, and c respectively. Before bundling, we have the general voltage equation similar to (4.152) but expanded to include all six wires.

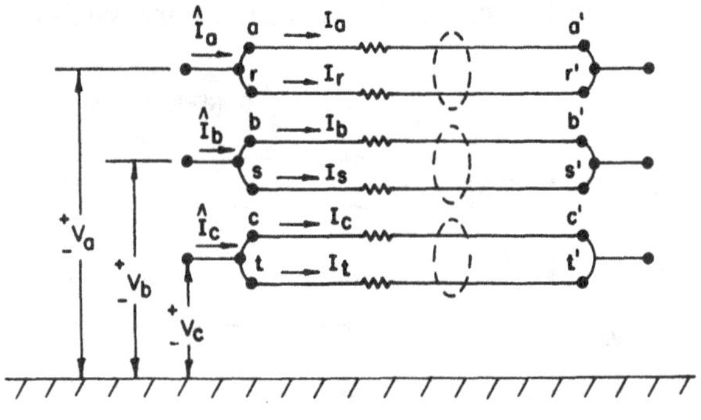

Fig. 4.17. Bundled conductors in all three phases.

$$\begin{bmatrix} V_a \\ V_b \\ V_c \\ V_r \\ V_s \\ V_t \end{bmatrix} = \begin{bmatrix} z_{aa} & z_{ab} & z_{ac} & z_{ar} & z_{as} & z_{at} \\ z_{ba} & z_{bb} & z_{bc} & z_{br} & z_{bs} & z_{bt} \\ z_{ca} & z_{cb} & z_{cc} & z_{cr} & z_{cs} & z_{ct} \\ z_{ra} & z_{rb} & z_{rc} & z_{rr} & z_{rs} & z_{rt} \\ z_{sa} & z_{sb} & z_{sc} & z_{sr} & z_{ss} & z_{st} \\ z_{ta} & z_{tb} & z_{tc} & z_{tr} & z_{ts} & z_{tt} \end{bmatrix} \begin{bmatrix} I_a \\ I_b \\ I_c \\ I_r \\ I_s \\ I_t \end{bmatrix} \text{ V/unit length} \tag{4.161}$$

After bundling, the constraining equations may be written,

$$V_r - V_a = 0, \quad V_s - V_b = 0, \quad V_t - V_c = 0 \text{ V} \tag{4.162}$$

and we define

$$\hat{I}_a = I_a + I_r, \quad \hat{I}_b = I_b + I_s, \quad \hat{I}_c = I_c + I_t \text{ A} \tag{4.163}$$

Then by the technique used before we alter (4.161) to write

$$\left[\begin{array}{c} V_a \\ V_b \\ V_c \\ \hline 0 \\ 0 \\ 0 \end{array}\right] = \left[\begin{array}{ccc|ccc} z_{aa} & z_{ab} & z_{ac} & (z_{ar} - z_{aa}) & (z_{as} - z_{ab}) & (z_{at} - z_{ac}) \\ z_{ba} & z_{bb} & z_{bc} & (z_{br} - z_{ba}) & (z_{bs} - z_{bb}) & (z_{bt} - z_{bc}) \\ z_{ca} & z_{cb} & z_{cc} & (z_{cr} - z_{ca}) & (z_{cs} - z_{cb}) & (z_{ct} - z_{cc}) \\ \hline (z_{ra} - z_{aa}) & (z_{rb} - z_{ab}) & (z_{rc} - z_{ac}) & \hat{z}_{rr} & \hat{z}_{rs} & \hat{z}_{rt} \\ (z_{sa} - z_{ba}) & (z_{sb} - z_{bb}) & (z_{sc} - z_{bc}) & \hat{z}_{sr} & \hat{z}_{ss} & \hat{z}_{st} \\ (z_{ta} - z_{ca}) & (z_{tb} - z_{cb}) & (z_{tc} - z_{cc}) & \hat{z}_{tr} & \hat{z}_{ts} & \hat{z}_{tt} \end{array}\right] \left[\begin{array}{c} \hat{I}_a \\ \hat{I}_b \\ \hat{I}_c \\ \hline I_r \\ I_s \\ I_t \end{array}\right] \text{ V/unit length} \tag{4.164}$$

where all elements in the lower right position, labeled $\hat{z}$, may be written from the formula

$$\begin{aligned} \hat{z}_{pq} &= z_{pq} - z_{iq} - z_{ph} + z_{ih} \\ i, h &= a, b, c \\ p, q &= r, s, t \end{aligned} \tag{4.165}$$

Then, following the partitioning of (4.164) we apply (4.158) to find the new impedance matrix from (4.159) or

$$\mathbf{z}_{new} = \mathbf{z}_1 - \mathbf{z}_2 \mathbf{z}_4^{-1} \mathbf{z}_3 \tag{4.166}$$

Here we must invert a 3×3 matrix $\mathbf{z}_4$, whereas this partition was a complex scalar in the previous case.

The above technique will permit the computation of the *a-b-c* impedance matrix of a bundled conductor line where each phase consists of two-wire bundles. This same result could be obtained by applying (4.158) with appropriate subscripts and adding the second wire to each phase one at a time. The amount of work involved is the same in either case.

Once the impedance matrix for the bundled conductor line is known, the sequence impedances are computed by similarity transformation.

$$\mathbf{z}_{012} = \mathbf{A}^{-1} \mathbf{z}_{new} \mathbf{A} \quad \Omega\text{/unit length} \tag{4.167}$$

Example 4.8

Compute the impedance matrices $\mathbf{Z}_{abc}$ and $\mathbf{Z}_{012}$ for the bundled conductor 345 kV line shown in Figure 4.18. There are two ACSR conductors in each phase bundle spaced 18 inches apart as shown in the insert of Figure 4.18. Ignore the

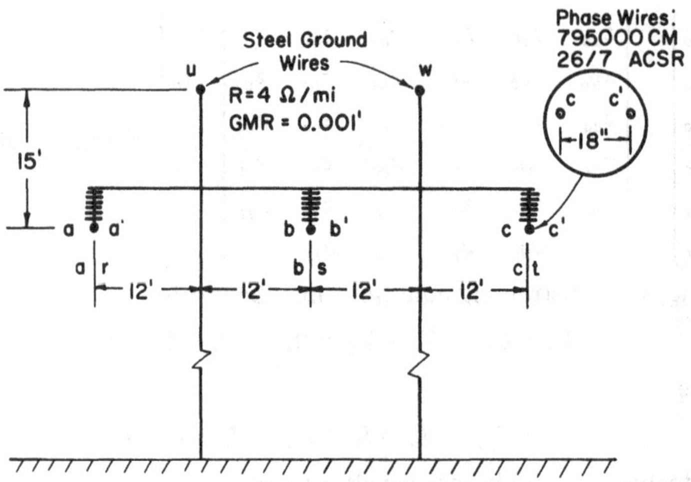

Fig. 4.18. Conductor arrangement for a 345 kV bundled-conductor line.

ground wires for this example. Let the line length be 40 miles so that results can be compared with previous computations.

Solution

From Appendix B we find the following wire data for 795000 CM ACSR:

$$r = 0.117 \quad \Omega/\text{mi at 60 Hz (small currents)}$$

$$D_s = 0.0375 \text{ ft}$$

From (4.161), using a', b', and c' for r, s, and t, we write

$$\begin{bmatrix} V_a \\ V_b \\ V_c \\ V_{a'} \\ V_{b'} \\ V_{c'} \end{bmatrix} = \begin{bmatrix} z_{aa} & z_{ab} & z_{ac} & z_{aa'} & z_{ab'} & z_{ac'} \\ z_{ba} & z_{bb} & z_{bc} & z_{ba'} & z_{bb'} & z_{bc'} \\ z_{ca} & z_{cb} & z_{cc} & z_{ca'} & z_{cb'} & z_{cc'} \\ z_{a'a} & z_{a'b} & z_{a'c} & z_{a'a'} & z_{a'b'} & z_{a'c'} \\ z_{b'a} & z_{b'b} & z_{b'c} & z_{b'a'} & z_{b'b'} & z_{b'c'} \\ z_{c'a} & z_{c'b} & z_{c'c} & z_{c'a'} & z_{c'b'} & z_{c'c'} \end{bmatrix} \begin{bmatrix} I_a \\ I_b \\ I_c \\ I_{a'} \\ I_{b'} \\ I_{c'} \end{bmatrix} \quad \Omega/\text{mi}$$

But, in our problem, the self impedances are, from (4.54) and (4.165)

$$z_{ii} = z_{pp} = (r_a + r_d) + \text{j}0.12134 \ln \frac{D_e}{D_{sp}} \quad \Omega/\text{mi}$$

$$p = a, b, c, a', b', c'$$

and since

$$r_a = r_b = r_c = r_{a'} = r_{b'} = r_{c'} = 0.117 \quad \Omega/\text{mi}$$

$$D_{sa} = D_{sb} = D_{sc} = D_{sa'} = D_{sb'} = D_{sc'} = 0.0375 \text{ ft}$$

we conclude that the diagonal terms of the **z** matrix are all equal. Furthermore, we may write, for any i and h or p and q, from (4.55) and (4.165)

$$z_{ih} = z_{pq} = r_d + \text{j}0.12134 \ln \frac{D_e}{D_{pq}} \quad \Omega/\text{mi}$$

$$p, q = a, b, c, a', b', c' \qquad p \neq q$$

and the off-diagonal terms are symmetric.

It is helpful to write the voltage matrix equation in partitioned form, taking advantage of symmetry in the way we write the matrix. Thus we have

$$\begin{bmatrix} \mathbf{V}_{abc} \\ \hline \mathbf{V}_{a'b'c'} \end{bmatrix} = \left[\begin{array}{c|c} \mathbf{z}_s & \mathbf{z}_m \\ \hline \mathbf{z}_m^t & \mathbf{z}_s \end{array}\right] \begin{bmatrix} \mathbf{I}_{abc} \\ \hline \mathbf{I}_{a'b'c'} \end{bmatrix}$$

where each partition of $\mathbf{z}_{abc}$ is 3×3. If the bundled conductor were made of two unlike conductors, the two $\mathbf{z}_s$ matrices would be different. From (4.164) we write

$$\begin{bmatrix} \mathbf{V}_{abc} \\ \hline 0 \end{bmatrix} = \left[\begin{array}{c|c} \mathbf{z}_s & (\mathbf{z}_m - \mathbf{z}_s) \\ \hline (\mathbf{z}_m^t - \mathbf{z}_s) & \mathbf{z}_k \end{array}\right] \begin{bmatrix} \mathbf{I}_{abc} + \mathbf{I}_{a'b'c'} \\ \hline \mathbf{I}_{a'b'c'} \end{bmatrix}$$

where each element of $\mathbf{z}_k$ is defined by (4.165).

Computing the various matrices for this example we have the following.

$$\mathbf{z}_s = \begin{bmatrix} (0.2123 + j1.3611) & (0.0953 + j0.5771) & (0.0953 + j0.4930) \\ (0.0953 + j0.5771) & (0.2123 + j1.3611) & (0.0953 + j0.5771) \\ (0.0953 + j0.4930) & (0.0953 + j0.5771) & (0.2123 + j1.3611) \end{bmatrix} \ \Omega/\text{mi}$$

$$\mathbf{z}_m = \begin{bmatrix} (0.0953 + j0.9135) & (0.0953 + j0.5697) & (0.0953 + j0.4892) \\ (0.0953 + j0.5849) & (0.0953 + j0.9135) & (0.0953 + j0.5697) \\ (0.0953 + j0.4968) & (0.0953 + j0.5849) & (0.0953 + j0.9135) \end{bmatrix} \ \Omega/\text{mi}$$

From these matrices the new matrices $(\mathbf{z}_m - \mathbf{z}_s)$ and $(\mathbf{z}_m^t - \mathbf{z}_s)$ can be easily computed. Then we compute

$$\mathbf{z}_k = (\mathbf{z}_s - \mathbf{z}_m) - (\mathbf{z}_m^t - \mathbf{z}_s)$$

$$= - \begin{bmatrix} (-0.2340 - j0.8952) & (0.0000 + j0.0005) & (0.0000 + j0.0001) \\ (0.0000 + j0.0005) & (-0.2340 - j0.8952) & (0.0000 + j0.0005) \\ (0.0000 + j0.0001) & (0.0000 + j0.0005) & (-0.2340 - j0.8952) \end{bmatrix} \ \Omega/\text{mi}$$

Inverting, we compute

$$z_k^{-1} = - \begin{bmatrix} (-0.2733 + j1.0456) & (-0.0003 + j0.0005) & (-0.0001 + j0.0001) \\ (-0.0003 + j0.0005) & (-0.2733 + j1.0456) & (-0.0003 + j0.0005) \\ (-0.0001 + j0.0001) & (-0.0003 + j0.0005) & (-0.2733 + j1.0456) \end{bmatrix} \ \text{mho} \cdot \text{mi}$$

To complete the matrix reduction we compute

$$\mathbf{z}_N = (\mathbf{z}_m - \mathbf{z}_s)\, \mathbf{z}_k^{-1} (\mathbf{z}_m^t - \mathbf{z}_s)$$

$$= \begin{bmatrix} (0.0585 + j0.2239) & (-0.0000 - j0.0001) & (0.0001 - j0.0001) \\ (-0.0000 - j0.0001) & (0.0585 + j0.2239) & (-0.0000 - j0.0001) \\ (0.0001 - j0.0001) & (-0.0000 - j0.0001) & (0.0585 + j0.2239) \end{bmatrix} \ \Omega/\text{mi}$$

Finally then, from (4.166)

$$\mathbf{z}_{\text{new}} = \mathbf{z}_s - \mathbf{z}_N$$

$$= \begin{bmatrix} (0.1538 + j1.1372) & (0.0953 + j0.5772) & (0.0952 + j0.4930) \\ (0.0953 + j0.5772) & (0.1538 + j1.1372) & (0.0953 + j0.5772) \\ (0.0953 + j0.4930) & (0.0953 + j0.5772) & (0.1538 + j1.1372) \end{bmatrix} \ \Omega/\text{mi}$$

From this result we readily compute

$$\mathbf{z}_{012} = \begin{bmatrix} (0.3444 + j2.2354) & (0.0243 - j0.0140) & (-0.0243 - j0.0140) \\ (-0.0243 - j0.0140) & (0.0585 + j0.5881) & (-0.0486 + j0.0281) \\ (0.0243 - j0.0140) & (0.0486 + j0.0280) & (0.0585 + j0.5881) \end{bmatrix} \ \Omega/\text{mi}$$

and

$$\mathbf{y}_{012} = \begin{bmatrix} (0.0675 - j0.4375) & (0.0224 - j0.0060) & (-0.0164 - j0.0164) \\ (-0.0164 - j0.0164) & (0.1725 - j1.6996) & (-0.1548 + j0.0517) \\ (0.0224 - j0.0060) & (0.1222 + j0.1082) & (0.1725 - j1.6996) \end{bmatrix} \ \text{mho} \cdot \text{mi}$$

Then the total impedance for the 40 miles of line is

$$\mathbf{Z}_{012} = \begin{bmatrix} (13.775 + j89.417) & (0.970 - j0.561) & (-0.971 - j0.560) \\ (-0.970 - j0.560) & (2.3409 + j23.523) & (-1.943 + j1.122) \\ (0.970 - j0.561) & (1.943 + j1.121) & (2.3409 + j23.523) \end{bmatrix} \ \Omega$$

We may also compute the unbalance factors two ways.

$$m_0 = \frac{y_{01}}{y_{11}} = \frac{0.0224 - j0.0060}{0.1725 - j1.6996} = 0.0136 \ \underline{/69.2^\circ}$$

$$\cong -\frac{z_{01}}{z_{00}} = -\frac{0.0243 - j0.0140}{0.3444 + j2.2354} = 0.0124 \ \underline{/68.8^\circ}$$

$$m_2 = \frac{y_{21}}{y_{11}} = \frac{0.1222 + j0.1082}{0.1725 - j1.6996} = 0.096 \ \underline{/125.7^\circ}$$

$$\cong -\frac{z_{21}}{z_{22}} = -\frac{0.0486 + j0.0280}{0.0585 + j0.5881} = 0.095 \ \underline{/125.6^\circ}$$

It is interesting to compare these results with those of the previous examples. First, comparing against the $\mathbf{Z}_{012}$ of the untransposed line of Example 4.4, we see that the off-diagonal terms are almost exactly the same for the two lines even though the conductor size and spacing are greatly different for the two cases. From this we observe that the coupling between sequences is not critically dependent on phase spacing, bundling, or wire size. The diagonal terms Z_{00}, Z_{11} and Z_{22} are reduced, both in the real and imaginary parts, by bundling.

The unbalance factors may be compared against those of a similar untransposed line in Example 4.7. Since the denominator terms in the unbalance equations, z_{00}, z_{22}, and y_{11} are all reduced by bundling, we would expect the unbalance factors to increase.

4.10 Sequence Impedance of Lines with One Ground Wire

In many physical transmission lines, wires are added above the phase wires to "shield" the line against direct lightning strokes. Determining the location of such wires is beyond the scope of our concern. We shall concentrate on the effect such ground wires have on the line impedances (also see [10, 11, and 24]).

Consider the line arrangement shown in Figure 4.19 where one ground wire, labeled w, is shown with each end connected solidly to the local ground point. Clearly, the voltage equation of this arrangement is exactly the same as that of Figure 4.15 given by equation (4.149) except that in this case $V_x = V_w = 0$. Thus

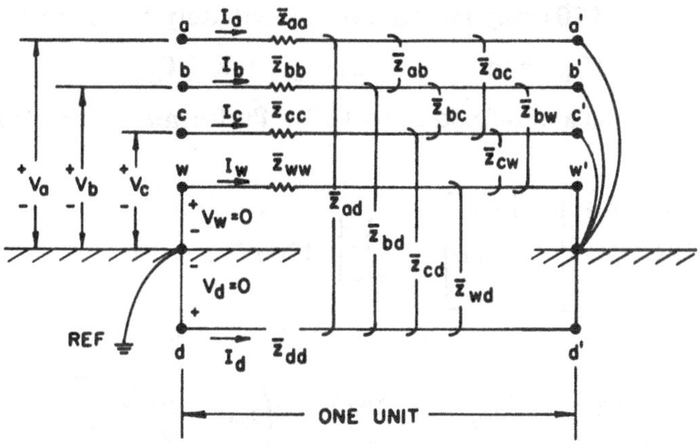

Fig. 4.19. A three-phase line with one ground wire.

we write the primitive equation

$$\begin{bmatrix} V_{aa'} \\ V_{bb'} \\ V_{cc'} \\ V_{ww'} \\ V_{dd'} \end{bmatrix} = \begin{bmatrix} V_a - V_{a'} \\ V_b - V_{b'} \\ V_c - V_{c'} \\ 0 - V_{w'} \\ 0 - V_{d'} \end{bmatrix} = \begin{bmatrix} \bar{z}_{aa} & \bar{z}_{ab} & \bar{z}_{ac} & \bar{z}_{aw} & \bar{z}_{ad} \\ \bar{z}_{ba} & \bar{z}_{bb} & \bar{z}_{bc} & \bar{z}_{bw} & \bar{z}_{bd} \\ \bar{z}_{ca} & \bar{z}_{cb} & \bar{z}_{cc} & \bar{z}_{cw} & \bar{z}_{cd} \\ \bar{z}_{wa} & \bar{z}_{wb} & \bar{z}_{wc} & \bar{z}_{ww} & \bar{z}_{wd} \\ \bar{z}_{da} & \bar{z}_{db} & \bar{z}_{dc} & \bar{z}_{dw} & \bar{z}_{dd} \end{bmatrix} \begin{bmatrix} I_a \\ I_b \\ I_c \\ I_w \\ I_d \end{bmatrix} \quad \text{V/unit length} \tag{4.168}$$

Now with ground wire w in parallel with the earth conductor d, the return current will divide between the two paths or $I_a + I_b + I_c = -(I_d + I_w)$. Rearranging, we write

$$I_d = -(I_a + I_b + I_c + I_w) \tag{4.169}$$

Using this result for I_d in (4.168), we subtract the $V_{dd'}$ equation from each of the others to compute

$$\left[\begin{array}{c} V_a \\ V_b \\ V_c \\ \hline V_w = 0 \end{array}\right] = \left[\begin{array}{ccc|c} z_{aa} & z_{ab} & z_{ac} & z_{aw} \\ z_{ba} & z_{bb} & z_{bc} & z_{bw} \\ z_{ca} & z_{cb} & z_{cc} & z_{cw} \\ \hline z_{wa} & z_{wb} & z_{wc} & z_{ww} \end{array}\right] \left[\begin{array}{c} I_a \\ I_b \\ I_c \\ \hline I_w \end{array}\right] \quad \text{V/unit length} \tag{4.170}$$

Where as before,

$$z_{pq} = \bar{z}_{pq} - \bar{z}_{pd} - \bar{z}_{dq} + \bar{z}_{dd}$$
$$p, q = a, b, c, w \tag{4.171}$$

Note that z_{pq} is defined to include r_a or r_w when $p = q$ but is purely imaginary when $p \neq q$. From (4.45)-(4.52) we compute

$$z_{pq} = (r_a + r_d) + j\omega k \ln \frac{D_e}{D_{pq}}, \quad p = q$$
$$= r_d + j\omega k \ln \frac{D_e}{D_{pq}}, \quad p \neq q \quad \Omega\text{/unit length} \tag{4.172}$$

Since $V_w = 0$, (4.170) may be reduced immediately to the form

$$\mathbf{V}_{abc} = (\mathbf{z}_1 - \mathbf{z}_2\mathbf{z}_4^{-1}\mathbf{z}_3)\,\mathbf{I}_{abc} = \hat{\mathbf{z}}_{abc}\mathbf{I}_{abc} \tag{4.173}$$

where the Z partitions are defined in (4.158). Performing the operation indicated in (4.173), we have

$$\hat{\mathbf{z}}_{abc} = \begin{bmatrix} z_{aa} & z_{ab} & z_{ac} \\ z_{ba} & z_{bb} & z_{bc} \\ z_{ca} & z_{cb} & z_{cc} \end{bmatrix} - \begin{bmatrix} z_{aw} \\ z_{bw} \\ z_{cw} \end{bmatrix} \begin{bmatrix} \dfrac{1}{z_{ww}} \end{bmatrix} [z_{wa}\; z_{wb}\; z_{wc}]$$

$$= \begin{array}{c} \\ a \\ b \\ c \end{array} \begin{array}{c} \begin{array}{ccc} a & b & c \end{array} \\ \begin{bmatrix} \left(z_{aa} - \dfrac{z_{aw}z_{wa}}{z_{ww}}\right) & \left(z_{ab} - \dfrac{z_{aw}z_{wb}}{z_{ww}}\right) & \left(z_{ac} - \dfrac{z_{aw}z_{wc}}{z_{ww}}\right) \\ \left(z_{ba} - \dfrac{z_{bw}z_{wa}}{z_{ww}}\right) & \left(z_{bb} - \dfrac{z_{bw}z_{wb}}{z_{ww}}\right) & \left(z_{bc} - \dfrac{z_{bw}z_{wc}}{z_{ww}}\right) \\ \left(z_{ac} - \dfrac{z_{cw}z_{wa}}{z_{ww}}\right) & \left(z_{cb} - \dfrac{z_{cw}z_{wb}}{z_{ww}}\right) & \left(z_{cc} - \dfrac{z_{cw}z_{wc}}{z_{ww}}\right) \end{bmatrix} \end{array} \;\Omega/\text{unit length} \tag{4.174}$$

or each element of the reduced matrix is of the form

$$\hat{z}_{pq} = z_{pq} - \frac{z_{pw}z_{wq}}{z_{ww}}$$

$$p, q \text{ (row, col)} = a, b, c \tag{4.175}$$

Each element of the matrix is smaller by a correction factor involving the mutual impedances to the ground wire w.

In most lines of interest we may assume that the three phase wires have equal self impedances or $z_{aa} = z_{bb} = z_{cc}$. This makes the leading terms on the diagonals equal, but the subtracted corrections are still unequal since the mutuals from w to a, b, and c are usually unequal.

4.10.1 Sequence impedance of an untransposed line with one ground wire

To compute the sequence impedances, we again use equations (2.46)–(2.48). Since the matrix (4.174) is symmetric but with unequal mutual terms, the sequences will, from (2.46), be coupled by nonreciprocal terms. The computation is straightforward, however, and involves only the application of (2.46)–(2.48) or $\mathbf{Z}_{012} = \mathbf{A}^{-1}\hat{\mathbf{Z}}_{abc}\mathbf{A}$. This result is too laborious to write out in detail. An example will illustrate the procedure.

Example 4.9

Compute the sequence impedance of the 40-mile line of Example 4.1, taking into account the presence of the ground wire. Assume that the ground wire has a resistance of 4 ohm/mile and a self GMD of 10^{-3} feet. There are no transpositions.

Solution

We are given that

$$r_w = 4.0 \;\Omega/\text{mi}, \qquad D_{sw} = D_{ww} = 10^{-3} \text{ ft}$$

Then the self impedance term is

$$z_{ww} = r_w + r_d + j\omega k \ln \frac{D_e}{D_{ww}} = 4.0 + 0.09528 + j(0.1213) \ln \frac{2790}{0.001}$$

$$= 4.095 + j1.800 \quad \Omega/\text{mi}$$

and for 40 miles of line $Z_{ww} = 163.80 + j72.00\ \Omega$.

To compute the mutuals, we need the distances

$$D_{aw} = D_{cw} = \sqrt{10^2 + 15^2} = 18.03 \text{ ft}$$

$$D_{bw} = 15.0 \text{ ft}$$

Then

$$z_{aw} = z_{cw} = r_d + j\omega k \ln \frac{D_e}{D_{aw}} = 0.09528 + j(0.1213) \ln \frac{2790}{18.03}$$

$$= 0.09528 + j0.612 \quad \Omega/\text{mi}$$

and

$$z_{bw} = 0.09528 + j(0.1213) \ln \frac{2790}{15.0} = 0.09528 + j0.634 \quad \Omega/\text{mi}$$

The total mutual impedances for the 40 mile line are

$$Z_{aw} = Z_{cw} = 3.812 + j24.470 \quad \Omega$$

$$Z_{bw} = 3.812 + j25.363 \quad \Omega$$

In matrix form

$$\mathbf{Z}_{abcw} = \left[\begin{array}{ccc|c} (14.932 + j58.376) & (3.812 + j27.332) & (3.812 + j23.968) & (3.812 + j24.470) \\ (3.812 + j27.332) & (14.932 + j58.376) & (3.812 + j27.332) & (3.812 + j25.363) \\ (3.812 + j23.968) & (3.812 + j27.332) & (14.932 + j58.376) & (3.812 + j24.470) \\ \hline (3.812 + j24.470) & (3.812 + j25.363) & (3.812 + j24.470) & (163.812 + j72.036) \end{array}\right] \Omega$$

Reducing along indicated partitioning, we compute

$$\hat{\mathbf{Z}}_{abc} = \begin{bmatrix} (17.500 + j56.107) & (6.484 + j24.996) & (6.380 + j21.698) \\ (6.484 + j24.996) & (17.712 + j55.972) & (6.484 + j24.996) \\ (6.380 + j21.698) & (6.484 + j24.996) & (17.500 + j56.107) \end{bmatrix} \Omega$$

Then by digital computer we find the sequence impedances

$$\mathbf{Z}_{012} = \begin{bmatrix} (30.470 + j103.85) & (0.860 - j0.618) & (-0.966 - j0.436) \\ (-0.966 - j0.436) & (11.121 + j32.165) & (-1.944 + j1.121) \\ (0.860 - j0.619) & (1.943 + j1.123) & (11.121 + j32.165) \end{bmatrix} \Omega$$

The unbalance factor may be easily computed as

$$m_0 = -\frac{z_{01}}{z_{00}} = -\frac{0.860 - j0.618}{30.470 + j103.85} = 0.00978 \underline{/70.65^\circ}$$

$$m_2 = -\frac{z_{21}}{z_{22}} = -\frac{1.943 + j1.123}{11.121 + j32.165} = 0.0659 \underline{/139.11^\circ}$$

Comparing this result with that of Example 4.4, we observe the following:

1. The positive and negative self impedances Z_{11} and Z_{22} are unchanged by the addition of the ground wire.
2. The zero sequence impedance is changed in a remarkable way, viz., (a) the real part R_{00} is *increased* by about 40% and (b) the imaginary part X_{00} has *decreased* by about 7%.
3. The off-diagonal terms are changed but not by a significant amount.
4. The zero sequence unbalance has been increased by the addition of the ground wire but the negative sequence unbalance is unchanged.

Example 4.10

Repeat the computations of $\mathbf{Z}_{012}$ made in Example 4.9 but with a ground wire of higher conductivity. Compare the results with those of Example 4.9.

Solution

For this example let us choose two different ground wires as follows:

1. 3/8-inch copperweld, 40% conductivity

$$D_s = 0.00497 \text{ ft}$$

$$r_w = 1.264 \quad \Omega/\text{mi}$$

$$z_{ww} = 1.359 + j0.12134 \ln \frac{2790}{0.00497} = 1.359 + j1.606 \quad \Omega/\text{mi}$$

2. 1F copperweld-copper

$$D_s = 0.00980 \text{ ft}$$

$$r_w = 0.705 \quad \Omega/\text{mi}$$

$$z_{ww} = 0.8003 + j0.12134 \ln \frac{2790}{0.00980} = 0.8003 + j1.524 \quad \Omega/\text{mi}$$

Then, by digital computer we compute the following sequence impedances for the 40 miles of line.

For the line with a 3/8-inch copperweld ground wire the impedance for 40 miles of line is given by

$$\mathbf{Z}_{012} = \begin{bmatrix} (31.207 + j90.148) & (0.706 - j0.568) & (-0.844 - j0.327) \\ (-0.844 - j0.327) & (11.122 + j32.163) & (-1.946 + j1.121) \\ (0.706 - j0.568) & (1.944 + j1.125) & (11.122 + j32.163) \end{bmatrix} \Omega$$

For the line with a 1F copperweld-copper ground wire the impedance for 40 miles is given by

$$\mathbf{Z}_{012} = \begin{bmatrix} (27.403 + j83.866) & (0.652 - j0.499) & (-0.758 - j0.316) \\ (-0.758 - j0.316) & (11.122 + j32.162) & (-1.947 + j1.122) \\ (0.652 - j0.499) & (1.945 + j1.125) & (11.122 + j32.162) \end{bmatrix} \Omega$$

It is apparent that the zero sequence reactance reduces appreciably as the ground wire impedance goes down. The zero sequence resistance remains high compared to the case with no ground wire.

It is also interesting to compare the unbalance factor for the three cases.

Table 4.7. Comparison of Zero Sequence Unbalance, m_0, in Percent, as a Function of Ground Wire Selection

	No Ground Wire	*3/8-inch Steel*	*3/8-inch Cw*	*1F Cw-Cu*
$\lvert m_0 \rvert$	1.054	0.978	0.950	0.931
angle	72.68°	70.65°	70.28°	70.67°

Obviously the negative sequence unbalance is unchanged (Why?). The zero sequence unbalance is tabulated for comparison in Table 4.7.

Example 4.11

Repeat the computation made in Example 4.9 except consider the configuration of Figure 4.20 where the ground wire is located at the same height as before but 5 feet to the right of center.

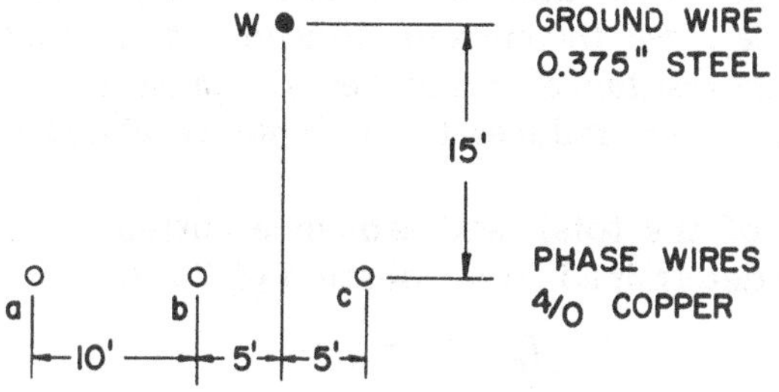

Fig. 4.20. Line configuration for Example 4.11.

Solution

We proceed exactly as in Example 4.10, the only difference being in the distance from w to a, b, and c. This makes the reactance terms different than before by a very small amount in the fourth row and column. Thus we compute

$$\mathbf{Z}_{abcw} = \left[\begin{array}{ccc|c} (14.931 + j58.375) & (3.812 + j27.331) & (3.812 + j23.966) & (3.812 + j23.680) \\ (3.812 + j27.331) & (14.931 + j58.375) & (3.812 + j27.331) & (3.812 + j25.107) \\ (3.812 + j23.966) & (3.812 + j27.331) & (14.931 + j58.375) & (3.812 + j25.107) \\ \hline (3.812 + j23.680) & (3.812 + j25.107) & (3.812 + j25.107) & (163.812 + j72.036) \end{array}\right] \Omega$$

Reducing to find $\mathbf{Z}_{abc}$, we compute

$$\mathbf{Z}_{abc} = \begin{bmatrix} (17.319 + j56.223) & (6.360 + j25.075) & (6.247 + j21.783) \\ (6.360 + j25.075) & (17.651 + j56.011) & (6.411 + j25.043) \\ (6.247 + j21.783) & (6.411 + j25.043) & (17.651 + j56.011) \end{bmatrix} \Omega$$

and converting to the 0-1-2 frame of reference we have

$$\mathbf{Z}_{012} = \begin{bmatrix} (30.374 + j103.92) & (0.804 - j0.454) & (-1.139 - j0.454) \\ (-1.139 - j0.454) & (11.123 + j32.165) & (-1.939 + j1.120) \\ (0.804 - j0.454) & (1.946 + j1.120) & (11.123 + j32.164) \end{bmatrix} \Omega$$

The unbalance factors for this case are

$$m_0 = -\frac{z_{01}}{z_{00}} = -\frac{0.804 - j0.454}{30.374 + j103.92} = 0.00853 \underline{/76.84^\circ}$$

$$m_2 = -\frac{z_{21}}{z_{22}} = -\frac{1.946 + j1.120}{11.123 + j32.164} = 0.0659\ \underline{/138.99^\circ}$$

Comparing results, we note that all terms of the $\mathbf{Z}_{abc}$ matrix have been changed by moving the ground wire, with the row and column for wire a seeing the greatest change. In the $\mathbf{Z}_{012}$ matrix the zero sequence row and column is affected. The zero sequence unbalance is reduced slightly, which may come as a surprise.

4.10.2 Current division between the ground wire and the earth

It is apparent that the addition of a ground wire to an untransposed line section reduces the zero sequence reactance x_{00} and increases the zero sequence resistance r_{00}. This is due in part to the relatively high impedance of the ground wire compared to the earth and in part to the fact that the ground wire is physically nearer the phase wires than the ground return path. Because the ground wire is near the phase wires, the current induced in this wire tends to be large and if the wire has a large resistance, this makes r_{00} appear to be great. The closeness of this coupling, however, reduces the inductance about in proportion to the current I_w.

The proportion of the total zero sequence current $3I_{a0}$ which flows in the ground wire may be determined from equation (4.168). If we set

$$I_a = I_b = I_c = I_{a0} \tag{4.176}$$

in (4.168) and subtract the d equation from the w equation

$$\begin{aligned} 0 = {} & [(\bar{z}_{wa} + \bar{z}_{wb} + \bar{z}_{wc}) - (\bar{z}_{da} + \bar{z}_{db} + \bar{z}_{dc})]\, I_{a0} \\ & + (\bar{z}_{ww} - \bar{z}_{dw})I_w + (\bar{z}_{wd} - \bar{z}_{dd})I_d \end{aligned} \tag{4.177}$$

Also, from (4.169) and (4.175)

$$I_d = -(3I_{a0} + I_w) \tag{4.178}$$

If we substitute (4.178) into (4.177), we compute

$$-\frac{I_w}{3I_{a0}} = \frac{(\bar{z}_{wa} + \bar{z}_{wb} + \bar{z}_{wc}) - (\bar{z}_{da} + \bar{z}_{db} + \bar{z}_{dc}) - 3\bar{z}_{wd} + 3\bar{z}_{dd}}{3(\bar{z}_{ww} - \bar{z}_{wd} - \bar{z}_{dw} + \bar{z}_{dd})} \tag{4.179}$$

The primitive impedances in this expression may be combined according to (4.171) to write

$$-\frac{I_w}{3I_{a0}} = \frac{z_{wa} + z_{wb} + z_{wc}}{3z_{ww}} \tag{4.180}$$

Also, from (4.178), we compute

$$-\frac{I_d}{3I_{a0}} = 1 - \frac{z_{wa} + z_{wb} + z_{wc}}{3z_{ww}} \tag{4.181}$$

The two foregoing expressions give the ratio of currents in the ground wire and earth respectively to the total zero sequence current. From (4.180) and (4.181) we compute the ratio of ground wire to earth current.

$$\frac{I_w}{I_d} = \frac{z_{wa} + z_{wb} + z_{wc}}{3z_{ww} - (z_{wa} + z_{wb} + z_{wc})}$$

Suppose we define

$$z_{ag} \triangleq (1/3)(z_{wa} + z_{wb} + z_{wc}) = r_d + j\omega k \ln \frac{D_e}{D_{ag}} \tag{4.182}$$

where

$$D_{ag} = (D_{wa} D_{wb} D_{wc})^{1/3} \tag{4.183}$$

is the GMD between the ground wire and the phase wires. Then

$$\frac{I_w}{I_d} = \frac{z_{ag}}{z_{ww} - z_{ag}} \tag{4.184}$$

Obviously as z_{ww} becomes smaller, the fraction of the total current flowing in the ground wire becomes large. Note that it is not possible for the denominator to go to zero. We have defined z_{ww} as

$$z_{ww} = r_w + r_d + j\omega k \ln \frac{D_e}{D_{ww}}$$

Therefore

$$z_{ww} - z_{ag} = r_w + j\omega k \ln \frac{D_{ag}}{D_{ww}} > 0$$

for any conceivable situation.

Obviously I_w can be greater than I_d. This occurs when the following conditions are met. If we compute the ratios of the phasor currents, we have

$$\frac{I_w}{I_d} = \frac{z_{ag}}{z_{ww} - z_{ag}} > 1 \tag{4.185}$$

then it is apparent that I_w is greater than I_d when

$$|z_{ww} - z_{ag}| < |z_{ag}| \tag{4.186}$$

Since the currents in (4.185) are both phasors, it may be more meaningful to take the absolute value of these expressions.

4.10.3 Sequence impedance of a transposed line with one ground wire

The mutual coupling between sequences may be reduced and even eliminated by transposing the line. Suppose that the line is transposed according to the rotation sequence in Figure 4.9. Let the three wire positions be identified as α, β, and γ such that the arrangement in Table 4.8 is observed. Then the impedance of each transposition section may be determined from (4.174) with the result shown in Table 4.9. The total impedance in ohms per unit length is always the sum of the three matrices of Table 4.9 or

$$\begin{bmatrix} \Sigma V_a \\ \Sigma V_b \\ \Sigma V_c \end{bmatrix} = s\,[f_1 \mathbf{z}_{f1} + f_2 \mathbf{z}_{f2} + f_3 \mathbf{z}_{f3}] \begin{bmatrix} I_a \\ I_b \\ I_c \end{bmatrix} \tag{4.187}$$

Table 4.8. Conductor Arrangement in Transposition Sections

Section	*Length*	*Position*	*Phasing*
1	$f_1 s$	α-β-γ	*a-b-c*
2	$f_2 s$	α-β-γ	*c-a-b*
3	$f_3 s$	α-β-γ	*b-c-a*

Table 4.9. Impedances of Transposition Sections

Section	*Impedance Matrix* (*Ω/unit length*)
s_1	$\mathbf{Z}_{f1} = \begin{bmatrix} \left(z_{\alpha\alpha} - \frac{z_{\alpha w} z_{w\alpha}}{z_{ww}}\right) & \left(z_{\alpha\beta} - \frac{z_{\alpha w} z_{w\beta}}{z_{ww}}\right) & \left(z_{\alpha\gamma} - \frac{z_{\alpha w} z_{w\gamma}}{z_{ww}}\right) \\ \left(z_{\beta\alpha} - \frac{z_{\beta w} z_{w\alpha}}{z_{ww}}\right) & \left(z_{\beta\beta} - \frac{z_{\beta w} z_{w\beta}}{z_{ww}}\right) & \left(z_{\beta\gamma} - \frac{z_{\beta w} z_{w\gamma}}{z_{ww}}\right) \\ \left(z_{\gamma\alpha} - \frac{z_{\gamma w} z_{w\alpha}}{z_{ww}}\right) & \left(z_{\gamma\beta} - \frac{z_{\gamma w} z_{w\beta}}{z_{ww}}\right) & \left(z_{\gamma\gamma} - \frac{z_{\gamma w} z_{w\gamma}}{z_{ww}}\right) \end{bmatrix}$
s_2	$\mathbf{Z}_{f2} = \begin{bmatrix} \left(z_{\beta\beta} - \frac{z_{\beta w} z_{w\beta}}{z_{ww}}\right) & \left(z_{\beta\gamma} - \frac{z_{\beta w} z_{w\gamma}}{z_{ww}}\right) & \left(z_{\beta\alpha} - \frac{z_{\beta w} z_{w\alpha}}{z_{ww}}\right) \\ \left(z_{\gamma\beta} - \frac{z_{\gamma w} z_{w\beta}}{z_{ww}}\right) & \left(z_{\gamma\gamma} - \frac{z_{\gamma w} z_{w\gamma}}{z_{ww}}\right) & \left(z_{\gamma\alpha} - \frac{z_{\gamma w} z_{w\alpha}}{z_{ww}}\right) \\ \left(z_{\alpha\beta} - \frac{z_{\alpha w} z_{w\beta}}{z_{ww}}\right) & \left(z_{\alpha\gamma} - \frac{z_{\alpha w} z_{w\gamma}}{z_{ww}}\right) & \left(z_{\alpha\alpha} - \frac{z_{\alpha w} z_{w\alpha}}{z_{ww}}\right) \end{bmatrix}$
s_3	$\mathbf{Z}_{f3} = \begin{bmatrix} \left(z_{\gamma\gamma} - \frac{z_{\gamma w} z_{w\gamma}}{z_{ww}}\right) & \left(z_{\gamma\alpha} - \frac{z_{\gamma w} z_{w\alpha}}{z_{ww}}\right) & \left(z_{\gamma\beta} - \frac{z_{\gamma w} z_{w\beta}}{z_{ww}}\right) \\ \left(z_{\alpha\gamma} - \frac{z_{\alpha w} z_{w\gamma}}{z_{ww}}\right) & \left(z_{\alpha\alpha} - \frac{z_{\alpha w} z_{w\alpha}}{z_{ww}}\right) & \left(z_{\alpha\beta} - \frac{z_{\alpha w} z_{w\beta}}{z_{ww}}\right) \\ \left(z_{\beta\gamma} - \frac{z_{\beta w} z_{w\gamma}}{z_{ww}}\right) & \left(z_{\beta\alpha} - \frac{z_{\beta w} z_{w\alpha}}{z_{ww}}\right) & \left(z_{\beta\beta} - \frac{z_{\beta w} z_{w\beta}}{z_{ww}}\right) \end{bmatrix}$

When the line is completely transposed, $f_1 = f_2 = f_3 = 1/3$, and some simplified equations for the sequence impedances may be derived. First we observe that the three diagonal terms are equal, thus we define z_s as

$$\begin{aligned} z_s &\triangleq z_{aa} = z_{bb} = z_{cc} \\ &= \frac{1}{3}(z_{\alpha\alpha} + z_{\beta\beta} + z_{\gamma\gamma}) - \frac{1}{3}\left(\frac{z_{\alpha w} z_{w\alpha}}{z_{ww}} + \frac{z_{\beta w} z_{w\beta}}{z_{ww}} + \frac{z_{\gamma w} z_{w\gamma}}{z_{ww}}\right) \ \Omega\text{/unit length} \end{aligned} \quad (4.188)$$

Similarly, all off-diagonal terms are equal such that we may define z_m as

$$\begin{aligned} z_m &\triangleq z_{ab} = z_{ba} = z_{bc} = z_{cb} = z_{ca} = z_{ac} \\ &= \frac{1}{3}(z_{\alpha\beta} + z_{\beta\gamma} + z_{\gamma\alpha}) - \frac{1}{3}\left(\frac{z_{\alpha w} z_{w\beta}}{z_{ww}} + \frac{z_{\beta w} z_{w\gamma}}{z_{ww}} + \frac{z_{\gamma w} z_{w\alpha}}{z_{ww}}\right) \ \Omega\text{/unit length} \end{aligned} \quad (4.189)$$

These relations are true since z_{pq} is reciprocal. Then the impedance to currents

$\mathbf{I}_{abc}$ may be written in matrix form as

$$\mathbf{z}_{abc} = \begin{bmatrix} z_s & z_m & z_m \\ z_m & z_s & z_m \\ z_m & z_m & z_s \end{bmatrix} \ \Omega\text{/unit length} \tag{4.190}$$

Because of the symmetry of (4.190) the sequence impedances are much simplified. From equations (2.47) and (2.48) we compute

$$z_{S0} = z_s, \qquad z_{S1} = z_{S2} = 0$$
$$z_{M0} = z_m, \qquad z_{M1} = z_{M2} = 0$$

Then

$$z_{11} = z_{22} = z_s - z_m, \qquad z_{00} = z_s + 2z_m \tag{4.191}$$

and all the off-diagonal terms in the $\mathbf{Z}_{012}$ matrix are zero.

From (4.191) the positive sequence impedance is found to be

$$z_{11} = \frac{1}{3}(z_{\alpha\alpha} + z_{\beta\beta} + z_{\gamma\gamma}) - \frac{1}{3}(z_{\alpha\beta} + z_{\beta\gamma} + z_{\gamma\alpha})$$
$$- \frac{1}{3}\left(\frac{z_{\alpha w} z_{w\alpha}}{z_{ww}} + \frac{z_{\beta w} z_{w\beta}}{z_{ww}} + \frac{z_{\gamma w} z_{w\gamma}}{z_{ww}}\right) + \frac{1}{3}\left(\frac{z_{\alpha w} z_{w\beta}}{z_{ww}} + \frac{z_{\beta w} z_{w\gamma}}{z_{ww}} + \frac{z_{\gamma w} z_{w\alpha}}{z_{ww}}\right)$$
$$\Omega\text{/unit length} \tag{4.192}$$

We examine this equation term by term. For the first term we compute

$$\frac{1}{3}(z_{\alpha\alpha} + z_{\beta\beta} + z_{\gamma\gamma}) = (r_a + r_d) + j\omega k \ln \frac{D_e}{D_s} \quad \Omega\text{/unit length} \tag{4.193}$$

where $D_s \triangleq D_{\alpha\alpha} = D_{\beta\beta} = D_{\gamma\gamma}$. For the second term we compute

$$\frac{1}{3}(z_{\alpha\beta} + z_{\beta\gamma} + z_{\gamma\alpha}) = r_d + j\omega k \ln \frac{D_e}{D_{eq}} \quad \Omega\text{/unit length} \tag{4.194}$$

where $D_{eq} \triangleq (D_{\alpha\beta} D_{\beta\gamma} D_{\gamma\alpha})^{1/3}$. The third and fourth terms may be combined to write

$$\frac{1}{3z_{ww}}\left[(z_{\alpha w} z_{w\alpha} + z_{\beta w} z_{w\beta} + z_{\gamma w} z_{w\gamma}) - (z_{\alpha w} z_{w\beta} + z_{\beta w} z_{w\gamma} + z_{\gamma w} z_{w\alpha})\right] = -\frac{\omega^2 k^2 M}{3z_{ww}} \tag{4.195}$$

where we let

$$M \triangleq \left(\ln \frac{D_e}{D_{\alpha w}}\right)^2 + \left(\ln \frac{D_e}{D_{\beta w}}\right)^2 + \left(\ln \frac{D_e}{D_{\gamma w}}\right)^2$$
$$- \left(\ln \frac{D_e}{D_{\alpha w}}\right)\left(\ln \frac{D_e}{D_{\beta w}}\right) - \left(\ln \frac{D_e}{D_{\beta w}}\right)\left(\ln \frac{D_e}{D_{\gamma w}}\right) - \left(\ln \frac{D_e}{D_{\gamma w}}\right)\left(\ln \frac{D_e}{D_{\alpha w}}\right) \tag{4.196}$$

and finally,

$$z_{ww} = r_w + r_d + j\omega k \ln \frac{D_e}{D_{ww}} \tag{4.197}$$

In most physical configurations $D_e >> D_{\alpha w}, D_{\beta w}, D_{\gamma w}$ so that $M \cong 0$. Then z_{11} is simply

$$z_{11} = (r_a + r_d) + j\omega k \ln \frac{D_e}{D_s} - r_d - j\omega k \ln \frac{D_e}{D_{eq}} = r_a + j\omega k \ln \frac{D_{eq}}{D_s} \quad \Omega\text{/unit length} \tag{4.198}$$

For convenience we define the quantities

$$x_a = \omega k \ln \frac{1}{D_s}, \qquad x_d = \omega k \ln D_{eq} \tag{4.199}$$

to write

$$z_{11} = r_a + j(x_a + x_d) \ \Omega\text{/unit length} \tag{4.200}$$

All quantities needed (r_a, x_a, and x_d) may be determined by the tables of Appendix B. Note also that equations (4.198) and (4.199) are identical to previous results for a completely transposed line, the only approximation necessary being that of $M \cong 0$.

For the zero sequence impedance we write, from (2.46),

$$\begin{aligned} z_{00} = z_s + 2z_m &= (1/3)(z_{\alpha\alpha} + z_{\beta\beta} + z_{\gamma\gamma}) + (2/3)(z_{\alpha\beta} + z_{\beta\gamma} + z_{\gamma\alpha}) \\ &- (1/3z_{ww})[(z_{\alpha w} z_{w\alpha} + z_{\beta w} z_{w\beta} + z_{\gamma w} z_{w\gamma}) + 2(z_{\alpha w} z_{w\beta} + z_{\beta w} z_{w\gamma} + z_{\gamma w} z_{w\alpha})] \end{aligned} \tag{4.201}$$

The first two terms are known from (4.193) and (4.194). The numerator of last term requires more detailed study. We know that $z_{xw} = z_{wx}$ for $x = \alpha, \beta, \gamma$. Thus the numerator is of the polynomial form

$$A^2 + B^2 + C^2 + 2AB + 2BC + 2CA = (A + B + C)^2 \tag{4.202}$$

or we may write the third term as

$$\begin{aligned} &\frac{1}{3z_{ww}} \left[\left(r_d + j\omega k \ln \frac{D_e}{D_{\alpha w}}\right) + \left(r_d + j\omega k \ln \frac{D_e}{D_{\beta w}}\right) + \left(r_d + j\omega k \ln \frac{D_e}{D_{\gamma w}}\right)\right]^2 \\ &= \frac{1}{3z_{ww}} \left(3r_d + j3\omega k \ln D_e + j3\omega k \ln \frac{1}{D_{ag}}\right)^2 \end{aligned} \tag{4.203}$$

where we define $D_{ag} = (D_{\alpha w} D_{\beta w} D_{\gamma w})^{1/3}$ which is the GMD between the phase wires and the ground wire.

Now for convenience we define the quantities shown in Table 4.10. These defined quantities may then be used to write all of (4.201). We define

Table 4.10. Definitions of Useful Quantities

Quantity	*Designation*	*Value in Ω/mi (f = 60 Hz, D_e = 2790 ft)*
$3r_d$	r_e	0.2858
$3\omega k \ln D_e$	x_e	2.888
$-\omega k \ln D_s$	x_a	...
$\omega k \ln D_{eq}$	x_d	...
$\omega k \ln D_{ag}$	x_{ag}	...
$-\omega k \ln D_{ww}$	x_g	...

$$\begin{aligned}
z_{0(a)} &\triangleq (1/3)(z_{\alpha\alpha} + z_{\beta\beta} + z_{\gamma\gamma}) + (2/3)(z_{\alpha\beta} + z_{\beta\gamma} + z_{\gamma\alpha}) \\
&= (r_a + 3r_d) + j\omega k \ln \frac{D_e^3}{D_s D_{eq}^2} \\
&= (r_a + r_e) + j(x_e + x_a - 2x_d)\ \Omega/\text{unit length} \qquad (4.204)
\end{aligned}$$

$$\begin{aligned}
z_{0(m)} &\triangleq [z_{\alpha w} z_{w\alpha} + z_{\beta w} z_{w\beta} + z_{\gamma w} z_{w\gamma} \\
&\quad + 2(z_{\alpha w} z_{w\beta} + z_{\beta w} z_{w\gamma} + z_{\gamma w} z_{w\alpha})]^{1/2} \\
&= 3r_d + j3\omega k \ln D_e - j3\omega k \ln D_{ag} \\
&= r_e + j(x_e - 3x_{ag}) \qquad (4.205)
\end{aligned}$$

$$z_{0(g)} \triangleq 3z_{ww} = (3r_w + r_e) + j(x_e + 3x_g)\ \Omega/\text{unit length} \qquad (4.206)$$

Using these definitions, we write

$$z_{00} = z_{0(a)} - \frac{z_{0(m)}^2}{z_{0(g)}} \quad \Omega/\text{unit length} \qquad (4.207)$$

This formula agrees with that of [14] for the case of one ground wire and is a commonly used expression for computing zero sequence impedances. It may be used when the following conditions are true.

1. The line is completely transposed.
2. The three phase wires are identical.
3. Either $D_e >> D_{\alpha w}$, $D_{\beta w}$, $D_{\gamma w}$, or $D_{\alpha w} \cong D_{\beta w} \cong D_{\gamma w}$ so that $M \cong 0$ in (4.196).
4. The phase wires are all the same height above the ground, i.e.,

$$D_{ae} = D_{be} = D_{ce} \triangleq D_e$$

4.11 Sequence Impedance of Lines with Two Ground Wires

A system of three phase wires and two ground wires is analyzed in exactly the same way as the case for one ground wire. Consider the system shown in Figure 4.21 where ground wires u and w in parallel with the phase wires are solidly

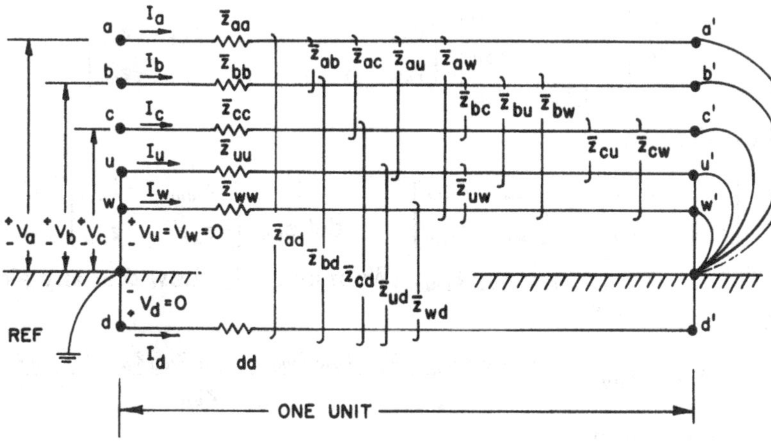

Fig. 4.21. A three-phase line with two ground wires.

grounded at each end of the line to the local ground point. We write the primitive voltage equations as follows:

$$\begin{bmatrix} V_{aa'} \\ V_{bb'} \\ V_{cc'} \\ V_{uu'} \\ V_{ww'} \\ V_{dd'} \end{bmatrix} = \begin{bmatrix} V_a - V_{a'} \\ V_b - V_{b'} \\ V_c - V_{c'} \\ 0 - V_{u'} \\ 0 - V_{w'} \\ 0 - V_{d'} \end{bmatrix} = \begin{bmatrix} \bar{z}_{aa} & \bar{z}_{ab} & \bar{z}_{ac} & \bar{z}_{au} & \bar{z}_{aw} & \bar{z}_{ad} \\ \bar{z}_{ba} & \bar{z}_{bb} & \bar{z}_{bc} & \bar{z}_{bu} & \bar{z}_{bw} & \bar{z}_{bd} \\ \bar{z}_{ca} & \bar{z}_{cb} & \bar{z}_{cc} & \bar{z}_{cu} & \bar{z}_{cw} & \bar{z}_{cd} \\ \bar{z}_{ua} & \bar{z}_{ub} & \bar{z}_{uc} & \bar{z}_{uu} & \bar{z}_{uw} & \bar{z}_{ud} \\ \bar{z}_{wa} & \bar{z}_{wb} & \bar{z}_{wc} & \bar{z}_{wu} & \bar{z}_{ww} & \bar{z}_{wd} \\ \bar{z}_{da} & \bar{z}_{db} & \bar{z}_{dc} & \bar{z}_{du} & \bar{z}_{dw} & \bar{z}_{dd} \end{bmatrix} \begin{bmatrix} I_a \\ I_b \\ I_c \\ I_u \\ I_w \\ I_d \end{bmatrix} \text{ V/unit length} \tag{4.208}$$

Since the return current divides between paths d, u, and w, we have

$$I_a + I_b + I_c = -(I_d + I_u + I_w)$$

or

$$I_d = -(I_a + I_b + I_c + I_u + I_w) \tag{4.209}$$

Making this substitution into (4.208) and subtracting $V_{dd'}$ from each of the other voltages, we compute

$$\begin{bmatrix} V_a \\ V_b \\ V_c \\ \hline V_u = 0 \\ V_w = 0 \end{bmatrix} = \left[\begin{array}{ccc|cc} z_{aa} & z_{ab} & z_{ac} & z_{au} & z_{aw} \\ z_{ba} & z_{bb} & z_{bc} & z_{bu} & z_{bw} \\ z_{ca} & z_{cb} & z_{cc} & z_{cu} & z_{cw} \\ \hline z_{ua} & z_{ub} & z_{uc} & z_{uu} & z_{uw} \\ z_{wa} & z_{wb} & z_{wc} & z_{wu} & z_{ww} \end{array}\right] \begin{bmatrix} I_a \\ I_b \\ I_c \\ \hline I_u \\ I_w \end{bmatrix} \text{ V/unit length} \tag{4.210}$$

where we define $z_{pq} = \bar{z}_{pq} - \bar{z}_{pd} - \bar{z}_{dq} + \bar{z}_{dd}$; p, $q = a, b, c, u, w$ exactly as in (4.171).

We immediately recognize that this matrix equation may be reduced to a third-order system in variables subscripted a, b, and c. Calling the resulting impedance matrix $\hat{\mathbf{Z}}_{abc}$, we have

$$\hat{\mathbf{Z}}_{abc} = \begin{bmatrix} z_{aa} & z_{ab} & z_{ac} \\ z_{ba} & z_{bb} & z_{bc} \\ z_{ca} & z_{cb} & z_{cc} \end{bmatrix} - \begin{bmatrix} z_{au} & z_{aw} \\ z_{bu} & z_{bw} \\ z_{cu} & z_{cw} \end{bmatrix} \begin{bmatrix} y_{uu} & y_{uw} \\ y_{wu} & y_{ww} \end{bmatrix} \begin{bmatrix} z_{ua} & z_{ub} & z_{uc} \\ z_{wa} & z_{wb} & z_{wc} \end{bmatrix} \ \Omega\text{/unit length} \tag{4.211}$$

where we have defined

$$\begin{bmatrix} y_{uu} & y_{uw} \\ y_{wu} & y_{ww} \end{bmatrix} = \begin{bmatrix} z_{uu} & z_{uw} \\ z_{wu} & z_{ww} \end{bmatrix}^{-1} = \frac{1}{\det \mathbf{z}_{uw}} \begin{bmatrix} z_{ww} & -z_{uw} \\ -z_{wu} & z_{uu} \end{bmatrix} \tag{4.212}$$

and where $\det \mathbf{z}_{uw} = z_{uu} z_{ww} - z_{uw} z_{wu}$. We may easily show that any element of (4.211) may be written as

$$\hat{z}_{pq} = z_{pq} - \frac{z_{pu} z_{ww} z_{uq} - z_{pu} z_{uw} z_{wq} - z_{pw} z_{wu} z_{uq} + z_{pw} z_{uu} z_{wq}}{z_{uu} z_{ww} - z_{uw}^2}$$

$$p, q \text{ (row, col)} = a, b, c \tag{4.213}$$

This result is similar to (4.174), the result for one ground wire.

4.11.1 Sequence impedance of an untransposed line with two ground wires

If the line is untransposed, we may prepare a table similar to Table 4.9 and use appropriate values of f_1, f_2, and f_3 to describe the line sections. In particular, for $f_2 = f_3 = 0$ we have a matrix of terms which may be computed from (4.213) with $p, q = \alpha, \beta, \gamma$ where we again use the Greek letters to indicate the wire positions. The resulting matrix would be exactly the same as (4.211), and a similarity transformation will transform this result to the symmetrical component impedances. Since this result may not be written in a more compact notation than (4.211), it will be left this way.

Example 4.12

Consider the bundled conductor line shown in Figure 4.18 of Example 4.8. Compute the impedance of this line, including the effect of the two ground wires, and compare the results with those of Example 4.8. The ground wires are 3/8-inch EBB steel.

Solution

The impedance matrix of the five conductors is computed first from (4.210), using $(\mathbf{z}_{\mathrm{new}})_{abc}$ as defined in Example 4.8 as the three equivalent single-conductor phase conductors.

$$\mathbf{z}_{abcuw} = \begin{array}{cc} & \begin{array}{ccccc} a & b & c & u & w \end{array} \\ \begin{array}{c} a \\ b \\ c \\ \\ u \\ w \end{array} & \left[\begin{array}{ccc:cc} (0.154 + j1.137) & (0.095 + j0.577) & (0.095 + j0.493) & (0.095 + j0.604) & (0.095 + j0.518) \\ (0.095 + j0.577) & (0.154 + j1.137) & (0.095 + j0.577) & (0.095 + j0.604) & (0.095 + j0.604) \\ (0.095 + j0.493) & (0.095 + j0.577) & (0.154 + j1.137) & (0.095 + j0.518) & (0.095 + j0.604) \\ \hdashline (0.095 + j0.604) & (0.095 + j0.604) & (0.095 + j0.518) & (4.095 + j1.801) & (0.095 + j0.577) \\ (0.095 + j0.518) & (0.095 + j0.604) & (0.095 + j0.604) & (0.095 + j0.577) & (4.095 + j1.801) \end{array}\right] \end{array} \Omega/\text{mi}$$

Reducing along the indicated partitions to reduce to the *a-b-c* matrix, we have

$$\hat{\mathbf{z}}_{abc} = \begin{array}{cc} & \begin{array}{ccc} a & b & c \end{array} \\ \begin{array}{c} a \\ b \\ c \end{array} & \left[\begin{array}{ccc} (0.243 + j1.036) & (0.192 + j0.470) & (0.183 + j0.392) \\ (0.192 + j0.470) & (0.259 + j1.023) & (0.192 + j0.470) \\ (0.183 + j0.392) & (0.192 + j0.470) & (0.243 + j1.036) \end{array}\right] \end{array} \Omega/\text{mi}$$

Finally, transforming and multiplying by 40 miles we have

$$\mathbf{Z}_{012} = \begin{bmatrix} (25.035 + j76.737) & (0.582 - j0.711) & (-0.906 - j0.149) \\ (-0.905 - j0.149) & (2.392 + j23.505) & (-1.923 + j1.139) \\ (0.581 - j0.709) & (1.948 + j1.096) & (2.392 + j23.505) \end{bmatrix} \Omega$$

The unbalance factors are

$$m_0 = -\frac{z_{01}}{z_{00}} = -\frac{0.582 - j0.711}{25.035 + j76.737} = 0.0114\underline{/57.37^\circ}$$

$$m_2 = -\frac{z_{21}}{z_{22}} = -\frac{1.948 + j1.096}{2.392 + j23.505} = 0.0946\underline{/125.17^\circ}$$

As in the case of one ground wire, the presence of two ground wires increases R_{00}

and reduces X_{00}. This results in a slight increase in the zero sequence unbalance factor.

4.11.2 Sequence impedance of a transposed line with two ground wires

If the line is transposed, the three matrices similar to those of Table 4.9 must be added to find the total impedance $\hat{\mathbf{Z}}_{abc}$. This result may then be transformed to $\mathbf{Z}_{012}$ by a similarity transformation.

When the line is completely transposed such that $f_1 = f_2 = f_3 = 1/3$, the result may be reduced to a more compact formula. Since this formula often appears in the literature, it will be derived.

For any transposed line we may write

$$\begin{bmatrix} \hat{z}_{aa} \\ \hat{z}_{bb} \\ \hat{z}_{cc} \end{bmatrix} = \begin{bmatrix} f_1 & f_2 & f_3 \\ f_3 & f_1 & f_2 \\ f_2 & f_3 & f_1 \end{bmatrix} \begin{bmatrix} \hat{z}_{\alpha\alpha} \\ \hat{z}_{\beta\beta} \\ \hat{z}_{\gamma\gamma} \end{bmatrix} \ \Omega/\text{unit length} \tag{4.214}$$

and

$$\begin{bmatrix} \hat{z}_{ab} \\ \hat{z}_{bc} \\ \hat{z}_{ca} \end{bmatrix} = \begin{bmatrix} f_1 & f_2 & f_3 \\ f_3 & f_1 & f_2 \\ f_2 & f_3 & f_1 \end{bmatrix} \begin{bmatrix} \hat{z}_{\alpha\beta} \\ \hat{z}_{\beta\gamma} \\ \hat{z}_{\gamma\alpha} \end{bmatrix} \ \Omega/\text{unit length} \tag{4.215}$$

where the Greek-subscripted quantities are defined by (4.213) with $p, q = \alpha, \beta, \gamma$.

In the special case where $f_1 = f_2 = f_3 = 1/3$, the three equations (4.214) are all equal so we define

$$z_s \triangleq \hat{z}_{aa} = \hat{z}_{bb} = \hat{z}_{cc} = (1/3)(\hat{z}_{\alpha\alpha} + \hat{z}_{\beta\beta} + \hat{z}_{\gamma\gamma})$$

Then from (4.213) with $p, q = \alpha, \beta, \gamma$

$$z_s = (1/3)(z_{\alpha\alpha} + z_{\beta\beta} + z_{\gamma\gamma}) - (1/3K)\, z_L \tag{4.216}$$

where we define

$$K \triangleq z_{uu} z_{ww} - z_{uw}^2 = \text{denominator of (4.213)} \tag{4.217}$$

and

$$\begin{aligned} z_L \triangleq\ & + z_{\alpha u} z_{ww} z_{u\alpha} - z_{\alpha u} z_{uw} z_{w\alpha} - z_{\alpha w} z_{wu} z_{u\alpha} + z_{\alpha w} z_{uu} z_{w\alpha} \\ & + z_{\beta u} z_{ww} z_{u\beta} - z_{\beta u} z_{uw} z_{w\beta} - z_{\beta w} z_{wu} z_{u\beta} + z_{\beta w} z_{uu} z_{w\beta} \\ & + z_{\gamma u} z_{ww} z_{u\gamma} - z_{\gamma u} z_{uw} z_{w\gamma} - z_{\gamma w} z_{wu} z_{u\gamma} + z_{\gamma w} z_{uu} z_{w\gamma} \end{aligned} \tag{4.218}$$

Moreover the three equations (4.215) are all equal such that we may define

$$\begin{aligned} z_m \triangleq \hat{z}_{ab} = \hat{z}_{bc} = \hat{z}_{ca} &= \frac{1}{3}(\hat{z}_{\alpha\beta} + \hat{z}_{\beta\gamma} + \hat{z}_{\gamma\alpha}) \\ &= (1/3)(z_{\alpha\beta} + z_{\beta\gamma} + z_{\gamma\alpha}) - (1/3K)\, z_M \end{aligned} \tag{4.219}$$

where K is defined in (4.217) and

$$\begin{aligned} z_M \triangleq\ & + z_{\alpha u} z_{ww} z_{u\beta} - z_{\alpha u} z_{uw} z_{w\beta} - z_{\alpha w} z_{wu} z_{u\beta} + z_{\alpha w} z_{uu} z_{w\beta} \\ & + z_{\beta u} z_{ww} z_{u\gamma} - z_{\beta u} z_{uw} z_{w\gamma} - z_{\beta w} z_{wu} z_{u\gamma} + z_{\beta w} z_{uu} z_{w\gamma} \\ & + z_{\gamma u} z_{ww} z_{u\alpha} - z_{\gamma u} z_{uw} z_{w\alpha} - z_{\gamma w} z_{wu} z_{u\alpha} + z_{\gamma w} z_{uu} z_{w\alpha} \end{aligned} \tag{4.220}$$

Then, by symmetry

$$z_{S0} = z_s, \qquad z_{S1} = z_{S2} = 0$$
$$z_{M0} = z_m, \quad z_{M1} = z_{M2} = 0 \qquad (4.221)$$

and the sequence impedances are

$$z_{11} = z_{22} = z_s - z_m, \qquad z_{00} = z_s + 2z_m \qquad (4.222)$$

as often noted before.

The positive and negative sequence impedance may be written in terms of (4.216) and (4.219) with the result

$$z_{11} = z_{22} = (1/3)(z_{\alpha\alpha} + z_{\beta\beta} + z_{\gamma\gamma}) - (1/3)(z_{\alpha\beta} + z_{\beta\gamma} + z_{\gamma\alpha})$$
$$-(1/3K)(z_L - z_M)\ \Omega\text{/unit length} \qquad (4.223)$$

The first two terms are known from (4.193) and (4.194) as follows:

$$\frac{1}{3}(z_{\alpha\alpha} + z_{\beta\beta} + z_{\gamma\gamma}) - \frac{1}{3}(z_{\alpha\beta} + z_{\beta\gamma} + z_{\gamma\alpha}) = r_a + \text{j}\omega\text{k}\frac{D_{eq}}{D_s} = r_a + \text{j}(x_a + x_d)$$

The third term of (4.223) may be expanded to compute

$$(1/3K)(z_L - z_M) = (1/3K)(-z_{ww} + z_{uw} + z_{wu} - z_{uu})(-\omega^2\text{k}^2M)$$

where M is defined by (4.196). In most overhead lines this term may be neglected such that we write

$$z_{11} = r_a + \text{j}(x_a + x_d)\ \Omega\text{/unit length} \qquad (4.224)$$

This is the same result as previously derived for the case of one ground wire in (4.200) and for no ground wires in (4.120).

The zero sequence impedance, also defined by (4.222), may also be rearranged into a familiar form. From (4.216) and (4.219) we compute

$$z_{00} = z_s + 2z_m = (1/3)(z_{\alpha\alpha} + z_{\beta\beta} + z_{\gamma\gamma})$$
$$+ (2/3)(z_{\alpha\beta} + z_{\beta\gamma} + z_{\gamma\alpha}) - (1/3K)(z_L + 2z_M)\ \Omega\text{/unit length} \qquad (4.225)$$

The first two terms are known from (4.204) and are identified as the quantity $z_{0(a)}$. The third term requires more detailed study. We may use (4.202) to compute

$$z_L + 2z_M = z_{uu}(z_{\alpha w} + z_{\beta w} + z_{\gamma w})^2 + z_{ww}(z_{\alpha u} + z_{\beta u} + z_{\gamma u})^2$$
$$- 2z_{uw}(z_{\alpha u} + z_{\beta u} + z_{\gamma u})(z_{\alpha w} + z_{\beta w} + z_{\gamma w}) \qquad (4.226)$$

If the two ground wires are symmetrically located with respect to the three phase wires, as is often the case, then $(z_{\alpha u} + z_{\beta u} + z_{\gamma u}) = (z_{\alpha w} + z_{\beta w} + z_{\gamma w})$ and (4.226) becomes $z_L + 2z_M = (z_{uu} + z_{ww} - 2z_{uw})(z_{\alpha w} + z_{\beta w} + z_{\gamma w})^2$. Moreover, if the ground wires are identical,

$$z_{uu} = z_{ww} \qquad (4.227)$$
$$z_L + 2z_M = 2(z_{ww} - z_{uw})(z_{\alpha w} + z_{\beta w} + z_{\gamma w})^2 \qquad (4.228)$$

We also evaluate $3K$ as $3K \triangleq 3(z_{uu}z_{ww} - z_{uw}^2)$. But with identical ground wires (4.227) applies, or

$$3K = 3(z_{ww}^2 - z_{uw}^2) = 3(z_{ww} + z_{uw})(z_{ww} - z_{uw})$$

Then the third term of (4.225) may be written as

$$\frac{1}{3K}(z_L + 2z_M) = \frac{(z_{\alpha w} + z_{\beta w} + z_{\gamma w})^2}{(3/2)(z_{ww} + z_{uw})} \triangleq \frac{z_{0(m)}^2}{z_{0(g)}} \tag{4.229}$$

Then

$$\begin{aligned} z_{0(m)} &= z_{\alpha w} + z_{\beta w} + z_{\gamma w} = 3r_d + j\omega k \ln \frac{D_e^3}{D_{\alpha w}D_{\beta w}D_{\gamma w}} \\ &= r_e + j(x_e - 3x_{ag})\ \Omega\text{/unit length} \end{aligned} \tag{4.230}$$

which is exactly the same as before. The remaining impedance $z_{0(g)}$ is defined as

$$\begin{aligned} z_{0(g)} &= \frac{3}{2}(z_{ww} + z_{uw}) = \left(\frac{3}{2}r_w + 3r_d\right) + j\frac{3}{2}\omega k \ln \frac{D_e^2}{D_{ww}D_{uw}} \\ &= \left(\frac{3}{2}r_w + r_e\right) + j\left(x_e + \frac{3}{2}x_g - \frac{3}{2}x_{dg}\right)\ \Omega\text{/unit length} \end{aligned} \tag{4.231}$$

where we define

$$x_{dg} = \omega k \ln \frac{1}{D_{uw}}\quad \Omega\text{/unit length} \tag{4.232}$$

Thus if we write

$$z_{00} = z_{0(a)} - \frac{z_{0(m)}^2}{z_{0(g)}}\quad \Omega\text{/unit length} \tag{4.233}$$

the only difference between one ground wire and two ground wires is in the term $z_{0(g)}$ as follows:

$$\begin{aligned} z_{0(g)} &= 3r_w + r_e + j(x_e + 3x_g) \quad \text{one ground wire} \\ &= (3/2)r_w + r_e + j[x_e + (3/2)x_g - (3/2)x_{dg}] \quad \text{two ground wires} \end{aligned} \tag{4.234}$$

Reviewing the above derivation, we note the following restrictions of the use of (4.233) in the computation of the zero sequence impedance where two ground wires are present.

1. The line is completely transposed, i.e., $f_1 = f_2 = f_3 = 1/3$.
2. The two ground wires are symmetrically located, or

$$D_{\alpha u}D_{\beta u}D_{\gamma u} = D_{\alpha w}D_{\beta w}D_{\gamma w}$$

3. The two ground wires are identical.
4. The three phase wires are identical and at the same height above the ground.

4.12 Sequence Impedance of Lines with *n* Ground Wires

Lines are seldom constructed with more than two ground wires, but the computation of line impedances may be extended to any number of ground wires by the methods used in the previous sections. The most general case is that in which n ground wires are used, where n is any integer. This problem is of academic interest only but will be examined briefly.

If there are n ground wires, the primitive impedance matrix will be of order $3 + n + 1$ and the earth conductor may be eliminated to reduce the system to order $3 + n$. Then we may reduce the impedance matrix to the usual third-order form by the equation

$$\hat{\mathbf{Z}}_{abc} = \mathbf{Z}_1 - \mathbf{Z}_2 \mathbf{Z}_4^{-1} \mathbf{Z}_3$$

$$\hat{\mathbf{Z}}_{abc} = \begin{bmatrix} z_{aa} & z_{ab} & z_{ac} \\ z_{ba} & z_{bb} & z_{bc} \\ z_{ca} & z_{cb} & z_{cc} \end{bmatrix} - \begin{bmatrix} z_{a1} & \cdots & z_{an} \\ z_{b1} & \cdots & z_{bn} \\ z_{c1} & \cdots & z_{cn} \end{bmatrix} \mathbf{Z}_n^{-1} \begin{bmatrix} z_{1a} & z_{1b} & z_{1c} \\ \cdots & \cdots & \cdots \\ z_{na} & z_{nb} & z_{nc} \end{bmatrix} \quad (4.235)$$

where

$$\mathbf{Z}_n = \begin{bmatrix} z_{11} & \cdots & z_{1n} \\ \cdots & \cdots & \cdots \\ z_{n1} & \cdots & z_{nn} \end{bmatrix} \quad (4.236)$$

is the matrix of ground wire self and mutual impedances and is surely nonsingular.

The result of this computation is certainly not obvious, even for a completely transposed line. When the line is completely transposed the result is given in [14] as

$$z_{0(g)} = \left(\frac{3r_w}{n} + r_e\right) + \mathrm{j}\left[x_e + \frac{3}{n}x_a - \frac{3(n-1)}{n}x_{dg}\right] \ \Omega/\text{unit length} \quad (4.237)$$

with $z_{0(a)}$ and $z_{0(m)}$ being defined as before. The proof of (4.237) is left as an exercise.

4.13 Zero Sequence Impedance of Transposed Lines with Ground Wires

The zero sequence impedance of a completely transposed line with ground wires may be computed by the method of sections 4.10–4.12. However, if the line is completely transposed the zero sequence impedance may be computed directly using the GMD method (see Wagner and Evans [10] and Stevenson [21] for a discussion of the GMD method). Since this method is often used, it will be derived below.

4.13.1 Single-phase circuit with earth return

First, consider the mutual impedance between circuits with common earth return as shown in Figure 4.22. Recall that the mutual inductance between two cir-

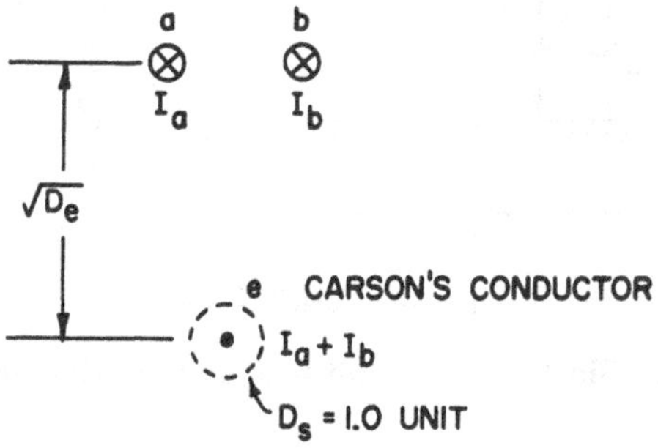

Fig. 4.22. Single-phase circuit with earth return.

cuits is defined in terms of the flux linkage λ. Thus if λ_{12} = flux linking circuit 1 due to I_2, then $M_{12} = \lambda_{12}/I_2$ H where (see [12])

$$\lambda_{12} = 2 \times 10^{-7}\, I_2 \ln \frac{D_2}{D_1} \quad \text{Wb turn/m} \tag{4.238}$$

Also, since the system is passive and linear, $M = M_{12} = M_{21}$.

In our case the flux linkage of circuit *ae* due to I_b in conductor *b* is

$$\lambda_{ae\text{-}b} = 2 \times 10^{-7} I_b \ln \frac{D_{be}}{D_{ba}} \quad \text{Wb turn/m} \tag{4.239}$$

where we have defined conductor 2 of (4.238) to be *b* and conductor 1 to be *a*. Similarly, the flux linkages of circuit *ae* due to current I_b in conductor *e* is

$$\lambda_{ae\text{-}e} = -2 \times 10^{-7} I_b \ln \frac{1}{D_{ae}} \quad \text{Wb turn/m} \tag{4.240}$$

Adding (4.239) and (4.240), we find the total flux linkages of circuit *ae* due to I_b as

$$\lambda_{ae} = 2 \times 10^{-7} I_b \ln \frac{D_{ae} D_{be}}{D_{ab}} \quad \text{Wb turn/m} \tag{4.241}$$

and if $D_{ae} \cong D_{be} \cong \sqrt{D_e}$ (the conductors are at equal heights),

$$\lambda_{ae} \cong 2 \times 10^{-7} I_b \ln \frac{D_e}{D_{ab}}$$

Then the mutual inductance between circuits *ae* and *be* is

$$m_{ae\text{-}be} = m_{ab} = \frac{\lambda_{ae}}{I_b} = 2 \times 10^{-7} \ln \frac{D_e}{D_{ab}} \quad \text{H/m} \tag{4.242}$$

or

$$m_{ab} = \text{k} \ln \frac{D_e}{D_{ab}} \quad \text{H/unit length} \tag{4.243}$$

In Figure 4.23 the circuit is shown with polarizing dots similar to Figure 4.2.

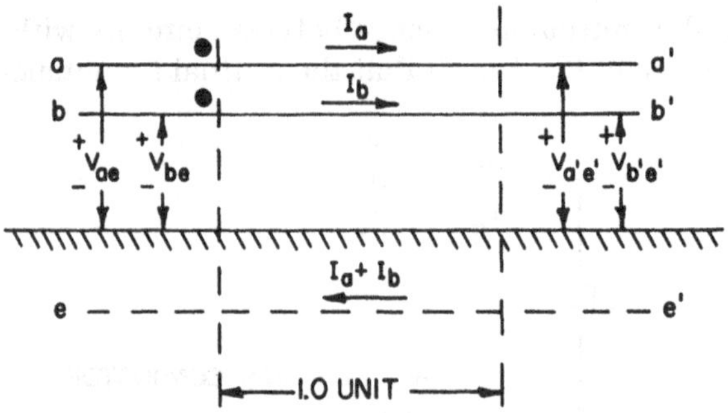

Fig. 4.23. Single-phase circuit with earth return, one unit long.

Current I_b, since it flows in the same direction as I_a, causes a *voltage drop* in circuit *ae* of

$$V_{ae\text{-}b} = (1/3) r_e I_b + j\omega m_{ab} I_b = z_{ab} I_b \quad \text{V/unit length} \tag{4.244}$$

where we have defined the mutual impedance

$$z_{ab} = (1/3)r_e + j\omega m_{ab} \quad \Omega/\text{unit length} \tag{4.245}$$

For example, if the unit of length is the mile

$$z_{ab} = (1.588 \times 10^{-3} f) + j(2\pi k) f \ln \frac{D_e}{D_{ab}} \quad \Omega/\text{mi} \tag{4.246}$$

Now consider a one unit length section of the two mutually coupled circuits a and b with earth return, as shown in Figure 4.23. The voltage drop in the direction I_a and I_b in one unit length of line and earth return is

$$V_{aa'} = V_{ae} - V_{a'e'}, \qquad V_{bb'} = V_{be} - V_{b'e'} \tag{4.247}$$

If we define

z_{aa} = self impedance of one unit length of circuit ae
z_{bb} = self impedance of one unit length of circuit be
$z_{ab} = z_{ba}$ = mutual impedance between ae and be for one unit length

then we may write

$$V_{aa'} = z_{aa} I_a + z_{ab} I_b, \qquad V_{bb'} = z_{ba} I_a + z_{bb} I_b \quad \text{V/unit length} \tag{4.248}$$

4.13.2 Single circuit with ground wire and earth return

Now modify the circuit of Figure 4.23 to that of a more useful configuration, as shown in Figure 4.24 where the following changes are noted: the circuit b has

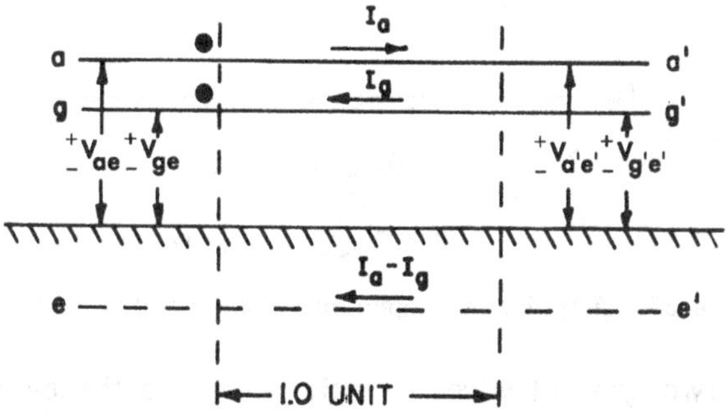

Fig. 4.24. Single circuit with ground wire and earth return.

been designated as the ground wire g and the current $I_b = -I_g$, or the return current is divided between the earth and the ground wire. Since circuit g is a ground circuit, this implies that its voltage drop to ground is zero. Mathematically, the changes made are

$$I_b = -I_g, \qquad V_{be} = V_{ge} = 0, \qquad V_{b'e'} = V_{g'e'} = 0 \tag{4.249}$$

We may now rewrite (4.248) as

$$V_{aa'} = z_{aa} I_a - z_{ag} I_g, \qquad 0 = -z_{ag} I_a + z_{gg} I_g \quad \text{V/unit length} \tag{4.250}$$

Solving (4.250) for I_a, we have

$$I_a = \frac{z_{gg} V_{aa'}}{z_{aa} z_{gg} - z_{ag}^2} \quad \text{A} \tag{4.251}$$

If we define the equivalent impedance of one unit length of circuit with ground wire and earth return as z_a', then by inspection of (4.251)

$$z'_a = z_{aa} - \frac{z_{ag}^2}{z_{gg}} \quad \Omega\text{/unit length} \tag{4.252}$$

where

$$z_{aa} = \left(r_a + \frac{r_e}{3}\right) + j\omega k \ln \frac{D_e}{D_{aa}} \quad \Omega\text{/unit length} \tag{4.253}$$

$$z_{ag} = \frac{r_e}{3} + j\omega k \ln \frac{D_e}{D_{ag}} \quad \Omega\text{/unit length} \tag{4.254}$$

$$z_{gg} = \left(r_g + \frac{r_e}{3}\right) + j\omega k \ln \frac{D_e}{D_{gg}} \quad \Omega\text{/unit length} \tag{4.255}$$

It is often convenient to interpret conductor a of Figure 4.24 as a composite consisting of all three phase wires and conductor g as a composite of n ground wires. In this case we modify these last three equations so that each of the three phase wires carries a current I_{a0} and the ground wires carry a total current I_g. The earth returns $3I_{a0} - I_g$ amperes. A typical arrangement is shown in Figure 4.25

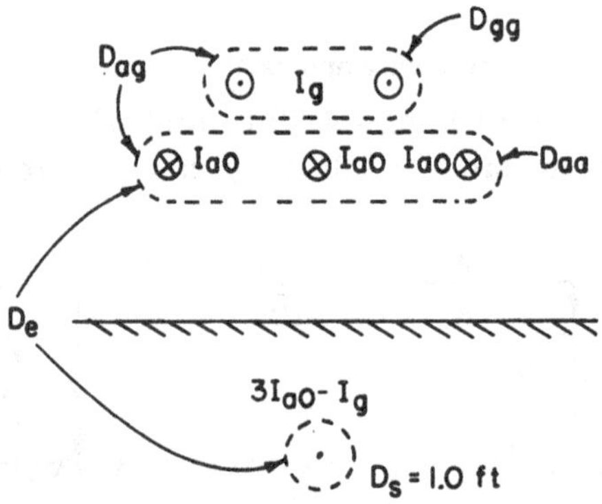

Fig. 4.25. A typical arrangement of phase and ground wires.

where there are two ground wires ($n = 2$). For the three-phase circuit with n ground wires we have (with I_a designating the current in composite conductor a)

$$z_0 = -\frac{V_{a0}}{I_{a0}} = \frac{V_{aa'}}{I_a/3} = 3z'_a$$

so that

$$z_{00} = 3\left(z_{aa} - \frac{z_{ag}^2}{z_{gg}}\right) \Omega\text{/unit length} \tag{4.256}$$

Note that this defines z_0 as the impedance seen by I_{a0}, which is consistent with our previous definition. We refer to this as "z_0 on an I_{a0} basis," as noted later. In equation (4.256) we require that the resistance terms be that of the composite conductors a or g. Thus we have

$$z_{aa} = \left(\frac{r_a}{3} + \frac{r_e}{3}\right) + j\omega k \ln \frac{D_e}{D_{aa}} \quad \Omega\text{/unit length}$$

$$z_{ag} = \frac{r_e}{3} + j\omega k \ln \frac{D_e}{D_{ag}} \quad \Omega\text{/unit length}$$

$$z_{gg} = \left(\frac{r_g}{n} + \frac{r_e}{3}\right) + j\omega k \ln \frac{D_e}{D_{gg}} \quad \Omega\text{/unit length} \tag{4.257}$$

4.13.3 The basis for zero sequence impedance: I_{a0} or $3I_{a0}$?

From the foregoing discussion of zero sequence impedance it is seen that there are two ways in which z_0 could be defined. The z'_a in (4.252) gives the total impedance of a circuit with earth and ground wire return and is the impedance seen by $(3I_{a0})$. Equation (4.256), on the other hand, gives the impedance seen by I_{a0}. We prefer the latter definition as one which is consistent with our definition of zero sequence impedance as "the impedance seen by the current I_{a0}" (see Chapter 2). Actually, either definition will suffice to compute the impedance. The problem arises at the end in knowing whether or not the result should be multiplied by three. Adding to the confusion is the fact that some references use a notation different from that used here such that the results appear to have the factor 3 missing. For example, in sections 4.10 and 4.11 the following expression was developed for computing the zero sequence impedance of a transposed line (see equations 4.207 and 4.233)

$$z_{00} = z_{0(a)} - (z^2_{0(m)}/z_{0(g)}) \tag{4.258}$$

where we have determined that

$$z_{0(a)} = (r_a + r_e) + j3\omega k \ln \frac{D_e}{D_s^{1/3} D_{eq}^{2/3}}$$

$$z_{0(m)} = r_e + j3\omega k \ln \frac{D_e}{D_{ag}}$$

$$z_{0(g)} = (3r_w + r_e) + j3\omega k \ln \frac{D_e}{D_{ww}} \quad \text{for one ground wire}$$

$$= \left(\frac{3}{2} r_w + r_e\right) + j3\omega k \ln \frac{D_e}{D_{ww} D_{uw}} \quad \text{for two ground wires} \tag{4.259}$$

Comparing (4.259) with (4.257), we verify that

$$z_{0(a)} = 3z_{aa}, \qquad z_{0(m)} = 3z_{ag}, \qquad z_{0(g)} = 3z_{gg}$$

and the two methods of computing z_{00} are identical.

Impedances $z_{0(a)}$, $z_{0(m)}$, and $z_{0(g)}$ are preferred for use over z_{aa}, z_{ag}, and z_{gg} since these latter quantities are described on a $3I_{a0}$ basis which is contrary to our definition of a zero sequence impedance.

4.14 Computations Involving Steel Conductors

It is common practice in the construction of high-voltage lines to use steel-cored aluminum phase conductors called ACSR[3] and either high-strength steel or a steel composite such as copperweld or alumoweld[4] for the ground wires. Both types of conductor contain magnetic material and therefore have a nonlinear flux linkage–current relationship and a nonlinear self inductance which is a function of the current in the conductor. In the case of ACSR phase conductors the change in permeability is not great, and the reactance is usually taken as a constant. Thus in Appendix B the conductor tables for ACSR show the inductive reactance at 1 foot

[3] ACSR is an abbreviation for "aluminum conductor, steel reinforced."
[4] See Appendix B for properties of these conductors.

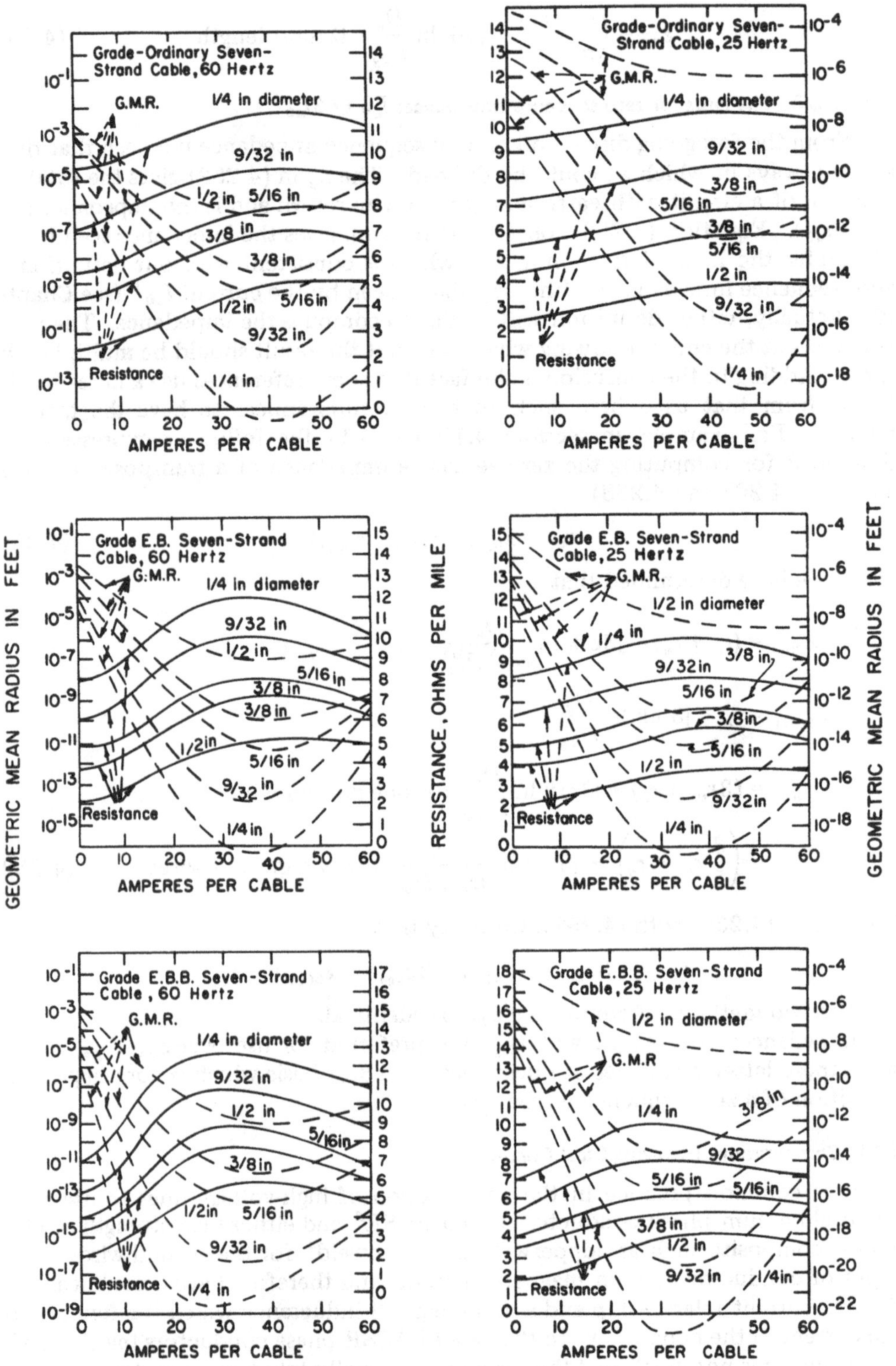

Fig. 4.26. Electrical characteristics of steel ground wires. (From *Symmetrical Components* by C. F. Wagner and R. D. Evans. Copyright McGraw-Hill, 1933. Used with permission of McGraw-Hill Book Co.)

spacing as a constant for all currents. This is because the current is largely confined to the aluminum strands which, of course, are nonmagnetic.

This is not the case, however, with steel conductors where both the resistance and the self inductance are strongly dependent upon the current magnitude. H. B. Dwight compiled data on the resistance and reactance of steel cables [31] in 1919. This data was converted by Wagner and Evans [10] to the more convenient form shown in Figure 4.26, where the resistance and D_s (or GMR) are plotted as a function of current. Since in any given fault computation the current is usually not known until the computation is completed, one possible procedure is to carry along two or three values of impedance for circuits with ground wires. Then, once the current is found, the problem as to the correct impedance may also be resolved. An example will illustrate the procedure.

Example 4.13

Compute the zero sequence impedance of the circuit of Example 4.1, assuming that the line is completely transposed and that the ground wire is 0.375-inch EBB steel.

Solution

From equation (4.207) we have for the zero sequence impedance of a line with one ground wire,

$$z_{00} = z_{0(a)} - \frac{z_{0(m)}^2}{z_{0(g)}} \quad \Omega/\text{mi}$$

The line is 40 miles long. To illustrate the procedure, we compute each quantity $z_{0(a)}$, $z_{0(m)}$, and $z_{0(g)}$, both by formula, and also by table where appropriate.

1. Zero sequence self impedance of the phase wire, $z_{0(a)}$. From (4.204)

$$z_{0(a)} = (r_a + r_e) + j\omega k \ln \frac{D_e^3}{D_s D_{eq}^2} \quad \Omega/\text{mi}$$
$$= (r_a + r_e) + j(x_e + x_a - 2x_d)\ \Omega/\text{mi}$$

where, from Table 4.1, $\omega k = 0.1213$ when $f = 60$ Hz. The first expression is sometimes written as

$$z_{0(a)} = (r_a + r_e) + j0.3639 \ln \frac{D_e}{D_{aa}} \quad \Omega/\text{mi}$$

where

$$D_{aa} = (D_s D_{eq}^2)^{1/3} = (D_{sa} D_{sb} D_{sc} D_{ab} D_{ac} D_{ba} D_{bc} D_{ca} D_{cb})^{1/9}$$
$$= [(0.01688)^3 (10)^4 (20)^2]^{1/9} = 1.393 \text{ ft}$$

Then with $D_e = 2790$, we compute

$$x_{0(a)} = 0.3639 \ln \frac{2790}{1.393} = 2.770 \quad \Omega/\text{mi}$$

From Tables B.1 and B.9 we also compute, with an equivalent spacing D_{eq} of $(10 \times 10 \times 20)^{1/3} = 12.6$ ft,

$$x_{0(a)} = x_e + x_a - 2x_d$$
$$= 2.888 + 0.497 - 2(0.3074) = 2.770 \quad \Omega/\text{mi}$$

The resistance is

$$r_{0(a)} = r_a + r_e = 0.278 + 0.286 = 0.564 \quad \Omega/\text{mi}$$

Then

$$z_{0(a)} = 0.564 + j2.770 \quad \Omega/\text{mi}$$

or

$$Z_{0(a)} = 22.56 + j110.8 \quad \Omega \text{ for 40 mi}$$

2. Zero sequence self impedance of the ground wire, $z_{0(g)}$. From equations (4.197) and (4.206)

$$z_{0(g)} = (3r_w + r_e) + j3\omega k \ln \frac{D_e}{D_{ww}} = (3r_w + r_e) + j(x_e + 3x_g)\ \Omega/\text{mi}$$

From Figure 4.26 for 3/8-inch = 0.375-inch EBB steel at 60 Hz we have

$$D_{ww} = \begin{bmatrix} 1 \times 10^{-3} \\ 5 \times 10^{-12} \\ 1.5 \times 10^{-10} \end{bmatrix} \text{ ft at } \begin{bmatrix} 1 \\ 30 \\ 60 \end{bmatrix} \text{ A}$$

Then, as a function of ground wire current, we have

$$x_{0(g)} = 0.3639 \ln \frac{2790}{D_{ww}} = 2.888 + 0.3639 \ln \frac{1}{D_{ww}}$$

$$= \begin{bmatrix} 5.40 \\ 12.29 \\ 11.10 \end{bmatrix} \Omega/\text{mi at } \begin{bmatrix} 1 \\ 30 \\ 60 \end{bmatrix} \text{ A}$$

The resistance is also a function of current. From Figure 4.26 we find

$$r_w = \begin{bmatrix} 3.5 \\ 7.8 \\ 6.0 \end{bmatrix} \Omega/\text{mi}$$

which is added to $r_e = 0.286$ to write, after rounding,

$$r_{0(g)} = \begin{bmatrix} 11.69 \\ 24.59 \\ 19.19 \end{bmatrix} \Omega/\text{mi at } \begin{bmatrix} 1 \\ 30 \\ 60 \end{bmatrix} \text{ A}$$

Then

$$z_{0(g)} = \begin{bmatrix} 11.69 \\ 24.59 \\ 19.19 \end{bmatrix} + j \begin{bmatrix} 5.40 \\ 12.29 \\ 11.10 \end{bmatrix} \Omega/\text{mi at } \begin{bmatrix} 1 \\ 30 \\ 60 \end{bmatrix} \text{ A}$$

For the 40 miles of line we have

$$R_{0(g)} = \begin{bmatrix} 467.6 \\ 983.6 \\ 767.6 \end{bmatrix} + j \begin{bmatrix} 216.0 \\ 491.6 \\ 444.0 \end{bmatrix} \Omega \text{ at } \begin{bmatrix} 1 \\ 30 \\ 60 \end{bmatrix} \text{ A}$$

or

$$Z_{0(g)} = \begin{bmatrix} 515.1\underline{/24.8^\circ} \\ 1099.6\underline{/26.6^\circ} \\ 886.8\underline{/30.0^\circ} \end{bmatrix} \text{ in polar form}$$

Here we note that the resistance is about one-half the reactance whereas for the copper phase wires the resistance was only about one-seventh the reactance.

3. Zero sequence mutual impedance, $z_{0(m)}$. From (4.205)

$$z_{0(m)} = r_e + j3\omega k \ln \frac{D_e}{D_{ag}} = r_e + j(x_e - 3x_{ag}) \ \Omega/\text{mi}$$

First, by slide rule we compute (see equation 4.203)

$$D_{ag} = (D_{aw} D_{bw} D_{cw})^{1/3} = [(18.03)(15.0)(18.03)]^{1/3} = 16.95 \text{ ft}$$

Then

$$z_{0(m)} = 0.286 + j0.3639 \ln \frac{2790}{16.95} = 0.286 + j1.855 \quad \Omega/\text{mi}$$

As a check we repeat this computation by table. From Table B.24(a) with $D_{ag} = 16.95$ we see that $x_{ag} = 0.3435\ \Omega/\text{mi}$.

Since $x_e = 2.888$, we compute

$$z_{0(m)} = r_e + j(x_e - 3x_{ag}) = 0.286 + j1.857 \quad \Omega/\text{mi}$$

which is an excellent check. The total impedance for the 40 mile line then is

$$Z_{0(m)} = 11.44 + j74.24 = 75.2\underline{/81.25^\circ} \quad \Omega$$

and $Z_{0(m)}^2 = 5660\underline{/162.5^\circ}$.

4. The total zero sequence impedance, Z_{00}. The total zero sequence Z_{00} may be computed from (4.207) as

$$Z_{00} = Z_{0(a)} - \frac{Z_{0(m)}^2}{Z_{0(g)}}$$

$$= \begin{bmatrix} 30.66 + j103.43 \\ 26.25 + j107.23 \\ 26.85 + j106.10 \end{bmatrix} \Omega \text{ at } \begin{matrix} 1 \\ 30 \\ 60 \end{matrix} \ \text{A}$$

4.15 Parallel Transposed and Untransposed Multicircuit Lines

A study of parallel untransposed multicircuit lines by Hesse [32, 33] has shown that the unbalance between parallel lines must be examined for two separate types: the net effect or "through unbalance" and the circulating currents due to the unbalanced voltages being out of phase in the parallel circuits. This research has revealed that the in-phase unbalanced currents which contribute to the overall through unbalance are usually quite low. The circulating currents, on the other hand, may be much greater. These latter currents are usually important where the parallel lines are bused at both ends such that the circulating current

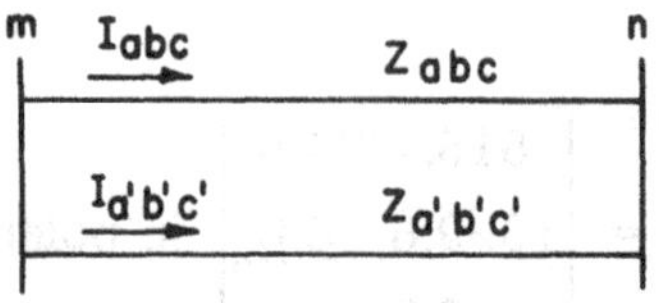

Fig. 4.27. A double circuit transmission line from m to n.

sees only the impedance of the parallel lines and is not affected by the impedance seen looking into the system. Any line compensation by series capacitors will have an adverse effect on these circulating currents and may dictate the relocation of such capacitors to another circuit. It is also important to examine the exact phase arrangement of untransposed parallel circuits since some configurations are significantly worse with respect to circulating current unbalance than others.

In performing the calculations of unbalanced circuits, it is convenient to neglect the Thevenin equivalent impedances, looking into the system at either end of the parallel line section, and consider only the line section between buses m and n. Also, we assume that the voltage drop between buses m and n is the same for each circuit since they have common termination and that these bus voltages are positive sequence (balanced) voltages. This is approximately true for physical systems.

With these assumptions, Hesse [33] gives a method of estimating the unbalance due to line currents and circulating currents. This can be done for all possible phasing arrangements. Making such estimates permits the elimination of the most undesirable phasing arrangements, leaving only a relatively few configurations for detailed study.

Consider the double circuit line shown schematically in Figure 4.27, where one circuit is designated a-b-c and the other a'-b'-c'. Each circuit may be untransposed or may have any number of rotations and twists in its length from m to n. Any line section, and therefore the total length, may be represented by a voltage equation expressing the voltage drop in the given length of line of the form

$$\begin{bmatrix} \Sigma V_a \\ \Sigma V_b \\ \Sigma V_c \\ \hline \Sigma V_{a'} \\ \Sigma V_{b'} \\ \Sigma V_{c'} \end{bmatrix} = \left[\begin{array}{ccc|ccc} Z_{aa} & Z_{ab} & Z_{ac} & Z_{aa'} & Z_{ab'} & Z_{ac'} \\ Z_{ba} & Z_{bb} & Z_{bc} & Z_{ba'} & Z_{bb'} & Z_{bc'} \\ Z_{ca} & Z_{cb} & Z_{cc} & Z_{ca'} & Z_{cb'} & Z_{cc'} \\ \hline Z_{a'a} & Z_{a'b} & Z_{a'c} & Z_{a'a'} & Z_{a'b'} & Z_{a'c'} \\ Z_{b'a} & Z_{b'b} & Z_{b'c} & Z_{b'a'} & Z_{b'b'} & Z_{b'c'} \\ Z_{c'a} & Z_{c'b} & Z_{c'c} & Z_{c'a'} & Z_{c'b'} & Z_{c'c'} \end{array}\right] \begin{bmatrix} I_a \\ I_b \\ I_c \\ \hline I_{a'} \\ I_{b'} \\ I_{c'} \end{bmatrix} \text{V} \tag{4.260}$$

If there are ground wires, it is assumed that the rows and columns of the matrix have been reduced to eliminate these equations so that (4.260) is a true representation of the actual voltage drop, including the effect of ground wires. If there are several transposition sections, the sum of all such voltage drops may be represented by equation (4.260).

Using matrix notation, we may write (4.260) as

$$\begin{bmatrix} \Sigma \mathbf{V}_{abc} \\ \Sigma \mathbf{V}_{a'b'c'} \end{bmatrix} = \begin{bmatrix} \mathbf{Z}_{aa} & \mathbf{Z}_{aa'} \\ \mathbf{Z}_{a'a} & \mathbf{Z}_{a'a'} \end{bmatrix} \begin{bmatrix} \mathbf{I}_{abc} \\ \mathbf{I}_{a'b'c'} \end{bmatrix} \text{V} \tag{4.261}$$

Since the circuit unbalance is expressed in terms of sequence current unbalance, we solve (4.261) for the currents.

$$\begin{bmatrix} \mathbf{I}_{abc} \\ \mathbf{I}_{a'b'c'} \end{bmatrix} = \begin{bmatrix} \mathbf{Y}_{aa} & \mathbf{Y}_{aa'} \\ \mathbf{Y}_{a'a} & \mathbf{Y}_{a'a'} \end{bmatrix} \begin{bmatrix} \Sigma \mathbf{V}_{abc} \\ \Sigma \mathbf{V}_{a'b'c'} \end{bmatrix} \quad \mathrm{A} \tag{4.262}$$

Since the **Z** matrix of (4.261) is symmetric, we can show that [7]

$$\begin{aligned} \mathbf{Y}_{aa} &= \mathbf{Z}_{aa}^{-1} + \mathbf{K}\mathbf{L}^{-1}\mathbf{K}^t, & \mathbf{Y}_{aa'} &= -\mathbf{K}\mathbf{L}^{-1} \\ \mathbf{Y}_{a'a} &= \mathbf{Y}_{aa'}^t, & \mathbf{Y}_{a'a'} &= \mathbf{L}^{-1} \end{aligned} \tag{4.263}$$

where

$$\mathbf{K} = \mathbf{Z}_{aa}^{-1}\mathbf{Z}_{aa'}, \quad \mathbf{K}^t = \mathbf{Z}_{a'a}\mathbf{Z}_{aa}^{-1}, \quad \mathbf{L} = \mathbf{Z}_{a'a'} - \mathbf{Z}_{a'a}\mathbf{Z}_{aa}^{-1}\mathbf{Z}_{aa'} \tag{4.264}$$

From (4.263) and (4.264) we observe that the 6×6 **Z** matrix may be inverted by performing two 3×3 inversions, one on $\mathbf{Z}_{aa}$ and one on **L**. The inversion is then completed by performing the indicated matrix multiplications.

Having computed the phase currents by (4.262), we need a method whereby these currents may be transformed to the 0-1-2 frame of reference. To do this, we introduce the transformation $\mathbf{A}_2$ defined by

$$\mathbf{A}_2 = \left[\begin{array}{c|c} \mathbf{A} & \mathbf{0} \\ \hline \mathbf{0} & \mathbf{A} \end{array}\right] \tag{4.265}$$

such that

$$\left[\begin{array}{c} \mathbf{I}_{abc} \\ \hline \mathbf{I}_{a'b'c'} \end{array}\right] \triangleq \left[\begin{array}{c|c} \mathbf{A} & \mathbf{0} \\ \hline \mathbf{0} & \mathbf{A} \end{array}\right] \left[\begin{array}{c} \mathbf{I}_{012} \\ \hline \mathbf{I}_{0'1'2'} \end{array}\right] \tag{4.266}$$

and where, obviously,

$$\mathbf{A}_2^{-1} = \begin{bmatrix} \mathbf{A}^{-1} & \mathbf{0} \\ \mathbf{0} & \mathbf{A}^{-1} \end{bmatrix} \tag{4.267}$$

Premultiplication of (4.262) by $\mathbf{A}_2^{-1}$ gives the result

$$\begin{bmatrix} \mathbf{I}_{012} \\ \mathbf{I}_{0'1'2'} \end{bmatrix} = \begin{bmatrix} \mathbf{A}^{-1}\mathbf{Y}_{aa}\mathbf{A} & \mathbf{A}^{-1}\mathbf{Y}_{aa'}\mathbf{A} \\ \mathbf{A}^{-1}\mathbf{Y}_{a'a}\mathbf{A} & \mathbf{A}^{-1}\mathbf{Y}_{a'a'}\mathbf{A} \end{bmatrix} \begin{bmatrix} \Sigma \mathbf{V}_{012} \\ \Sigma \mathbf{V}_{0'1'2'} \end{bmatrix} \tag{4.268}$$

We now observe that the sequence voltage at each end of the line is a positive sequence voltage by definition. Therefore the sequence voltage *drops* expressed in (4.268) are entirely positive sequence or

$$\left[\begin{array}{c} \Sigma \mathbf{V}_{012} \\ \hline \Sigma \mathbf{V}_{0'1'2'} \end{array}\right] = \left[\begin{array}{c} 0 \\ \Sigma V_{a1} \\ 0 \\ \hline 0 \\ \Sigma V_{a'1} \\ 0 \end{array}\right] = \left[\begin{array}{c} 0 \\ 1 \\ 0 \\ \hline 0 \\ 1 \\ 0 \end{array}\right] \Sigma V_{a1} \tag{4.269}$$

where we note also that $V_{a1} = V_{a'1}$ since the lines are bused at each end.

Expanding (4.268) to the full 6 × 6 array and recognizing (4.269), we may write

$$\begin{bmatrix} I_{a0} \\ I_{a1} \\ I_{a2} \\ \hline I_{a'0} \\ I_{a'1} \\ I_{a'2} \end{bmatrix} = \left[\begin{array}{ccc|ccc} Y_{00} & Y_{01} & Y_{02} & Y_{00'} & Y_{01'} & Y_{02'} \\ Y_{10} & Y_{11} & Y_{12} & Y_{10'} & Y_{11'} & Y_{12'} \\ Y_{20} & Y_{21} & Y_{22} & Y_{20'} & Y_{21'} & Y_{22'} \\ \hline Y_{0'0} & Y_{0'1} & Y_{0'2} & Y_{0'0'} & Y_{0'1'} & Y_{0'2'} \\ Y_{1'0} & Y_{1'1} & Y_{1'2} & Y_{1'0'} & Y_{1'1'} & Y_{1'2'} \\ Y_{2'0} & Y_{2'1} & Y_{2'2} & Y_{2'0'} & Y_{2'1'} & Y_{2'2'} \end{array}\right] \begin{bmatrix} 0 \\ 1 \\ 0 \\ \hline 0 \\ 1 \\ 0 \end{bmatrix} \Sigma V_{a1} \tag{4.270}$$

Having done this, we easily compute

$$\begin{bmatrix} I_{a0} \\ I_{a1} \\ I_{a2} \\ I_{a'0} \\ I_{a'1} \\ I_{a'2} \end{bmatrix} = \begin{bmatrix} Y_{01} + Y_{01'} \\ Y_{11} + Y_{11'} \\ Y_{21} + Y_{21'} \\ Y_{0'1} + Y_{0'1'} \\ Y_{1'1} + Y_{1'1'} \\ Y_{2'1} + Y_{2'1'} \end{bmatrix} \Sigma V_{a1} \tag{4.271}$$

and we note that only the second and fifth columns of the Y matrix are used.

Then we define the net *through* unbalances as

$$m_{0t} \triangleq \frac{I_{a0} + I_{a'0}}{I_{a1} + I_{a'1}} \text{ pu}, \qquad m_{2t} \triangleq \frac{I_{a2} + I_{a'2}}{I_{a1} + I_{a'1}} \text{ pu} \tag{4.272}$$

and the net circulating current unbalance as

$$m_{0c} = \frac{I_{a0} - I_{a'0}}{I_{a1} + I_{a'1}} \text{ pu}, \qquad m_{2c} = \frac{I_{a2} - I_{a'2}}{I_{a1} + I_{a'1}} \text{ pu} \tag{4.273}$$

or substituting from these equations, we may write

$$\begin{bmatrix} m_{0t} \\ m_{2t} \\ m_{0c} \\ m_{2c} \end{bmatrix} = \begin{bmatrix} Y_{01} + Y_{01'} + Y_{0'1} + Y_{0'1'} \\ Y_{21} + Y_{21'} + Y_{2'1} + Y_{2'1'} \\ Y_{01} + Y_{01'} - Y_{0'1} - Y_{0'1'} \\ Y_{21} + Y_{21'} - Y_{2'1} - Y_{2'1'} \end{bmatrix} \frac{1}{Y_1}$$

where

$$Y_1 = Y_{11} + Y_{11'} + Y_{1'1} + Y_{1'1'} \tag{4.274}$$

Hesse [33] shows that, for parallel lines in flat configuration, $m_{0c} >> m_{0t}$ and $m_{2t} >> m_{2c}$ often by about a factor of ten or so in each case. This is not obvious from an inspection of (4.274) because it is hard to appreciate the vast difference in phase angle of the admittances.

Example 4.14

Compute the through and circulating current unbalances for the double circuit 345 kV untransposed line shown in Figure 4.28.

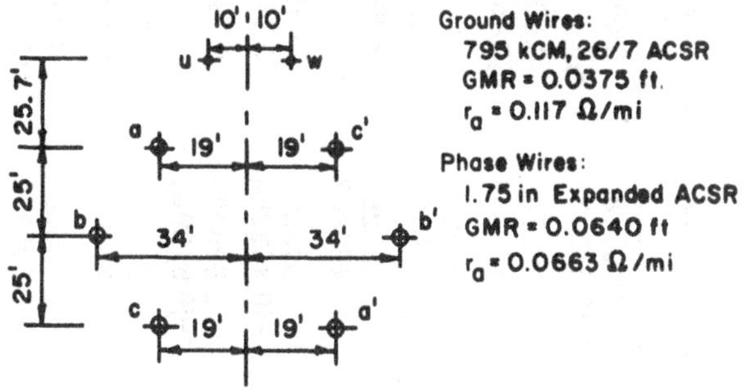

Fig. 4.28. A 345 kV double circuit configuration.

Solution

First we must determine the impedance matrix of self and mutual impedances defined by equation (4.79). These 64 impedances are given in Table 4.11, and the 36 impedances which result after reducing this matrix to eliminate rows u and w are given in Table 4.12.

The inverse of the matrix given in Table 4.12 is computed by digital computer with the result

$$\mathbf{Y}_{aa} = \begin{bmatrix} (0.1057 - j1.1573) & (-0.0214 + j0.2131) & (-0.0056 + j0.1195) \\ (-0.0214 + j0.2131) & (0.0983 - j1.1178) & (-0.0245 + j0.2348) \\ (-0.0056 + j0.1195) & (-0.0245 + j0.2348) & (0.0954 - j1.1006) \end{bmatrix}$$

$$\mathbf{Y}_{aa'} = \begin{bmatrix} (-0.0011 + j0.0877) & (-0.0011 + j0.0894) & (-0.0087 + j0.1416) \\ (-0.0039 + j0.1099) & (-0.0000 + j0.0827) & (-0.0011 + j0.0894) \\ (-0.0179 + j0.1950) & (-0.0039 + j0.1099) & (-0.0011 + j0.0877) \end{bmatrix}$$

$$\mathbf{Y}_{a'a} = \mathbf{Y}_{aa'}^{t}$$

$$\mathbf{Y}_{a'a'} = \begin{bmatrix} (0.0954 - j1.1006) & (-0.0245 + j0.2348) & (-0.0056 + j0.1195) \\ (-0.0245 + j0.2348) & (0.0983 - j1.1178) & (-0.0214 + j0.2131) \\ (-0.0056 + j0.1195) & (-0.0214 + j0.2131) & (0.1057 - j1.1573) \end{bmatrix}$$

Each partition above is now transformed to the 0-1-2 frame of reference as specified by (4.268), with the following results

$$\mathbf{A}^{-1}\mathbf{Y}_{aa}\mathbf{A} = \begin{bmatrix} (0.0655 - j0.7469) & (0.0287 - j0.0351) & (-0.0154 - j0.0426) \\ (-0.0155 - j0.0421) & (0.1170 - j1.3156) & (-0.0634 + j0.0200) \\ (0.0286 - j0.0346) & (0.0546 + j0.0400) & (0.1171 - j1.3156) \end{bmatrix}$$

$$\mathbf{A}^{-1}\mathbf{Y}_{a'a}\mathbf{A} = \begin{bmatrix} (-0.0129 + j0.3311) & (-0.0309 + j0.0113) & (0.0329 + j0.0010) \\ (0.0057 + j0.0323) & (0.0018 - j0.0376) & (0.0417 - j0.0161) \\ (-0.0156 + j0.0288) & (-0.0348 - j0.0281) & (0.0018 - j0.0355) \end{bmatrix}$$

$$\mathbf{A}^{-1}\mathbf{Y}_{aa'}\mathbf{A} = \begin{bmatrix} (-0.0179 + j0.3311) & (-0.0156 + j0.0290) & (0.0056 + j0.0325) \\ (0.0330 + j0.0013) & (0.0089 - j0.0355) & (0.0417 - j0.0162) \\ (-0.0310 - j0.0116) & (-0.0348 - j0.0281) & (0.0018 - j0.0376) \end{bmatrix}$$

$$\mathbf{A}^{-1}\mathbf{Y}_{a'a'}\mathbf{A} = \begin{bmatrix} (0.0655 - j0.7469) & (0.0446 + j0.0079) & (-0.0448 - j0.0072) \\ (-0.0449 - j0.0067) & (0.1170 - j1.3156) & (-0.0617 + 0.0278) \\ (0.0446 + j0.0085) & (0.0487 + j0.0455) & (0.1171 - j1.3156) \end{bmatrix}$$

Table 4.11. Impedance Tabulation for the Circuit of Figure 4.27
(8 × 8 matrix of phase wire and ground wire impedances)

	a	*b*	*c*	*a'*	*b'*	*c'*	*u*	*w*
a	0.16158 +j1.296235	0.09528 +j0.553453	0.09528 +j0.488002	0.09528 +j0.460342	0.09528 +j0.468743	0.09528 +j0.521302	0.09528 +j0.561739	0.09528 +j0.518934
b	0.09528 +j0.553453	0.16158 +j1.296235	0.09528 +j0.553453	0.09528 +j0.468743	0.09528 +j0.450692	0.09528 +j0.468743	0.09528 +j0.474048	0.09528 +j0.452254
c	0.09528 +j0.488002	0.09528 +j0.553453	0.16158 +j1.296235	0.09528 +j0.521302	0.09528 +j0.468743	0.09528 +j0.460342	0.09528 +j0.436824	0.09528 +j0.429368
a'	0.09528 +j0.460346	0.09528 +j0.468743	0.09528 +j0.521302	0.16158 +j1.296235	0.09528 +j0.553453	0.09528 +j0.488002	0.09528 +j0.429368	0.09528 +j0.436824
b'	0.09528 +j0.468743	0.09528 +j0.450692	0.09528 +j0.468743	0.09528 +j0.553453	0.16158 +j1.296235	0.09528 +j0.553453	0.09528 +j0.452254	0.09528 +j0.474048
c'	0.09528 +j0.521302	0.09528 +j0.468743	0.09528 +j0.460342	0.09528 +j0.488002	0.09528 +j0.553453	0.16158 +j1.296235	0.09528 +j0.518934	0.09528 +j0.561739
u	0.09528 +j0.561739	0.09528 +j0.474048	0.09528 +j0.436824	0.09528 +j0.429368	0.09528 +j0.452254	0.09528 +j0.518934	0.21228 +j1.361096	0.09528 +j0.599185
w	0.09528 +j0.518934	0.09528 +j0.452254	0.09528 +j0.429368	0.09528 +j0.436824	0.09528 +j0.474048	0.09528 +j0.561739	0.09528 +j0.599185	0.21228 +j1.361096

Table 4.12. Impedance Tabulation for the Circuit of Figure 4.27 (6 X 6 matrix of phase impedances after matrix reduction)

	a	*b*	*c*	*a'*	*b'*	*c'*
a	0.1034 +j0.9973	0.0379 +j0.2978	0.0381 +j0.2493	0.0380 +j0.2220	0.0377 +j0.2142	0.0368 +j0.2247
b	0.0379 +j0.2978	0.1058 +j1.0775	0.0402 +j0.3494	0.0402 +j0.2648	0.0395 +j0.2327	0.0377 +j0.2142
c	0.0381 +j0.2493	0.0402 +j0.3494	0.1073 +j1.1055	0.0410 +j0.3307	0.0402 +j0.2648	0.0380 +j0.2220
a'	0.0380 +j0.2220	0.0402 +j0.2648	0.0410 +j0.3307	0.1073 +j1.1055	0.0402 +j0.3494	0.0381 +j0.2493
b'	0.0377 +j0.2142	0.0395 +j0.2327	0.0402 +j0.2648	0.0402 +j0.3494	0.1058 +j1.0775	0.0378 +j0.2978
c'	0.0368 +j0.2247	0.0377 +j0.2142	0.0380 +j0.2220	0.0381 +j0.2493	0.0378 +j0.2978	0.1034 +j0.9973

Then from (4.271) we combine these results to write

$$\begin{bmatrix} I_{a0} \\ I_{a1} \\ I_{a2} \\ I_{a'0} \\ I_{a'1} \\ I_{a'2} \end{bmatrix} = \begin{bmatrix} (0.0286 - j0.0351) + (-0.0349 + j0.0133) \\ (0.1169 - j1.3144) + (0.0059 - j0.0386) \\ (0.0545 + j0.0396) + (-0.0541 - j0.0437) \\ (-0.0116 + j0.0044) + (0.0446 + j0.0079) \\ (0.0047 - j0.0345) + (0.1169 - j1.3144) \\ (-0.0154 - j0.0122) + (0.0487 + j0.0451) \end{bmatrix} \Sigma V_{a1} = \begin{bmatrix} -0.0131 - j0.0061 \\ 0.1259 - j1.3511 \\ 0.0197 + j0.0119 \\ 0.0137 - j0.0034 \\ 0.1188 - j1.3532 \\ 0.0139 + j0.0175 \end{bmatrix} \Sigma V_{a1}$$

Then, from (4.272) we compute the "through unbalances."

$$m_{0t} = \frac{I_{a0} + I_{a'0}}{I_{a1} + I_{a'1}} = \frac{0.0268 - j0.0095}{0.2447 - j2.7043} = 0.0104\underline{/65.2^\circ}$$

$$m_{2t} = \frac{I_{a2} + I_{a'2}}{I_{a1} + I_{a'1}} = \frac{0.0336 + j0.0294}{0.2447 - j2.7043} = 0.0163\underline{/126.0^\circ}$$

Similarly, from (4.273) we find the "circulatory unbalances."

$$m_{0c} = \frac{I_{a0} - I_{a'0}}{I_{a1} + I_{a'1}} = -\frac{0.0006 - j0.0027}{0.2447 - j2.7043} = 0.00102\underline{/-17.7^\circ}$$

$$m_{2c} = \frac{I_{a2} - I_{a'2}}{I_{a1} + I_{a'1}} = +\frac{0.0058 - j0.0209}{0.2447 - j2.7043} = 0.00296\underline{/40.8^\circ}$$

For the vertical configuration of this problem the negative sequence unbalances are about equal but not much greater than m_{0c}.

4.16 Optimizing a Parallel Circuit for Minimum Unbalance

In the previous section each parallel circuit line was treated in terms of its total impedance characteristic from one terminal to the other. Obviously, it would be possible to write separate equations for each line for each section of like configuration and then apply the transposition matrices $\mathbf{R}_\phi$ and $\mathbf{T}_\phi$ to each circuit to reorder the equations in an *a-b-c-a'-b'-c'* sequence. Once each section has been so arranged, the matrices may be added to find the total impedance of each individual line.

In many practical situations the lines are not transposed at all. This is because the transpositions are costly and, perhaps even more important, they are often the cause of circuit failure due to either mechanical or electrical weakness at the transposition structures [34, 35]. The general problem of the effect of circuit unbalance due to lack of transpositions has been studied in detail by E. T. B. Gross [28, 29, 30, 34, 35, 36] and is an excellent resource for the interested reader.

In double circuit lines which are not transposed, Hesse [33] has shown that there is an optimum conductor arrangement which will minimize the unbalance factors. There is value, therefore, in developing a method whereby the unbalance factors can be determined in a straightforward way, while making sure that all possible configurations are examined. We shall do this for lines which are themselves untransposed but which can have any desired phase identification.

From equation (4.268) we have

$$\begin{bmatrix} \mathbf{I}_{012} \\ \mathbf{I}_{0'1'2'} \end{bmatrix} = \begin{bmatrix} \mathbf{A}^{-1}\mathbf{Y}_{aa}\mathbf{A} & \mathbf{A}^{-1}\mathbf{Y}_{aa'}\mathbf{A} \\ \mathbf{A}^{-1}\mathbf{Y}_{a'a}\mathbf{A} & \mathbf{A}^{-1}\mathbf{Y}_{a'a'}\mathbf{A} \end{bmatrix} \begin{bmatrix} \mathbf{V}_{012} \\ \mathbf{V}_{0'1'2'} \end{bmatrix}$$

$$= \begin{bmatrix} \mathbf{Y}_{00} & \mathbf{Y}_{00'} \\ \mathbf{Y}_{0'0} & \mathbf{Y}_{0'0'} \end{bmatrix} \begin{bmatrix} \mathbf{V}_{012} \\ \mathbf{V}_{0'1'2'} \end{bmatrix} \begin{matrix} \text{phased } a\text{-}b\text{-}c \\ \text{phased } a'\text{-}b'\text{-}c' \end{matrix} \tag{4.275}$$

where we have defined new admittance submatrices $\mathbf{Y}_{00}, \mathbf{Y}_{00'}, \mathbf{Y}_{0'0}$ and $\mathbf{Y}_{0'0'}$ as a matter of convenience. We have also indicated the phase designation quite arbitrarily as *a-b-c* and *a'-b'-c'* for the configuration used in writing these equations.

Suppose, however, that the second (primed) circuit had been phased *b'-c'-a'* (on wires 1-2-3). This means that the voltage (or current) equation must be premultiplied by $\mathbf{R}_\phi$ to rearrange the elements to the desired *a-b-c* order. However, since equation (4.275) is already in the 0-1-2 frame of reference, we may use the rotation matrix $\mathbf{R}_{012}$ defined by equation (4.90). Similarly, multiplication by $\mathbf{R}_{012}^{-1}$ defines a rotation in the opposite direction.

These rotations of only the primed circuit may be performed mathematically by transformations

$$\begin{bmatrix} \mathbf{U} & \mathbf{O} \\ \mathbf{O} & \mathbf{R}_{012} \end{bmatrix} \begin{matrix} a\text{-}b\text{-}c \\ b'\text{-}c'\text{-}a' \end{matrix} \tag{4.276}$$

and

$$\begin{bmatrix} \mathbf{U} & \mathbf{O} \\ \mathbf{O} & \mathbf{R}_{012}^{-1} \end{bmatrix} \begin{matrix} a\text{-}b\text{-}c \\ c'\text{-}a'\text{-}b' \end{matrix} \tag{4.277}$$

For example, premultiplying both sides of (4.275) by (4.276) gives

$$\begin{bmatrix} \mathbf{I}_{012} \\ \hline \mathbf{R}_{012}\mathbf{I}_{0'1'2'} \end{bmatrix} = \left[\begin{array}{c|c} \mathbf{Y}_{00} & \mathbf{Y}_{00'}\mathbf{R}_{012}^{-1} \\ \hline \mathbf{R}_{012}\mathbf{Y}_{0'0} & \mathbf{R}_{012}\mathbf{Y}_{0'0'}\mathbf{R}_{012}^{-1} \end{array}\right] \begin{bmatrix} \mathbf{V}_{012} \\ \hline \mathbf{R}_{012}\mathbf{V}_{0'1'2'} \end{bmatrix} \begin{matrix} a\text{-}b\text{-}c \\ b'\text{-}c'\text{-}a' \end{matrix} \tag{4.278}$$

This operation is equivalent to multiplying every element in the fifth row and the sixth column of (4.270) by a and multiplying elements of the sixth row and fifth column by a^2. This result, with only positive sequence voltages applied, is

$$\begin{bmatrix} I_{a0} \\ I_{a1} \\ I_{a2} \\ \hline I_{a'0} \\ I_{a'1} \\ I_{a'2} \end{bmatrix} = \begin{bmatrix} Y_{01} + a^2 Y_{01'} \\ Y_{11} + a^2 Y_{11'} \\ Y_{21} + a^2 Y_{21'} \\ \hline Y_{0'1} + a^2 Y_{0'1'} \\ a(Y_{1'1} + a^2 Y_{1'1'}) \\ a^2 (Y_{2'1} + a^2 Y_{2'1'}) \end{bmatrix} \begin{matrix} \text{phased } a\text{-}b\text{-}c \\ \Sigma V_{a1} \\ \text{phased } b'\text{-}c'\text{-}a' \end{matrix} \tag{4.279}$$

Another rotation of the second circuit would give a similar result, but with operators a and a^2 interchanged.

If instead the second circuit is transposed according to transformation T_{012}, the result is equivalent to interchanging the fifth and sixth rows of (4.270) and then interchanging the fifth and sixth columns. The result is

$$\begin{bmatrix} I_{a0} \\ I_{a1} \\ I_{a2} \\ \hline I_{a'0} \\ I_{a'1} \\ I_{a'2} \end{bmatrix} = \begin{bmatrix} Y_{01} + Y_{02'} \\ Y_{11} + Y_{12'} \\ Y_{21} + Y_{22'} \\ \hline Y_{0'1} + Y_{0'2'} \\ Y_{2'1} + Y_{2'2'} \\ Y_{1'1} + Y_{1'2'} \end{bmatrix} \begin{matrix} \text{phased } a\text{-}b\text{-}c \\ \Sigma V_{a1} \\ \text{phased } a'\text{-}c'\text{-}b' \end{matrix} \tag{4.280}$$

If this arrangement is then changed by a rotation of the second circuit to a-b-c:b'-a'-c', the Y elements of (4.280) are multiplied by a and a^2 exactly as in (4.279). In all, six arrangements are possible, with the first circuit remaining unchanged. These are

$$1\begin{bmatrix} U & a\text{-}b\text{-}c \\ U & a'\text{-}b'\text{-}c' \end{bmatrix} \quad 4\begin{bmatrix} U & a\text{-}b\text{-}c \\ T_{012} & a'\text{-}c'\text{-}b' \end{bmatrix}$$
$$2\begin{bmatrix} U & a\text{-}b\text{-}c \\ R_{012} & b'\text{-}c'\text{-}a' \end{bmatrix} \quad 5\begin{bmatrix} U & a\text{-}b\text{-}c \\ R_{012}T_{012} & b'\text{-}a'\text{-}c' \end{bmatrix}$$
$$3\begin{bmatrix} U & a\text{-}b\text{-}c \\ R_{012}^2 & c'\text{-}a'\text{-}b' \end{bmatrix} \quad 6\begin{bmatrix} U & a\text{-}b\text{-}c \\ R_{012}^2T_{012} & c'\text{-}b'\text{-}a' \end{bmatrix} \tag{4.281}$$

and another six relationships may be obtained with the U for the first circuit replaced by T_{012}. These are the only unique arrangements—a fact that may require a little careful thought. This is due to the fact that *which* circuit is unbalanced with respect to its neighbor is immaterial. Thus 12 arrangements in all must be examined. In general, it is possible to have $6^n/3$ significantly different phase arrangements for an n-circuit system.

Problems

Some of the problems suggested here should be solved by digital computer to provide high accuracy, to minimize labor, and to correctly display the desired impedance characteristics. Some helpful programs are available in Appendix A.

4.1. Compute the per mile positive and negative sequence impedance for the line configuration shown in Figure P4.1 where the conductor is 4/0, 7-strand copper. Assume that the

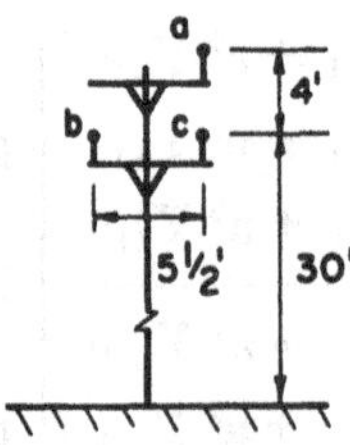

Fig. P4.1.

line is transposed such that the methods of section 4.1 are applicable. Compute x_1 by equation (4.3) and use equation (4.10) as a check.

4.2. Compute the per mile positive and negative sequence impedance for the line configuration of Figure P4.2 where the conductor is 3/0 ACSR. Use the method of section 4.1.

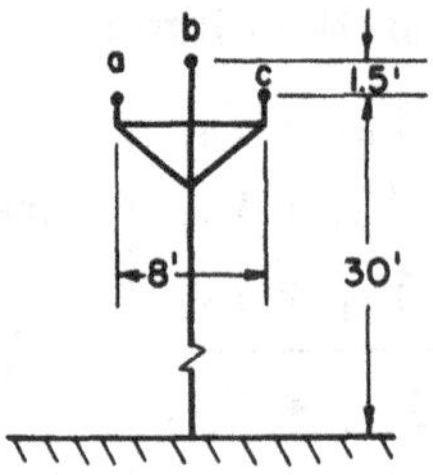

Fig. P4.2.

4.3. Compute the per mile positive and negative sequence impedance for the line configuration of Figure P4.3 where the conductor is 500,000 CM, 61% conductivity, 37-strand, hard-drawn aluminum.

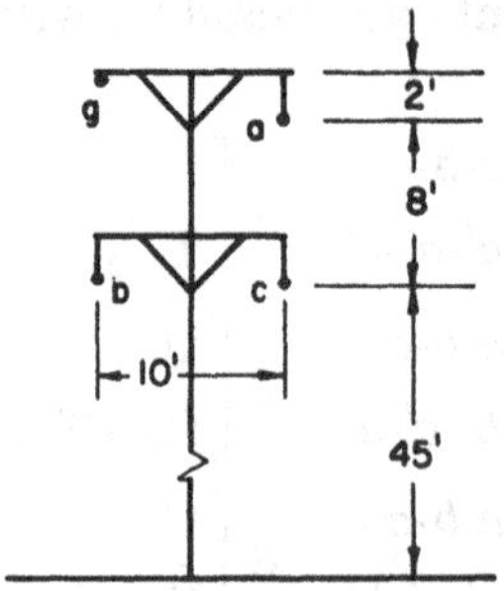

Fig. P4.3.

4.4. Compute the per mile positive and negative sequence impedance for the line configuration of Figure P4.4 where the conductor is 336,400 CM, 26/7-strand ACSR.

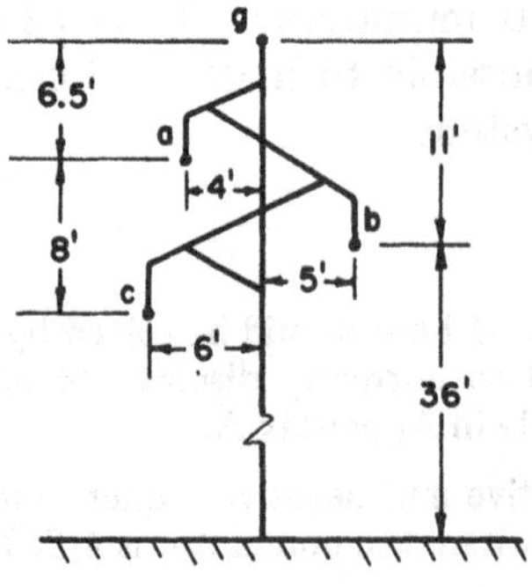

Fig. P4.4.

4.5. Compute the per mile positive and negative sequence impedance for the line configuration of Figure P4.5 where the conductor is 397,500 CM, 26/7-strand ACSR.

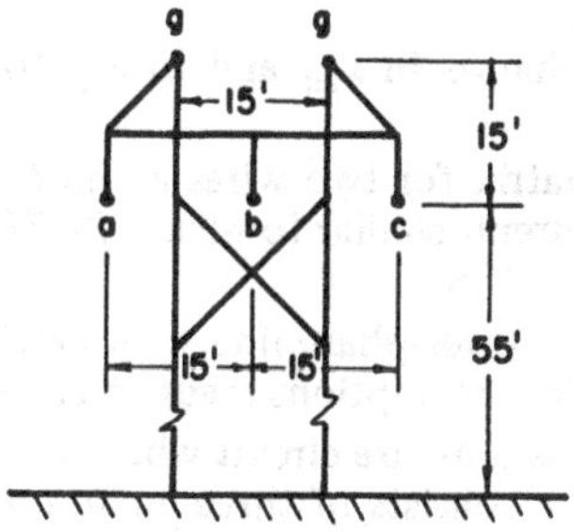

Fig. P4.5.

4.6. Write the phasor equations for a two-winding transformer with self inductances L_p and L_s in primary and secondary windings respectively and mutual inductance M. Compare these equations with equation (4.14) for two mutually coupled, parallel wires. Review the concept of placing polarity markings on the transformer coils.

4.7. Verify that the four-terminal equivalent circuits shown in Figure P4.7a and Figure 4.7b accurately represent the equations for two mutually coupled wires given by equation (4.14). Circuit *b* is due to Starr [37].

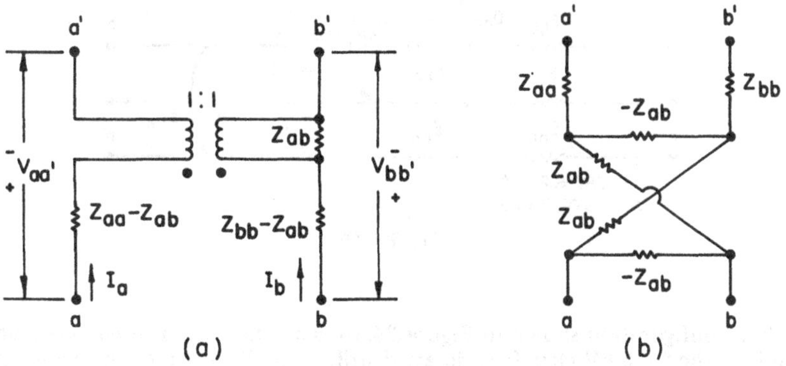

Fig. P4.7.

4.8. Find the internal inductance in millihenrys/mile of 10^6 circular mil, all aluminum (61% conductivity) conductor when the operating frequency is 60 Hz.

4.9. Find the internal inductance in millihenrys/mile of a certain iron wire having a dc resistance of 2.0 ohms/mile and a relative permeability of 100. The frequency is 60 Hz and the wire is cylindrical with a diameter of 0.394 inches (4/0 SWG pure iron wire).

4.10. Verify the impedance for Carson's line, given by equations (4.35)–(4.40).

4.11. Compute the voltage drop V_{mn} across the circuit of Figure P4.11 and compare with equation (4.35). Neglect the resistance of the windings.

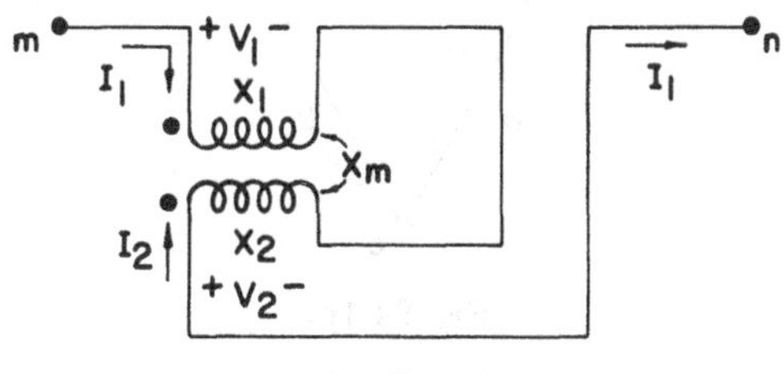

Fig. P4.11.

4.12. Consider a single conductor plus earth return circuit. The conductor is size 4/0, 7-strand hard-drawn copper at 50 C suspended 30 ft above the earth; $f = 60$ Hz.

(a) Find the self impedance of the circuit, $r_{aa} + jx_{aa}$, in ohms/mile, for (1) $\rho = 10\ \Omega \cdot \text{m}$ and (2) $\rho = 1000\ \Omega \cdot \text{m}$

(b) What is the percent change in x_{aa} and in z_{aa} for this 100-fold increase in earth resistivity?

4.13. Compute the impedance matrix for two wires a and b with earth return d and sketch an equivalent circuit for this circuit similar to Figure P4.7(a) where the effect of the earth is included in the circuit parameters.

4.14. Verify equation (4.47) for a three-phase line with earth return and compute all self and mutual impedances where the assumptions made in equation (4.53) are *not* true.

4.15. Illustrated in Figure P4.15 is a 4-wire circuit which is used to serve a three-phase, 4-wire, Y-connected load. The line consists of three phase conductors a-a', b-b', and c-c' separated by distances D_{ab}, D_{bc}, and D_{ca} as shown. Also shown is the neutral conductor n-n' separated from the phase conductors by distances D_{an}, D_{bn}, and D_{cn}. The neutral conductor is *not* connected to the earth, and its impedance does not involve the earth in any way. One end of the line is short circuited in order to facilitate finding the line impedances. Find z_0, the impedance of this line to the flow of zero sequence currents. Assume that the four conductors are identical with resistance r ohms per unit length and a GMR of D_s.

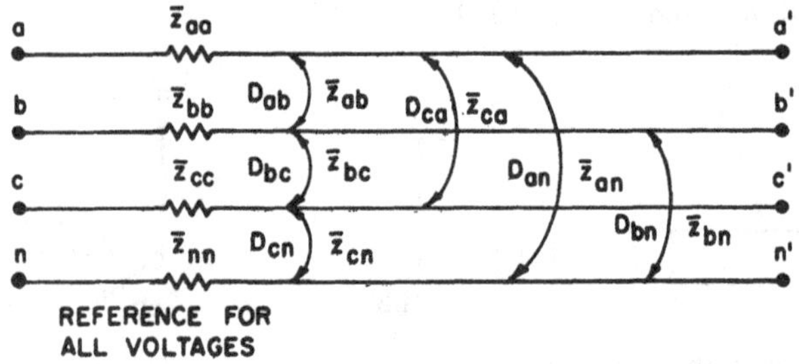

Fig. P4.15.

4.16. The line configuration shown in Figure P4.16 is similar to that used in certain overhead circuits in the 5–15 kV class for primary distribution. Phases a, b, and c are insulated but not shielded cables. The neutral conductor n is a grounded bare messenger which supports all wires mechanically and also carries any unbalanced neutral current ($3I_{a0}$). The frequency is 60 Hz.

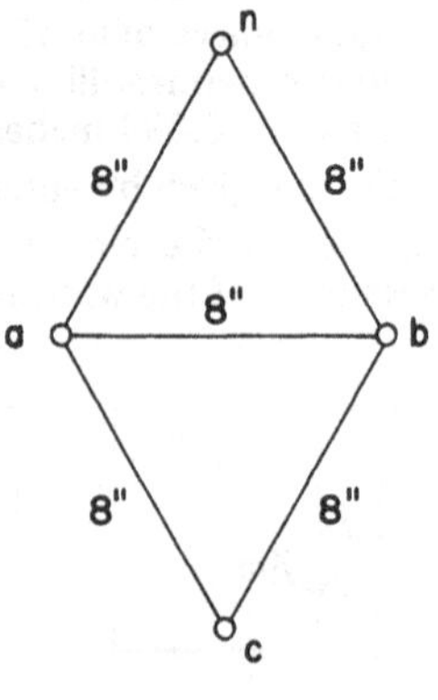

Fig. P4.16.

Phase conductors:

3/0 hard-drawn aluminum, 19 strand
$D_s = 0.01483$ ft = 0.178 inches
$r = 0.558\ \Omega$/mi @ 25 C

Neutral conductor:

1/0 ACSR, 6/1 strand
$D_s = 0.00446$ ft = 0.0535 inches
$r = 0.888\ \Omega$/mi @ 25 C

(a) Ignore the ground wire and determine the phase impedance matrix $\mathbf{Z}_{abc}$ for this circuit.
(b) Assume balanced applied voltages and unbalanced loading as follows:

$V_a = 8000\underline{/0^\circ}$ line-to-neutral $\quad I_a = 200\underline{/-30}$ A
$V_b = a^2 V_a \quad I_b = 200\underline{/-150}$ A
$V_c = a\, V_a$

Let I_c be variable with the following values:

Case	$\lvert I_c \rvert$	*Angle of* I_c	*Case*	$\lvert I_c \rvert$	*Angle of* I_c
1	100	60°	6	200	60°
2	100	90°	7	200	90°
3	100	120°	8	200	120°
4	100	150°	9	200	150°
5	100	180°	10	200	180°

Find the line-to-neutral voltages two miles from the sending end of the line for cases 1-10.
(c) Tabulate the results of part (b) and also the power factor of phase c and the percent voltage drop in each phase.
(d) Plot the percent voltage drop in each phase versus the phase c power factor: (1) for $\lvert I_c \rvert = 100$ A and (2) for $\lvert I_c \rvert = 200$ A. Identify the one point which represents a balanced load.

4.17. Compute the phase impedance matrix $\mathbf{Z}_{abc}$ for the line described in problem 4.1. Assume that the line is 25 miles long and is not transposed.
4.18. Compute the phase impedance matrix $\mathbf{Z}_{abc}$ for the line described in problem 4.2. Assume that the line is 35 miles long and is not transposed.
4.19. Compute the phase impedance matrix $\mathbf{Z}_{abc}$ for the line described in problem 4.3. Assume that the line is 50 miles long and is not transposed. Ignore the ground wire.
4.20. Compute the phase impedance matrix $\mathbf{Z}_{abc}$ for the line described in problem 4.4. Assume that the line is 50 miles long and is not transposed. Ignore the ground wire.
4.21. Compute the phase impedance matrix $\mathbf{Z}_{abc}$ for the line described in problem 4.5. Assume that the line is 60 miles long and is not transposed. Ignore the ground wires.
4.22. Compute the total impedance matrix $\mathbf{Z}_{abc}$ for the lines of problems 4.17–4.21 with the following transposition arrangements.

	Fraction	*Configuration*
(a)	$f_1 = 0.20$	*a-b-c*
	$f_2 = 0.80$	*b-c-a*
	$f_3 = 0.00$	*c-a-b*
(b)	$f_1 = 0.25$	*a-b-c*
	$f_2 = 0.35$	*b-c-a*
	$f_3 = 0.40$	*b-a-c*
(c)	$f_1 = 0.30$	*a-b-c*
	$f_2 = 0.60$	*c-a-b*
	$f_3 = 0.10$	*c-b-a*

(d) $f_1 = 0.33$ *a-b-c*
$f_2 = 0.33$ *c-a-b*
$f_3 = 0.33$ *b-c-a*

4.23. Verify the results given in Table 4.3 by performing the indicated operations.

4.24. (a) Repeat problem 4.22(a) for any given configuration and let f_1 be variable from 0 to 1 in steps of 0.2. Plot X_{ab} as a function of f_1.
(b) Repeat problem 4.22(a) with $f_3 = 0.5$ and let f_1 be a variable from 0 to 0.5 in steps of 0.1. Plot X_{ab} as a function of f_1 and compare results with part (a).

4.25. Verify equations (4.78) and (4.82) for a transposed line.

4.26. Compute the sequence impedance matrix $\mathbf{Z}_{012}$ for each line of problem 4.22.

4.27. Compute the sequence impedances for the study specified in problem 4.24 and determine the way in which the sequences are coupled together as a function of f_1.

4.28. Verify the twist equation (4.110).

4.29. Compute the unbalance factors of each line of problem 4.22,
(a) Using the exact formula (4.144).
(b) Using the approximate formula (4.146).

4.30. Write the matrix equations for the flux linkages in each section of the transposed line of Figure 4.9.

4.31. Consider the four circuits *a*, *b*, *c*, and *d* shown in Figure 4.5.
(a) Write the instantaneous voltage equations of the form $v = ri + d\lambda/dt$ for each circuit.
(b) Apply the phasor transformation (1.50) to the equations of (a).
(c) Compute the phase impedance matrix $\mathbf{Z}_{abc}$ from the result of (b).

4.32. Verify the formula for unbalance factors given by (4.144) by performing the computation implied.

4.33. Consider the line configuration of problem 4.3 shown in Figure P4.3. A ground wire of 1/0 K copperweld-copper is present but is used as a second conductor for one of the phase wires (assume here that the ground wire is fully insulated). Compute the phase impedance matrix $\mathbf{Z}_{abc}$ if this extra wire is connected as follows:
(a) In parallel with phase *a*.
(b) In parallel with phase *b*.
(c) In parallel with phase *c*.

4.34. Consider the line configuration shown in Figure P4.4. Instead of using a single conductor of 336,400 CM ACSR in each phase, with current-carrying capacity of 530 amperes, suppose that each phase consists of a two-conductor bundle of two 3/0 ACSR conductors with capacity of 300 amperes/conductor. Let the two conductors of each bundle be separated by 1.0 ft vertically.
(a) Compute the phase impedance matrix $\mathbf{Z}_{abc}$ for the bundled conductor configuration and compare with the previous solution (problem 4.20).
(b) Compute the sequence impedance matrix for both the new and old conductor arrangements and compare.

4.35. Let the circuit described in problem 4.5 and Figure P4.5 be altered to consist of a two-conductor horizontal bundle of 397,500 CM ACSR phase wires. Compute phase impedance matrix $\mathbf{Z}_{abc}$ and sequence impedance matrix $\mathbf{Z}_{012}$. Bundles with 1.2 ft spacing.

4.36. Verify equation (4.164) for a bundled conductor line. Then show that the necessary matrix operations are all 3 × 3 matrix operations as suggested in Example 4.8.

4.37. Verify equation (4.174) for the phase impedance matrix $\hat{\mathbf{z}}_{abc}$ of a line with one ground wire. Then extend this analysis to the case of a partially transposed line where $f_1 \neq f_2 \neq f_3$.

4.38. Verify the computations which result in equations (4.198) and (4.201) for a completely transposed line with one ground wire.

4.39. Consider an untransposed line described in problem 4.3 and Figure P4.3. Let the ground wire be 1/0 K copperweld-copper and recalculate the phase impedance matrix $\mathbf{Z}_{abc}$, the sequence impedance matrix $\mathbf{Z}_{012}$, and the unbalance factors m_0 and m_2. Compare with previous results from problem 4.19 for the same line without the ground wire.

4.40. Consider an untransposed line described in problem 4.4 and Figure P4.4. Let the ground wire be 1/0 ACSR and recalculate the phase impedance matrix $\mathbf{Z}_{abc}$, the sequence impedance matrix $\mathbf{Z}_{012}$, and the unbalance factors. Compare with previous results from problem 4.20 for the same line without the ground wire.

4.41. Consider an untransposed line described in problem 4.5 and Figure P4.5 with two ground wires of 2/0 ACSR. Compute the phase impedance matrix $\mathbf{Z}_{abc}$, the sequence impedance matrix $\mathbf{Z}_{012}$, and the unbalance factors m_0 and m_2. Compare with the results of problem 4.21 for the same line without ground wires.

4.42. Verify carefully the result given by equation (4.230)-(4.234) for impedance of a transposed line with two ground wires.

4.43. Derive the general expression for a completely transposed line with n ground wires in such a form that (4.233) may be used.

4.44. Repeat the computation of the sequence impedances for problem 4.40 if the ground wire is taken to be 1/2-inch EBB steel messenger. Perform this computation for 1, 30, and 60 amperes in the ground wire.

4.45. Repeat problem 4.44 for the case where the line is completely transposed.

4.46. Compute the sequence impedance matrix for the line described in problem 4.4 and Figure P4.4. Let the ground wire have a GMR of 0.001 feet and let the resistance of the ground wire vary from 0 to 5 ohms/mile. Plot the ratio of I_w to $3I_{a0}$ and to I_d as a function of r_w for fixed GMR.

4.47. Repeat problem 4.46 for a range of GMR's from 0.001 to 0.10 and plot the current ratios I_w/I_d and $I_w/3I_{a0}$ for the entire family.

4.48. Equation (4.184), giving the ratio I_w/I_d, indicates that if $z_{ww} = z_{ag}$ then $I_w/I_d \longrightarrow \infty$. Show that this is impossible for the range of r_w and GMR encountered in physical situations. How high does this ratio become as a practical matter?

4.49. Suppose that the lines shown in Figure P4.4 and P4.5 (with ground wires specified in problems 4.40 and 4.41 respectively) are located on the same right-of-way and separated by a horizontal center-to-center distance of 50 feet.
   (a) Compute the matrix of phase impedances including all ground wires and reduce this matrix to a 6×6 phase impedance matrix similar to equation (4.260).
   (b) Invert and transform the matrix of (a) to compute the four unbalance factors m_{0t}, m_{2t}, m_{0c}, and m_{2c}.
   (c) Recalculate (b) for all 12 possible conductor arrangements for the 2 circuits.

4.50. Prove that there are $6^n/3$ unique arrangements of n parallel circuits which must be examined to determine the optimum phase configuration.

4.51. Repeat the computations of Example 4.14 for all configurations specified in equation (4.281).

chapter 5

Sequence Capacitance of Transmission Lines

This chapter focuses upon the shunt admittance of the transmission line. In overhead lines this shunt admittance is a pure susceptance since the conduction current between wires or between wires and ground is negligible. Furthermore, this susceptance is purely capacitive.

We begin with the computation of the capacitance to neutral of an isolated transposed line and review the method, adequate for most problems, which is used for this situation. In later sections, we examine capacitance in greater detail and find the capacitance between wires (mutual capacitance) and the capacitance unbalance in untransposed lines.

5.1 Positive and Negative Sequence Capacitance of Transposed Lines

The capacitance to ground of each phase of a transposed line is computed by finding the ratio of the linear charge density to the voltage, averaged for each section of the transposition. From field theory we recall that the voltage drop from point 1 to point 2, both external to a linear charge density q_x, is given by

$$V_{12} = \frac{q_x}{2\pi\epsilon} \ln \frac{D_{x2}}{D_{x1}} \quad \text{V} \tag{5.1}$$

Note that we can keep the subscripts straight if we follow this notation carefully, always interpreting the distance from a charged line to itself, i.e., D_{xx} as the conductor radius. Figure 5.1 illustrates the distances expressed in equation (5.1). Note that if $q_x > 0$, and $D_{x2} > D_{x1}$, then V_{12} is a positive drop in potential of polarity indicated.

Since a transmission line is passive, the capacitance is the same for positive and negative sequence systems because the physical parameters do not change with a change in sequence of the applied voltage. Stevenson [9] uses a convenient "modified GMD method" for finding the capacitance to neutral of transposed three-phase lines with the result,

$$c_n = \frac{k'}{\ln(D_m/r)} \quad \text{nF/unit length, to neutral} \tag{5.2}$$

where D_m is the same GMD computed in equation (4.4), r is the radius of the phase conductor, and k' is a constant depending on units of length given by Table 5.1. Note the similarity in the logarithm term to the inductance equations where

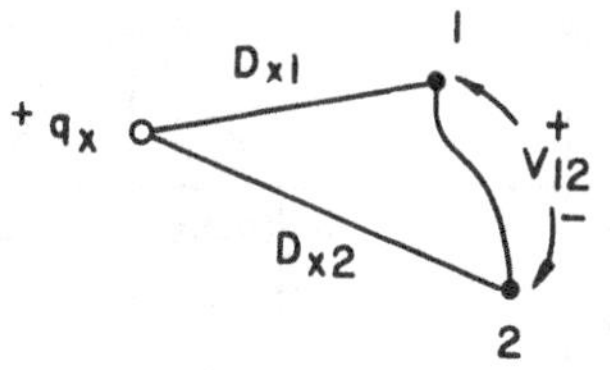

Fig. 5.1. Configuration for equation (5.1).

Table 5.1. Capacitance Multiplying Constants for *C* in nF/Unit Length*

Constant		*Unit of Length*	*Natural Logarithm* (ln)	*Base 10 Logarithm* ($\log_{10}$)
	k′	km	55.630	24.159
		mi	89.525	38.880
	(1/3)k′	km	18.543	8.053
		mi	29.842	12.960
f = 50 Hz				
	fk′	km	2781.49	1207.97
		mi	4476.24	1943.99
	ωk′	km	17476.57	7589.90
		mi	28125.04	12214.42
f = 60 Hz				
	fk′	km	3337.78	1449.57
		mi	5317.49	2309.33
	ωk′	km	20971.89	9107.88
		mi	33750.07	14657.32

*n = nano = 10^{-9}; multiply table values by 10^{-3} if C is desired in microfarads; multiply table values by 10^{-9} if C is desired in farads; 1.0 mi = 1.6093 km; f = 50 Hz, ω = 314.195 rad/sec; f = 60 Hz, ω = 376.991 rad/sec; $\epsilon_0 = (1/\mu_o c^2) = 8.854 \times 10^{-12}$ F/m.

we wrote $\ln(D_m/D_s)$. Here the D_m is the same, but D_s has been replaced by the conductor radius since all charge on a conductor resides on its surface. If we interpret the D_s of each phase to be based on the radius r instead of D_s for each wire, we may generalize the equation to write

$$c_n = \frac{k'}{\ln(D_m/D'_s)} \quad \text{nF/unit length} \tag{5.3}$$

and $b_n = \omega c_n$ nmho/unit length where $D'_s = (D'_{s1}D'_{s2}D'_{s3})^{1/3}$ and D'_{si} = the geometric mean of distances from conductor center to outside radius in part i of a transposition. Usually b_n is in the range 4.8 to 5.5 micromho/mile for single circuit, 60 Hz overhead lines. This can best be illustrated by means of an example.

Example 5.1

Find the capacitance to neutral of the double circuit line shown in Figure 5.2, where the notation in parenthesis, (a,c,b) for example, indicates that this position is occupied in turn by phases a, c, and then b in the three parts of the transposition cycle. All conductors are 477,000 circular mil ACSR, 26/7 strand. It is assumed that $f_1 = f_2 = f_3$, or each transposition section is one-third the total line length.

Solution

Since the line of Figure 5.2 is a double circuit line, we assume that conductors a and a' are connected together at each end and form a parallel phase a conductor,

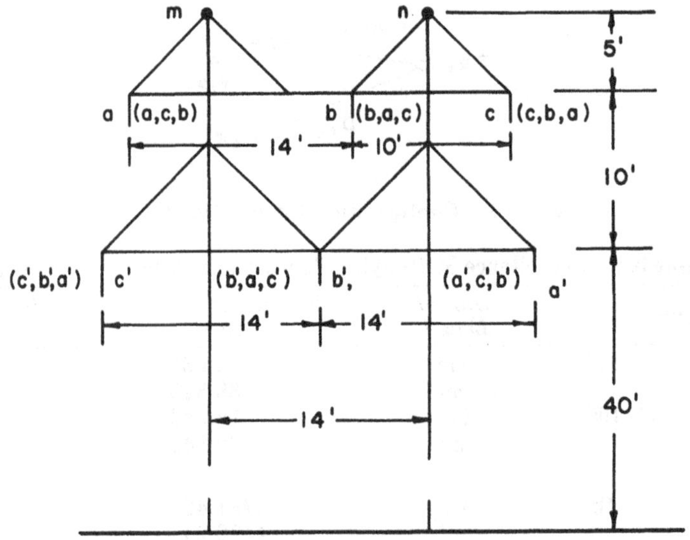

Fig. 5.2. A double circuit line configuration.

and similarly for phases b and c. Then $D_s' = (D_{s1}' D_{s2}' D_{s3}')^{1/3}$, where

$$D_{s1}' = (r_a D_{aa'} r_{a'} D_{a'a})^{1/4}$$
$$D_{s2}' = (r_b D_{bb'} r_{b'} D_{b'b})^{1/4}$$
$$D_{s3}' = (r_c D_{cc'} r_{c'} D_{c'c})^{1/4}$$

We note that $D_{aa'} = D_{cc'}$ and, assuming that all wires have the same radius, we have

$$D_s' = (r^{3/2} D_{aa'} D_{bb'}^{1/2})^{1/3} = r^{1/2} D_{aa'}^{1/3} D_{bb'}^{1/6}$$

The GMD may be computed as (for position 1, shown in Figure 5.2)

$$D_m = (\overline{D}_{ab} \overline{D}_{bc} \overline{D}_{ca})^{1/3}$$

where

$$\overline{D}_{ab} = (D_{ab} D_{ab'} D_{a'b} D_{a'b'})^{1/4}$$
$$\overline{D}_{bc} = (D_{bc} D_{bc'} D_{b'c} D_{b'c'})^{1/4}$$
$$\overline{D}_{ca} = (D_{ca} D_{ca'} D_{c'a} D_{c'a'})^{1/4}$$

But

$$D_{ab} = D_{a'b'} = D_{b'c'}, \qquad D_{ab'} = D_{b'c} = D_{a'b}, \qquad D_{c'a} = D_{ca'}$$

Then

$$D_m = (D_{ab} D_{ab'})^{1/4} (D_{bc} D_{bc'} D_{ca} D_{c'a'})^{1/12} D_{ca'}^{1/6}$$

From Figure 5.2 we compute

$$D_{aa'} = (10^2 + 26^2)^{1/2} = 27.8 \text{ ft}, \qquad D_{bb'} = (10^2 + 2^2)^{1/2} = 10.2 \text{ ft} = D_{ca'}$$
$$D_{ab} = 14 \text{ ft}, \qquad D_{ab'} = (10^2 + 12^2)^{1/2} = 15.65 \text{ ft}$$
$$D_{bc} = 10 \text{ ft}, \qquad D_{bc'} = (10^2 + 16^2)^{1/2} = 18.9 \text{ ft}$$

$$D_{ca} = 24 \text{ ft}, \qquad D_{c'a'} = 28 \text{ ft}$$

$$r = \frac{0.858}{2 \times 12} \text{ (from table B.8)} = 0.0357 \text{ ft}$$

Substituting, we compute

$$D_s' = (0.0357)^{1/2}(27.8)^{1/3}(10.2)^{1/6} = (0.189)(3.029)(1.473) = 0.843 \text{ ft}$$

$$D_m = (14 \times 15.65)^{1/4}(10 \times 18.9 \times 24 \times 28)^{1/12}(10.2)^{1/6}$$
$$= (3.847)(2.663)(1.473) = 15.086 \text{ ft}$$

Then from (5.3) the capacitance per phase (or per equivalent conductor) to neutral is

$$c_n = \frac{89.5}{\ln(15.086/0.843)} = \frac{89.5}{2.88} = 31.03 \text{ nF/mi}$$

and we may compute the 60 Hz susceptance as

$$b_c = 2\pi(60)C_n = 11.70 \quad \mu\text{mho/mi/phase}$$

The preceding computations ignore the effect of the conductor height above the ground. Stevenson [9] shows that (5.3) is modified slightly when conductor height is taken into account. If conductor heights are measured to the image conductors as shown in Figure 5.3, we have

$$c_n = \frac{k'}{\ln(D_m/D_s') - \ln(H_{12}H_{23}H_{31}/H_1H_2H_3)^{1/3}} \text{ nF/unit length} \tag{5.4}$$

From this equation we observe that the effect of taking conductor height into consideration amounts to subtracting from the denominator the term

$$\ln\left(\frac{H_{12}H_{23}H_{31}}{H_1H_2H_3}\right)^{1/3} \tag{5.5}$$

If the conductor height is large compared to the spacing between wires, this term is nearly zero. It is therefore often omitted.

Usually, the actual height of attachment to poles or towers is modified for these calculations. A figure often used is the attachment height minus one-third the sag [38].

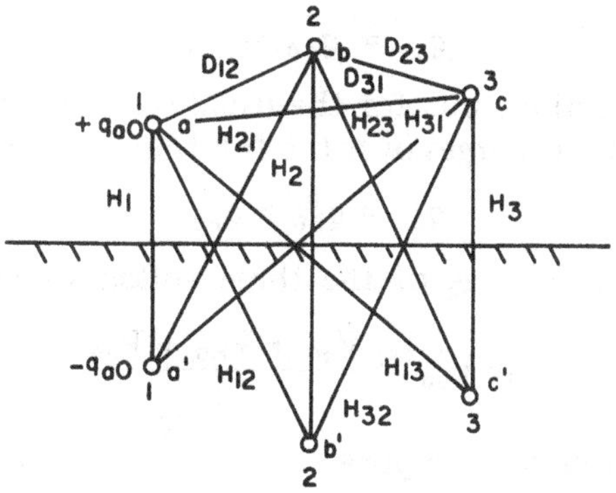

Fig. 5.3. Three phase conductors and their images.

5.2 Zero Sequence Capacitance of Transposed Lines

The shunt admittance between line connections and the zero-potential bus in the zero sequence network depends upon the capacitance to neutral as seen by zero sequence voltages. To compute the zero sequence susceptance, we use the method of images as applied to zero sequence charges and compute the capacitance of a transposed line. Using this method, we consider first a line without ground wires and later add the ground wire. (Also see Stevenson [21].)

Capacitance with no ground wires. Consider the system shown in Figure 5.3 where charges q_{a0}, q_{b0}, and q_{c0} reside on conductors a, b, and c respectively, while the negative of these charges resides on the image conductors. Then the voltage drop from wire a to neutral is half the voltage drop $V_{aa'}$ or, if ϵ is the permittivity (see [12]),

$$V_{a0} = (1/2)V_{aa'} = (1/2)\sum_i \int_{D_1}^{D_2} \frac{q_i}{2\pi\epsilon x}\,dx = \sum_i \frac{q_i}{4\pi\epsilon} \ln\frac{D_2}{D_1} \tag{5.6}$$

$$= \frac{1}{4\pi\epsilon}\left(q_{a0}\ln\frac{H_1}{r_a} + q_{b0}\ln\frac{H_{12}}{D_{12}} + q_{c0}\ln\frac{H_{31}}{D_{31}} - q_{a0}\ln\frac{r_a}{H_1} - q_{b0}\ln\frac{D_{12}}{H_{12}} - q_{c0}\ln\frac{D_{31}}{H_{31}}\right) \tag{5.7}$$

which simplifies to

$$V_{a0} = \frac{1}{2\pi\epsilon}\left(q_{a0}\ln\frac{H_1}{r_a} + q_{b0}\ln\frac{H_{12}}{D_{12}} + q_{c0}\ln\frac{H_{31}}{D_{31}}\right)\ \text{V} \tag{5.8}$$

Similarly,

$$V_{b0} = \frac{1}{2\pi\epsilon}\left(q_{a0}\ln\frac{H_{12}}{D_{12}} + q_{b0}\ln\frac{H_2}{r_b} + q_{c0}\ln\frac{H_{32}}{D_{32}}\right)\ \text{V} \tag{5.9}$$

$$V_{c0} = \frac{1}{2\pi\epsilon}\left(q_{a0}\ln\frac{H_{13}}{D_{13}} + q_{b0}\ln\frac{H_{23}}{D_{23}} + q_{c0}\ln\frac{H_3}{r_c}\right)\ \text{V} \tag{5.10}$$

By definition we know that

$$V_{a0} = V_{b0} = V_{c0} \tag{5.11}$$

so we conclude from (5.8)-(5.10) that

$$q_{a0} \neq q_{b0} \neq q_{c0} \tag{5.12}$$

in any transposition section, but for the usual spacings these charges are nearly equal. If we assume equal charges over the full transposition cycle, or

$$q_{a0} = q_{b0} = q_{c0}$$

the voltage becomes an average of the three voltages given by (5.8)-(5.10), or

$$\text{av } V_{a0} \cong \frac{V_{a0} + V_{b0} + V_{c0}}{3}$$

Performing this operation, we compute

$$V_{a0} = \frac{3q_{a0}}{2\pi\epsilon}\ln\left[\frac{H_1H_2H_3(H_{12}H_{23}H_{31})^2}{r_ar_br_c(D_{12}D_{23}D_{31})^2}\right]^{1/9}\ \text{V} \tag{5.13}$$

or we observe that

$$V_{a0} = \frac{3q_{a0}}{2\pi\epsilon} \ln \frac{H_{aa}}{D_{aa}} \tag{5.14}$$

where

H_{aa} = GMD between the three conductors and their images
D_{aa} = self GMD of the overhead conductors as a composite group, but with D_s of each wire taken as its radius

Then

$$c_0 = \frac{q_{a0}}{V_{a0}} = \frac{2\pi\epsilon}{3\ \ln(H_{aa}/D_{aa})} \quad \text{F/m/phase}$$

$$= \frac{(1/3)\text{k}'}{\ln\,(H_{aa}/D_{aa})} \quad \text{nF/unit length} \tag{5.15}$$

As a rule of thumb b_{c0} has values in the range of 2.5–3.5 μmho/mile for single circuit 60 Hz overhead lines.

Capacitance with ground wires. To solve the capacitance problem with ground wires, we first analyze the case of a single conductor a with one ground wire g as shown in Figure 5.4. This may later be extended to a more general situation.

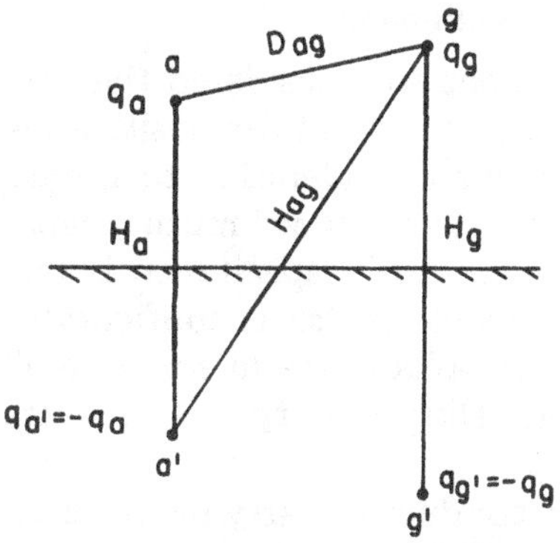

Fig. 5.4. One conductor with one ground wire.

Referring to Figure 5.4, we write

$$V_a = \frac{1}{2}\,V_{aa'} = \frac{1}{2\pi\epsilon}\left(q_a \ln \frac{H_a}{r_a} + q_g \ln \frac{H_{ag}}{D_{ag}}\right) \tag{5.16}$$

Likewise,

$$V_g = 0 = \frac{1}{2}\,V_{gg'} = \frac{1}{2\pi\epsilon}\left(q_a \ln \frac{H_{ag}}{D_{ag}} + q_g \ln \frac{H_g}{r_g}\right) \tag{5.17}$$

Solving (5.16) and (5.17) for q_a, we have

$$q_a = \frac{2\pi\epsilon\ V_a \ln(H_g/r_g)}{\ln \dfrac{H_a}{r_a} \ln \dfrac{H_g}{r_g} - \left(\ln \dfrac{H_{ag}}{D_{ag}}\right)^2} \tag{5.18}$$

Now suppose that conductor a is the composite of three phase wires and conductor g is the composite of n ground wires. Then q_a is the total charge of all three conductors or, if all wires have the same charge,

$$q_{a0} = (1/3)\, q_a \tag{5.19}$$

Also, since $V_{a0} = V_{b0} = V_{c0}$, we have

$$V_{a0} = V_a \tag{5.20}$$

From (5.18)-(5.20) we compute

$$c_0 = \frac{(1/3)k' \ln(H_{gg}/D_{gg})}{\ln \dfrac{H_{aa}}{D_{aa}} \ln \dfrac{H_{gg}}{D_{gg}} - \left(\ln \dfrac{H_{ag}}{D_{ag}}\right)^2} \quad \text{nF/unit length/phase} \tag{5.21}$$

where

D_{aa} = the same as in equation (5.14)
D_{gg} = self GMD of ground wires with $D_s = r_g$
D_{ag} = GMD between phase wires and ground wires
H_{aa} = GMD between phase wires and their images
H_{gg} = GMD between ground wires and their images
H_{ag} = GMD between phase wires and images of ground wires

5.3 Mutual Capacitance of Transmission Lines

In the preceding paragraphs we considered the calculation of capacitance between a charged conductor and ground for positive, negative, and zero sequence charges. We now expand our consideration to the capacitance between nearby conductors and shall refer to the result as "mutual capacitance."

The subject of self and mutual capacitance is treated in many references.[1] Rather than treat every possible physical configuration, we will investigate the transposed single circuit three-phase line in some detail and then consider briefly the effect of ground wires. Dissymmetry due to untransposed lines will be covered in section 5.8.

Consider a group of n conductors carrying linear charge densities $q_a, q_b, \ldots, q_n$, located above the ground plane as shown in Figure 5.5. Using the method of images (see [12]), we compute the voltage drop between two points 1 and 2 by superposition (since the system is linear). From (5.1) we write

$$V_{12} = \frac{1}{2\pi\epsilon}\left(q_a \ln\frac{D_{a2}}{D_{a1}} + q_b \ln\frac{D_{b2}}{D_{b1}} + \cdots + q_n \ln\frac{D_{n2}}{D_{n1}} - q_a \ln\frac{H_{a2}}{H_{a1}} - q_b \ln\frac{H_{b2}}{H_{b1}} - \cdots - q_n \ln\frac{H_{n2}}{H_{n1}}\right) \text{V} \tag{5.22}$$

where the D's are distances between conductors or points above the ground and the H's are distances between conductors or points in space and image charges. In this way we compute the voltages of all conductors to ground by noting that

[1] The interested reader is referred to Clarke [11], Wagner and Evans [10], and Lyon [39] for additional reading on the subject. Our approach follows Calabrese [24] which is recommended for its clarity and complete coverage of the subject.

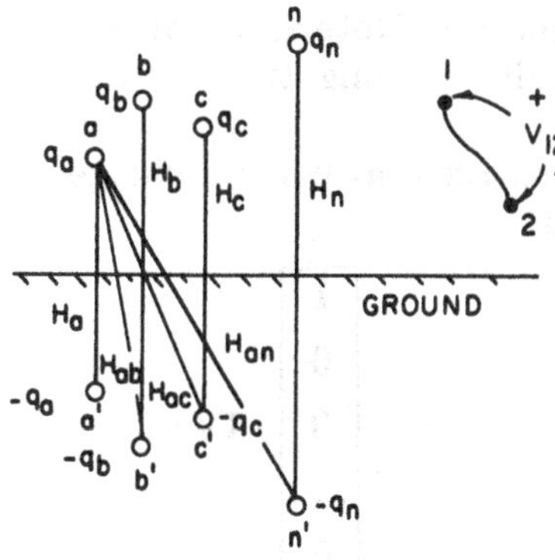

Fig. 5.5. Group of n charged lines and their images.

$$
\begin{aligned}
V_a &= \frac{1}{2} V_{aa'} \\
V_b &= \frac{1}{2} V_{bb'} \\
&\cdots\cdots\cdots\cdots \\
V_n &= \frac{1}{2} V_{nn'}
\end{aligned}
\tag{5.23}
$$

Then, for example,

$$
\begin{aligned}
V_a = \frac{1}{2} V_{aa'} = \frac{1}{4\pi\epsilon} \bigg(& q_a \ln \frac{H_a}{r_a} + q_b \ln \frac{H_{ab}}{D_{ab}} + \cdots + q_n \ln \frac{H_{an}}{D_{an}} \\
& - q_a \ln \frac{r_a}{H_a} - q_b \ln \frac{D_{ab}}{H_{ab}} - \cdots - q_n \ln \frac{D_{an}}{H_{an}} \bigg)
\end{aligned}
\tag{5.24}
$$

and combining terms,

$$
V_a = \frac{1}{2\pi\epsilon} \left(q_a \ln \frac{H_a}{r_a} + q_b \ln \frac{H_{ab}}{D_{ab}} + \cdots + q_n \ln \frac{H_{an}}{D_{an}} \right) \text{ V} \tag{5.25}
$$

Similarly,

$$
\begin{aligned}
V_b &= \frac{1}{2\pi\epsilon} \left(q_a \ln \frac{H_{ab}}{D_{ab}} + q_b \ln \frac{H_b}{r_b} + \cdots + q_n \ln \frac{H_{bn}}{D_{bn}} \right) \text{ V} \\
&\cdots\cdots\cdots\cdots\cdots\cdots\cdots\cdots\cdots\cdots \\
V_n &= \frac{1}{2\pi\epsilon} \left(q_a \ln \frac{H_{an}}{D_{an}} + q_b \ln \frac{H_{bn}}{D_{bn}} + \cdots + q_n \ln \frac{H_n}{r_n} \right) \text{ V}
\end{aligned}
\tag{5.26}
$$

Equations (5.25) and (5.26) are often written in matrix form as

$$
\mathbf{V} = \mathbf{P}\mathbf{q} \tag{5.27}
$$

where $\mathbf{V}$ is the voltage vector, $\mathbf{q}$ is the charge vector, and $\mathbf{P}$ is a matrix of coefficients called "potential coefficients" [24] where

$$
\begin{aligned}
p_{ij} &= \frac{1}{2\pi\epsilon} \ln \frac{H_i}{r_i}, \; i = j \\
&= \frac{1}{2\pi\epsilon} \ln \frac{H_{ij}}{D_{ij}}, \; i \neq j \qquad \text{F}^{-1}\text{m}
\end{aligned}
\tag{5.28}
$$

and we note that **P** is symmetric. Note also that **V** and **q** are not restricted in any way but may be sinusoidally varying time domain quantities or transformed quantities such as phasors.

One can think of the matrix **P** as being a chargeless or "zero charge matrix." For example, suppose we set

$$\mathbf{q} = \begin{bmatrix} 1 \\ 0 \\ 0 \\ \cdots \\ 0 \end{bmatrix} \text{ C/m} \tag{5.29}$$

i.e., q_a is the only charge present and its value is 1 coulomb/meter. Then

$$\begin{bmatrix} V_a \\ V_b \\ \cdots \\ V_n \end{bmatrix} = \begin{bmatrix} p_{aa} \\ p_{ba} \\ \cdots \\ p_{na} \end{bmatrix} = \frac{1}{2\pi\epsilon} \begin{bmatrix} \ln \dfrac{H_a}{r_a} \\ \ln \dfrac{H_{ab}}{D_{ab}} \\ \cdots \\ \ln \dfrac{H_{an}}{D_{an}} \end{bmatrix} \text{ V} \tag{5.30}$$

or p_{ka} is the potential in volts assumed by wire k due to linear charge on wire a alone (of 1 coulomb/meter) *with all other charges zero*. All these coefficients are positive quantities since $H_{an} > D_{an}$, these being equal only if charge a is on the ground plane. With charge q_a acting alone as in (5.29) and (5.30), the potential V_a is the greatest of all potentials in (5.30), but *all* are positive.

In a similar way we can define each column of **P** as a set of potentials derived by setting the charges equal to 1 coulomb/meter and with all other charges equal to zero. Thus each column of **P** is independent of all others depending only upon the geometry of the system, and the columns of **P** comprise a basis for an n-dimensional vector space [40]. This being the case, **P** is nonsingular and has an inverse $\mathbf{P}^{-1}$. Thus from (5.27) we may compute

$$\mathbf{q} = \mathbf{cV} \tag{5.31}$$

where[2]

$$\mathbf{c} = \mathbf{P}^{-1} \tag{5.32}$$

The elements of the matrix **c** are called "Maxwell's coefficients" or more specifically are called "capacitance coefficients" when referring to diagonal terms and "coefficients of electrostatic induction" when referring to off-diagonal terms. These coefficients may be thought of as short circuit parameters which relate the charges to the system voltages. For example, with

[2] Obviously, this is not the same matrix **C** as defined in Chapter 2. There will be no ambiguity, however, as we usually use $\mathbf{A}^{-1}$ for the symmetrical component matrix. Here the **c** is written lowercase for capacitance per unit length. The uppercase **C** is for total capacitance of the line.

$$\mathbf{V} = \begin{bmatrix} 1 \\ 0 \\ 0 \\ \cdots \\ 0 \end{bmatrix} \tag{5.33}$$

i.e., with $V_a = 1$ and all other voltages zero (short circuited), we find the first column of **c** to be

$$\begin{bmatrix} q_a \\ q_b \\ \cdots \\ q_n \end{bmatrix} = \begin{bmatrix} c_{aa} \\ -c_{ba} \\ \cdots \\ -c_{na} \end{bmatrix} \text{ C/m} \tag{5.34}$$

The negative signs from the off-diagonal terms arise due to the fact that the coefficients of **P** in (5.27) are all positive. Then the inverse of **P** is

$$\mathbf{c} = \mathbf{P}^{-1} = \frac{\text{adj}\,\mathbf{P}}{\det \mathbf{P}} = \frac{[P_{ij}]^t}{\det \mathbf{P}} \tag{5.35}$$

where P_{ij} is the cofactor of element p_{ij}. But

$$P_{ij} = (-1)^{i+j} M_{ij} \tag{5.36}$$

where M_{ij} is the minor of element p_{ij}. We set [24]

$$\begin{aligned} c_{ij} &= \frac{M_{ij}}{\det \mathbf{P}} \text{ for } i = j \text{ or } (i + j) \text{ even} \\ &= -\frac{M_{ij}}{\det \mathbf{P}} \text{ for } i \neq j \text{ or } (i + j) \text{ odd} \end{aligned} \tag{5.37}$$

Then

$$\mathbf{c} = \begin{bmatrix} +c_{aa} & -c_{ab} & \cdots & -c_{an} \\ -c_{ba} & +c_{bb} & \cdots & -c_{bn} \\ \cdots & \cdots & \cdots & \cdots \\ -c_{na} & -c_{nb} & \cdots & +c_{nn} \end{bmatrix} \text{ F/m} \tag{5.38}$$

and only the diagonal terms have positive signs. The coefficients (c_{ij}) themselves are all positive quantities and are all capacitances. Note that all coefficients may be found by superposition, i.e., by applying 1 volt to wire *a*, *b*, etc., always with every other wire shorted to ground. The negative sign really means, then, that applying a positive potential to one wire induces a negative charge on the other wires. This is intuitively correct since the application of a dc voltage to a capacitor makes one terminal (or plate) of the capacitor positive but also makes the other terminal (plate) negative.

Equation (5.31) is the desired relationship for self and mutual capacitances of an *n* wire system. It is not in the best form for physical interpretation, however. To make this clearer, suppose that both **q** and **V** are sinusoidal time varying quantities. Specifically, let one of the charge densities be

$$q = Q_m \cos \omega t \tag{5.39}$$

Then, since current is the time rate of change of charge, we compute

$$i = dq/dt = -\omega Q_m \sin \omega t \tag{5.40}$$

Now transform both (5.39) and (5.40) into phasors. Then the phasor charge density Q and phasor charging current I are defined as

$$Q = \frac{Q_m}{\sqrt{2}} e^{j0} \tag{5.41}$$

and

$$I = \frac{\omega Q_m}{\sqrt{2}} e^{j\pi/2} \tag{5.42}$$

or

$$I = \omega Q e^{j\pi/2} = j\omega Q \tag{5.43}$$

If we write (5.31) in terms of phasor charge density and voltage **Q** and **V**, we have

$$\mathbf{Q} = \mathbf{CV} \tag{5.44}$$

which we change to a current equation by multiplying by $j\omega$ as in (5.43), or

$$\mathbf{I} = j\omega \mathbf{Q} = j\omega \mathbf{CV} \tag{5.45}$$

But, from circuit theory we write the charging current as

$$\mathbf{I} = \mathbf{YV} \tag{5.46}$$

or the phasor admittances are

$$\mathbf{Y} = j\omega \mathbf{C} \tag{5.47}$$

For example, from (5.46) we interpret that

$$\begin{aligned} Y_{kk} &= \text{sum of all admittances connected to } k \\ &= j\omega c_{kk}, \text{ a capacitive susceptance} \end{aligned} \tag{5.48}$$

$$\begin{aligned} Y_{km} &= \text{the } \textit{negative} \text{ of all admittance connected between } k \text{ and } m \\ &= -j\omega c_{km} \end{aligned} \tag{5.49}$$

or the actual admittance between k and m is (using lowercase for the actual quantity and uppercase for the matrix element)

$$y_{km} = -Y_{km} = g_{km} + jb_{km} = +j\omega c_{km} \tag{5.50}$$

which is again a capacitive susceptance. From (5.47)-(5.50) we visualize an equivalent circuit as shown in Figure 5.6. Since Y_{kk} is defined in (5.48), the capacitances to ground are

$$\begin{aligned} c_{ag} &= c_{aa} - c_{ab} - c_{ac} - \cdots - c_{an} \quad \text{F/m} \\ c_{bg} &= -c_{ba} + c_{bb} - c_{bc} - \cdots - c_{bn} \quad \text{F/m} \\ &\cdots\cdots\cdots\cdots\cdots\cdots \\ c_{ng} &= -c_{na} - c_{nb} - c_{nc} - \cdots + c_{nn} \quad \text{F/m} \end{aligned} \tag{5.51}$$

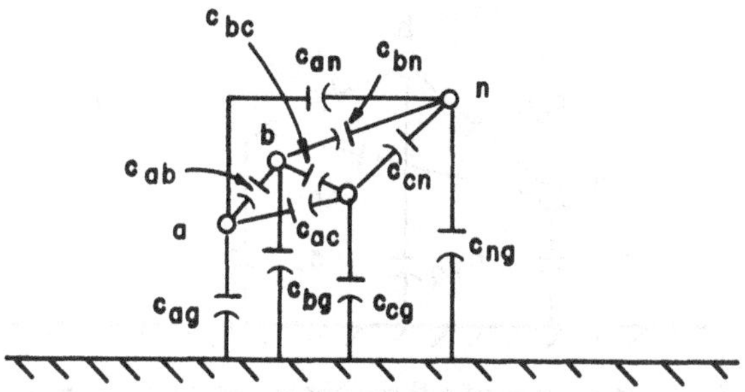

Fig. 5.6. Self and mutual capacitances of an n-phase system.

With capacitances so defined, it is easy to see that the sum of admittances connected to node k is indeed Y_{kk}. The equivalent circuit of Figure 5.6 can also be obtained by algebraic manipulation of (5.47). (See problem 5.2.)

5.4 Mutual Capacitance of Three-Phase Lines without Ground Wires

We now consider a special case of the general mutual capacitance problem where there are only the three charged conductors of a transposed three-phase line with no ground wires.

From (5.25) and (5.26) the potential equations are, in terms of potential coefficients, $\mathbf{V} = \mathbf{Pq}$, or

$$\begin{bmatrix} V_a \\ V_b \\ V_c \end{bmatrix} = \begin{bmatrix} p_{aa} & p_{ab} & p_{ac} \\ p_{ba} & p_{bb} & p_{bc} \\ p_{ca} & p_{cb} & p_{cc} \end{bmatrix} \begin{bmatrix} q_a \\ q_b \\ q_c \end{bmatrix} \text{ V} \tag{5.52}$$

where the elements of $\mathbf{P}$ are given by (5.28). The charge equation is found by solving (5.52) for $\mathbf{q}$. Thus, in terms of Maxwell's coefficients, $\mathbf{q} = \mathbf{cV}$, or

$$\begin{bmatrix} q_a \\ q_b \\ q_c \end{bmatrix} = \begin{bmatrix} c_{aa} & -c_{ab} & -c_{ac} \\ -c_{ba} & c_{bb} & -c_{bc} \\ -c_{ca} & -c_{cb} & c_{cc} \end{bmatrix} \begin{bmatrix} V_a \\ V_b \\ V_c \end{bmatrix} \text{ C/m} \tag{5.53}$$

where the elements of $\mathbf{c}$ are given by (5.37). Let

$$\det \mathbf{P} = p_{aa}(p_{bb}p_{cc} - p_{bc}^2) - p_{ab}^2 p_{cc} + 2p_{ab}p_{bc}p_{ac} - p_{ac}^2 p_{bb}$$

Then compute the minors (noting that $p_{ik} = p_{ki}$)

$$\begin{aligned} M_{aa} &= p_{bb}p_{cc} - p_{bc}^2 & M_{ab} &= p_{ab}p_{cc} - p_{ac}p_{bc} \\ M_{bb} &= p_{aa}p_{cc} - p_{ac}^2 & M_{ac} &= p_{ab}p_{bc} - p_{bb}p_{ac} \\ M_{cc} &= p_{aa}p_{bb} - p_{ab}^2 & M_{bc} &= p_{aa}p_{bc} - p_{ab}p_{ac} \end{aligned} \tag{5.54}$$

Then the elements of $\mathbf{c}$ are computed from (5.37). For the capacitance to ground we compute from (5.51)

$$\begin{aligned} c_{ag} &= c_{aa} - c_{ab} - c_{ac} \text{ F/m} \\ c_{bg} &= c_{bb} - c_{ab} - c_{bc} \text{ F/m} \\ c_{cg} &= c_{cc} - c_{ac} - c_{bc} \text{ F/m} \end{aligned} \tag{5.55}$$

These various capacitances may be viewed symbolically as shown in Figure 5.7 where the various capacitances are shown as circuit elements.

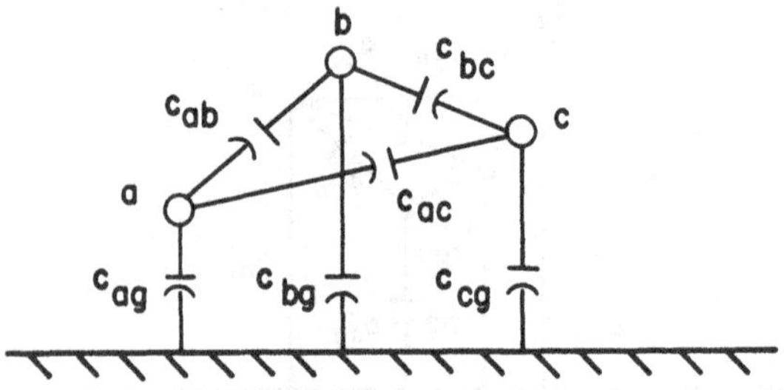

Fig. 5.7. Capacitances of a three-phase line with no ground wires.

If the line is transposed in sections of per unit length f_1, f_2, and f_3 the voltage equation (5.52) may be written using the notation of Chapter 4. Thus we write

$$\mathbf{q}_{abc} = (f_1 \mathbf{C}_{123} + f_2 \mathbf{C}_{231} + f_3 \mathbf{C}_{312})\, \mathbf{V}_{abc} \quad \text{c/m} \tag{5.56}$$

where

$$\mathbf{C}_{231} = \mathbf{R}_\phi^{-1} \mathbf{C}_{123} \mathbf{R}_\phi, \qquad \mathbf{C}_{312} = \mathbf{R}_\phi \mathbf{C}_{123} \mathbf{R}_\phi^{-1}$$

and $\mathbf{C}_{123}$ is defined to be the Maxwell's coefficient matrix for section f_1. The potential matrix is then found by inverting the entire matrix expression in (5.56).

If the line is completely transposed, each phase occupies each position for one-third the total line length. Since each section is one-third the total length, the capacitance per phase per meter consists of one-third meter of each of c_{ab}, c_{bc}, and c_{ac} or

$$c_{M0} = (1/3)(c_{ab} + c_{bc} + c_{ac}) \text{ F/m} \tag{5.57}$$

Similarly, for the capacitance to ground

$$c_{g0} = (1/3)(c_{ag} + c_{bg} + c_{cg}) = c_{S0} - 2c_{M0} \quad \text{F/m} \tag{5.58}$$

where

$$c_{S0} = (1/3)(c_{aa} + c_{bb} + c_{cc}) \text{ F/m} \tag{5.59}$$

Actually, these "transposed capacitances" are just averages of the capacitance seen by each phase in each transposition section. The capacitance to ground can be thought of as a combination of a self capacitance c_S and the mutual capacitance c_M.

Example 5.2

Find the mutual capacitance and capacitance to ground of the lower circuit only (ignore upper circuit and ground wires) of Figure 5.2. Compare the capacitance to ground computed in this way to that obtained by application of (5.3).

Solution

First, we form the **P** matrix, the coefficients of which are entirely dependent upon geometry. In our case (see Figure 5.2)

$$D_{ab} = D_{bc} = 14.0 \text{ ft}, \qquad D_{ca} = 28.0 \text{ ft}$$

$$H_a = H_b = H_c = 80.0 \text{ ft}, \qquad H_{ab} = H_{bc} = (80^2 + 14^2)^{1/2} = 81.2 \text{ ft}$$

$$H_{ac} = (80^2 + 28^2)^{1/2} = 84.8 \text{ ft}, \qquad r = 0.0357 \text{ ft}$$

Then from (5.28)

$$p_{aa} = p_{bb} = p_{cc} = \frac{1}{2\pi\epsilon} \ln \frac{H_a}{r_a}$$

and if we let

$$\epsilon = \epsilon_0 \kappa = \kappa/(36\pi \times 10^9)$$

where $\kappa = 1$ for air dielectric, we have

$$\begin{aligned} p_{aa} &= 18 \times 10^{+9} \ln \frac{H_a}{r_a} \quad \text{F}^{-1}\,\text{m} \\ &= 11.185 \ln \frac{H_a}{r_a} \quad \text{MF}^{-1}\,\text{mi} \\ &= 11.185 \ln \frac{80}{0.0357} = 86.28 \quad \text{MF}^{-1}\,\text{mi} \end{aligned}$$

$$p_{ab} = p_{bc} = 11.185 \ln \frac{H_{ab}}{D_{ab}} = 11.185 \ln \frac{81.3}{14.0} = 19.66 \quad \text{MF}^{-1}\,\text{mi}$$

$$p_{ac} = 11.185 \ln \frac{H_{ac}}{D_{ac}} = 11.185 \ln \frac{84.6}{28.0} = 12.39 \quad \text{MF}^{-1}\,\text{mi}$$

Then

$$\mathbf{P} = \begin{bmatrix} 86.3 & 19.7 & 12.4 \\ 19.7 & 86.3 & 19.7 \\ 12.4 & 19.7 & 86.3 \end{bmatrix} \text{MF}^{-1}\text{mi}$$

Direct inversion of **P** by digital computer (see Appendix A) gives the result

$$\mathbf{c} = \begin{bmatrix} 12.34 & -2.54 & -1.19 \\ -2.54 & 12.75 & -2.54 \\ -1.19 & -2.54 & 12.34 \end{bmatrix} \quad \text{nF/mi}$$

Since a digital computer is not always available, we confirm this result by hand computation. From (5.54) we compute,

$$\begin{aligned} \Delta = \det \mathbf{P} &= p_{aa}p_{bb}p_{cc} - p_{aa}p_{bc}^2 - p_{ab}^2 p_{cc} + 2p_{ab}p_{ca}p_{bc} - p_{ac}^2 p_{bb} \\ &= 86.3^3 - 86.3(19.7)^2 - (19.7)^2(86.3) + 2(19.7)^2(12.4) - (12.4)^2(86.3) \\ &\cong 647{,}500 - 33{,}600 - 33{,}600 + 8{,}330 - 13{,}300 \\ &= 575{,}330 \end{aligned}$$

Also from (5.54),

$$\begin{aligned} M_{aa} &= 7091 \qquad M_{cc} = 7091 \qquad M_{ac} = -685 \\ M_{bb} &= 7326 \qquad M_{ab} = 1459 \qquad M_{bc} = 1459 \end{aligned}$$

from which we compute trom (5.37),

$$\begin{aligned} c_{aa} &= 12.31 \text{ nF/mi} \qquad c_{cc} = 12.31 \text{ nF/mi} \qquad c_{ac} = 1.190 \text{ nF/mi} \\ c_{bb} &= 12.72 \text{ nF/mi} \qquad c_{ab} = 2.535 \text{ nF/mi} \qquad c_{bc} = 2.535 \text{ nF/mi} \end{aligned}$$

These values are seen to be very close indeed to the digital computer results. Since the more accurate digital computer solution is available, we use it as a basis for computations which follow. The capacitances to ground are, from (5.55),

$$c_{ag} = c_{aa} - c_{ab} - c_{ac} = 8.606 \quad \text{nF/mi}$$
$$c_{bg} = c_{bb} - c_{ab} - c_{bc} = 7.667 \quad \text{nF/mi}$$
$$c_{cg} = c_{cc} - c_{ac} - c_{bc} = 8.606 \quad \text{nF/mi}$$

For a transposed line the capacitance to ground is the average of these three, i.e., from (5.58),

$$c_{g0} = (1/3)(c_{ag} + c_{bg} + c_{cg}) = 8.293 \quad \text{nF/mi}$$

The average (transposed) mutual capacitance is, from (5.57),

$$c_{M0} = (1/3)(c_{ab} + c_{bc} + c_{ac}) = 2.091 \quad \text{nF/mi}$$

The average (transposed) self capacitance is, from (5.59),

$$c_{S0} = (1/3)(c_{aa} + c_{bb} + c_{cc}) = 12.475 \quad \text{nF/mi}$$

Now compare c_{g0} computed above to the capacitance to neutral computed by equation (5.3) which neglects the effect of the earth and is computed under the assumption of equal charge density in each part of a transposition cycle [9]. From (5.3)

$$c_n = \frac{89.5}{\ln(D_m/D_s')} \quad \text{nF/mi}$$

where

$$D_m = D_{eq} = (D_{ab}D_{bc}D_{ca})^{1/3} = 17.5 \text{ ft}$$
$$D_s' = r = 0.0357 \text{ ft}$$

Then $c_n = 14.4$ nF/mi. How do we reconcile this difference? If we imagine an equivalent circuit similar to Figure 5.7, the Δ-connected mutual capacitors all have a value of $c_{M0} = 2.091$ nF/mi. Convert this Δ to a Y. Then each capacitor in the Y has a value $c_y = 6.273$ nF/mi. Since this is in parallel with c_{g0}, the total capacitance to neutral per phase is the sum $c_n = c_{g0} + c_y = 14.566$ nF/mi. If the c_n computed by (5.3) is corrected by its height above the ground, these values will be in quite close agreement.

5.5 Sequence Capacitance of a Transposed Line without Ground Wires

Consider a three-phase line, the capacitance of which is described in terms of its Maxwell coefficients as $\mathbf{q} = \mathbf{CV}$, where $\mathbf{q}$ and $\mathbf{V}$ are phasors. Then by (5.45), $\mathbf{I} = j\omega\mathbf{CV}$ or, to emphasize the *a-b-c* coordinate system,

$$\mathbf{I}_{abc} = j\omega\mathbf{C}\mathbf{V}_{abc} \tag{5.60}$$

But this is easily changed to a 0-1-2 coordinate system by a similarity transformation. Thus

$$\mathbf{I}_{012} = j\omega\mathbf{A}^{-1}\mathbf{C}\mathbf{A}\,\mathbf{V}_{012} \tag{5.61}$$

or

$$\mathbf{I}_{012} = j\omega\mathbf{C}_{012}\,\mathbf{V}_{012} \tag{5.62}$$

where we have defined

$$\mathbf{C}_{012} = \mathbf{A}^{-1}\mathbf{C}\,\mathbf{A} = \begin{bmatrix} c_{00} & c_{01} & c_{02} \\ c_{10} & c_{11} & c_{12} \\ c_{20} & c_{21} & c_{22} \end{bmatrix} \tag{5.63}$$

Performing the indicated transformation, we compute

$$\mathbf{C}_{012} = \begin{matrix} \\ 0 \\ 1 \\ 2 \end{matrix} \begin{matrix} \begin{matrix} 0 & \qquad\qquad 1 & \qquad\qquad 2 \end{matrix} \\ \begin{bmatrix} (c_{S0} - 2c_{M0}) & (c_{S2} + c_{M2}) & (c_{S1} + c_{M1}) \\ (c_{S1} + c_{M1}) & (c_{S0} + c_{M0}) & (c_{S2} - 2c_{M2}) \\ (c_{S2} + c_{M2}) & (c_{S1} - 2c_{M1}) & (c_{S0} + c_{M0}) \end{bmatrix} \end{matrix} \tag{5.64}$$

where we define c_{S0} and c_{M0} as in (5.59) and (5.57) respectively[3] and where

$$\begin{aligned} c_{S1} &= (1/3)(c_{aa} + \mathrm{a}c_{bb} + \mathrm{a}^2 c_{cc}) \\ c_{S2} &= (1/3)(c_{aa} + \mathrm{a}^2 c_{bb} + \mathrm{a}c_{cc}) \\ c_{M1} &= (1/3)(c_{bc} + \mathrm{a}c_{ac} + \mathrm{a}^2 c_{ab}) \\ c_{M2} &= (1/3)(c_{bc} + \mathrm{a}^2 c_{ac} + \mathrm{a}c_{ab}) \end{aligned} \tag{5.65}$$

If the line is transposed, the values in (5.65) are all zero because each phase occupies each position for an equal distance, thereby acquiring a multiplier of $(1 + \mathrm{a} + \mathrm{a}^2)$ for each capacitance term. With these sequence mutuals all zero, (5.64) becomes

$$\mathbf{C}_{012} = \begin{bmatrix} c_{S0} - 2c_{M0} & 0 & 0 \\ 0 & c_{S0} + c_{M0} & 0 \\ 0 & 0 & c_{S0} + c_{M0} \end{bmatrix} \tag{5.66}$$

and the mutual coupling between sequence networks is eliminated.

Note from (5.66) that the zero sequence capacitance is much less than the positive and negative sequence capacitances. Also note that the positive and negative sequence capacitance to neutral is given by

$$c_{11} = c_{22} = c_{S0} + c_{M0} \tag{5.67}$$

which should agree closely with (5.3). Similarly, for the zero sequence

$$c_{00} = c_{S0} - 2c_{M0} \tag{5.68}$$

which should check with (5.15).

Example 5.3

Compute the matrix $\mathbf{C}_{012}$ for the transposed line of Figure 5.2, data for which is computed in Example 5.2. Compare with results computed from (5.3) and (5.15).

[3]Note that the signs in $\mathbf{C}_{012}$ are not in the same pattern as in the $\mathbf{Z}_{012}$ matrix of the last chapter. This is because of the negative signs in the $\mathbf{C}_{abc}$ matrix of equation (5.53).

Solution

From Example 5.2 we have

$$c_{S0} = 12.475 \quad \text{nF/mi}, \qquad c_{M0} = 2.091 \quad \text{nF/mi}$$

Then $c_{11} = c_{22} = c_{S0} + c_{M0} = 14.566$ nF/mi which checks exactly with the c_n computed in Example 5.2. Also $c_{00} = 8.293$ nF/mi. For the zero sequence capacitance we have from (5.15),

$$c_0 = \frac{29.8}{\ln(H_{aa}/D_{aa})} \quad \text{nF/mi}$$

where

$$\begin{aligned} H_{aa} &= [H_a H_b H_c (H_{ab} H_{bc} H_{ac})^2]^{1/9} \\ &= (80)^{1/3}(81.3 \times 81.3 \times 84.6)^{2/9} = (4.3)(18.9) = 81.3 \text{ ft} \\ D_{aa} &= [r^3 (D_{ab} D_{bc} D_{ca})^2]^{1/9} \\ &= (0.0357)^{1/3}(14 \times 14 \times 28)^{2/9} = (0.33)(6.75) = 2.22 \text{ ft} \end{aligned}$$

Thus

$$c_0 = \frac{29.8}{\ln(81.3/2.22)} = 8.28 \quad \text{nF/mi}$$

which checks very well with c_{00}.

5.6 Mutual Capacitance of Three-Phase Lines with Ground Wires

Having computed the self and mutual capacitance of a circuit without ground wires, we now consider the additional complication added by the ground wire and study its effect upon capacitance of a line. As before, we consider the line as being transposed, leaving the problem of unequal phase capacitances for a later section.

Using the subscript n to denote the ground wire, we may write the equation of potential coefficients for the four-wire system shown in Figure 5.8. We write $\mathbf{V} = \mathbf{Pq}$ or, in more detail and with $V_n = 0$,

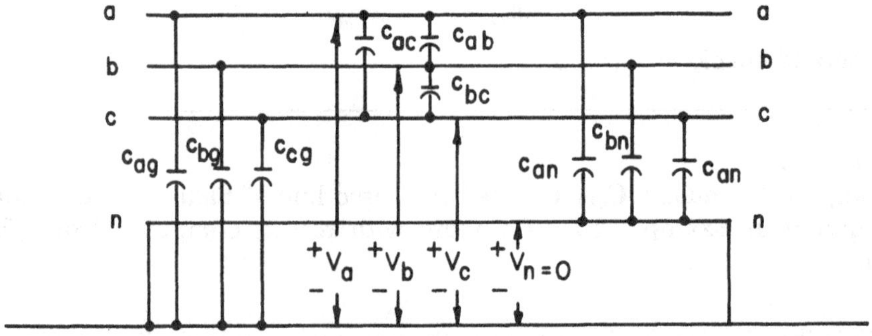

Fig. 5.8. A three-phase line with one ground wire n.

$$\begin{bmatrix} V_a \\ V_b \\ V_c \\ \hline 0 \end{bmatrix} = \left[\begin{array}{ccc|c} p_{aa} & p_{ab} & p_{ac} & p_{an} \\ p_{ba} & p_{bb} & p_{bc} & p_{bn} \\ p_{ca} & p_{cb} & p_{cc} & p_{cn} \\ \hline p_{na} & p_{nb} & p_{nc} & p_{nn} \end{array}\right] \begin{bmatrix} q_a \\ q_b \\ q_c \\ \hline q_n \end{bmatrix} \tag{5.69}$$

But this matrix can be reduced to three equations by eliminating the fourth row and column. Solving the last equation and substituting back, we have

$$\begin{bmatrix} V_a \\ V_b \\ V_c \end{bmatrix} = \begin{bmatrix} \left(p_{aa} - \dfrac{p_{an}p_{na}}{p_{nn}}\right) & \left(p_{ab} - \dfrac{p_{an}p_{nb}}{p_{nn}}\right) & \left(p_{ac} - \dfrac{p_{an}p_{nc}}{p_{nn}}\right) \\ \left(p_{ba} - \dfrac{p_{bn}p_{na}}{p_{nn}}\right) & \left(p_{bb} - \dfrac{p_{bn}p_{nb}}{p_{nn}}\right) & \left(p_{bc} - \dfrac{p_{bn}p_{nc}}{p_{nn}}\right) \\ \left(p_{ca} - \dfrac{p_{cn}p_{na}}{p_{nn}}\right) & \left(p_{cb} - \dfrac{p_{cn}p_{nb}}{p_{nn}}\right) & \left(p_{cc} - \dfrac{p_{cn}p_{nc}}{p_{nn}}\right) \end{bmatrix} \begin{bmatrix} q_a \\ q_b \\ q_c \end{bmatrix} \tag{5.70}$$

This could be simplified slightly by taking advantage of the fact that **P** is symmetric.

Since the elements of **P** are all positive, the new **P** of (5.70) contains elements all of which are smaller than corresponding elements for the same line with no ground wires. Note that the "correction factor" for each element is a positive quantity which depends only upon the geometry of the system (see (5.28)). If the ground wire is moved toward infinity, this correction factor approaches zero since the entire row and column associated with the ground wire vanish.

Using the corrected **P** matrix of (5.70), we again find the capacitance matrix of Maxwell's coefficients by matrix inversion, i.e., $\mathbf{q} = \mathbf{C}_{abc}\mathbf{V}$, where $\mathbf{C}_{abc}$ is the inverse of the 3×3 matrix of (5.70). Since the presence of the ground wire makes the elements of **P** smaller, we would expect the elements of **C** to be larger than the case of the same system with no ground wires.

The sequence capacitance matrix $\mathbf{C}_{012}$ is again found by a similarity transformation, exactly as in the case of no ground wire, by (5.63).

If it is desirable to consider the capacitance to ground separately from the capacitance to neutral, as shown in Figure 5.8, this can be done by inverting the 4×4 **P** matrix of (5.69). This would permit separate identification of elements such as c_{an}, c_{bn}, and c_{cn} as shown in Figure 5.8. Since these capacitances are in parallel with c_{ag}, c_{bg}, and c_{cg}, it is apparent that the capacitances are increased by the ground wire.

If there are two ground wires m and n, the procedure is exactly the same. In this case the equations involving potential coefficients are

$$\begin{bmatrix} V_a \\ V_b \\ V_c \\ \hline 0 \\ 0 \end{bmatrix} = \left[\begin{array}{ccc|cc} p_{aa} & p_{ab} & p_{ac} & p_{am} & p_{an} \\ p_{ba} & p_{bb} & p_{bc} & p_{bm} & p_{bn} \\ p_{ca} & p_{cb} & p_{cc} & p_{cm} & p_{cn} \\ \hline p_{ma} & p_{mb} & p_{mc} & p_{mm} & p_{mn} \\ p_{na} & p_{nb} & p_{nc} & p_{nm} & p_{nn} \end{array}\right] \begin{bmatrix} q_a \\ q_b \\ q_c \\ \hline q_m \\ q_n \end{bmatrix} \tag{5.71}$$

which we rewrite as

$$\begin{bmatrix} \mathbf{V}_{abc} \\ \hline 0 \end{bmatrix} = \left[\begin{array}{c|c} \mathbf{P}_1 & \mathbf{P}_2 \\ \hline \mathbf{P}_3 & \mathbf{P}_4 \end{array}\right] \begin{bmatrix} \mathbf{q}_{abc} \\ \hline \mathbf{q}_{mn} \end{bmatrix} \tag{5.72}$$

where the matrices $\mathbf{P}_1$, $\mathbf{P}_2$, $\mathbf{P}_3$, and $\mathbf{P}_4$ are defined according to the partitioning of (5.71). Solving for $\mathbf{V}_{abc}$ we have

$$\mathbf{V}_{abc} = (\mathbf{P}_1 - \mathbf{P}_2\mathbf{P}_4^{-1}\ \mathbf{P}_3)\, \mathbf{q}_{abc} = \mathbf{P}_{abc}\mathbf{q}_{abc} \tag{5.73}$$

where

$$\mathbf{P}_{abc} = \mathbf{P}_1 - \mathbf{P}_2\mathbf{P}_4^{-1}\mathbf{P}_3 \tag{5.74}$$

Thus, $\mathbf{P}_2\mathbf{P}_4^{-1}\mathbf{P}_3$ may be thought of as a correction due to ground wires. The Maxwell coefficients are found by inverting (5.74), i.e.,

$$\mathbf{C}_{abc} = \mathbf{P}_{abc}^{-1} \tag{5.75}$$

and the sequence capacitances are found by applying the similarity transformation (5.63) to $\mathbf{C}_{abc}$.

Example 5.4

Repeat the capacitance calculation of Example 5.2, this time including the effect of the ground wires shown in Figure 5.2. Assume the ground wire is 3/8-inch EBB steel.

Solution

First we compute the potential coefficients for the ground wires. Thus

$$p_{mm} = p_{nn} = 11.185 \ln \frac{H_m}{r_m}$$

where

$$r_m = (1/2)\,(3/8)\,(1/12) = 0.01562 \text{ ft}, \qquad H_m = 2(55) = 110 \text{ ft}$$

or

$$p_{mm} = p_{nn} = 11.185 \ln \frac{110}{0.01562} = 99.096 \quad \text{MF}^{-1}\,\text{mi}$$

Also,

$$p_{mn} = p_{nm} = 11.185 \ln \frac{H_{mn}}{D_{mn}}$$

where

$$H_{mn} = (110^2 + 14^2)^{1/2} = 110.89 \text{ ft}, \qquad D_{mn} = 14 \text{ ft}$$

or

$$p_{mn} = p_{nm} = 11.185 \ln (111/14) = 23.147 \quad \text{MF}^{-1}\,\text{mi}$$

We also compute

$$p_{an} = p_{bn} = p_{bm} = p_{cm} = 11.185 \ln \frac{H_{an}}{D_{an}}$$

and

$$p_{am} = p_{cn} = 11.185 \ln \frac{H_{am}}{D_{am}}$$

where

$$H_{an} = (95^2 + 7^2)^{1/2} \cong 95.26 \text{ ft}, \qquad H_{am} = (95^2 + 21^2)^{1/2} = 97.29 \text{ ft}$$
$$D_{an} = (15^2 + 7^2)^{1/2} = 16.55 \text{ ft}, \qquad D_{am} = (15^2 + 21^2)^{1/2} = 25.81 \text{ ft}$$

Then

$$p_{an} = 11.185 \ln \frac{95.26}{16.55} = 19.574 \quad \text{MF}^{-1}\,\text{mi}$$

and

$$p_{am} = 11.185 \ln \frac{97.29}{25.81} = 14.844 \quad \text{MF}^{-1}\,\text{mi}$$

Thus

$$\mathbf{P} = \begin{array}{c} \\ a \\ b \\ c \\ m \\ n \end{array}\begin{array}{c} \begin{array}{ccccc} a & b & c & m & n \end{array} \\ \left[\begin{array}{ccc|cc} 86.3 & 19.7 & 12.4 & 14.8 & 19.6 \\ 19.7 & 86.3 & 19.7 & 19.6 & 19.6 \\ 12.4 & 19.7 & 86.3 & 19.6 & 14.8 \\ \hline 14.8 & 19.6 & 19.6 & 99.1 & 23.1 \\ 19.6 & 19.6 & 14.8 & 23.1 & 99.1 \end{array}\right] \end{array} \text{MF}^{-1}\,\text{mi} = \begin{bmatrix} \mathbf{P}_1 & \mathbf{P}_2 \\ \mathbf{P}_3 & \mathbf{P}_4 \end{bmatrix}$$

Now $\mathbf{P}_4^{-1} = \text{adj}\,\mathbf{P}_4/\det \mathbf{P}_4$, so we compute

$$\det \mathbf{P}_4 = 99.1^2 - 23.1^2 = 9821 - 534 = 9287$$

$$\mathbf{P}_4^{-1} = \frac{1}{9287}\begin{bmatrix} 99.1 & -23.1 \\ -23.1 & 99.1 \end{bmatrix} = \begin{bmatrix} 0.0107 & -0.0025 \\ -0.0025 & 0.0107 \end{bmatrix}$$

Then

$$\mathbf{P}_2\mathbf{P}_4^{-1}\mathbf{P}_3 = \begin{bmatrix} 14.8 & 19.6 \\ 19.6 & 19.6 \\ 19.6 & 14.8 \end{bmatrix}\begin{bmatrix} 0.0107 & -0.0025 \\ -0.0025 & 0.0107 \end{bmatrix}\begin{bmatrix} 14.8 & 19.6 & 19.6 \\ 19.6 & 19.6 & 14.8 \end{bmatrix}$$

$$= \begin{bmatrix} 4.99 & 5.51 & 4.70 \\ 5.51 & 6.27 & 5.51 \\ 4.70 & 5.51 & 4.99 \end{bmatrix}$$

and

$$\mathbf{P}_{abc} = \mathbf{P}_1 - \mathbf{P}_2\mathbf{P}_4^{-1}\mathbf{P}_3 = \begin{bmatrix} 86.3 & 19.7 & 12.4 \\ 19.7 & 86.3 & 19.7 \\ 12.4 & 19.7 & 86.3 \end{bmatrix} - \begin{bmatrix} 4.99 & 5.51 & 4.70 \\ 5.51 & 6.27 & 5.51 \\ 4.70 & 5.51 & 4.99 \end{bmatrix}$$

$$= \begin{bmatrix} 81.30 & 14.15 & 7.69 \\ 14.15 & 80.02 & 14.15 \\ 7.69 & 14.15 & 81.30 \end{bmatrix}$$

Finally, $\mathbf{c}_{abc} = \mathbf{P}_{abc}^{-1}$, which we compute by digital computer to be

$$\mathbf{c}_{abc} = \begin{bmatrix} 12.747 & -2.106 & -0.839 \\ -2.106 & 13.242 & -2.106 \\ -0.839 & -2.106 & 12.747 \end{bmatrix} \quad \text{nF/mi}$$

The capacitances to ground are

$$c_{ag} = c_{aa} - c_{ab} - c_{ac} = 9.802 \quad \text{nF/mi}$$
$$c_{bg} = c_{bb} - c_{ab} - c_{bc} = 9.030 \quad \text{nF/mi}$$
$$c_{cg} = c_{cc} - c_{ac} - c_{bc} = 9.802 \quad \text{nF/mi}$$

For the transposed line the capacitance to ground in each phase is the average of these three values, i.e.,

$$c_{g0} = (1/3)\,(c_{ag} + c_{bg} + c_{cg}) = 9.545 \quad \text{nF/mi}$$

which is considerably larger than the 8.293 nF/mi for this same line without ground wires.

The average (transposed) mutual capacitance is

$$c_{M0} = (1/3)\,(c_{ab} + c_{bc} + c_{ac}) = 1.683 \quad \text{nF/mi}$$

which is smaller than the 2.091 nF/mi for the same line without ground wires.

The average (transposed) self capacitance is

$$c_{S0} = (1/3)\,(c_{aa} + c_{bb} + c_{cc}) = 12.912 \quad \text{nF/mi}$$

which is greater than the 12.475 nF/mi for the same line without ground wires.

The sequence capacitances are $c_{00} = c_{S0} - 2c_{M0} = 9.546$ nF/mi and

$$c_{11} = c_{22} = c_{S0} + c_{M0} = 14.595 \quad \text{nF/mi}$$

These values are both greater than the 8.293 and 14.566 computed for the same line without ground wires.

5.7 Capacitance of Double Circuit Lines

For the case of a double circuit line with or without ground wires, the problem becomes more complicated because of the presence of so many charges. Consider the configuration of Figure 5.9 where the distances from one conductor c' to all other conductors and to all images are shown. Using the method of section 5.4, we may write the voltage equations in terms of potential coefficients as $\mathbf{V} = \mathbf{P}\mathbf{q}$ or

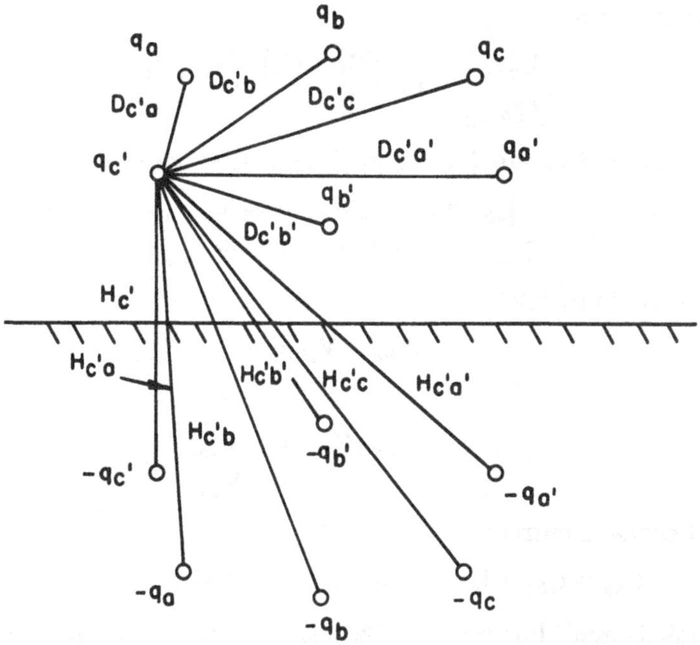

Fig. 5.9. Configuration of a double circuit line.

$$\begin{bmatrix} V_a \\ V_b \\ V_c \\ \hline V_{a'} \\ V_{b'} \\ V_{c'} \end{bmatrix} = \left[\begin{array}{ccc|ccc} p_{aa} & p_{ab} & p_{ac} & p_{aa'} & p_{ab'} & p_{ac'} \\ p_{ba} & p_{bb} & p_{bc} & p_{ba'} & p_{bb'} & p_{bc'} \\ p_{ca} & p_{cb} & p_{cc} & p_{ca'} & p_{cb'} & p_{cc'} \\ \hline p_{a'a} & p_{a'b} & p_{a'c} & p_{a'a'} & p_{a'b'} & p_{a'c'} \\ p_{b'a} & p_{b'b} & p_{b'c} & p_{b'a'} & p_{b'b'} & p_{b'c'} \\ p_{c'a} & p_{c'b} & p_{c'c} & p_{c'a'} & p_{c'b'} & p_{c'c'} \end{array}\right] \begin{bmatrix} q_a \\ q_b \\ q_c \\ \hline q_{a'} \\ q_{b'} \\ q_{c'} \end{bmatrix} \tag{5.76}$$

which we may write as

$$\begin{bmatrix} \mathbf{V}_{abc} \\ \mathbf{V}_{a'b'c'} \end{bmatrix} = \begin{bmatrix} \mathbf{P}_{11} & \mathbf{P}_{12} \\ \mathbf{P}_{21} & \mathbf{P}_{22} \end{bmatrix} \begin{bmatrix} \mathbf{q}_{abc} \\ \mathbf{q}_{a'b'c'} \end{bmatrix} \tag{5.77}$$

The capacitance matrix $\mathbf{C}$ is the inverse of $\mathbf{P}$, which may be computed according to the partitioning of (5.77) as (see [7])

$$\mathbf{C} = \mathbf{P}^{-1} = \begin{bmatrix} (\mathbf{P}_{11}^{-1} + \mathbf{F}\,\mathbf{E}^{-1}\mathbf{F}^t) & (-\mathbf{F}\,\mathbf{E}^{-1}) \\ (-\mathbf{F}\,\mathbf{E}^{-1})^t & (\mathbf{E}^{-1}) \end{bmatrix} \tag{5.78}$$

where we define

$$\mathbf{F} = \mathbf{P}_{11}^{-1}\mathbf{P}_{12} \tag{5.79}$$

so that $\mathbf{F}^t = \mathbf{P}_{21}\mathbf{P}_{11}^{-1}$ since $\mathbf{P}$ is symmetric, and

$$\mathbf{E} = \mathbf{P}_{22} - \mathbf{P}_{21}\mathbf{P}_{11}^{-1}\mathbf{P}_{12} = \mathbf{P}_{22} - \mathbf{P}_{21}\mathbf{F} \tag{5.80}$$

Once $\mathbf{C}$ is determined, we may write $\mathbf{I} = j\omega\,\mathbf{C}\,\mathbf{V}$, where both $\mathbf{I}$ and $\mathbf{V}$ are 6×1. It will be helpful to write

$$\begin{bmatrix} \mathbf{I}_{abc} \\ \mathbf{I}_{a'b'c'} \end{bmatrix} = j\omega \begin{bmatrix} \mathbf{C}_1 & \mathbf{C}_2 \\ \mathbf{C}_3 & \mathbf{C}_4 \end{bmatrix} \begin{bmatrix} \mathbf{V}_{abc} \\ \mathbf{V}_{a'b'c'} \end{bmatrix} \tag{5.81}$$

where $\mathbf{C}_1$, $\mathbf{C}_2$, $\mathbf{C}_3$, and $\mathbf{C}_4$ are defined in (5.78). Then we compute

$$\begin{aligned} \mathbf{I}_{abc} &= j\omega (\mathbf{C}_1 \mathbf{V}_{abc} + \mathbf{C}_2 \mathbf{V}_{a'b'c'}) \\ \mathbf{I}_{a'b'c'} &= j\omega (\mathbf{C}_3 \mathbf{V}_{abc} + \mathbf{C}_4 \mathbf{V}_{a'b'c'}) \end{aligned} \tag{5.82}$$

Since the lines are in parallel

$$\mathbf{V}_{abc} = \mathbf{V}_{a'b'c'} \tag{5.83}$$

or

$$\begin{aligned} \mathbf{I}_{abc} &= j\omega (\mathbf{C}_1 + \mathbf{C}_2) \mathbf{V}_{abc} \\ \mathbf{I}_{a'b'c'} &= j\omega (\mathbf{C}_3 + \mathbf{C}_4) \mathbf{V}_{abc} \end{aligned} \tag{5.84}$$

and the total charging current is

$$\mathbf{I}_{\text{chg}} = \mathbf{I}_{abc} + \mathbf{I}_{a'b'c'} = j\omega (\mathbf{C}_1 + \mathbf{C}_2 + \mathbf{C}_3 + \mathbf{C}_4) \mathbf{V}_{abc} \tag{5.85}$$

Then the parallel circuit line behaves like a single circuit line with capacitance $\mathbf{C}_{eq}$, where

$$\mathbf{C}_{eq} = \mathbf{C}_1 + \mathbf{C}_2 + \mathbf{C}_3 + \mathbf{C}_4 \tag{5.86}$$

The sequence capacitance may be found for each circuit or for the parallel equivalent by similarity transformation, i.e.,

$$\mathbf{C}_{012eq} = \mathbf{A}^{-1} \mathbf{C}_{eq} \mathbf{A} \tag{5.87}$$

Also, from (5.86) and (5.78) we compute

$$\mathbf{C}_{eq} = \mathbf{P}_{11}^{-1} + (\mathbf{F} - \mathbf{U}) \mathbf{E}^{-1} (\mathbf{F} - \mathbf{U})^t \tag{5.88}$$

where $\mathbf{U}$ is the unit matrix. Thus we see that the effect of the second circuit on the first depends strongly upon the matrix $(\mathbf{F} - \mathbf{U})$. Since $\mathbf{P}_{11} \neq \mathbf{P}_{12}$, we write

$$\mathbf{F} - \mathbf{U} = \mathbf{P}_{11}^{-1} \mathbf{P}_{12} - \mathbf{U} \tag{5.89}$$

If $\mathbf{F} - \mathbf{U}$ is nearly zero, $\mathbf{C}_{eq}$ becomes nearly $\mathbf{P}_{11}^{-1}$, which is the value for $\mathbf{C}_{abc}$ alone and would indicate no effect at all due to the second circuit. This will be the case if $\mathbf{P}_{12} \cong \mathbf{P}_{11}$ such that $\mathbf{F} - \mathbf{U} = \mathbf{P}_{11}^{-1} \mathbf{P}_{11} - \mathbf{U} = \mathbf{U} - \mathbf{U} = \mathbf{0}$. This will be approximately true if the height is large compared to the distance between conductors and if the conductors are arranged in vertical phase configuration such that $H_{ab} \cong H_{ab'}$, etc. The matrices $\mathbf{P}_{11}$ and $\mathbf{P}_{12}$ will never be exactly equal since this would require the wire radius to equal the distance between phases, or $r = D_{aa'}$ such that $p_{aa} = p_{aa'}$.

For the case of double circuits with ground wires, the equation involving $\mathbf{P}$ will be similar to (5.76) but will have added rows and columns for the ground wires. These can be eliminated by matrix reduction since the voltage of the ground wires is zero. The result then can always be reduced to a 6×6 $\mathbf{P}$ matrix equivalent.

Example 5.5

Compute the capacitance matrix for the double circuit line of Example 5.1 and Figure 5.2, including the effect of the ground wires which are assumed to be 3/8-inch EBB steel. Examine only one transposition section.

Solution

First we compute the **P** matrix. If we arrange as shown in (5.76), we know the lower right partition from Example 5.4, which involved only the lower a'-b'-c' circuit and the ground wires.

For the upper circuit

$$D_{ab} = 14 \text{ ft}, \qquad D_{bc} = 10 \text{ ft}, \qquad D_{ca} = 24 \text{ ft}$$

and

$$\begin{aligned} H_a &= H_b = H_c = 100 \text{ ft} \\ H_{ab} &= (100^2 + 14^2)^{1/2} = 100.98 \text{ ft} \\ H_{bc} &= (100^2 + 10^2)^{1/2} = 100.50 \text{ ft} \\ H_{ca} &= (100^2 + 24^2)^{1/2} = 102.84 \text{ ft} \\ r &= 0.0357 \text{ ft} \end{aligned}$$

Then

$$p_{aa} = p_{bb} = p_{cc} = 11.185 \ln \frac{H_a}{r_a} = 11.185 \ln \frac{100}{0.0357} = 88.784 \quad \text{MF}^{-1}\,\text{mi}$$

$$p_{ab} = 11.185 \ln \frac{100.98}{14.0} = 22.100 \quad \text{MF}^{-1}\,\text{mi}$$

$$p_{bc} = 11.185 \ln \frac{100.50}{10.0} = 25.810 \quad \text{MF}^{-1}\,\text{mi}$$

$$p_{ca} = 11.185 \ln \frac{102.84}{24.0} = 16.276 \quad \text{MF}^{-1}\,\text{mi}$$

Between the a-b-c circuit and the ground wires we compute

$$\begin{aligned} D_{am} &= 7.07 \text{ ft} & D_{bm} &= 10.30 \text{ ft} & D_{cm} &= 19.65 \text{ ft} \\ D_{an} &= 19.65 \text{ ft} & D_{bn} &= 7.07 \text{ ft} & D_{cn} &= 7.07 \text{ ft} \\ H_{am} &= 105.12 \text{ ft} & H_{bm} &= 105.39 \text{ ft} & H_{cm} &= 106.71 \text{ ft} \\ H_{an} &= 106.71 \text{ ft} & H_{bn} &= 105.12 \text{ ft} & H_{cn} &= 105.12 \text{ ft} \end{aligned}$$

such that

$$p_{am} = p_{cn} = p_{bn} = 11.185 \ln \frac{105.12}{7.07} = 30.189 \quad \text{MF}^{-1}\,\text{mi}$$

$$p_{an} = p_{cm} = 11.185 \ln \frac{106.71}{19.65} = 18.927 \quad \text{MF}^{-1}\,\text{mi}$$

$$p_{bm} = 11.185 \ln \frac{105.12}{10.30} = 26.015 \quad \text{MF}^{-1}\,\text{mi}$$

Between the a-b-c circuit and the a'-b'-c' circuit we compute

$$D_{aa'} = 27.86 \text{ ft} \quad D_{ba'} = 15.62 \text{ ft} \quad D_{ca'} = 10.20 \text{ ft}$$
$$D_{ab'} = 15.62 \text{ ft} \quad D_{bb'} = 10.20 \text{ ft} \quad D_{cb'} = 15.62 \text{ ft}$$
$$D_{ac'} = 10.20 \text{ ft} \quad D_{bc'} = 18.87 \text{ ft} \quad D_{cc'} = 27.86 \text{ ft}$$
$$H_{aa'} = 93.68 \text{ ft} \quad H_{ba'} = 90.80 \text{ ft} \quad H_{ca'} = 90.02 \text{ ft}$$
$$H_{ab'} = 90.80 \text{ ft} \quad H_{bb'} = 90.02 \text{ ft} \quad H_{cb'} = 90.80 \text{ ft}$$
$$H_{ac'} = 90.02 \text{ ft} \quad H_{bc'} = 91.41 \text{ ft} \quad H_{cc'} = 93.68 \text{ ft}$$

such that

$$p_{aa'} = p_{cc'} = 11.185 \ln \frac{93.68}{27.86} = 13.565 \quad \text{MF}^{-1} \text{ mi}$$

$$p_{ab'} = p_{ba'} = p_{cb'} = 11.185 \ln \frac{90.80}{15.62} = 19.686 \quad \text{MF}^{-1} \text{ mi}$$

$$p_{ac'} = p_{bb'} = p_{ca'} = 11.185 \ln \frac{90.02}{10.20} = 24.359 \quad \text{MF}^{-1} \text{ mi}$$

$$p_{bc'} = 11.185 \ln \frac{91.41}{18.87} = 17.649 \quad \text{MF}^{-1} \text{ mi}$$

Rounding to the nearest 0.1, we write

$$\mathbf{P} = \begin{array}{c} a \\ b \\ c \\ a' \\ b' \\ c' \\ m \\ n \end{array} \left[\begin{array}{ccc|ccc|cc} 88.8 & 22.1 & 16.3 & 13.6 & 19.7 & 24.4 & 30.2 & 18.9 \\ 22.1 & 88.8 & 25.8 & 19.7 & 24.4 & 17.6 & 26.0 & 30.2 \\ 16.3 & 25.8 & 88.8 & 24.4 & 19.7 & 13.6 & 18.9 & 30.2 \\ \hline 13.6 & 19.7 & 24.4 & 86.3 & 19.7 & 12.4 & 14.8 & 19.6 \\ 19.7 & 24.4 & 19.7 & 19.7 & 86.3 & 19.7 & 19.6 & 19.6 \\ 24.4 & 17.6 & 13.6 & 12.4 & 19.7 & 86.3 & 19.6 & 14.8 \\ \hline 30.2 & 26.0 & 18.9 & 14.8 & 19.6 & 19.6 & 99.1 & 23.1 \\ 18.9 & 30.2 & 30.2 & 19.6 & 19.6 & 14.8 & 23.1 & 99.1 \end{array} \right] \text{MF}^{-1} \text{ mi}$$

(columns: a, b, c, a', b', c', m, n)

We eliminate the two "outside" rows and columns delineated by the solid partition lines to compute a "correction matrix" $\mathbf{P}_c$, where we define $\mathbf{P}_c$ such that the 6×6 matrix is $\mathbf{P}_{abc} = \mathbf{P} - \mathbf{P}_c$.

$$\mathbf{P}_c = \left[\begin{array}{ccc|ccc} 10.70 & 10.98 & 9.03 & 6.56 & 7.86 & 7.27 \\ 10.98 & 13.04 & 11.60 & 8.04 & 9.00 & 7.78 \\ 9.03 & 11.60 & 10.70 & 7.27 & 7.86 & 6.56 \\ \hline 6.56 & 8.04 & 7.27 & 4.99 & 5.51 & 4.70 \\ 7.86 & 9.00 & 7.86 & 5.51 & 6.27 & 5.51 \\ 7.27 & 7.78 & 6.56 & 4.70 & 5.51 & 4.99 \end{array} \right]$$

Finally,

$$\mathbf{P}_{abc} = \begin{array}{c} \\ a \\ b \\ c \\ a' \\ b' \\ c' \end{array} \begin{array}{c} \begin{array}{cccccc} a & b & c & a' & b' & c' \end{array} \\ \left[\begin{array}{ccc|ccc} 78.08 & 11.12 & 7.24 & 7.00 & 11.82 & 17.09 \\ 11.12 & 75.75 & 14.21 & 11.64 & 15.36 & 9.87 \\ 7.24 & 14.21 & 78.08 & 17.09 & 11.82 & 7.00 \\ \hline 7.00 & 11.64 & 17.09 & 81.30 & 14.15 & 7.69 \\ 11.82 & 15.36 & 11.82 & 14.15 & 80.02 & 14.15 \\ 17.09 & 9.87 & 7.00 & 7.69 & 14.15 & 81.30 \end{array}\right] \end{array} \text{MF}^{-1}\,\text{mi}$$

which is inverted by digital computer to find

$$\mathbf{C} = \begin{array}{c} \\ a \\ b \\ c \\ a' \\ b' \\ c' \end{array} \begin{array}{c} \begin{array}{cccccc} a & b & c & a' & b' & c' \end{array} \\ \left[\begin{array}{ccc|ccc} 13.80 & -1.29 & -0.55 & -0.45 & -1.20 & -2.45 \\ -1.29 & 14.43 & -1.88 & -1.14 & -1.95 & -0.87 \\ -0.55 & -1.88 & 13.93 & -2.38 & -1.12 & -0.44 \\ \hline -0.45 & -1.14 & -2.38 & 13.34 & -1.63 & -0.54 \\ -1.20 & -1.95 & -1.12 & -1.63 & 13.79 & -1.66 \\ -2.45 & -0.87 & -0.44 & -0.54 & -1.66 & 13.30 \end{array}\right] \end{array} \text{nF/mi}$$

We also compute

$$\mathbf{P}_{11}^{-1}\mathbf{P}_{12} = \begin{bmatrix} 0.0526 & 0.1177 & 0.2010 \\ 0.1091 & 0.1648 & 0.0905 \\ 0.1939 & 0.1105 & 0.0546 \end{bmatrix} = \mathbf{F}$$

and

$$\mathbf{F} - \mathbf{U} = \begin{bmatrix} -0.9438 & 0.1177 & 0.2010 \\ 0.1091 & -0.8352 & 0.0905 \\ 0.1939 & 0.1105 & -0.9454 \end{bmatrix}$$

Since **F** - **U** is quite different from **0**, we conclude that the second circuit has a substantial effect upon the capacitance of the first. This is also evident if one compares the *a-b-c* partition of **C** with the **C** matrix of Example 5.4.

5.8 Electrostatic Unbalance of Untransposed Lines

If transmission lines are left untransposed, a practice which is becoming relatively common, an electrostatic unbalance exists in addition to the electromagnetic unbalance studied in Chapter 4. Any unbalance in transmission line charging currents results in the flow of neutral "residual" current in solidly grounded systems, and this current flows at all times, independent of load current. If the unbalance is great and these residual currents are large, they could possibly affect system relaying or cause the voltages to become unbalanced.

This problem has been studied extensively and methods have been developed for computing the amount of unbalance in a given situation (see [28–30, 34–36,

41–43]). Having established a definition of the "unbalance factor," different line configurations may be examined in detail to optimize the line design.

Ground displacement of lines. In the early 1950s, Gross and others [34–36] developed a definition for the electrostatic unbalance of a line. This definition is established with reference to a line supplied from a Y-connected transformer bank as shown in Figure 5.10 where we recognize the presence of capaoitance between

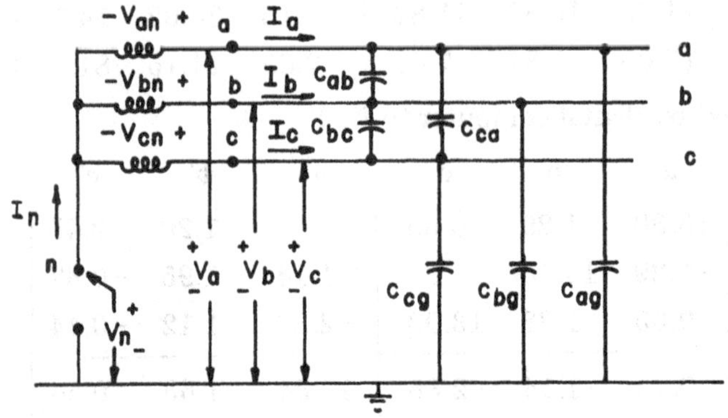

Fig. 5.10. Transmission line supplied from a Y-connected transformer.

wires and capacitance to ground. The neutral connection may be closed (grounded) or open, but in many modern systems it is grounded. The unbalance factor is defined differently for each connection, i.e., for the system neutral either grounded or ungrounded.

The system shown in Figure 5.10 is conveniently defined electrostatically by (5.27), i.e., $\mathbf{V}_{abc} = \mathbf{P}\mathbf{q}_{abc}$, where we assume that the effect of ground wires is included according to the method discussed in section 5.6. Then the charging currents flowing at no load are, from (5.60),

$$\mathbf{I}_{abc} = j\omega\mathbf{C}\mathbf{V}_{abc} = j\mathbf{B}\mathbf{V}_{abc} \tag{5.90}$$

where $\mathbf{B}$ is the shunt susceptance matrix. Also from (5.62)

$$\mathbf{I}_{012} = j\mathbf{B}_{012}\mathbf{V}_{012} \tag{5.91}$$

where

$$\mathbf{B}_{012} = \mathbf{A}^{-1}\mathbf{B}\mathbf{A} = \mathbf{A}^{-1}(\omega\mathbf{C})\mathbf{A} \tag{5.92}$$

Since the applied transformer voltages are balanced, positive sequence voltages, we write

$$\mathbf{V} = \begin{bmatrix} V_{an} \\ V_{bn} \\ V_{cn} \end{bmatrix} = \begin{bmatrix} V_a - V_n \\ V_b - V_n \\ V_c - V_n \end{bmatrix} = \mathbf{V}_{abc} - \mathbf{V}_n$$

then $\mathbf{V}_{abc} = \mathbf{V} + \mathbf{V}_n$ and

$$\mathbf{V}_{012} = \mathbf{A}^{-1}\mathbf{V}_{abc} = \mathbf{A}^{-1}(\mathbf{V} + \mathbf{V}_n) \tag{5.93}$$

where $\mathbf{V}$ is strictly positive sequence. Expanding (5.93), we compute

$$\mathbf{V}_{012} = \begin{bmatrix} V_n \\ V_{an} \\ 0 \end{bmatrix} \tag{5.94}$$

or V_n is a zero sequence voltage, V_{an} is positive sequence, and $V_{a2} = 0$. Then (5.91) may be written as

$$\begin{aligned} I_{a0} &= \mathrm{j}(B_{00} V_{a0} + B_{01} V_{a1}) \\ I_{a1} &= \mathrm{j}(B_{10} V_{a0} + B_{11} V_{a1}) \\ I_{a2} &= \mathrm{j}(B_{20} V_{a0} + B_{21} V_{a1}) \end{aligned} \tag{5.95}$$

Neutral ungrounded. If the system neutral is not grounded, the neutral voltage V_n will usually be nonzero and the neutral current will be zero, or $I_{a0} = 0$. Then from (5.95), $B_{00} V_{a0} + B_{01} V_{a1} = 0$, and we define the neutral "displacement" or unbalance as

$$d_0 = V_{a0}/V_{a1} = -B_{01}/B_{00} = -c_{01}/c_{00} \tag{5.96}$$

From equation (5.64) we also write

$$d_0 = \frac{-(c_{S2} + c_{M2})}{c_{S0} - 2c_{M0}} \tag{5.97}$$

Also from (5.58)

$$d_0 = \frac{-(c_{S2} + c_{M2})}{c_{g0}} \tag{5.98}$$

or, writing the numerator in terms of capacitances to ground,

$$d_0 = \frac{c_{ag} + \mathrm{a}^2 c_{bg} + \mathrm{a} c_{cg}}{c_{ag} + c_{bg} + c_{cg}} = \frac{c_{ag} + \mathrm{a}^2 c_{bg} + \mathrm{a} c_{cg}}{3c_{g0}} \tag{5.99}$$

Neutral grounded. If the system neutral is grounded, $V_n = V_{a0} = 0$ and we write from (5.95),

$$I_{a0} = \mathrm{j}B_{01} V_{a1}, \quad I_{a1} = \mathrm{j}B_{11} V_{a1}, \quad I_{a2} = \mathrm{j}B_{21} V_{a1} \tag{5.100}$$

In this case we define the displacement or unbalance as

$$d_0 = I_{a0}/I_{a1} = B_{01}/B_{11} = c_{01}/c_{11} \tag{5.101}$$

Then from (5.64)

$$\begin{aligned} d_0 &= \frac{c_{S2} + c_{M2}}{c_{S0} + c_{M0}} = \frac{c_{S2} + c_{M2}}{c_{g0} + 3c_{M0}} \\ &= \frac{c_{ag} + \mathrm{a}^2 c_{bg} + \mathrm{a} c_{cg}}{(c_{ag} + c_{bg} + c_{cg}) + 3(c_{ab} + c_{bc} + c_{ca})} = \frac{c_{ag} + \mathrm{a}^2 c_{bg} + \mathrm{a} c_{cg}}{3(c_{g0} + 3c_{M0})} \end{aligned} \tag{5.102}$$

This expression may be simplified to neglect the capacitance between conductors since they are considerably smaller than the capacitance to ground. If this is done, we write

$$d_0 \cong \frac{c_{ag} + \mathrm{a}^2 c_{bg} + \mathrm{a} c_{cg}}{3c_{g0}} \tag{5.103}$$

which is exactly the same as (5.99). Thus we have a convenient expression for the electrostatic unbalance which is independent of the system neutral grounding.

Example 5.6

Compute the displacement or unbalance of the lower circuit of Figure 5.2, using computed values of Example 5.4 where possible, where we now assume the line to be untransposed.

Solution

From Example 5.4 we have

$$c_{g0} = 9.545 \text{ nF/mi} \qquad c_{ab} = c_{bc} = 2.106 \text{ nF/mi}$$
$$c_{M0} = 1.683 \text{ nF/mi} \qquad c_{ac} = 0.839 \text{ nF/mi}$$
$$c_{S0} = 12.912 \text{ nF/mi} \qquad c_{ag} = c_{cg} = 9.802 \text{ nF/mi}$$
$$c_{bg} = 9.030 \text{ nF/mi}$$

Then

$$d_0 \cong \frac{c_{ag} + a^2 c_{bg} + a c_{cg}}{3c_{g0}} = \frac{(1+a)\, c_{ag} + a^2 c_{bg}}{3c_{g0}} = -\frac{a^2 (c_{ag} - c_{bg})}{3c_{g0}}$$
$$= -\frac{a^2 (0.772)}{3(9.545)} = 0.0269\underline{/-60^\circ} \text{ or } 2.69\%$$

If the system had a grounded neutral and the capacitance between phases is not neglected, we compute from (5.102)

$$d_0 = -\frac{a^2 (c_{ag} - c_{bg})}{3(9.545 + 5.049)} = -\frac{a^2 (0.772)}{3(14.594)} = 0.0177\underline{/-60^\circ} \text{ or } 1.77\%$$

Obviously, in this case the use of the approximate equation gives a very pessimistic result.

Reference [34] gives examples of similar computations using (5.99) with various wire sizes, spacing, and conductor heights and also shows the effect of ground wires. The results may be summarized as follows:

1. Electrostatic unbalance may be reduced by the addition of ground wires and by increasing the spacing between wires.
2. Electrostatic unbalance may be reduced by changing the arrangement of phase and ground wires, e.g., by lowering the middle conductor of a flat configuration or by arranging the wires *a-c-b*, rather than *a-b-c*, in a vertical configuration.

Note that from (5.95) we may also define a negative sequence unbalance. In the case of the grounded neutral system we write

$$d_2 = I_{a2}/I_{a1} = B_{21}/B_{11} = -\, c_{21}/c_{11}$$

or

$$d_2 = \frac{c_{M1} - c_{S1}}{c_{S0} + c_{M0}} \tag{5.104}$$

In the case of flat horizontal spacing with wire b in the center, this reduces to

$$d_2 = \frac{a(c_{bb} - c_{aa} + c_{bc} - c_{ac})}{3(c_{S0} + c_{M0})} \tag{5.105}$$

This unbalance factor is small and is usually ignored.

Problems

5.1. Compute the positive and negative sequence capacitance to neutral for the line configuration indicated below. Then compute the 60 Hz susceptance and the charging kVA per mile, assuming the line to operate at the nominal voltage indicated (neglect the effect of conductor height).
(a) Configuration of Figure P4.1, 34.5 kV.
(b) Configuration of Figure P4.2, 34.5 kV.
(c) Configuration of Figure P4.3, 69 kV.
(d) Configuration of Figure P4.4, 69 kV.
(e) Configuration of Figure P4.5, 161 kV.

5.2. Derive the equivalent circuit for self and mutual capacitances shown in Figure 5.6 by algebraic manipulation of (5.47).

5.3. Compute the change in capacitance in Example 5.1 if the height above the ground is considered.

5.4. Verify (5.57)-(5.59) by using (5.56) as a starting point.

5.5. Show that the total capacitance to neutral in the positive sequence network may be computed by converting the mutual capacitance Δ to a Y and adding the per phase capacitance to c_{g0} (see section 5.4).

5.6. Compute the positive and negative sequence capacitance for the circuit of Figure 4.6, using the method of section 5.1.

5.7. Compute the zero sequence capacitance for the circuits of Figure 4.6, using the methods of section 5.2.

5.8. Repeat the computation of positive and negative sequence capacitance of problem 5.6, this time taking into account the height of the conductor above the ground.

5.9. Examine the circuit of Figure 4.6 and compute the following (neglecting ground wires).
(a) The matrix **P** of potential coefficients.
(b) The matrix **C** of Maxwell coefficients. (Use of a digital computer is recommended for this step, but manual methods may be used.)
(c) The matrix $\mathbf{C}_{012}$ of sequence capacitances for a transposed line.

5.10. Repeat problem 5.9 for the circuit of Figure 4.18. Assume wire height is 70 ft.

5.11. Repeat problem 5.9 for the upper circuit of Figure 5.2, ignoring ground wires.

5.12. Examine the circuit of Figure 4.6 and compute the following, including the effect of ground wires.
(a) The matrix **P** of potential coefficients.
(b) The matrix **C** of Maxwell coefficients.
(c) The matrix $\mathbf{C}_{012}$ of sequence capacitances for a transposed line.

5.13. Repeat problem 5.12 for the circuit in Figure 4.18. Assume the phase wires to be 70 ft above the ground.

5.14. Repeat problem 5.12 for the upper circuit in Figure 5.2, including the effect of ground wires.

5.15. Repeat the computations of Example 5.4 by inverting the 5 × 5 **P** matrix to obtain a new 5 × 5 **C** matrix. Explain the meaning of each term of **C** and label these capacitances on a sketch.

5.16. Repeat the computations of Example 5.5, omitting the effect of the ground wires. Compare results of the two computations and justify the change in capacitance by physical reasoning.

5.17. Compute the capacitance (*a-b-c* coordinate system) for a double circuit line consisting of two identical lines like that of Figure 4.6 and separated by a distance of 25 ft, assuming the two circuits operate in parallel at 66 kV. Suggestion: Set up the **P** matrix by hand computation but use a digital computer to invert the matrices (see Appendix A).

5.18. Compute the electrostatic unbalance factor d_0 for the upper circuit of Figure 5.2.
(a) Neglecting the ground wires.
(b) Including the effect of the ground wires.

chapter 6

Sequence Impedance of Machines

An important problem in the determination of sequence impedances of a power system is concerned with machines. This problem is especially difficult since machines are complex devices to describe mathematically, requiring that many assumptions must be made in deriving expressions for impedances. For example, the speed, degree of saturation, linearity of the magnetic circuit, and other phenomena must be considered. Our discussion here is divided into two parts; synchronous machines and induction machines. In these devices the several circuits are coupled inductively and are therefore related by differential equations. Having established the appropriate equations, however, we will immediately assume that the load is constant but unbalanced. Thus we will again be concerned with algebraic equations, and phasor notation will be used. This treatment should not be considered exhaustive by any means, and the interested reader should consult the many excellent books on the subject.[1]

I. SYNCHRONOUS MACHINE IMPEDANCES

6.1 General Considerations

A synchronous machine is sometimes called a "dynamic circuit" because it consists of circuits which are moving with respect to each other, and therefore the impedance seen by currents entering or leaving the terminals is continually changing. There are several complications here. First, there is the problem of changing flux linkages in circuits where the mutual inductances change with time (i.e., with changing rotor position). There is also the problem of dc offset when a fault occurs. This is due to the shift in the ac envelope required, since there can be no discontinuity in the current wave of an inductive circuit. Since the normal (prefault) currents differ in phase by 120°, each phase current experiences a different dc offset. The concept of constant flux linkages over the period from just prior until just after the fault also requires certain fast reactions in the coupled circuits which generate large but rapidly decaying alternating currents whose magnitudes must be estimated. There is also the consideration of the machine speed. Since the machine sees a faulted condition, the load (active power) that it can deliver is changed suddenly. The prime mover requires a finite time to sense this change in load, so the rotating mass responds by changing its speed and allowing energy to be taken from or supplied to its inertia. Thus a suddenly ap-

[1] For example, see [10, 11, 14, 19, 39, 44 and 45] and examine the excellent list of references given by Kimbark [19].

plied fault sets up a dynamic response in a machine. This response must be estimated to permit computation of fault currents. All these questions require elaboration.

6.1.1 Machine dynamics

First, consider the problem of the change in speed of the generator due to a sudden change in load. Consider a single machine which is supplying a passive load when a three-phase fault is applied at its terminals. Since the voltage of all three phases becomes zero, the three-phase electrical power leaving the machine suddenly becomes zero. But the input power supplied by the prime mover is the same as before the fault. Thus all the input (mechanical) power is available to accelerate the machine.

Note there can be no discontinuity in the angle of the machine rotor. If this angle is θ, write

$$\theta = \omega_1 t + \delta + \pi/2 \tag{6.1}$$

where ω_1 is the synchronous angular frequency, δ is the torque angle, and the constant $\pi/2$ is added to conform with the usual convention, as noted later. Thus the angle advances linearly with time up to the time t_0 when the fault is applied. During this prefault period the speed of the machine is a constant, and the torque angle δ is a constant or

$$\dot{\theta} = \omega_1 \quad \text{rad/sec} \qquad t < t_0 \tag{6.2}$$

After the fault occurs, the speed changes to a new value ω which is usually not constant,

$$\dot{\theta} \triangleq \omega = \omega_1 + \dot{\delta} \qquad t > t_0 \tag{6.3}$$

and the angle advances at a new rate given by (6.1) as the shaft accelerates at a rate

$$\ddot{\theta} = \dot{\omega} = \ddot{\delta} \tag{6.4}$$

This problem of solving the differential equations of a machine following a disturbance, even a balanced three-phase disturbance, is a formidable problem in itself and involves the solution of differential equations of the machine, the prime mover, and its control system. As explained in Chapter 1, our goal is to simplify the solution of a faulted system to an assumed steady state condition such that algebraic equations involving phasor quantities may be used. The application of a fault near a machine is obviously *not* a steady state condition and requires rationalization.

Actually the *severity* of the fault is the key to this problem. If the fault is not severe but is an unbalanced load or other permanent condition, we compute the fault voltages and currents, using phasor notation, *after* all transients have died away and the system is in a steady (ω = a constant) condition. However, if the fault is severe (such as a short circuit), we assume the fault will be removed *before* the frequency has changed appreciably. We will, however, include all known machine circuit responses required to maintain constant flux linkages over the discontinuity. Thus we will solve a fictitious circuit problem in which we replace the generator by a Thevenin equivalent wherein both the voltage and impedance are arbitrary quantities intended to represent the (worst) condition immediately after the fault occurs. We then proceed with an algebraic solution.

The assumption that this procedure will give usable results to compute the settings of relays has been established through years of experience.

6.1.2 Direct current

At the instant a fault occurs, the generator currents change to new values which depend on the new value of impedance seen at the generator terminals. However, since the circuit is inductive, there cannot be a discontinuity in the current. That is, the current just prior to t_0 equals the current just after t_0 in each phase or, mathematically,

$$\mathbf{i}_{abc}(t_0^-) = \mathbf{i}_{abc}(t_0^+) \tag{6.5}$$

The current just after t_0 is composed of two components, a dc component which dies out exponentially and an ac component. The rms value of the ac component after the fault is different from that before the fault.

The exact amount of dc offset depends on the exact time in the current (or voltage) wave at which the disturbance appears and on the angle of the impedance seen by the generator. A typical offset for the three phase currents is shown in Figure 6.1 where the response to a 3ϕ fault is illustrated. Note that the sum of the dc components in the three phases is zero. Kimbark [19] shows that the amount of the dc offset can be found by taking $\sqrt{2}$ times the projection of the negative of the phasor fault currents on the real axis in the complex plane (see problem 6.2).

6.1.3 Initial value of fault currents—the flux linkage equations

As seen in Figure 6.1, the initial value of the fault current has both a dc and an ac component, with the dc component decaying to zero in a short time and the

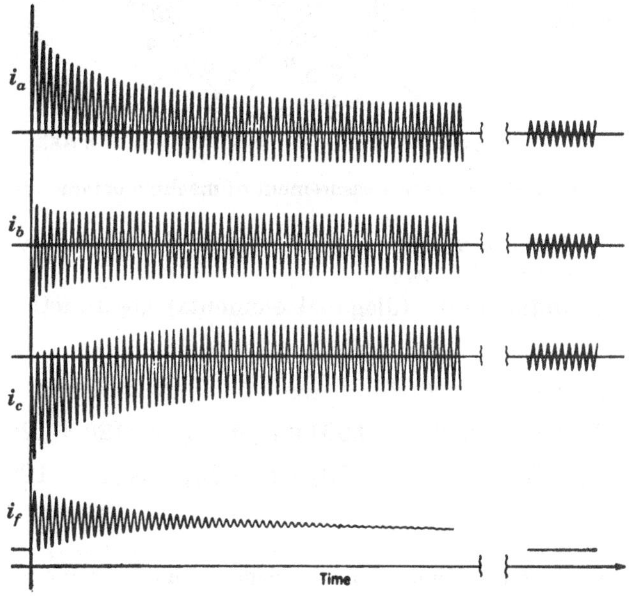

Fig. 6.1. Short circuit currents of a synchronous generator. Armature currents i_a, i_b, and i_c; field current i_f. (From Kimbark [19]. Used with permission.)

ac component decaying slowly to a much lower steady state value. This phenomenon can be explained by the principle of constant flux linkages (see [19]). Assume that prior to the fault the generator field is energized but the machine is unloaded (i = 0). It helps to visualize this situation as that of six separate but mutually coupled circuits consisting of three phase windings (a, b, c), a field winding (F), and two equivalent damper windings (D and Q) for which we may write the flux linkage equation[2]

$$\begin{bmatrix} \lambda_a \\ \lambda_b \\ \lambda_c \\ \hline \lambda_F \\ \lambda_D \\ \lambda_Q \end{bmatrix} = \left[\begin{array}{ccc|ccc} L_{aa} & L_{ab} & L_{ac} & L_{aF} & L_{aD} & L_{aQ} \\ L_{ba} & L_{bb} & L_{bc} & L_{bF} & L_{bD} & L_{bQ} \\ L_{ca} & L_{cb} & L_{cc} & L_{cF} & L_{cD} & L_{cQ} \\ \hline L_{Fa} & L_{Fb} & L_{Fc} & L_{FF} & L_{FD} & L_{FQ} \\ L_{Da} & L_{Db} & L_{Dc} & L_{DF} & L_{DD} & L_{DQ} \\ L_{Qa} & L_{Qb} & L_{Qc} & L_{QF} & L_{QD} & L_{QQ} \end{array}\right] \begin{bmatrix} i_a \\ i_b \\ i_c \\ \hline i_F \\ i_D \\ i_Q \end{bmatrix} \text{ Wb turns} \quad (6.6)$$

The notation adopted here is to use lowercase subscripts for stator quantities and uppercase subscripts for rotor quantities. Most of the elements of the inductance matrix are functions of the rotor angle θ. These inductances may be computed by carefully examining Figure 6.2 which has been suggested by the IEEE as the

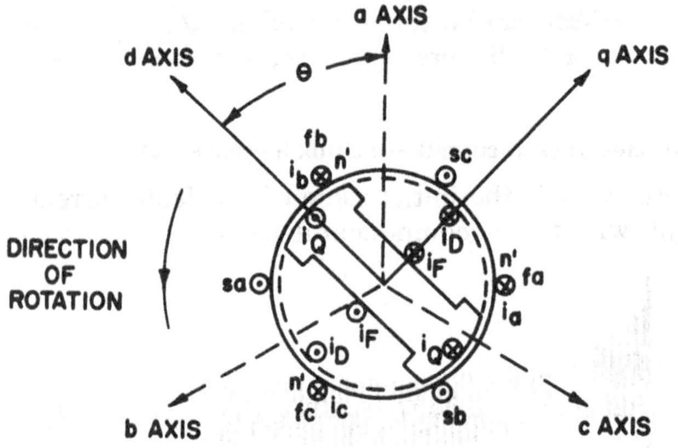

Fig. 6.2. Reference for measurement of machine parameters.

standard definition for the several physical parameters involved. These inductances may be computed as follows.

The *stator self inductances* (diagonal elements) are functions of twice the angle θ.[3]

$$\begin{aligned} L_{aa} &= L_s + L_m \cos 2\theta \text{ H} \\ L_{bb} &= L_s + L_m \cos 2(\theta - 120) = L_s + L_m \cos(2\theta + 120) \text{ H} \\ L_{cc} &= L_s + L_m \cos 2(\theta + 120) = L_s + L_m \cos(2\theta - 120) \text{ H} \end{aligned} \quad (6.7)$$

where $L_s > L_m$.

[2] We use the symbol λ for flux linkage in accordance with the American National Standard, ANSI Y10.5, 1968.

[3] Here we adopt the useful convention of designating a *constant* self or mutual inductance by a single subscript.

The *stator-to-stator mutual inductances* are also functions of 2θ.

$$\begin{aligned} L_{ab} = L_{ba} &= -M_s - L_m \cos 2(\theta + 30) = -M_s + L_m \cos(2\theta - 120) \text{ H} \\ L_{bc} = L_{cb} &= -M_s - L_m \cos 2(\theta - 90) = -M_s + L_m \cos 2\theta \text{ H} \\ L_{ca} = L_{ac} &= -M_s - L_m \cos 2(\theta + 150) = -M_s + L_m \cos(2\theta + 120) \text{ H} \end{aligned} \tag{6.8}$$

where $|M_s| > L_m$.

The *rotor self inductances* are all constants, so we redefine these quantities to have a single subscript.

$$\begin{aligned} L_{FF} &= L_F \text{ H} \\ L_{DD} &= L_D \text{ H} \\ L_{QQ} &= L_Q \text{ H} \end{aligned} \tag{6.9}$$

The *rotor mutual inductances* are also constants.

$$\begin{aligned} L_{FD} &= L_{DF} = M_R \text{ H} \\ L_{FQ} &= L_{QF} = 0 \text{ H} \\ L_{DQ} &= L_{QD} = 0 \text{ H} \end{aligned} \tag{6.10}$$

Finally, the *stator-to-rotor mutual inductances* are functions of the rotor position θ.

$$\begin{aligned} L_{aF} &= L_{Fa} = M_F \cos\theta \text{ H} \\ L_{bF} &= L_{Fb} = M_F \cos(\theta - 120) \text{ H} \\ L_{cF} &= L_{Fc} = M_F \cos(\theta + 120) \text{ H} \end{aligned} \tag{6.11}$$

$$\begin{aligned} L_{aD} &= L_{Da} = M_D \cos\theta \text{ H} \\ L_{bD} &= L_{Db} = M_D \cos(\theta - 120) \text{ H} \\ L_{cD} &= L_{Dc} = M_D \cos(\theta + 120) \text{ H} \end{aligned} \tag{6.12}$$

$$\begin{aligned} L_{aQ} &= L_{Qa} = M_Q \sin\theta \text{ H} \\ L_{bQ} &= L_{Qb} = M_Q \sin(\theta - 120) \text{ H} \\ L_{cQ} &= L_{Qc} = M_Q \sin(\theta + 120) \text{ H} \end{aligned} \tag{6.13}$$

Now consider the problem of a generator with negligible load currents (compared to the fault currents) which is to be faulted symmetrically on all three phases at $t = 0$. At $t = 0^-$ the currents are

$$i_a = i_b = i_c \cong 0, \qquad i_D = i_Q = 0$$

and the flux linkages are computed to be

$$\begin{bmatrix} \lambda_a \\ \lambda_b \\ \lambda_c \\ \lambda_F \\ \lambda_D \\ \lambda_Q \end{bmatrix} = \begin{bmatrix} L_{aF} \\ L_{bF} \\ L_{cF} \\ L_{FF} \\ L_{DF} \\ L_{QF} \end{bmatrix} i_F = \begin{bmatrix} M_F \cos\theta \\ M_F \cos(\theta - 120) \\ M_F \cos(\theta + 120) \\ L_F \\ M_R \\ 0 \end{bmatrix} i_F \tag{6.14}$$

But at $t = 0$, $\theta = \delta_0 + \pi/2$, and the flux linkages are functions of the torque angle.

Consider now the sequence of events associated with a three-phase fault on the unloaded generator. Before the fault occurs, the field with flux linkages λ_F produces an air gap flux ϕ_{ag}, leaving the N pole of the field and entering the armature, thereby establishing an armature S pole which moves with respect to the armature and produces time varying flux linkages expressed by (6.6). At the instant $t = 0$, the fault is applied, at which time the flux linkages are given by (6.14) and (by the principle of constant flux linkages) must remain at this value (at least for an instant). Thus exactly the same flux ϕ_{ag} which entered the armature S pole at $t = 0^-$ continues to do so, and this S pole remains fixed at the exact (stationary) location it occupied at $t = 0^-$, even though the field winding continues to rotate. The field N pole similarly continues to produce an air gap flux ϕ_{ag} which emerges from the field winding as before and is fixed with respect to the field winding. Obviously, similar statements could be made concerning the field S pole and armature N pole but they will be omitted here.

We summarize the flux condition at the time of the fault as follows:

1. At the field N pole—(a) ϕ_{ag} = a constant, leaving N, and (b) N fixed with respect to the field winding.
2. At the armature S pole—(a) ϕ_{ag} = a constant, entering S, and (b) S fixed with respect to the armature winding.

As Kimbark puts it, it is "as if the poles had stamped their images upon the armature" at $t = 0$.

Now both the armature and the field windings will react by inducing currents to maintain the flux linkages of (6.14) as described above. They do so as follows:

1. To maintain the stationary armature poles and force the flux ϕ_{ag} to enter S as described requires a dc component of current flow in the armature circuits as shown in Figure 6.1, with a different dc magnitude in each phase winding, depending on the rotor position θ corresponding to $t = 0$.
2. To counteract the production of armature flux linkages due to the spinning rotor field requires that alternating currents flow in the armature windings. These currents are positive sequence armature currents which are of a magnitude sufficient to hold the armature flux linkages at the prefault value specified by (6.14). The MMF produced by these currents rotates with synchronous speed, is stationary with respect to the field winding, and opposes the field MMF. The field winding reacts in turn with an increased current, the two forces balancing each other such that the flux linkages remain constant.
3. The stationary armature field, viewed from the field winding, appears as an alternating field and induces an alternating current in the field winding. Such an alternating current produces a pulsating MMF wave which is stationary with respect to the rotor. If one thinks of this pulsating wave as being composed of two moving MMF waves, one going forward and one backward, the backward wave will be such as to oppose the stationary armature field. The forward wave moves at twice synchronous speed with respect to the armature and induces a second harmonic current in the armature circuit. In machines with damper windings this second harmonic induction is small.

If the flux linkages of the machine were to be constant for all time, the situation would remain exactly as just described. However, as noted in Figure 6.1, the induced currents decay to new, lower values at different time constants. These time constants will be described in more detail in section 6.5.

6.2 Positive Sequence Impedance

The occurrence of a fault on a synchronous machine causes many different responses in the machine windings, each circuit responding to physical laws in known ways. The result is a response which is difficult to express mathematically so that sequence impedances can be defined in the usual way. For passive networks we write

$$\mathbf{Z}_{012} = \mathbf{A}^{-1}\mathbf{Z}_{abc}\mathbf{A} \tag{6.15}$$

where we assume that the elements of $\mathbf{Z}_{abc}$ are complex numbers and the problem is stated algebraically in phasor notation. This is not possible for a synchronous machine, as will be shown. Instead, we usually work problems either in the time domain, as in the case of stability problems, or make certain bold assumptions and use the phasor domain with the impedances assumed constant. This practice results in a whole family of positive sequence impedances, depending upon the exact condition under study, but only one negative and one zero sequence impedance. Different time constants are associated with the different impedances. Since all these quantities are in common use, their definitions are reviewed here. For a more thorough treatment see Kimbark [19] and Prentice [46].

6.2.1 Park's transformation

It has been shown that the equations for a synchronous machine can be greatly simplified if all variables are transformed to a new coordinate system defined by the transformation

$$\mathbf{i}_{0dq} \triangleq \mathbf{P}\,\mathbf{i}_{abc} \tag{6.16}$$

where

$$\mathbf{P} = \sqrt{\frac{2}{3}}\begin{bmatrix} 1/\sqrt{2} & 1/\sqrt{2} & 1/\sqrt{2} \\ \cos\theta & \cos(\theta - 120) & \cos(\theta + 120) \\ \sin\theta & \sin(\theta - 120) & \sin(\theta + 120) \end{bmatrix} \tag{6.17}$$

We call this transformation the *d-q* transformation or Park's transformation, named for its early proponent R. H. Park [47, 48]. The transformation (6.17) is different from the one used by Park in the constant $\sqrt{2/3}$, used to make **P** orthogonal.

When this transformation is used, the current i_d may be regarded as the current in a fictitious armature winding which rotates with the field winding and with its MMF axis aligned with the field *d* axis. The MMF thus produced is the same as that produced by the actual phase currents flowing in their actual armature windings. The *q* axis current i_q is similarly interpreted except that its fictitious winding is aligned with the *q* axis (see Fig. 6.2). The third component (usually called i_0) is really i_{a0}, the zero sequence current, expressed in instantaneous rather than phasor form.

It is not difficult to show that the transformation **P** is unique and that its inverse is given by

$$\mathbf{i}_{abc} = \mathbf{P}^{-1}\,\mathbf{i}_{0dq} \tag{6.18}$$

where

$$\mathbf{P}^{-1} = \mathbf{P}^{t} = \sqrt{\frac{2}{3}}\begin{bmatrix} 1/\sqrt{2} & \cos\theta & \sin\theta \\ 1/\sqrt{2} & \cos(\theta - 120) & \sin(\theta - 120) \\ 1/\sqrt{2} & \cos(\theta + 120) & \sin(\theta + 120) \end{bmatrix} \tag{6.19}$$

Park's transformation is used to simplify the usual expressions for the phase voltages of a synchronous machine as follows. From Figure 6.3 we write an equation for each phase-to-neutral voltage of a Y-connected synchronous generator in the form $v = -ri - \dot{\lambda} + v_n$.

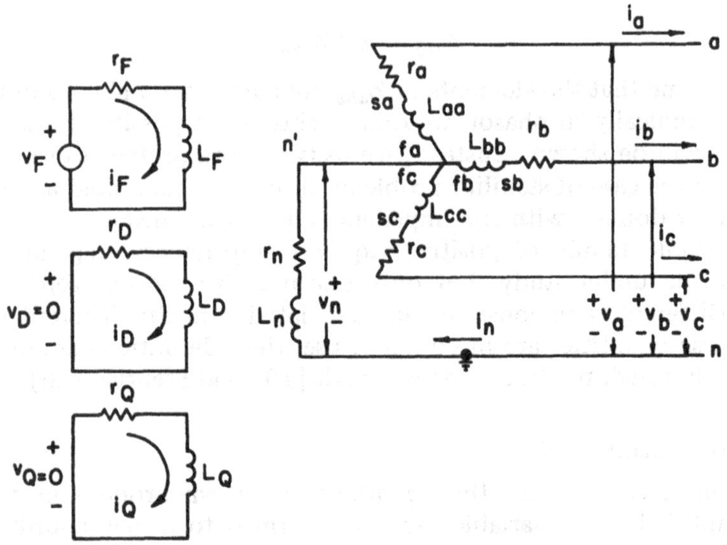

Fig. 6.3. Schematic diagram of a synchronous generator.

The voltage equation for the six coupled circuits of Figure 6.3 may be written in matrix form as follows, with $r_a = r_b = r_c = r$.

$$\begin{array}{l} \text{stator} \\ \\ \text{rotor} \end{array}\left[\begin{array}{c} v_a \\ v_b \\ v_c \\ \hline -v_F \\ -v_D = 0 \\ -v_Q = 0 \end{array}\right] = -\left[\begin{array}{ccc|ccc} r & 0 & 0 & & & \\ 0 & r & 0 & & 0 & \\ 0 & 0 & r & & & \\ \hline & & & r_F & 0 & 0 \\ & 0 & & 0 & r_D & 0 \\ & & & 0 & 0 & r_Q \end{array}\right]\left[\begin{array}{c} i_a \\ i_b \\ i_c \\ \hline i_F \\ i_D \\ i_Q \end{array}\right] - \left[\begin{array}{c} \dot{\lambda}_a \\ \dot{\lambda}_b \\ \dot{\lambda}_c \\ \hline \dot{\lambda}_F \\ \dot{\lambda}_D \\ \dot{\lambda}_Q \end{array}\right] + \left[\begin{array}{c} \mathbf{v}_n \\ \hline \mathbf{0} \end{array}\right] \tag{6.20}$$

where

$$\mathbf{v}_n = -r_n\begin{bmatrix} 1 & 1 & 1 \\ 1 & 1 & 1 \\ 1 & 1 & 1 \end{bmatrix}\begin{bmatrix} i_a \\ i_b \\ i_c \end{bmatrix} - L_n\begin{bmatrix} 1 & 1 & 1 \\ 1 & 1 & 1 \\ 1 & 1 & 1 \end{bmatrix}\begin{bmatrix} \dot{i}_a \\ \dot{i}_b \\ \dot{i}_c \end{bmatrix} \tag{6.21}$$

$$\triangleq -\mathbf{R}_n\mathbf{i}_{abc} - \mathbf{L}_n\dot{\mathbf{i}}_{abc}$$

This equation could be in volts or in pu. If we define

$$\mathbf{v}_{FDQ} = \begin{bmatrix} -v_F \\ 0 \\ 0 \end{bmatrix}, \quad \mathbf{R}_S = r\,\mathbf{U}, \quad \mathbf{R}_R = \begin{bmatrix} r_F & 0 & 0 \\ 0 & r_D & 0 \\ 0 & 0 & r_Q \end{bmatrix} \tag{6.22}$$

then (6.20) may be written in matrix form as

$$\begin{bmatrix} \mathbf{v}_{abc} \\ \mathbf{v}_{FDQ} \end{bmatrix} = -\begin{bmatrix} \mathbf{R}_S & \mathbf{0} \\ \mathbf{0} & \mathbf{R}_R \end{bmatrix} \begin{bmatrix} \mathbf{i}_{abc} \\ \mathbf{i}_{FDQ} \end{bmatrix} - \begin{bmatrix} \dot{\boldsymbol{\lambda}}_{abc} \\ \dot{\boldsymbol{\lambda}}_{FDQ} \end{bmatrix} + \begin{bmatrix} \mathbf{v}_n \\ \mathbf{0} \end{bmatrix} \tag{6.23}$$

It is convenient to transform the *a-b-c* partition of (6.23) to the 0-*d*-*q* frame of reference by the transformation **P** as in (6.16). This greatly simplifies the flux linkage equation (6.6) since it removes all time varying inductances. To do this we premultiply both sides of (6.23) by the transformation

$$\begin{bmatrix} \mathbf{P} & \mathbf{0} \\ \mathbf{0} & \mathbf{U} \end{bmatrix}$$

which by definition transforms the left side to 0-*d*-*q* voltages. Thus

$$\begin{bmatrix} \mathbf{P} & \mathbf{0} \\ \mathbf{0} & \mathbf{U} \end{bmatrix} \begin{bmatrix} \mathbf{v}_{abc} \\ \mathbf{v}_{FDQ} \end{bmatrix} = \begin{bmatrix} \mathbf{v}_{0dq} \\ \mathbf{v}_{FDQ} \end{bmatrix} \tag{6.24}$$

The resistance term becomes

$$\begin{bmatrix} \mathbf{P} & \mathbf{0} \\ \mathbf{0} & \mathbf{U} \end{bmatrix} \begin{bmatrix} \mathbf{R}_S & \mathbf{0} \\ \mathbf{0} & \mathbf{R}_R \end{bmatrix} \begin{bmatrix} \mathbf{i}_{abc} \\ \mathbf{i}_{FDQ} \end{bmatrix} = \begin{bmatrix} \mathbf{R}_S & \mathbf{0} \\ \mathbf{0} & \mathbf{R}_R \end{bmatrix} \begin{bmatrix} \mathbf{i}_{0dq} \\ \mathbf{i}_{FDQ} \end{bmatrix} \tag{6.25}$$

The flux linkage term requires careful study. Applying the transformation, we write

$$\begin{bmatrix} \mathbf{P} & \mathbf{0} \\ \mathbf{0} & \mathbf{U} \end{bmatrix} \begin{bmatrix} \dot{\boldsymbol{\lambda}}_{abc} \\ \dot{\boldsymbol{\lambda}}_{FDQ} \end{bmatrix} = \begin{bmatrix} \mathbf{P}\dot{\boldsymbol{\lambda}}_{abc} \\ \dot{\boldsymbol{\lambda}}_{FDQ} \end{bmatrix} \tag{6.26}$$

We can evaluate $\mathbf{P}\dot{\boldsymbol{\lambda}}_{abc}$ as follows. From the definition (6.16) we write $\boldsymbol{\lambda}_{0dq} = \mathbf{P}\boldsymbol{\lambda}_{abc}$. Then, taking the derivative, $\dot{\boldsymbol{\lambda}}_{0dq} = \dot{\mathbf{P}}\boldsymbol{\lambda}_{abc} + \mathbf{P}\dot{\boldsymbol{\lambda}}_{abc}$ from which we find

$$\mathbf{P}\dot{\boldsymbol{\lambda}}_{abc} = \dot{\boldsymbol{\lambda}}_{0dq} - \dot{\mathbf{P}}\boldsymbol{\lambda}_{abc} = \dot{\boldsymbol{\lambda}}_{0dq} - \dot{\mathbf{P}}\mathbf{P}^{-1}\boldsymbol{\lambda}_{0dq} \tag{6.27}$$

We easily show that

$$\dot{\mathbf{P}}\mathbf{P}^{-1} = \begin{bmatrix} 0 & 0 & 0 \\ 0 & 0 & -\omega \\ 0 & \omega & 0 \end{bmatrix}$$

so that the last term of (6.27), which we shall call **s**, becomes

$$\mathbf{s} \triangleq \dot{\mathbf{P}}\mathbf{P}^{-1}\boldsymbol{\lambda}_{0dq} = \begin{bmatrix} 0 \\ -\omega\lambda_q \\ +\omega\lambda_d \end{bmatrix} \tag{6.28}$$

and this is recognized to be a speed voltage term.

The last term of (6.23) is transformed as follows.

$$\begin{bmatrix} \mathbf{P} & \mathbf{0} \\ \mathbf{0} & \mathbf{U} \end{bmatrix} \begin{bmatrix} \mathbf{v}_n \\ \mathbf{0} \end{bmatrix} = \begin{bmatrix} \mathbf{P}\,\mathbf{v}_n \\ \mathbf{0} \end{bmatrix} \triangleq \begin{bmatrix} \mathbf{n}_{0dq} \\ \mathbf{0} \end{bmatrix}$$

where we have defined the voltage

$$\mathbf{n}_{0dq} \triangleq \mathbf{P}\,\mathbf{v}_n = -\begin{bmatrix} 3r_n i_0 \\ 0 \\ 0 \end{bmatrix} - \begin{bmatrix} 3L_n \dot{i}_0 \\ 0 \\ 0 \end{bmatrix} \tag{6.29}$$

Substituting all transformed quantities into (6.23) we have the new voltage equations,

$$\begin{bmatrix} \mathbf{v}_{0dq} \\ \mathbf{v}_{FDQ} \end{bmatrix} = -\begin{bmatrix} \mathbf{R}_S & \mathbf{0} \\ \mathbf{0} & \mathbf{R}_R \end{bmatrix} \begin{bmatrix} \mathbf{i}_{0dq} \\ \mathbf{i}_{FDQ} \end{bmatrix} - \begin{bmatrix} \dot{\boldsymbol{\lambda}}_{0dq} \\ \dot{\boldsymbol{\lambda}}_{FDQ} \end{bmatrix} + \begin{bmatrix} \mathbf{s} \\ \mathbf{0} \end{bmatrix} + \begin{bmatrix} \mathbf{n}_{0dq} \\ \mathbf{0} \end{bmatrix} \tag{6.30}$$

We shall now show that this is much simpler than the original voltage equation which included many time varying coefficients (L_{aa}, L_{ab}, etc.).

It is convenient to think of the 0-*d*-*q* partition of (6.30) as the stator voltages referred to or seen from the rotor. Since fundamental frequency ac quantities in the stator appear to the rotor to be dc quantities, we would expect these voltages to be constants under steady state operation.

Because of the transformation of stator quantities to the rotor or 0-*d*-*q* quantities, two remarkable things happen. First, the inductances which were so complicated and time varying in (6.6) have been transformed into constants. Second, a speed voltage term s has appeared, which adds voltage components to v_d and v_q proportional to ω, the rotor angular velocity. Looking at this result another way, we have replaced a linear system with time varying coefficients by a nonlinear system (because of s) with constant coefficients. Since under most conditions the speed ω is nearly constant, the nonlinearity is of little concern. The important change is in the inductances.

If we examine the transformation of the flux linkage equation, we have the following. By definition of the P transformation,

$$\begin{bmatrix} \boldsymbol{\lambda}_{0dq} \\ \hline \boldsymbol{\lambda}_{FDQ} \end{bmatrix} = \left[\begin{array}{c|c} \mathbf{P} & \mathbf{0} \\ \hline \mathbf{0} & \mathbf{U} \end{array}\right] \begin{bmatrix} \boldsymbol{\lambda}_{abc} \\ \hline \boldsymbol{\lambda}_{FDQ} \end{bmatrix}$$

$$= \left[\begin{array}{c|c} \mathbf{P} & \mathbf{0} \\ \hline \mathbf{0} & \mathbf{U} \end{array}\right] \left[\begin{array}{c|c} \mathbf{L}_{abc} & \mathbf{L}_{aR} \\ \hline \mathbf{L}_{Ra} & \mathbf{L}_{RR} \end{array}\right] \left[\begin{array}{c|c} \mathbf{P} & \mathbf{0} \\ \hline \mathbf{0} & \mathbf{U} \end{array}\right]^{-1} \left[\begin{array}{c|c} \mathbf{P} & \mathbf{0} \\ \hline \mathbf{0} & \mathbf{U} \end{array}\right] \begin{bmatrix} \mathbf{i}_{abc} \\ \hline \mathbf{i}_{FDQ} \end{bmatrix}$$

where the partitions of the inductance matrix are defined in (6.6). Then

$$\begin{bmatrix} \boldsymbol{\lambda}_{0dq} \\ \hline \boldsymbol{\lambda}_{FDQ} \end{bmatrix} = \left[\begin{array}{c|c} \mathbf{P}\,\mathbf{L}_{abc}\,\mathbf{P}^{-1} & \mathbf{P}\,\mathbf{L}_{aR} \\ \hline \mathbf{L}_{Ra}\,\mathbf{P}^{-1} & \mathbf{L}_{RR} \end{array}\right] \begin{bmatrix} \mathbf{i}_{0dq} \\ \hline \mathbf{i}_{FDQ} \end{bmatrix} \tag{6.31}$$

By straightforward computation of the partitions of (6.31) we may show that

$$\mathbf{P}\,\mathbf{L}_{abc}\,\mathbf{P}^{-1} = \begin{bmatrix} L_0 & 0 & 0 \\ 0 & L_d & 0 \\ 0 & 0 & L_q \end{bmatrix} \triangleq \mathbf{L}_{0dq}$$

where

$$L_0 = L_s - 2M_s, \quad L_d = L_s + M_s + (3/2)\,L_m, \quad L_q = L_s + M_s - (3/2)\,L_m$$

and these inductances are all constants. We also compute

$$\mathbf{P\,L}_{aR} = \begin{bmatrix} 0 & 0 & 0 \\ \sqrt{\frac{3}{2}}M_F & \sqrt{\frac{3}{2}}M_D & 0 \\ 0 & 0 & \sqrt{\frac{3}{2}}M_Q \end{bmatrix} \triangleq \mathbf{L}_m$$

and

$$\mathbf{L}_{Ra}\mathbf{P}^{-1} = \begin{bmatrix} 0 & \sqrt{\frac{3}{2}}M_F & 0 \\ 0 & \sqrt{\frac{3}{2}}M_D & 0 \\ 0 & 0 & \sqrt{\frac{3}{2}}M_Q \end{bmatrix} = \mathbf{L}_m^t$$

Thus we may write the transformed flux linkage equation as

$$\begin{bmatrix} \lambda_0 \\ \lambda_d \\ \lambda_q \\ \hline \lambda_F \\ \lambda_D \\ \lambda_Q \end{bmatrix} = \left[\begin{array}{ccc|ccc} L_0 & 0 & 0 & 0 & 0 & 0 \\ 0 & L_d & 0 & \mathrm{k}M_F & \mathrm{k}M_D & 0 \\ 0 & 0 & L_q & 0 & 0 & \mathrm{k}M_Q \\ \hline 0 & \mathrm{k}M_F & 0 & L_F & M_R & 0 \\ 0 & \mathrm{k}M_D & 0 & M_R & L_D & 0 \\ 0 & 0 & \mathrm{k}M_Q & 0 & 0 & L_Q \end{array}\right] \begin{bmatrix} i_0 \\ i_d \\ i_q \\ \hline i_F \\ i_D \\ i_Q \end{bmatrix} \tag{6.32}$$

where for convenience we set $\mathrm{k} \triangleq \sqrt{3/2}$. Observe now that every element in this inductance matrix is constant. Furthermore, the λ_0 equation is completely uncoupled from the other equations and may be discarded when balanced conditions are under study. Since every inductance in (6.32) is constant, the time derivative of this equation, required in (6.30), is easily found.

If the voltage (6.30) is now written in expanded notation, it is instructive to see the way in which the transformed equations are coupled.

$$\begin{bmatrix} v_0 \\ v_d \\ v_q \\ \hline -v_F \\ 0 \\ 0 \end{bmatrix} = -\left[\begin{array}{ccc|ccc} r & 0 & 0 & & & \\ 0 & r & 0 & & \mathbf{0} & \\ 0 & 0 & r & & & \\ \hline & & & r_F & 0 & 0 \\ & \mathbf{0} & & 0 & r_D & 0 \\ & & & 0 & 0 & r_Q \end{array}\right] \begin{bmatrix} i_0 \\ i_d \\ i_q \\ \hline i_F \\ i_D \\ i_Q \end{bmatrix}$$

$$
-\left[\begin{array}{ccc|ccc}
L_0 & 0 & 0 & 0 & 0 & 0 \\
0 & L_d & 0 & kM_F & kM_D & 0 \\
0 & 0 & L_q & 0 & 0 & kM_Q \\
\hline
0 & kM_F & 0 & L_F & M_R & 0 \\
0 & kM_D & 0 & M_R & L_D & 0 \\
0 & 0 & kM_Q & 0 & 0 & L_Q
\end{array}\right]
\left[\begin{array}{c}
\dot{i}_0 \\ \dot{i}_d \\ \dot{i}_q \\ \hline \dot{i}_F \\ \dot{i}_D \\ \dot{i}_Q
\end{array}\right]
+\left[\begin{array}{c}
0 \\
-\omega(L_q i_q + kM_Q i_Q) \\
\omega(L_d i_d + kM_F i_F + kM_D i_D) \\
\hline
0 \\ 0 \\ 0
\end{array}\right]
+\left[\begin{array}{c}
-3r_n i_0 - 3L_n \dot{i}_0 \\
0 \\
0 \\
\hline
0 \\ 0 \\ 0
\end{array}\right]
\tag{6.33}
$$

Or, written in a more compact form

$$
\left[\begin{array}{c}
v_0 \\ v_d \\ v_q \\ \hline -v_F \\ 0 \\ 0
\end{array}\right]
= -\left[\begin{array}{ccc|ccc}
r + 3r_n & 0 & 0 & 0 & 0 & 0 \\
0 & r & \omega L_q & 0 & 0 & k\omega M_Q \\
0 & -\omega L_d & r & -k\omega M_F & -k\omega M_D & 0 \\
\hline
0 & 0 & 0 & r_F & 0 & 0 \\
0 & 0 & 0 & 0 & r_D & 0 \\
0 & 0 & 0 & 0 & 0 & r_Q
\end{array}\right]
\left[\begin{array}{c}
i_0 \\ i_d \\ i_q \\ \hline i_F \\ i_D \\ i_Q
\end{array}\right]
$$

$$
-\left[\begin{array}{ccc|ccc}
L_0 + 3L_n & 0 & 0 & 0 & 0 & 0 \\
0 & L_d & 0 & kM_F & kM_D & 0 \\
0 & 0 & L_q & 0 & 0 & kM_Q \\
\hline
0 & kM_F & 0 & L_F & M_R & 0 \\
0 & kM_D & 0 & M_R & L_D & 0 \\
0 & 0 & kM_Q & 0 & 0 & L_Q
\end{array}\right]
\left[\begin{array}{c}
\dot{i}_0 \\ \dot{i}_d \\ \dot{i}_q \\ \hline \dot{i}_F \\ \dot{i}_D \\ \dot{i}_Q
\end{array}\right]
\tag{6.34}
$$

By proper choice of rotor and stator base quantities, all the foregoing equations may be written in exactly the same way in pu or in system quantities (volt, ohm, ampere, etc.).

Equation (6.34) is very unusual as a network equation because the "resistance" matrix is not symmetric. This is because the network is an active one which contains the controlled source terms due to speed voltages.

We now investigate the meaning of the newly defined inductances of (6.32) and some related quantities.

6.2.2 Direct axis synchronous inductance, L_d

If we apply positive sequence currents to the armature of a synchronous machine with the field circuit open and the field winding rotated at synchronous speed with the d axis aligned with the rotating MMF wave as shown in Figure 6.4a,

$$L_d = \lambda_a/i_a = \lambda_b/i_b = \lambda_c/i_c = L_s + M_s + (3/2)L_m \tag{6.35}$$

where we require that

$$\begin{bmatrix} i_a \\ i_b \\ i_c \end{bmatrix} = \begin{bmatrix} \sqrt{2}I\cos\theta \\ \sqrt{2}I\cos(\theta - 120) \\ \sqrt{2}I\cos(\theta + 120) \end{bmatrix}$$

Under these conditions only d axis current exists or

$$\begin{bmatrix} i_0 \\ i_d \\ i_q \end{bmatrix} = \begin{bmatrix} 0 \\ \sqrt{3}I \\ 0 \end{bmatrix}$$

where the $\sqrt{3}$ factor is due to the arbitrary constant multiplier chosen to make the **P** transformation orthogonal.

It is known that positive sequence currents flowing in the armature produce a space MMF wave which travels at synchronous speed. However, the flux produced by this rotating MMF wave depends upon the reluctance of the magnetic circuit. The reluctance is greatly influenced by the air gap. In cylindrical machines, therefore, the reactance seen by positive sequence currents is almost a constant irrespective of rotor position. In salient pole machines this is not true since the reluctance, and therefore the flux, depends strongly upon the relative position of the MMF wave and the protruding pole faces of the rotor. This is shown in Figures 6.4a and 6.4b. Thus we would expect the inductance of a salient pole machine to be a function of rotor position. This is true not only of the self inductance but the mutual inductance as well. The inductance will be a maximum when the rotor position is as shown in Figure 6.4a, and this maximum inductance is called L_d, the d axis synchronous inductance.

The d axis synchronous *reactance* is defined as

$$x_d = \omega_1 L_d \tag{6.36}$$

where ω is restricted to be the synchronous speed ω_1 since the concept of reactance is meaningless otherwise.

Kimbark [19] points out that (6.35) is valid whether instantaneous, maximum, or effective values are used for the flux linkage and current. This flux linkage appears to the armature as a sinusoidally varying linkage and induces an EMF in quadrature with the flux and current, the ratio of this induced voltage to i_a also being equal to x_d.

6.2.3 Quadrature axis synchronous inductance, L_q

Proceeding as before, we may determine the q axis synchronous inductance by applying positive sequence currents to the armature with the field circuit open and the field winding turning at synchronous speed with the rotor q axis aligned

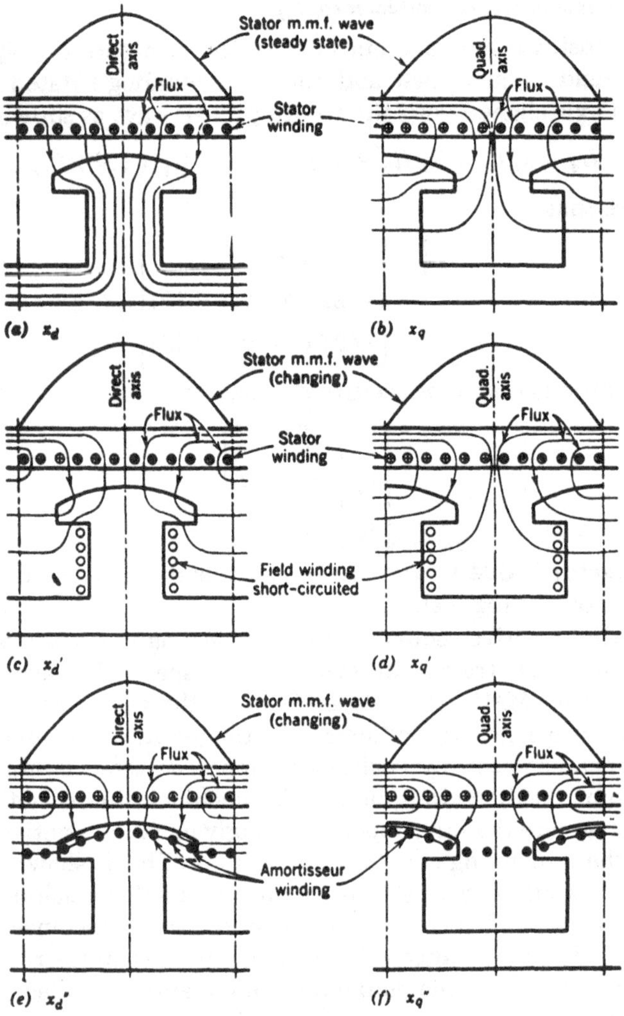

Fig. 6.4. Flux patterns under synchronous, transient, and subtransient conditions. (From Prentice [46].)

with the rotating MMF wave. Under these conditions the currents are shifted by 90° in phase or

$$\begin{bmatrix} i_a \\ i_b \\ i_c \end{bmatrix} = \begin{bmatrix} \sqrt{2}I \cos(\theta - 90) \\ \sqrt{2}I \cos(\theta - 90 - 120) \\ \sqrt{2}I \cos(\theta - 90 + 120) \end{bmatrix}$$

This makes i_0 and i_d go to zero, so that only $i_q = \sqrt{3}I$ exists. Then

$$L_q = \lambda_a/i_a = \lambda_b/i_b = \lambda_c/i_c = L_s + M_s - (3/2)L_m \tag{6.37}$$

With the rotor in the position specified, the synchronous inductance will assume its lowest value, since the reluctance of the flux path is a maximum, as seen in Figure 6.4b. Corresponding to this rotor position we also define the inductive reactance

$$x_q = \omega_1 L_q \tag{6.38}$$

In cylindrical rotor machines

$$x_d \cong x_q \tag{6.39}$$

but for all machines

$$x_d > x_q \tag{6.40}$$

with the inequality being much more pronounced in salient pole machines.

6.2.4 Direct axis subtransient inductance, L_d''

The d axis transient and subtransient inductances are defined with the field circuits shorted, i.e.,

$$\begin{bmatrix} v_F \\ v_D \\ v_Q \end{bmatrix} = \begin{bmatrix} 0 \\ 0 \\ 0 \end{bmatrix} \tag{6.41}$$

and with positive sequence voltages applied suddenly at $t = 0$ to the stator. Thus if $u(t)$ is the unit step function,

$$\begin{bmatrix} v_a \\ v_b \\ v_c \end{bmatrix} = \begin{bmatrix} \sqrt{2}\,V\cos\theta \\ \sqrt{2}\,V\cos(\theta - 120) \\ \sqrt{2}\,V\cos(\theta + 120) \end{bmatrix} u(t) \tag{6.42}$$

or

$$\begin{bmatrix} v_0 \\ v_d \\ v_q \end{bmatrix} = \begin{bmatrix} 0 \\ \sqrt{3}\,Vu(t) \\ 0 \end{bmatrix} \tag{6.43}$$

and only v_d exists after $t = 0$ and it jumps suddenly from zero to $\sqrt{3}V$. Similarly, i_d is the only current component, but it must build up more slowly according to the time constant of the d axis circuit.

Since all currents are zero at $t = 0^-$, the flux linkages are also zero and by the law of constant flux linkages must remain zero at $t = 0^+$. At this instant we may write

$$\lambda_F = 0 = \mathrm{k}M_F i_d + L_F i_F + M_R i_D, \qquad \lambda_D = 0 = \mathrm{k}M_D i_d + M_R i_F + L_D i_D \tag{6.44}$$

from which we may find i_F and i_D as a function of i_d with the result

$$i_F = -\frac{\mathrm{k}(L_D M_F - M_D M_R)}{L_F L_D - M_R^2}\, i_d, \qquad i_D = -\frac{\mathrm{k}(L_F M_D - M_F M_R)}{L_F L_D - M_R^2}\, i_d \tag{6.45}$$

Substituting these currents into the flux linkage equation for λ_d, we have

$$\lambda_d = \left[L_d - \frac{\mathrm{k}^2}{L_F L_D - M_R^2}\left(L_D M_F^2 + L_F M_D^2 - 2M_F M_D M_R\right)\right] i_d \triangleq L_d'' i_d \tag{6.46}$$

Thus

$$L_d'' = L_d - \frac{k^2}{L_F L_D - M_R^2}(L_D M_F^2 + L_F M_D^2 - 2M_F M_D M_R) \tag{6.47}$$

which we call the subtransient inductance.

Often a subtransient *reactance* is used and this is defined as

$$x_d'' = \omega_1 L_d'' \tag{6.48}$$

As illustrated in Figure 6.4e, very little flux is established initially in the field winding and the subtransient inductance is due largely to the damper windings. These windings have a very small time constant, and the subtransient currents die away fast leaving only the so-called transient currents.

Since the air gap flux prior to applying the stator voltage was zero, the damper winding currents try to maintain this no-flux condition. The result is that the flux path established is a high-reluctance air gap path, as in Figure 6.4e, and L_d'' is very small.

6.2.5 Direct axis transient inductance, L_d'

If we examine the transient situation just described only a few cycles after the stator voltages are applied, the damper winding currents have decayed to zero and we are in the so-called transient period where induced currents in the field winding are important. This situation also exists if there are no damper winding so that the air gap flux links the field winding as shown in Figure 6.4c.

To compute the d axis inductance seen under this condition, we can let $i_D = 0$ in (6.44) to compute

$$i_F = -\frac{kM_F}{L_F} i_d \tag{6.49}$$

Then from (6.32) we may compute

$$\lambda_d = \left(L_d - \frac{k^2 M_F^2}{L_F}\right) i_d \triangleq L_d' i_d$$

or

$$L_d' = L_d - \frac{k^2 M_F^2}{L_F} \tag{6.50}$$

and

$$x_d' = \omega_1 L_d' \tag{6.51}$$

This transient inductance is determined under the same rotor condition as the d axis synchronous inductance, the only difference being in the fact that it is measured immediately after the sudden application of the three-phase (positive sequence) voltages. The sudden establishment of flux across the air gap is opposed by establishing a current in the field winding, tending to hold λ_F at zero. Thus, as shown in Figure 6.4c, the only flux established is that which does *not* link the field winding, and this is a small flux. Hence L_d' is small but is greater than L_d''.

6.2.6 Quadrature axis subtransient and transient inductances, L_q'' and L_q'

The q axis subtransient and transient inductances are defined in a similar way to the d axis inductances except that the rotor in this case is positioned with its q

axis opposite the spatial MMF wave of the stator as shown in Figures 6.4d and 6.4f. The flux in this case is not much different than the synchronous case for the salient pole machine pictured in Figure 6.4. In round rotor machines the synchronous case results in a greater flux similar to that pictured in Figure 6.4 for the d axis. Thus we see the need to carefully distinguish between salient pole and round rotor machines in determining q axis inductances.

If we consider the rotor as spinning with the correct alignment and positive sequence voltages applied suddenly, we have a situation similar to equations (6.42) and (6.43) except for a 90° phase lag in the applied voltages to get the proper alignment with the q axis. Thus

$$\begin{bmatrix} v_a \\ v_b \\ v_c \end{bmatrix} = \begin{bmatrix} \sqrt{2}V \sin\theta \\ \sqrt{2}V \sin(\theta - 120) \\ \sqrt{2}V \sin(\theta + 120) \end{bmatrix} u(t) \tag{6.52}$$

or

$$\begin{bmatrix} v_0 \\ v_d \\ v_q \end{bmatrix} = \begin{bmatrix} 0 \\ 0 \\ \sqrt{3}Vu(t) \end{bmatrix} \tag{6.53}$$

Since the flux linkages in the q axis are zero both before and after the voltage is applied, we compute $\lambda_Q = 0 = \mathrm{k}M_Q i_q + L_Q i_Q$ or

$$i_Q = -\frac{\mathrm{k}M_Q}{L_Q} i_q \tag{6.54}$$

Then

$$\lambda_q = L_q i_q - \frac{\mathrm{k}^2 M_Q^2}{L_Q} i_q \triangleq L_q'' i_q$$

where we define

$$L_q'' = L_q - \mathrm{k}^2 M_Q^2 / L_Q \tag{6.55}$$

and

$$x_q'' = \omega_1 L_q'' \tag{6.56}$$

In round rotor machines the same effect as just described is also noted in the transient period because of the rotor iron acting much like a field winding. Thus we often assume for *round rotor machines* that

$$L_q >> L_q' \cong L_d' > L_q'' \cong L_d'' \tag{6.57}$$

In salient pole machines the presence of a damper winding makes a great deal of difference. Thus for *salient pole machines* we estimate that

$$\begin{aligned} &\text{with dampers:} && L_q = L_q' > L_q'' > L_d'' \\ &\text{without dampers:} && L_q = L_q' = L_q'' \end{aligned} \tag{6.58}$$

6.3 Negative Sequence Impedance

If negative sequence voltages (sequence a-c-b) are applied to the stator windings of a synchronous machine with the field winding shorted and the rotor spinning forward at synchronous speed, the currents in the stator see the negative sequence impedance of the machine. Mathematically, the boundary conditions are

$$\mathbf{i}_{abc} = \begin{bmatrix} \sqrt{2}I \cos \theta \\ \sqrt{2}I \cos (\theta + 120) \\ \sqrt{2}I \cos (\theta - 120) \end{bmatrix} \tag{6.59}$$

and $v_F = 0$. Then by Park's transformation we compute [19]

$$\mathbf{i}_{0dq} = \begin{bmatrix} 0 \\ \sqrt{3}I \cos 2\theta \\ \sqrt{3}I \sin 2\theta \end{bmatrix} \tag{6.60}$$

and observe that both i_d and i_q are second harmonic currents. Since i_d acts only on the d axis of the rotor, it induces a second harmonic voltage in the field winding. If we assume that at double frequency the field reactance is much greater than its resistance, we have $v_F = 0 = r_F i_F + \dot{\lambda}_F \cong \dot{\lambda}_F$ which requires that λ_F be a constant, or in the steady state

$$\lambda_F = 0 \tag{6.61}$$

But this is exactly the transient condition described in section 6.2.5. Thus (6.49) applies, and the flux linkages are found from (6.32) to be

$$\begin{bmatrix} \lambda_0 \\ \lambda_d \\ \lambda_q \\ \hline \lambda_F \\ \lambda_D \\ \lambda_Q \end{bmatrix} = \begin{bmatrix} 0 \\ L'_d i_d \\ L_q i_q \\ \hline 0 \\ 0 \\ 0 \end{bmatrix} \tag{6.62}$$

where L'_d is defined in (6.50). Then the flux linkage of phase a, λ_a, is computed as

$$\lambda_a = \sqrt{2}\, I \left[\frac{L'_d + L_q}{2} \cos \theta + \frac{L'_d - L_q}{2} \cos (3\theta + 2\alpha) \right] \tag{6.63}$$

which is observed to have both a fundamental and a third harmonic component, and λ_a / i_a is not a constant. Kimbark [19] defines the fundamental frequency component ratio as the negative sequence inductance, i.e., by definition

$$L_2 = \frac{L'_d + L_q}{2} = \frac{\lambda_a \text{ (fundamental)}}{i_a \text{ (fundamental)}} \tag{6.64}$$

and this definition applies for the case of *no damper windings*.

If damper windings exist on the rotor, we compute

$$\begin{bmatrix} \lambda_0 \\ \lambda_d \\ \lambda_q \end{bmatrix} = \begin{bmatrix} 0 \\ L''_d i_d \\ L''_q i_q \end{bmatrix}$$

and

$$L_2 = (L''_d + L''_q)/2 \tag{6.65}$$

We also define the negative sequence reactance as

$$x_2 = \omega_1 L_2 = (x''_d + x''_q)/2 \tag{6.66}$$

The relationship between x_2 and the subtransient reactances x''_d and x''_q depends strongly upon the presence of damper windings, as shown in Figure 6.5 where measurements are recorded with the rotor blocked or stationary.

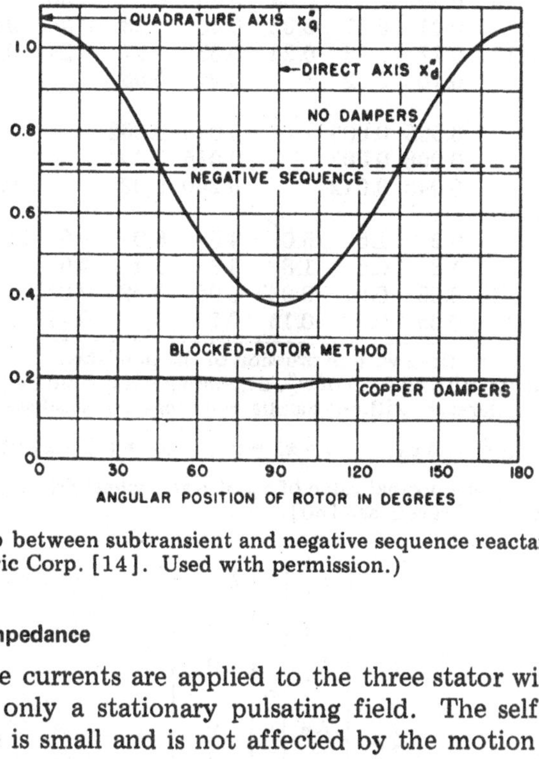

Fig. 6.5. Relationship between subtransient and negative sequence reactances. (From Westinghouse Electric Corp. [14]. Used with permission.)

6.4 Zero Sequence Impedance

If zero sequence currents are applied to the three stator windings, there is no rotating MMF but only a stationary pulsating field. The self inductance or reactance in this case is small and is not affected by the motion of the rotor. The pulsating field is opposed by currents induced in the rotor circuits, and very little air gap flux is established. Thus L_0 is very small, generally smaller than L''_d.

The boundary conditions for this situation are

$$\mathbf{i}_{abc} = \begin{bmatrix} \sqrt{2}I\cos\theta \\ \sqrt{2}I\cos\theta \\ \sqrt{2}I\cos\theta \end{bmatrix} \tag{6.67}$$

where $i_F = i_D = i_Q = 0$, and θ, the rotor angle, may be taken as any value. Then by Park's transformation

$$\mathbf{i}_{0dq} = \begin{bmatrix} \sqrt{6}I\cos\theta \\ 0 \\ 0 \end{bmatrix} \tag{6.68}$$

and

Table 6.1. Typical Synchronous Machine Constants

	Turbo-generators (solid rotor)			*Water-Wheel Generators (with dampers)*†			*Synchronous Condensers*			*Synchronous Motors (general purpose)*		
	Low	*Avg.*	*High*	*Low*	*Avg.*	*High*	*Low*	*Avg.*	*High*	*Low*	*Avg.*	*High*
Reactances in pu												
x_d	0.95	1.10	1.45	0.60	1.15	1.45	1.50	1.80	2.20	0.80	1.20	1.50
x_q	0.92	1.08	1.42	0.40	0.75	1.00	0.95	1.15	1.40	0.60	0.90	1.10
x_d'	0.12	0.23	0.28	0.20	0.37	0.50‡	0.30	0.40	0.60	0.25	0.35	0.45
x_q'	0.12	0.23	0.28	0.40	0.75	1.00	0.95	1.15	1.40	0.60	0.90	1.10
x_d''	0.07	0.12	0.17	0.13	0.24	0.35	0.18	0.25	0.38	0.20	0.30	0.40
x_q''	0.10	0.15	0.20	0.23	0.34	0.45	0.23	0.30	0.43	0.30	0.40	0.50
x_p	0.07	0.14	0.21	0.17	0.32	0.40	0.23	0.34	0.45			
x_2	0.07	0.12	0.17	0.13	0.24	0.35	0.17	0.24	0.37	0.25	0.35	0.45
x_0*	0.01		0.10	0.02		0.21	0.03		0.15	0.04		0.27
Resistances in pu												
r_a(dc)	0.0015		0.005	0.003		0.020	0.002		0.015			
r(ac)	0.003		0.008	0.003		0.015	0.004		0.010			
r_2	0.025		0.045	0.012		0.200	0.025		0.070			
Time constants in seconds												
T_{d0}'	2.8	5.6	9.2	1.5	5.6	9.5	6.0	9.0	11.5			
T_d'	0.4	1.1	1.8	0.5	1.8	3.3	1.2	2.0	2.8			
$T_d'' = T_q''$	0.02	0.035	0.05	0.01	0.035	0.05	0.02	0.035	0.05			
T_a	0.04	0.16	0.35	0.03	0.15	0.25	0.1	0.17	0.3			

Source: Kimbark [19]. Used with permission of the publisher.
*x_0 varies from about 0.15 to 0.60 of x_d'', depending upon winding pitch.
†For water-wheel generators without damper windings, x_0 is as listed and

$$x_d'' = 0.85 x_d', \quad x_q'' = x_q' = x_q, \quad x_2 = (x_d' + x_q)/2$$

‡For curves showing the normal value of x_d' of water-wheel-driven generators as a function of kilovolt-ampere rating and speed, see [50].

$$\begin{bmatrix} \lambda_0 \\ \lambda_d \\ \lambda_q \\ \hline \lambda_F \\ \lambda_D \\ \lambda_Q \end{bmatrix} = \begin{bmatrix} L_0 i_0 \\ 0 \\ 0 \\ \hline 0 \\ 0 \\ 0 \end{bmatrix} \tag{6.69}$$

From (6.69) we compute

$$\lambda_{abc} = \begin{bmatrix} 1 \\ 1 \\ 1 \end{bmatrix} L_0 i_0 / \sqrt{3} \tag{6.70}$$

where, from (6.32)

$$L_0 = \lambda_a / i_a = L_s - 2M_s \tag{6.71}$$

and L_0 is small compared to L_d and L_q. We also define

$$x_0 = \omega_1 L_0 \tag{6.72}$$

The actual value of x_0 depends upon the pitch of the windings but is usually in the range [19] of $0.15\, x_d'' < x_0 < 0.60\, x_d''$. Typical values of synchronous machine reactances are given in Table 6.1 which also gives values of the various time constants of the machine circuits defined in section 6.5.

6.5 Time Constants

If a 3ϕ fault is applied to an unloaded synchronous machine, an oscillogram of the phase current appears as in Figure 6.1. If we replot the current in one phase with the dc component removed, the result is shown in Figure 6.6. A careful examination of this damped exponential reveals that the current envelope has an unusually high initial value 0-*c* and that it decays in a few cycles to a lower rate of decrement. The current then continues to decay at this lower rate until it finally reaches its steady state value, represented by the peak value 0-*a*. The ac component of current in the field circuit decays over a very long period of time, as noted in Figure 6.1. We investigate these various decrements in greater detail since the time constants are of direct interest in faulted systems. In particular, we need to establish the approximate point in time on Figure 6.6 (or 6.1) at which the circuit relays will be likely to open. This is the value of fault current we would like to compute if possible or, alternatively, we seek a value which will give results to provide relaying margins on the safe side.

There are several time constants associated with the behavior noted in Figures 6.1 and 6.6. Since these are often quoted in the literature, they are reviewed here briefly.

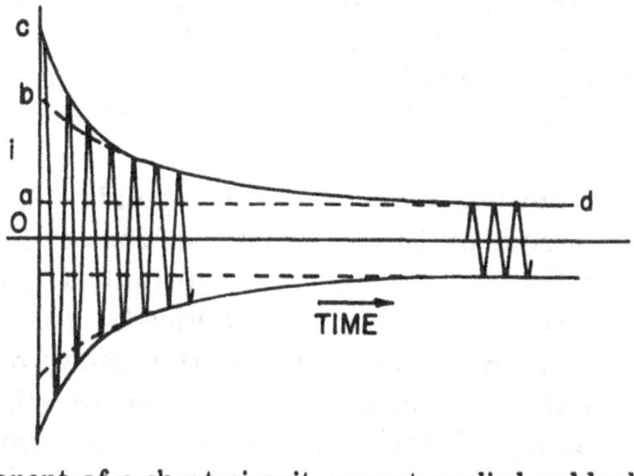

Fig. 6.6. The ac component of a short circuit current applied suddenly to a synchronous generator. (From *Elements of Power System Analysis*, by William D. Stevenson, Jr. Copyright McGraw-Hill, 1962. Used with permission of McGraw-Hill Book Co.)

6.5.1 Direct axis transient open circuit time constant, T'_{d0}

Consider a synchronous machine with no damper windings which is operating with the armature circuits open. Then a step change in the voltage applied to the field is unaffected by any other circuit, and the response is a function of the field resistance and inductance only. From (6.34) with $i_d = i_D = 0$ we write

$$v_F = r_F i_F + L_F \dot{i}_F \tag{6.73}$$

If i_F is initially zero and we let $v_F = \text{k}u(t)$, where k is a constant and $u(t)$ is the unit step function, the result is

$$i_F = \frac{\text{k}}{r_F}(1 - \text{e}^{-r_F t/L_F})\, u(t) \tag{6.74}$$

Thus the time constant of this circuit, defined as T'_{d0}, is

$$T'_{d0} = L_F/r_F \quad \text{sec} \tag{6.75}$$

Since the open circuit armature voltage varies directly with i_F, this voltage changes at the same rate as the field current. Typical values of T'_{d0} are quoted [19] as from 2 to 11 seconds, with 6 seconds an average value. The large size of T'_{d0} is due to the large inductance of the field.

6.5.2 Direct axis transient short circuit time constant, T'_d

A synchronous machine operating with both the field and armature circuits closed is different from the preceding case. If currents flow in both windings, each induces voltages in the other, which in turn cause current responses. If the rotor were stationary, the coupled circuits would behave as a transformer with currents of frequency f in one circuit inducing currents of this same frequency in the other. In the case of a machine, however, the frequency of induced currents is different because of the rotation of the rotor. Thus a direct current in the rotor causes a positive sequence current in the armature, whereas a direct current in the armature is associated with an alternating current in the field winding. This alternating component is clearly shown in Figure 6.1. Each of these induced currents changes at a different time constant depending upon the resistance and inductance each current sees. We call the time constant which governs the rate of change of direct current in the field the *field time constant*. But this must correspond to the rate of change of the amplitude (or current envelope) in the armature, which we like to call the *direct axis transient time constant*. We call the time constant which governs the rate of change of the direct current in the armature the *armature time constant*, and this is the same as the time constant of the envelope of alternating currents in the field.

The field or d axis transient time constant depends upon the inductance seen by i_F, which in turn depends on the impedance of the armature circuit. With the armature open we have found that this time constant is T'_{d0}. If the armature is shorted, the inductance seen by i_F is greatly reduced. We have already computed the required values. From (6.35) we compute the armature inductance to be L_d with the field open. With the field shorted as noted by (6.50), the armature inductance is computed as L'_d. Clearly then, shorting the armature as viewed from the field will change the inductance seen by the ratio L'_d/L_d, and the d axis transient (short circuit) time constant T'_d is defined by

$$T'_d = (L'_d/L_d)\,T'_{d0} \quad \text{sec} \tag{6.76}$$

Kimbark [19] notes that the resistance seen in the two cases is practically the same since with the armature shorted its resistance is negligible compared to its inductance. Typically, T'_d is about 1/4 that of T'_{d0}, or approximately 1.5 seconds.

Thus T'_{d0} and T'_d are the two extremes for the field time constant. With the machine loaded normally or faulted through a finite impedance, the time constant will be somewhere between these extremes. If a known external inductance L_e exists between the machine and the fault point, this inductance may be added to both L_d and L'_d in (6.76).

6.5.3 Armature time constant, T_a

As explained above, the armature time constant applies to the rate of change of direct currents in the armature or to the envelope of alternating currents in the field winding. It is equal to the ratio of armature inductance to armature resistance under the given condition. To determine the armature inductance we note that this situation, with rotor currents of frequency f, is analogous to the case of negative sequence armature currents (and inductance L_2) which produced field currents of frequency $2f$. In both cases the rotor flux linkages are constant, and the flux is largely leakage flux. Thus with alternating currents in the rotor, the stator inductance is essentially L_2, and the armature time constant T_a is given by

$$T_a = L_2/r \quad \text{sec} \tag{6.77}$$

A typical value of T_a is 0.15 seconds for a fault on the machine terminals.

6.5.4 Direct axis subtransient time constants, T''_{d0} and T''_d

In a machine with damper windings there is an additional coupled circuit in the d axis of the rotor, namely, the damper winding. This is a low impedance winding, and currents induced in its conductors may be large but decay rapidly to zero. Viewed from the armature, the direct currents in both the field and damper windings appear to the armature as positive sequence currents whose magnitudes reflect the coupling to *both* rotor circuits. Thus, as shown in Figure 6.6, there are *two* distinct time constants apparent in the alternating current wave, one with a time constant much shorter than the other.

The shorter of these time constants is due to the damper winding and is identified by double-primed notation such as T''_d. Here, as in the case of no damper windings, T''_{d0} applies with the armature circuits open, whereas T''_d applies with the armature shorted. Typical values are T''_{d0} = 0.125 sec (7.5 cycles) and T''_d = 0.035 sec (2 cycles).

6.5.5 Quadrature axis time constants, T'_{q0}, T'_q, T''_{q0}, and T''_q

We can identify time constants associated with the q axis in a way similar to the treatment of the d axis except that there is no field winding on the q axis. It is also important to recognize the significantly different structure of the q axis between salient pole and cylindrical rotor machines. Thus in salient pole machines, T'_q has no meaning since there is no quadrature rotor winding; but with damper windings present, a time constant of $T''_q \cong T''_d$ is used.

In cylindrical rotor machines the q axis flux has a lower reluctance path than in salient pole machines, and currents may be established in the steel which decay at various time constants depending upon the impedance of the current path. It is usually observed that this situation may be approximated by representing the armature current as the sum of two exponentials, one with time constants T'_{q0} and T'_q corresponding to a reactance x'_q and another with time constants T''_{q0} and T''_q corresponding to x''_q. Typical values are $T''_q = T''_d$ = 0.035 sec and T'_q = 0.8 sec.

6.6 Synchronous Generator Equivalent Circuits

Having carefully defined the parameters of a synchronous generator, we now consider the construction of an equivalent circuit. Referring to (6.33), we note first of all that the zero sequence equation is uncoupled from the others. A simple passive R-L network will satisfy this equation. The remaining five equations are more difficult and require further study. From (6.34) we write by rearranging,

$$
\begin{bmatrix} v_d \\ -v_F \\ -v_D = 0 \\ \hline v_q \\ -v_Q = 0 \end{bmatrix}
= -
\left[\begin{array}{ccc|cc} r & 0 & 0 & & \\ 0 & r_F & 0 & & 0 \\ 0 & 0 & r_D & & \\ \hline & 0 & & r & 0 \\ & & & 0 & r_Q \end{array}\right]
\begin{bmatrix} i_d \\ i_F \\ i_D \\ \hline i_q \\ i_Q \end{bmatrix}
+
\begin{bmatrix} -\omega\lambda_q \\ 0 \\ 0 \\ \hline \omega\lambda_d \\ 0 \end{bmatrix}
$$

$$
-
\left[\begin{array}{ccc|cc} L_d & kM_F & kM_D & & \\ kM_F & L_F & M_R & & 0 \\ kM_D & M_R & L_D & & \\ \hline & & & L_q & kM_Q \\ & 0 & & kM_Q & L_Q \end{array}\right]
\begin{bmatrix} \dot{i}_d \\ \dot{i}_F \\ \dot{i}_D \\ \hline \dot{i}_q \\ \dot{i}_Q \end{bmatrix}
\qquad (6.78)
$$

These equations represent a reciprocal set of coupled d circuits and coupled q circuits with controlled sources $\omega\lambda_q$ and $\omega\lambda_d$ as shown in Figure 6.7, where the controlled source terms are speed voltages and depend on currents in the other circuit. Note that the zero sequence circuit is completely uncoupled and is passive.

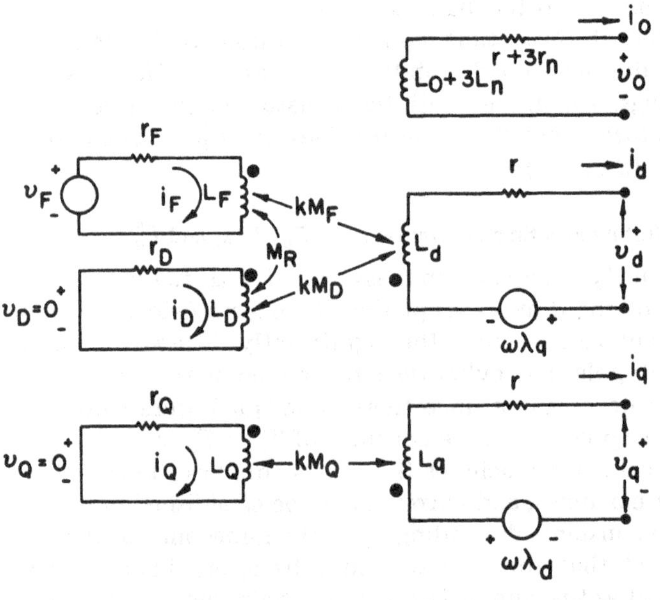

Fig. 6.7. Synchronous generator equivalent circuit.

Lewis [48] shows that the d and q equivalents may be greatly simplified if a T circuit is used to represent the mutual coupling. This requires that (1) all circuits in the equivalent be represented in pu on the same time basis, (2) all circuits in the equivalent be represented in pu on the same voltage basis (base voltage), and (3) all circuits in the equivalent have the same voltampere base. If we assume that these requirements are all met, the equivalent circuit may be redrawn as in Figure 6.8 where, because of the restrictions in the choice of base quantities, the off-diagonal mutual inductances are equal. Then we define

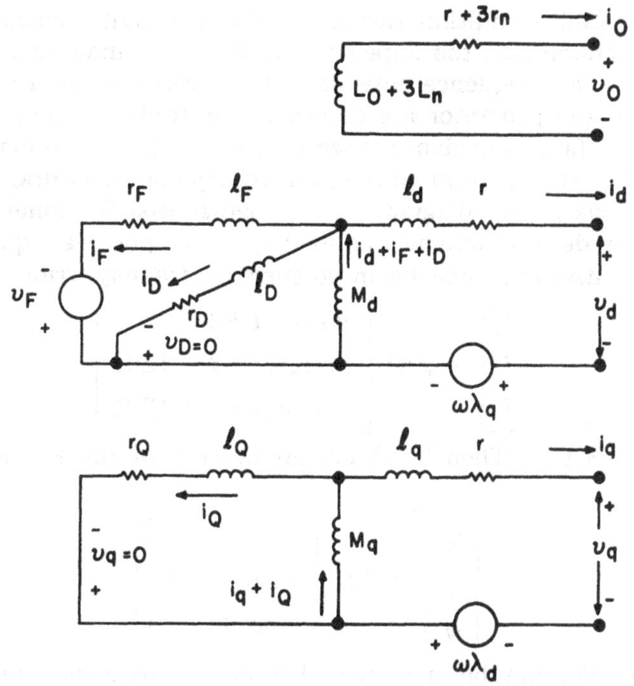

Fig. 6.8. Equivalent T circuits for a synchronous generator.

$$M_d \triangleq kM_F = kM_D = M_R \quad \text{pu}, \qquad M_q \triangleq kM_Q \quad \text{pu} \tag{6.79}$$

We also define the leakage inductances, designated by script ℓ, according to the equation

$$\begin{aligned} M_d &= L_d - \ell_d = L_F - \ell_F = L_D - \ell_D \quad \text{pu} \\ M_q &= L_q - \ell_q = L_Q - \ell_Q \quad \text{pu} \end{aligned} \tag{6.80}$$

The equivalent circuit of Figure 6.8 is often used in computation because of its simplicity. It is also convenient for visualizing transient and subtransient conditions. For example, the subtransient impedance and time constant in the d and q axes is that seen looking into the d and q circuits from the terminals on the right and with all voltage sources shorted. The transient condition is determined in the same way but with the damper circuits open.

The concept of leakage inductance is also useful since these inductances are linear. Thus in Figure 6.8 only the mutual inductance M_d would normally become saturated and it should be considered nonlinear.

6.7 Phasor Diagram of a Synchronous Generator

In fault studies we prefer to work with phasor quantities since phasor solutions require less work than solving the nonlinear differential equations for a synchronous machine. But phasors usually are assumed to represent a steady state condition. We therefore derive the phasor diagram for a generator operating in the steady state. We may then examine this diagram to determine the operating condition immediately following a fault.

A three-phase fault or load may be studied by making the appropriate impedance connection to the positive sequence network only. But we may also view any other fault condition this way, adding a fault impedance at the fault point. In the general case the impedance to be added may be a combination of the negative and zero sequence networks. In all cases the positive sequence currents flowing in the generator are determined entirely by the positive sequence network and this fault impedance if we neglect the load currents. Thus we limit our consideration at this point to the positive sequence network. This permits us to derive only one phasor diagram. We recognize that for some kinds of unbalance the diagram derived here may apply only to the positive sequence quantities. If we assume positive sequence, balanced currents, we may write

$$\begin{bmatrix} i_a \\ i_b \\ i_c \end{bmatrix} = \sqrt{2}I \begin{bmatrix} \cos(\omega_1 t + \phi) \\ \cos(\omega_1 t + \phi - 120) \\ \cos(\omega_1 t + \phi + 120) \end{bmatrix} \tag{6.81}$$

and $\theta = \omega_1 t + \delta + 90$. Then by direct application of the **P** transformation we compute

$$\begin{bmatrix} i_0 \\ i_d \\ i_q \end{bmatrix} = \sqrt{3}I \begin{bmatrix} 0 \\ -\sin(\delta - \phi) \\ \cos(\delta - \phi) \end{bmatrix} \tag{6.82}$$

Equation (6.82) may be manipulated to define the angle ϕ in terms of i_d and i_q with the result

$$\cos\phi = \frac{-i_d \sin\delta + i_q \cos\delta}{\sqrt{3}I}$$

$$\sin\phi = \frac{i_q \sin\delta + i_d \cos\delta}{\sqrt{3}I} \tag{6.83}$$

These values may be substituted into (6.81) to write

$$i_a = \sqrt{2/3}\,[i_d \cos(\omega_1 t + \delta + 90) + i_q \cos(\omega_1 t + \delta)] \tag{6.84}$$

and, since the phase currents are balanced, i_b and i_c are known also. Applying the phasor transformation (1.50), we could write (6.84) immediately as a phasor quantity. Before doing this, however, we digress to examine the reference systems which might be convenient to use.

6.7.1 Phasor frames of reference for synchronous machines

In synchronous machines there are at least two convenient and widely used frames of reference. One we shall call the *arbitrary* reference frame.[4] The second is the *d-q* reference frame discussed in the preceding sections.

Using an arbitrary reference frame we can write a phasor current $\tilde{I}_a$ as

$$\begin{aligned} \tilde{I}_a &= \tilde{I}_{ar} + \tilde{I}_{ax} = I_a e^{j\phi} \\ &= I_a \cos\phi + jI_a \sin\phi = I_{ar} + jI_{ax} \end{aligned} \tag{6.85}$$

[4] The "arbitrary reference" used here is a *phasor reference* and should not be confused with the arbitrary reference frame defined by Krause and Thomas [49] which is a rotating reference that revolves at an arbitrary angular velocity and is used in the study of induction motors.

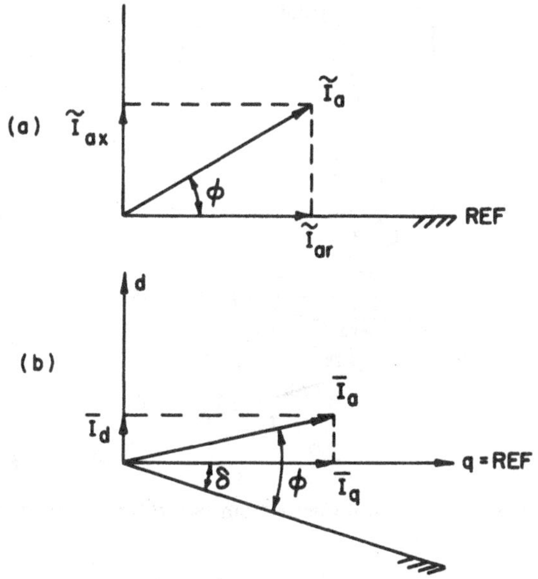

Fig. 6.9. Two frames of reference for phasor quantities: (a) arbitrary reference, (b) *d-q* reference.

where this phasor is pictured in Figure 6.9a. Note that we use the tilde over the phasor quantity symbol to indicate that this quantity is based on the arbitrary reference frame. Usually such attention to detail is unnecessary, but here we need to clearly distinguish between two reference systems. From (6.85) the components of $\tilde{I}_a$ are

$$\tilde{I}_{ar} = I_{ar} = I_a \cos\phi, \qquad \tilde{I}_{ax} = jI_{ax} = jI_a \sin\phi \tag{6.86}$$

Note carefully that the quantity I_a (without the tilde) is a scalar quantity.

The *d-q* frame of reference is pictured in Figure 6.9b. Here we use a bar over the phasor symbol to indicate that this phasor is based on the *d-q* reference frame. Thus

$$\bar{I}_a = \bar{I}_q + \bar{I}_d = I_q + jI_d = I_a \cos(\phi - \delta) + jI_a \sin(\phi - \delta) \tag{6.87}$$

Then

$$\bar{I}_q = I_q = I_a \cos(\phi - \delta), \qquad \bar{I}_d = jI_d = jI_a \sin(\phi - \delta) \tag{6.88}$$

It is now possible to relate a phasor expressed in one reference frame to the same quantity expressed on the second basis by the formula[5]

$$\tilde{I}_a = \bar{I}_a e^{j\delta} \tag{6.89}$$

From this formula we compute

$$I_{ar} + jI_{ax} = (I_q \cos\delta - I_d \sin\delta) + j(I_d \cos\delta + I_q \sin\delta)$$

[5] For a more detailed discussion of the subject of transformation of bases, see Hueslman [5].

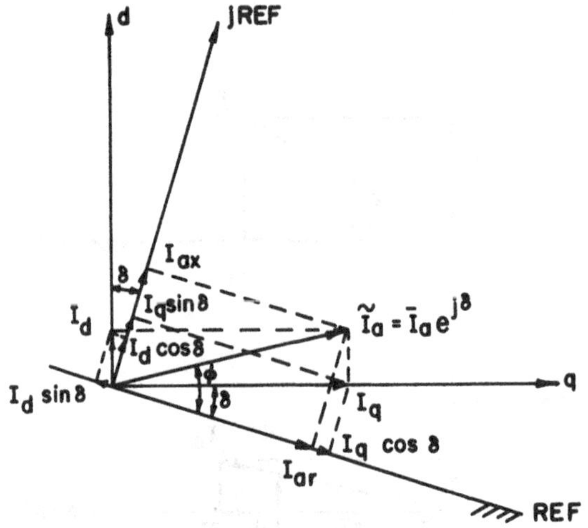

Fig. 6.10. $\tilde{I}_a$ components from two reference frames.

or

$$I_{ar} = I_q \cos\delta - I_d \sin\delta, \qquad I_{ax} = I_d \cos\delta + I_q \sin\delta \tag{6.90}$$

These quantities are shown in Figure 6.10.

6.7.2 Phasor transformation of generator quantities

From (6.84) we have the generator current i_a expressed as a time domain quantity in terms of the magnitudes i_d and i_q, which are rotor equivalents of the instantaneous stator currents. We now define rms stator equivalents of these rotor-referenced magnitudes as

$$I_d = i_d/\sqrt{3}, \qquad I_q = i_q/\sqrt{3} \quad \text{rms A or rms pu} \tag{6.91}$$

where the $\sqrt{3}$ comes from the way the **P** transformation was defined and therefore arises as a scale factor for i_d and i_q. Using (6.91), we write (6.84) as

$$i_a = \sqrt{2}\, I_d \cos(\omega_1 t + \delta + 90) + \sqrt{2}\, I_q \cos(\omega_1 t + \delta)$$

Applying the phasor transformation (1.50), we compute

$$\tilde{I}_a = I_d e^{j(\delta+90)} + I_q e^{j\delta} = (I_q + jI_d)e^{j\delta} = \bar{I}_a e^{j\delta} \tag{6.92}$$

as given by (6.89).

The magnitude of I_a should be the same in any frame of reference. Thus

$$I_a = |\tilde{I}_a| = |\bar{I}_a| = \sqrt{I_q^2 + I_d^2} = \frac{1}{\sqrt{3}}\sqrt{i_q^2 + i_d^2} \tag{6.93}$$

The magnitude could also be computed from the components of $\tilde{I}_a$, but these values are not usually specified.

The foregoing permits us to represent a machine in the convenient d-q frame of reference and then easily transfer to a more general reference which is used for all machines in the system. Knowing the d-q voltages and currents, we may by (6.92) write these quantities as phasors.

6.7.3 The steady state phasor diagram

To derive the steady state phasor diagram, we write the d-q voltage equations for the balanced case and then transform them to the complex domain. Since we are considering the balanced, positive sequence case, the zero sequence voltage and current are both zero, i.e.,

$$v_0 = i_0 = 0 \tag{6.94}$$

Also, since the system is in steady state, the speed is constant and the damper currents and all current derivatives are zero, i.e.,

$$\begin{aligned} \omega &= \omega_1 \\ i_D &= i_Q = 0 \\ \dot{i}_d &= \dot{i}_q = \dot{i}_F = \dot{i}_D = \dot{i}_Q = 0 \end{aligned} \tag{6.95}$$

From (6.34) the d axis voltage is $v_d = -ri_d - \omega L_q i_q - \mathrm{k}\omega M_Q i_Q - \dot{\lambda}_d$ and, incorporating (6.95), this voltage may be written as

$$v_d = -ri_d - \omega_1 L_q i_q \tag{6.96}$$

Also from (6.34) the q axis voltage is

$$v_q = -ri_q + \omega L_d i_d + \mathrm{k}\omega M_F i_F + \mathrm{k}\omega M_D i_D - \dot{\lambda}_q$$

which may be written as

$$v_q = -ri_q + \omega_1 L_d i_d + \mathrm{k}\omega_1 M_F i_F \tag{6.97}$$

We now define an rms stator equivalent of the rotor field current to be

$$I_F = i_F/\sqrt{3} \tag{6.98}$$

Then, using this definition and definition (6.91) of d and q axis currents, we define rms equivalent d and q axis voltages by dividing (6.96) and (6.97) by $\sqrt{3}$, with the result

$$\begin{aligned} V_d &= v_d/\sqrt{3} = -rI_d - x_q I_q \\ V_q &= v_q/\sqrt{3} = -rI_q + x_d I_d + \mathrm{k}x_{mF} I_F \end{aligned} \tag{6.99}$$

Here we recognize the previously defined reactances x_d and x_q and define the mutual reactance

$$x_{mF} = \omega_1 M_F \tag{6.100}$$

Using (6.92), we may compute the rms phase a terminal voltage on an arbitrary reference basis to be

$$\tilde{V}_a = (V_q + \mathrm{j}V_d)\,\mathrm{e}^{\mathrm{j}\delta} = [(-rI_q + x_d I_d + \mathrm{k}x_{mF} I_F) + \mathrm{j}(-rI_d - x_q I_q)]\,\mathrm{e}^{\mathrm{j}\delta}$$

We recognize the quantity in brackets to be V_a on a d-q basis, according to (6.89). Thus using the bar notation for this basis and rearranging, we have

$$\begin{aligned} \overline{V}_a &= V_q + \mathrm{j}V_d = -r(I_q + \mathrm{j}I_d) - \mathrm{j}x_q I_q + x_d I_d + \mathrm{k}x_{mF} I_F \\ &= -r\overline{I}_a - \mathrm{j}x_q I_q + x_d I_d + \mathrm{k}x_{mF} I_F \end{aligned} \tag{6.101}$$

It is convenient to define a field excitation voltage

$$\overline{E}_F = \overline{E}_q + \overline{E}_d = E_q + \mathrm{j}E_d = \mathrm{k}x_{mF} I_F \tag{6.102}$$

such that

$$\overline{E}_q = E_q = kx_{mF}I_F, \qquad \overline{E}_d = jE_d = 0 \tag{6.103}$$

The reason $\overline{E}_d$ is zero is that there is no field winding on the q axis. Then we may write (6.101) as

$$\overline{E}_q = \overline{V}_a + r\overline{I}_a + jx_qI_q - x_dI_d \tag{6.104}$$

which is still an equation of mixed phasor and scalar notation. We clarify this by applying (6.88), where we see that

$$I_q = \overline{I}_q, \qquad I_d = -j\overline{I}_d$$

Then (6.104) in d-q phasor notation becomes

$$\overline{E}_q = \overline{V}_a + r\overline{I}_a + jx_q\overline{I}_q + jx_d\overline{I}_d \tag{6.105}$$

or, multiplying by $e^{j\delta}$, we can transform this equation to an arbitrary frame of reference.

Equation (6.105) defines the phasor diagram of a synchronous machine when operating in the steady state. Two typical phasor diagrams, one for a leading power factor and one for a lagging power factor, are shown in Figure 6.11. In

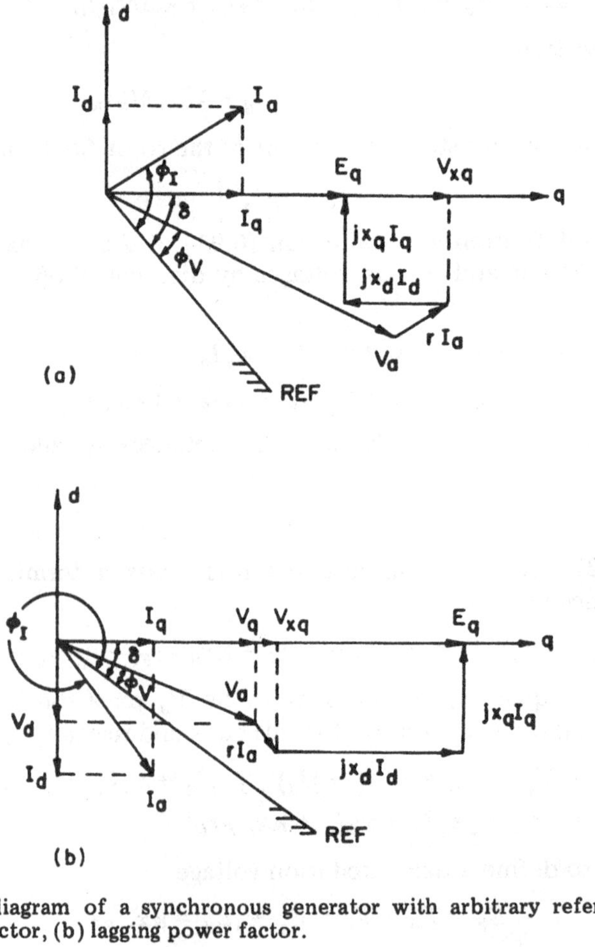

Fig. 6.11. Phasor diagram of a synchronous generator with arbitrary reference: (a) leading power factor, (b) lagging power factor.

either case we can define

$$F_p = \text{power factor} = \cos\theta \tag{6.106}$$

where θ = angle by which $\tilde{V}_a$ leads $\tilde{I}_a = \phi_V - \phi_I$. The voltage $\tilde{V}_{xq}$ shown in Figure 6.11 is defined to be

$$\tilde{V}_{xq} = \tilde{E}_q - jx_d\tilde{I}_d \tag{6.107}$$

This voltage will be needed in a later development.

In many references the phasor diagram of a synchronous generator is drawn in a slightly different way than shown in Figure 6.11. If a quantity $jx_q\tilde{I}_d$ is added and subtracted from (6.105), we can develop

$$\tilde{E}_q = \tilde{V}_a + r\tilde{I}_a + jx_q\tilde{I}_a + j(x_d - x_q)\tilde{I}_d \tag{6.108}$$

This form is interesting because the last term is approximately zero for a round rotor generator since $x_d \cong x_q$ for these machines (see Table 6.1). A phasor diagram which combines (6.105) and (6.108) is shown in Figure 6.12.

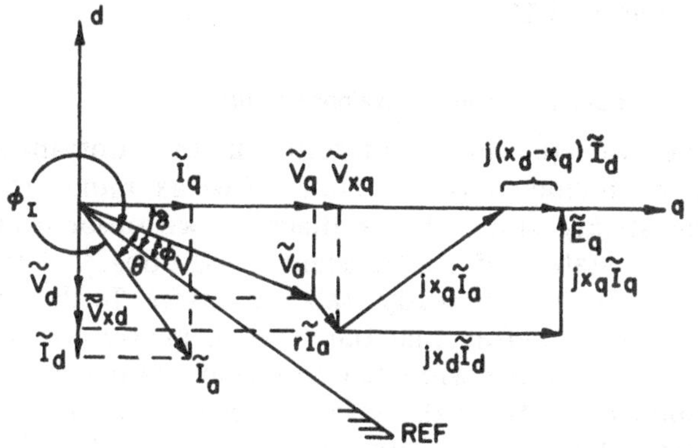

Fig. 6.12. Another form of phasor diagram of a synchronous generator.

The condition described in (6.105) and (6.108) or by the phasor diagrams of Figures 6.11 and 6.12 can also be visualized in terms of the synchronous generator equivalent circuits of Figure 6.7 and 6.8. Under the steady state condition described above, the equivalent circuit reduces to that of Figure 6.13 where all

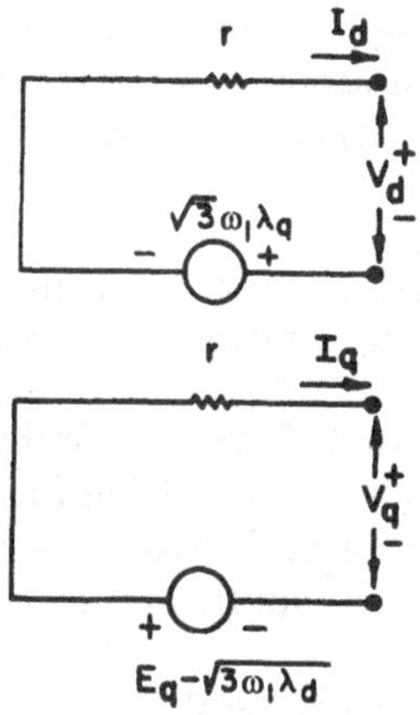

Fig. 6.13. The *d*-*q* equivalent circuits in the steady state case.

voltages and currents have been divided by $\sqrt{3}$ to use the equivalent rms stator (uppercase) variables as in (6.99).

After combining the d and q equations to obtain (6.108), we can draw the simple equivalent circuit shown in Figure 6.14, where an EMF due to saliency of

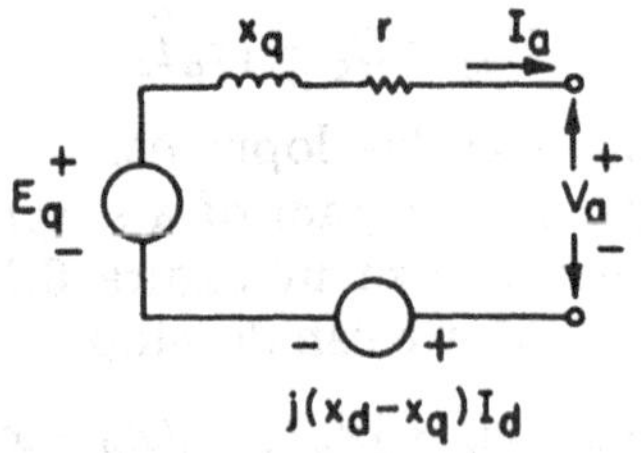

Fig. 6.14. Equivalent circuit of a synchronous generator.

$j(x_d - x_q)I_d$ is shown as a separate voltage. Since the quantity $(x_d - x_q)$ is relatively small, this EMF is small and is always in phase with E_q as indicated by the phasor diagram of Figure 6.12.

6.8 Subtransient Phasor Diagram and Equivalent Circuit

In fault studies we are usually interested in the subtransient period immediately following the incidence of the fault. If we examine the assumptions used in developing the steady state phasor diagram, we observe certain differences between the steady state and subtransient conditions. Equation (6.95) is a mathematical statement of the steady state condition. In the subtransient period we can usually assume as before that the speed is nearly constant. However, we cannot assume that the damper winding currents are zero as in (6.95).

The third condition, setting the current derivatives to zero, is equivalent to stating that the rate of change of flux linkages is zero. In other words, the flux linkages are the same immediately following the fault as they were just before the fault occurred. This is a good assumption to make for the subtransient period and follows from Lenz's law in that the currents do not change instantaneously.

The solution of the voltage equations (6.34) for a transient condition depends upon the initial condition (at $t = 0^-$) and the nature of the change or driving function which is the fault in our problem. A complete solution also requires the solution of a torque or inertia equation to find the speed ω as a function of time. Even if we assume balanced conditions and constant speed, we are faced with the simultaneous solution of six equations (0, d, q, F, D, and Q). Then the complete solution may be written as

$$\text{complete solution} = \text{final steady state solution} + \text{complimentary or transient solution}$$

In our case we are usually satisfied if we can find only the subtransient solution, i.e., the solution at $t = 0^+$. This is not as difficult as solving for the complete solution as a function of time, but it still involves a consideration of the same five equations. The desired subtransient solution may be written in words as

$$\text{subtransient solution at } t = 0^+ = \text{initial condition at } t = 0^- + \text{delta condition or change from } t = 0^- \text{ to } t = 0^+ \tag{6.109}$$

The initial or prefault condition is the steady state solution discussed in the previous section, where the voltages and currents depend on the machine load.

These prefault ($t = 0^-$) conditions are:

1. v_d, v_q, i_d, i_q are usually nonzero and are a function of load.
2. All current derivatives are zero.
3. All flux linkage derivatives are zero due to item 2.
4. All damper currents are zero, $i_D = i_Q = 0$.
5. The field current is constant, i_F = a constant.
6. All zero sequence quantities are zero.
7. The speed is constant, $\omega = \omega_1$.

Immediately after the fault ($t = 0^+$) we assume:

1. The fault is represented by a step change in i_d and i_q.
2. The current derivatives are nonzero (an impulse) at $t = 0$ but are zero at $t = 0^+$.
3. The flux linkage derivatives are zero at $t = 0^+$.
4. The damper currents are nonzero at $t = 0^+$.
5. The field current changes to a new value at $t = 0^+$.
6. The zero sequence quantities are zero.
7. The speed is constant, $\omega = \omega_1$.

We shall represent all variables by denoting the initial condition plus the change as indicated in (6.109), using the notation

$$i = i_0 + i_\Delta \tag{6.110}$$

where

$$\begin{aligned} i_0 &= \text{initial condition, } t = 0^- \\ i_\Delta &= \text{change from } t = 0^- \text{ to } t = 0^+ \end{aligned} \tag{6.111}$$

At $t = 0^+$ we may write the flux linkages from (6.32) in matrix form as $\boldsymbol{\lambda} = \mathbf{Li}$, or

$$\boldsymbol{\lambda}_0 + \boldsymbol{\lambda}_\Delta = \mathbf{L}(\mathbf{i}_0 + \mathbf{i}_\Delta) \tag{6.112}$$

Then the initial condition establishes the prefault flux linkages as

$$\begin{bmatrix} \lambda_{d0} \\ \lambda_{q0} \\ \lambda_{F0} \\ \lambda_{D0} \\ \lambda_{Q0} \end{bmatrix} = \begin{bmatrix} L_d i_{d0} + kM_F i_{F0} \\ L_q i_{q0} \\ kM_F i_{d0} + L_F i_{F0} \\ kM_D i_{d0} + M_R i_{F0} \\ kM_Q i_{q0} \end{bmatrix} \tag{6.113}$$

Now, following a change in currents $i_{d\Delta}$ and $i_{q\Delta}$ the currents in the coupled circuits will adjust themselves so that the flux linkages remain constant, in the F, D, and Q circuits, i.e., from (6.32) and (6.112)

$$\begin{bmatrix} \lambda_{d\Delta} \\ \lambda_{q\Delta} \\ \lambda_{F\Delta} \\ \lambda_{D\Delta} \\ \lambda_{Q\Delta} \end{bmatrix} = \begin{bmatrix} L_d i_{d\Delta} + kM_F i_{F\Delta} + kM_D i_{D\Delta} \\ L_q i_{q\Delta} + kM_Q i_{Q\Delta} \\ kM_F i_{d\Delta} + L_F i_{F\Delta} + M_R i_{D\Delta} \\ kM_D i_{d\Delta} + M_R i_{F\Delta} + L_D i_{D\Delta} \\ kM_Q i_{q\Delta} + L_Q i_{Q\Delta} \end{bmatrix} = \begin{bmatrix} ? \\ ? \\ 0 \\ 0 \\ 0 \end{bmatrix} \tag{6.114}$$

This equation sets up a constraint among the current variables, for we have from (6.114)

$$L_F i_{F\Delta} + M_R i_{D\Delta} = -\mathrm{k}M_F i_{d\Delta}, \qquad M_R i_{F\Delta} + L_D i_{D\Delta} = -\mathrm{k}M_D i_{d\Delta}$$

which we can solve for $i_{F\Delta}$ and $i_{D\Delta}$ with the result

$$\begin{aligned} i_{F\Delta} &= -\frac{\mathrm{k}L_D M_F + \mathrm{k}M_D M_R}{L_F L_D - M_R^2}\, i_{d\Delta} \\ i_{D\Delta} &= -\frac{\mathrm{k}L_F M_D + \mathrm{k}M_F M_R}{L_F L_D - M_R^2}\, i_{d\Delta} \end{aligned} \tag{6.115}$$

Also, from the last equation in (6.114) we compute

$$i_{Q\Delta} = -\frac{\mathrm{k}M_Q}{L_Q}\, i_{q\Delta} \tag{6.116}$$

Currents from (6.115) and (6.116) may now be used in (6.114) to compute

$$\lambda_{d\Delta} = \left[L_d - \frac{\mathrm{k}^2(L_D M_F^2 + L_F M_D^2 - 2M_F M_D M_R)}{L_F L_D - M_R^2} \right] i_{d\Delta} \triangleq L_d'' i_{d\Delta} \tag{6.117}$$

where L_d'' is defined in (6.47) to be the quantity in the brackets. Similarly, we compute

$$\lambda_{q\Delta} = \left(L_q - \frac{\mathrm{k}^2 M_Q^2}{L_Q} \right) i_{q\Delta} \triangleq L_q'' i_{q\Delta} \tag{6.118}$$

Thus we have identified the changes in d and q circuit flux linkages in terms of the change in currents.

The voltage equations may also be written in terms of initial-plus-change notation. For the d axis voltage we write from (6.34)

$$v_d = v_{d0} + v_{d\Delta} = -r(i_{d0} + i_{d\Delta}) - \omega_1 L_q (i_{q0} + i_{q\Delta}) - \mathrm{k}\omega_1 M_Q (i_{Q0} + i_{Q\Delta}) - \dot{\lambda}_d$$

Since this equation must represent the solution at $t = 0^+$, we note that $\dot{\lambda}_d = 0$. (Why?) Also, by definition the initial condition is steady state so that $i_{Q0} = 0$. Finally, we note that $i_{Q\Delta}$ may be written in terms of $i_{q\Delta}$ from equation (6.116). Making these changes, we write $v_{d0} = -ri_{d0} - \omega_1 L_q i_{q0}$ and

$$v_{d\Delta} = -ri_{d\Delta} - \omega_1 L_q i_{q\Delta} + \frac{\omega_1 \mathrm{k}^2 M_Q^2}{L_Q} i_{q\Delta} = -ri_{d\Delta} - \omega_1 L_q'' i_{q\Delta}$$

or, adding the two equations, $v_d = -ri_d - \omega_1 L_q i_{q0} - \omega_1 L_q'' i_{q\Delta}$.

Adding and subtracting a quantity $\omega_1 L_q'' i_{q0}$ gives the result

$$v_d = -ri_d - \omega_1 L_q'' i_q + e_d'' \tag{6.119}$$

where we define

$$e_d'' \triangleq e_{d0} + e_{d\Delta} = -\omega_1 (L_q - L_q'')\, i_{q0} \tag{6.120}$$

From (6.96) it is apparent that $e_{d0} = 0$ since this component does not exist in the steady state.

For the q axis we similarly compute from (6.34)

$$\begin{aligned} v_q = v_{q0} + v_{q\Delta} &= -r(i_{q0} + i_{q\Delta}) + \omega_1 L_d (i_{d0} + i_{d\Delta}) + \mathrm{k}\omega_1 M_F (i_{F0} + i_{F\Delta}) \\ &\quad + \mathrm{k}\omega_1 M_D (i_{D0} + i_{D\Delta}) - \dot{\lambda}_q \end{aligned}$$

In this case i_{F0} = a constant, $i_{D0} = 0$, $\dot{\lambda}_q = 0$, and $i_{F\Delta}$ and $i_{D\Delta}$ are given by (6.115). Then

$$v_{q0} = -ri_{q0} + \omega_1 L_d i_{d0} + e_{q0} \tag{6.121}$$

where from (6.103)

$$e_{q0} = k\omega_1 M_F i_{F0} \tag{6.122}$$

Making substitutions of $i_{F\Delta}$ and $i_{D\Delta}$, we compute

$$v_{q\Delta} = -ri_{q\Delta} + \omega_1 L_d'' i_{d\Delta} \tag{6.123}$$

Equations (6.121) and (6.123) are combined to write

$$v_q = -ri_q + \omega_1 L_d'' i_d + e_q'' \tag{6.124}$$

where we define

$$e_q'' = e_{q0} + e_{q\Delta} \tag{6.125}$$

and

$$e_{q\Delta} = \omega_1 (L_d - L_d'')\, i_{d0} \tag{6.126}$$

It is interesting to note that the definitions for e_d'' and e_q'' parallel similar definitions for the steady state condition. In particular, we compute from (6.103) and (6.126) $e_q'' - e_{q0} = \omega_1 (L_d - L_d'')\, i_{d0}$, or

$$v_{xq} = e_{q0} + x_d i_{d0} = e_q'' + x_d'' i_{d0} \tag{6.127}$$

We may also compute $e_{d0} - e_d'' = \omega_1 (L_q - L_q'')\, i_{q0}$ or

$$v_{xd} = -x_q i_{q0} = e_d'' - x_q'' i_{q0} \tag{6.128}$$

since $e_{d0} = 0$.

To construct a phasor diagram for the subtransient case, we divide the v_d and v_q equations by $\sqrt{3}$ to rewrite the equations as rms stator equivalents. Thus

$$V_d = -rI_d - x_q'' I_q + E_d'', \qquad V_q = -rI_q + x_d'' I_d + E_q'' \tag{6.129}$$

where we note the striking similarity to the steady state equations (6.99), the only difference being the addition of the double primes ($''$) and the appearance of the voltage E_d''. As before, we compute on a *d-q* reference,

$$\overline{V}_a = V_q + jV_d = -r\overline{I}_a + x_d'' I_d - jx_q'' I_q + \overline{E}'' \tag{6.130}$$

where we define

$$\overline{E}'' = E_q'' + jE_d'' \tag{6.131}$$

Equation (6.130) may be changed to the form

$$\overline{E}'' = \overline{V}_a + r\overline{I}_a + jx_d''\overline{I}_a + j(x_q'' - x_d'')\,\overline{I}_q = E_q'' + jE_d'' \tag{6.132}$$

This form is convenient since the quantity $(x_q'' - x_d'')$ is positive on both round rotor and salient pole machines but is usually quite small.

The phasor diagram for the subtransient condition is constructed according to (6.132) in Figure 6.15 where the condition of Figure 6.12 is used as an initial condition. Several observations are in order concerning this important phasor diagram.

We note that the new (fault) current is larger than the initial current, and it

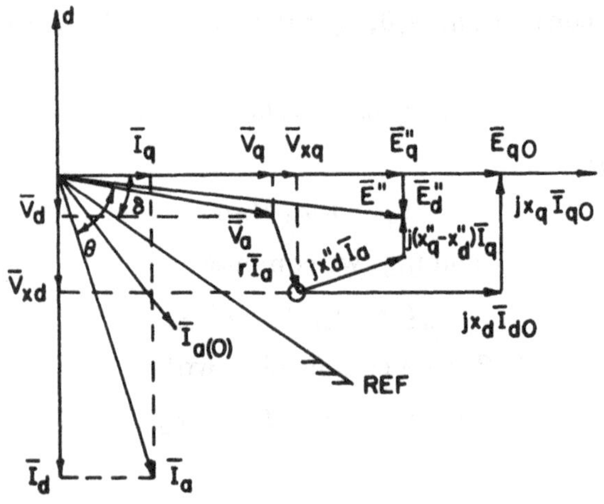

Fig. 6.15. Subtransient phasor diagram of a synchronous generator.

lags the terminal voltage by a larger angle θ, which is nearly 90° for a fault on the machine terminals. This means that I_q is small and I_d is large. Since the speed is assumed constant, δ has not changed.

An important point on the diagram is the encircled point located at

$$\bar{V}_{xq} + \bar{V}_{xd} = \bar{V}_a + r\bar{I}_a \tag{6.133}$$

But from (6.127) and (6.128)

$$\bar{V}_{xq} + \bar{V}_{xd} = (\bar{E}_{q0} - \mathrm{j}x_d\bar{I}_{d0}) + (-x_q\bar{I}_{q0}) = (\bar{E}''_q - \mathrm{j}x''_d\bar{I}_{d0}) + (\bar{E}''_d - x''_q\bar{I}_{q0}) \tag{6.134}$$

or this voltage is a constant and depends only upon the prefault or initial condition. Therefore, the location of the circled point is unchanged from $t = 0^-$ to $t = 0^+$ and forms a kind of pivot about which the other phasors change. Thus

$$\bar{V}_{xq} + \bar{V}_{xd} = \bar{V}_a + r\bar{I}_a = \bar{V}_{a(0)} + r\bar{I}_{a(0)} = \text{a constant} \tag{6.135}$$

Referring again to the phasor diagram, we note that the phasor $\mathrm{j}(x''_q - x''_d)\,\bar{I}_q$ is very small since from Table 6.1 $(x''_q - x''_d)$ is small, and I_q is usually small under fault conditions. This quantity might be neglected as an approximation, or set

$$\mathrm{j}(x''_q - x''_d)\,I_q \cong 0 \tag{6.136}$$

We observe that

$$E''_q >> E''_d \tag{6.137}$$

or approximately,

$$\bar{E}'' \cong \bar{E}''_q \tag{6.138}$$

This is quite apparent from the phasor diagram but could also be concluded from the definitions of E''_q and E''_d, where E_{q0} will be the dominant component.

If approximations (6.136) and (6.138) are allowed, the voltage equation (6.132) becomes

$$\bar{E}'' \cong \bar{E}''_q \cong \bar{V}_a + r\bar{I}_a + \mathrm{j}x''_d\bar{I}_a \tag{6.139}$$

Both the exact equation (6.132) and the approximate equation (6.139) can be represented by equivalent circuits, as shown in Figure 6.16. Note that E'' and E_q'' are constants since from (6.125) and (6.126) they depend only on the initial conditions.

The circuit of Figure 6.16b is widely used for fault calculations, often with the resistance neglected since it is small. We may think of this circuit as a sub-

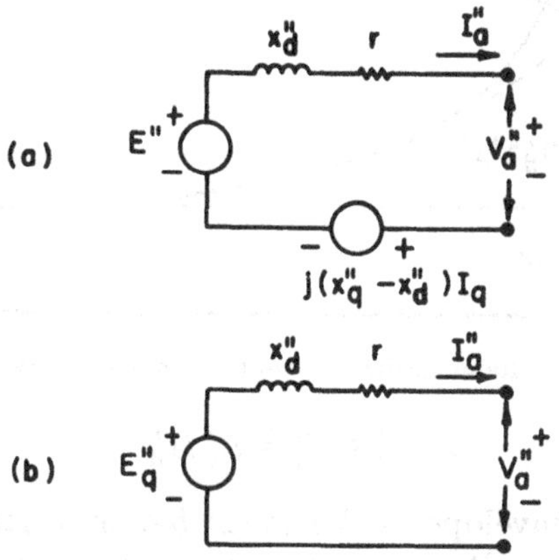

Fig. 6.16. Equivalent circuits of the subtransient condition: (a) exact equivalent, (b) approximate equivalent.

transient Thevenin equivalent since it is clearly a constant voltage behind a "constant" impedance. Obviously, it is valid only at the instant $t = 0^+$, but this is the time of greatest interest in fault calculations.

6.9 Armature Current Envelope

It is convenient to have a mathematical expression for the envelope of the armature currents of Figures 6.1 and 6.6. The envelope is the sum of several terms, each with its own time constant, which combine to determine the peak value or envelope of the current as a function of time.

The final value of the current in Figure 6.6 has a peak value 0-*a*. It is called the *synchronous component* of the envelope and may be plotted alone as the line *a-d*. Since the current of interest is a fault current, I_q will be small and

$$I_a \cong I_d \tag{6.140}$$

Making this substitution into (6.108), we compute for r negligible,

$$\bar{E}_q = \bar{V}_a + jx_d\bar{I}_d \tag{6.141}$$

or for a three-phase fault on the machine terminals ($V_a = 0$) the current magnitude I_d shown in Figure 6.17 is given by

$$I_a \cong I_d = E_q/x_d \tag{6.142}$$

Superimposed on the synchronous component of current is the *transient component* which is given by the quantity $(i_d' - i_d)$ in Figure 6.17. Equations for the transient condition have not been derived here but are identical to the results of section 6.8 if the double-prime notation is replaced with single primes. The result for a 3ϕ fault with negligible resistance can be computed from (6.139), with the result

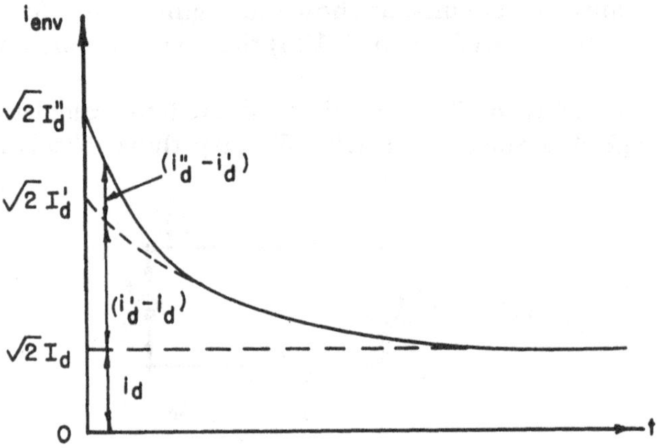

Fig. 6.17. The armature current ac envelope, excluding i_{dc}.

$$I'_a \cong I'_d = E'_q / x'_d \tag{6.143}$$

The transient current envelope is the locus *b-d* in Figure 6.6 and includes the synchronous component. It is usually convenient to write the *transient current envelope* as

$$i'_d - i_d = \sqrt{2}(I'_d - I_d)\, e^{-t/T'_d} \tag{6.144}$$

and this envelope decays with time constant T'_d (1 to 2 seconds).

The subtransient component of fault current is superimposed on the previously defined components. Its initial rms value for a 3ϕ fault on the generator terminals is found from (6.139) with r neglected. Thus

$$I''_a \cong I''_d = E''_q / x''_d \tag{6.145}$$

The *subtransient component envelope* is written as

$$i''_d - i'_d = \sqrt{2}\,(I''_d - I'_d)\, e^{-t/T''_d} \tag{6.146}$$

and this component decays very fast with a time constant of T''_d (2 or 3 cycles). This component determines the initial rms current if there is no dc component.

The ac component of the current envelope is the sum of the synchronous, transient, and subtransient components

$$i_{env} = \sqrt{2}\,[I_d + (I'_d - I_d)\, e^{-t/T'_d} + (I''_d - I'_d)\, e^{-t/T''_d}] \tag{6.147}$$

or the rms value of the current as a function of time is $1/\sqrt{2}$ times this amount, i.e., $I_{ac} = i_{env}/\sqrt{2}$. In terms of the machine reactances the rms value of alternating current as a function of time may be estimated from (6.142), (6.143), and (6.145) with the assumption that $E_q \cong E'_q \cong E''_q$ or

$$I_{ac} \cong E_q \left[\frac{1}{x_d} + \left(\frac{1}{x'_d} - \frac{1}{x_d}\right) e^{-t/T'_d} + \left(\frac{1}{x''_d} - \frac{1}{x'_d}\right) e^{-t/T''_d}\right] \tag{6.148}$$

Usually E_q is assumed to be in the range of 1.0 to 1.10 pu for fault computations.

The total current also has a dc component which depends on the switching angle α (radians) given by (see [19])

$$I_{dc} = \frac{\sqrt{2}\, E_q \cos\alpha}{x''_d}\, e^{-t/T_a} \tag{6.149}$$

Then the effective value of armature current is

$$I = (I_{ac}^2 + I_{dc}^2)^{1/2} \tag{6.150}$$

which has a maximum value at $t = 0$ corresponding to $\alpha = 0$ of

$$I_{max} = \sqrt{3}E_q/x_d'' \tag{6.151}$$

Values of x_d, x_d', x_d'', T_d', T_d'', and T_a may all be found from an oscillograph of armature currents similar to Figure 6.1. The technique for doing this is explained fully in [19] and will not be repeated here. These values, expressed in pu, tend to be nearly constant for a particular type of machine and are often tabulated for use in cases where the actual machine parameters are unavailable. Reference values are given in Table 6.1.

Since our motivation is to perform fault calculations, we are often concerned with fault analysis under extremum conditions; i.e., we are trying to determine either the maximum or minimum value of fault current seen by a particular relay or protective device. For maximum current we usually take the current I_d'' as the rms value of fault current, or we consider that the generator has reactance x_d'' in the positive sequence network, with x_2 and x_0 in the other networks. For other than maximum fault conditions the value of reactance used may be changed to a value which will give the correct current at the time the protective device operates. Thus the time may be substituted into (6.148), and the current and hence the equivalent reactance found.

6.10 Momentary Currents

The equivalent circuit developed in section 6.8 will permit the solution of the subtransient current I_d'' corresponding to the highest point in the current envelope of Figure 6.17. This current is the symmetrical current, i.e., it includes no dc offset at all and corresponds to the highest rms value of Figure 6.6. If our purpose in computing the fault current is the selection of a circuit breaker to interrupt I_d'' either at the generator or some other location, we should include the dc offset. Thus the total effective value of the current must be computed as in (6.150). If this is done, for the worst case the result is that of (6.151) for 100% dc offset in a given phase, or

$$I_{d\,max}'' = \sqrt{3}E_q''/x_d'' = 1.732\,I_{d\,sym}'' \tag{6.152}$$

where $I_{d\,sym}''$ = the symmetrical part of the total current. Actually, the current to be interrupted by the circuit breaker will never be this great since this would require zero fault detection time (relay time) and zero breaker operating time.

The IEEE, recognizing that some consideration should be given to the inclusion of dc offset in the selection of breakers, has formulated a recommended method for computing the value of current to be used in these computations. The computed current, called the *momentary current*, depends upon the time after initiation of the fault at which the circuit will be likely to be interrupted. Thus the natural decrement of the dc offset, which occurs at time constant T_a, is accounted for approximately and provides a reasonable compromise between the limits of $I_{d\,sym}''$ and $1.732\,I_{d\,sym}''$. In the recommended method of computation, the momentary duty of the breaker is based upon the symmetrical current $I_{d\,sym}''$, with multipliers to be applied according to the breaker speed. These multipliers are given in Table 6.2 which is taken from [51]. These multipliers determine the so-called "interrupting duty" of the circuit breaker. Another rating called the

Table 6.2. Current Multiplying Factors for Computing Circuit Breaker Interrupting Rating

Breaker Speed or Location	$I''_{d\,sym}$ *Multiplier M*
8 cycle or slower	1.0
5 cycle	1.1
3 cycle	1.2
2 cycle	1.4
Located on generator bus	add 0.1 to multiplier
If sym fault MVA > 500	add 0.1 to multiplier

"momentary duty" is found by multiplying the symmetrical current by 1.6, and this may be checked against a published breaker momentary rating. (The present practice is to use only the symmetrical interrupting duty, and breakers are so rated.) In summary the two breaker ratings which must be checked are

$$\text{momentary rating} > 1.6\, I''_{d\,\text{sym}}$$
$$\text{interrupting rating} > M\, I''_{d\,\text{sym}} \tag{6.153}$$

where M is found from Table 6.2.

II. INDUCTION MOTOR IMPEDANCES

6.11 General Considerations

The striking difference between induction and synchronous machines, insofar as their response to faults is concerned, is in the method of machine excitation. A synchronous machine obtains its excitation from a separate dc source which is virtually unaffected by the fault. Thus as the prime mover continues to drive the synchronous generator, excited at its prefault level, it responds by forcing large transient currents toward the fault. The induction machine, on the other hand, receives its excitation from the line. If the line voltage drops, the machine excitation is reduced and its ability to drive the mechanical load is greatly impaired. If a 3ϕ fault occurs at the induction motor terminals, the excitation is completely lost; but because of the need for constant flux linkages, the machine's residual excitation will force currents into the fault for one or two cycles. During these first few cycles following a fault the contribution of induction motors to the total fault current should not be neglected. However, it is somewhat unusual to find an induction motor large enough to make a significant difference in the total current. For example, if the base MVA is chosen to be close to the size of the average synchronous generator, an induction motor would have to be larger than about 1% of base MVA to be of any consequence. When induction motors are included in the computation, a subtransient reactance of about 25% (on the motor base) is often used. The transient reactance is infinite.

6.12 Induction Motor Equivalent Circuit

The induction motor is usually represented by a "transformer equivalent," i.e., a T equivalent where separate series branches represent the stator (primary) circuit and the rotor (secondary) circuit, with a shunt branch to represent iron losses and excitation. At standstill or locked rotor the induction machine is indeed a transformer. With the rotor turning, however, the equivalent impedance of the secondary or rotor circuit is seen to be a function of the slip s as noted in Figure 6.18. If the rotor induced EMF is E_r at standstill, it is sE_r when the rotor is turning at

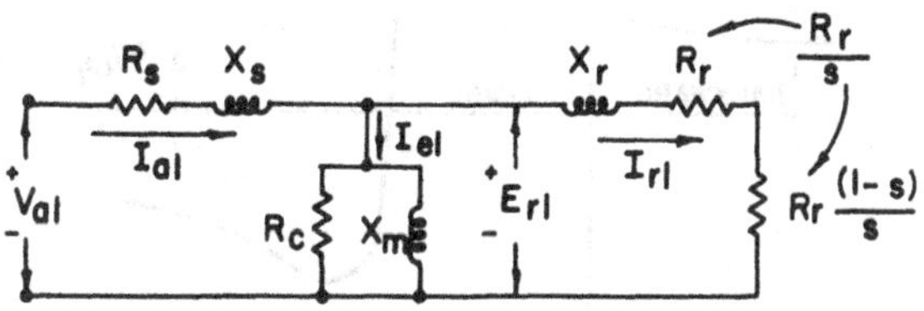

Fig. 6.18. Positive sequence induction motor equivalent circuit.

slip s. Similarly, the rotor apparent inductance is a function of slip. Since induced rotor currents have frequency sf where f is the stator frequency, the reactance of the rotor will be sX_r where X_r is the rotor reactance at standstill. Then, we compute at any slip s, $I_{r1} = sE_{r1}/(R_r + jsX_r)$ or

$$I_{r1} = \frac{E_{r1}}{(R_r/s) + jX_r} \tag{6.154}$$

Thus the rotor impedance is

$$Z_{r1} = (R_r/s) + jX_r \tag{6.155}$$

as noted in Figure 6.18. We usually write (6.155) as

$$Z_{r1} = (R_r + jX_r) + \frac{1-s}{s}R_r$$

where the first term is the impedance at standstill and is not a function of s. The second term represents the shaft load to which the power delivered is

$$P_{m1} = I_{r1}^2 R_r \frac{1-s}{s} \quad \text{W/phase} \tag{6.156}$$

The shaft or mechanical torque is

$$T_{m1} = \frac{P_{m1}}{\omega} = \frac{P_{m1}}{\omega_1(1-s)} \quad \text{Nm/phase} \tag{6.157}$$

$$= \frac{I_{r1}^2 R_r}{\omega_1 s} \quad \text{Nm/phase} \tag{6.158}$$

where

ω = shaft speed in rad/sec
ω_1 = synchronous speed in rad/sec
s = slip in pu

If the mechanical torque is expressed in pu, we select the base torque in terms of base power (base voltamperes) and base speed ω_1. Then from (6.157)

$$\text{pu } T_{m1} = \frac{P_{m1}/\omega_1(1-s)}{S_B/\omega_1} = \frac{P_{m1u}}{1-s} = \frac{I_{r1}^2 R_r}{s} \quad \text{pu} \tag{6.159}$$

which has units of pu power/pu slip or pu torque. Some authors state pu T_{m1} in

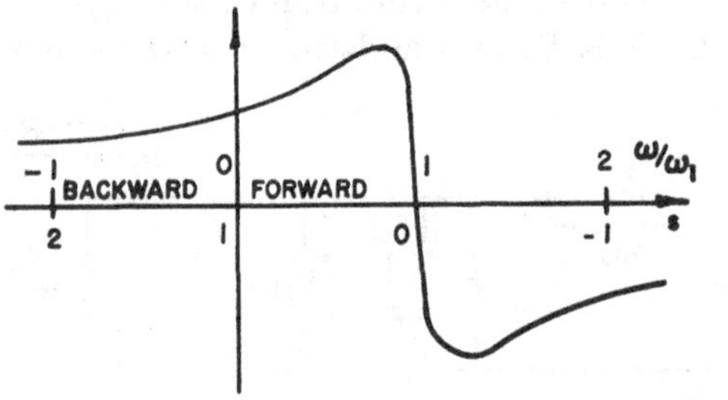

Fig. 6.19. Torque of an induction motor as a function of speed or slip. (From Kimbark [19]. Used with permission.)

"synchronous watts" (see [19]). The average torque-speed characteristic of an induction motor is shown in Figure 6.19.

If negative sequence voltages are applied to the induction motor, a revolving MMF wave is established in the machine air gap which is rotating backwards, or with a slip of 2.0 pu. Then the slip of the *rotor* with respect to the negative sequence field is $2 - s$ when the rotor is moving with forward rotation. Since s is small, the approximation is sometimes made that this negative sequence slip is 2.0 rather than $2 - s$. If s_2 is the negative sequence slip, $s_2 = 2 - s \cong 2$.

The equivalent circuit for negative sequence currents is the same as that of Figure 6.18 with s replaced by s_2 as shown in Figure 6.20. The mechanical power

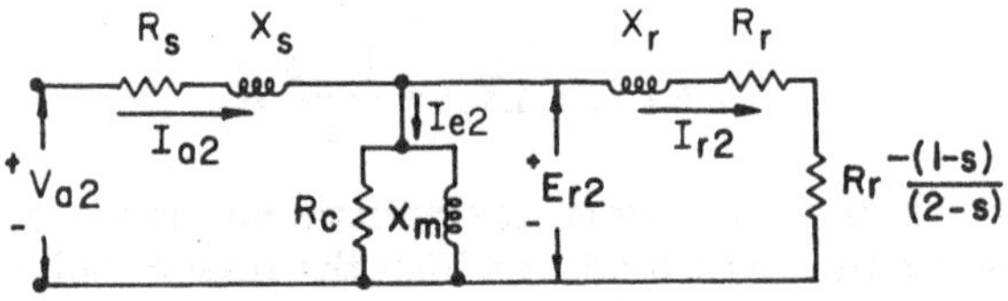

Fig. 6.20. Negative sequence induction motor equivalent circuit.

associated with the negative sequence rotor current I_{r2} is

$$P_{m2} = -I_{r2}^2 \, R_r \frac{(1-s)}{(2-s)} \quad \text{W/phase} \tag{6.160}$$

the negative sign indicating that this would cause a retarding torque T_{m2} where

$$T_{m2} = \frac{I_{r2}^2 \, R_r}{\omega_1 (2-s)} \quad \text{Nm/phase} \tag{6.161}$$

Then the net torque per phase is

$$T_m = T_{m1} + T_{m2} = \frac{R_r}{\omega_1}\left(\frac{I_{r1}^2}{s} - \frac{I_{r2}^2}{2-s}\right) \quad \text{Nm/phase} \tag{6.162}$$

which is a smaller torque than that present when only positive sequence voltages are applied. The net effect of unbalanced voltages applied to an induction motor is to reduce the mechanical torque. Since the speed of the motor depends upon the simultaneous matching of motor and load torque-speed characteristics, the effect of the unbalance depends on the type of load served by the motor. Should the motor applied voltage change but remain balanced, the torque T_{m1} will vary as the square of the applied voltage since I_{r1} varies directly with V_{a1}. Depending

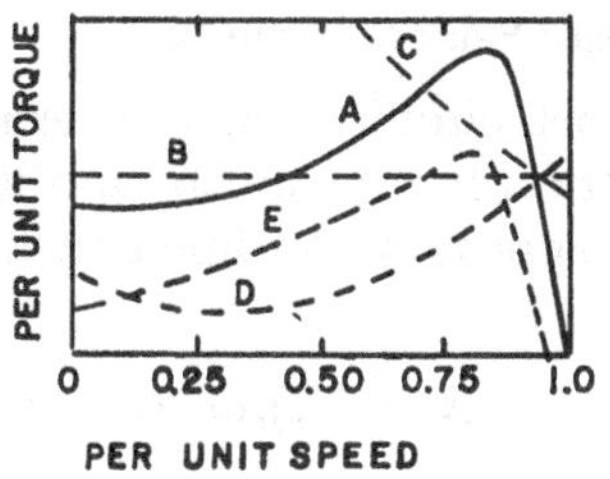

Fig. 6.21. Torque-speed relationships for motor and load: (A) motor torque-speed curve at rated voltage, (B) constant torque load, (C) constant power load, (D) typical load torque-speed curve, (E) motor torque-speed curve with unbalanced applied voltages. (From Clarke [11, vol. 2]. Used with permission.)

on the load characteristic, a 3ϕ fault near the motor or a reduced voltage may cause the motor to stall. A system disturbance which unbalances the voltage will also cause the torque of the motor to be reduced, as shown by curve E of Figure 6.21. In this case the motor will slow down and seek a new operating point on the torque-speed characteristic with the load. If the load is constant torque such as a conveyor or is constant power such as a regulated dc generator driving a constant impedance load, the motor may stall if it is unable to deliver the necessary torque at the lower speed. For a load such as a fan or pump the torque varies about as the square of the speed, such as curve D of Figure 6.21. Such a load would continue to be served at a reduced speed if the motor torque is reduced due to any cause, unless the reduction in motor torque drops practically to zero. By the use of equivalent circuits for the positive and negative sequence networks, these torques can be determined and the motor performance may be evaluated if the torque-speed characteristic of the shaft load is known or can be estimated.

Since induction motors are usually wound either for Δ or ungrounded Y connection, the zero sequence currents in the motor are always zero and there is no need for a zero sequence equivalent circuit.

Approximate values for induction motor equivalent circuits are given by Clarke [11, vol. 2] and credited to several references. This data is given in Table 6.3. Clarke suggests that the positive sequence impedance for any load may also be determined by dividing rated voltage by the assumed pu current or by computing the machine driving-point impedance. However, as the actual current would seldom be known, Figures 6.18 and 6.20 may be used for the equivalent circuits of positive and negative sequence networks respectively.

Table 6.3. Approximate Constants for Three-Phase Induction Motors

Rating	*Full Load Efficiency*	*Full Load Power Factor*	*Full Load Slip*	*R and X in per Unit**			
				$X_s + X_r^{\dagger}$	X_m	R_s	R_r
(HP)	*(%)*	*(%)*	*(%)*	*(pu)*	*(pu)*	*(pu)*	*(pu)*
Up to 5	75-80	75-85	3.0-5.0	0.10-0.14	1.6-2.2	0.040-0.06	0.040-0.06
5-25	80-88	82-90	2.5-4.0	0.12-0.16	2.0-2.8	0.035-0.05	0.035-0.05
25-200	86-92	84-91	2.0-3.0	0.15-0.17	2.2-3.2	0.030-0.04	0.030-0.04
200-1000	91-93	85-92	1.5-2.5	0.15-0.17	2.4-3.6	0.025-0.03	0.020-0.03
Over 1000	93-94	88-93	~1.0	0.15-0.17	2.6-4.0	0.015-0.02	0.015-0.025

Source: Clarke [11, vol. 2]. Used with permission of the publisher.
*Based on full load kVA rating and rated voltage.
†Assume that $X_s = X_r$ for constructing the equivalent circuit.

6.13 Induction Motor Subtransient Fault Contribution

The application of a short circuit near the terminals of an induction motor removes the source of excitation for the motor and its field collapses very rapidly. Clarke [11, vol. 2] gives the approximate time constant of the decay of rotor flux as

$$T_r = (X_s + X_r)/\omega_1 R_r \quad \text{sec} \tag{6.163}$$

where the quantities used in the equation are defined in Figure 6.18 and are assumed to be in ohms. However, since the ratio of X to R is used in (6.163), these quantities may be in pu. From Table 6.3, if we take 0.16 and 0.035 as average values of $X_s + X_r$ and R_r respectively, we compute $t_r = 0.0121$ sec for a 60 Hz motor which is less than one cycle (1 cycle = 0.01667 sec if $f = 60$ Hz). The current to be interrupted by the circuit breaker in transmission systems is that which exists 2–4 cycles after the fault occurs. In this case the current contribution from induction motors may be neglected. In industrial plants, however, low-voltage, instantaneous, air circuit breakers are often used which clear faults in about one cycle. In such cases the fault contribution from induction motors should not be neglected.

However, if the maximum value of current is required for computing fuse melting or circuit breaker mechanical stresses, the contribution from an induction motor may be found by treating it during this subtransient period as a synchronous machine. That is, we will compute a generated EMF behind a reactance, and this will constitute the subtransient equivalent circuit as shown in Figure 6.22.

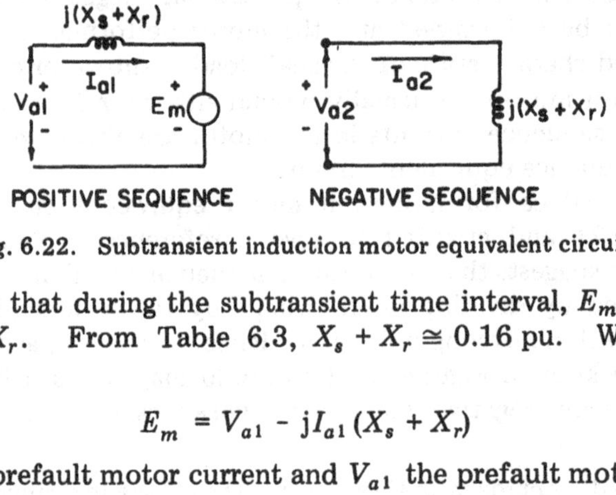

Fig. 6.22. Subtransient induction motor equivalent circuits.

Here, we assume that during the subtransient time interval, E_m acts through the reactance $X_s + X_r$. From Table 6.3, $X_s + X_r \cong 0.16$ pu. We compute from Figure 6.22,

$$E_m = V_{a1} - jI_{a1}(X_s + X_r) \tag{6.164}$$

where I_{a1} is the prefault motor current and V_{a1} the prefault motor voltage.

Example 6.1

Approximate the subtransient value of E_m for a 100 hp induction motor, operating before the fault at rated voltage and drawing rated current at 88% power factor. Then find the subtransient contribution of the motor to a 3ϕ fault at the motor terminals.

Solution

From (6.164), taking V_{a1} as the reference phasor,

$$E_m = 1.0\underline{/0^\circ} - j(1.0\underline{/-28.5^\circ})(.16) = 0.9235 - j0.1406 = 0.935\underline{/-9.35^\circ} \text{ pu}$$

Therefore, for a 3ϕ fault on the motor terminals,

$$I_m = \frac{0.935\angle -9.35^\circ}{0.16\angle 90^\circ} = 5.85\angle -99.35^\circ \quad \text{pu}$$

where the current is in pu on the rated base kVA of the motor.

6.14 Operation with One Phase Open

If one of the three leads supplying an induction motor should become open, the resulting series unbalance may be evaluated by symmetrical components. In such a case the boundary conditions are

$$I_a = 0, \qquad I_b = -I_c, \qquad V_{bc} = V_b - V_c \tag{6.165}$$

From the first two boundary conditions we compute $\mathbf{I}_{012} = \mathbf{A}^{-1}\,\mathbf{I}_{abc}$ or (let h = 1)

$$\begin{bmatrix} I_{a0} \\ I_{a1} \\ I_{a2} \end{bmatrix} = \frac{1}{3}\begin{bmatrix} 1 & 1 & 1 \\ 1 & a & a^2 \\ 1 & a^2 & a \end{bmatrix}\begin{bmatrix} 0 \\ I_b \\ -I_b \end{bmatrix} = \begin{bmatrix} 0 \\ a - a^2 \\ a^2 - a \end{bmatrix}\frac{I_b}{3}$$

$$= \begin{bmatrix} 0 \\ 1 \\ -1 \end{bmatrix} \mathrm{j}\frac{I_b}{\sqrt{3}} \tag{6.166}$$

from which we see that $I_{a1} = -I_{a2}$. This means that the positive and negative sequence networks must be connected as shown in Figure 6.23.

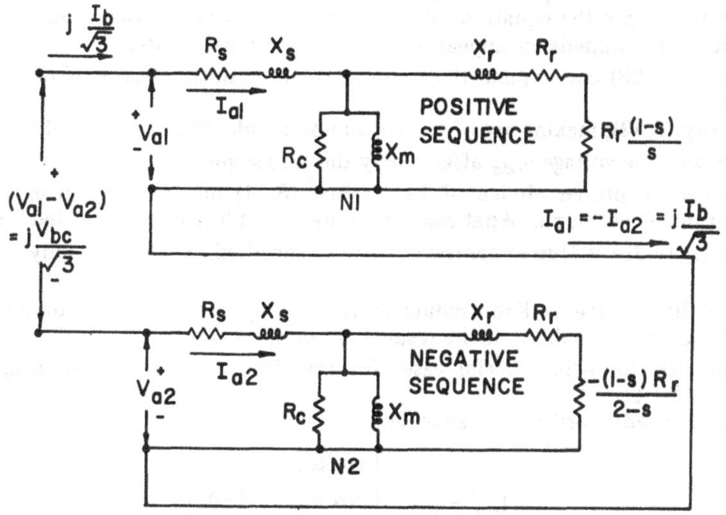

Fig. 6.23. Sequence network connections for an induction motor with line *a* open.

The only known voltage is, from (6.165), with h = 1

$$V_{bc} = V_b - V_c = (V_{a0} + a^2 V_{a1} + aV_{a2}) - (V_{a0} + aV_{a1} + a^2 V_{a2})$$
$$= -\,\mathrm{j}\sqrt{3}(V_{a1} - V_{a2}) \tag{6.167}$$

Summarizing from (6.166) and (6.167)

$$I_{a1} = -I_{a2} = j\frac{I_b}{\sqrt{3}}, \qquad V_{a1} - V_{a2} = j\frac{V_{bc}}{\sqrt{3}} \tag{6.168}$$

and if V_{bc} is known, I_b may be determined from Figure 6.23. This completely solves the sequence networks (assuming that s is known), and the net torque may be computed from (6.162). Wagner and Evans [10] show that the net torque in this situation remains positive as long as the slip is small. This means that the shaft load must not be too large, or the motor will stall. Clarke [11, vol. 2] discusses the case of an open conductor in the supply to an induction motor where a capacitor bank is in parallel with the motor. In this case a negative torque may result which will cause the induction motor to reverse direction of rotation. This condition will not be discussed here.

Problems

6.1. Consider a series R-L circuit supplied from an ac voltage source $v(t) = V_m \sin(\omega t + \alpha)$ and assume that initially the current is zero because of an open switch in the circuit. If the switch is closed at $t = 0$, find the current as a function of the series impedance magnitude and angle, $Z = |Z| \angle\theta$. Sketch the resulting current for $\alpha - \theta = 0, 45^\circ$, and 90°.

6.2. Express the voltage and final value of current for problem 6.1 as phasors. Show how the dc offset can be found by projecting the current phasor on the real axis.

6.3. Verify the inductances given by (6.7)-(6.13).

6.4. Show that $\mathbf{P}^{-1}$ of (6.19) is indeed the inverse of $\mathbf{P}$ given in (6.17).

6.5. In Figure 6.3 consider that all self and mutual inductances are constants instead of the time varying inductances given by (6.6)-(6.14). Write the equations for the synchronous machine voltages, then transform by the phasor transformation, and transform again to the 0-1-2 coordinate system.

6.6. Repeat 6.4 with an impedance $Z_n = r_n + j\omega L_n$ in the neutral and a current I_n entering the neutral. Write the equations of phase voltage to *ground* rather than to neutral. How does the neutral impedance appear in the 0-1-2 coordinate system?

6.7. (a) Verify (6.28) and explain the meaning of the results. What kind of induced voltage is s?
(b) Verify (6.32), making use of the trigonometric identities of Appendix C.

6.8. Explain why the voltage $\mathbf{n}_{0dq}$ affects only the zero sequence.

6.9. Show that by proper choice of base values (6.34) may be written exactly the same whether in volts or in pu. What restriction does this place on the selected base values?

6.10. Prove that (6.35) is true under the conditions specified. Do this for only one phase, e.g., λ_b/i_b.

6.11. Given (6.20), reduce by Kron reduction to write $\mathbf{v}_{abc}$ in terms of v_f only by eliminating the rotor current variables. Is the result Laplace transformable?

6.12. Examine the following special cases for the Park's transformation $\mathbf{i}_{0dq} = \mathbf{P}\mathbf{i}_{abc}$ with $\theta = \omega_1 t + \delta + 90$.
(a) $\mathbf{i}_{abc}$ is strictly positive sequence, i.e.,

$$\mathbf{i}_{abc} = \sqrt{2}I \begin{bmatrix} \cos\omega_1 t \\ \cos(\omega_1 t - 120) \\ \cos(\omega_1 t + 120) \end{bmatrix}$$

find $\mathbf{i}_{0dq}$.
(b) $\mathbf{i}_{abc}$ is strictly negative sequence, i.e.,

$$\mathbf{i}_{abc} = \sqrt{2}I \begin{bmatrix} \cos\omega_1 t \\ \cos(\omega_1 t + 120) \\ \cos(\omega_1 t - 120) \end{bmatrix}$$

find $\mathbf{i}_{0dq}$.

(c) $\mathbf{i}_{abc}$ is strictly zero sequence, i.e.,

$$\mathbf{i}_{abc} = \sqrt{2} I \cos \omega_1 t \begin{bmatrix} 1 \\ 1 \\ 1 \end{bmatrix}$$

find $\mathbf{i}_{0dq}$.

6.13. Given the unbalanced set of currents

$$\mathbf{i}_{abc} = \begin{bmatrix} I_{am} \cos(\omega_1 t + \alpha) \\ I_{bm} \cos(\omega_1 t - 120 + \beta) \\ I_{cm} \cos(\omega_1 t + 120 + \gamma) \end{bmatrix}$$

compute $\mathbf{i}_{0dq}$.

6.14. Verify equation (6.46) and explain the meaning of this result in two or three sentences.

6.15. Compute $\mathbf{v}_{0dq}$ from equation (6.34) under the condition that (a) $v_F = 0$ and (b) $v_F - r_F i_F = 0$.

6.16. Compute $\mathbf{v}_{0dq}$ by making a Park's transformation of the result of problem 6.11.

6.17. Verify that the equivalent T circuit of Figure 6.8 can be derived from the coupled circuits of Figure 6.7.

6.18. Show how (6.79) can be satisfied by proper choice of base.

6.19. Construct equivalent T circuits for a typical synchronous machine, using values from Table 6.1 where the machine is (a) round rotor and (b) salient pole.

6.20. Find the impedance seen looking into the generator d and q terminals of the equivalent T circuit under (a) steady state conditions, (b) transient conditions, and (c) subtransient conditions

6.21. Compute numerical values for the impedances found in problem 6.20 using the data from problem 6.19.

6.22. Find the time constant of the impedance seen looking into the generator terminals under (a) steady state conditions, (b) transient conditions, and (c) subtransient conditions. Use data from problem 6.19.

6.23. A synchronous generator is operating in the steady state and supplying a lagging power factor load. Assume that the machine is connected to a "solid" bus with several other generators. Explain, using before and after phasor diagrams, what happens if the operator increases the field voltage (at constant power) (a) gradually and (b) suddenly.

6.24. Repeat problem 6.23, but this time assume that the load is increased while the excitation remains constant.

6.25. Verify (6.82) by performing the $\mathbf{P}$ transformation.

6.26. Find I_a if $i_d = 0.5$, $i_q = 0.8$, and $i_0 = 0$. Sketch a phasor diagram. Find I_{ar} and I_{ax} if $\delta = 60°$.

6.27. Explain in words how we can interpret $\mathrm{k}x_{mF} I_F$ as an induced EMF in equation (6.104).

6.28. Use data from Table 6.1 for a synchronous machine loaded such that on an arbitrary basis we have

$$\tilde{V}_a = 1.0\underline{/30°}, \qquad \tilde{I}_a = 0.8\underline{/-15°}$$

Then compute $\tilde{E}_q$ and construct a phasor diagram similar to Figure 6.12 if the machine is (a) round rotor and (b) salient pole.

6.29. Construct the equivalent circuit similar to Figure 6.13 for the machines of problem 6.28.

6.30. Using the condition of problem 6.28 as the initial condition for a 3ϕ fault, compute the initial flux linkages at time $t = 0^+$.

6.31. Compute all necessary quantities to sketch the phasor diagram immediately following a 3ϕ fault on the generator terminals, using the condition of problem 6.28 as the prefault condition.

6.32. Compute the maximum current two cycles after the fault of problem 6.31 is applied. Repeat for 70 cycles and 150 cycles.

6.33. Estimate the interrupting and momentary rating of the circuit breaker required to interrupt a 3ϕ fault on the terminals of the following round rotor generators: (a) 150 MVA, (b) 600 MVA, and (c) 900 MVA.

6.34. Justify the approximations specified in equations (6.57) and (6.58).

6.35. A cylindrical rotor, synchronous machine having (average) machine constants given in Table 6.1 is operating at rated current (1.0 pu) and 90% lagging power factor when a 3ϕ fault occurs on the machine terminals. Compute:
(a) The voltage E behind the synchronous impedance.
(b) The voltage E'' behind the subtransient impedance.
(c) The initial symmetrical subtransient fault current.
(d) The peak symmetrical current after 5 cycles and 10 cycles.
(e) The maximum asymmetrical current after 5 cycles and 10 cycles.

6.36. Repeat problem 6.35 for a salient pole machine.

6.37. A 1000 hp induction motor operates at a slip of 2.0% and with 93% efficiency while driving a rated shaft load. Use average values from Table 6.3 to construct a positive sequence equivalent circuit for this machine when the base MVA is the machine base. The motor is rated at 4 kV.

6.38. If the motor of problem 6.37 operates in a large system which is under study, recompute the equivalent circuit if the base for the system study is 100 MVA, 200 MVA, and 500 MVA.

6.39. Suppose that the voltage supplying the motor of problem 6.37 has a 5% negative sequence component. Compute the positive, negative, and net torque in pu.

6.40. Compute the time constant for the decay of rotor flux for the motor of problem 6.37 by applying (6.163). What is the time constant in cycles at a frequency of 60 Hz?

6.41. Suppose that line a of the supply to the motor of problem 6.37 is opened. Find the sequence voltages and currents if V_{bc} is 1.0 pu on a *line-to-line* basis.

chapter 7

Sequence Impedance of Transformers

An examination of the transformer completes our study of the major components of the power system. Insofar as the study of faulted networks is concerned, we need to solve two kinds of problems. First, given a particular transformation, we must determine the sequence network representation of the physical device. This requires a knowledge of transformer equivalents, standard terminal markings, and the various kinds of transformer connections. The second problem is the estimation of reasonable transformer parameters for installations still in the planning stage. In this case a knowledge of average impedance values is required, and the way in which the various transformer connections affect the sequence networks is helpful. Since the type of transformer connection often influences the type of protective scheme, these concepts must be studied in system planning to assure workability and reliability in the overall design.

I. SINGLE-PHASE TRANSFORMERS

Single-phase transformers, connected to form three-phase banks, were in wide general use for transforming both high and low voltages in the first half of the twentieth century. One reason for this was that a fourth transformer could be purchased and installed with a three-phase bank for later connection in the event one single-phase transformer failed. Recently the trend has been toward the use of three-phase transformers because they are cheaper and more efficient and are generally regarded as very reliable. Still, however, many single-phase units are in service, and large transformers are often single phase because of limitations in shipping sizes and weights.

7.1 Single-Phase Transformer Equivalents

The equivalent circuit often used for a single-phase transformer is shown in Figure 7.1a where the high and low voltage leakage impedances Z_H and Z_X are given in ohms. The transformer core losses are assumed to vary as the square of the H-winding voltage and are represented by R_c. The rms value of magnetizing current is represented by the reactance X_m. The turns ratio is defined as

$$n = n_H / n_X \tag{7.1}$$

where

n_H = number of turns in winding H
n_X = number of turns in winding X

We usually eliminate the magnetization branch since $I_e << I_H$ and also transfer the impedance Z_X to the H side of the circuit as shown in Figure 7.1b. Thus at open circuit

$$n = V_H / V_X \tag{7.2}$$

and, since the ampere-turns of the transformer windings are equal except for the excitation MMF $n_H I_e$, we have $n_H I_H = n_X I_X$ or

$$n = n_H / n_X = I_X / I_H \tag{7.3}$$

Modern substation transformers usually have I_e of less than 1% in sizes up to about 10 MVA and high-voltage ratings up to about 69 kV.

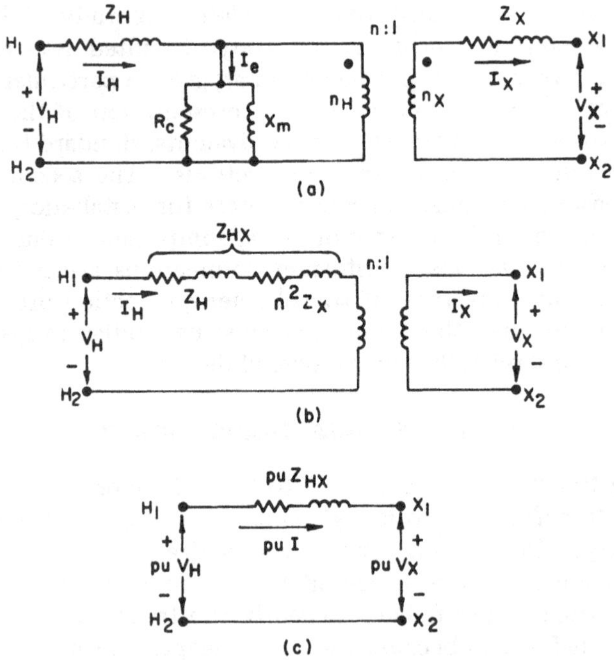

Fig. 7.1. Single-phase, two-winding transformer equivalents: (a) equivalent in system quantities, (b) simplified equivalent with all series impedance referred to the H winding, (c) pu equivalent.

Any X-winding current I_X is seen by the primary side as I_X / n and the impedance Z_X appears to I_H to be $n^2 Z_X$. Thus the total transformer impedance as seen by the H-winding current is

$$Z_{HX} = Z_H + n^2 Z_X \tag{7.4}$$

which is shown in Figure 7.1b as the total transformer impedance where, since I_e is small, the shunt branch is eliminated.

It is convenient to change Figure 7.1b to a pu equivalent circuit. We arbitrarily select the base quantities as

$$S_B = \text{transformer VA rating}$$
$$V_{HB} = \text{rated } V_H$$
$$V_{XB} = \text{rated } V_X \tag{7.5}$$

Then

$$I_{HB} = S_B / V_{HB} \qquad I_{XB} = S_B / V_{XB} \tag{7.6}$$

and

$$Z_{HB} = V_{HB}^2 / S_B \qquad Z_{XB} = V_{XB}^2 / S_B \tag{7.7}$$

From (7.2) and (7.5) we compute in pu *at no load*, pu V_X = pu V_H, and the pu turns ratio is

$$\text{pu } n = 1.0 \tag{7.8}$$

Also, from (7.7) we compute the total transformer impedance as viewed from the H winding to be

$$\text{pu } Z_{HX} = (Z_H + n^2 Z_X)/Z_{HB} \tag{7.9}$$

Finally, from (7.3) we compute

$$\text{pu } I_H = \text{pu } I_X \tag{7.10}$$

From (7.8)-(7.10) we establish the equivalent circuit of Figure 7.1c and this is the equivalent we will almost always use for load flow, fault, and stability studies. (In the study of ferroresonance, switching surges, traveling waves, harmonics, etc., the magnetizing reactance should not be ignored.)

7.2 Transformer Impedances

Transformer impedances are nearly always given in pu (or percent) based on the transformer rating. Thus from (7.9)

$$\text{pu } Z_{HX} = (Z_H + n^2 Z_X)/Z_{HB} = \text{pu } Z_H + n^2 Z_X / Z_{HB} = \text{pu } Z_H + \text{pu } Z_X \tag{7.11}$$

This result may be verified by dividing Z_{HB} by Z_{XB} to compute

$$Z_{HB}/Z_{XB} = (V_{HB}/V_{XB})^2 = n^2 \tag{7.12}$$

Values of pu Z_{HX} are nearly constant for transformers of a given size and design. Tables of average values are available from the manufacturers and may be used when actual nameplate data are not known. Table 7.1 gives typical values for two-winding distribution transformers rated 500 kVA and below. Note that in the larger sizes the impedance is almost entirely inductive reactance. In substation applications where these larger sizes are sometimes used, the resistance is often neglected entirely. Usually the resistance is considered only when losses, efficiency, or economic studies are being considered.

Table 7.2 gives typical values for large two-winding power transformers. This table is convenient for estimating impedances of transformers where different methods of cooling may be under consideration. Note that a range of impedances is specified for each voltage class. If the transformer is self cooled (OA), the impedance should be taken from the lower end of the range specified. Forced cooling allows a transformer of a given size to dissipate heat faster and operate at

Table 7.1. Distribution Transformer Impedances, Standard Reactances and Impedances for Ratings 500 kVA and below (for 60 Hz transformers)

	Rated-Voltage Class in kV							
	2.5		*15*		*25*		*69*	
*Single-Phase kVA Rating**	*Average Reactance %*	*Average Impedance %*	*Average Reactance %*	*Average Impedance %*	*Average Reactance %*	*Average Impedance %*	*Average Reactance %*	*Average Impedance %*
3	1.1	2.2	0.8	2.8				
10	1.5	2.2	1.3	2.4	4.4	5.2		
25	2.0	2.5	1.7	2.3	4.8	5.2		
50	2.1	2.4	2.1	2.5	4.9	5.2	6.3	6.5
100	3.1	3.3	2.9	3.2	5.0	5.2	6.3	6.5
500	4.7	4.8	4.9	5.0	5.1	5.2	6.4	6.5

Source: Westinghouse Electric Corp. [14]. Used with permission.
*For three-phase transformers use 1/3 of the three-phase kVA rating, and enter table with rated line-to-line voltages.

a greater voltampere loading and therefore should be represented by a greater impedance than a similar rating in a self cooled unit. Resistance is usually neglected in power transformers.

7.3 Transformer Polarity and Terminal Markings

The terminals of a single-phase transformer manufactured in the United States are marked according to specifications published by the American National Standards Institute [52], often called ASA Standards. These ASA Standards specify that the highest voltage winding be designated *HV* or *H* and that numbered subscripts be used to identify the terminals, e.g., H_1 and H_2. The low-voltage winding is designated *LV* or *X* and is subscripted in a similar way. If there are more than two windings, the others are designated *Y* and *Z* with appropriate subscripts.

This same marking scheme for windings permits the identification of taps in a winding. Thus in a given winding the subscripts $1, 2, 3, \ldots, n$ may be used to identify all terminals, with one and n marking the full winding and the intermediate numbers $2, 3, \ldots, n-1$ marking the fractional windings or taps. These numbers are arranged so that when terminal $i+1$ is positive (or negative) with respect to terminal i, the i is also positive (negative) with respect to $i-1$. The specifications further require that if H_1 and X_1 are tied together and the H winding is energized, the voltage between the highest numbered H winding and the highest numbered X winding shall be less than the voltage across the H winding.

The standards also specify the relative location of the numbered terminals on the transformer tank or enclosure. The two possibilities are shown in Figure 7.2. Examples of transformers with tapped windings are shown in [14] and [52]. We have avoided use of the terms "primary" and "secondary" here since these terms refer to the direction of energy flow and this is not specified when deriving general results.

Figure 7.2 also illustrates what is meant by the terms "additive" and "sub-

Table 7.2. Standard Range in Impedances for Two-Winding Power Transformers Rated at 65 C Rise (Both 25- and 60-Hz transformers)

High-Voltage Winding Insulation Class kV	*Low-Voltage Winding Insulation Class kV*	*Impedance Limit in Percent*			
		Class OA OW OA/FA OA/FA/FOA**		*Class FOA FOW*	
		Min	*Max*	*Min*	*Max*
15	15	4.5	7.0	6.75	10.5
25	15	5.5	8.0	8.25	12.0
34.5	15	6.0	8.0	9.0	12.0
	25	6.5	9.0	9.75	13.5
46	25	6.5	9.0	9.75	13.5
	34.5	7.0	10.0	10.5	15.0
69	34.5	7.0	10.0	10.5	15.0
	46	8.0	11.0	12.0	16.5
92	34.5	7.5	10.5	11.25	15.75
	69	8.5	12.5	12.75	18.75
115	34.5	8.0	12.0	12.0	18.0
	69	9.0	14.0	13.5	21.0
	92	10.0	15.0	15.0	23.25
138	34.5	8.5	13.0	12.75	19.5
	69	9.5	15.0	14.25	22.5
	115	10.5	17.0	15.75	25.5
161	46	9.5	15.0	13.5	21.0
	92	10.5	16.0	15.75	24.0
	138	11.5	18.0	17.25	27.0
196	46	10.0	15.0	15.0	22.5
	92	11.5	17.0	17.25	25.5
	161	12.5	19.0	18.75	28.5
230	46	11.0	16.0	16.5	24.0
	92	12.5	18.0	18.75	27.0
	161	14.0	20.0	21.0	30.0

Source: Westinghouse Electric Corp. [14]. Used with permission.
*The impedances are expressed in percent on the self-cooled rating of OA/FA and OA/FA/FOA.
Definition of transformer classes:
OA—Oil-immersed, self-cooled OW—Oil-immersed, water-cooled.
OA/FA—Oil-immersed, self-cooled/forced-air-cooled.
OA/FA/FOA—Oil-immersed, self-cooled/forced-air-cooled/forced-oil-cooled.
FOA—Oil-immersed, forced-oil-cooled with forced-air cooler.
FOW—Oil-immersed, forced-oil-cooled with water cooler.
Note: The through impedance of a two-winding autotransformer can be estimated knowing rated circuit voltages, by multiplying impedance obtained from this table by the factor ($HV - LV/HV$).

tractive" polarity. The lower of the three sets of drawings shows a core segment with voltage polarities and terminals labeled. Obviously there are two ways to orient two windings on a core and these possibilities are shown in Figures 7.2a and 7.2b. If I_X is increasing in the circuit on the left, this would cause a flux to tend to increase in the upward direction. But by Lenz's law a current I_H will flow to oppose this change in core flux and hold the flux linkages constant. This establishes H_1 as a positive terminal. Since the H_1 and X_1 bushings have the

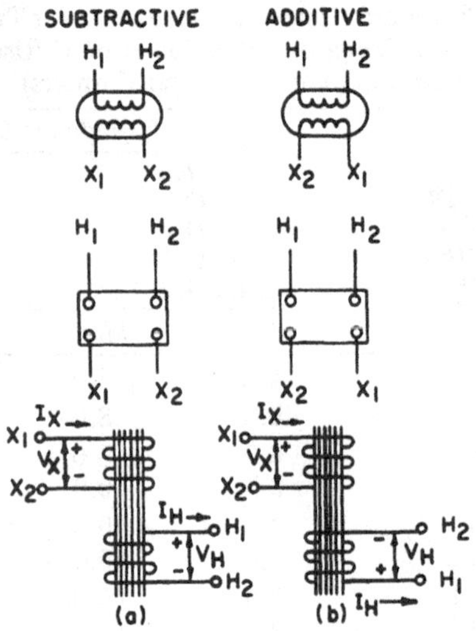

Fig. 7.2. Standard polarity markings for two-winding transformers: (a) subtractive, (b) additive. (From Westinghouse Electric Corp. [14]. Used with permission.)

same relative position on the tank, as noted on the left-hand set of drawings, this is *by definition* a "subtractive polarity." Thus for the transformer on the left the polarity may be determined by connecting *adjacent* terminals together (viz., H_1 and X_1), applying a voltage between H_1 and H_2, and then checking that the voltage between *adjacent* connections H_2 and X_2 is *less than* the applied voltage. If so, the polarity is subtractive. Obviously, the reverse is true on additive polarity transformers. Additive polarity is standard for single-phase transformers of 500 kVA and smaller when the *HV* winding is rated 8660 volts and below [14]. All other transformers are normally subtractive polarity, although the nameplate should be checked before connecting any transformers in a three-phase bank.

From the viewpoint of circuit analysis, we often identify the coils of a coupled circuit by polarity markings or dots. By this convention we easily establish that if H_1 is a dotted terminal, X_1 is likewise dotted as shown in Figure 7.3. This fact follows from the definition of additive polarity where we establish that the instant H_1 is positive with respect to H_2, then X_1 is likewise positive with respect to X_2. If we choose I_H as entering the dotted terminal (H_1) and I_X as leaving the dotted terminal (X_1), these currents are in phase if exciting current is negligible. Equations (7.2) to (7.10) are written on this basis.

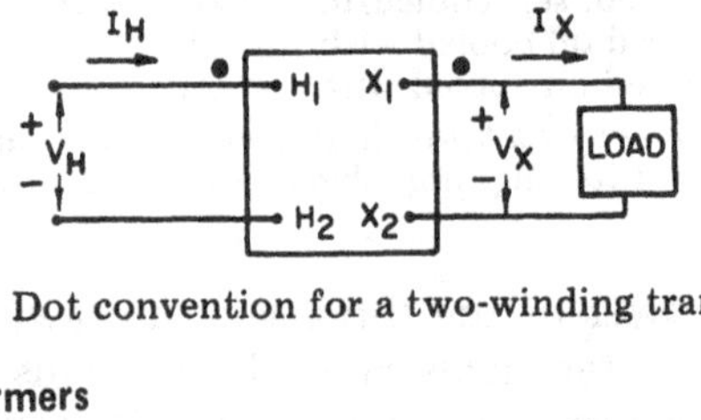

Fig. 7.3. Dot convention for a two-winding transformer.

7.4 Three-Winding Transformers

It is quite common in power systems to utilize transformers with more than two windings.[2] This is especially true in large transmission substations where volt-

[2] This analysis follows closely that of [14] to which the interested reader is referred for additional information.

ages are transformed from high-voltage transmission levels to intermediate subtransmission levels. In such cases a third voltage level is often established for local distribution, for application of power factor correcting capacitors or reactors, or perhaps simply to establish a Δ connection to provide a path for zero sequence currents. Although such transmission substations are usually equipped with three-phase transformers, the theory of the three-winding transformer is more easily understood by examining a single-phase unit. These results may later be used on a per phase basis in the study of three-phase applications.

The windings of a three-winding transformer are designated as H for the highest voltage, X for the intermediate voltage, and Y for the lowest voltage. We also assume that exciting currents are negligible and will use equivalent circuits similar to Figure 7.1b and 7.1c, where the excitation branch is omitted.

Consider the winding diagram of a three-winding transformer shown in Figure 7.4a where the windings H, X, and Y are shown to have turns n_H, n_X, and

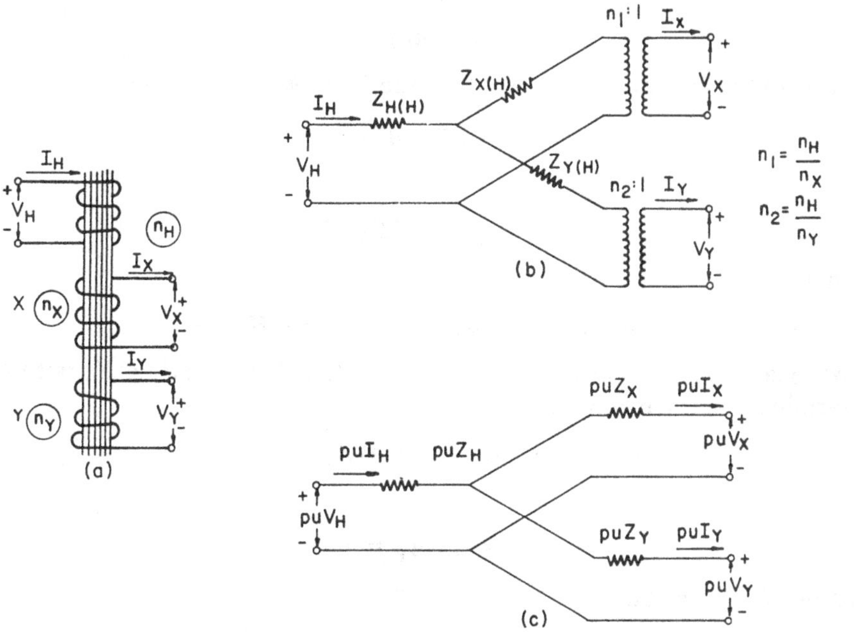

Fig. 7.4. Three-winding transformer: (a) winding diagram, (b) equivalent circuit in ohms, (c) equivalent circuit in pu. (From Westinghouse Electric Corp. [14]. Used with permission.)

n_Y respectively. In testing such a transformer to determine its impedance we make short circuit tests between pairs of windings. This is done with the third winding open. Applying such a test between the H and X windings with Y open gives the ohmic impedance Z_{HX}. Following the procedure of (7.4) with instrumentation on the H winding and the X winding shorted, we have

$$Z_{HX} = Z_H + n_1^2 Z_X \quad \Omega \tag{7.13}$$

or, in pu based on the H winding voltage and voltampere capacity we write

$$Z_{HX(H)} = Z_{H(H)} + Z_{X(H)} \quad \text{pu} \tag{7.14}$$

where the basis is indicated by the notation (H). We shall refer all impedances arbitrarily to the H winding where

$$
\begin{aligned}
S_{HB} &= \text{VA rating of winding } H \\
V_{HB} &= \text{rated voltage of winding } H
\end{aligned}
\tag{7.15}
$$

Then

$$Z_{HB} = V_{HB}^2/S_{HB} \quad \Omega \tag{7.16}$$

and if we multiply (7.14) by Z_{HB}, we have

$$Z_{HX(H)} = Z_{H(H)} + Z_{X(H)} \quad \Omega \text{ on } H \text{ base} \tag{7.17}$$

where $Z_{H(H)}$ is the same Z_H as in (7.13) but

$$Z_{X(H)} = n_1^2 Z_X \quad \Omega \tag{7.18}$$

where

$$n_1 = n_H/n_X \tag{7.19}$$

To complete the equivalent circuit of Figure 7.4b, we also define

$$n_2 = n_H/n_Y \tag{7.20}$$

and

$$Z_{Y(H)} = n_2^2 Z_Y \quad \Omega \tag{7.21}$$

such that

$$Z_{HY(H)} = Z_{H(H)} + Z_{Y(H)} \quad \Omega \text{ on } H \text{ base} \tag{7.22}$$

A third measurement may be made from winding X to winding Y, with H open, to determine Z_{XY}. We compute

$$Z_{XY} = Z_X + n_3^2 Z_Y \quad \Omega \tag{7.23}$$

where

$$n_3 = n_X/n_Y = n_2/n_1 \tag{7.24}$$

Rewriting (7.23), we have

$$
\begin{aligned}
Z_{XY(X)} &= \frac{n_1^2 Z_X + n_2^2 Z_Y}{n_1^2} \\
&= \frac{Z_{X(H)} + Z_{Y(H)}}{n_1^2} \quad \Omega \text{ on } X \text{ base}
\end{aligned}
\tag{7.25}
$$

We have now established three equations in three unknowns—viz., (7.17), (7.22), and (7.25)—which we rewrite as

$$
\begin{bmatrix} Z_{HX(H)} \\ Z_{HY(H)} \\ Z_{XY(X)} \end{bmatrix} = \begin{bmatrix} 1 & 1 & 0 \\ 1 & 0 & 1 \\ 0 & 1/n_1^2 & 1/n_1^2 \end{bmatrix} \begin{bmatrix} Z_{H(H)} \\ Z_{X(H)} \\ Z_{Y(H)} \end{bmatrix} \quad \Omega \text{ on mixed } X \text{ and } H \text{ base}
\tag{7.26}
$$

Note that the column vector on the left is known or can be determined by tests on the transformer. The column vector on the right contains the desired quantities for the equivalent circuit. Since the determinant of the coefficient matrix is nonzero (det = $-2/n_1^2$), a unique solution exists,

$$\begin{bmatrix} Z_{H(H)} \\ Z_{X(H)} \\ Z_{Y(H)} \end{bmatrix} = 1/2 \begin{bmatrix} 1 & 1 & -n_1^2 \\ 1 & -1 & n_1^2 \\ -1 & 1 & n_1^2 \end{bmatrix} \begin{bmatrix} Z_{HX(H)} \\ Z_{HY(H)} \\ Z_{XY(X)} \end{bmatrix} \ \Omega \text{ on mixed } X \text{ and } H \text{ base} \tag{7.27}$$

These values are inserted into the equivalent circuit of Figure 7.4b.

We convert the equivalent circuit to a pu equivalent by dividing (7.27) by the base ohms of the H winding. We also note that

$$n_1^2 = (n_H/n_X)^2 = (V_{HB}/V_{XB})^2 = S_{HB} Z_{HB}/S_{XB} Z_{XB} \tag{7.28}$$

Incorporating (7.28), we divide (7.27) by Z_{HB} and find the pu expression

$$\begin{bmatrix} Z_H \\ Z_X \\ Z_Y \end{bmatrix} = 1/2 \begin{bmatrix} 1 & 1 & -1 \\ 1 & -1 & 1 \\ -1 & 1 & 1 \end{bmatrix} \begin{bmatrix} Z_{HX(H)} \\ Z_{HY(H)} \\ Z_{XY(H)} \end{bmatrix} \text{ pu} \tag{7.29}$$

where

$$Z_{XY(H)} = (S_{HB}/S_{XB})\, Z_{XY(X)}$$

These values are used in the pu (or percent) equivalent circuit of Figure 7.4c. Note carefully that these values are computed from pu winding-to-winding impedances where Z_{HX} and Z_{HY} are on an H base and Z_{XY} is on an X base (both voltampere and voltage).

The above procedure is necessary in transformers with more than one winding since the pu impedances must be carefully identified as to the base used in specifying each value. The values of equation (7.29) could be based upon the rating of any winding, but the H winding is often selected. It is important to realize that when fully loaded the H winding divides its capacity between X and Y. It is reasonable then that the ratio S_{HB}/S_{XB} affects the impedance Z_{XY} which in turn affects the equivalent circuit. Note also that the effect of S_{HB}/S_{XB} on $Z_{XY(X)}$ is to convert this quantity to an H-based pu impedance.

Observe that the "T equivalent" of Figure 7.4 has a node at the junction of Z_H, Z_X, and Z_Y which is fictitious and has no physical significance whatever. The voltage at this point is not usually computed. This node can be eliminated by changing the T to a Δ. Equation (7.29) contains negative signs and the computed results are not necessarily positive. Indeed, it is rather common for one leg of a T equivalent to be negative or zero.

Equivalent circuits are developed in the literature (e.g., see [11, vol. 2] and [14]) for transformers with four or more circuits. Since these devices are not in common use, they will not be studied here.

7.5 Autotransformer Equivalents

In the modern power system a large number of major transformations are accomplished by autotransformers rather than separate-winding transformers.[3] One reason for this is the lower cost of autotransformers. There are, however, other

[3] This development follows closely that of [14] but [10] and [11, vol. 2] are also recommended for further reading.

significant differences [14]. The impedance of an autotransformer is smaller than that of a conventional transformer of the same rating. This is both an advantage and a disadvantage. The lower impedance means lower regulation, but it also means lower fault limiting capability and sometimes necessitates the inclusion of external impedance. Autotransformers also have lower losses and smaller exciting currents than two-winding transformers of the same rating. They are physically smaller and generally have higher efficiencies.

Consider the autotransformer circuit of Figure 7.5 where we denote the ratio

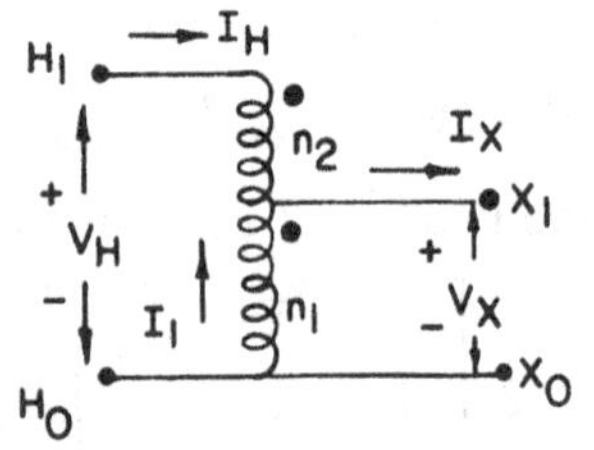

Fig. 7.5. Circuit for a two-winding autotransformer.

of H to X voltages by the turns ratio n, i.e.,

$$V_H/V_X = (n_1 + n_2)/n_1 = n \tag{7.30}$$

Then, according to this definition $n_2/n_1 = n - 1$, and the ratio of turns in the two windings is *not* equal to the no-load voltage ratio n.

When the transformer is loaded, the MMF of the two windings must be the same (if ideal), i.e.,

$$I_1 n_1 = I_H n_2 \tag{7.31}$$

and since

$$I_X = I_1 + I_H \tag{7.32}$$

we compute

$$I_X = \left(\frac{n_2}{n_1} + 1\right) I_H = nI_H \tag{7.33}$$

Comparing (7.30) and (7.33) with similar equations (7.2) and (7.3) for a two-winding transformer, we see that the autotransformer looks exactly like a two-winding transformer with turns ratio n where n is defined by (7.30).

From transformer theory we know that the total circuit voltamperes in a transformer must be the same in both the input and output (for an ideal transformer). For the autotransformer we compute, using subscript c for "circuit" voltamperes,

$$S_c = V_H I_H = V_X I_X \tag{7.34}$$

but the winding voltamperes are

$$\begin{aligned} S_1 &= V_X I_1 = V_X I_H (n_2/n_1) \text{ for winding 1} \\ S_2 &= (V_H - V_X) I_H = V_X I_H (n_2/n_1) = S_1 \text{ for winding 2} \end{aligned} \tag{7.35}$$

The ratio of winding to circuit voltamperes is

$$S_1/S_c = n_2/(n_1 + n_2) = (n - 1)/n \tag{7.36}$$

This amounts to a considerable "savings" in winding capacity. For example, if the transformation is two to one, i.e., $n = 2$, the ratio (7.36) is one-half, or the autotransformer windings have twice the voltampere capacity of the equivalent two-winding transformer.

The autotransformer impedance may be measured, as in the case of the two-winding transformer, by making a short circuit test. The circuit for performing this test is shown in Figure 7.6 where we compute the leakage impedance

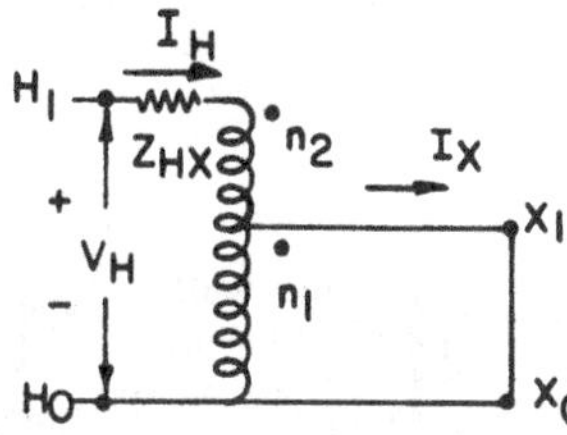

Fig. 7.6. Short circuit test circuit.

$$Z_{HX} = (V_H/I_H)_{V_X=0} \quad \Omega \text{ referred to } H \tag{7.37}$$

This is the impedance shown in the equivalent of the autotransformer of Figure 7.7 where we depict the autotransformer as a *two-winding* transformer with turns ratio n and impedance Z_{HX}.

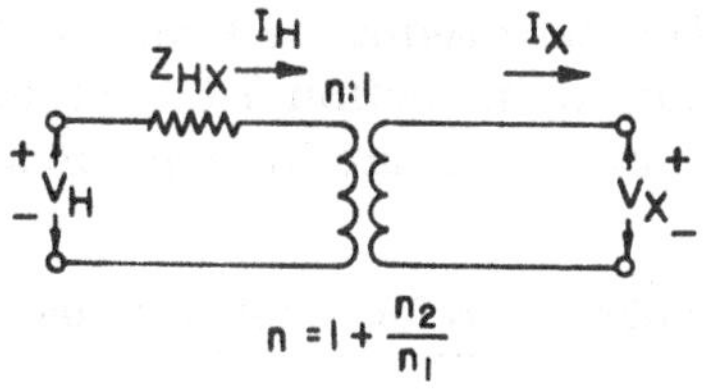

Fig. 7.7. Conventional equivalent circuit for an autotransformer.

7.5.1 Three-winding autotransformers

Three-winding autotransformers are also widely used. Because of the physical arrangement of autotransformer circuits, three transformer windings are usually connected Y-Y with grounded neutral. Since the Y-Y connection contains no path for third-harmonic exciting currents, a third winding is provided in each phase, which can be Δ connected. This winding, called the Δ-tertiary, provides a path for exciting currents and can be used to transform power as well.

In three-winding autotransformers it is necessary to recognize clearly whether one is computing impedances on a *circuit* basis or on a *winding* basis. Care must also be used in applying formulas given in the literature since there does not appear to be a commonly accepted practice (e.g., [14] uses a circuit basis whereas [11] uses a winding basis).

Wagner and Evans [10] give an excellent treatment of the comparison between the impedances of a three-winding autotransformer computed on a circuit basis and a winding basis. Three-winding autotransformers usually consist of two physical windings, one of which is tapped to form two subwindings which are separately identified. In the following analysis we shall call the *windings* s, c, and t (for series, common, and tertiary) and the *circuits* H, X, and Y (sometimes called

H, M, L or H, L, T by other authors). These circuit and winding configurations are identified in Figure 7.8 where the three-terminal equivalent circuits are also

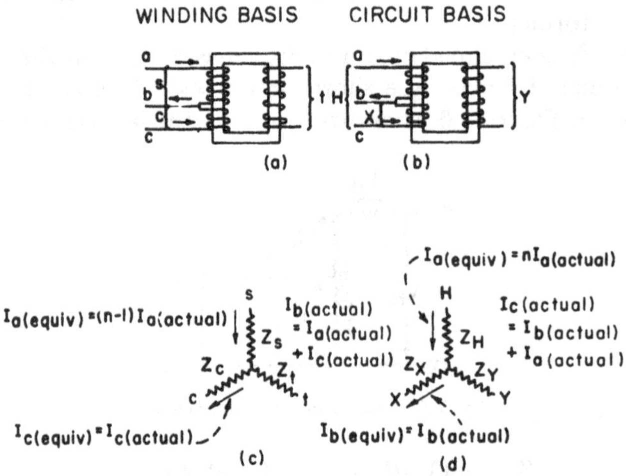

Fig. 7.8. Autotransformer connections and equivalents for analysis on a winding or circuit basis. (From *Symmetrical Components* by C. F. Wagner and R. D. Evans. Copyright McGraw-Hill, 1933. Used with permission of McGraw-Hill Book Co.)

shown for each case. These are the equivalent circuits for the positive and negative sequence networks. The zero sequence equivalent depends upon the nature of the connection, as discussed in section 7.6. Note, however, that since three circuits to which external connections are to be made are present in either case (a)

Table 7.3. Conversion Formulas for Autotransformer with Tertiary Winding*

	Winding Basis		*Circuit Basis*
(a)	$Z_t = 1/2(Z_{tc} + Z_{ts} - Z_{cs})$	(b)	$Z_Y = 1/2(Z_{YX} + Z_{YH} - Z_{XH})$
	$Z_c = 1/2(Z_{ct} + Z_{cs} - Z_{st})$		$Z_X = 1/2(Z_{XY} + Z_{XH} - Z_{YH})$
	$Z_s = 1/2(Z_{st} + Z_{sc} - Z_{ct})$		$Z_H = 1/2(Z_{HY} + Z_{HX} - Z_{YX})$
(c)	$Z_{XY} = Z_{ct}$	(d)	$Z_{ct} = Z_{XY}$
	$Z_{HX} = \left(\frac{n-1}{n}\right)^2 Z_{sc}$		$Z_{sc} = \left(\frac{n}{n-1}\right)^2 Z_{HX}$
	$Z_{HY} = \frac{Z_{tc}}{n} + \frac{n-1}{n} Z_{ts} - \frac{n-1}{n^2} Z_{cs}$		$Z_{ts} = \frac{n}{n-1} Z_{HY} - \frac{1}{n-1} Z_{XY} + \frac{n}{(n-1)^2} Z_{HX}$
(e)	$Z_c = \frac{n}{n-1} Z_X$	(f)	$Z_X = \frac{n-1}{n} Z_c$
	$Z_t = Z_Y - \frac{Z_X}{n-1}$		$Z_Y = Z_t + \frac{Z_c}{n}$
	$Z_s = \left(\frac{n}{n-1}\right)^2 Z_H + \frac{n}{(n-1)^2} Z_X$		$Z_H = \left(\frac{n-1}{n}\right)^2 Z_s - \frac{n-1}{n^2} Z_c$

Source: Wagner and Evans [10]. Used with permission.
*c-winding or X circuit used as voltage base. For pu use arbitrary VA base. $n = V_H/V_X$

or (b) of Figure 7.8, the equivalent circuit must have three terminals as in (c) and (d). Note also that in Figure 7.8 we use the ratio

$$n = V_H/V_X \tag{7.38}$$

which agrees with (7.30) for the two-winding autotransformers.

Table 7.3 gives the equations which convert from one basis to another with all values based on either the c winding or the X circuit voltages, and where all impedances are in pu on these bases, with an arbitrary voltampere base.

Then equations (a) and (b) are the same as (7.29) and the various conversion equations of (c), (d), (e), and (f) are all in pu on these same bases. The only exception to this is that the current relationships given in the Figure 7.8 equivalent circuits (c) and (d) are ampere relations. The equations shown are developed in [10]. Table 7.4 gives typical values of autotransformer impedances.

Table 7.4. Typical Characteristics of Autotransformers

	Reactances†*				*Losses, Percent on Rated kVA Base†*			
	High to low		*Low to tertiary*	*High to tertiary*	*No load losses*		*Total losses*	
Ratings	*Range*	*Recommended average*			*Range*	*Recommended average*	*Range*	*Recommended average*
Three-phase: 300–600 MVA								
Three single-phase: 600–1200 MVA								
345–138 kV	10–13	12	11–19	24–34	0.07–0.10	0.09	0.25–0.40	0.33
345–161 kV	8–11	10	20–25	10–20	0.06–0.10	0.08	0.20–0.35	0.30
345–230 kV	5–9	7	...	...	0.06–0.10	0.07	0.15–0.30	0.26
500–138 kV	15–20	17	5–15	20–30	0.07–0.11	0.09	0.30–0.50	0.38
500–161 kV	12–18	15	7–18	20–30	0.06–0.10	0.08	0.25–0.45	0.35
500–230 kV	10–13	12	15–25	20–30	0.06–0.10	0.08	0.20–0.40	0.30
500–345 kV	6–10	8	23–25	30–40	0.06–0.10	0.07	0.15–0.35	0.26
Three single-phase: 300–1200 MVA								
700–138 kV	20–30	25	...	...	...	0.10	...	0.55
700–161 kV	18–28	23	...	...	...	0.09	...	0.48
700–230 kV	15–25	20	...	...	...	0.09	...	0.37
700–345 kV	13–21	17	...	...	...	0.08	...	0.28
700–500 kV	10–20	15	...	...	...	0.07	...	0.22

Source: Federal Power Commission [53].

*All reactances in percent on rated voltage and kVA base.

†The above transformer reactances and loss data are based on 1050 kV BIL for 345 kV transformer windings, 1550 kV BIL for 500 kV transformer windings, and 2175 kV BIL for 700 kV transformer windings. Increasing or decreasing the BIL values results in a corresponding increase or decrease in transformer reactances and losses of up to approximately 10% of the suggested values.

7.6 Three-Phase Banks of Single-Phase Units

Since our concern is primarily with three-phase power systems, we are naturally interested in the application of single-phase transformers connected to form three-phase banks.[4] In this study we will limit our consideration to three *iden-*

[4] This discussion follows closely that of Clarke [11, vol. 2] to which the interested reader is referred for further information.

tical single-phase units and will develop the sequence network representations for commonly used connections. The use of dissimilar units introduces a series unbalance which may be treated separately as such once the basic sequence network configurations are known.

One problem in the computation of transformer impedances is the selection of the base for converting ohmic values to pu. For transformers it is common to choose the voltampere rating of a single-phase unit as the base or, in the case of three-winding transformers, we refer all impedances to the voltampere rating of the one winding. If the windings are Y connected, the LN voltage is taken as the base voltage, and this is the winding rated voltage. If windings are Δ connected the LN voltage is $1/\sqrt{3}$ the winding rated voltage, and this should be computed as the base value. But as noted in (1.16) and (1.20), one may choose the voltampere base as the total bank rating, or three times the rating of each unit, and use the LL voltage as base voltage. Note that once a voltage base is selected for one winding, the base is automatically fixed, through the turns ratio, for the other windings. If these values do not coincide with the arbitrarily chosen base voltages, the off-nominal ratio must be accounted for. This may be done by inserting an ideal transformer with a turns ratio equal to the off-nominal ratio of these voltages. Another method for doing this will be given in Part III of this chapter.

It should be emphasized here that all Δ-connected transformer banks will be represented as an equivalent (ungrounded) Y insofar as the positive and negative sequence networks are concerned. For the zero sequence network, however, we must also recognize that the Δ connection permits zero sequence currents to flow, whereas the ungrounded Y does not. This concept of thinking in terms of an equivalent Y is quite acceptable insofar as the relationship between phase quantities is concerned. Thus the total current entering the bank is correct and phase relationships are correct among quantities on either side of the transformer. If we wish to compare the phase relationship of quantities *across* a Y-Δ bank, this must be done as a separate computation since the equivalent Y representation destroys this relationship.

Using this equivalent Y idea, we see that the transformer equivalent of Figure 7.1c is the (per phase) equivalent for the positive and negative sequence networks for *any* transformer connection of two-winding transformers. For three-winding transformers the T circuit of Figure 7.4c is adequate. Thus with the base quantities chosen as discussed above, the positive and negative sequence representations are identical and are the same as the single-phase transformer equivalent.

The zero sequence equivalent circuit depends upon the type of connection and the method of grounding. In a Δ-connected winding, zero sequence currents can flow, but since they are equal in each leg of the Δ, they do not leave the transformer terminals. Thus the Δ winding as viewed from the external circuit appears as an impedance with both ends shorted to ground and an open circuit to the external circuit. In a grounded Y circuit the three currents I_{a0}, I_{b0}, and I_{c0} add to produce $3I_{a0}$ flowing to (or from) ground. Since the zero sequence network represents only one phase with current I_{a0}, we include any neutral impedance Z_n as $3Z_n$ in the sequence network so that the neutral voltage will be computed correctly. The ungrounded Y appears open to zero sequence currents since the three-phase currents add to zero, from which we conclude that I_{a0} is zero.

These concepts are illustrated in the zero sequence equivalent circuits of two- and three-winding transformers shown in Figure 7.9 where the letters P, Q, and R

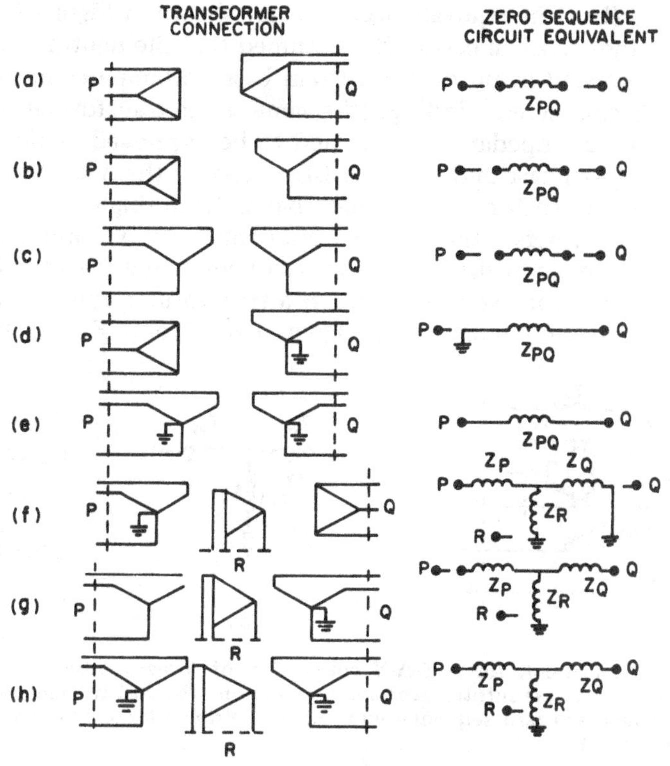

Fig. 7.9. Zero sequence equivalent circuits for transformer banks of three identical single-phase units—exciting currents neglected. (From Clarke [11, vol. 1]. Used with permission.)

designate the transformer circuits. Once the voltage levels are specified, these designations may be replaced by H, X, and Y for the appropriate windings. In this figure the ground symbols represent the zero potential bus N_0.

Two additional special cases of zero sequence equivalent circuits are shown in Figure 7.10 where a neutral impedance is present in (a) and a corner of the Δ im-

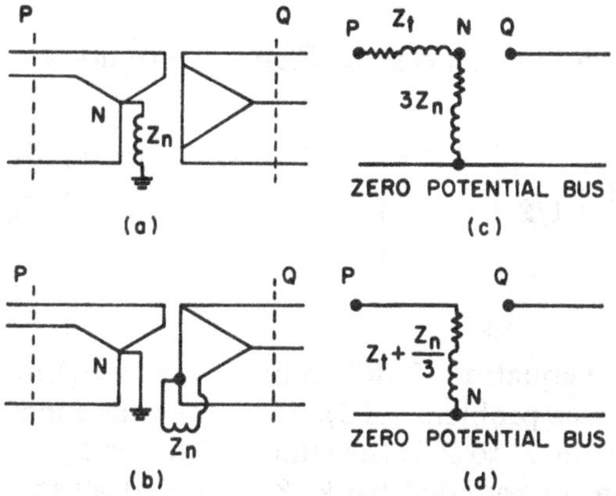

Fig. 7.10. Y-Δ bank with (a) neutral grounded through Z_n, (b) Z_n in a corner of the Δ, (c) and (d) zero sequence equivalent circuits for (a) and (b) respectively, where Z_t = leakage impedance between windings. (From Clarke [11, vol. 2]. Used with permission.)

pedance exists in (b). The equivalent circuits are shown in Figure 7.10c and 7.10d respectively for these special cases. It is assumed that the neutral impedance Z_n in (a) and (c) is expressed in pu on the system base voltamperes per phase and LN voltage of the Y-connected windings (the same as the transformer impedance Z_t). The corner of the Δ impedance is assumed to be expressed in pu on the system base voltamperes per phase and the base LL voltage of the Δ circuit.

It is also possible to form three-phase banks from single-phase, two- or three-winding autotransformers. Such banks are commonly Y connected, often with the neutral grounded, and with a third (tertiary) winding connected in Δ to supply third-harmonic MMF for excitation. Such a transformer connection is shown in Figure 7.11a where the circuits are designated H, X, and Y. Clarke [11, vol. 2]

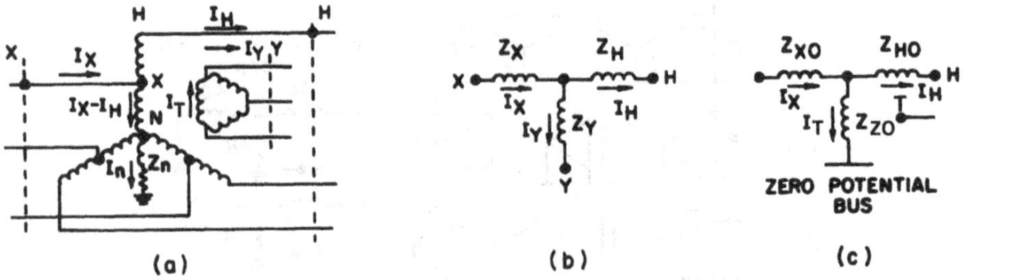

Fig. 7.11. Equivalent circuit for a Y-Δ-Y connection of three identical three-winding autotransformers: (a) autotransformer connection, (b) positive and negative sequence equivalent, (c) zero sequence equivalent. (From Clarke [11, vol. 2]. Used with permission.)

shows that the equivalent circuit for the positive and negative sequence networks is that shown in Figure 7.11b where the T circuit parameters are taken from Table 7.3(b) to be

$$\begin{bmatrix} Z_H \\ Z_X \\ Z_Y \end{bmatrix} = 1/2 \begin{bmatrix} 1 & 1 & -1 \\ 1 & -1 & 1 \\ -1 & 1 & 1 \end{bmatrix} \begin{bmatrix} Z_{HX} \\ Z_{HY} \\ Z_{XY} \end{bmatrix} \text{ pu} \tag{7.39}$$

which is seen to be exactly the same as (7.29). Here all impedances are in pu on the same base.

The zero sequence impedances for Figure 7.11c are also given by Clarke [11, vol. 2] to be

$$\begin{bmatrix} Z_{X0} \\ Z_{H0} \\ Z_{n0} \end{bmatrix} = 1/2 \begin{bmatrix} 1 & -1 & 1 & (n-1)/n \\ 1 & 1 & -1 & -(n-1)/n^2 \\ -1 & 1 & 1 & 1/n \end{bmatrix} \begin{bmatrix} Z_{HX} \\ Z_{HY} \\ Z_{XY} \\ 6Z_n \end{bmatrix} \text{ pu} \tag{7.40}$$

where n is defined by equation (7.30) to be $n = (n_1 + n_2)/n_1$. The proof of (7.40) is left as an exercise (see problem 7.13). If the tertiary winding is missing, $Z_{n0} = \infty$ and the impedance from H to X is the sum of Z_{X0} and Z_{H0}.

In the case of an ungrounded bank, $Z_n = \infty$ and all the impedances in (7.40) go to infinity. One way to solve this problem is to convert the Y equivalent (consisting of Z_{X0}, Z_{H0}, and Z_{n0}) to a Δ equivalent and then take the limit as Z_{n0} approaches infinity. If we call these Δ impedances $Z_{HX\Delta}$, $Z_{HY\Delta}$, and $Z_{XY\Delta}$, we compute the so-called "resonant Δ" equivalent to be

$$\begin{bmatrix} Z_{HX\Delta} \\ Z_{HY\Delta} \\ Z_{XY\Delta} \end{bmatrix} = \begin{bmatrix} (n-1)^2/n \\ -(n-1) \\ (n-1)/n \end{bmatrix} Z_{st} \text{ pu} \tag{7.41}$$

where Z_{st} is the series-tertiary impedance defined (winding basis) in Figure 7.8.

II. THREE-PHASE TRANSFORMERS

Much of the information needed to perform studies of a faulted system and to properly represent the transformers in such studies is given in the preceding sections. It is often of little concern whether the transformer bank consists of three single-phase units or one three-phase unit. There are, however, certain differences which may be important under some conditions. Among these are the standard practice of marking the terminals of three-phase units, the impedance differences due to different core designs, and the phase shift across transformer banks.

7.7 Three-Phase Transformer Terminal Markings

The terminals of three-phase transformers are labeled H, X, Y, etc., exactly as for the single-phase transformers discussed in section 7.3. In this case, however,

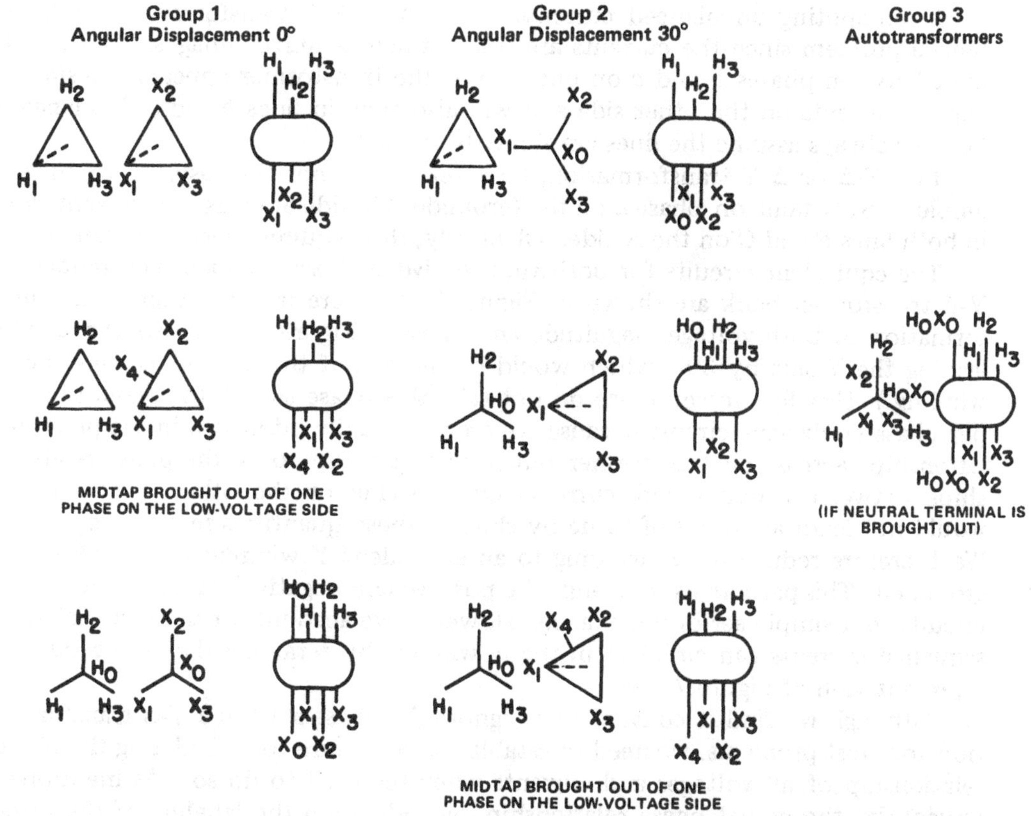

Fig. 7.12. Angular displacement and terminal markings for three-phase transformers. Dashed lines show angular displacement between high- and low-voltage windings. Angular displacement is the angle between a line drawn from neutral to H_1 and a line drawn from neutral to X_1. (Courtesy USAS Institute [52].)

each voltage level will have three terminals, one for each phase, and may have four if the neutral is brought out as an external connection. These are labeled by subscripts 1, 2, and 3, with 0 designating the neutral. The subscripts are intended to designate the time order of the instantaneous voltages if connected in a logical manner. Thus if the phase sequence is *a-b-c* and these phases are connected to H_1, H_2, and H_3, then X_1, X_2, and X_3 will be of phase sequence *a-b-c*.

The USAS [52] also specify a definite phase relationship between windings of a three-phase transformer. For Y-Y and Δ-Δ connections the phase angle between like-numbered terminals is 0°, whereas for Y-Δ or Δ-Y connections the *H*-terminal voltage leads the corresponding *X*-terminal voltage by 30°. This is shown in Figure 7.12 which also shows the USAS arrangement for terminal positions on the transformer tank.

From Figure 7.12 we observe that the voltage associated with phase *a* on one side of the transformer may have many different phase relations with the phase *a* voltage on the other side, depending on the way the phases are labeled. Thus, for Y-Y or Δ-Δ transformers this phase relationship may be 0°, +120°, or - 120°. In Y-Δ or Δ-Y connections, the relation between phase *a* windings may be ±30°, ±150°, or ±90°, again depending on labeling. The transformer itself, if of standard design and marking, is consistent in the relationships shown in Figure 7.12.

7.8 Phase Shift in Y-Δ Transformers

In computing unbalanced currents, the Δ-Δ or Y-Y transformer presents no special problem since the currents are transformed as mirror images. Thus, a LL fault between phases *b* and *c* on one side of the transformer appears as a similar pair of currents on the other side and will also flow in lines *b* and *c* if so labeled (and we always assume the lines are similarly labeled).

In a Y-Δ or Δ-Y transformation, however, this is not the case. Here, for example, a SLG fault on phase *a* on the (grounded) Y side appears as a current flow in both lines *B* and *C* on the Δ side. Obviously, this requires special treatment.

The equivalent circuits for positive, negative, and zero sequence currents of a Y-Δ transformer bank are shown in Figure 7.13, where we note there is a transformation in both voltage magnitude and phase. (Figure 7.13 shows the Δ side leading the Y side by 30°, which would be the case if the Δ winding were the *H* winding.) Usually, however, we do not take this phase shift into account in our per phase equivalent circuit because we are not usually interested in the phase relationships across the transformer but need to examine only the phase relationships between voltages and currents on one side or the other. Indeed, we would not learn anything of value by shifting these quantities in phase by ±30°. We therefore reduce the Δ winding to an equivalent *Y* winding with neutral ungrounded. This permits us to change the positive and negative sequence equivalent circuits to a simple series impedance. However, we account for the fact that zero sequence currents can circulate in the Δ winding by retaining the zero sequence representation of Figure 7.13d.

Although we find it convenient to ignore the phase shift in a Y-Δ transformation for most problems, we need to establish a procedure for calculating the phase relationship of all voltages and currents when required to do so. As mentioned previously, the actual phase relationship depends upon the labeling of the three phases and may be ±30°, ±150°, or ±90° as shown in Figure 7.14. However, there is no need to exhaustively analyze all six arrangements shown in Figure 7.14.

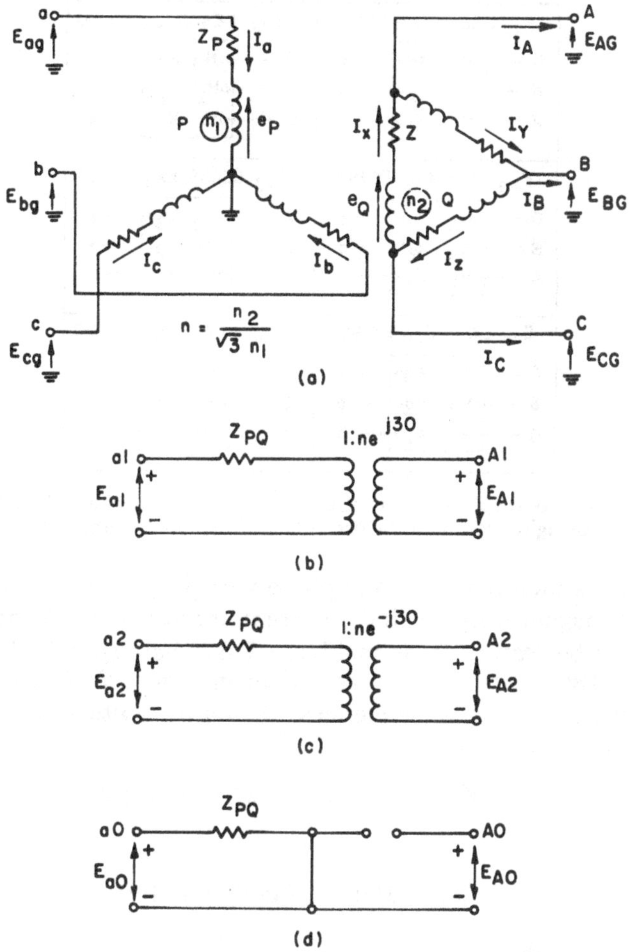

Fig. 7.13. Equivalent circuits of a Y-Δ transformer bank: (a) schematic diagram, (b) positive sequence, (c) negative sequence, (d) zero sequence. (From Westinghouse Electric Corp. [14]. Used with permission.)

Indeed, if we analyze only one, we may make any desired computation and can then relabel the results to fit any of the six possible configurations. The problem is choosing arbitrarily the most convenient of the six phase-shift arrangements.

Stevenson [9, 21] has analyzed the 90° phase shift. This is very convenient to work with and will be chosen as the basis for all computations here. The basic arrangement is shown in Figure 7.15 where several important features should be noticed. First, note that the phase labeled *a* is connected to H_1 and that labeled *A* is connected to X_3. Since by definition the positive sequence voltage is *a-b-c*, once we choose to connect *b* to H_2, we *must* connect *B* and *C* to X_1 and X_2 respectively in order that the phase sequence be *a-b-c* (*A-B-C*) on each side of the transformation. Comparing with Figure 7.14, we see that this is a 90° connection (or labeling) of terminals. This is verified in the positive sequence phasor diagram in Figure 7.15b where we see that V_{a1} lags V_{A1} by 90°. Figure 7.15 also follows the convention that windings drawn in parallel (e.g., *bn* and *CA*) are linked magnetically and that labels are arranged such that *a-n* on the Y side is linked to *B-C* on the Δ side. Lowercase letters are used on one side, uppercase on the other.

E_{ag} LAGS E_{AG} BY 30°	E_{ag} LEADS E_{AG} BY 30°
C —oH_3 X_3o— c B —oH_2 X_2o— b A —oH_1 X_1o— a	A —oH_3 X_3o— a B —oH_2 X_2o— b C —oH_1 X_1o— c

E_{ag} LAGS E_{AG} BY 150°	E_{ag} LEADS E_{AG} BY 150°
C —oH_3 X_3o— b B —oH_2 X_2o— a A —oH_1 X_1o— c	A —oH_3 X_3o— b B —oH_2 X_2o— c C —oH_1 X_1o— a

E_{ag} LAGS E_{AG} BY 90°	E_{ag} LEADS E_{AG} BY 90°
A —oH_3 X_3o— c B —oH_2 X_2o— a C —oH_1 X_1o— b	C —oH_3 X_3o— a B —oH_2 X_2o— c A —oH_1 X_1o— b

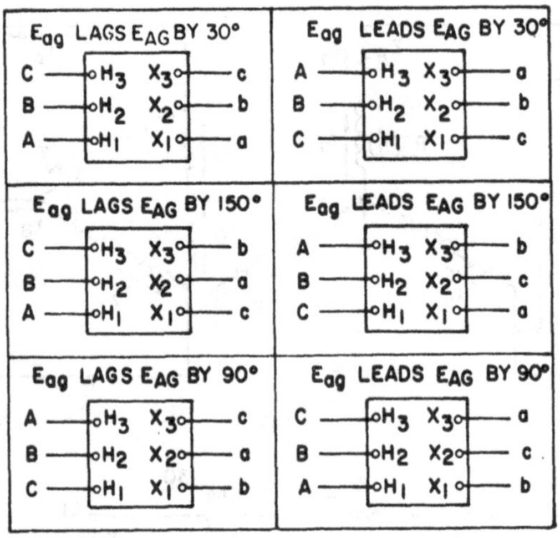

Fig. 7.14. Angular phase displacements obtainable with three-phase Y-Δ transformer units. (From Westinghouse Electric Corp. [14]. Used with permission.)

Our notation is such that V_{a1} is in phase with V_{BC1}. This being the case, these coils are linked magnetically so that a current *entering a* would produce a current I_{BC} which would be *leaving B*, as in Figure 7.3 (where $I_H = I_a$, $I_X = -I_{BC}$). Thus I_{an} and I_{BC} are 180° out of phase, as shown in Figure 7.16, and this is true for both the positive and negative sequences. From the voltage and current relationships of Figure 7.15 we write

$$V_{A1} = +jV_{a1}, \qquad V_{A2} = -jV_{a2} \text{ pu} \qquad (7.42)$$

and

$$I_{A1} = +jI_{a1}, \qquad I_{A2} = -jI_{a2} \text{ pu} \qquad (7.43)$$

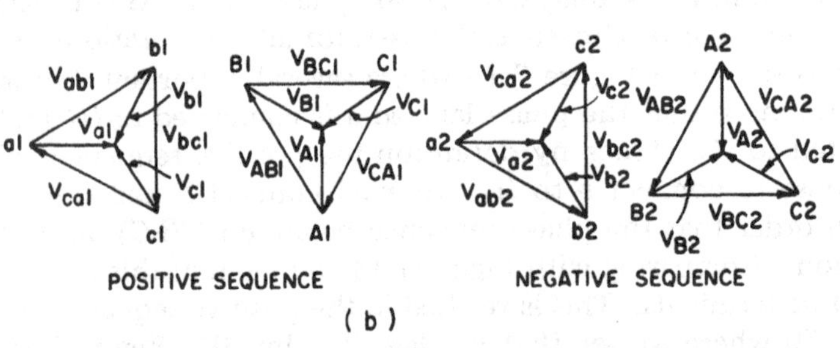

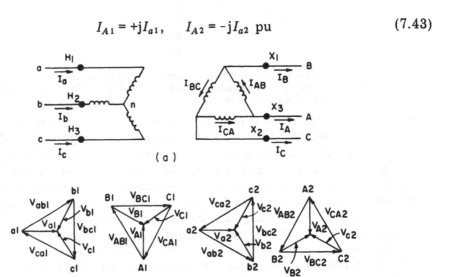

Fig. 7.15. Wiring diagram and voltage phasors for a three-phase transformer connected Y-Δ: (a) wiring diagram, (b) voltage components. (From *Elements of Power System Analysis* by William D. Stevenson, Jr. Copyright McGraw-Hill, 1962. Used with permission of McGraw-Hill Book Co.)

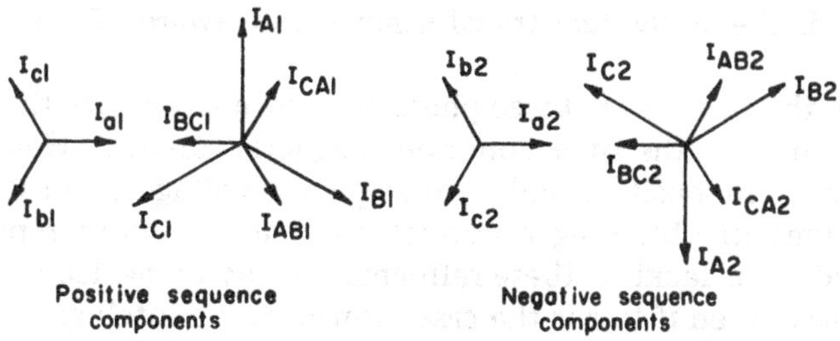

Fig. 7.16. Current phasors of a three-phase transformer connected Y-Δ. (From *Elements of Power System Analysis* by William D. Stevenson, Jr. Copyright McGraw-Hill, 1962. Used with permission of McGraw-Hill Book Co.)

Obviously, this tells us nothing about the phase relationship between sequence quantities since this depends entirely on the boundary conditions at the fault.

Example 7.1

As an example of the use of (7.42) and (7.43) we compute the currents flowing on both sides of transformer T1 in Example 3.1, where we have already established that the total fault current is $I_a = 3I_{a1} = 2.34\underline{/-42.8^\circ}$ pu and $I_b = I_c = 0$.

Solution

We easily compute from

$$I_{a0} = I_{a1} = I_{a2} = 0.78\underline{/-42.8^\circ} = 0.572 - j0.530 \text{ pu}$$

that

$$I_{A1} = j0.78\underline{/-42.8^\circ} = 0.78\underline{/+47.2^\circ} = 0.530 + j0.572 \text{ pu}$$

and

$$I_{A2} = -j0.78\underline{/-42.8^\circ} = 0.78\underline{/-132.8^\circ} = -0.530 - j0.572 \text{ pu}$$

and by inspection, $I_{A0} = 0$. Thus

$$\begin{aligned} I_A &= I_{A0} + I_{A1} + I_{A2} = 0 \\ I_B &= I_{A0} + a^2 I_{A1} + aI_{A2} = 0 + (0.230 - j0.745) + (0.760 - j0.173) \\ &= 0.990 - j0.918 = 1.35\underline{/-42.8^\circ} \text{ pu} \end{aligned}$$

and

$$\begin{aligned} I_C &= I_{A0} + aI_{A1} + a^2 I_{A2} = 0 + (-0.760 + j0.173) + (-0.230 + j0.745) \\ &= -0.990 + j0.918 = 1.35\underline{/180^\circ - 42.8^\circ} \text{ pu} \end{aligned}$$

Thus the generator views the SLG fault as a fault in lines *B* and *C* and sees no zero sequence current at all. Voltages at the generator bus can be calculated by a similar procedure and this is left as an exercise for the interested reader (see problem 7.16).

7.9 Zero Sequence Impedance of Three-Phase Transformers

In analyzing the sequence impedances of three-phase transformer banks consisting of three single-phase units, we used the same impedance for the zero sequence network as for the positive and negative sequence networks. This is because the individual transformers present a given impedance to any applied voltage. Indeed, the individual transformer is not aware of the sequence of applied voltages.

This is not the case with three-phase transformers where the fluxes of each phase winding share paths of a common magnetic circuit. Here the device responds differently to positive and zero sequence voltages, and these differences may be important in obtaining a correct sequence network representation. If precise information is lacking, these refinements may be neglected (and often are) in system studies. Even if this is the case, however, the intelligent engineer should be able to estimate the possible error in making a simplifying assumption.

There are two basic designs commonly used for three-phase transformers—the *core-type* design and the *shell-type* design. These core configurations are shown in Figure 7.17 where fluxes are shown due to positive sequence applied voltages.

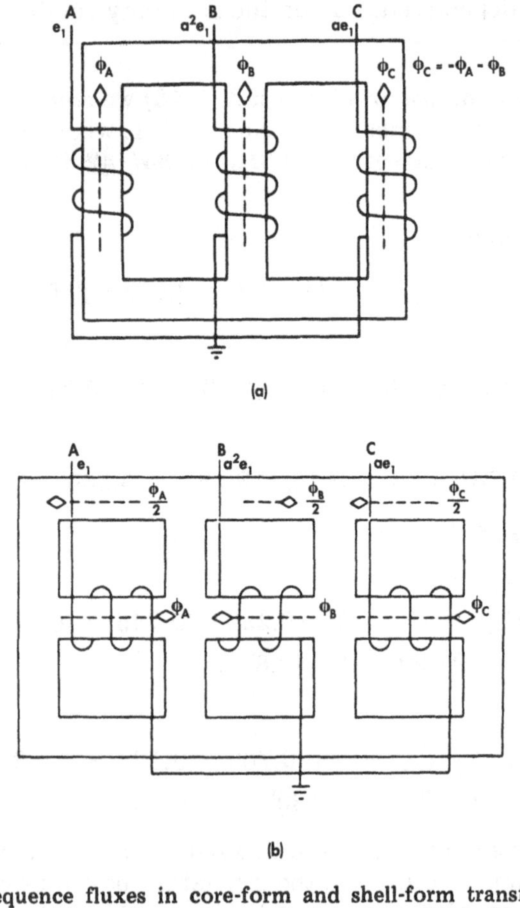

Fig. 7.17. Positive sequence fluxes in core-form and shell-form transformers: (a) core-form, (b) shell-form. (From Westinghouse Electric Corp. [54]. Used with permission.)

Note that the core-form windings (only the primary windings are shown) are all wound in the same sense. In the shell-form transformer the center leg is wound in the opposite sense of the other two legs to reduce the flux in the core sections between windings. Since nearly all flux is confined to iron paths, the excitation current is low and the shunt excitation branch is usually omitted from the positive and negative sequence transformer equivalent circuits for either core-type or shell-type designs.

The zero sequence impedance of a three-phase transformer may be found by performing open circuit and short circuit tests with zero sequence voltages applied. If this is done, the short circuit test determines leakage impedances which are nearly the same as positive sequence impedances, providing the test does not saturate the core. The open circuit test, however, reveals a substantial difference in the excitation (shunt) branch of the zero sequence equivalent for core-type and shell-type units. This is due to the different flux patterns inherent in the two designs, as shown in Figure 7.18 where zero sequence voltages are applied to

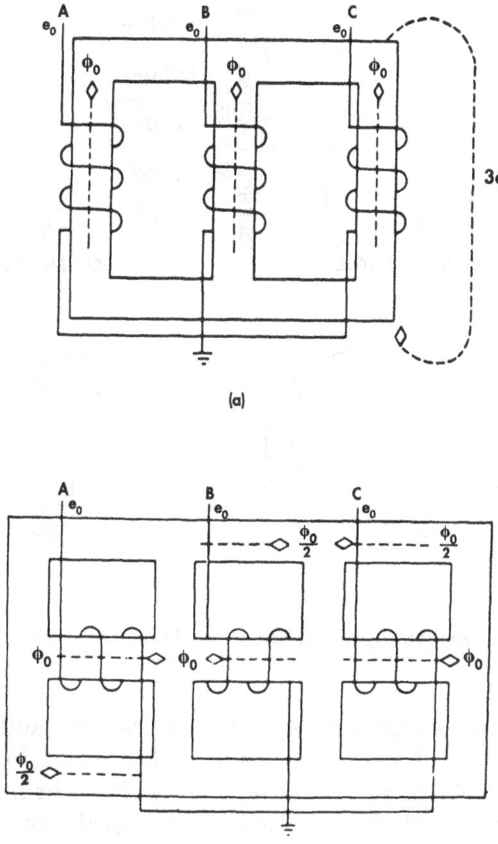

Fig. 7.18. Zero sequence fluxes in core-form and shell-form transformers: (a) core-form, (b) shell-form. (From Westinghouse Electric Corp. [54]. Used with permission.)

establish zero sequence fluxes. In the *core-type* design the flux in the three legs does not add to zero as in the positive sequence case. Instead, the sum $3\phi_0$ must seek a path through the air (or oil) or through the transformer tank, either of which presents a high reluctance. The result is a low zero sequence excitation impedance, so low that it should not be neglected in the equivalent circuit if high precision is required in computations.

The *shell-type* design may also present a problem if the legs between windings become saturated. Usually, however, the excitation impedance of the shell-type design is neglected.

The excitation of either the core-type or shell-type design is dependent upon the magnitude of the applied zero sequence voltage, as shown in Figure 7.19, but the shell-type design is much more variable than the core-type due to saturation of the shell-type core by zero sequence fluxes. Figure 7.19a also shows how the transformer tank acts as a flux path for zero sequence fluxes in core-type transformers and is sometimes treated as a fictitious Δ tertiary winding of high impedance [10].

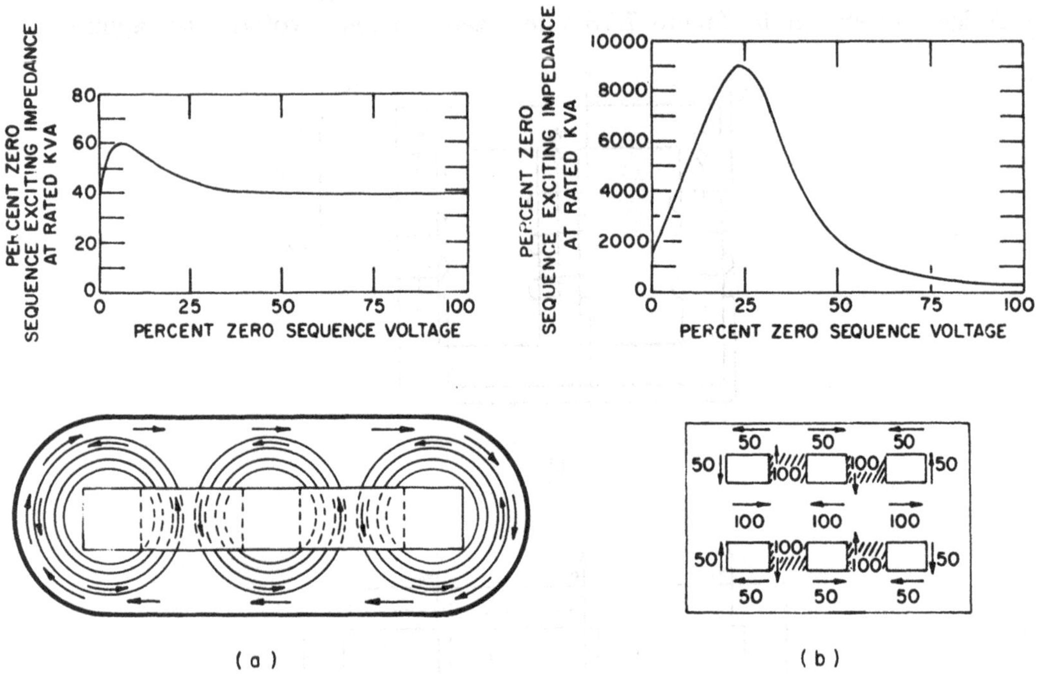

Fig. 7.19. Typical zero sequence open circuit impedances: (a) core-type, (b) shell-type. (From Clarke [11, vol. 2]. Used with permission.)

In any situation where high precision is required in fault computation, the transformer manufacturer should be consulted for exact information on zero sequence impedances. Where precise data is unavailable or high precision is not required (as is often the case), the zero sequence impedance is taken to be equal to the positive sequence impedance and the zero sequence equivalent is taken to be the same as that developed for three single-phase units.

This problem of finite excitation impedance is more pronounced in small units than in large three-phase designs. Three-phase distribution transformers, rated below 500 kVA and below 79 kV, are always of core-type design. Many but not all of the large power transformers are of shell-type design. There is also a five-legged core-type design which has an excitation impedance value between the three-legged core-type and the shell-type.

This treatment of the subject of excitation of three-phase transformers is certainly not exhaustive, and the interested reader is referred to the many references available, particularly [11, vol. 2], [14], and [55]. Reference [20] gives data for representation of three-phase, three-legged, core-type transformers as shown in Figures 7.20 and 7.21, where the notation $Q//N$ indicates that terminals Q and N would be connected together, or $Z_{PQ//N}$ is the impedance between P and the par-

TRANSFORMER CONNECTION	ZERO SEQUENCE EQUIVALENT CIRCUIT	APPROXIMATE ZERO SEQUENCE REACTANCE
1		$Z_{PQ-O} = Z_{PQ-I}$ $Z_{PN-O} = 5 Z_{PQ-I}$ $Z_{QN-O} = 6 Z_{PQ-I}$
2		$Z_{PN-O} = 5 Z_{PQ-I}$
3		$Z_{PQ-O} = \infty$
4		$Z_{PQ//N-O} = .85 Z_{PQ-I}$
5		
6		

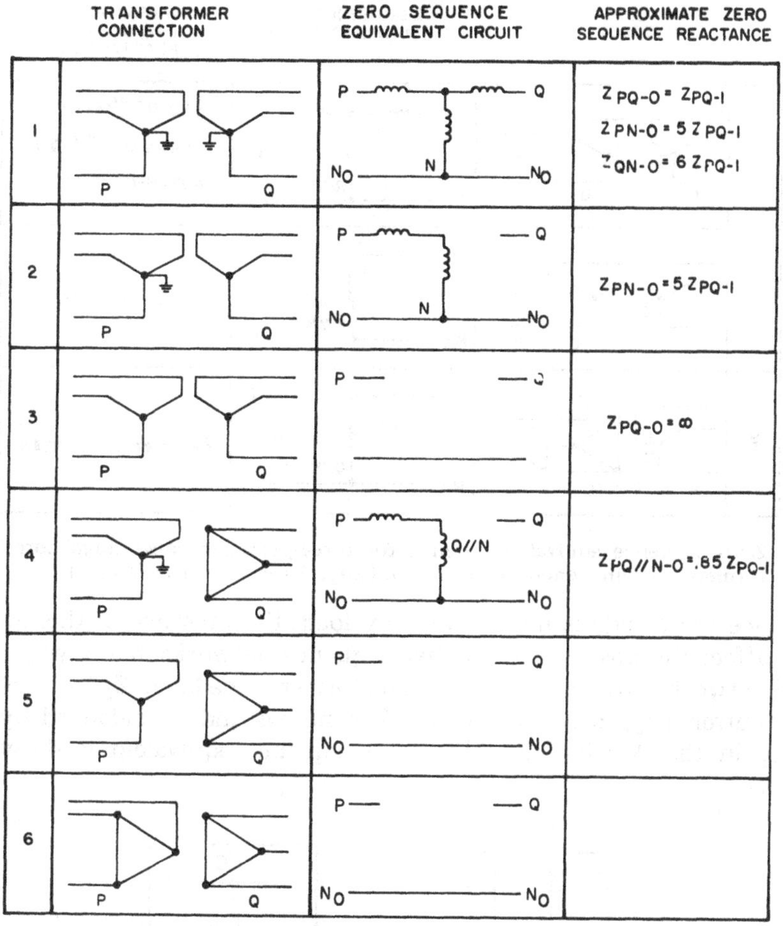

Fig. 7.20. Zero sequence equivalent circuits for three-phase, two-winding, core-type transformers. (From General Electric Co. [20]. Used with permission.)

allel combination of Q and N. Note that these data apply for *core-type* units only. Transformer circuits are labeled P, Q, and R and these labels may be replaced by H, X, and Y when voltage levels are known.

7.10 Grounding Transformers

Grounding transformers are sometimes used in systems which are ungrounded or which have high-impedance ground connections. Such units serve as a source of zero sequence currents for polarizing ground relays and for limiting overvoltages. These transformers must have some connection to ground, and this is usually through some sort of Y connection. As a system component the grounding transformer carries no load and does not affect the *normal* system behavior. When unbalances occur, the grounding transformer provides a low impedance in the zero sequence network. Two kinds of grounding transformers are used, the Y-Δ and the zigzag designs. These will be discussed separately.

The Y-Δ grounding transformer is an ordinary Y-Δ transformer connection but with the Δ winding isolated as shown in Figure 7.22a. Viewed from the Y side, the impedance is the excitation impedance which is usually taken to be in-

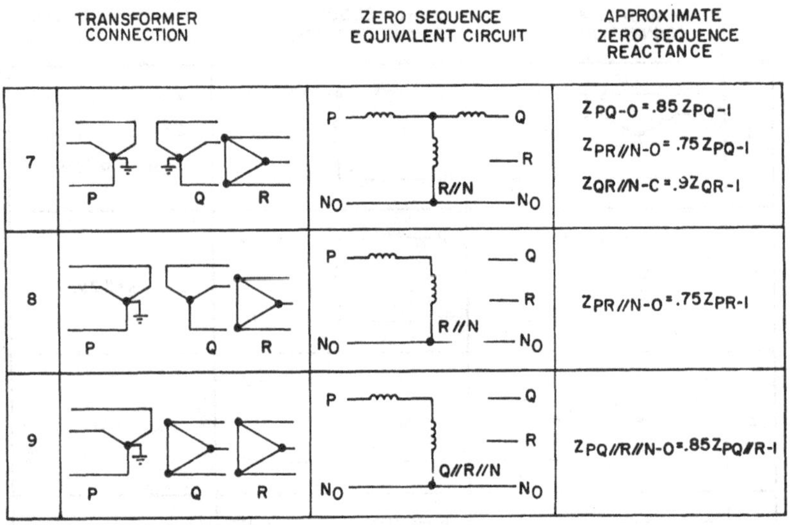

Fig. 7.21. Zero sequence equivalent circuits for three-phase, three-winding, core-type transformers. (From General Electric Co. [20]. Used with permission.)

finite. Since the Δ side is not serving any load, the presence of this transformer does not affect the positive or negative sequence networks in any way. The zero sequence network, however, sees the transformer impedance Z_t from point P_0 to N_0 since currents I_{a0} may flow in all Y windings and be balanced by currents circulating in the Δ winding. The zero sequence equivalent is shown in Figure 7.22b.

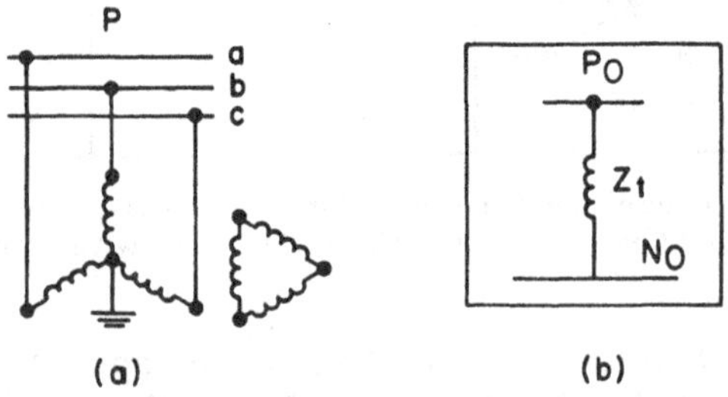

Fig. 7.22. Y-Δ grounding transformer: (a) connection at P, (b) zero sequence network representation.

The zigzag grounding transformer is a connection of 1:1 autotransformer windings, where primary and secondary windings are interconnected as shown in Figure 7.23. When positive or negative sequence voltages are applied to this connection, the impedance seen is the excitation impedance which is usually considered to be infinite. Thus the positive and negative sequence networks are unaffected by the grounding transformer. When zero sequence currents are applied, it is noted that the currents are all in phase and are connected so that the MMF produced in each coil is opposed by an equal MMF from another phase winding. These ideas are expressed graphically in Figure 7.24 where (b) shows the normal positive sequence condition and (a) shows the opposing sense of the coil connections. (Note that Figure 7.24a is valid for three single-phase units or one three-phase unit such as Figure 7.23.) Thus it appears to winding a_1 that it is "loaded" in winding a_2 by "load current" I_c which is equal to I_a. The impedance seen by

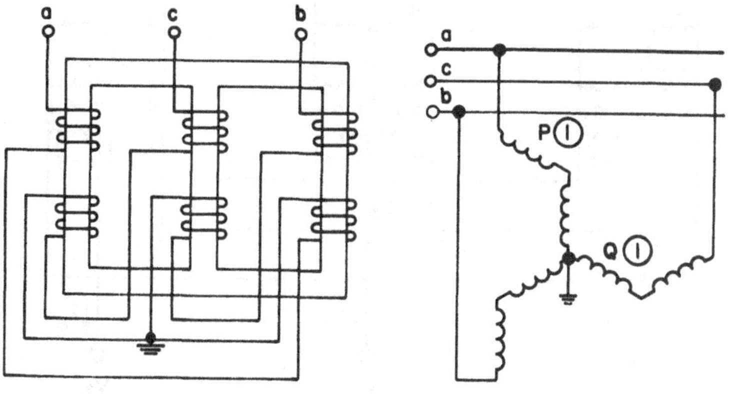

(a) (b)

Fig. 7.23. Zigzag grounding transformer connections: (a) winding arrangement on a core-form magnetic circuit, (b) schematic arrangement where parallel windings P and Q share core 1. (From Westinghouse Electric Corp. [14]. Used with permission.)

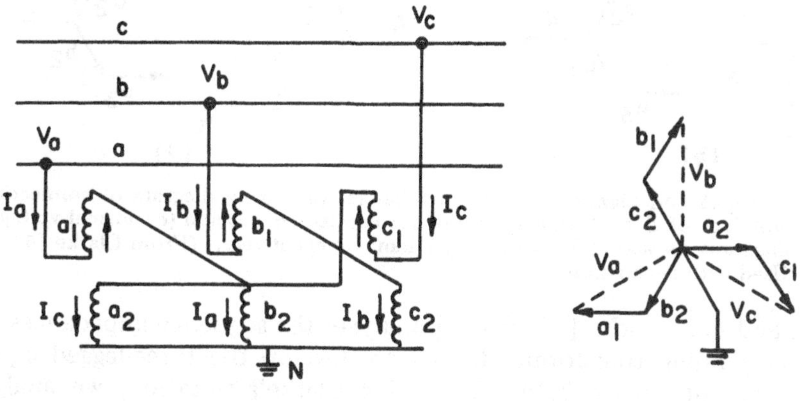

(a) (b)

Fig. 7.24. Zigzag grounding transformer: (a) wiring diagram, (b) normal voltage phasor diagram. (From Clarke [11, vol. 2]. Used with permission.)

I_a, then, is the leakage impedance between a_1 and a_2, or Z_t per phase. Thus the zero sequence representation is exactly the same as Figure 7.22b.

7.11 The Zigzag-Δ Power Transformer

If another winding connected in Δ is wound on the same core as the zigzag-connected windings, a transformer connection exists which is capable of power transmission with grounding and with no phase shift.[5] Such a transformation might be useful to parallel an existing Δ-Δ bank and supply grounding at the same time. If the added winding is Y connected the phase shift of the bank is 30°, exactly the same as for the Y-Δ connection. Both connections are shown in Figure 7.25 where windings subscripted 2 and 3 are connected in zigzag. The windings subscripted 1 are not connected in the figure, but parts (b) and (c) show the voltage phasor diagrams which result from Δ and Y connections respectively

[5] Some authors refer to this connection as the "interconnected star-delta" connection. We reject this label in preference for the shorter "zigzag-Δ" or "zigzag-Y" name.

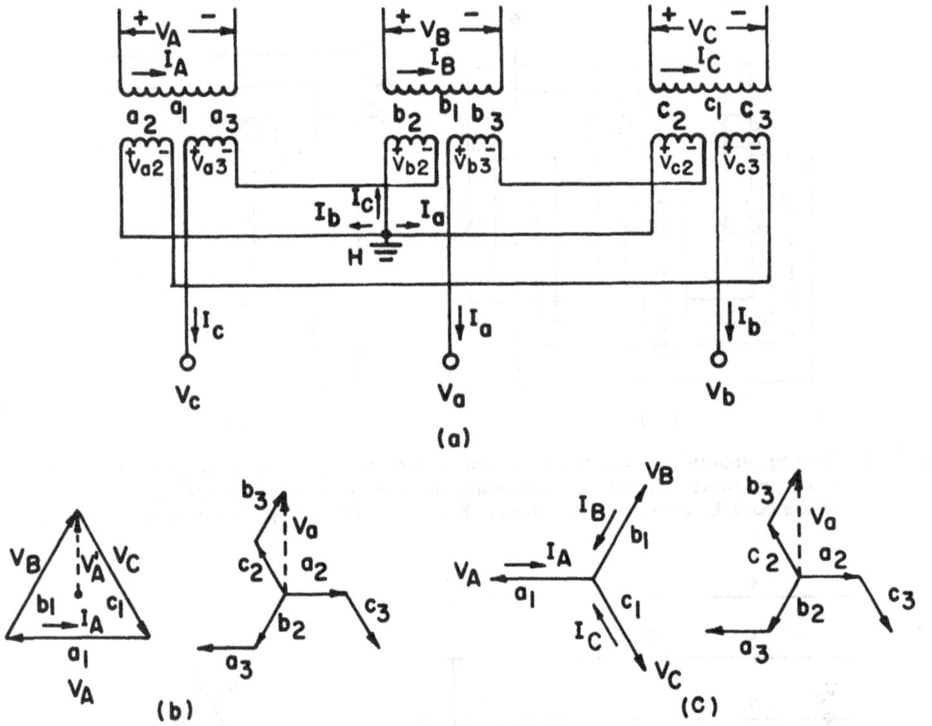

Fig. 7.25. Zigzag-Δ and zigzag-Y transformer banks: (a) wiring diagram of connections with windings of Δ or Y indicated but not connected, (b) and (c) normal voltage phasor diagrams of zigzag-Δ and zigzag-Y banks respectively. (From Clarke [11, vol. 2]. Used with permission.)

of a_1, b_1, and c_1. Clarke [11, vol. 2] analyzes the sequence impedances according to three-winding transformer theory to develop the three-legged equivalent similar to that of Figure 7.4c. Following Clarke's notation, we analyze the zigzag-Δ connection by defining the three-winding leakage impedances for core a as

$$\begin{aligned} Z_{12} &= \text{leakage impedance between } a_1 \text{ and } a_2 \\ Z_{13} &= \text{leakage impedance between } a_1 \text{ and } a_3 \\ Z_{23} &= \text{leakage impedance between } a_2 \text{ and } a_3 \end{aligned} \tag{7.44}$$

where all impedances are in pu based on rated voltamperes per phase and rated voltage of the windings. We take rated voltage of a_1 to be the Δ LL voltages, and for a_2 and a_3 (which have the same number of turns) we use $1/\sqrt{3}$ times the base LN voltage of the zigzag side. Since all impedances are in pu on the same voltampere base, we compute from (7.29),

$$\begin{bmatrix} Z_x \\ Z_y \\ Z_z \end{bmatrix} = 1/2 \begin{bmatrix} 1 & 1 & -1 \\ 1 & -1 & 1 \\ -1 & 1 & 1 \end{bmatrix} \begin{bmatrix} Z_{12} \\ Z_{13} \\ Z_{23} \end{bmatrix} \text{ pu} \tag{7.45}$$

where Z_x, Z_y, and Z_z are the impedances of the three-legged equivalent as shown in Figure 7.26a, and with similar results for the b and c windings of Figure 7.25 given by Figure 7.26b and 7.26c respectively. The notation and current directions

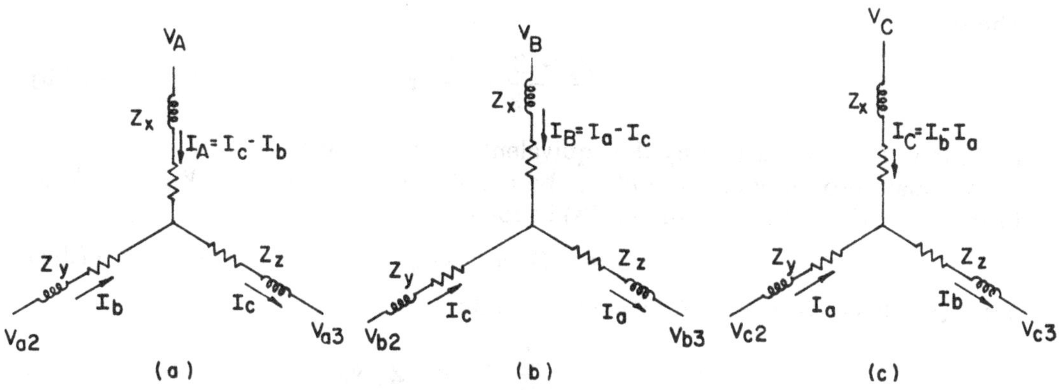

Fig. 7.26. Identical equivalent circuits to replace each of the three-winding transformers, with currents and voltages indicated. All values in pu. (From Clarke [11, vol. 2]. Used with permission.)

for Figure 7.26 correspond to those of Figure 7.25. From Figure 7.25a

$$V_a = V_{b3} - V_{c2} \tag{7.46}$$

But from Figure 7.26b and 7.26c

$$\begin{aligned} V_{b3} &= V_B - I_B Z_x - I_a Z_z \\ V_{c2} &= V_C - I_C Z_x + I_a Z_y \end{aligned} \tag{7.47}$$

combining (7.46) and (7.47), we compute

$$V_a = V_B - V_C - I_a(2Z_x + Z_y + Z_z) + (I_b + I_c) Z_x \tag{7.48}$$

Now let V_A, V_B, and V_C be a positive sequence set of voltages, V_{A1}, V_{B1}, and V_{C1}. Then $V_{B1} - V_{C1} = -\mathrm{j}\sqrt{3}\, V_{A1}$, and we also note that

$$V_a = V_{a1}, \qquad I_a = I_{a1}, \qquad I_b + I_c = I_{b1} + I_{c1} = -I_{a1}$$

and (7.48) becomes

$$V_{a1} = -\mathrm{j}\sqrt{3}\, V_{A1} - (3Z_x + Z_y + Z_z) I_{a1} \tag{7.49}$$

But this is computed for V_{a1} based on $1/\sqrt{3}$ times the rated LN voltage (or based on winding a_2 or a_3 voltage). Based on the system LN voltage, V_{a1} would be $1/\sqrt{3}$ times this amount. Similarly, I_{a1} is based on rated voltamperes and rated a_2 voltage; and if we choose a new base voltage $\sqrt{3}$ times as large as the old base, the new pu current is $\sqrt{3}$ times as large as the old pu current. Using a bar to signify the new pu value we have

$$\overline{V}_{a1} = \frac{V_{a1}}{\sqrt{3}}, \qquad \overline{I}_{a1} = \sqrt{3}\, I_{a1} \tag{7.50}$$

which changes (7.49) to

$$\overline{V}_{a1} = -\mathrm{j}\, V_{A1} - \frac{3Z_x + Z_y + Z_z}{3} \overline{I}_{a1} \tag{7.51}$$

or, rearranging,

$$-\mathrm{j}\, V_{A1} = \overline{V}_{a1} + Z_1 \overline{I}_{a1} \tag{7.52}$$

where

$$Z_1 = \frac{3Z_x + Z_y + Z_z}{3} \text{ pu} \tag{7.53}$$

Equation (7.52) is satisfied by the equivalent circuit of Figure 7.27a.

We may also let V_A, V_B, and V_C be a zero sequence set V_{A0}, V_{B0}, and V_{C0}. In this case, $V_{B0} - V_{C0} = 0$ and (7.48) becomes

$$V_{a0} = -(Z_y + Z_z) I_{a0} \tag{7.54}$$

After the change of base in (7.50) we compute

$$\overline{V}_{a0} = -\frac{Z_y + Z_z}{3} \overline{I}_{a0} = -Z_0 \overline{I}_{a0} \tag{7.55}$$

where

$$Z_0 = \frac{Z_y + Z_z}{3} = \frac{Z_{23}}{3} \text{ pu} \tag{7.56}$$

Equation (7.55) is satisfied by the circuit of Figure 7.27b.

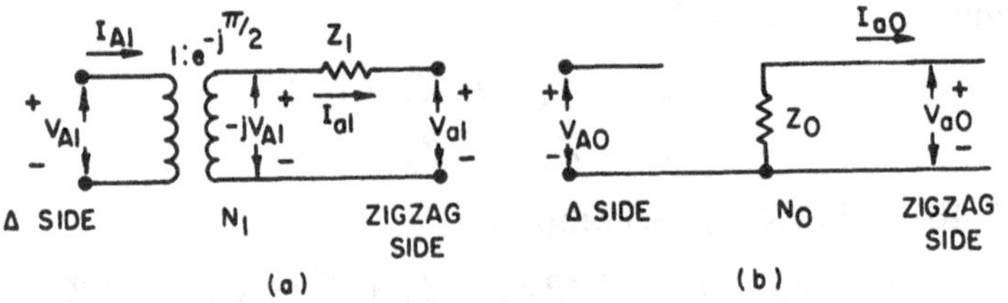

Fig. 7.27. Equivalent circuits of a zigzag-Δ transformer: (a) positive sequence, (b) zero sequence.

III. TRANSFORMERS IN SYSTEM STUDIES

7.12 Off-Nominal Turns Ratios

In the study of very small radial systems there is no problem in representing the transformer, following the guidelines previously presented, and the chosen base voltages are conveniently taken to be the rated transformer voltages. This was the case in Example 1.2. In multiply interconnected systems involving two or more voltage levels, however, it is not always possible to choose the base voltage as the transformer rated voltage because the transformer voltages are not always rated the same.

Consider, for example, the simple system shown in Figure 7.28 where three systems $S1$, $S2$, and $S3$ are interconnected as shown and where the three transformers may have different voltage ratings. Yet the three transformers operate essentially in parallel, interconnecting systems nominally rated 69 kV and 161 kV. Such a connection of transformers presents two problems, an "operating" problem and a "mathematical" problem. If the transformers are of different turns ratio, even though fairly close to the 69–161 kV nominal transformation ratio, an interconnection like that of Figure 7.28 will cause currents to circulate and reactive power to circulate in the interconnected loops. This is the operating problem, which is often ignored for fault computation, although we realize that an off-

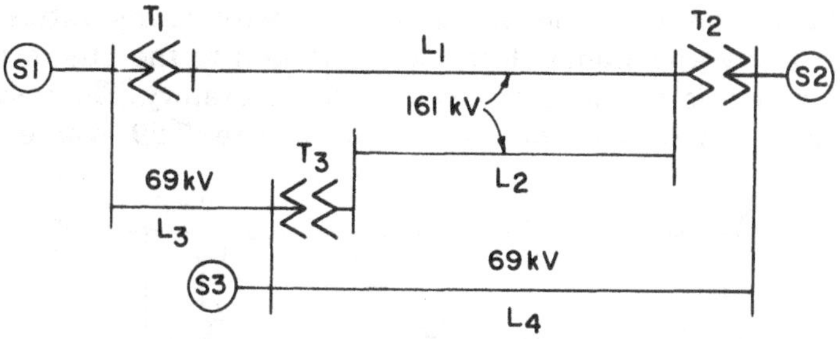

Fig. 7.28. System with unmatched transformations.

nominal transformer tap ratio may be employed, either in fixed taps or in load tap changing equipment, to eliminate any circulating currents. The mathematical problem is that of deriving a correct representation for a system such as that in Figure 7.28. This problem is of interest since it involves the correct system representation for an assumed operating condition. In other words, if our chosen base voltages do not coincide with the transformation ratio of the transformers, how do we compensate for this difference? We illustrate this problem by an example.

Example 7.2

Given the following data for the system of Figure 7.28, examine the transformer representations for arbitrarily chosen base voltages of 69 kV and 161 kV and for a base MVA of 100 if the transformers are rated as follows:

T1: X = 10%, 50 MVA, 161 (grd Y)-69 (Δ) kV
T2: X = 10%, 40 MVA, 161 (grd Y)-66 (Δ) kV
T3: X = 10%, 40 MVA, 154 (grd Y)-69 (Δ) kV

Discussion

We compute the voltage transformation ratios as follows:

T1: Ratio = 161/69 = 2.33 (nominal)
T2: Ratio = 161/66 = 2.44 (11% high)
T3: Ratio = 154/69 = 2.23 (10% low)

If we view each transformer from the 161 kV system only, we compute the following impedances.

$$\text{T1: } X = (0.1)(100/50)(161/161)^2 = 0.20 \text{ pu}$$

$$\text{T2: } X = (0.1)(100/40)(161/161)^2 = 0.25 \text{ pu}$$

$$\text{T3: } X = (0.1)(100/40)(154/161)^2 = 0.229 \text{ pu}$$

Suppose now that we have exactly 1.0 pu voltage on the 161 kV system with all transformer low-voltage connections open. Then the voltages on the transformer low-voltage buses would be as follows:

T1: V = 69/69 = 1.0 pu
T2: V = 66/69 = 0.957 pu
T3: V = (161/154)(69/69) = 1.045 pu

Thus we could correct for the off-nominal turns ratio by inserting an *ideal trans-*

former on the low-voltage side of T2 and T3 with turns ratios 0.957 : 1 and 1.045 : 1 respectively and would close the connection to the bus in this way. This is exactly the way this problem is solved in the laboratory. The resulting positive and zero sequence networks are shown in Figure 7.29 where the intercon-

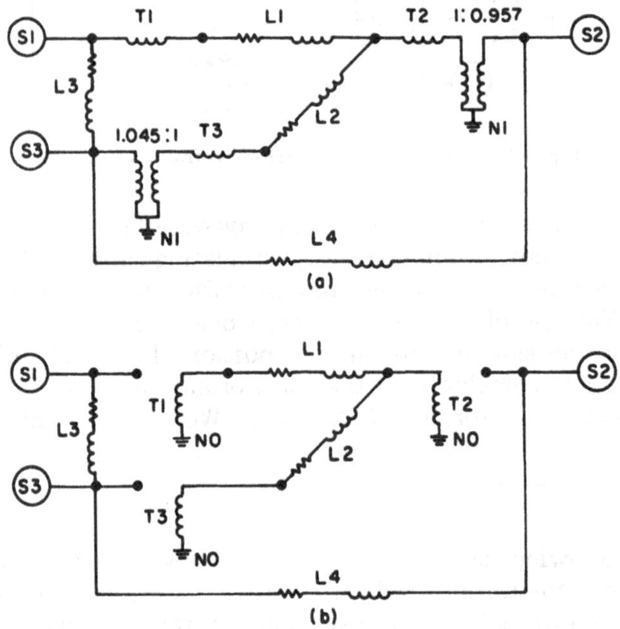

Fig. 7.29. **Positive and zero sequence networks for the system of Example 7.2: (a) positive sequence, (b) zero sequence.**

necting systems $S1$, $S2$, and $S3$ are not shown in detail. The negative sequence network is exactly like the positive sequence network in this problem since there are no generators in the system under consideration. If the tap changing mechanisms were located on the high side (the 161 kV side) on the actual transformers, the transformer impedances should be computed on the low-voltage base and the ideal transformer shown on the side corresponding to the physical tap changer. Actually, in a Y-Δ transformer it is quite likely that the tap changer will be located near the neutral end of the Y side.

In analytical work it is somewhat awkward to deal with an ideal transformer. Instead, a passive network equivalent is sometimes used where the transformer tap changer (ideal transformer) and series impedance are replaced by a pi equivalent circuit. Such an equivalent circuit may be derived with reference to Figure 7.30 where an LTC (load tap changing) transformer is represented between nodes j and k and with tap changing equipment on the j side. Note also that the tap ratio is indicated as from k toward j, e.g., a tap ratio $n > 1.0$ would indicate that j is in a boost (step-up) position with respect to k. The admittance Y is the inverse of the usual transformer impedance Z and is used as a matter of convenience.

With currents defined as in Figure 7.30 we compute

$$I_k = Y(V_m - V_k) \tag{7.57}$$

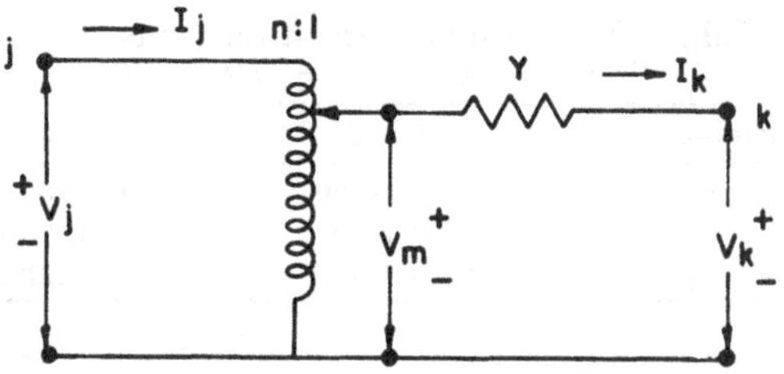

Fig. 7.30. Equivalent circuit of an LTC transformer with tap changer on node j side.

But for the ideal transformer we have

$$I_k = nI_j, \qquad V_j = nV_m \tag{7.58}$$

and we eliminate V_m from (7.57) by substitution to compute

$$I_k = Y\left(\frac{1}{n} V_j - V_k\right) \tag{7.59}$$

Multiplying by $1/n$, we have

$$I_j = \frac{Y}{n}\left(\frac{1}{n} V_j - V_k\right) \tag{7.60}$$

We now establish similar equations for the pi equivalent circuit of Figure 7.31

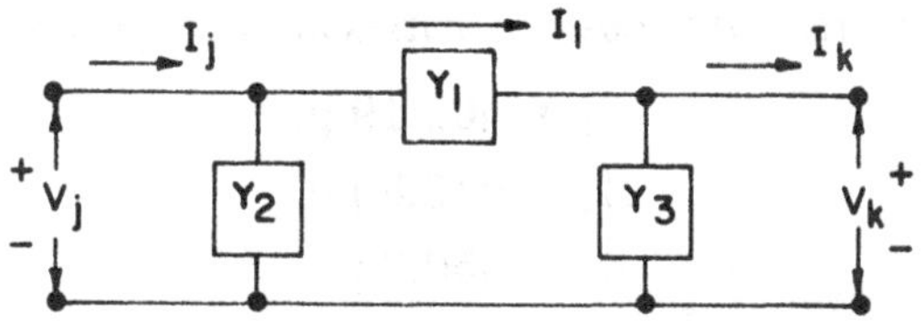

Fig. 7.31. Pi equivalent circuit for Figure 7.30.

where we write by inspection

$$I_1 = Y_1(V_j - V_k), \qquad I_j = Y_2 V_j + I_1, \qquad I_1 = Y_3 V_k + I_k \tag{7.61}$$

From (7.61) we compute

$$I_k = Y_1 V_j - (Y_1 + Y_3) V_k \tag{7.62}$$

and

$$I_j = (Y_1 + Y_2) V_j - Y_1 V_k \tag{7.63}$$

We now compare (7.59) and (7.60) with (7.62) and (7.63) respectively to write

$$Y_1 = \frac{1}{n} Y, \qquad Y_2 = \frac{1 - n}{n^2} Y, \qquad Y_3 = \frac{n - 1}{n} Y \tag{7.64}$$

Note that all three of the admittances (7.64) are functions of the turns ratio n. Furthermore, the sign associated with the shunt components Y_2 and Y_3 are always opposite so that Y_2 and Y_3 are either inductive or capacitive for Y representing a pure inductance, depending entirely on n. This is shown in Table 7.5 and is further illustrated by Example 7.3.

Table 7.5. Nature of Circuit Elements of the Pi Equivalent for $Y = -jB, B > 0$.

Element	$n > 1$	$n < 1$
Y_1	inductance	inductance
Y_2	inductance	capacitance
Y_3	capacitance	inductance

Example 7.3

Compute the pi equivalent representation of transformer T3 of Example 7.2.

Solution

If the node which corresponds to node j of Figure 7.30 is arbitrarily set at 1.0 pu voltage (this is the $S3$ node of Figure 7.29), we set $n = 0.957$ pu. Then compute

$$Y = -j\frac{1}{X} = -j\frac{1}{0.229} = -j4.36 \text{ pu}$$

$$1/n = 1.045$$

$$Y_1 = (1.045)(-j4.36) = -j4.56 \text{ pu (ind)}$$

$$Y_2 = (1.045)^2\,(0.043)(-j4.36) = -j0.0205 \text{ pu (ind)}$$

$$Y_3 = (-0.043)(-j4.36) = +j0.01965 \text{ pu (cap)}$$

As impedances the inverses of these quantities are computed as

$$Z_1 = +j0.219 \text{ pu}$$

$$Z_2 = +j48.8 \text{ pu}$$

$$Z_3 = -j50.9 \text{ pu}$$

Since we usually work with impedances rather than admittances and where Z_t is the transformer impedance, we compute from (7.64)

$$Z_t = 1/Y, \qquad Z_1 = nZ_t$$

$$Z_2 = \frac{n^2}{1-n}Z_t, \qquad Z_3 = \frac{n}{n-1}Z_t \tag{7.65}$$

The above conclusions are summarized in Figure 7.32.

Note that as n approaches unity, the shunt branches of the pi equivalent both approach an infinite impedance and the series branch approaches Z_t. Also note that if transformer resistance is included in Z_t, one of the shunt impedances has a

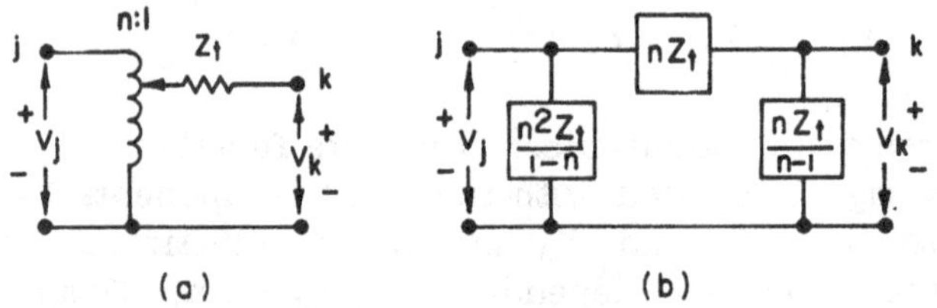

Fig. 7.32. The transformer pi equivalent: (a) transformer equivalent, (b) pi equivalent.

negative real part. This is of no concern in analytical work but would be difficult to simulate on a network analyzer.

7.13 Three-Winding Off-Nominal Transformers

A three-winding transformer is usually represented by a Y equivalent similar to Figure 7.4c. If, however, one or more of the three windings differs from the chosen base voltage or if a winding operates at a tap position which causes that winding to differ from base voltage, that leg of the Y equivalent must be changed to account for this voltage error.

Suppose that the equivalent circuit of a transformer is computed in pu on the transformer base or, given Z_{HX}, Z_{HY}, and Z_{XY}, we compute (on the H winding base voltamperes and voltage) from (7.29)

$$\begin{bmatrix} Z_H \\ Z_X \\ Z_Y \end{bmatrix} = 1/2 \begin{bmatrix} 1 & 1 & -1 \\ 1 & -1 & 1 \\ -1 & 1 & 1 \end{bmatrix} \begin{bmatrix} Z_{HX} \\ Z_{HY} \\ Z_{XY} \end{bmatrix} \text{ pu}$$

Now suppose that one of the windings, say winding Y, has a voltage rating different from the chosen base voltage for the Y voltage system. Then to match the transformer to the system, it is necessary to add an ideal transformer of voltage ratio n where n is defined as

$$n = \frac{\text{Base } V_Y}{\text{transformer rated } V_Y} \tag{7.66}$$

This connection is shown in Figure 7.33 where the Y impedances are taken from (7.29).

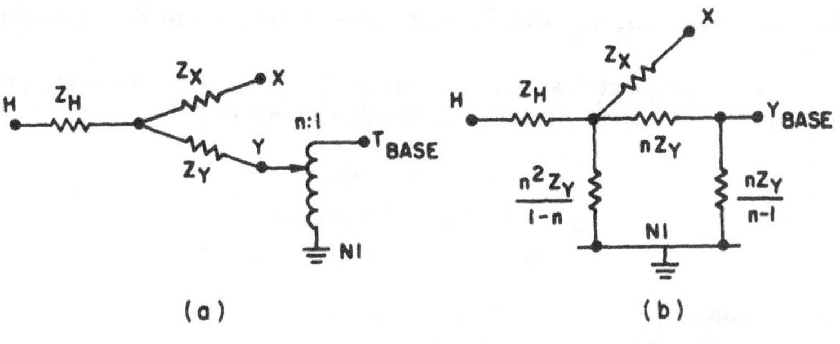

Fig. 7.33. Three-winding transformer equivalent with off-nominal Y winding: (a) ideal transformer, (b) pi equivalent.

Problems

7.1. A single-phase distribution transformer is rated 50 kVA, 60 Hz, 2400-240 V and has impedances as follows:

$$Z_H = 0.7 + j1.0 \ \Omega$$
$$Z_X = 7 + j10 \ \text{m}\Omega$$
$$Y_m = 1/Z_m = 3 - j20 \text{ mmho referred to LV winding}$$

Compute all parameters needed to represent this transformer according to equivalent circuits in Figures 7.1a, 7.1b, and 7.1c. Compare your solution to the values given in Table 7.1.

7.2. For the transformer of Figure P7.2 determine the terminal markings X_1 and X_2 for the right-hand winding.

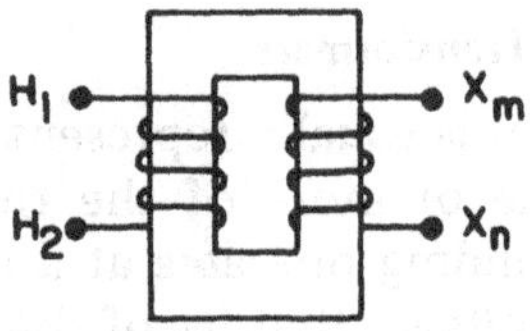

Fig. P7.2.

7.3. The terminals of the transformer of Figure P7.3 are assumed to be unlabeled and the polarity is not known. As a test, terminal H_1 is connected to X_n and 200 V are applied between H_1 and H_2. Under this test condition a voltmeter connected from H_2 to X_m reads 220 V. What is the transformer polarity and what is its turns ratio?

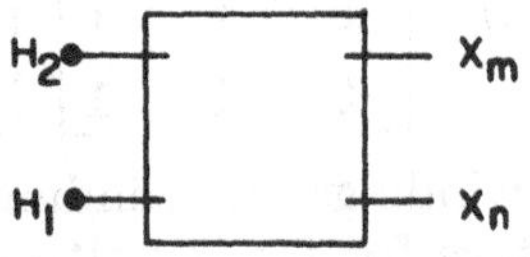

Fig. P7.3.

7.4. Verify (7.23)-(7.25).

7.5. The following is taken from the nameplate of a three-winding transformer.

$$\text{Rated } V_H = 161 \text{ kV}, \quad \text{Rated } V_X = 69 \text{ kV}, \quad \text{Rated } V_Y = 13.8 \text{ kV}$$
$$Z_{HX} = 8.69\%, \quad Z_{HY} = 5.33\%, \quad Z_{XY} = 1.92\%$$
$$S_H = 30 \text{ MVA}, \quad S_X = 10.5 \text{ MVA}, \quad S_Y = 10.5 \text{ MVA}$$

Compute the values Z_H, Z_X, and Z_Y in percent to be used in the equivalent circuit of Figure 7.4c.

7.6. Given a three-winding autotransformer whose H, X, and Y windings are rated 200 kV, 100 kV, and 10 kV respectively and with circuit impedances of

$$Z_{HX} = 10\% \text{ on a 30 MVA base}$$
$$Z_{XY} = 9\% \text{ on a 10 MVA base}$$
$$Z_{HY} = 15\% \text{ on a 30 MVA base}$$

Compute the following in pu on a 100 MVA base.
(a) Equivalent circuit impedances Z_H, Z_X, and Z_Y.
(b) Equivalent circuit impedances Z_c, Z_t, and Z_s.
(c) Circuit impedance Z_{ct}, Z_{sc}, and Z_{ts}.

7.7. Verify equations (a) and (b) of Table 7.3.

7.8. Verify equation (c) of Table 7.3.

7.9. Verify equation (d) of Table 7.3.

7.10. Verify equations (e) and (f) of Table 7.3.

7.11. Three 10 MVA, 100-15 kV transformers have nameplate impedances of 10% and are connected Δ-Y with the high voltage side Δ. Find the zero sequence equivalent circuit.
(a) If the neutral is ungrounded.
(b) If the neutral is grounded solidly.
(c) If the neutral is grounded through a 5 Ω resistance.
(d) If the neutral is grounded through 5000 μF capacitance.

7.12. Find the resistance in ohms which must be placed in the corner of the Δ of the transformer bank of problem 7.11 to result in exactly the same zero sequence circuit as in problem 7.11a-7.11d.

7.13. Verify (7.40).

7.14. Suppose that three identical autotransformers are connected Y-Δ-Y as shown in Figure 7.11a. If the transformers have the impedances specified in problem 7.6, find the equivalent circuits: (a) For positive and negative sequence networks, corresponding to Figure 7.11b. (Let $Z_n = 0.1$ pu on an X base.) (b) For the zero sequence network, corresponding to Figure 7.11c.

7.15. Reconnect the transformer windings of Figure 7.15 so that $V_{A1} = -jV_{a1}$. Is this an acceptable phase relationship with which to perform all other computations? Draw phasor diagrams to show all sequence voltages and currents.

7.16. Extend the computation begun in Example 7.1 to include the voltages on both sides of transformer T1. Note that this requires the computation of the voltage drop from bus B to the fault point.

7.17. Show that third-harmonic exciting currents can be supplied in the zigzag grounding transformer connection of Figure 7.22.

7.18. Consider a bank of 3-winding autotransformers connected zigzag-Δ and connected in a system with voltages of 34.5 kV ungrounded and 69 kV grounded. The transformer is to be connected to provide a ground to the 34.5 kV system. The transformer impedances and ratings are (for each of the three units),

$$Z_{12} = j0.15 \text{ pu on a 5 MVA base}$$
$$Z_{13} = j0.15 \text{ pu on a 5 MVA base}$$
$$Z_{23} = j0.10 \text{ pu on a 2.5 MVA base}$$

Find the positive and zero sequence impedances for a problem in which the Base MVA of the study is to be 25 MVA.

7.19. Compute the pi equivalent representation of transformer T2 of Example 7.2.

7.20. Verify the circuit of Figure 7.33b.

7.21. Given a 3-phase, wye-connected autotransformer with delta tertiary winding and with ratings and known data as follows, find the positive and zero sequence equivalents for this transformer:

Winding H: S_{HB} V_{HB} given $Z_{HX(X)}$ in per unit based on S_{HB} and V_{HB}

Winding X: S_{XB} V_{XB} given $Z_{HY(H)}$ in per unit based on S_{HB} and V_{HB}

Winding Y: S_{YB} V_{YB} given $Z_{XY(X)}$ in per unit based on S_{XB} and V_{XB}

7.22. Figure P7.22 shows three single-phase transformers connected Y-Δ with the transformer terminals marked with both the IEEE H-X markings and the a-b-c phase designations.

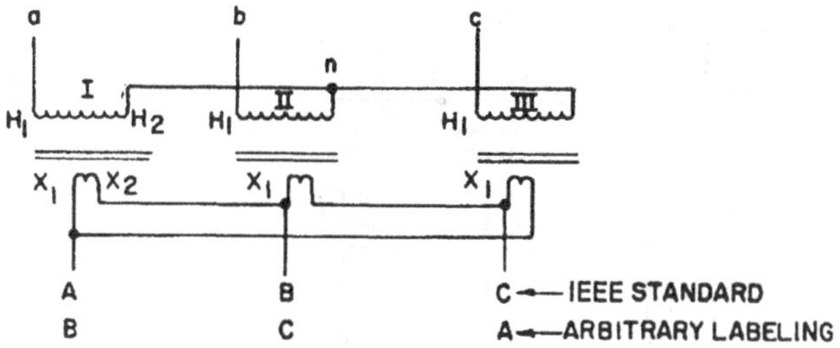

Fig. P7.22.

Note in particular the Δ winding where an arbitrary labeling is specified which does not agree with the IEEE terminal marking.

(a) Sketch two phasor diagrams of the H-winding voltages, assuming the applied voltages to be first positive then negative sequences. Let V_{an} be the reference voltage.

(b) Sketch phasor diagrams of the X-winding positive and negative sequence voltages, using the IEEE labeling. Show that HV_{LN} leads LV_{LN} by 30° in the positive sequence case but lags by 30° in the negative sequence.

(c) Sketch phasor diagrams of the X-winding positive and negative sequence voltages, using the arbitrary labeling. Show that HV_{LN} lags LV_{LN} by 90° in the positive sequence but leads by 90° in the negative sequence.

7.23. In the three-phase transformer connection shown in Figure P7.23 all three windings of each transformer are assumed to be in use but at essentially no load. Given that $V_{AB} = 69{,}000\underline{/0^\circ}$ V and the sequence is A-B-C on the H winding side. Given that $n = n_H/n_X = 144$, $n' = n_H/n_Y = 16.5$.

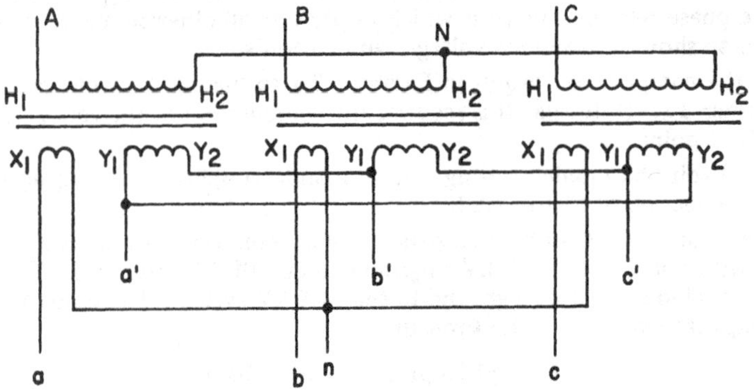

Fig. P7.23.

(a) Sketch three voltage phasor diagrams, one each showing all six (3 LL and 3 LN) voltages for the three voltages designated H, X, and Y. Label each diagram to show all voltage magnitudes, phase angles, and the A-B-C, a-b-c, or a'-b'-c' phase designations. Orient the three diagrams to show the phase shifts between V_{AN}, V_{an}, and $V_{a'n'}$. (Where is n'?)

(b) Assume that the H-windings are energized with sinusoidal LL balanced three-phase voltages. List the odd harmonic orders through $n = 11$ which exist in
   (1) The H-winding LN voltages.
   (2) The X-winding LN voltages.
   (3) The X-winding LL voltages.
   (4) The Y-winding voltages.
   (5) The H-winding currents.
   (6) The Y-winding currents.

(c) Repeat (b) but with the Y-winding open.

(d) Repeat (b) with the Y-windings open and a high-voltage neutral wire connecting N to the high-voltage source.

(e) Are any of the connections (b), (c), or (d) unsuited for practical application? Why?

7.24. In the transformer connection of Figure P7.24 each single-phase transformer has a turns

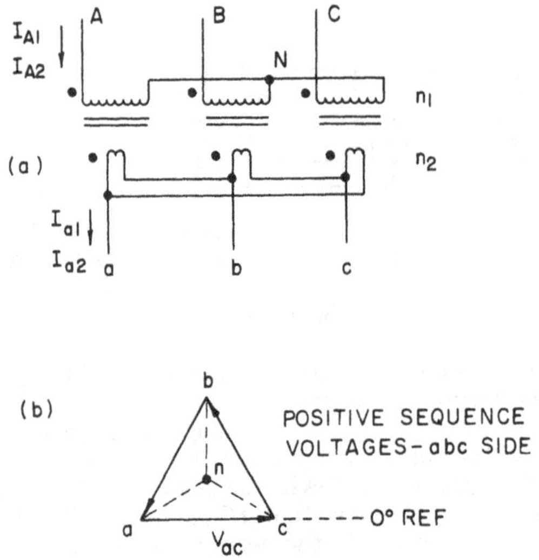

Fig. P7.24.

ratio of $n = n_1/n_2 = 10$. The low-voltage side carries an unbalanced load with sequence currents given as $I_{a1} = -I_{a2} = 100\underline{/-30^\circ}$ A. Find the sequence currents I_{A1} and I_{A2} as phasors.

7.25. Consider the single-phase three-winding transformer shown in Figure P7.25, with the windings labeled arbitrarily as *s-c-t* and with ratings given for each winding. The following partial short circuit test data is known.

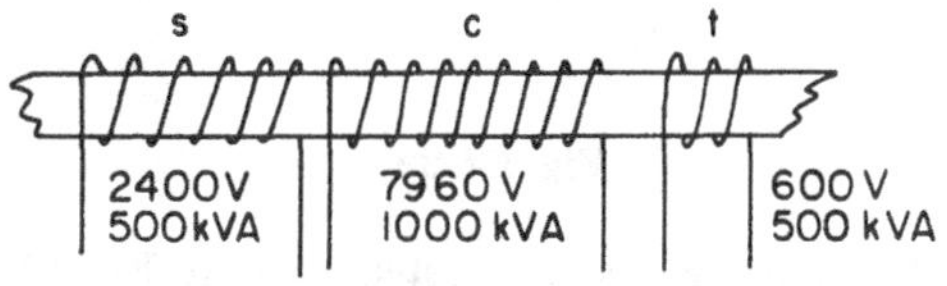

Fig. P7.25.

Excited Winding	*Shorted Winding*	*Applied Volts*	*Amperes in Excited Winding*
c	*s*	252	62.7
c	*t*	770	62.7
s	*t*	217	208.0

Note the absence of wattmeter data for copper-loss measurements. Assume that winding resistances are negligible.

Consider a three-phase connection of identical units with windings *c-s-t* connected Y-Y-Δ with both Y's ungrounded. Let the base kVA be 3000 kVA.

(a) Determine the equivalent circuit parameters $Z_{H(H)}$, $Z_{X(H)}$, and $Z_{Y(H)}$ in ohms referred to the *H*-winding.

(b) Determine Z_H, Z_X, and Z_Y in pu.

7.26. Consider a three-phase bank of transformers identical to those of problem 7.25 but connected as autotransformers (Figure P7.26) and again connected Y-Y-Δ with the *t*-windings connected in Δ and the *c*-windings connected to a common neutral.

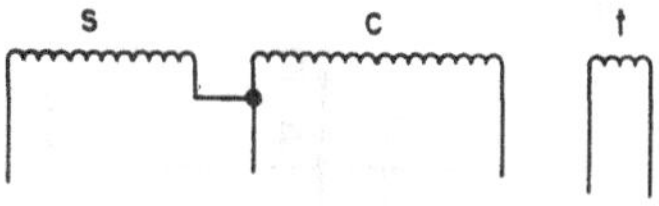

Fig. P7.26.

(a) Determine the *s-c-t* parameters of the *winding base* equivalent circuit for positive and negative sequences, viz., values $Z_{s(c)}$, $Z_{c(c)}$, and $Z_{t(c)}$ in ohms referred to the *c*-winding.

(b) Convert the *s-c-t* parameters to pu on a 3.0 MVA base.

7.27. Consider the autotransformer connection of problem 7.26.

(a) Find the parameters of the *H-X-Y* equivalent circuit on a *circuit basis* for positive and negative sequences; viz., find the values $Z_{H(X)}$, $Z_{X(X)}$, and $Z_{Y(X)}$ in ohms referred to the *X* circuit.

(b) Convert the *H-X-Y* parameters to pu on a 3.0 MVA base.

7.28. Consider the three-phase autotransformer connection of Problem 7.26 with the neutral grounded through an impedance $Z_N = 0 + j1.0\ \Omega$. Determine the T circuit parameters in pu for the zero sequence network.

7.29. Let the three-phase transformer bank T1 in Figure P7.29 be the bank described in Problem 7.28. Transformer bank T2 has reactances $X_1 = X_2 = 0.10$ pu based on the transformer rating of 79.6/138-17.95 kV and 30 MVA. Thevenin impedances $Z_1 = Z_2$ for the 138 kV supply system are such that the three-phase fault MVA at bus 1 is 600 MVA.

(a) Draw and label the three sequence networks and mark the pu values of all impedances. Assume the Thevenin equivalent voltage is 1.0 pu.

(b) Compute the symmetrical SLG and 3ϕ fault currents in amperes for a fault on bus 3.

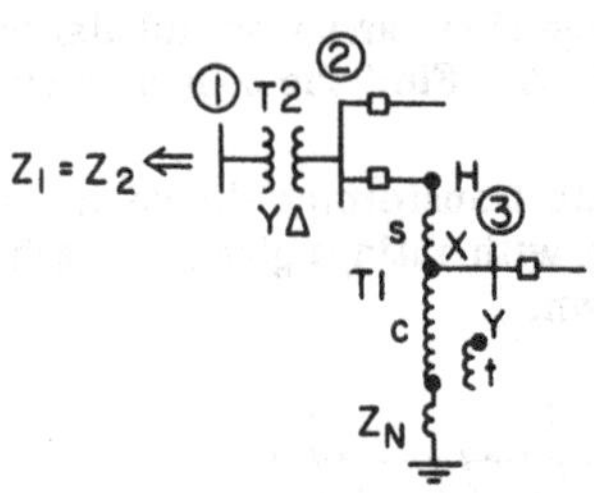

Fig. P7.29.

7.30. A three-phase 34.6 kV subtransmission system (Figure P7.30) supplies several distribution substations (which are not shown) all of which are Δ connected on the high-voltage side.

(a) Sketch the zero sequence network for this system and note the position of the Peterson coil, labeled PC.

(b) The purpose of the Peterson coil is to reduce the SLG fault currents to small, harmless values and to avoid service outages when such faults occur. Describe how this could be possible and suggest any problems you can think of which might limit the effectiveness of this device.

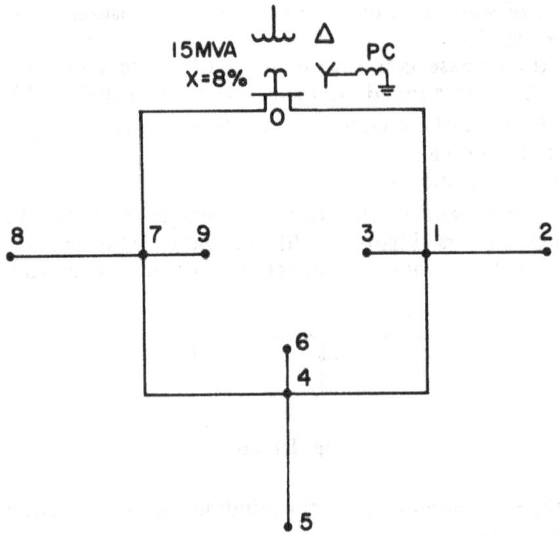

Fig. P7.30.

7.31. Let the line configuration for the subtransmission network of problem 7.30 be that given in Figure P7.31, where the line data is given as follows:

Line Section	*Conductor*				*Section Length*	*Ground Wire*	
	Size	*Type*	*STR*	*Radius*		*Size*	*Type*
0-1	266.8	ACSR	26/7	0.0268 ft	10 mi	1/2″	SM steel
1-2	2/0	ACSR	6/1	0.0186	15	none	
1-3	2/0	ACSR	6/1	0.0186	5	none	
1-4	266.8	ACSR	26/7	0.0268	10	none	
4-5	2/0	ACSR	6/1	0.0186	15	none	
4-6	2/0	ACSR	6/1	0.0186	5	none	
0-7	266.8	ACSR	26/7	0.0268	10	1/2″	SM steel
7-8	2/0	ACSR	6/1	0.0186	15	none	
7-9	2/0	ACSR	6/1	0.0186	5	none	
7-4	266.8	ACSR	26/7	0.0268	10	none	

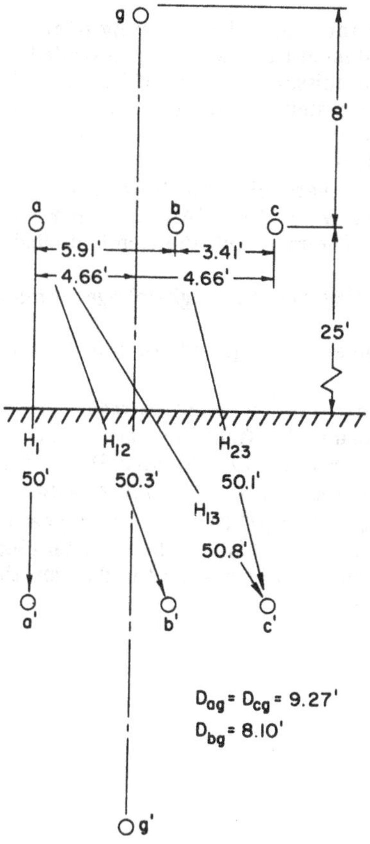

Fig. P7.31.

(a) Compute the zero sequence capacitance and 60 Hz susceptance per phase per mile for all line sections.

(b) Specify the following parameters for the Peterson coil:

(1) inductance
(2) reactance
(3) continuous current rating
(4) continuous kVA rating
(5) continuous voltage rating

7.32. Consider the subtransmission bus shown in Figure P7.32 where the secondary winding of the supply transformer is connected in Δ. A grounding transformer bank made up of three identical single-phase transformers is connected to the bus to provide a ground connection for the Peterson coil. Assume that these single-phase transformers have a leakage reactance of 6% based on their ratings.

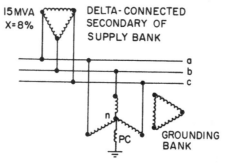

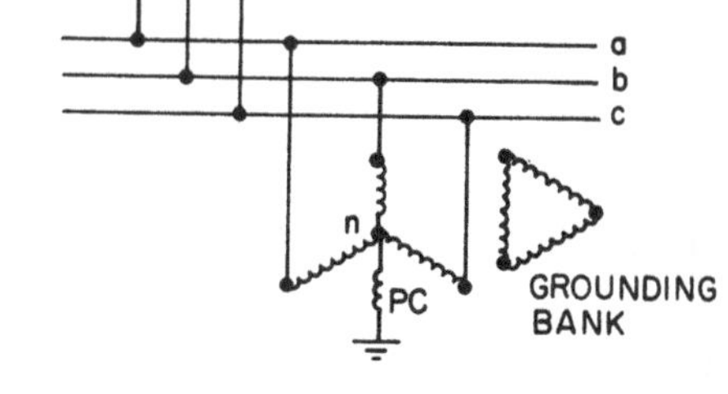

Fig. P7.32.

We wish to determine the required kVA rating of the single-phase grounding transformers from the list of standard available sizes provided below. Assume that the bus *a-b-c* is to serve the subtransmission network of Figure P7.30 in place of the Δ-Y bank originally specified for that system. Assume that a SLG fault may persist indefinitely to fix the grounding transformer ratings. Available single-phase transformer sizes (kVA) are 50, 75, 100, 167, 250, 333, and 500.

(a) Draw the three sequence networks and show the network interconnections to represent the SLG fault. Show the 15 MVA transformer reactances, the grounding transformer reactances, the Peterson coil, the terminal labeled *n*, and the zero sequence susceptance.

(b) Compute the kVA rating and the high-voltage terminal rating for the grounding transformer bank.

(c) Find the reactance value now required for the Peterson coil and compute the PC rating.

7.33. In Problem 7.32 assume that the low-voltage Δ rating of each grounding transformer is 240 volts. Find the fundamental frequency circulating current in the Δ-connected winding when the system (a) is normal or (b) has a SLG fault.

7.34. Assume that the prefault voltages in the system of problem 7.32 are balanced, positive sequence voltages with $V_{an} = 20.0\underline{/0^\circ}$ kV. Sketch a phasor diagram to show the bus voltages when phase *a* sustains a SLG fault. Show the terminal *n* on your diagram as well as the ground point. Calculate and show on the diagram the phasor voltage V_{nF0}, the voltage across the Peterson coil.

chapter 8

Changes in Symmetry

In previous chapters we introduced the concept of solving an unsymmetrical system by a symmetrical per phase technique known as symmetrical components. We followed this by examining the system components, transmission lines, machines, and transformers to determine the impedance of these components to the flow of sequence currents. We should therefore be capable of representing any power system and analyzing all elementary series and shunt unbalances. We now examine other interesting and challenging problems.

8.1 Creating Symmetry by Labeling

It has been carefully noted that our calculations of symmetrical component currents and voltages are always to be in terms of phase a as the symmetrical phase. Should an unbalance occur which is not symmetrical with respect to phase a, such as a SLG fault on phase b, we have suggested that the phases of the system simply be relabeled. For example, if phase b is the phase of symmetry, let (a, b, c) be labeled (C, A, B), and proceed in the usual manner. In simple cases of one unbalance this is often the best and easiest way to proceed.

In some cases, however, it may be desirable to compute an unbalanced fault where phase a is not the symmetrical phase. There are at least two ways to do this and both will be explored. The first method is due to Atabekov [56] and is based on the construction of a generalized fault diagram. The second method is an analytical one based on the work of Kron [57] and others.

8.2 Generalized Fault Diagrams for Shunt (Transverse) Faults

A systematic way of expressing fault information which is perfectly general insofar as phase symmetry is concerned is through the "generalized fault diagram" of Atabekov [56]. For shunt or transverse faults we begin with the general condition at the fault point shown in Figure 8.1 where, depending upon the values assumed by Z_a, Z_b, Z_c, and Z_g (including 0 and ∞), the fault could be any of the types usually considered.

We write by inspection

$$\begin{bmatrix} V_a \\ V_b \\ V_c \end{bmatrix} = \begin{bmatrix} Z_a + Z_g & Z_g & Z_g \\ Z_g & Z_b + Z_g & Z_g \\ Z_g & Z_g & Z_c + Z_g \end{bmatrix} \begin{bmatrix} I_a \\ I_b \\ I_c \end{bmatrix} \tag{8.1}$$

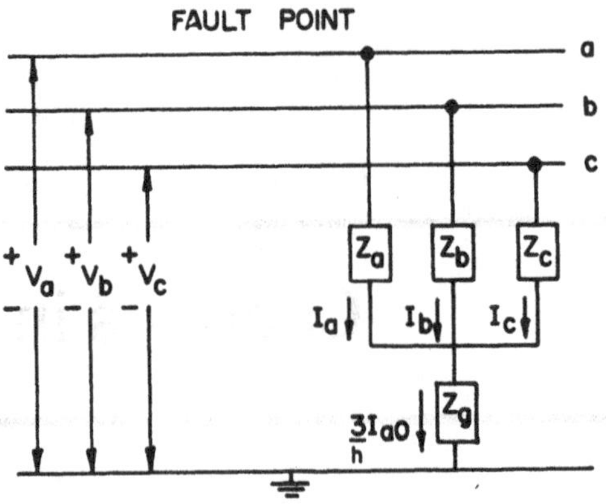

Fig. 8.1. A general shunt (transverse) fault condition at F.

which is easily transformed to the symmetrical component domain by premultiplying both sides of (8.1) by $\mathbf{A}^{-1}$. In matrix notation

$$\mathbf{V}_{abc} = \mathbf{Z}_{abc}\mathbf{I}_{abc} \tag{8.2}$$

Then

$$\begin{aligned}\mathbf{V}_{012} &= \mathbf{A}^{-1}\mathbf{V}_{abc} \\ &= \mathbf{A}^{-1}\mathbf{Z}_{abc}\mathbf{A}\,\mathbf{A}^{-1}\mathbf{I}_{abc} = \mathbf{Z}_{012}\mathbf{I}_{012}\end{aligned} \tag{8.3}$$

Then we may define

$$\mathbf{Z}_{012} = \mathbf{A}^{-1}\mathbf{Z}_{abc}\mathbf{A} \triangleq \begin{bmatrix} Z_{00} & Z_{01} & Z_{02} \\ Z_{10} & Z_{11} & Z_{12} \\ Z_{20} & Z_{21} & Z_{22} \end{bmatrix} \tag{8.4}$$

where we compute the elements of (8.4) as

$$\begin{aligned}
Z_{00} &= (1/3)(Z_a + Z_b + Z_c) + 3Z_g & Z_{10} &= (1/3)(Z_a + \mathrm{a}Z_b + \mathrm{a}^2 Z_c) \\
Z_{01} &= (1/3)(Z_a + \mathrm{a}^2 Z_b + \mathrm{a}Z_c) & Z_{11} &= (1/3)(Z_a + Z_b + Z_c) \\
Z_{02} &= (1/3)(Z_a + \mathrm{a}Z_b + \mathrm{a}^2 Z_c) & Z_{12} &= (1/3)(Z_a + \mathrm{a}^2 Z_b + \mathrm{a}Z_c)
\end{aligned}$$

$$\begin{aligned}
Z_{20} &= (1/3)(Z_a + \mathrm{a}^2 Z_b + \mathrm{a}Z_c) \\
Z_{21} &= (1/3)(Z_a + \mathrm{a}Z_b + \mathrm{a}^2 Z_c) \\
Z_{22} &= (1/3)(Z_a + Z_b + Z_c)
\end{aligned} \tag{8.5}$$

By adding the rows of (8.1), we may write (8.3) in a different form. Adding the three rows directly, we have

$$V_{a0} - 3Z_g I_{a0} = \frac{\mathrm{h}}{3} Z_a I_a + \frac{\mathrm{h}}{3} Z_b I_b + \frac{\mathrm{h}}{3} Z_c I_c \tag{8.6}$$

Adding row 1, a times row 2 and a^2 times row 3 gives

$$V_{a1} = \frac{\mathrm{h}}{3} Z_a I_a + \frac{\mathrm{ha}}{3} Z_b I_b + \frac{\mathrm{ha}^2}{3} Z_c I_c \tag{8.7}$$

Finally, adding row 1, a^2 times row 2 and a times row 3 gives

$$V_{a2} = \frac{h}{3} Z_a I_a + \frac{ha^2}{3} Z_b I_b + \frac{ha}{3} Z_c I_c \tag{8.8}$$

These three equations (8.6)-(8.8) may be used to construct a diagram which completely defines the condition described in Figure 8.1. To do so, however, requires the use of a phase shifting device which may be thought of as a transformer with complex turns ratio of (1, a, or a^2). Such a device rotates both the current *and* voltage by a given phase angle (0° or ± 120°) without changing the magnitude of either quantity. We need not think of this as a physical device but merely as a symbolic phase shifter (we will have no occasion to be concerned with the physical realizability of the network we are constructing). The phase shifter is characterized by the equations

$$V_2 = n\, V_1, \qquad I_2 = n\, I_1 \tag{8.9}$$

where $n = e^{j\theta}$ and where a shift of θ radians is apparent. The circuit is shown in Figure 8.2 and is called the generalized fault diagram for shunt faults. It is easily verified by noting that the points Q, Q_0, Q_1, and Q_2 are all at the same potential (and they could be connected together). The voltages at P_0, P_1, and P_2 are

$$V_{p0} = (1/3)Z_a I_a, \qquad V_{p1} = (1/3)Z_b I_b, \qquad V_{p2} = (1/3)Z_c I_c$$

and these voltages appear across the top, center, and lower transformers respectively, viewed from the right side. Taking the transformer phase shifts into account gives the voltages on the right side of (8.6), (8.7), and (8.8) for the zero, positive, and negative sequence networks respectively.

Usually we are interested in two types of unbalanced faults, a SLG fault and a 2LG fault (where we may consider the LL fault as a special case of the 2LG). Atabekov [56] examines these two cases in detail and develops special, simplified, generalized fault diagrams for these two cases.

SLG fault. For a SLG fault on phase *a* we set the network conditions as

$$Z_a = 0, \qquad Z_b = Z_c = \infty, \qquad Z_g \neq 0 \tag{8.10}$$

and we easily show that since

$$V_a = Z_g I_a = \frac{3}{h} Z_g I_{a0}, \qquad I_b = I_c = 0 \tag{8.11}$$

then

$$I_{a0} = I_{a1} = I_{a2} = \frac{h}{3} I_a$$

$$V_{a0} + V_{a1} + V_{a2} = 3 I_{a0} Z_g \tag{8.12}$$

Through the boundary conditions (8.11) we establish a simpler form of generalized fault diagram for the SLG fault as shown in Figure 8.3. The phase shifts of the transformers (specified as n_0, n_1, and n_2) are given by

$$n_0 = 1, \qquad n_1 = 1, \qquad n_2 = 1 \tag{8.13}$$

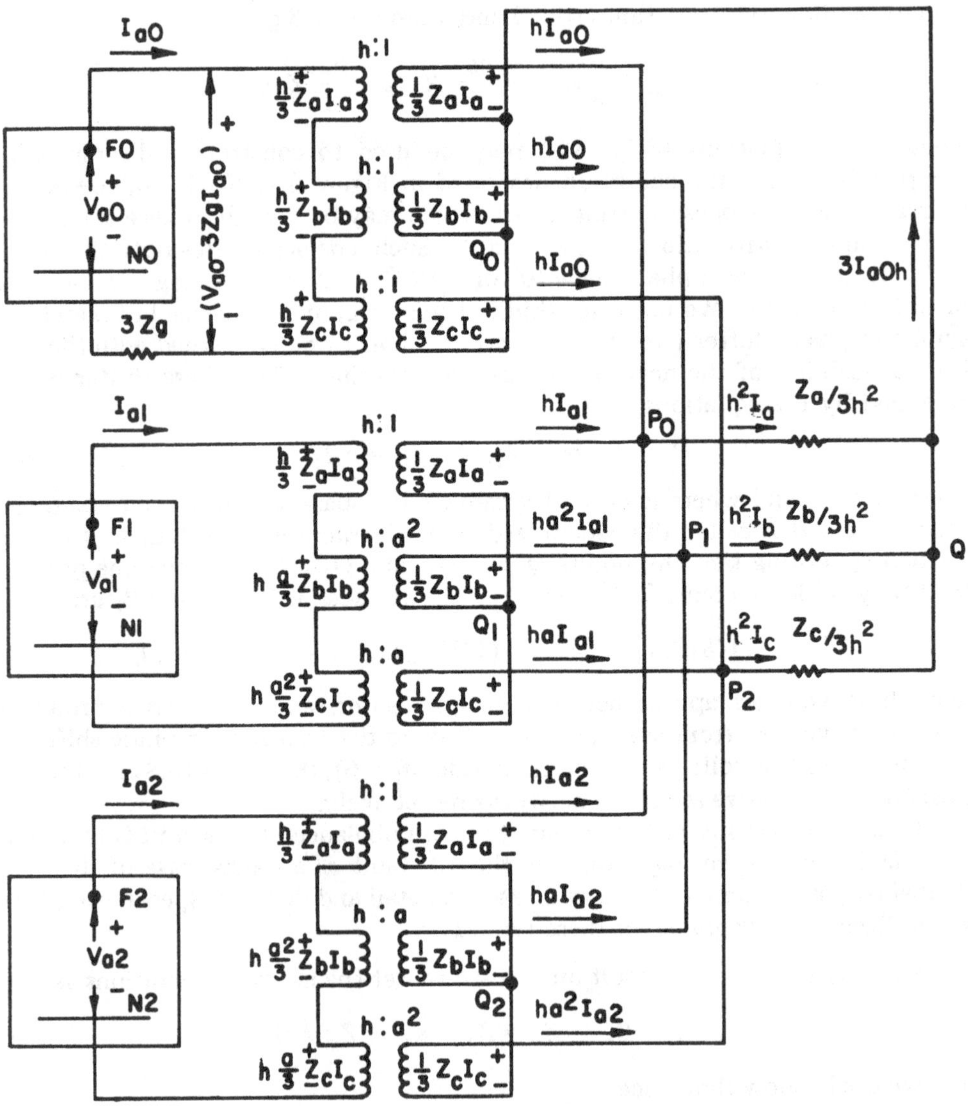

Fig. 8.2. Generalized fault diagram for shunt (transverse) faults.

for a SLG fault on phase a. Phase shifts for faults on phases b or c may be easily verified. Note that the factor h is not required in Figure 8.3 since equation (8.12) is valid for any h, but a factor h/3 is required if phase current is used.

Example 8.1

Verify the phase shifts given in Figure 8.3 for a SLG fault on phase b.

Solution

For a SLG fault on phase b we have the network conditions

$$Z_b = 0, \qquad Z_a = Z_c = \infty, \qquad Z_g \neq 0$$

giving boundary conditions

$$V_b = Z_g I_b, \qquad I_a = I_c = 0$$

Then

$$\frac{1}{h}(I_{a0} + I_{a1} + I_{a2}) = \frac{1}{h}(I_{a0} + aI_{a1} + a^2 I_{a2}) = 0$$

from which we compute $I_{a0} = a^2 I_{a1} = aI_{a2}$. This verifies that if $n_0 = 1$, then for any h

$$n_1 = a^2, \qquad n_2 = a$$

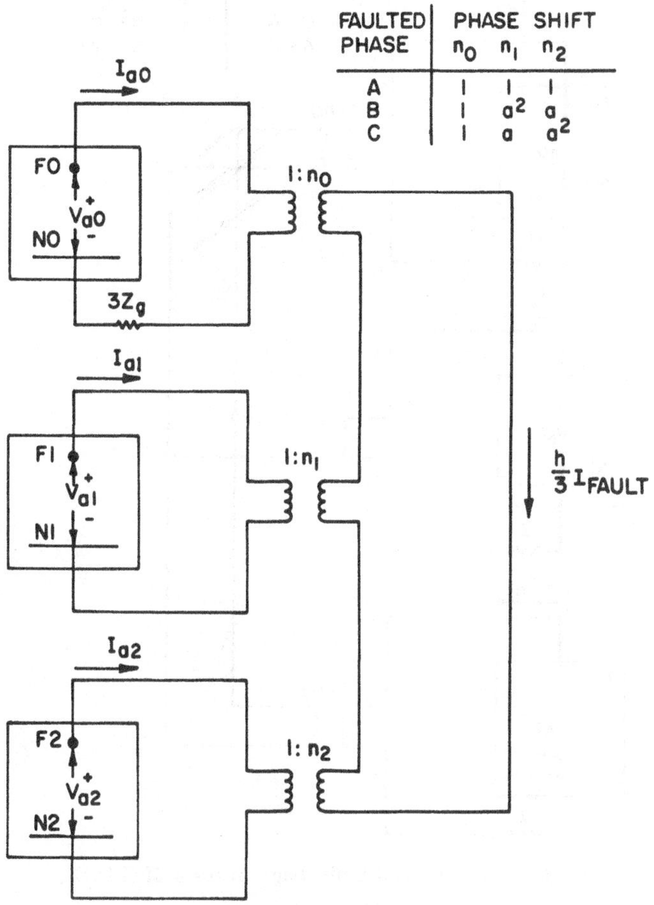

Fig. 8.3. Generalized fault diagram for a SLG fault.

2LG fault. For a 2LG fault on phases b and c we set the network conditions to be

$$Z_b = Z_c = Z, \qquad Z_a = \infty, \qquad Z_g \neq 0 \tag{8.14}$$

The LL fault will be a special case with $Z_g = \infty$ and will give an impedance $2Z$ between lines.

The boundary conditions corresponding to (8.14) are

$$I_a = 0, \qquad V_b = ZI_b + 3Z_g I_{a0}/h, \qquad V_c = ZI_c + 3Z_g I_{a0}/h \tag{8.15}$$

In this case the generalized fault diagram reduces to the configuration shown in

Figure 8.4 where phase shifts are all 1:1. If phases b or c are not the faulted phases, other turns ratios are required as noted in Figure 8.4.

The two diagrams given in Figure 8.3 and 8.4 enable the engineer to solve all common shunt fault configurations for any faulted phase arrangement.

FAULTED PHASE	PHASE SHIFT n_0	n_1	n_2
B–C	1	1	1
C–A	1	a^2	a
A–B	1	a	a^2

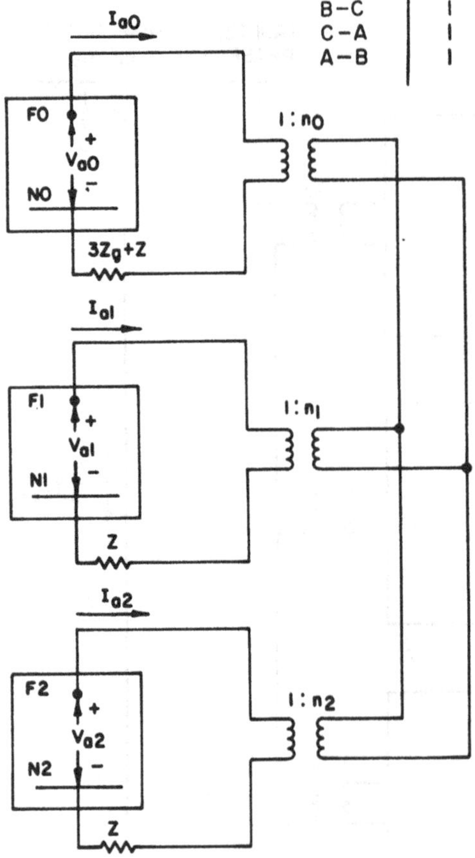

Fig. 8.4. Generalized fault diagram for a 2LG fault.

8.3 Generalized Fault Diagrams for Series (Longitudinal) Faults

We may denote a general series or longitudinal fault by the set of unbalanced series impedances of Figure 8.5, where we conceive of three impedances connecting network points f and m with currents defined as flowing in the f-m direction.

The impedances Z_a, Z_b, and Z_c are free to take on any value including 0 and ∞, thereby representing all common series fault conditions such as open lines. This network condition is described by

$$\mathbf{V}_{f\text{-}abc} - \mathbf{V}_{m\text{-}abc} = \mathbf{V}_{fm\text{-}abc} = \mathbf{Z}_{fm\text{-}abc}\mathbf{I}_{abc} \tag{8.16}$$

which is easily transformed to the symmetrical component equation

$$\mathbf{V}_{fm\text{-}012} = \mathbf{Z}_{fm\text{-}012}\mathbf{I}_{012} \tag{8.17}$$

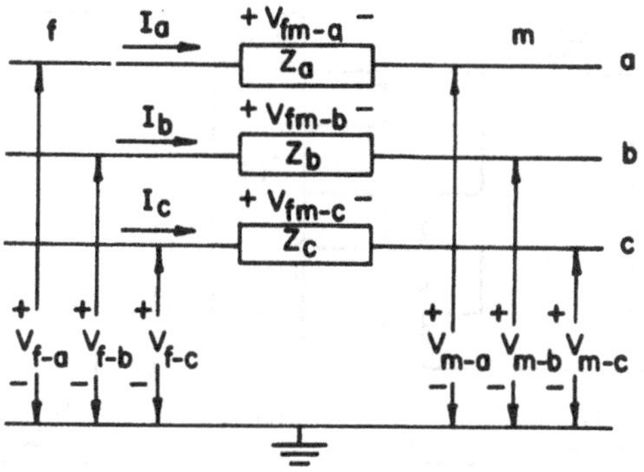

Fig. 8.5. General series (longitudinal) fault condition between f and m.

This was done in section 2.5 for the general case where $\mathbf{Z}_{abc}$ includes mutual impedances. In our case $\mathbf{Z}_{mn\text{-}012}$ is a circulant matrix,

$$\mathbf{Z}_{fm\text{-}012} = \begin{bmatrix} Z_{s0} & Z_{s2} & Z_{s1} \\ Z_{s1} & Z_{s0} & Z_{s2} \\ Z_{s2} & Z_{s1} & Z_{s0} \end{bmatrix}$$

where Z_{s0}, Z_{s1}, and Z_{s2} are given by (2.47). If we write voltage equation (8.17) in terms of Z_a, Z_b, and Z_c, the result is similar in form to (8.6)–(8.8), the only difference being the absence of the Z_g term. Thus these equations may be represented by a generalized fault diagram very similar to Figure 8.2. This diagram is given in Figure 8.6 and may be verified by writing (8.17) in the form

$$\begin{aligned} V_{a0} &= \frac{\mathrm{h}}{3} Z_a I_a + \frac{\mathrm{h}}{3} Z_b I_b + \frac{\mathrm{h}}{3} Z_c I_c \\ V_{a1} &= \frac{\mathrm{h}}{3} Z_a I_a + \frac{\mathrm{ha}}{3} Z_b I_b + \frac{\mathrm{ha}^2}{3} Z_c I_c \\ V_{a2} &= \frac{\mathrm{h}}{3} Z_a I_a + \frac{\mathrm{ha}^2}{3} Z_b I_b + \frac{\mathrm{ha}}{3} Z_c I_c \end{aligned} \tag{8.18}$$

The striking similarity between the generalized fault diagrams of Figures 8.2 and 8.6 is apparent. Note that in each diagram, the symmetrical components of fault current *leave* the networks at F and return at N or M. We have defined the sequence voltages in the two cases as a voltage *drop* through the symmetrical (unfaulted) network from F to N or M. Thus F is labeled positive in both cases.

Two cases of series or longitudinal faults are generally of greatest interest, one line open and two lines open. It is easy to show that these two cases reduce the generalized fault diagram of Figure 8.6 to the diagrams of Figure 8.7 for two lines open and to Figure 8.8 for one line open. These two diagrams should be compared immediately with Figures 8.3 and 8.4 for shunt (transverse) faults. The similarity is obvious. In fact, we can generalize the results for the two most common shunt and series fault connections by only two diagrams, as long as we make a proper interpretation as to the physical meaning of points F and M in each case. The phase shifters required to represent these faults on the various phases are the same in the sense that the phase shift depends only on the choice of *symmetrical phase*. By this we mean the phase that is symmetrical with respect

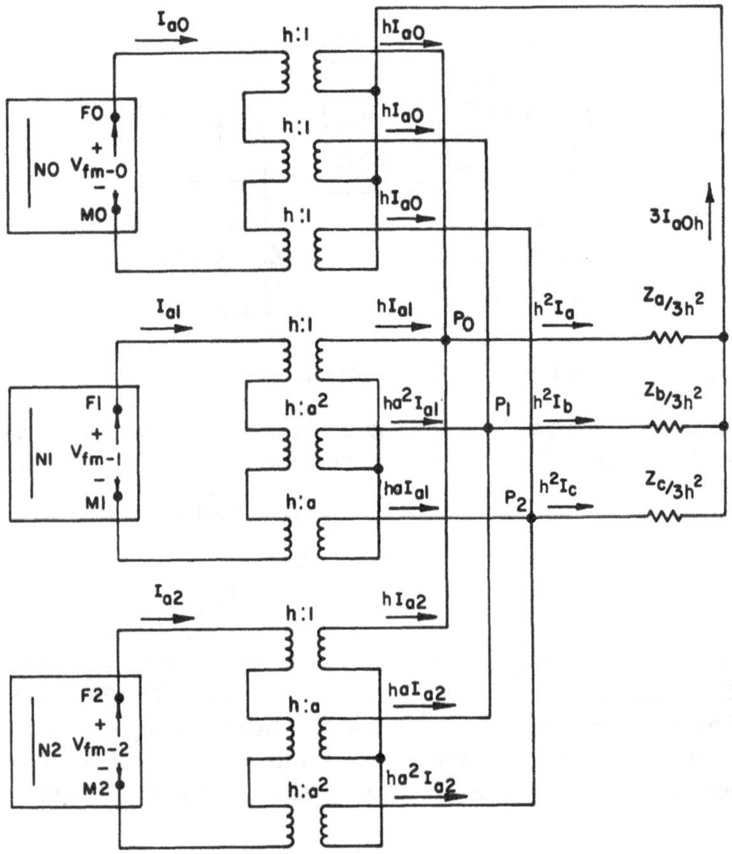

Fig. 8.6. Generalized fault diagram for series (longitudinal) faults.

to the other two, as shown in Table 8.1. Thus if only one phase is involved in the fault, that phase is symmetrical. If two phases are faulted, the third (unfaulted) phase would be considered symmetrical.

In summary, then, two sequence network connections can be used to represent all common faults. The *series* connections of Figure 8.3 and 8.7 represent the SLG fault and two lines open respectively. The *parallel* connections of Figure 8.4 and 8.8 represent the 2LG fault and one line open respectively. These ideas are summarized in Figure 8.9, which may be used in the computation of all common faults irrespective of symmetry. Note that in using Figure 8.9 in computations involving longitudinal faults, Z_g is not involved and should be set to zero in the diagrams. The impedance Z is used both for series line impedance and for the impedance from line to Z_g in the case of the 2LG fault.

8.4 Computation of Fault Currents and Voltages

It is useful to establish a definite procedure for the computation of fault currents for series or shunt faults which are determined by series or parallel connections of the sequence networks. (Note the dual use of the word "series." The meaning is usually clear from the context of use, however.) For both the series and parallel network connections of Figure 8.9 we will view the system as that of three one-port networks, one active and two passive. Then taking the

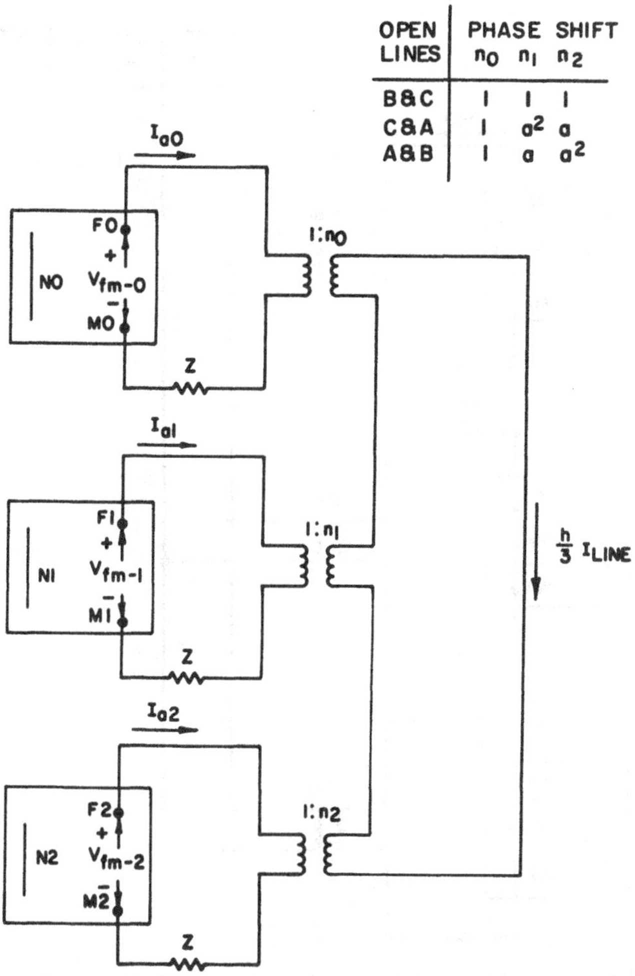

Fig. 8.7. Generalized fault diagram for two lines open and an impedance Z in the sound line.

external transformations and terminal connections into account, we may derive equations for the fault currents or voltages.

Referring again to Figure 8.9, note that a consistent identity has been established for all port voltages and currents external to the transformers. We arbitrarily define voltages V_0, V_1, and V_2 and prescribe that currents I_0, I_1, and I_2 agree in direction with I_{a0}, I_{a1}, and I_{a2} and enter the + terminal. We also identify voltage V between terminals ϕ-ϕ and specify current I to enter the $+V$ node.

Since the transformers terminating the networks are considered ideal phase shifters, we write

$$n_i = I_i/I_{ai} = V_i/V_{Ki}, \qquad i = 0, 1, 2 \tag{8.19}$$

This defines a new voltage V_{Ki} for each sequence network, which is a voltage drop in the direction of I_{ai} and includes the drop across the fault impedance and across the sequence network. The values of n_i are always 1, a, or a^2 as specified in Figure 8.9 and depend only on the fault symmetry. We easily compute V_{Ki} by

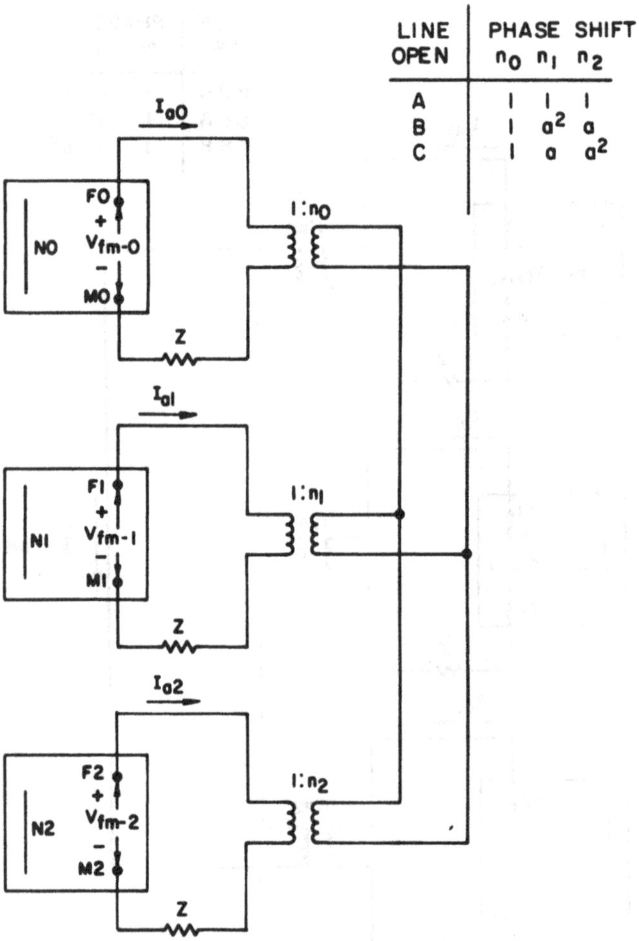

Fig. 8.8. Generalized fault diagram for one line open and an impedance Z in the two sound lines.

inspection to be

$$\begin{bmatrix} V_{K0} \\ V_{K1} \\ V_{K2} \end{bmatrix} = \begin{bmatrix} (3Z_g + Z) & 0 & 0 \\ 0 & Z & 0 \\ 0 & 0 & Z \end{bmatrix} \begin{bmatrix} I_{a0} \\ I_{a1} \\ I_{a2} \end{bmatrix} - \begin{bmatrix} V_{a0} \\ V_{a1} \\ V_{a2} \end{bmatrix} \quad (8.20)$$

The voltage vector $[V_{a0} \ V_{a1} \ V_{a2}]^t$ depends on the kind of fault, i.e., whether a shunt or series unbalance. In every case, however, the zero and negative sequence networks consist of passive impedances Z_0 and Z_2, and the positive sequence network consists of an EMF E in series with an impedance Z_1. Thus it is

Table 8.1. Definition of the Symmetrical Phase

		Phase Shift		
Fault Location	*Symmetrical Phase*	n_0	n_1	n_2
a or b-c	a	1	1	1
b or c-a	b	1	a^2	a
c or a-b	c	1	a	a^2

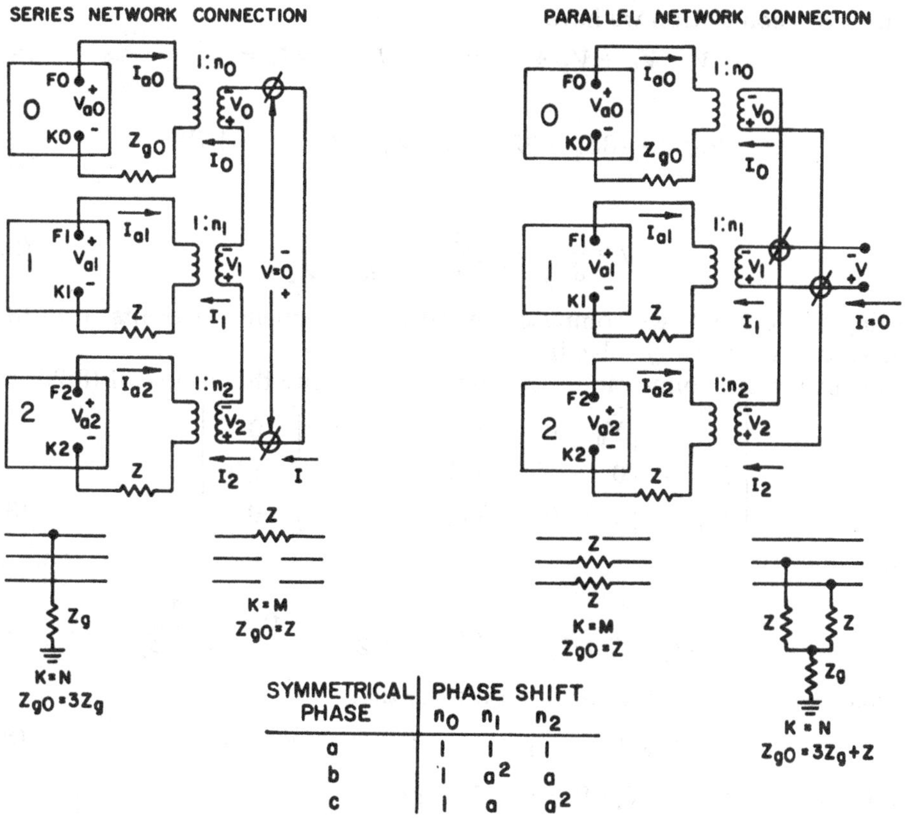

SYMMETRICAL PHASE	PHASE SHIFT n_0	n_1	n_2
a	1	1	1
b	1	a^2	a
c	1	a	a^2

Fig. 8.9. Summary of network connections for common fault conditions.

always possible to write

$$\begin{bmatrix} V_{a0} \\ V_{a1} \\ V_{a2} \end{bmatrix} = \begin{bmatrix} 0 \\ hE \\ 0 \end{bmatrix} - \begin{bmatrix} Z_0 & 0 & 0 \\ 0 & Z_1 & 0 \\ 0 & 0 & Z_2 \end{bmatrix} \begin{bmatrix} I_{a0} \\ I_{a1} \\ I_{a2} \end{bmatrix} \tag{8.21}$$

and for any fault we define E to be the value of V_{a1} when $I_{a1} = 0$. Obviously, the values of Z_0, Z_1, and Z_2 depend on the fault type and location but are always the driving point impedance between terminals Fi and Ki for $i = 0, 1, 2$.

Incorporating (8.21) into (8.20), we write

$$\begin{bmatrix} V_{K0} \\ V_{K1} \\ V_{K2} \end{bmatrix} = \begin{bmatrix} (3Z_g + Z + Z_0) & 0 & 0 \\ 0 & (Z + Z_1) & 0 \\ 0 & 0 & (Z + Z_2) \end{bmatrix} \begin{bmatrix} I_{a0} \\ I_{a_1} \\ I_{a2} \end{bmatrix} - \begin{bmatrix} 0 \\ hE \\ 0 \end{bmatrix} \tag{8.22}$$

Premultiplying (8.22) by a diagonal matrix, diag (n_0, n_1, n_2), and using (8.19), we have

$$\begin{bmatrix} V_0 \\ V_1 \\ V_2 \end{bmatrix} = \begin{bmatrix} n_0 V_{K0} \\ n_1 V_{K1} \\ n_2 V_{K2} \end{bmatrix} = \begin{bmatrix} (3Z_g + Z + Z_0) & 0 & 0 \\ 0 & (Z + Z_1) & 0 \\ 0 & 0 & (Z + Z_2) \end{bmatrix} \begin{bmatrix} I_0 \\ I_1 \\ I_2 \end{bmatrix} - \begin{bmatrix} 0 \\ n_1 hE \\ 0 \end{bmatrix} \tag{8.23}$$

For the series connection we have

$$V = V_0 + V_1 + V_2 = 0, \qquad I = I_0 = I_1 = I_2 \tag{8.24}$$

or

$$V = 0 = (3Z_g + 3Z + Z_0 + Z_1 + Z_2)I - n_1 \mathrm{h} E$$

and

$$I = \frac{n_1 \mathrm{h} E}{3Z_g + 3Z + Z_0 + Z_1 + Z_2} \tag{8.25}$$

Knowing this current, we compute the individual sequence currents from (8.19) and may then completely solve the networks.

For the parallel connection it is convenient to write the inverse of (8.23) or

$$\begin{bmatrix} I_0 \\ I_1 \\ I_2 \end{bmatrix} = \begin{bmatrix} \hat{Y}_0 & 0 & 0 \\ 0 & \hat{Y}_1 & 0 \\ 0 & 0 & \hat{Y}_2 \end{bmatrix} \begin{bmatrix} V_0 \\ V_1 \\ V_2 \end{bmatrix} + \begin{bmatrix} 0 \\ \hat{Y}_1 n_1 \mathrm{h} E \\ 0 \end{bmatrix} \tag{8.26}$$

where

$$\hat{Y}_0 = \frac{1}{3Z_g + Z + Z_0}, \qquad \hat{Y}_1 = \frac{1}{Z + Z_1}, \qquad \hat{Y}_2 = \frac{1}{Z + Z_2} \tag{8.27}$$

For this connection we note from Figure 8.9 that

$$I = I_0 + I_1 + I_2 = 0, \qquad V = V_0 = V_1 = V_2 \tag{8.28}$$

Then $I = 0 = (\hat{Y}_0 + \hat{Y}_1 + \hat{Y}_2)V + \hat{Y}_1 n_1 \mathrm{h} E$ and

$$V = \frac{-\hat{Y}_1 n_1 \mathrm{h} E}{\hat{Y}_0 + \hat{Y}_1 + \hat{Y}_2} \tag{8.29}$$

Knowing this voltage, we may compute the individual sequence voltages from (8.19) and completely solve the networks.

The methods described above, culminating in the diagrams of Figure 8.9, are extremely useful in computing common faults with any possible symmetry. We now explore an entirely different method based almost exclusively on matrix algebra. This method is due originally to Kron [57] who used tensor analysis to solve a wide variety of network problems including those where the circuits are in motion, such as machines. He treated stationary networks as a special case in his general scheme of analysis but still used tensor notation. It is our view that the methods of tensor analysis, although entirely adequate for network problems, are not required and the more familiar matrix notation will be of greater value to engineers. Several authors have adapted portions of Kron's work to network problems using matrix algebra. One such work is LeCorbeiller [58], which is recommended reading. Our approach follows more closely that of Lewis and Pryce [13], which is more directly aimed at our problem. Other works from this area which are worth investigating include Hancock [59] and Stigant [60].

8.5 A Fundamental Result: The Invariance of Power

Let $\mathbf{x}_1$ be an n vector and let $\mathbf{A}$ and $\mathbf{B}$ be $n \times n$ matrices, where these quantities are related by the homogeneous equation

$$(\mathbf{A} - \mathbf{B})\,\mathbf{x}_1 = \mathbf{0} \tag{8.30}$$

But from matrix theory we learn that the system (8.30) has a nontrivial solution if and only if the coefficient matrix $\mathbf{A} - \mathbf{B}$ is singular, i.e.,

$$\det(\mathbf{A} - \mathbf{B}) = 0 \tag{8.31}$$

Note that (8.30) does *not* imply that $\mathbf{A} = \mathbf{B}$.

Suppose that we now find additional $\mathbf{x}$ vectors $\mathbf{x}_2, \mathbf{x}_3, \ldots, \mathbf{x}_n$ such that the n $\mathbf{x}$ vectors are linearly independent (in which case we say that they "span" an n-dimensional vector space) so that we have the set of equations

$$\begin{aligned} (\mathbf{A} - \mathbf{B})\,\mathbf{x}_1 &= 0 \\ (\mathbf{A} - \mathbf{B})\,\mathbf{x}_2 &= 0 \\ \cdots \quad & \quad \cdots \\ (\mathbf{A} - \mathbf{B})\,\mathbf{x}_n &= 0 \end{aligned} \tag{8.32}$$

We may write (8.32) in matrix form by defining the matrix $\mathbf{X}$ where

$$\mathbf{X} = [\mathbf{x}_1\, \mathbf{x}_2 \ldots \mathbf{x}_n] \tag{8.33}$$

so that (8.32) becomes

$$(\mathbf{A} - \mathbf{B})\,\mathbf{X} = \mathbf{0} \tag{8.34}$$

Since $\mathbf{X}$ is composed of linearly independent columns, $\det \mathbf{X} \neq 0$ and $\mathbf{X}^{-1}$ exists. If we postmultiply (8.34) by $\mathbf{X}^{-1}$, we have $\mathbf{A} - \mathbf{B} = \mathbf{0}$ or

$$\mathbf{A} = \mathbf{B} \tag{8.35}$$

Note that the foregoing argument is not changed if $\mathbf{A}$ and $\mathbf{B}$ are row vectors of n elements.

Suppose now that a network is under consideration for which we have a voltage equation (in matrix form)

$$\mathbf{V} = \mathbf{Z}\,\mathbf{I} \tag{8.36}$$

but for some reason we require that the currents be transformed to a new set of currents $\mathbf{I}'$ which are linearly related to $\mathbf{I}$ by the transformation

$$\mathbf{I} = \mathbf{K}\,\mathbf{I}' \tag{8.37}$$

We now seek a voltage transformation such that we may write

$$\mathbf{V}' = \mathbf{Z}'\mathbf{I}' \tag{8.38}$$

where this transformation is subject to the restriction that the *power* is *invariant*, i.e.,

$$P + jQ = \mathbf{V}^t\mathbf{I}^* = \mathbf{V}'^t\mathbf{I}'^* \tag{8.39}$$

This restriction will assure us that the law of conservation of energy is preserved.

Substituting (8.37) into (8.39), we have $\mathbf{V}^t\mathbf{K}^*\mathbf{I}'^* = \mathbf{V}'^t\mathbf{I}'^*$ or

$$(\mathbf{V}^t\mathbf{K}^* - \mathbf{V}'^t)\,\mathbf{I}'^* = 0 \tag{8.40}$$

and since this is valid for all current vectors $\mathbf{I}'$, we conclude from (8.35) that

$$\mathbf{V}^t\mathbf{K}^* - \mathbf{V}'^t = 0 \tag{8.41}$$

Transposing and rearranging, we have

$$\mathbf{V}' = \mathbf{K}^{*t}\mathbf{V} \tag{8.42}$$

as the voltage constraint which must accompany the current transformation (8.37) if power is to be invariant.

Example 8.2

When the phase currents are transformed into symmetrical component currents according to the equation $\mathbf{I}_{abc} = \mathbf{A}\,\mathbf{I}_{012}$ determine the corresponding voltage transformation if power is to be invariant.

Solution

From (8.37) we write $\mathbf{I} = \mathbf{K}\,\mathbf{I}'$ where

$$\mathbf{I} = \mathbf{I}_{abc}, \qquad \mathbf{K} = \mathbf{A}, \qquad \mathbf{I}' = \mathbf{I}_{012}$$

Then from (8.42) $\mathbf{V}_{012} = \mathbf{K}^{*t}\mathbf{V}_{abc} = \mathbf{A}^{*t}\mathbf{V}_{abc}$.

Since $\mathbf{A}$ is symmetric, $\mathbf{A} = \mathbf{A}^t$ and $\mathbf{V}_{012} = \mathbf{A}^*\mathbf{V}_{abc}$. Since

$$\mathbf{A} = \frac{1}{h}\begin{bmatrix} 1 & 1 & 1 \\ 1 & a^2 & a \\ 1 & a & a^2 \end{bmatrix}$$

$$\mathbf{A}^* = \frac{1}{h}\begin{bmatrix} 1 & 1 & 1 \\ 1 & a & a^2 \\ 1 & a^2 & a \end{bmatrix} = \frac{3}{h^2}\,\mathbf{A}^{-1}$$

Thus, we have $\mathbf{V}_{012} = (3/h^2)\,\mathbf{A}^{-1}\,\mathbf{V}_{abc}$ which agrees with the conclusions of Chapter 2 derived by a different method.

8.6 Constraint Matrix K

It is appropriate to recognize that the operation implied by equation $\mathbf{I} = \mathbf{K}\,\mathbf{I}'$ is not a new concept to the student of circuit theory. This is just a way of rearranging the available information into a different and perhaps more useful form. For example, if $\mathbf{I}$ represents the branch currents in a network, then $\mathbf{I}'$ might represent the loop or mesh currents. In such a case $\mathbf{K}$ would be rectangular, and the transformation (8.37) is not unique since $\mathbf{K}$ is singular. This simply means that the meshes are not always defined in the same way.

The matrix $\mathbf{K}$ which expresses the relationship between the old and new current vector is called the *constraint matrix* or the *connection matrix*. It is also referred to as Kron's transformation matrix. It determines exactly how $\mathbf{I}$ and $\mathbf{I}'$ are related and is usually composed only of the *real* elements +1, −1, and 0. (How would this simplify (8.42)?) In the literature one often finds the Kron transformation designated by the letter C. We prefer $\mathbf{K}$ here to avoid confusion with the C used for the capacitance matrix of Chapter 5.

If (8.37) is written in elementary form, we have in the case where I represents branch currents and I' the mesh currents,

$$I_p = \sum_{q=1}^{M} K_{pq} I'_q \qquad (p = 1, 2, \ldots, B) \tag{8.43}$$

where

K_{pq} = +1 if branch p is contained in mesh q and has the same sense
= -1 if branch p is contained in mesh q and has the opposite sense
= 0 if branch p is not contained in mesh q
B = the total number of elements in **I**, i.e., the number of branches
M = the total number of elements in **I**′, i.e., the number of meshes

If there are N nodes, we know that [61]

$$M = B - N + 1 \tag{8.44}$$

so that $M < B$, sometimes by quite a margin, which means that **K** is rectangular of dimension $B \times M$.

The transformation may also be used to change from one set of mesh currents to another since mesh currents are arbitrarily defined. In this case **K** is nonsingular and the transformation is unique.

To gain experience with the constraint matrix we solve a simple example.

Example 8.3

Given a network with $N = 4$ nodes, $B = 6$ branches, and connected as shown in Figure 8.10, find the constraint matrix **K** and determine the voltage relation which guarantees the invariance of power. Note the sense markings on all branches.

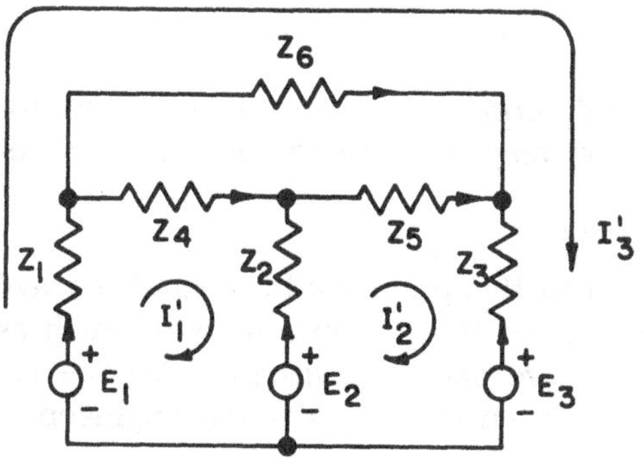

Fig. 8.10. Circuit for Example 8.3.

Solution

By inspection of Figure 8.10 we write **I** = **K I**′

$$\begin{bmatrix} I_1 \\ I_2 \\ I_3 \\ I_4 \\ I_5 \\ I_6 \end{bmatrix} = \begin{matrix} & 1' & 2' & 3' \\ 1 & 1 & 0 & 1 \\ 2 & -1 & 1 & 0 \\ 3 & 0 & -1 & -1 \\ 4 & 1 & 0 & 0 \\ 5 & 0 & 1 & 0 \\ 6 & 0 & 0 & 1 \end{matrix} \begin{bmatrix} I'_1 \\ I'_2 \\ I'_3 \end{bmatrix}$$

Then the voltage equation is

$$V' = K^{*t}V = \begin{matrix} & 1 & 2 & 3 & 4 & 5 & 6 \\ 1' & 1 & -1 & 0 & 1 & 0 & 0 \\ 2' & 0 & 1 & -1 & 0 & 1 & 0 \\ 3' & 1 & 0 & -1 & 0 & 0 & 1 \end{matrix} \; V$$

If we can say that the original currents I and voltages V in a network are related by an impedance matrix Z, we write $V = Z\,I$, and for the transformed case we have $V' = Z'I'$. Since $V' = K^{*t}V$, we have $V' = K^{*t}Z\,I$ and substituting $I = K\,I'$,

$$V' = K^{*t}Z\,K\,I' \tag{8.45}$$

Then, apparently,

$$Z' = K^{*t}Z\,K \tag{8.46}$$

expresses the relationship between Z and Z′. Synge [61] proves that Z′ is nonsingular by considering the network to be dissipative. If $(Z')^{-1}$ exists, (and it always will) we may write $I' = (Z')^{-1}V'$ which we premultiply by K to obtain

$$I = K\,(K^{*t}Z\,K)^{-1}\,V'$$

Finally, we eliminate V′ to compute I as

$$I = K\,(K^{*t}Z\,K)^{-1}\,K^{*t}V \tag{8.47}$$

In many cases K is simply constructed, and the computation of (8.47) is straightforward. Using (8.47), we may compute the branch currents directly.

8.7 Kron's Primitive Network

In our discussion up to this point we have carefully avoided defining exactly what we mean by the vectors V and I in a network such as that of Figure 8.10. The quantities V′ and I′ are easier to visualize for the engineer accustomed to solving networks by mesh currents, and most engineers could write Z′ by inspection.

The real problem lies in identifying what we mean by the equation $V = Z\,I$ in a physical network. Kron visualized a collection of individual branches for which he could write branch voltage equations. The collection of these equations he called the primitive equations and the disconnected "network" he termed the "primitive network" shown in Figure 8.11. The primitive equations are easily written as

$$V = Z\,I - E \tag{8.48}$$

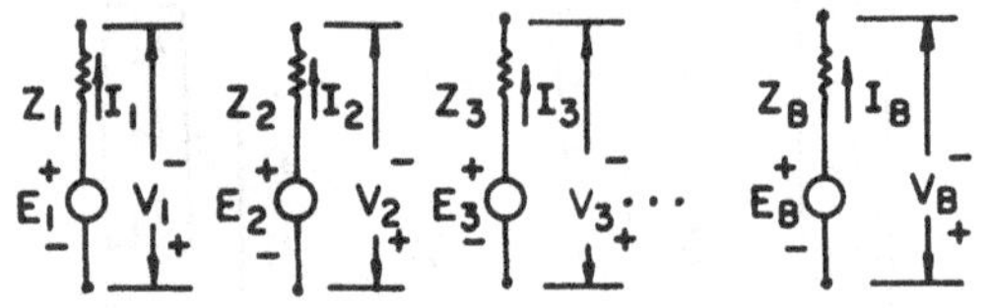

Fig. 8.11. Kron's primitive network for a *B*-branched circuit.

Since the matrix K specifies the assembly of the primitive branches into a network, it is apparent from Kirchhoff's voltage law that $\mathbf{K}^{*t}\mathbf{V} = \mathbf{0}$, which simply says that the total voltage around any mesh specified by K is zero. Then (8.48) may be written as if the primitive network were shorted, i.e.,

$$\mathbf{E} = \mathbf{Z}\,\mathbf{I} \tag{8.49}$$

with I the shorted current. Since the branch impedances here are considered to be independent, Z is diagonal, i.e.,

$$\mathbf{Z} = \begin{array}{c} \\ 1 \\ 2 \\ \\ B \end{array}\!\!\begin{array}{c} \begin{array}{cccc} 1 & 2 & & B \end{array} \\ \begin{bmatrix} Z_1 & 0 & \dots & 0 \\ 0 & Z_2 & \dots & 0 \\ \dots & \dots & \dots & \dots \\ 0 & 0 & \dots & Z_B \end{bmatrix} \end{array}$$

Furthermore, E and I are B-dimensioned vectors of branch EMF's and branch currents, with all branch currents defined as leaving the positive terminal of each voltage source. In the remainder of this chapter we use E for the EMF vector.

This is a remarkable way to solve a network. The equation (8.49) which relates branch EMF's to branch currents in the individual shorted branches does not describe any network at all! That is to say, the primitive network equation gives absolutely no information as to how the individual branches are interconnected to form a useful network. This "connection" information is conveyed exclusively through the constraint matrix K. An example will illustrate the method.

Example 8.4

The circuit in Figure 8.12 has three branches and two nodes and is labeled as specified in Figure 8.11. Find the K matrix which will specify the connection indicated and transform branch currents to mesh currents.

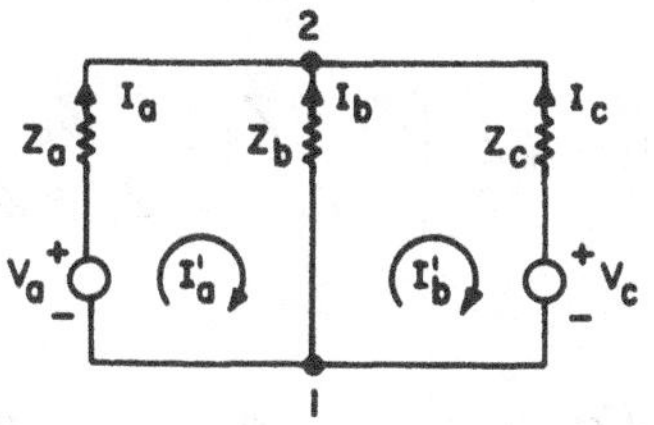

Fig. 8.12. Network with three branches and two nodes.

Solution

We write the primitive network equation from (8.48) as $\mathbf{E} = \mathbf{Z}\,\mathbf{I}$ where

$$\mathbf{E} = \begin{bmatrix} V_a \\ 0 \\ V_c \end{bmatrix}, \quad \mathbf{I} = \begin{bmatrix} I_a \\ I_b \\ I_c \end{bmatrix}, \quad \mathbf{Z} = \begin{bmatrix} Z_a & 0 & 0 \\ 0 & Z_b & 0 \\ 0 & 0 & Z_c \end{bmatrix}$$

From (8.37) $\mathbf{I} = \mathbf{K}\mathbf{I}'$ where, by inspection

$$\begin{bmatrix} I_a \\ I_b \\ I_c \end{bmatrix} = \begin{array}{c} \\ a \\ b \\ c \end{array} \overset{\begin{array}{cc} a' & b' \end{array}}{\begin{bmatrix} 1 & 0 \\ -1 & 1 \\ 0 & -1 \end{bmatrix}} \begin{bmatrix} I'_a \\ I'_b \end{bmatrix}$$

or

$$\mathbf{K} = \begin{bmatrix} 1 & 0 \\ -1 & 1 \\ 0 & -1 \end{bmatrix}$$

Then from (8.46) $\mathbf{Z}' = \mathbf{K}^{*t}\mathbf{Z}\mathbf{K}$ or

$$\mathbf{Z}' = \begin{bmatrix} 1 & -1 & 0 \\ 0 & 1 & -1 \end{bmatrix} \begin{bmatrix} Z_a & 0 & 0 \\ 0 & Z_b & 0 \\ 0 & 0 & Z_c \end{bmatrix} \begin{bmatrix} 1 & 0 \\ -1 & 1 \\ 0 & -1 \end{bmatrix} = \begin{bmatrix} Z_a + Z_b & -Z_b \\ -Z_b & Z_b + Z_c \end{bmatrix}$$

We also write $\mathbf{V}' = \mathbf{Z}'\mathbf{I}'$ or

$$\begin{bmatrix} V'_a \\ V'_b \end{bmatrix} = \begin{bmatrix} Z_a + Z_b & -Z_b \\ -Z_b & Z_b + Z_c \end{bmatrix} \begin{bmatrix} I'_a \\ I'_b \end{bmatrix}$$

which the engineer familiar with writing mesh current equations will verify by inspection of Figure 8.12 to be correct. We also note that from (8.44)

$$M = B - N + 1 = 3 - 2 + 1 = 2 \text{ meshes}$$

which agrees with our solution.

We extend this example by one additional step to compute the branch currents. From (8.47) we compute

$$\mathbf{I} = \mathbf{K}(\mathbf{Z}')^{-1}\mathbf{K}^{*t}\mathbf{E}$$

We define $\Delta = (Z_a + Z_b)(Z_b + Z_c) - Z_b^2 = \det \mathbf{Z}'$ to write

$$(\mathbf{Z}')^{-1} = \frac{1}{\Delta} \begin{bmatrix} Z_b + Z_c & Z_b \\ Z_b & Z_a + Z_b \end{bmatrix}$$

and

$$\begin{bmatrix} I_a \\ I_b \\ I_c \end{bmatrix} = \frac{1}{\Delta} \begin{bmatrix} Z_b + Z_c & -Z_c & -Z_b \\ -Z_c & Z_a + Z_c & -Z_a \\ -Z_b & -Z_a & Z_a + Z_b \end{bmatrix} \begin{bmatrix} V_a \\ 0 \\ V_c \end{bmatrix}$$

Finally, we note that we can verify by inspection that $\mathbf{V}' = \mathbf{K}^{*t}\mathbf{E} = \mathbf{K}^t\mathbf{E}$ or

$$\begin{bmatrix} V'_a \\ V'_b \end{bmatrix} = \begin{bmatrix} 1 & -1 & 0 \\ 0 & 1 & -1 \end{bmatrix} \begin{bmatrix} V_a \\ 0 \\ V_c \end{bmatrix} = \begin{bmatrix} V_a \\ -V_c \end{bmatrix}$$

8.8 Other Useful Transformations

In section 8.7, through the mechanism of the primitive network, we see the way in which branch currents of the primitive network may be transformed into mesh currents of the connected network. This transformation is accomplished through the constraint matrix **K**. We now investigate other applications of the idea of a transformation of voltage and current under the constraint of power invariance.

The first transformation we consider is one which takes one set of mesh currents into a new set of mesh currents. This is possible since mesh currents are arbitrarily defined and the transformation technique permits us to discard one set of network equations in favor of another, presumedly more useful, set of equations. We will illustrate by example.

Example 8.5

The network shown in Figure 8.13 is correctly described by the mesh current equations

$$\begin{bmatrix} V_a \\ V_b \\ V_c \end{bmatrix} = \begin{bmatrix} -V_{s1} \\ V_{s1} \\ -V_{s2} \end{bmatrix} = \begin{bmatrix} (Z_1+Z_2+Z_5) & -Z_5 & -Z_2 \\ -Z_5 & (Z_4+Z_5+Z_6) & -Z_6 \\ -Z_2 & -Z_6 & (Z_2+Z_3+Z_6) \end{bmatrix} \begin{bmatrix} I_a \\ I_b \\ I_c \end{bmatrix}$$

in terms of I_a, I_b, and I_c. Change the current variables to the new mesh currents I_1, I_2, and I_3 and find the new mesh equations.

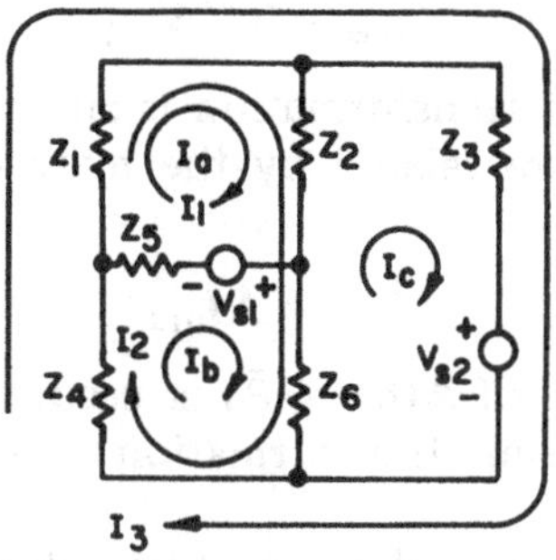

Fig. 8.13. Network for Example 8.5.

Solution

By inspection of Figure 8.13 find the relationship between the old currents I (in this case I_a, I_b, and I_c) and the new currents I' which are I_1, I_2, and I_3. By selecting an appropriate tree, we relate the mesh currents by the equations

$$I_a = I_1 + I_2 + I_3, \qquad I_b = I_2 + I_3, \qquad I_c = I_3$$

or, in matrix form $\mathbf{I} = \mathbf{K}\mathbf{I}'$ or

$$\begin{bmatrix} I_a \\ I_b \\ I_c \end{bmatrix} = \begin{bmatrix} 1 & 1 & 1 \\ 0 & 1 & 1 \\ 0 & 0 & 1 \end{bmatrix} \begin{bmatrix} I_1 \\ I_2 \\ I_3 \end{bmatrix}$$

The new voltage vector is computed by

$$\mathbf{V}' = \mathbf{K}^{*t}\mathbf{E} = \begin{bmatrix} 1 & 0 & 0 \\ 1 & 1 & 0 \\ 1 & 1 & 1 \end{bmatrix} \begin{bmatrix} -V_{s1} \\ V_{s1} \\ -V_{s2} \end{bmatrix} = \begin{bmatrix} -V_{s1} \\ 0 \\ -V_{s2} \end{bmatrix}$$

which is easily verified by inspection.

The new impedance matrix is

$$\mathbf{Z}' = \mathbf{K}^{*t}\mathbf{Z}\mathbf{K} = \begin{bmatrix} (Z_1 + Z_2 + Z_5) & (Z_1 + Z_2) & Z_1 \\ (Z_1 + Z_2) & (Z_1 + Z_2 + Z_4 + Z_6) & (Z_1 + Z_4) \\ Z_1 & (Z_1 + Z_4) & (Z_1 + Z_3 + Z_4) \end{bmatrix}$$

which is also verified by inspection. Finally, we write $\mathbf{V}' = \mathbf{Z}'\mathbf{I}'$ and the transformation is complete. Note that in this case $\mathbf{K}$ is nonsingular and the transformation is unique. We could therefore reverse the process, transforming $\mathbf{V}'$, $\mathbf{I}'$, and $\mathbf{Z}'$ back to $\mathbf{E}$, $\mathbf{I}$, and $\mathbf{Z}$ through the matrix $\mathbf{K}^{-1}$.

Observe that just as in the transformation from branch currents to mesh currents, the process described above is based upon the idea

$$\mathbf{I}_{\text{old}} = \mathbf{K}\mathbf{I}_{\text{new}} \tag{8.50}$$

or $\mathbf{I} = \mathbf{K}\mathbf{I}'$. Furthermore, if the power is to be invariant, the old and new voltages are related by

$$\mathbf{V}_{\text{new}} = \mathbf{K}^{*t}\mathbf{V}_{\text{old}} \tag{8.51}$$

The symmetrical component transformation is an arbitrary transformation where the new and old currents are related by the matrix $\mathbf{K} = \mathbf{A}$ or from (8.50) with $\mathbf{I}_{\text{old}} = \mathbf{I}_{abc}$ and $\mathbf{I}_{\text{new}} = \mathbf{I}_{012}$

$$\mathbf{I}_{abc} = \mathbf{A}\mathbf{I}_{012} \tag{8.52}$$

If the power invariant form of $\mathbf{A}(h = \sqrt{3})$ is used, the voltage transformation is given by (8.51). The Fortescue transformation ($h = 1$) requires special treatment as noted in Example 8.2.

Other transformations can be devised which relate currents of one network to those of a second network, thus specifying mathematically the connection or constraint $\mathbf{K}$ between the networks. An example will illustrate this procedure.

Example 8.6

Consider the two networks of Figure 8.14 where the network mesh equations are given by

$$\begin{bmatrix} V_1 \\ V_2 \\ V_3 \end{bmatrix} = \begin{bmatrix} Z_1 & 0 & 0 \\ 0 & Z_2 & 0 \\ 0 & 0 & Z_3 \end{bmatrix} \begin{bmatrix} I_1 \\ I_2 \\ I_3 \end{bmatrix}$$

and $V_n = Z_n I_n$. The two networks are to be combined with the constraints

$$I_1 = I_1', \qquad I_2 = I_2'$$
$$I_3 = I_3', \qquad I_n = I_1' + I_2' + I_3'$$

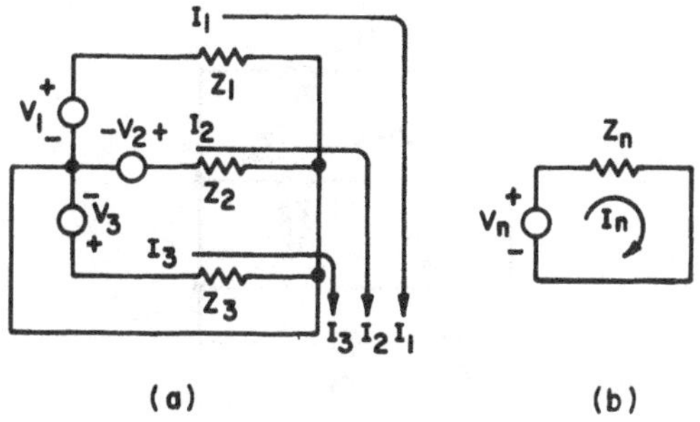

Fig. 8.14. Networks for Example 8.6: (a) three-mesh circuit, (b) one-mesh circuit.

Find the network connection and examine the special case where $V_n = 0$.

Solution

The relationship between the "old" voltages and currents is $\mathbf{E} = \mathbf{ZI}$ or

$$\begin{bmatrix} V_1 \\ V_2 \\ V_3 \\ V_n \end{bmatrix} = \begin{bmatrix} Z_1 & 0 & 0 & 0 \\ 0 & Z_2 & 0 & 0 \\ 0 & 0 & Z_3 & 0 \\ 0 & 0 & 0 & Z_n \end{bmatrix} \begin{bmatrix} I_1 \\ I_2 \\ I_3 \\ I_n \end{bmatrix}$$

The constraints between old and new currents completely specify $\mathbf{K}$, i.e.,

$$\mathbf{I}_{\text{old}} = \mathbf{K}\mathbf{I}_{\text{new}}, \qquad \mathbf{I} = \mathbf{KI}'$$

or

$$\begin{bmatrix} I_1 \\ I_2 \\ I_3 \\ I_n \end{bmatrix} = \begin{bmatrix} 1 & 0 & 0 \\ 0 & 1 & 0 \\ 0 & 0 & 1 \\ 1 & 1 & 1 \end{bmatrix} \begin{bmatrix} I_1' \\ I_2' \\ I_3' \end{bmatrix}$$

or

$$\mathbf{K} = \begin{bmatrix} 1 & 0 & 0 \\ 0 & 1 & 0 \\ 0 & 0 & 1 \\ 1 & 1 & 1 \end{bmatrix}$$

The voltages are related by $\mathbf{V}' = \mathbf{K}^{*t}\mathbf{E}$ or

$$\begin{bmatrix} V_1' \\ V_2' \\ V_3' \end{bmatrix} = \begin{bmatrix} 1 & 0 & 0 & 1 \\ 0 & 1 & 0 & 1 \\ 0 & 0 & 1 & 1 \end{bmatrix} \begin{bmatrix} V_1 \\ V_2 \\ V_3 \\ V_n \end{bmatrix} = \begin{bmatrix} V_1 + V_n \\ V_2 + V_n \\ V_3 + V_n \end{bmatrix}$$

The new network is shown in Figure 8.15. In the special case where $V_n = 0$, we note that $\mathbf{V} = \mathbf{V}'$ and the new connection is that of a three-phase generator with phase impedances Z_1, Z_2, and Z_3 and with neutral impedance Z_n. Since the original network of Figure 8.14a was that of a three-phase generator with zero neutral impedance, we have added an arbitrary neutral impedance through the choice of a particular constraint $\mathbf{K}$.

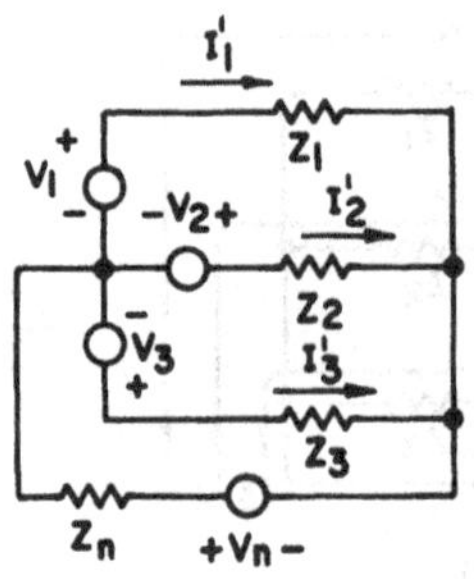

Fig. 8.15. Solution for Example 8.6.

8.9 Shunt Fault Transformations

We now consider the application of Kron's technique to the solution of faulted networks. Suppose that the network is to be faulted at a fault point F and that the positive, negative, and zero sequence Thevenin equivalents are known. These equivalents, shown in Figure 8.16, are described by the sequence network

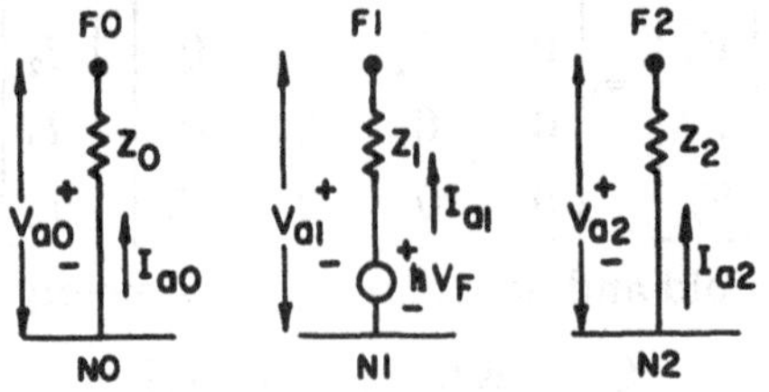

Fig. 8.16. Sequence networks.

equations

$$\begin{bmatrix} V_{a0} \\ V_{a1} \\ V_{a2} \end{bmatrix} = \begin{bmatrix} 0 \\ hV_F \\ 0 \end{bmatrix} - \begin{bmatrix} Z_0 & 0 & 0 \\ 0 & Z_1 & 0 \\ 0 & 0 & Z_2 \end{bmatrix} \begin{bmatrix} I_{a0} \\ I_{a1} \\ I_{a2} \end{bmatrix} \tag{8.53}$$

But by the method of Kron we need to consider the sequence networks as primitive ones. Then we may specify the connection of these primitive networks by developing a constraint matrix K which describes the fault condition under study. From Figure 8.11 and (8.49) we see that the primitive sequence networks are those in which

$$V_{a0} = V_{a1} = V_{a2} = 0 \tag{8.54}$$

as shown in Figure 8.17. From (8.53) and (8.54) we have the primitive network equation E =ZI or

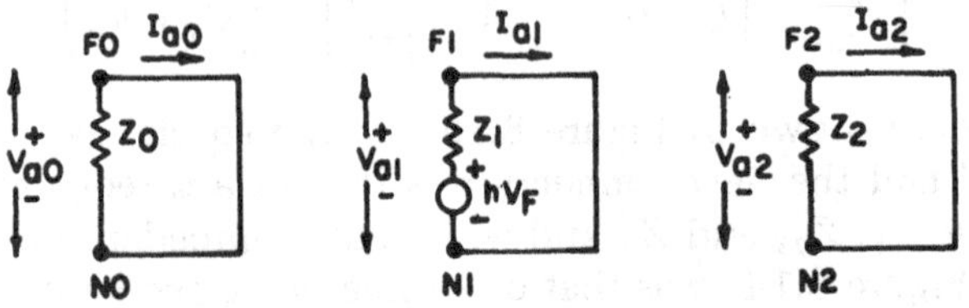

Fig. 8.17. Primitive sequence networks.

$$\begin{bmatrix} 0 \\ \mathrm{h}V_F \\ 0 \end{bmatrix} = \begin{bmatrix} Z_0 & 0 & 0 \\ 0 & Z_1 & 0 \\ 0 & 0 & Z_2 \end{bmatrix} \begin{bmatrix} I_{a0} \\ I_{a1} \\ I_{a2} \end{bmatrix} \tag{8.55}$$

The application of (8.55) to specific fault conditions requires the specification of **K**.

SLG faults. For a SLG fault on phase a we have the condition

$$I'_{a0} = I'_{a1} = I'_{a2} \tag{8.56}$$

where we use the prime notation to indicate sequence currents in the connected sequence networks or where $\mathbf{I} = \mathbf{K}\mathbf{I}'$ with

I = sequence currents of primitive networks
I' = sequence currents of interconnected sequence networks

For the SLG fault on phase a we require the networks to be connected such that

$$\begin{bmatrix} I_{a0} \\ I_{a1} \\ I_{a2} \end{bmatrix} = \begin{bmatrix} I'_{a0} \\ I'_{a0} \\ I'_{a0} \end{bmatrix} = \begin{bmatrix} 1 \\ 1 \\ 1 \end{bmatrix} I'_{a0} \tag{8.57}$$

Then

$$\mathbf{K} = \begin{bmatrix} 1 \\ 1 \\ 1 \end{bmatrix}$$

and $\mathbf{I}' = I'_{a0}$, a scalar. We also compute

$$\mathbf{V}' = \mathbf{K}^{*t}\mathbf{E} = [1 \quad 1 \quad 1] \begin{bmatrix} 0 \\ \mathrm{h}V_F \\ 0 \end{bmatrix} = \mathrm{h}V_F \tag{8.58}$$

Thus both $\mathbf{I}'$ and $\mathbf{V}'$ are scalars, i.e., 1×1 matrices. For the connected network we write

$$\mathbf{V}' = \mathbf{Z}'\mathbf{I}' \tag{8.59}$$

so we must also compute $\mathbf{Z}'$ where

$$\mathbf{Z}' = \mathbf{K}^{*t}\mathbf{Z}\mathbf{K} = [1 \quad 1 \quad 1] \begin{bmatrix} Z_0 & 0 & 0 \\ 0 & Z_1 & 0 \\ 0 & 0 & Z_2 \end{bmatrix} \begin{bmatrix} 1 \\ 1 \\ 1 \end{bmatrix}$$
$$= (Z_0 + Z_1 + Z_2)$$

also a scalar. Then (8.59) may be written as

$$\mathrm{h}V_F = (Z_0 + Z_1 + Z_2)\, I'_{a0} \tag{8.60}$$

which is obviously correct for the case where the fault impedance is either zero or is included as a part of Z_0, i.e., $Z_0 = Z_{0\,(\mathrm{system})} + 3Z_f$. We will examine the fault impedance problem in greater detail later.

For a SLG fault on phase b we have a different constraint among the sequence currents. In this case $I_a = I_c = 0$, and we compute

$$\mathbf{I}_{012} = \mathbf{A}^{-1}\mathbf{I}_{abc} = \frac{h}{3}\begin{bmatrix} 1 \\ a \\ a^2 \end{bmatrix} I_b \tag{8.61}$$

In other words the relationship among sequence currents corresponding to (8.56) in the previous case is from (8.61),

$$I'_{a0} = a^2 I'_{a1} = a I'_{a2} \tag{8.62}$$

Note that this agrees with the solution obtained from the network of Figure 8.3. From (8.62) and Figure 8.3 we equate primitive to connected currents to write

$$\begin{bmatrix} I_{a0} \\ I_{a1} \\ I_{a2} \end{bmatrix} = \begin{bmatrix} I'_{a0} \\ I'_{a1} \\ I'_{a2} \end{bmatrix} = \begin{bmatrix} I'_{a0} \\ aI'_{a0} \\ a^2 I'_{a0} \end{bmatrix} = \begin{bmatrix} 1 \\ a \\ a^2 \end{bmatrix} I'_{a0} \tag{8.63}$$

or

$$\mathbf{K} = \begin{bmatrix} 1 \\ a \\ a^2 \end{bmatrix} \tag{8.64}$$

Note that since $a = e^{j2\pi/3}$ then $a^* = e^{-j2\pi/3} = a^2$ and, similarly, $(a^2)^* = a$. Then $\mathbf{K}^{*t} = [1 \quad a^2 \quad a]$ and

$$\mathbf{V}' = [1 \quad a^2 \quad a]\begin{bmatrix} 0 \\ hV_F \\ 0 \end{bmatrix} = a^2 h V_F$$

Finally,

$$\mathbf{Z}' = \mathbf{K}^{*t}\mathbf{Z}\,\mathbf{K} = [1 \quad a^2 \quad a]\begin{bmatrix} Z_0 & 0 & 0 \\ 0 & Z_1 & 0 \\ 0 & 0 & Z_2 \end{bmatrix}\begin{bmatrix} 1 \\ a \\ a^2 \end{bmatrix}$$

$$= Z_0 + Z_1 + Z_2 \tag{8.65}$$

To compute the fault current, we write $\mathbf{V}' = \mathbf{Z}'\mathbf{I}'$ or $a^2 h V_F = (Z_0 + Z_1 + Z_2)\, I'_{a0}$ and

$$I'_{a0} = \frac{a^2 h V_F}{Z_0 + Z_1 + Z_2} \tag{8.66}$$

which is seen to be the same as the current computed from (8.60) except the result is rotated by +240° as it should be for a fault on phase b.

2LG faults. As another example of the Kron method applied to faulted networks we examine a 2LG fault on phases b and c with zero fault impedance. For this condition we recall that the sequence networks are connected in parallel as shown in Figure 3.15. With zero fault impedance we have, for the connected network, $I_a = 0$, or

$$I'_{a0} + I'_{a1} + I'_{a2} = 0 \tag{8.67}$$

Then we determine the constraint matrix K from

$$\begin{bmatrix} I_{a0} \\ I_{a1} \\ I_{a2} \end{bmatrix} = \begin{bmatrix} I'_{a0} \\ I'_{a1} \\ I'_{a2} \end{bmatrix} = \begin{bmatrix} -I'_{a1} - I'_{a2} \\ I'_{a1} \\ I'_{a2} \end{bmatrix} = \begin{bmatrix} -1 & -1 \\ 1 & 0 \\ 0 & 1 \end{bmatrix} \begin{bmatrix} I'_{a1} \\ I'_{a2} \end{bmatrix} \tag{8.68}$$

or

$$\mathbf{K} = \begin{bmatrix} -1 & -1 \\ 1 & 0 \\ 0 & 1 \end{bmatrix}$$

The voltage equation is

$$\mathbf{V}' = \mathbf{K}^{*t}\mathbf{E} = \begin{bmatrix} -1 & 1 & 0 \\ -1 & 0 & 1 \end{bmatrix} \begin{bmatrix} 0 \\ \mathrm{h}V_F \\ 0 \end{bmatrix} = \begin{bmatrix} \mathrm{h}V_F \\ 0 \end{bmatrix} \tag{8.69}$$

The transformed impedance is

$$\mathbf{Z}' = \mathbf{K}^{*t}\,\mathbf{Z}\,\mathbf{K} = \begin{bmatrix} -1 & 1 & 0 \\ -1 & 0 & 1 \end{bmatrix} \begin{bmatrix} Z_0 & 0 & 0 \\ 0 & Z_1 & 0 \\ 0 & 0 & Z_2 \end{bmatrix} \begin{bmatrix} -1 & -1 \\ 1 & 0 \\ 0 & 1 \end{bmatrix} = \begin{bmatrix} Z_0 + Z_1 & Z_0 \\ Z_0 & Z_0 + Z_2 \end{bmatrix} \tag{8.70}$$

Then the connected network may be solved for I'_{a1} and I'_{a2} from $\mathbf{V}' = \mathbf{Z}'\mathbf{I}'$ or

$$\begin{bmatrix} \mathrm{h}V_F \\ 0 \end{bmatrix} = \begin{bmatrix} Z_0 + Z_1 & Z_0 \\ Z_0 & Z_0 + Z_2 \end{bmatrix} \begin{bmatrix} I'_{a1} \\ I'_{a2} \end{bmatrix} \tag{8.71}$$

Solving for the currents, we have

$$\begin{aligned} I'_{a1} &= \frac{\mathrm{h}V_F}{Z_1 + Z_0 Z_2/(Z_0 + Z_2)} \\ I'_{a2} &= -\frac{Z_0}{Z_0 + Z_2}\, I'_{a1} \end{aligned} \tag{8.72}$$

which from (3.22) are obviously correct. Then I'_{a0} may be computed from (8.67) and $\mathbf{I}_{abc}$ computed in the usual way. If the 2LG fault is on phases other than b and c, the K matrix becomes complex. For example, if the fault is on phases a and b, then $I_c = 0$, or

$$I'_{a0} + aI'_{a1} + a^2 I'_{a2} = 0 \tag{8.73}$$

Then

$$\begin{bmatrix} I_{a0} \\ I_{a1} \\ I_{a2} \end{bmatrix} = \begin{bmatrix} -aI'_{a1} - a^2 I'_{a2} \\ I'_{a1} \\ I'_{a2} \end{bmatrix} = \begin{bmatrix} -a & -a^2 \\ 1 & 0 \\ 0 & 1 \end{bmatrix} \begin{bmatrix} I'_{a1} \\ I'_{a2} \end{bmatrix} \tag{8.74}$$

and

$$K = \begin{bmatrix} -a & -a^2 \\ 1 & 0 \\ 0 & 1 \end{bmatrix} \tag{8.75}$$

which is complex.

From the foregoing brief experience in solving faulted networks by Kron's method, we note the following features concerning the formation of K.

1. In the case of SLG faults it is always possible to write two current constraints, e.g., for a SLG fault on phase b,

$$I'_{a0} = a^2 I'_{a1}, \qquad I'_{a0} = aI'_{a2}$$

   In such cases when we write $\mathbf{I} = \mathbf{K}\,\mathbf{I}'$, there are two constraints among the three variables of $\mathbf{I}'$ and we can always eliminate two currents. This means that K will always be 3×1.
2. In the case of 2LG faults it is possible to write only one current constraint. For example, if the fault involves lines a and b, we have

$$I'_{a0} + aI'_{a1} + a^2 I'_{a2} = 0$$

   Here we have *one* constraint for *three* variables and can eliminate only one variable. Thus the K matrix is always 3×2.
3. In all cases if the fault is symmetric with respect to phase a, then K is real and $K^* = K$. If any other symmetry is presented, K is complex but it involves only the complex parameters 1, a, and a^2.
4. In cases where *two current constraints* exist, the K matrix may be written such that any *one* of the currents I'_{a0}, I'_{a1}, or I'_{a2} is explicitly in the solution. In the SLG cases above we solved for I'_{a0}, but we could have chosen I'_{a1} or I'_{a2}.
5. In cases where only one current constraint exists, the K matrix may be written such that any two currents are explicitly in the solution. In the 2LG faults above we solved for I'_{a1} and I'_{a2}, but we could have chosen any two currents.

8.10 Transformations for Shunt Faults with Impedance

In Figure 8.1 a generalized shunt fault is shown with arbitrary values of fault impedance. The equations describing this situation are given in (8.1) in terms of phase quantities and in (8.3)-(8.5) in terms of sequence quantities. The various shunt fault conditions of interest are then specified by prescribing values to the impedances Z_a, Z_b, Z_c, and Z_g. This in turn gives certain constraints among the sequence voltages and currents which form the boundary conditions for the different fault situations.

We may take advantage of this analysis and adapt it to the Kron analytical technique by enlarging our view of the primitive network to include certain fault impedances as well as the sequence impedances for the normal (symmetrical) portion of the network.

Since we are usually concerned with only the SLG, LL, and 2LG faults, with appropriate impedance, we focus our attention on the following special cases:

$$\text{SLG fault:} \quad Z_g \neq 0$$
$$Z_a = Z_b = Z_c = 0$$

LL fault: $Z_g = \infty$

any two of Z_a, Z_b, and $Z_c = Z$

2LG fault: $Z_g \neq 0$

any two of Z_a, Z_b, and $Z_c = Z$

These are exactly the special cases studied in section 8.2 where we developed the sequence network connections given in Figure 8.3 for the SLG fault and in Figure 8.4 for the LL and 2LG faults. The LL and 2LG faults differ only in the specification of Z_g, and we note that $I_{a0} = 0$ for the LL fault such that the zero sequence network is isolated from the other networks (see Figure 8.4). For any symmetry, Figures 8.3 and 8.4 provide us with the desired constraints for determining the K matrix if we interpret the Kron primitive networks as those of Figure 8.18.

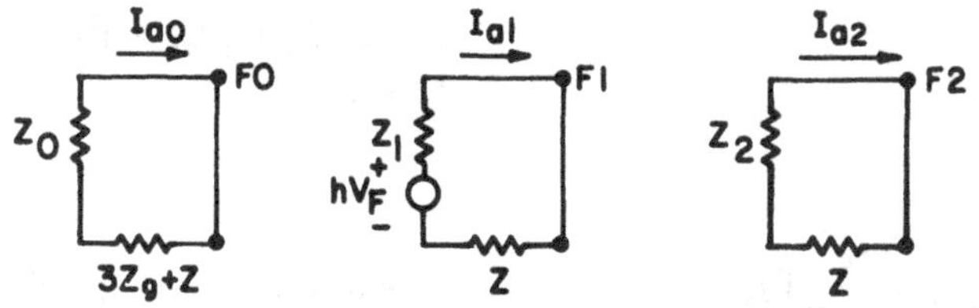

Fig. 8.18. Primitive sequence networks for shunt faults with impedance.

We must also decide, as discussed in section 8.9, which sequence currents in the connected network are to be specified. The choice here is arbitrary. It seems reasonable, however, that in any fault involving the ground the ground current $3I'_{a0}$ is of direct interest. Hence I'_{a0} should be the current to find in both SLG and 2LG faults. In LL faults, which are a special case of 2LG, either I'_{a1} or I'_{a2} is required. Let us arbitrarily choose I'_{a1}. Then we may specify I'_{a0} and I'_{a1} for the 2LG fault and can easily identify the LL case as that for which $I'_{a0} = 0$.

For any fault we have the primitive equation $\mathbf{E} = \mathbf{Z}\,\mathbf{I}$ or

$$\begin{bmatrix} 0 \\ hV_F \\ 0 \end{bmatrix} = \begin{bmatrix} (Z_0 + 3Z_g + Z) & 0 & 0 \\ 0 & (Z_1 + Z) & 0 \\ 0 & 0 & (Z_2 + Z) \end{bmatrix} \begin{bmatrix} I_{a0} \\ I_{a1} \\ I_{a2} \end{bmatrix} \tag{8.76}$$

which is transformed by the current constraint $\mathbf{I} = \mathbf{K}\,\mathbf{I}'$ to a new equation

$$\mathbf{V}' = \mathbf{Z}'\mathbf{I}' \tag{8.77}$$

The constraint matrix may be determined from the connections of Figures 8.3 and 8.4 which are tabulated in Table 8.2. Note that it is unnecessary to provide a separate tabulation for LL faults since the constraint matrix in this case is immediately available from the 2LG results by eliminating the column of K corresponding to I'_{a0}.

To solve for the sequence currents, we must solve (8.77) for the current $\mathbf{I}'$ given in the second column of Table 8.2. This requires two additional items $\mathbf{V}'$ and $\mathbf{Z}'$, which are readily computed once K is known. These results are given in Table 8.3.

We illustrate the use of Tables 8.2 and 8.3 by the following examples.

Table 8.2. The Constraint Matrix K in Terms of I_{a0} and I_{a1}

Type Fault	Solution Equation	Symmetrical Phase	Current Constraints: System	Current Constraints: Sequence	Constraint Matrix K
SLG or two lines open	$\mathbf{I} = \mathbf{K}I'_{a0}$	a	$I_b = 0$, $I_c = 0$	$I'_{a0} = I'_{a1} = I'_{a2}$	$\begin{array}{c\|c} & 0' \\ \hline 0 & 1 \\ 1 & 1 \\ 2 & 1 \end{array}$
		b	$I_a = 0$, $I_c = 0$	$I'_{a0} = a^2 I'_{a1} = a I'_{a2}$	$\begin{array}{c\|c} & 0' \\ \hline 0 & 1 \\ 1 & a \\ 2 & a^2 \end{array}$
		c	$I_a = 0$, $I_b = 0$	$I'_{a0} = a I'_{a1} = a^2 I'_{a2}$	$\begin{array}{c\|c} & 0' \\ \hline 0 & 1 \\ 1 & a^2 \\ 2 & a \end{array}$
2LG or one line open	$\mathbf{I} = \mathbf{K}\begin{bmatrix} I'_{a0} \\ I'_{a1} \end{bmatrix}$	a (b-c fault)	$I_a = 0$	$I'_{a0} + I'_{a1} + I'_{a2} = 0$	$\begin{array}{c\|cc} & 0' & 1' \\ \hline 0 & 1 & 0 \\ 1 & 0 & 1 \\ 2 & -1 & -1 \end{array}$
		b (a-c fault)	$I_b = 0$	$I'_{a0} + a^2 I'_{a1} + a I'_{a2} = 0$	$\begin{array}{c\|cc} & 0' & 1' \\ \hline 0 & 1 & 0 \\ 1 & 0 & 1 \\ 2 & -a^2 & -a \end{array}$
		c (a-b fault)	$I_c = 0$	$I'_{a0} + a I'_{a1} + a^2 I'_{a2} = 0$	$\begin{array}{c\|cc} & 0' & 1' \\ \hline 0 & 1 & 0 \\ 1 & 0 & 1 \\ 2 & -a & -a^2 \end{array}$
LL	$\mathbf{I} = \mathbf{K}I'_{a1}$	...	...	same as 2LG but with $I'_{a0} = 0$	use 2nd column of 2LG results

Example 8.7

Consider a SLG fault on a system for which the sequence networks are given as those of Figure 8.18. Since the fault is SLG, we let $Z = 0$ and $Z_g \neq 0$. Find fault currents for a SLG fault on phase a, then on phase b. Compare with results of similar computations from section 8.9.

Solution

For a SLG fault on phase a we have from Table 8.2

$$\mathbf{K} = \begin{bmatrix} 1 \\ 1 \\ 1 \end{bmatrix}$$

Table 8.3. Transformed Voltages and Impedances Corresponding to Constraint Matrices of Table 8.2 *

Type Fault	*Symmetrical Phase*	$\mathbf{V}' = \mathbf{K}^{*t}\mathrm{h}\mathbf{E}$	$\mathbf{Z}' = \mathbf{K}^{*t}\,\mathbf{Z}\,\mathbf{K}$†	k
SLG or two lines open	a	$V'_{a0} = \mathrm{h}V_F$	$Z' = Z_0 + Z_1 + Z_2 + 3Z_{gg}$	does not apply
	b	$V'_{a0} = \mathrm{a}^2\mathrm{h}V_F$	same	
	c	$V'_{a0} = \mathrm{ah}V_F$	same	
2LG or one line open	a (b-c fault)	$\begin{bmatrix} V'_{a0} \\ V'_{a1} \end{bmatrix} = \begin{bmatrix} 0 \\ \mathrm{h}V_F \end{bmatrix}$	$\mathbf{Z}' = \begin{bmatrix} Z_{00} & \mathrm{k}Z_{22} \\ \mathrm{k}^2 Z_{22} & Z_{11} \end{bmatrix}$	1
	b (a-c fault)	same	change k only	a^2
	c (a-b fault)	same	change k only	a
LL	any	$V'_{a1} = \mathrm{h}V_F$	$\mathbf{Z}' = Z_{11}$	. . .

*Values for $\mathbf{V}'$ and $\mathbf{Z}'$ in this table correspond to solution for currents I_{a0} and I_{a1}.

†$\mathbf{Z}'$ components are defined as $Z_{00} = Z_0 + Z_2 + 3Z_g + 2Z$, $Z_{11} = Z_1 + Z_2 + 2Z$, $Z_{22} = Z_2 + Z$, and $Z_{gg} = Z_g$ (SLG) $= Z$ (2LO).

and $\mathbf{I}' = I'_{a0}$, a scalar.

Also from Table 8.3, $\mathbf{V}' = \mathrm{h}V_F$, a scalar, and $\mathbf{Z}' = Z_0 + Z_1 + Z_2, + 3Z_g$. Then, since $\mathbf{V}' = \mathbf{Z}'\mathbf{I}'$, we have $\mathrm{h}V_F = (Z_0 + Z_1 + Z_2 + 3Z_g)\, I'_{a0}$ and

$$I'_{a0} = \frac{\mathrm{h}V_F}{Z_0 + Z_1 + Z_2 + 3Z_g}$$

This is exactly the same result given in equation (3.4).

By symmetry we know that a SLG fault on phase b will be the same as on phase a except rotated by $-120°$ (or $+240°$). We check this by taking values from the tables. From Table 8.2

$$\mathbf{K} = \begin{bmatrix} 1 \\ \mathrm{a} \\ \mathrm{a}^2 \end{bmatrix}$$

and from Table 8.3

$$\mathbf{V}' = \mathrm{a}^2\mathrm{h}V_F, \qquad \mathbf{Z}' = Z_0 + Z_1 + Z_2 + 3Z_g$$

so that

$$I'_{a0} = \frac{\mathrm{a}^2\mathrm{h}V_F}{Z_0 + Z_1 + Z_2 + 3Z_g}$$

which is obviously correct. In both cases for faults on phase a and phase b, these results agree with equations (8.60) and (8.66) except that the impedance has been increased by the amount $3Z_g$.

Example 8.8

Consider a 2LG fault on a system for which the sequence networks are given as those of Figure 8.18 where $Z \neq 0$ and $Z_g \neq 0$. Find the sequence current I'_{a1} and compare with equation (8.72) for a 2LG fault on phases b and c.

Solution

From Table 8.2 we have

$$\mathbf{K} = \begin{bmatrix} 1 & 0 \\ 0 & 1 \\ -1 & -1 \end{bmatrix}$$

and

$$\mathbf{I}' = \begin{bmatrix} I'_{a0} \\ I'_{a1} \end{bmatrix}$$

From Table 8.3 we find that

$$\mathbf{V}' = \begin{bmatrix} 0 \\ \mathrm{h} V_F \end{bmatrix}$$

and

$$\mathbf{Z}' = \begin{bmatrix} Z_{00} & Z_{22} \\ Z_{22} & Z_{11} \end{bmatrix}$$

Since $\mathbf{V}' = \mathbf{Z}'\mathbf{I}'$ we compute

$$\begin{bmatrix} 0 \\ \mathrm{h} V_F \end{bmatrix} = \begin{bmatrix} Z_{00} & Z_{22} \\ Z_{22} & Z_{11} \end{bmatrix} \begin{bmatrix} I'_{a0} \\ I'_{a1} \end{bmatrix}$$

Then

$$I'_{a1} = \frac{\det \begin{bmatrix} Z_{00} & 0 \\ Z_{22} & \mathrm{h}V_F \end{bmatrix}}{\det Z'} = \frac{Z_{00}\,\mathrm{h}V_F}{Z_{00}Z_{11} - Z_{22}^2}$$

But, from Table 8.3

$$Z_{00} = Z_0 + Z_2 + 3Z_g + 2Z, \qquad Z_{11} = Z_1 + Z_2 + 2Z, \qquad Z_{22} = Z_2 + Z$$

and

$$\begin{aligned} I'_{a1} &= \frac{\mathrm{h}V_F}{Z_{11} - Z_{22}^2/Z_{00}} \\ &= \frac{\mathrm{h}V_F}{(Z_1 + Z_2 + 2Z) - Z_{22}^2/(Z_0 + Z_2 + 3Z_g + 2Z)} \end{aligned}$$

If $Z = Z_g = 0$, this reduces to (8.72) exactly.

Example 8.9

Repeat Example 8.8 except that this time let the fault be applied to phases a and c. Then simplify to the case of a LL fault, also on phases a and c.

Solution

For a 2LG fault on phases a and c we have from Table 8.2

$$\mathbf{K} = \begin{bmatrix} 1 & 0 \\ 0 & 1 \\ -a^2 & -a \end{bmatrix}$$

and

$$\mathbf{I}' = \begin{bmatrix} I'_{a0} \\ I'_{a1} \end{bmatrix}$$

Then from Table 8.3

$$\mathbf{V}' = \begin{bmatrix} V'_{a0} \\ V'_{a1} \end{bmatrix} = \begin{bmatrix} 0 \\ hV_F \end{bmatrix}$$

and

$$\mathbf{Z}' = \begin{bmatrix} Z_{00} & a^2 Z_{22} \\ aZ_{22} & Z_{11} \end{bmatrix}$$

where

$$Z_{00} = Z_0 + Z_2 + 3Z_g + 2Z, \qquad Z_{11} = Z_1 + Z_2 + 2Z, \qquad Z_{22} = Z_2 + Z$$

We compute the sequence currents from

$$\begin{bmatrix} 0 \\ hV_F \end{bmatrix} = \begin{bmatrix} Z_{00} & a^2 Z_{22} \\ aZ_{22} & Z_{11} \end{bmatrix} \begin{bmatrix} I'_{a0} \\ I'_{a1} \end{bmatrix}$$

Then

$$I'_{a0} = \frac{1}{\det \mathbf{Z}'} \det \begin{bmatrix} 0 & a^2 Z_{22} \\ hV_F & Z_{11} \end{bmatrix} = -\frac{a^2 Z_{22} hV_F}{\det \mathbf{Z}'}$$

and

$$I'_{a1} = \frac{1}{\det \mathbf{Z}'} \det \begin{bmatrix} Z_{00} & 0 \\ aZ_{22} & hV_F \end{bmatrix} = \frac{Z_{00} hV_F}{\det \mathbf{Z}'}$$

From Table 8.2 we have the sequence current constraint $I'_{a0} + a^2 I'_{a1} + aI'_{a2} = 0$ or

$$I'_{a2} = -a^2 I'_{a0} - aI'_{a1} = \frac{a(Z_{22} - Z_{00})\, hV_F}{\det \mathbf{Z}'} = \frac{-a\,(Z_0 + 3Z_g + Z)\, hV_F}{\det \mathbf{Z}'}$$

All currents may be found from I'_{a0}, I'_{a1}, and I'_{a2}.

In the case of a LL fault, $I'_{a0} = 0$ or

$$\mathbf{I}' = \begin{bmatrix} 0 \\ I'_{a1} \end{bmatrix}$$

and we have

$$\begin{bmatrix} 0 \\ hV_F \end{bmatrix} = \begin{bmatrix} Z_{00} & a^2 Z_2 \\ aZ_2 & Z_{11} \end{bmatrix} \begin{bmatrix} 0 \\ I'_{a1} \end{bmatrix}$$

or

$$I'_{a1} = \mathrm{h}V_F/Z_{11} = \mathrm{h}V_F/(Z_1 + Z_2 + 2Z)$$

From Table 8.2 with $I'_{a0} = 0$, we also compute $\mathrm{a}^2 I'_{a1} + \mathrm{a}I'_{a2} = 0$ or $I'_{a2} = -\mathrm{a}I'_{a1}$. Again, by knowing the sequence currents, we can solve for all currents.

8.11 Series Fault Transformations

It is apparent that a one-to-one relationship exists between the generalized fault diagram of Figure 8.9 and Kron's analytical technique of section 8.10. We recognize this relationship in developing an analytical technique for series faults which will be based upon the network unbalance described in Figure 8.5 and (8.17).

The similarity between series faults involving one or two open lines and shunt faults of the 2LG and SLG conditions is noted in Figure 8.9. In the case of series faults the impedance Z_g is always taken as zero but the impedance Z is not zero in general. Thus we may take the primitive sequence networks to be those of Figure 8.19 except that $Z_g = 0$.

Two lines open. In the case of two lines open we may write two current constraints, or if the open lines are b and c,

$$I_b = 0, \qquad I_c = 0$$

Since we have two constraints, we need to solve for only one of the three sequence currents. Let us arbitrarily choose I'_{a0}. Then this case is mathematically identical to that of a SLG fault or phase a, and the entries in Tables 8.2 and 8.3 apply. Similar statements could also be made for two lines open involving other phases.

One line open. If only one line is open, there is only one current constraint which we can write, and we must therefore solve for two sequence currents. If we arbitrarily choose I'_{a0} and I'_{a1}, the entries under the 2LG fault of Tables 8.2 and 8.3, with Z_g always taken as zero, apply.

Because of the unique similarity in the symmetry of shunt and series faults there is no need to develop separate tables for the series fault situation. In using Tables 8.2 and 8.3, the only caution is that Z_g must always be set to zero for a series fault. Obviously, $\mathrm{h}V_F$ has a different meaning in the two situations also.

8.12 Summary

We have presented two methods for dealing with symmetries different from phase a symmetry. The first method is valuable since it places emphasis on circuit concepts with which most electrical engineers are familiar. It results in a quite general pair of circuit diagrams given in Figure 8.9, from which the engineer can solve nearly any fault which may be of interest.

The second method, due to Kron, is more analytical than the first in the sense that it is based upon topology and matrix techniques. It is readily adapted to computer analysis because of its orderly method of computation and because it does *not* rely on the drawing of a circuit diagram.

Both methods have their unique advantages, and both can be extended to help us solve even more difficult problems.

Problems

8.1. Verify the network connections of Figure 8.3 by network reduction of Figure 8.2 for the condition of a SLG fault on phase *a*.

8.2. Verify the transformer phase shift of Figure 8.3 for the case of a SLG fault on phase *c*.

8.3. Verify the network connections of Figure 8.4 by network reduction of Figure 8.2 for the condition of a 2LG fault on phases *b* and *c*.

8.4. Verify the transformer phase shift of Figure 8.4 for the case of a 2LG fault on phases *c* and *a*. Repeat for a 2LG fault on phases *a* and *b*.

8.5. Given a system with Thevenin equivalent sequence network representations shown in Figure P8.5, sketch the generalized fault diagram for a SLG fault on phase *b* at *F* and compute the fault currents in all phases. Let the fault impedance be j0.2 pu.

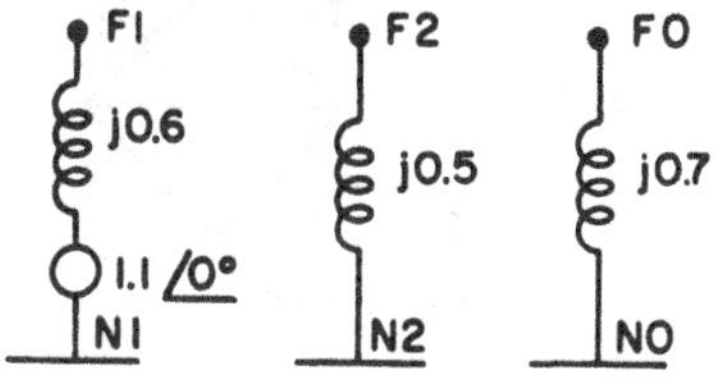

Fig. P8.5.

8.6. Using the sequence networks of problem 8.5, sketch the generalized fault diagram for a 2LG fault between phases *a* and *b* with fault impedance of j0.1 in each line (i.e., $Z = j0.1$) and grounding impedance Z_g of j0.1. Compute the total fault current in each phase.

8.7. Given a system with two-port Thevenin equivalents at points *F* and *M* as shown in Figure P8.7, sketch the generalized fault diagram for the case of two lines *a* and *b* open between *F* and *M*, and line *c* having an impedance of j0.1 pu. Compute the three line currents.

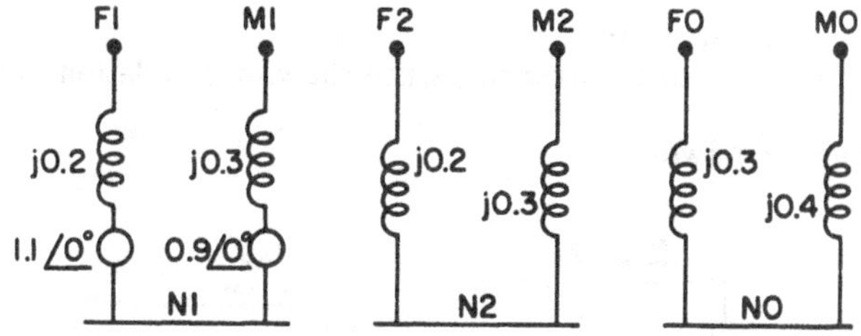

Fig. P8.7.

8.8. Using the sequence networks of problem 8.7, sketch the generalized fault diagram for the case of line *b* open and with lines *a* and *c* each having an impedance of j0.2 pu.

8.9. Find the constraint matrix K if the mesh of I_3' in Example 8.2 is taken to be the mesh $Z_6, -Z_5, -Z_4$. (This is the mesh in the "window" defined in the clockwise sense.)

8.10. Given a connected network with *B* branches and *N* nodes (see [57]),
(a) Show that there are $N(N-1)$ possible voltages between nodes, taking polarity into account.
(b) Show that of these $N(N-1)$ voltages only $N-1$ are independent.
(c) Show that if there are *S* subnetworks, the number of independent voltages is $N-S$, where *N* is the total number of nodes in all networks.

8.11. Given a connected network with *B* branches and *N* nodes and $P = N - 1$ independent node voltages (see problem 8.10),
(a) How many independent node equations can we write?
(b) How many independent branch voltage equations can we write?
(c) How many independent mesh current equations can we write? Call this number *M*.

(d) How many independent mesh current equations can we write if there are S subnetworks?

8.12. Given the circuit of Figure P8.12 with branch currents, in the directions indicated,
(a) Write the primitive network equation $\mathbf{E} = \mathbf{Z}\,\mathbf{I}$.
(b) Transform the primitive network equation into a mesh equation where the meshes are defined to be 1-2-4, 2-3-6, and 4-5-6 in the order indicated.
(c) Transform the mesh equations of (b) to a new set of equations, viz., 1-2-4, 2-3-6, and 2-3-5-4.
(d) Compute I from (8.47). In both parts (b) and (c) carefully define the constraint matrix K. Note that in part (c), K should be nonsingular. What does this mean?

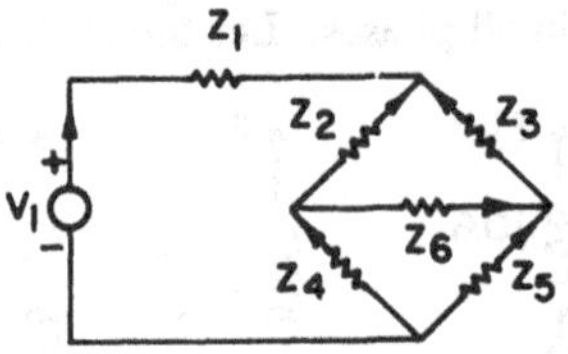

Fig. P8.12.

8.13. Repeat problem 8.12 where a mutual impedance Z_{35} exists between branches 3 and 5.
(a) Define a new Z matrix which includes Z_{35}. Consider the mutual impedance voltage drop as positive and explain what this implies.
(b), (c), (d). Transform from branch to mesh currents and compute branch currents as in 8.12.

8.14. Given the circuits shown in Figure P8.14 with mesh currents defined as flowing through impedances (a) *a-n*, *b-n*, *c-n* and (b) *d-e* and *f-e*,
(a) Write the mesh equations for circuit (a) and circuit (b).
(b) Define a new network by combining (a) and (b) according to the constraint

$$I_a' = I_{a1} = I_{a2}, \quad I_b' = I_{b1}, \quad I_c' = I_{c1} = I_{c2}$$

Write the constraint matrix.
(c) Assuming power is to be invariant, write the voltage relationship between new and old networks.
(d) Sketch the new network.

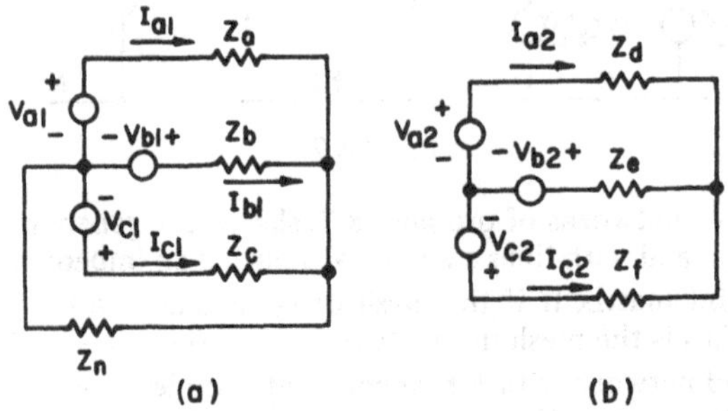

Fig. P8.14.

8.15. (a) Write the mesh current equations for the circuit of Figure P8.15.
(b) Consider the circuits of Figures P8.14a and P8.15 with the current constraint given in problem 8.14b. Find the voltage relationship between the given networks and the combined networks.
(c) Find the combined impedance network, i.e., $\mathbf{Z}' = \mathbf{K}^{*t}\mathbf{Z}\,\mathbf{K}$.
(d) Sketch the new network.

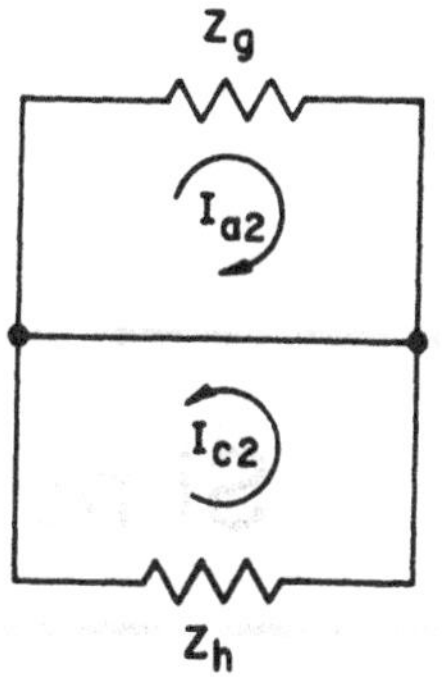

Fig. P8.15.

8.16. What fault condition does Figure 8.17 and (8.55) describe? Give proof for your answer.

8.17. Find the constraint matrix K which will transform the primitive sequence currents into SLG fault on phase c currents, where I'_{a1} is to be solved explicitly.

8.18. Find the constraint matrix K which will transform the primitive sequence currents into 2LG fault currents on phases a and c where I'_{a0} and I'_{a2} are to be found explicitly.

8.19. Find the constraint matrix K which will transform the primitive sequence currents into LL fault currents where I'_{a2} is to be found explicitly and where the faulted phases are (a) b and c, (b) a and c, and (c) a and b.

8.20. Repeat problem 8.5, using Kron's analytical technique and Tables 8.2 and 8.3.

8.21. Repeat problem 8.6, using Kron's analytical technique and Tables 8.2 and 8.3.

8.22. Repeat problem 8.7, using Kron's analytical technique and Tables 8.2 and 8.3.

8.23. Repeat problem 8.8, using Kron's analytical technique and Tables 8.2 and 8.3.

chapter 9

Simultaneous Faults

One of the most difficult problems in the solution of faulted networks is that involving two or more faults which occur simultaneously. Such an occurrence may be the result of some event such as a stroke of lightning or a man-caused accident, which forces more than one fault at a single location. On the other hand, a simultaneous fault situation may arise from the chance occurrence of two fault conditions at two (or more) remote points. Usually only two simultaneous faults are considered. This is a practical limitation because the joint probability of even two simultaneous faults is quite low since it is computed as the product of the two single-event probabilities. In a well-designed circuit the likelihood of a single fault is fairly remote even in stormy weather and may occur only once every 5–10 years per 100 miles of line.

In the case of two simultaneous faults there are four cases of interest. For faults occurring at points A and B these four cases are:

1. A shunt fault at A and a shunt fault at B.
2. A shunt fault at A and a series fault at B.
3. A series fault at A and a series fault at B.
4. A series fault at A and a shunt fault at B.

Actually, we can view this as only three *different* fault configurations since situations 2 and 4 require exactly the same computation scheme.

We will develop a method for computing two simultaneous faults by two methods, each of which will permit us to deal with the three situations noted for common fault configurations. The first method is oriented toward the generalized network configurations, and the second is an extension of Kron's analytical technique. Both methods are valuable but either will suffice.

I. SIMULTANEOUS FAULTS BY TWO-PORT NETWORK THEORY

9.1 Two-Port Networks

Before proceeding to a consideration of the simultaneous fault problem we engage in a brief review of two-port network theory. This will enable us to lay the groundwork for subsequent applications of the two-port method and will introduce the two-port conventions and notations to be used later. Our treatment of the subject will be brief, and the interested reader is advised to consult some of the excellent references on the subject such as Hueslman [5] or Guillemin [62].

A two-port network, as the term is to be used here, is one with two pairs of terminals emerging as shown on the network of Figure 9.1. We make a special

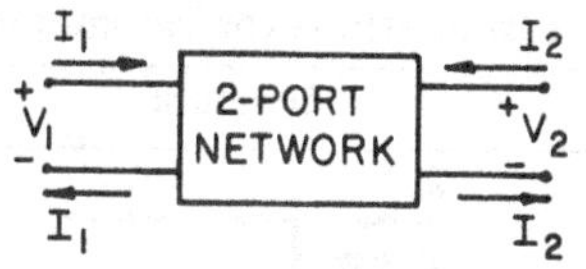

Fig. 9.1 A two-port network.

distinction between a "port" and a "pair of terminals" by requiring that in a port the same current must leave one of the terminals (the bottom terminal) as enters the other (the top terminal). This is not a restriction on the network itself but on the external connections made to the network. Thus we must always be certain that no matter how complicated the connections among a group of two-port networks, this current requirement is always preserved. Often we guarantee this by placing 1:1 transformers on the ports before making any external connections (see [5] for a discussion of this technique). We note that two-port networks are sometimes called "four-terminal networks" or "two terminal-pair networks," but we reject these names as being inadequate for the reasons discussed above.

Passive two-port networks are commonly specified in terms of the network parameters defined in Table 9.1. Huelsman [5] reviews the method for finding all parameters by appropriate tests on the networks. In every case the tests consist of placing appropriate terminations on the ports, preserving the port current criteria and leaving the network unchanged as test conditions are changed. That is, if a port is shorted for the first test, it must be shorted for subsequent tests as well. The port terminations used are reviewed in Table 9.2. Note especially the

Table 9.1. Two-Port Network Parameters

Designation	*Equation*
Z (impedance) parameters	$\begin{bmatrix} V_1 \\ V_2 \end{bmatrix} = \begin{bmatrix} z_{11} & z_{12} \\ z_{21} & z_{22} \end{bmatrix} \begin{bmatrix} I_1 \\ I_2 \end{bmatrix}$
Y (admittance) parameters	$\begin{bmatrix} I_1 \\ I_2 \end{bmatrix} = \begin{bmatrix} y_{11} & y_{12} \\ y_{21} & y_{22} \end{bmatrix} \begin{bmatrix} V_1 \\ V_2 \end{bmatrix}$
H (hybrid) parameters*	$\begin{bmatrix} V_1 \\ I_2 \end{bmatrix} = \begin{bmatrix} h_{11} & h_{12} \\ h_{21} & h_{22} \end{bmatrix} \begin{bmatrix} I_1 \\ V_2 \end{bmatrix}$
G (inverse hybrid) parameters	$\begin{bmatrix} I_1 \\ V_2 \end{bmatrix} = \begin{bmatrix} g_{11} & g_{12} \\ g_{21} & g_{22} \end{bmatrix} \begin{bmatrix} V_1 \\ I_2 \end{bmatrix}$
A (transmission) parameters†	$\begin{bmatrix} V_1 \\ I_1 \end{bmatrix} = \begin{bmatrix} A_{11} & A_{12} \\ A_{21} & A_{22} \end{bmatrix} \begin{bmatrix} V_2 \\ I_2 \end{bmatrix}$
$\mathfrak{A}$ (inverse transmission) parameters‡	$\begin{bmatrix} V_2 \\ I_2 \end{bmatrix} = \begin{bmatrix} \mathfrak{A}_{11} & \mathfrak{A}_{12} \\ \mathfrak{A}_{21} & \mathfrak{A}_{22} \end{bmatrix} \begin{bmatrix} V_1 \\ I_1 \end{bmatrix}$

*The hybrid parameters used here are similar but not equal to the D parameters used by Atabekov [56].

†These parameters are usually designated $ABCD$ parameters where $A = A_{11}$, $B = -A_{12}$, $C = A_{21}$ and $D = -A_{22}$ (see [5]).

‡These parameters are usually designated $\mathfrak{A}\,\mathfrak{B}\,\mathfrak{C}\,\mathfrak{D}$ parameters (see [5]).

Table 9.2. Port Terminations for Parameter Determination

Type	*Description*	*Diagram*	*Termination Impedance*
1	voltage source	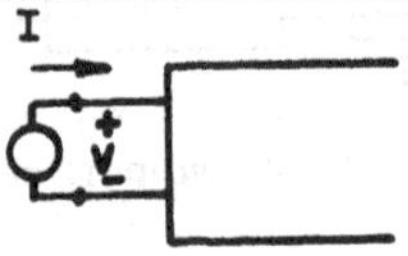	zero
2	current source	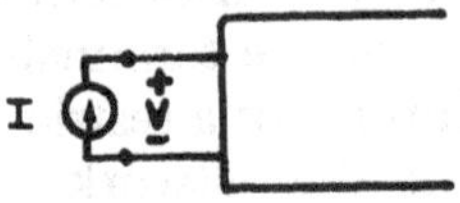	infinite
3	short circuit	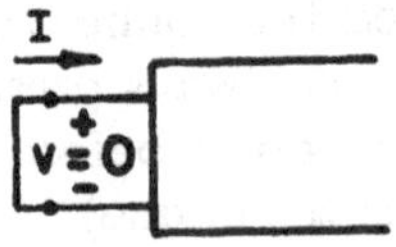	zero
4	open circuit	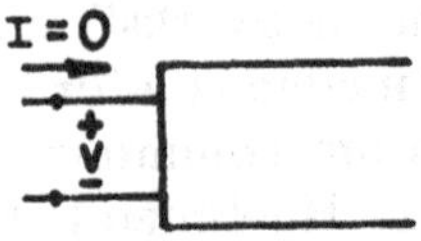	infinite

specified termination impedance. Thus if a port has a current source attached for one test, its impedance is infinite, and either a current source or an open circuit must be used at that port in subsequent tests. An example will illustrate the procedure.

Example 9.1

Compute the Z parameters for the network shown in Figure 9.2.

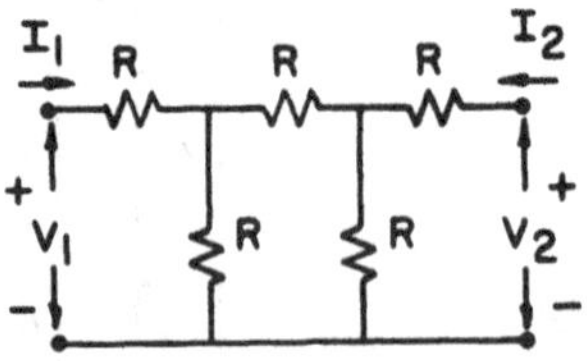

Fig. 9.2. Network for which Z parameters are required.

Solution

It is apparent from the way the Z parameters are defined in Table 9.1,

$$V_1 = z_{11} I_1 + z_{12} I_2, \qquad V_2 = z_{21} I_1 + z_{22} I_2$$

that

$$z_{11} = (V_1/I_1)_{I_2=0} \quad \text{and} \quad z_{21} = (V_2/I_1)_{I_2=0} \tag{9.1}$$

and that

$$z_{12} = (V_1/I_2)_{I_1=0} \quad \text{and} \quad z_{22} = (V_2/I_2)_{I_1=0} \tag{9.2}$$

Note that conditions (9.1) require that port 2 be open ($I_2 = 0$) and conditions (9.2) require that port 1 be open ($I_1 = 0$). This explains why the Z parameters are sometimes called the "open circuit Z parameters." That is, these parameters are determined with both ports terminated in an infinite impedance.

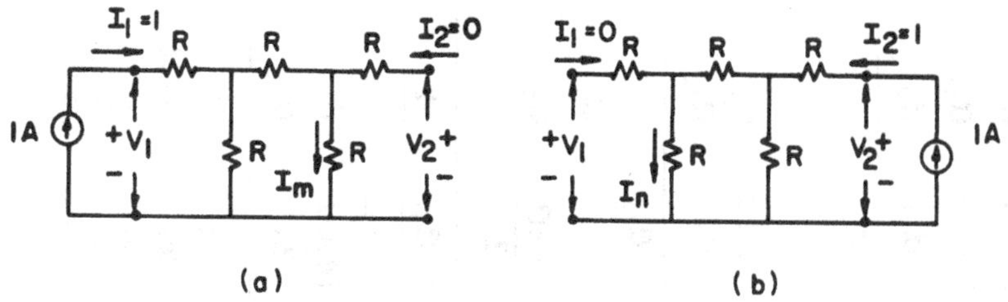

Fig. 9.3. Network test conditions for Z parameters: (a) network condition for (9.1), (b) network condition for (9.2).

To satisfy (9.1), we visualize the network as shown in Figure 9.3a, where we arbitrarily set $I_1 = 1$ ampere (or 1 pu). Then from (9.1) with $I_1 = 1$ the voltage V_1 is numerically equal to z_{11} or

$$z_{11} = V_1 = R + 2R^2/3R = 5R/3$$

and, similarly,

$$z_{21} = V_2 = RI_m = R(1/3)$$

In a similar way we use the network of Figure 9.3b to compute the second column of the Z matrix. Here we set $I_2 = 1$ arbitrarily and compute

$$z_{12} = V_1 = RI_n = R(1/3)$$

and

$$z_{22} = V_2 = R + 2R^2/3R = 5R/3$$

Thus we have the Z matrix

$$\mathbf{Z} = \begin{bmatrix} \frac{5R}{3} & \frac{R}{3} \\ \frac{R}{3} & \frac{5R}{3} \end{bmatrix}$$

which we observe is symmetric. This is always the case in a reciprocal network and will always be the case in a power system.

In complicated networks it is often helpful to sketch figures similar to Figure 9.3 as an aid in computation and as a visual check that the port impedance is the same for both tests.

All two-port network descriptions given in Table 9.1 show the relationship between pairs of variables. In each case there are two equations in two unknowns. Thus it is apparent that the equations can be rearranged to determine unique relationships among the various port parameter sets. These relationships are given in Table 9.3. Note that some sets of parameters may not exist for a given network. For example, if $z_{22} = 0$ (see Table 9.3, row H, column Z), the H matrix for that network is infinite in all its terms and does not exist. The use of Table 9.3 will be illustrated by means of an example.

Example 9.2

Compute the Y and H parameters from the Z parameters found in Example 9.1.

Table 9.3. Relationships among Two-Port Parameters

	Z		Y		H		G		A		$\mathcal{A}$	
Z	z_{11}	z_{12}	$\frac{y_{22}}{\det Y}$	$-\frac{y_{12}}{\det Y}$	$\frac{\det H}{h_{22}}$	$\frac{h_{12}}{h_{22}}$	$\frac{1}{g_{11}}$	$-\frac{g_{12}}{g_{11}}$	$\frac{A_{11}}{A_{21}}$	$-\frac{\det A}{A_{21}}$	$-\frac{\mathcal{A}_{22}}{\mathcal{A}_{21}}$	$\frac{1}{\mathcal{A}_{21}}$
	z_{21}	z_{22}	$-\frac{y_{21}}{\det Y}$	$\frac{y_{11}}{\det Y}$	$-\frac{h_{21}}{h_{22}}$	$\frac{1}{h_{22}}$	$\frac{g_{21}}{g_{11}}$	$\frac{\det G}{g_{11}}$	$\frac{1}{A_{21}}$	$-\frac{A_{22}}{A_{21}}$	$-\frac{\det \mathcal{A}}{\mathcal{A}_{21}}$	$\frac{\mathcal{A}_{11}}{\mathcal{A}_{21}}$
Y	$\frac{z_{22}}{\det Z}$	$-\frac{z_{12}}{\det Z}$	y_{11}	y_{12}	$\frac{1}{h_{11}}$	$-\frac{h_{12}}{h_{11}}$	$\frac{\det G}{g_{22}}$	$\frac{g_{12}}{g_{22}}$	$\frac{A_{22}}{A_{12}}$	$-\frac{\det A}{A_{12}}$	$-\frac{\mathcal{A}_{11}}{\mathcal{A}_{12}}$	$\frac{1}{\mathcal{A}_{12}}$
	$-\frac{z_{21}}{\det Z}$	$\frac{z_{11}}{\det Z}$	y_{21}	y_{22}	$\frac{h_{21}}{h_{11}}$	$\frac{\det H}{h_{11}}$	$-\frac{g_{21}}{g_{22}}$	$\frac{1}{g_{22}}$	$\frac{1}{A_{12}}$	$-\frac{A_{11}}{A_{12}}$	$-\frac{\det \mathcal{A}}{\mathcal{A}_{12}}$	$\frac{\mathcal{A}_{22}}{\mathcal{A}_{12}}$
H	$\frac{\det Z}{z_{22}}$	$\frac{z_{12}}{z_{22}}$	$\frac{1}{y_{11}}$	$-\frac{y_{12}}{y_{11}}$	h_{11}	h_{12}	$\frac{g_{22}}{\det G}$	$-\frac{g_{12}}{\det G}$	$\frac{A_{12}}{A_{22}}$	$\frac{\det A}{A_{22}}$	$-\frac{\mathcal{A}_{12}}{\mathcal{A}_{11}}$	$\frac{1}{\mathcal{A}_{11}}$
	$-\frac{z_{21}}{z_{22}}$	$\frac{1}{z_{22}}$	$\frac{y_{21}}{y_{11}}$	$\frac{\det Y}{y_{11}}$	h_{21}	h_{22}	$-\frac{g_{21}}{\det G}$	$\frac{g_{11}}{\det G}$	$\frac{1}{A_{22}}$	$-\frac{A_{21}}{A_{22}}$	$\frac{\det \mathcal{A}}{\mathcal{A}_{11}}$	$\frac{\mathcal{A}_{21}}{\mathcal{A}_{11}}$
G	$\frac{1}{z_{11}}$	$-\frac{z_{12}}{z_{11}}$	$\frac{\det Y}{y_{22}}$	$\frac{y_{12}}{y_{22}}$	$\frac{h_{22}}{\det H}$	$-\frac{h_{12}}{\det H}$	g_{11}	g_{12}	$\frac{A_{21}}{A_{11}}$	$\frac{\det A}{A_{11}}$	$\frac{\mathcal{A}_{21}}{\mathcal{A}_{22}}$	$\frac{1}{\mathcal{A}_{22}}$
	$\frac{z_{21}}{z_{11}}$	$\frac{\det Z}{z_{11}}$	$-\frac{y_{21}}{y_{22}}$	$\frac{1}{y_{22}}$	$-\frac{h_{21}}{\det H}$	$\frac{h_{11}}{\det H}$	g_{21}	g_{22}	$\frac{1}{A_{11}}$	$-\frac{A_{12}}{A_{11}}$	$\frac{\det \mathcal{A}}{\mathcal{A}_{22}}$	$\frac{\mathcal{A}_{12}}{\mathcal{A}_{22}}$
A	$\frac{z_{11}}{z_{21}}$	$-\frac{\det Z}{z_{21}}$	$-\frac{y_{22}}{y_{21}}$	$\frac{1}{y_{21}}$	$-\frac{\det H}{h_{21}}$	$\frac{h_{11}}{h_{21}}$	$\frac{1}{g_{21}}$	$-\frac{g_{22}}{g_{21}}$	A_{11}	A_{12}	$\frac{\mathcal{A}_{22}}{\det \mathcal{A}}$	$-\frac{\mathcal{A}_{12}}{\det \mathcal{A}}$
	$\frac{1}{z_{21}}$	$-\frac{z_{22}}{z_{21}}$	$-\frac{\det Y}{y_{21}}$	$\frac{y_{11}}{y_{21}}$	$-\frac{h_{22}}{h_{21}}$	$\frac{1}{h_{21}}$	$\frac{g_{11}}{g_{21}}$	$-\frac{\det G}{g_{21}}$	A_{21}	A_{22}	$-\frac{\mathcal{A}_{21}}{\det \mathcal{A}}$	$\frac{\mathcal{A}_{11}}{\det \mathcal{A}}$
$\mathcal{A}$	$\frac{z_{22}}{z_{12}}$	$-\frac{\det Z}{z_{12}}$	$-\frac{y_{11}}{y_{12}}$	$\frac{1}{y_{12}}$	$\frac{1}{h_{12}}$	$-\frac{h_{11}}{h_{12}}$	$-\frac{\det G}{g_{12}}$	$\frac{g_{22}}{g_{12}}$	$\frac{A_{22}}{\det A}$	$-\frac{A_{12}}{\det A}$	$\mathcal{A}_{11}$	$\mathcal{A}_{12}$
	$\frac{1}{z_{12}}$	$-\frac{z_{11}}{z_{12}}$	$-\frac{\det Y}{y_{12}}$	$\frac{y_{22}}{y_{12}}$	$\frac{h_{22}}{h_{12}}$	$-\frac{\det H}{h_{12}}$	$-\frac{g_{11}}{g_{12}}$	$\frac{1}{g_{12}}$	$-\frac{A_{21}}{\det A}$	$\frac{A_{11}}{\det A}$	$\mathcal{A}_{21}$	$\mathcal{A}_{22}$

Solution

From Example 9.1 we have

$$\mathbf{Z} = \begin{bmatrix} \dfrac{5R}{3} & \dfrac{R}{3} \\ \dfrac{R}{3} & \dfrac{5R}{3} \end{bmatrix}$$

Then we compute det **Z** = $(25R^2/9) - (R^2/9) = (24R^2/9)$ and from Table 9.3

$$\mathbf{Y} = \frac{1}{\det \mathbf{Z}} \begin{bmatrix} z_{22} & -z_{12} \\ -z_{21} & z_{11} \end{bmatrix} = \frac{9}{24R^2} \begin{bmatrix} \dfrac{5R}{3} & -\dfrac{R}{3} \\ -\dfrac{R}{3} & \dfrac{5R}{3} \end{bmatrix} = \begin{bmatrix} \dfrac{5}{8R} & -\dfrac{1}{8R} \\ -\dfrac{1}{8R} & \dfrac{5}{8R} \end{bmatrix}$$

We also compute

$$\mathbf{H} = \begin{bmatrix} \dfrac{\det \mathbf{Z}}{z_{22}} & \dfrac{z_{12}}{z_{22}} \\ -\dfrac{z_{21}}{z_{22}} & \dfrac{1}{z_{22}} \end{bmatrix} = \begin{bmatrix} \dfrac{8R}{5} & \dfrac{1}{5} \\ -\dfrac{1}{5} & \dfrac{3}{5R} \end{bmatrix}$$

which can be checked by computing again from the **Y** parameters.

Two-port networks with internal sources. As a final consideration in the description of two-port networks we consider the effect of internal sources. Internal sources can be of two kinds, dependent or independent. Dependent or controlled sources develop an output voltage or current which depends on some other network voltage or current. For example, the *d* axis voltage in a synchronous machine depends upon the *q* axis current. Controlled sources affect the matrix description of the two-port network by altering one or more of the elements. Usually if the matrix is initially symmetric, as with the **Z** or **Y** matrix of a passive network, the controlled source causes the matrix to be unsymmetric. This will be illustrated by means of an example.

Example 9.3

The two-port network shown in Figure 9.4 is identical with the network of Example 9.1 except that a controlled current source $I = kV_1$ has been added. Find the **Z** parameters of this two-port network.

Solution

We solve the network by node voltages, applying test conditions as before. In doing so, note that any impedance in series with a current source is negligible com-

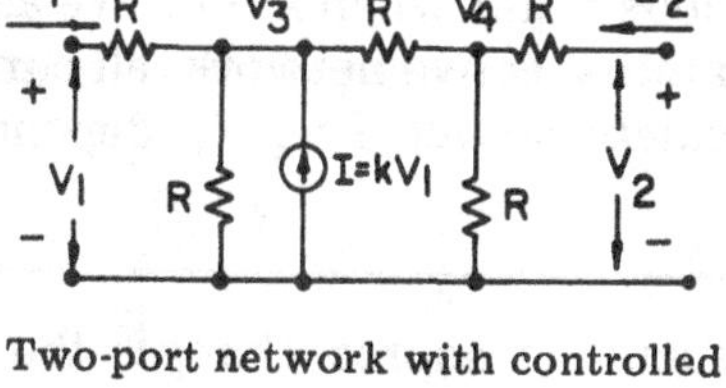

Fig. 9.4. Two-port network with controlled source.

pared to the infinite internal impedance of the source. Thus with $I_1 = 1$ ampere, $I_2 = 0$ we have the nodal equations

$$\frac{2}{R} V_3 - \frac{1}{R} V_4 = 1 + kV_1, \qquad -\frac{1}{R} V_3 + \frac{2}{R} V_4 = 0$$

and we note that $V_1 = R + V_3$, $V_2 = V_4$. Solving, we find that

$$V_3 = \frac{2R\,(1 + kR)}{3 - 2kR}$$

and

$$V_4 = \frac{V_3}{2} = \frac{R\,(1 + kR)}{3 - 2kR}$$

Then

$$z_{11} = V_1 = R + V_3 = \frac{5R}{3 - 2kR}$$

and

$$z_{21} = V_2 = V_4 = \frac{R\,(1 + kR)}{3 - 2kR}$$

Similarly, with $I_2 = 1$, $I_1 = 0$ we write

$$\frac{2}{R} V_3 - \frac{1}{R} V_4 = kV_1 = kV_3, \qquad -\frac{1}{R} V_3 + \frac{2}{R} V_4 = 1$$

from which we find that

$$V_3 = \frac{R}{3 - 2kR}, \qquad V_4 = \frac{R\,(2 - kR)}{3 - 2kR}$$

Then

$$z_{12} = V_1 = V_3 = \frac{R}{3 - 2kR}$$

and

$$z_{22} = V_2 = V_4 + R = \frac{R\,(5 - 3kR)}{3 - 2kR}$$

Finally then, we write

$$\mathbf{Z} = \frac{1}{3 - 2kR} \begin{bmatrix} 5R & R \\ R\,(1 + kR) & R\,(5 - 3kR) \end{bmatrix}$$

which reduces to the previous result when $k = 0$. We also note that if the network is not connected to any external active network, all currents and voltages are zero. This is because the dependent source $I = kV_1$ depends upon the existence of a voltage V_1.

The presence of a dependent source changes the two-port network description. As seen in Example 9.3, the source parameter k is present in all matrix elements. We now consider an *independent* source located within the two-port network. Such a source will hold its output to a specified value irrespective of external connections to the two-port network. Indeed, if the two-port network is completely isolated, the internal independent source will cause measurable open circuit voltages or short circuit currents to appear at the two ports. Thus for independent sources within the network we write in the case of **Z** parameters,

$$\begin{bmatrix} V_1 \\ V_2 \end{bmatrix} = \begin{bmatrix} z_{11} & z_{12} \\ z_{21} & z_{22} \end{bmatrix} \begin{bmatrix} I_1 \\ I_2 \end{bmatrix} + \begin{bmatrix} V_{z1} \\ V_{z2} \end{bmatrix} \tag{9.3}$$

and we interpret the equation as follows.

1. All port voltages and currents are defined as before.
2. When $I_1 = I_2 = 0$, the port voltages are

$$\begin{bmatrix} V_1 \\ V_2 \end{bmatrix} = \begin{bmatrix} V_{z1} \\ V_{z2} \end{bmatrix} \tag{9.4}$$

   and these open circuit voltages are due to the internal source or sources.
3. We may determine the **Z** matrix of (9.3) by removing the internal independent sources where we are careful to (a) short out all independent voltage sources and (b) open all independent current sources. Having done this, equation (9.3) reduces to

$$\begin{bmatrix} V_1 \\ V_2 \end{bmatrix} = \begin{bmatrix} z_{11} & z_{12} \\ z_{21} & z_{22} \end{bmatrix} \begin{bmatrix} I_1 \\ I_2 \end{bmatrix} \tag{9.5}$$

In a similar way we can establish equations similar to (9.3) for networks described by any other set of two-port parameters. In general we must add to each equation a source term of the same *dimension* as that equation, and we determine the size of the source term by testing at the appropriate terminal with all ports properly terminated. The "proper" termination depends upon the equations but will be either short or open circuits. Mathematically, we write

$$\mathbf{U} = \mathbf{P}\,\mathbf{W} + \mathbf{U}_s \tag{9.6}$$

where

$\mathbf{U}$, $\mathbf{W}$ are vectors containing port voltages and/or currents
$\mathbf{P}$ is the 2×2 matrix of two-port parameters
$\mathbf{U}_s$ is the independent source term, each element having the same dimension as $\mathbf{U}$

Then we find $\mathbf{U}_s$ by the test

$$\mathbf{U} = \mathbf{U}_s \quad \text{when} \quad \mathbf{W} = 0 \tag{9.7}$$

and we compute $\mathbf{P}$ by the test

$$\mathbf{U} = \mathbf{P}\,\mathbf{W} \quad \text{when} \quad \mathbf{U}_s = 0 \tag{9.8}$$

In (9.8) we set $\mathbf{U}_s = 0$ by properly removing the internal sources. We note also that test (9.7) will ordinarily be impossible for transmission parameters (**A** or $\mathcal{A}$) since it is not usually possible to set $I_2 = V_2 = 0$ simultaneously. Thus a two-port description including the effect of an independent source is limited to the **Z**, **Y**,

Table 9.4. Equations of Two-Port Networks with Independent Sources

Designation	*Equation*	*Test for Source Term*
Z	$\begin{bmatrix} V_1 \\ V_2 \end{bmatrix} = \begin{bmatrix} z_{11} & z_{12} \\ z_{21} & z_{22} \end{bmatrix} \begin{bmatrix} I_1 \\ I_2 \end{bmatrix} + \begin{bmatrix} V_{z1} \\ V_{z2} \end{bmatrix}$	
Y	$\begin{bmatrix} I_1 \\ I_2 \end{bmatrix} = \begin{bmatrix} y_{11} & y_{12} \\ y_{21} & y_{22} \end{bmatrix} \begin{bmatrix} V_1 \\ V_2 \end{bmatrix} + \begin{bmatrix} I_{y1} \\ I_{y2} \end{bmatrix}$	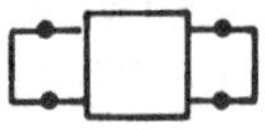
H	$\begin{bmatrix} V_1 \\ I_2 \end{bmatrix} = \begin{bmatrix} h_{11} & h_{12} \\ h_{21} & h_{22} \end{bmatrix} \begin{bmatrix} I_1 \\ V_2 \end{bmatrix} + \begin{bmatrix} V_{h1} \\ I_{h2} \end{bmatrix}$	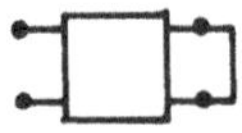
G	$\begin{bmatrix} I_1 \\ V_2 \end{bmatrix} = \begin{bmatrix} g_{11} & g_{12} \\ g_{21} & g_{22} \end{bmatrix} \begin{bmatrix} V_1 \\ I_2 \end{bmatrix} + \begin{bmatrix} I_{g1} \\ V_{g2} \end{bmatrix}$	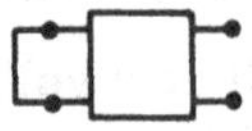

H, and G descriptions. These are tabulated in Table 9.4. The computation of the independent source term is illustrated by an example.

Example 9.4

Consider the current source in Figure 9.4 to be an independent source I_s and compute the independent source term associated with the Z parameter description of the two-port network.

Solution

As indicated in Table 9.4, to determine V_{z1} and V_{z2} we test the two-port network with $I_1 = I_2 = 0$, or with both ports open. Then we compute V_1 and V_2, which result from the presence of current source I_s. By inspection of Figure 9.4 it is apparent that the source current sees two paths, R (to the left) and $2R$ (to the right). This divides the current inversely as the impedance or, with $I_1 = I_2 = 0$,

$$I_L = (2/3)\, I_s, \qquad I_R = (1/3)\, I_s$$

and we compute, with $I_1 = 0$ and $I_2 = 0$ simultaneously,

$$\begin{bmatrix} V_{z1} \\ V_{z2} \end{bmatrix} = \begin{bmatrix} V_1 \\ V_2 \end{bmatrix} = \begin{bmatrix} RI_L \\ RI_R \end{bmatrix} = \begin{bmatrix} (2/3)\, RI_s \\ (1/3)\, RI_s \end{bmatrix}$$

Thus the complete description for the two-port network with independent current source I_s is

$$\begin{bmatrix} V_1 \\ V_2 \end{bmatrix} = \begin{bmatrix} \frac{5R}{3} & \frac{R}{3} \\ \frac{R}{3} & \frac{5R}{3} \end{bmatrix} \begin{bmatrix} I_1 \\ I_2 \end{bmatrix} + \begin{bmatrix} \frac{2}{3} RI_s \\ \frac{1}{3} RI_s \end{bmatrix}$$

The two-port parameters are related to one another in a definite way as noted in Table 9.3. It is not surprising, therefore, that the independent source terms of Table 9.4 are also related. If we write the Z and Y parameter equations in matrix form, we have

$$\mathbf{V} = \mathbf{Z}\mathbf{I} + \mathbf{V}_z \tag{9.9}$$

$$\mathbf{I} = \mathbf{Y}\mathbf{V} + \mathbf{I}_y \tag{9.10}$$

For convenience we also write

$$\mathbf{M} = \mathbf{H}\mathbf{N} + \mathbf{M}_h \tag{9.11}$$

and

$$\mathbf{N} = \mathbf{G}\mathbf{M} + \mathbf{N}_g \tag{9.12}$$

where we arbitrarily define

$$\mathbf{M} = \begin{bmatrix} V_1 \\ I_2 \end{bmatrix}, \quad \mathbf{M}_h = \begin{bmatrix} V_{h1} \\ I_{h2} \end{bmatrix} \tag{9.13}$$

and

$$\mathbf{N} = \begin{bmatrix} I_1 \\ V_2 \end{bmatrix}, \quad \mathbf{N}_g = \begin{bmatrix} I_{g1} \\ V_{g2} \end{bmatrix} \tag{9.14}$$

Then straightforward algebraic manipulation reveals the relationships among the source parameters. For example, we write

$$\mathbf{V}_z = \mathbf{U}\mathbf{V}_z = -\mathbf{Y}^{-1}\mathbf{I}_y = \begin{bmatrix} 1 & -\dfrac{h_{12}}{h_{22}} \\ 0 & -\dfrac{1}{h_{22}} \end{bmatrix} \mathbf{M}_h = \begin{bmatrix} -\dfrac{1}{g_{11}} & 0 \\ -\dfrac{g_{21}}{g_{11}} & 1 \end{bmatrix} \mathbf{N}_g \tag{9.15}$$

Table 9.5. Relationship among Source Parameters*

	$\mathbf{V}_z$		$\mathbf{I}_y$		$\mathbf{M}_h$		$\mathbf{N}_g$	
$V_z = \begin{bmatrix} V_{z1} \\ V_{z2} \end{bmatrix}$	1	0	$-\dfrac{y_{22}}{\det \mathbf{Y}}$	$\dfrac{y_{12}}{\det \mathbf{Y}}$	1	$-\dfrac{h_{12}}{h_{22}}$	$-\dfrac{1}{g_{11}}$	0
	0	1	$\dfrac{y_{21}}{\det \mathbf{Y}}$	$-\dfrac{y_{11}}{\det \mathbf{Y}}$	0	$-\dfrac{1}{h_{22}}$	$-\dfrac{g_{21}}{g_{11}}$	1
$I_y = \begin{bmatrix} I_{y1} \\ I_{y2} \end{bmatrix}$	$-\dfrac{z_{22}}{\det \mathbf{Z}}$	$\dfrac{z_{12}}{\det \mathbf{Z}}$	1	0	$-\dfrac{1}{h_{11}}$	0	1	$-\dfrac{g_{12}}{g_{22}}$
	$\dfrac{z_{21}}{\det \mathbf{Z}}$	$-\dfrac{z_{11}}{\det \mathbf{Z}}$	0	1	$-\dfrac{h_{21}}{h_{11}}$	1	0	$-\dfrac{1}{g_{22}}$
$M_h = \begin{bmatrix} V_{h1} \\ I_{h2} \end{bmatrix}$	1	$-\dfrac{z_{12}}{z_{22}}$	$-\dfrac{1}{y_{11}}$	0	1	0	$-\dfrac{g_{22}}{\det \mathbf{G}}$	$\dfrac{g_{12}}{\det \mathbf{G}}$
	0	$-\dfrac{1}{z_{22}}$	$-\dfrac{y_{21}}{y_{11}}$	1	0	1	$\dfrac{g_{21}}{\det \mathbf{G}}$	$-\dfrac{g_{11}}{\det \mathbf{G}}$
$N_g = \begin{bmatrix} I_{g1} \\ V_{g2} \end{bmatrix}$	$-\dfrac{1}{z_{11}}$	0	1	$-\dfrac{y_{12}}{y_{22}}$	$-\dfrac{h_{22}}{\det \mathbf{H}}$	$\dfrac{h_{12}}{\det \mathbf{H}}$	1	0
	$-\dfrac{z_{21}}{z_{11}}$	1	0	$-\dfrac{1}{y_{22}}$	$\dfrac{h_{21}}{\det \mathbf{H}}$	$-\dfrac{h_{11}}{\det \mathbf{H}}$	0	1

*(Vector in first column) = (matrix from table) × (vector above table entry).

where U = the unit matrix. Exploring all possible combinations, we may develop the matrices of Table 9.5. We illustrate the use of Table 9.5 by means of an example.

Example 9.5

Use the value of $\mathbf{V}_z$ computed in Example 9.4 and the Z parameters found in Example 9.1 to determine the H parameter equation for the active network of Example 9.4.

Solution

From Example 9.4 we have

$$\mathbf{V}_z = \frac{RI_s}{3}\begin{bmatrix} 2 \\ 1 \end{bmatrix}$$

and from Example 9.1

$$\mathbf{Z} = \frac{R}{3}\begin{bmatrix} 5 & 1 \\ 1 & 5 \end{bmatrix}$$

We desire the H parameter description so, using Table 9.3, we have

$$\mathbf{H} = \frac{1}{z_{22}}\begin{bmatrix} \det \mathbf{Z} & z_{12} \\ -z_{21} & 1 \end{bmatrix}$$

and since det $\mathbf{Z} = (25R^2/9) - (R^2/9) = 24R^2/9$, we compute

$$\mathbf{H} = \begin{bmatrix} \frac{8R}{5} & \frac{1}{5} \\ -\frac{1}{5} & \frac{3}{5R} \end{bmatrix}$$

Also from Table 9.5 we have

$$\mathbf{M}_h = \begin{bmatrix} 1 & -\frac{z_{12}}{z_{22}} \\ 0 & -\frac{1}{z_{22}} \end{bmatrix} \mathbf{V}_z$$

$$= \begin{bmatrix} 1 & -\frac{1}{5} \\ 0 & -\frac{3}{5R} \end{bmatrix}\begin{bmatrix} \frac{2}{3}RI_s \\ \frac{1}{3}RI_s \end{bmatrix} = \begin{bmatrix} \frac{3}{5}RI_s \\ -\frac{I_s}{5} \end{bmatrix}$$

These results may be easily verified by computing the H equation directly. The result is

$$\begin{bmatrix} V_1 \\ I_2 \end{bmatrix} = \begin{bmatrix} \frac{8R}{5} & \frac{1}{5} \\ -\frac{1}{5} & \frac{3}{5R} \end{bmatrix}\begin{bmatrix} I_1 \\ V_2 \end{bmatrix} + \begin{bmatrix} \frac{3}{5}RI_s \\ -\frac{I_s}{5} \end{bmatrix}$$

9.2 Interconnection of Two-Port Networks

One of the principal reasons for describing two-port networks so carefully is to simplify the problem of interconnecting two-port networks to form larger, more complicated systems. Furthermore, in doing this we preserve a well-defined notation in the interconnected system. There are several ways two-port networks may be interconnected. These will be examined briefly here, but the interested reader should consult other references such as Huelsman [5] on the subject.

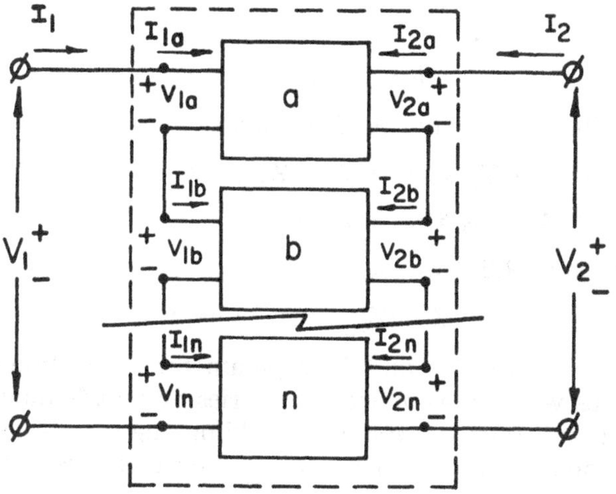

Fig. 9.5. Series connection of n two-port networks.

Series connection. Consider a group of n two-port networks, labeled a, $b, \ldots, n$, which are connected in *series* as shown in Figure 9.5 where we define

$$\mathbf{V}_k = \begin{bmatrix} V_{1k} \\ V_{2k} \end{bmatrix} \qquad k = a, b, \ldots, n \tag{9.16}$$

$$\mathbf{I}_k = \begin{bmatrix} I_{1k} \\ I_{2k} \end{bmatrix} \qquad k = a, b, \ldots, n \tag{9.17}$$

and

$$\mathbf{Z}_k = \begin{bmatrix} z_{11k} & z_{12k} \\ z_{21k} & z_{22k} \end{bmatrix} \qquad k = a, b, \ldots, n \tag{9.18}$$

Then we write for each two-port network

$$\mathbf{V}_k = \mathbf{Z}_k \mathbf{I}_k \qquad k = a, b, \ldots, n \tag{9.19}$$

We now seek a description of the interconnected group of two-port networks designated in Figure 9.5 by the terminal symbol ϕ and defined by

$$\mathbf{V} = \mathbf{Z}\,\mathbf{I} \tag{9.20}$$

where

$$\mathbf{V} = \begin{bmatrix} V_1 \\ V_2 \end{bmatrix}, \qquad \mathbf{I} = \begin{bmatrix} I_1 \\ I_2 \end{bmatrix} \tag{9.21}$$

and where

$$\mathbf{Z} = \begin{bmatrix} z_{11} & z_{12} \\ z_{21} & z_{22} \end{bmatrix} \tag{9.22}$$

describes the relationship between these "external" quantities. Since the networks are in series, we observe that

$$\mathbf{V} = \mathbf{V}_a + \mathbf{V}_b + \cdots + \mathbf{V}_n \tag{9.23}$$

and

$$\mathbf{I} = \mathbf{I}_a = \mathbf{I}_b = \cdots = \mathbf{I}_n \tag{9.24}$$

From (9.19), (9.20), (9.23), and (9.24) we write

$$\begin{aligned} \mathbf{V} = \mathbf{V}_a + \mathbf{V}_b + \cdots + \mathbf{V}_n &= \mathbf{Z}_a\mathbf{I}_a + \mathbf{Z}_b\mathbf{I}_b + \cdots + \mathbf{Z}_n\mathbf{I}_n \\ &= (\mathbf{Z}_a + \mathbf{Z}_b + \cdots + \mathbf{Z}_n)\,\mathbf{I} \end{aligned} \tag{9.25}$$

Then from (9.20) and (9.25) we see that

$$\mathbf{Z} = (\mathbf{Z}_a + \mathbf{Z}_b + \cdots + \mathbf{Z}_n) \tag{9.26}$$

This means that we can easily find the Z parameter description of any number of two-port networks which are connected in series by first finding the Z parameters of the individual two-port networks and adding these together as in (9.26). It must be understood, however, that port descriptions must be preserved in the interconnection.

Parallel connection. Consider a group of n two-port networks, labeled $a, b, \ldots, n$, connected in parallel as shown in Figure 9.6. We again define the port voltage and current vectors as in (9.16) and (9.17) respectively. For the parallel connection, however, we do not use the Z parameter description but instead define

$$\mathbf{Y}_k = \begin{bmatrix} y_{11k} & y_{12k} \\ y_{21k} & y_{22k} \end{bmatrix} \qquad k = a, b, \ldots, n \tag{9.27}$$

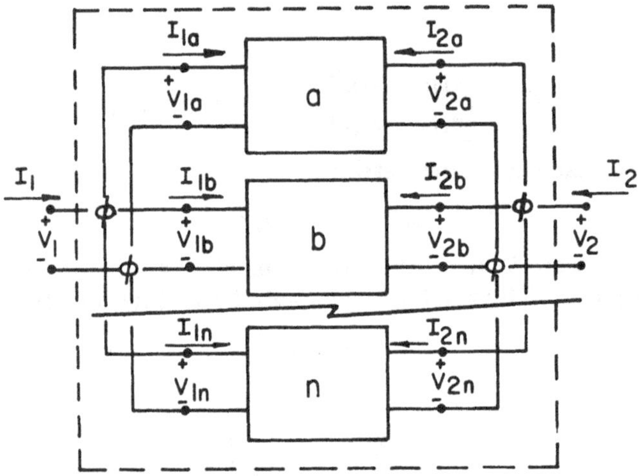

Fig. 9.6. Parallel connection of two-port networks.

such that we may write for any two-port network,

$$\mathbf{I}_k = \mathbf{Y}_k \mathbf{V}_k \qquad k = a, b, \ldots, n \tag{9.28}$$

If we again define voltage and current vectors as (9.21) for the ϕ-ϕ network, we may write

$$\mathbf{I} = \mathbf{Y}\,\mathbf{V} \tag{9.29}$$

where

$$\mathbf{Y} = \begin{bmatrix} y_{11} & y_{12} \\ y_{21} & y_{22} \end{bmatrix} \tag{9.30}$$

describes the relationship between **V** and **I**. For the parallel connection we observe that

$$\mathbf{I} = \mathbf{I}_a + \mathbf{I}_b + \cdots + \mathbf{I}_n \tag{9.31}$$

and

$$\mathbf{V} = \mathbf{V}_a = \mathbf{V}_b = \cdots = \mathbf{V}_n \tag{9.32}$$

Combining these last two equations with (9.28), we write

$$\begin{aligned} \mathbf{I} &= \mathbf{I}_a + \mathbf{I}_b + \cdots + \mathbf{I}_n = \mathbf{Y}_a\mathbf{V}_a + \mathbf{Y}_b\mathbf{V}_b + \cdots + \mathbf{Y}_n\mathbf{V}_n \\ &= (\mathbf{Y}_a + \mathbf{Y}_b + \cdots + \mathbf{Y}_n)\,\mathbf{V} \end{aligned} \tag{9.33}$$

Comparing this result with (9.29), we see that

$$\mathbf{Y} = \mathbf{Y}_a + \mathbf{Y}_b + \cdots + \mathbf{Y}_n \tag{9.34}$$

and the two-port description for this connection is easily found by adding the **Y** parameters of the individual networks. We again require that two-port descriptions are preserved in the interconnection. Note that it is *not* convenient to find a **Z** parameter description for the parallel connection. It is preferable to first convert the **Z** parameters of the individual two-port networks to **Y** parameters and add these as in (9.34).

Hybrid connection. Consider a group of n two-port networks, labeled $a, b, \ldots, n$, connected in series at one port and in parallel at the other as shown in Figure 9.7. We refer to this as the *hybrid* connection. Here, instead of the voltage and current parameters of (9.16) and (9.17), we define the port hybrid voltage-current vectors

$$\mathbf{M}_k = \begin{bmatrix} V_{1k} \\ I_{2k} \end{bmatrix} \qquad k = a, b, \ldots, n \tag{9.35}$$

$$\mathbf{N}_k = \begin{bmatrix} I_{1k} \\ V_{2k} \end{bmatrix} \qquad k = a, b, \ldots, n \tag{9.36}$$

and the hybrid matrix

$$\mathbf{H}_k = \begin{bmatrix} h_{11k} & h_{12k} \\ h_{21k} & h_{22k} \end{bmatrix} \qquad k = a, b, \ldots, n \tag{9.37}$$

Then from Table 9.1 we write

$$\mathbf{M}_k = \mathbf{H}_k \mathbf{N}_k \qquad k = a, b, \ldots, n \tag{9.38}$$

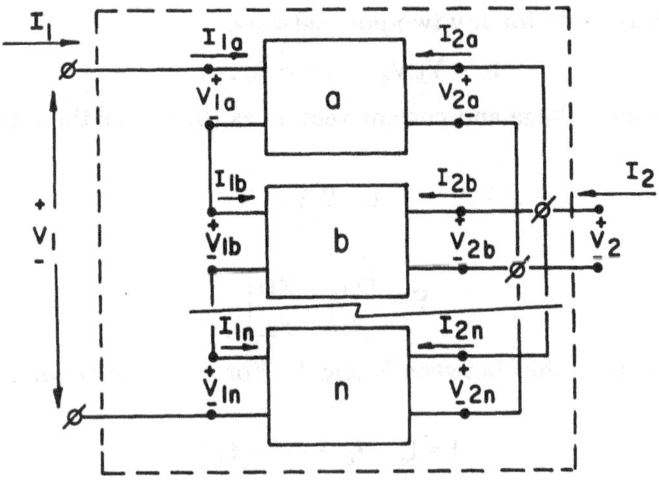

Fig. 9.7. Hybrid (series-parallel) connection of two-port networks.

We now describe the interconnected system according to the equation

$$\mathbf{M} = \mathbf{H}\,\mathbf{N} \tag{9.39}$$

where M and N are defined according to "external" quantities,

$$\mathbf{M} = \begin{bmatrix} V_1 \\ I_2 \end{bmatrix} \tag{9.40}$$

and

$$\mathbf{N} = \begin{bmatrix} I_1 \\ V_2 \end{bmatrix} \tag{9.41}$$

and where

$$\mathbf{H} = \begin{bmatrix} h_{11} & h_{12} \\ h_{21} & h_{22} \end{bmatrix} \tag{9.42}$$

is the hybrid parameter matrix for the interconnection.

From Figure 9.7 we observe that

$$\mathbf{M} = \mathbf{M}_a + \mathbf{M}_b + \cdots + \mathbf{M}_n \tag{9.43}$$

and

$$\mathbf{N} = \mathbf{N}_a = \mathbf{N}_b = \cdots = \mathbf{N}_n \tag{9.44}$$

Combining these observed results with (9.38) we write

$$\begin{aligned} \mathbf{M} = \mathbf{M}_a + \mathbf{M}_b + \cdots + \mathbf{M}_n &= \mathbf{H}_a\mathbf{N}_a + \mathbf{H}_b\mathbf{N}_b + \cdots + \mathbf{H}_n\mathbf{N}_n \\ &= (\mathbf{H}_a + \mathbf{H}_b + \cdots + \mathbf{H}_n)\,\mathbf{N} \end{aligned} \tag{9.45}$$

Then comparing (9.45) and (9.39), we have

$$\mathbf{H} = \mathbf{H}_a + \mathbf{H}_b + \cdots + \mathbf{H}_n \tag{9.46}$$

Again we require that port descriptions be preserved.

Note that a similar result could be obtained by reversing the series-parallel connection of Figure 9.7 to form a parallel-series connection. In such a case the G parameters will be found to add and a description corresponding to N = G M derived. This description is redundant in a sense since we are at liberty to number the ports any way we choose, and if we would exchange the 1 and 2 subscripts, we would have a parallel-series (G matrix) hybrid. Thus, the hybrid connection of Figure 9.7 will suffice for either situation requiring a series and a parallel connection.

Cascade connection. Finally, we review the cascade connection of two-port networks as shown in Figure 9.8 where we define

$$\begin{bmatrix} V_{1k} \\ I_{1k} \end{bmatrix} = \begin{bmatrix} A_{11k} & A_{12k} \\ A_{21k} & A_{22k} \end{bmatrix} \begin{bmatrix} V_{2k} \\ I_{2k} \end{bmatrix} \qquad k = a, b, \ldots, n \tag{9.47}$$

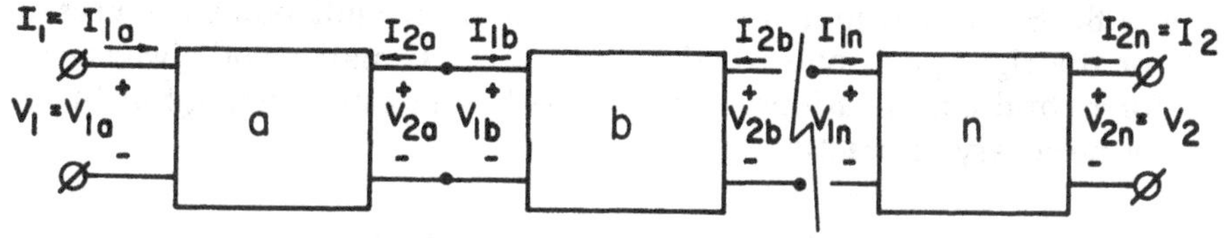

Fig. 9.8. Cascade connection of two-port networks.

We seek a description

$$\begin{bmatrix} V_1 \\ I_1 \end{bmatrix} = \begin{bmatrix} A_{11} & A_{12} \\ A_{21} & A_{22} \end{bmatrix} \begin{bmatrix} V_2 \\ I_2 \end{bmatrix} \tag{9.48}$$

which describes the connection. If we observe the constraints of the network interconnection, we write

$$\begin{bmatrix} V_1 \\ I_1 \end{bmatrix} = \begin{bmatrix} V_{1a} \\ I_{1a} \end{bmatrix}, \begin{bmatrix} V_{2a} \\ -I_{2a} \end{bmatrix} = \begin{bmatrix} V_{1b} \\ I_{1b} \end{bmatrix}, \ldots \begin{bmatrix} V_{2n} \\ I_{2n} \end{bmatrix} = \begin{bmatrix} V_2 \\ I_2 \end{bmatrix} \tag{9.49}$$

Combining with (9.47), we have

$$\begin{aligned} \begin{bmatrix} V_1 \\ I_1 \end{bmatrix} &= \begin{bmatrix} A_{11a} & -A_{12a} \\ A_{21a} & -A_{22a} \end{bmatrix} \begin{bmatrix} V_{2a} \\ -I_{2a} \end{bmatrix} = \begin{bmatrix} A_{11a} & -A_{12a} \\ A_{21a} & -A_{22a} \end{bmatrix} \begin{bmatrix} A_{11b} & -A_{12b} \\ A_{21b} & -A_{22b} \end{bmatrix} \begin{bmatrix} V_{2b} \\ -I_{2b} \end{bmatrix} \\ &= \begin{bmatrix} A_{11a} & -A_{12a} \\ A_{21a} & -A_{22a} \end{bmatrix} \begin{bmatrix} A_{11b} & -A_{12b} \\ A_{21b} & -A_{22b} \end{bmatrix} \cdots \begin{bmatrix} A_{11n} & -A_{12n} \\ A_{21n} & -A_{22n} \end{bmatrix} \begin{bmatrix} V_2 \\ -I_2 \end{bmatrix} \end{aligned} \tag{9.50}$$

Comparing (9.50) with (9.48), we see that the composite A matrix is the product of the individual A matrices with the second column of each changed in sign. These negative signs arise because the currents between two-port networks flow in opposite directions. Using the *ABCD* parameter description, with I_1 directed *in* and I_2 *out* of the two-port network, the cascade connection is simply a matrix product with no sign changes. We define the currents as entering the network to conform with established usage [5, 63].

9.3 Simultaneous Fault Connection of Sequence Networks

In Chapter 8 we developed the interconnection of sequence networks for all common shunt and series faults. These results, summarized in Figure 8.9, show the sequence networks as *one-port networks*, with the current always entering the network at K and leaving at F. We further defined K to be the zero potential bus N for shunt faults and the current-receiving terminal M for series faults. (See Figures 8.3, 8.4, 8.6, and 8.7.) The point F is always defined to be the "fault point." We note also that the positive sequence network is an active one-port network since it contains independent sources, but the negative and zero sequence networks are passive one-port networks.

Suppose that two faults occur simultaneously on a system. Using the two-port network theory developed in sections 9.1 and 9.2, we can represent the two faults by finding the two-port parameters of the positive, negative, and zero sequence networks and then combining them in a way which will properly represent the common fault configurations of Figure 8.9 [63]. Considering the four common types of faults which can occur at two different locations, there are a total of 16 possible problems which require our attention. These are summarized in Table 9.6. Since it is immaterial whether a given fault point is represented as the left or the right port, one could eliminate the **G** matrix and always use the **H** matrix for hybrid combinations. We will do this in our analysis, but this restriction is not necessary in practice.

Table 9.6. Two-Port Fault Combinations

			Right Port (primed)			
			Shunt fault ($K = N$)		*Series fault* ($K = M$)	
	Fault Type	*Kind*	SLG *(series)*	2LG *(parallel)*	1LO *(parallel)*	2LO *(series)*
Left Port (unprimed)	*Shunt fault* ($K = N$)	SLG *(series)*	$\mathbf{Z}_{NN}$	$\mathbf{H}_{NN}$	$\mathbf{H}_{NM}$	$\mathbf{Z}_{NM}$
		2LG *(parallel)*	$\mathbf{G}_{NN}$	$\mathbf{Y}_{NN}$	$\mathbf{Y}_{NM}$	$\mathbf{G}_{NM}$
	Series fault ($K = M$)	1LO *(parallel)*	$\mathbf{G}_{MN}$	$\mathbf{Y}_{MN}$	$\mathbf{Y}_{MM}$	$\mathbf{G}_{MM}$
		2LO *(series)*	$\mathbf{Z}_{MN}$	$\mathbf{H}_{MN}$	$\mathbf{H}_{MM}$	$\mathbf{Z}_{MM}$

Reference to Figure 8.9, which shows the various port connections, reveals the several pertinent facts concerning fault analysis. First, note that each sequence network is always terminated in an ideal transformer so that port currents and voltages are preserved for *any* network interconnection. This means that the two-port network theory of sections 9.1 and 9.2 will always apply. Note also that the added impedances Z and Z_g, used to represent impedance at the fault, are considered here to be a part of the network and must be added to the usual Thevenin equivalent sequence impedance. We also recognize that for simultaneous faults the sequence networks will have two fault points and two-port networks will be defined for each sequence network, with points F and K being defined according to the fault location and type.

The simultaneous fault problems to be solved can be generalized by three forms: (1) a series-series fault, (2) a shunt-shunt fault, and (3) a series-shunt fault, where the terms "series" and "shunt" refer to the type of network interconnection shown in Figure 8.9. If we view the sequence network interconnec-

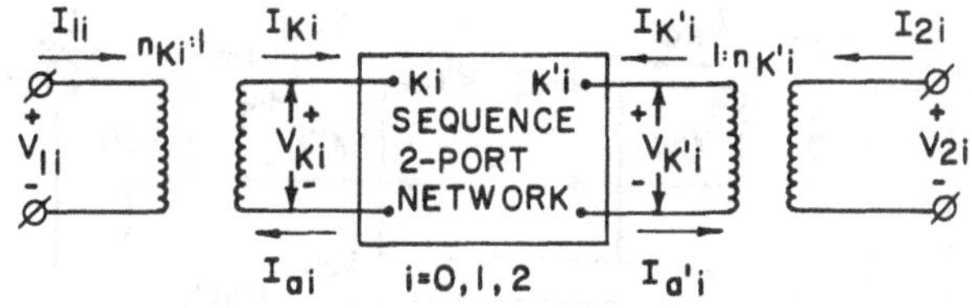

Fig. 9.9. Two-port network representation for simultaneous faults at F and F'.

tion from the external ϕ-ϕ connections, the network is shown in Figure 9.9 where the network voltages and currents are defined for simultaneous faults at F and F'. (Note that there should be no confusion between I_1 and I_2 as defined in Figure 9.9 and the sequence currents I_{a1} and I_{a2}). This figure will be used to define the current directions and voltage polarities to be considered in subsequent development. The transformations shown are composites of the actual transformers or phase shifters required for various faults. Since these phase shifters are ideal, we define the transformation ratios

$$n_{Ki} = V_{1i}/V_{Ki} = I_{1i}/I_{Ki}, \qquad n_{K'i} = V_{2i}/V_{K'i} = I_{2i}/I_{K'i} \tag{9.51}$$

These quantities will always be subscripted with i equal to 0, 1, or 2 for the various sequences and will take a value given in Figure 8.9, for a particular fault type and symmetry.

9.4 Series-Series Connection (Z-Type Faults)

A series-series connection of two-port sequence networks is required to represent (see Table 9.6):

1. Simultaneous SLG faults at F and F' ($\mathbf{Z}_{NN}$).
2. A SLG fault at F and two lines open at F' ($\mathbf{Z}_{NM}$).
3. Two lines open at F and a SLG fault at F' ($\mathbf{Z}_{MN}$).
4. Two lines open at F and two lines open at F' ($\mathbf{Z}_{MM}$).

The sequence network connection is shown in Figure 9.10. For the positive sequence network we write

$$\begin{bmatrix} V_{K1} \\ V_{K'1} \end{bmatrix} = \begin{bmatrix} z_{11(1)} & z_{12(1)} \\ z_{21(1)} & z_{22(1)} \end{bmatrix} \begin{bmatrix} I_{K1} \\ I_{K'1} \end{bmatrix} + \begin{bmatrix} V_{z1} \\ V_{z2} \end{bmatrix} \tag{9.52}$$

where

$$\begin{bmatrix} V_{z1} \\ V_{z2} \end{bmatrix}$$

is the independent source term viewed from the $K1$ - $K'1$ terminals (inside the transformers). But from (9.51) with $i = 1$

$$n_{K1} = V_{11}/V_{K1} = I_{11}/I_{K1}, \qquad n_{K'1} = V_{21}/V_{K'1} = I_{21}/I_{K'1} \tag{9.53}$$

Premultiplying (9.52) by

$$\begin{bmatrix} n_{K1} & 0 \\ 0 & n_{K'1} \end{bmatrix}$$

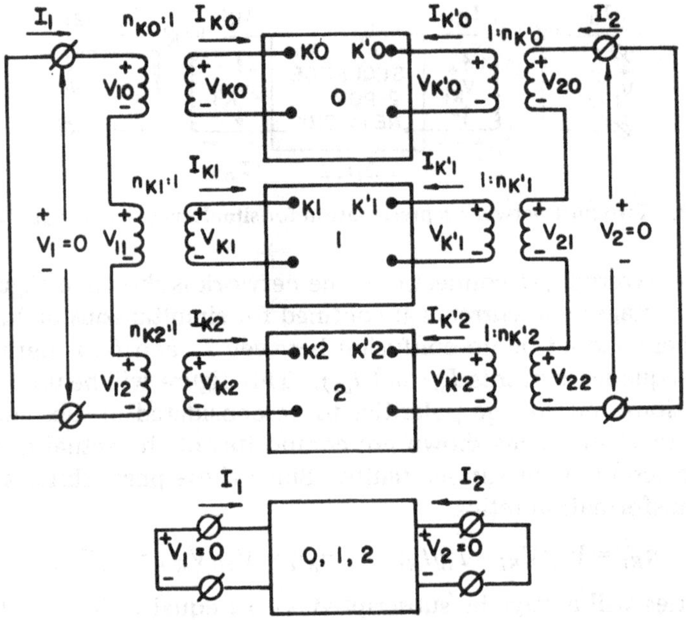

Fig. 9.10. Sequence network connection for simultaneous Z-type faults.

we get

$$\begin{bmatrix} V_{11} \\ V_{21} \end{bmatrix} = \begin{bmatrix} z_{11(1)} & (n_{K1}/n_{K'1})\, z_{12(1)} \\ (n_{K'1}/n_{K1})\, z_{21(1)} & z_{22(1)} \end{bmatrix} \begin{bmatrix} I_{11} \\ I_{21} \end{bmatrix} + \begin{bmatrix} n_{K1}\, V_{z1} \\ n_{K'1}\, V_{z2} \end{bmatrix} \tag{9.54}$$

For the negative sequence network we write

$$\begin{bmatrix} V_{K2} \\ V_{K'2} \end{bmatrix} = \begin{bmatrix} z_{11(2)} & z_{12(2)} \\ z_{21(2)} & z_{22(2)} \end{bmatrix} \begin{bmatrix} I_{K2} \\ I_{K'2} \end{bmatrix} \tag{9.55}$$

which we premultiply by

$$\begin{bmatrix} n_{K2} & 0 \\ 0 & n_{K'2} \end{bmatrix}$$

to obtain

$$\begin{bmatrix} V_{12} \\ V_{22} \end{bmatrix} = \begin{bmatrix} z_{11(2)} & (n_{K2}/n_{K'2})\, z_{12(2)} \\ (n_{K'2}/n_{K2})\, z_{21(2)} & z_{22(2)} \end{bmatrix} \begin{bmatrix} I_{12} \\ I_{22} \end{bmatrix} \tag{9.56}$$

And for the zero sequence network we write

$$\begin{bmatrix} V_{K0} \\ V_{K'0} \end{bmatrix} = \begin{bmatrix} z_{11(0)} & z_{12(0)} \\ z_{21(0)} & z_{22(0)} \end{bmatrix} \begin{bmatrix} I_{K0} \\ I_{K'0} \end{bmatrix} \tag{9.57}$$

which transforms (since $n_{K0} = n_{K'0} = 1$) to

$$\begin{bmatrix} V_{10} \\ V_{20} \end{bmatrix} = \begin{bmatrix} z_{11(0)} & z_{12(0)} \\ z_{21(0)} & z_{22(0)} \end{bmatrix} \begin{bmatrix} I_{10} \\ I_{20} \end{bmatrix} \tag{9.58}$$

But from Figure 9.10 we observe that for a series-series connection,

$$\begin{bmatrix} V_1 \\ V_2 \end{bmatrix} = \begin{bmatrix} V_{11} \\ V_{21} \end{bmatrix} + \begin{bmatrix} V_{12} \\ V_{22} \end{bmatrix} + \begin{bmatrix} V_{10} \\ V_{20} \end{bmatrix} = \begin{bmatrix} 0 \\ 0 \end{bmatrix} \tag{9.59}$$

and

$$\begin{bmatrix} I_1 \\ I_2 \end{bmatrix} = \begin{bmatrix} I_{11} \\ I_{21} \end{bmatrix} = \begin{bmatrix} I_{12} \\ I_{22} \end{bmatrix} = \begin{bmatrix} I_{10} \\ I_{20} \end{bmatrix} \tag{9.60}$$

Performing the addition indicated in (9.59) and making the substitution (9.60), we write

$$\begin{bmatrix} V_1 \\ V_2 \end{bmatrix} = \begin{bmatrix} 0 \\ 0 \end{bmatrix} = \begin{bmatrix} z_{11} & z_{12} \\ z_{21} & z_{22} \end{bmatrix} \begin{bmatrix} I_1 \\ I_2 \end{bmatrix} + \begin{bmatrix} n_{K1} V_{z1} \\ n_{K'1} V_{z2} \end{bmatrix} \tag{9.61}$$

where

$$\begin{aligned} z_{11} &= z_{11(0)} + z_{11(1)} + z_{11(2)} \\ z_{12} &= z_{12(0)} + (n_{K1}/n_{K'1})\, z_{12(1)} + (n_{K2}/n_{K'2})\, z_{12(2)} \\ z_{21} &= z_{21(0)} + (n_{K'1}/n_{K1})\, z_{21(1)} + (n_{K'2}/n_{K2})\, z_{21(2)} \\ z_{22} &= z_{22(0)} + z_{22(1)} + z_{22(2)} \end{aligned} \tag{9.62}$$

Equation (9.61) may be written as

$$\mathbf{V} = \mathbf{0} = \mathbf{ZI} + \mathbf{V}_z \tag{9.63}$$

Then

$$\mathbf{I} = -\mathbf{Z}^{-1}\mathbf{V}_z \tag{9.64}$$

is the solution of the series-series connection. Knowing this current, we know all currents in the outside transformer windings, and we may compute all voltages from (9.54), (9.56), and (9.58). Then the individual sequence network currents and voltages may be found by applying Kirchhoff's laws to the individual networks.

We may write out the solution (9.63) as

$$\begin{bmatrix} I_1 \\ I_2 \end{bmatrix} = \frac{-1}{\det \mathbf{Z}} \begin{bmatrix} z_{22} & -z_{12} \\ -z_{21} & z_{11} \end{bmatrix} \begin{bmatrix} n_{K1} V_{z1} \\ n_{K'1} V_{z2} \end{bmatrix} \tag{9.65}$$

In this form, we see that the current at each fault depends upon the independent source voltages at *both* fault locations. We illustrate this method by means of an example.

Example 9.6

A simple power system is shown in Figure 9.11 with simultaneous faults indicated by X's at fault points F and F'. The following system data is known:

Generator L: $Z_1'' = Z_2 = j0.12$, $Z_0 = j0.10$, $E_L = 1.1\underline{/30^\circ}$

Generator R: $Z_1'' = Z_2 = j0.15$, $Z_0 = j0.13$, $E_R = 1.0\underline{/0^\circ}$

Transformer AB: $Z_1 = Z_2 = Z_0 = j0.10$

Transmission line BC: $Z_1 = Z_2 = j0.50$, $Z_0 = j1.00$

Transformer CD: $Z_1 = Z_2 = Z_0 = j0.12$

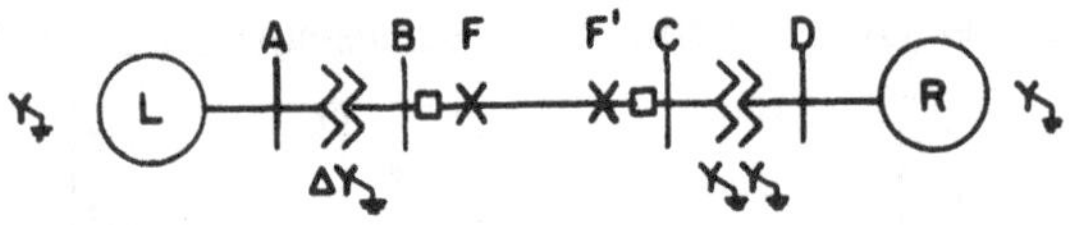

Fig. 9.11. Power system for Example 9.6.

We assume that faults are applied as follows:

Fault F: lines a and b open
Fault F': SLG fault on phase b

Solve the system for the currents and voltages at the circuit breaker positions at B and C. Assume that all fault impedances are zero.

Solution

First, we note that the two faults specified require a *series* connection of sequence networks. Thus a series-series two-port description is applicable. Referring to Table 9.6, we must determine the Z parameters Z_{MN} for each sequence network.

The sequence networks are sketched in Figure 9.12, where it is noted that the series fault requires the insertion of an *open* section adjacent to breaker B. We arbitrarily define the current as positive in the BC direction, (I_K) since this would constitute positive tripping current for that breaker, as noted at F. We also identify the positive tripping current direction for breaker C as away from bus C and toward the faulted line.

To solve the system, we must first find the Z parameters. We easily compute the following.

$$\begin{bmatrix} V_{K0} \\ V_{K'0} \end{bmatrix} = \begin{bmatrix} j1.35 & -j0.25 \\ -j0.25 & j0.25 \end{bmatrix} \begin{bmatrix} I_{K0} \\ I_{K'0} \end{bmatrix}$$

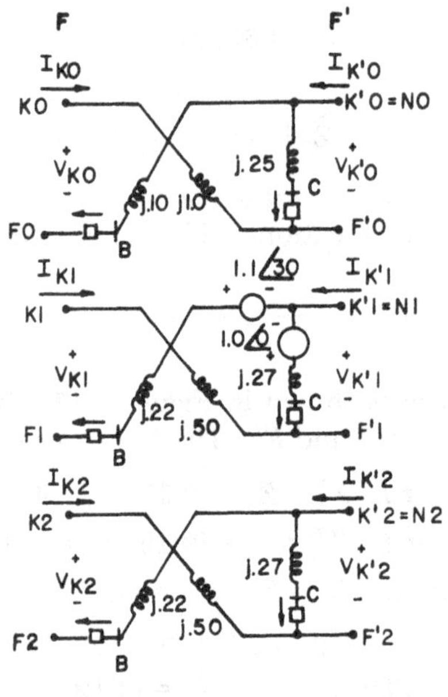

Fig. 9.12. Sequence two-port networks for Example 9.6.

$$\begin{bmatrix} V_{K1} \\ V_{K'1} \end{bmatrix} = \begin{bmatrix} j0.99 & -j0.27 \\ -j0.27 & j0.27 \end{bmatrix} \begin{bmatrix} I_{K1} \\ I_{K'1} \end{bmatrix} + \begin{bmatrix} 0.047 - j0.550 \\ -1.0\underline{/0^\circ} \end{bmatrix}$$

$$\begin{bmatrix} V_{K2} \\ V_{K'2} \end{bmatrix} = \begin{bmatrix} j0.99 & -j0.27 \\ -j0.27 & j0.27 \end{bmatrix} \begin{bmatrix} I_{K2} \\ I_{K'2} \end{bmatrix}$$

Fault symmetry is the next consideration. From Figure 8.9 we confirm the following:

Fault F: two lines open on a and b

$$n_{K0} = 1, \quad n_{K1} = \text{a}, \quad n_{K2} = \text{a}^2$$

Fault F': SLG fault on b

$$n_{K'0} = 1, \quad n_{K'1} = \text{a}^2, \quad n_{K'2} = \text{a}$$

The impedance matrix is then computed from (9.62) with the result

$$\mathbf{Z} = \begin{bmatrix} j3.33 & j0.02 \\ j0.02 & j0.79 \end{bmatrix}$$

Finally, from (9.65) we compute

$$\begin{bmatrix} I_1 \\ I_2 \end{bmatrix} = \frac{-1}{\det \mathbf{Z}} \begin{bmatrix} (j0.79)\,\text{a}V_{K1} - (j0.02)\,\text{a}^2 V_{K'1} \\ -(j0.02)\,\text{a}V_{K1} + (j3.33)\,\text{a}^2 V_{K'1} \end{bmatrix}$$

$$= \begin{bmatrix} -0.088 + j0.1221 \\ -1.094 + j0.630 \end{bmatrix} = \begin{bmatrix} 0.1589\underline{/123.76} \\ 1.2622\underline{/150.08} \end{bmatrix}$$

and we note that

$$\begin{bmatrix} I_1 \\ I_2 \end{bmatrix} = \begin{bmatrix} I_{10} \\ I_{20} \end{bmatrix} = \begin{bmatrix} I_{11} \\ I_{21} \end{bmatrix} = \begin{bmatrix} I_{12} \\ I_{22} \end{bmatrix}$$

From (9.51) we write

$$\begin{bmatrix} I_{K0} \\ I_{K'0} \end{bmatrix} = \begin{bmatrix} (1/n_{K0})I_{10} \\ (1/n_{K'0})I_{20} \end{bmatrix} = \begin{bmatrix} I_1 \\ I_2 \end{bmatrix}$$

$$\begin{bmatrix} I_{K1} \\ I_{K'1} \end{bmatrix} = \begin{bmatrix} (1/n_{K1})I_{11} \\ (1/n_{K'1})I_{21} \end{bmatrix} = \begin{bmatrix} \text{a}^2 I_1 \\ \text{a}I_2 \end{bmatrix} = \begin{bmatrix} 0.1589\underline{/3.76^\circ} \\ 1.262\underline{/270.08^\circ} \end{bmatrix}$$

$$\begin{bmatrix} I_{K2} \\ I_{K'2} \end{bmatrix} = \begin{bmatrix} (1/n_{K2})I_{12} \\ (1/n_{K'2})I_{22} \end{bmatrix} = \begin{bmatrix} \text{a}I_1 \\ \text{a}^2 I_2 \end{bmatrix} = \begin{bmatrix} 0.1589\underline{/243.76^\circ} \\ 1.262\underline{/30.08^\circ} \end{bmatrix}$$

Currents and voltages at breaker B:

$$I_{a0} = I_{K0} = 0.1615\underline{/123.75^\circ}$$
$$I_{a1} = I_{K1} = \text{a}^2 I_{a0}$$
$$I_{a2} = I_{K2} = \text{a}I_{a0}$$
$$I_a = I_{a0} + I_{a1} + I_{a2} = 0$$
$$I_b = I_{a0} + \text{a}^2 I_{a1} + \text{a}I_{a2} = 0$$
$$I_c = 3I_{a0} = 0.477\underline{/123.76^\circ}$$

$$V_{a0} = -(j0.10)\, I_{K0} = 0.0134 + j0.00897$$
$$V_{a1} = 1.1\underline{/30^\circ} - (j0.22)\, I_{K1} = 0.954 + j0.514$$
$$V_{a2} = -(j0.22)\, I_{K2} = -0.0381 + j0.0157$$
$$V_a = V_{a0} + V_{a1} + V_{a2} = 1.08\underline{/29.9^\circ}$$
$$V_b = V_{a0} + a^2 V_{a1} + aV_{a2} = 1.11\underline{/269.1^\circ}$$
$$V_c = V_{a0} + aV_{a1} + a^2 V_{a2} = 1.06\underline{/145.8^\circ}$$

Currents and voltages at breaker C:

$$I_{a0} = I_{K'0} - I_{K0} = -1.094 + j0.630$$
$$I_{a1} = I_{K'1} - I_{K1} = -0.002 - j1.262$$
$$I_{a2} = I_{K'2} - I_{K2} = 1.092 + j0.633$$

Then

$$\begin{bmatrix} I_a \\ I_b \\ I_c \end{bmatrix} = \begin{bmatrix} 1 & 1 & 1 \\ 1 & a^2 & a \\ 1 & a & a^2 \end{bmatrix} \begin{bmatrix} -1.094 + j0.630 \\ -0.002 - j1.262 \\ 1.092 + j0.633 \end{bmatrix} = \begin{bmatrix} 0 + j0 \\ 3.79\underline{/150.1^\circ} \\ 0 + j0 \end{bmatrix}$$

We also compute

$$V_{a0} = -I_{a0}\,(j0.25) = 0.124 + j0.251$$
$$V_{a1} = 1.0\underline{/0^\circ} - I_{a1}\,(j0.27) = 0.656 + j0.042$$
$$V_{a2} = -I_{a2}\,(j0.27) = 0.209 - j0.314$$

so that at breaker C

$$\begin{bmatrix} V_a \\ V_b \\ V_c \end{bmatrix} = \mathbf{AV}_{012} = \begin{bmatrix} 0.990\underline{/-1.16^\circ} \\ 0 + j0 \\ 0.990\underline{/128.5^\circ} \end{bmatrix}$$

9.5 Parallel-Parallel Connection (Y-Type Faults)

A parallel-parallel connection of two-port networks is required to represent the following simultaneous fault conditions (see Table 9.6):

1. Simultaneous 2LG faults at F and F' ($\mathbf{Y}_{NN}$).
2. A 2LG fault at F and one line open at F' ($\mathbf{Y}_{NM}$).
3. One line open at F and a 2LG fault at F' ($\mathbf{Y}_{MN}$).
4. One line open at F and one line open at F' ($\mathbf{Y}_{MM}$).

The sequence network connection with parallel-parallel termination is shown in Figure 9.13 where the ideal transformations are phase shifters with voltage and current relations as given by (9.51). We write the two-port equations for the three sequence networks as follows. For the positive sequence network

$$\begin{bmatrix} I_{K1} \\ I_{K'1} \end{bmatrix} = \begin{bmatrix} y_{11(1)} & y_{12(1)} \\ y_{21(1)} & y_{22(1)} \end{bmatrix} \begin{bmatrix} V_{K1} \\ V_{K'1} \end{bmatrix} + \begin{bmatrix} I_{y1} \\ I_{y2} \end{bmatrix} \tag{9.66}$$

where $\mathbf{I}_y$ is the independent source term viewed from the $K1$ and $K'1$ ports. Simi-

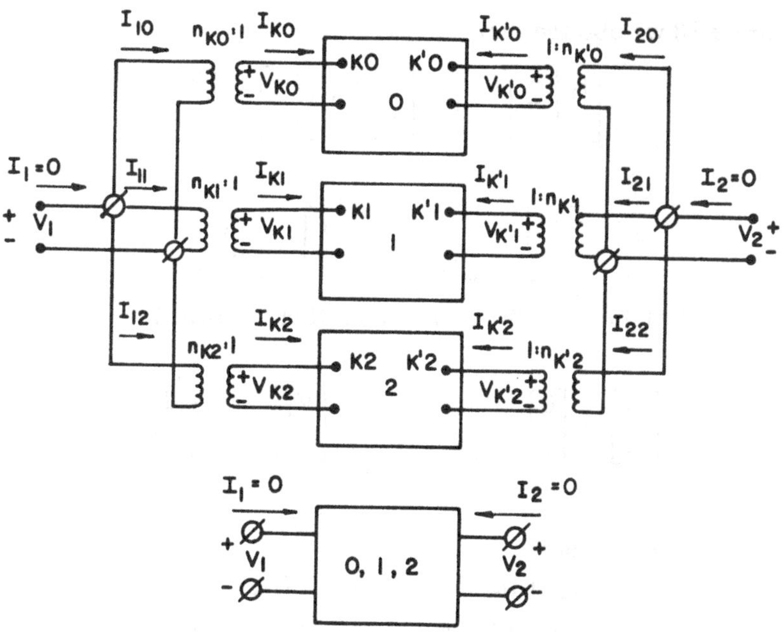

Fig. 9.13. Sequence network connection for simultaneous Y-type faults.

larly, for the passive sequence networks we have

$$\begin{bmatrix} I_{K0} \\ I_{K'0} \end{bmatrix} = \begin{bmatrix} y_{11(0)} & y_{12(0)} \\ y_{21(0)} & y_{22(0)} \end{bmatrix} \begin{bmatrix} V_{K0} \\ V_{K'0} \end{bmatrix} \tag{9.67}$$

$$\begin{bmatrix} I_{K2} \\ I_{K'2} \end{bmatrix} = \begin{bmatrix} y_{11(2)} & y_{12(2)} \\ y_{21(2)} & y_{22(2)} \end{bmatrix} \begin{bmatrix} V_{K2} \\ V_{K'2} \end{bmatrix} \tag{9.68}$$

Using (9.51) for each sequence, we reflect the two-port expressions across the transformations to obtain new two-port equations which include the transformers. This is done by premultiplying (9.66)-(9.68) by

$$\begin{bmatrix} n_{Ki} & 0 \\ 0 & n_{K'i} \end{bmatrix}$$

for $i = 1, 0$, and 2 respectively. The result for the positive sequence network is

$$\begin{bmatrix} I_{11} \\ I_{21} \end{bmatrix} = \begin{bmatrix} y_{11(1)} & (n_{K1}/n_{K'1})y_{12(1)} \\ (n_{K'1}/n_{K1})y_{21(1)} & y_{22(1)} \end{bmatrix} \begin{bmatrix} V_{11} \\ V_{21} \end{bmatrix} + \begin{bmatrix} n_{K1} I_{y1} \\ n_{K'1} I_{y2} \end{bmatrix} \tag{9.69}$$

Similarly, since $n_{K0} = n_{K'0} = 1$ in all cases, for the zero sequence we have

$$\begin{bmatrix} I_{10} \\ I_{20} \end{bmatrix} = \begin{bmatrix} y_{11(0)} & y_{12(0)} \\ y_{21(0)} & y_{22(0)} \end{bmatrix} \begin{bmatrix} V_{10} \\ V_{20} \end{bmatrix} \tag{9.70}$$

And for the negative sequence network we compute

$$\begin{bmatrix} I_{12} \\ I_{22} \end{bmatrix} = \begin{bmatrix} y_{11(2)} & (n_{K2}/n_{K'2})y_{12(2)} \\ (n_{K'2}/n_{K2})y_{21(2)} & y_{22(2)} \end{bmatrix} \begin{bmatrix} V_{12} \\ V_{22} \end{bmatrix} \tag{9.71}$$

From Figure 9.13 we observe that

$$\begin{bmatrix} I_1 \\ I_2 \end{bmatrix} = \begin{bmatrix} I_{10} \\ I_{20} \end{bmatrix} + \begin{bmatrix} I_{11} \\ I_{21} \end{bmatrix} + \begin{bmatrix} I_{12} \\ I_{22} \end{bmatrix} = \begin{bmatrix} 0 \\ 0 \end{bmatrix} \tag{9.72}$$

and

$$\begin{bmatrix} V_1 \\ V_2 \end{bmatrix} = \begin{bmatrix} V_{10} \\ V_{20} \end{bmatrix} = \begin{bmatrix} V_{11} \\ V_{21} \end{bmatrix} = \begin{bmatrix} V_{12} \\ V_{22} \end{bmatrix} \tag{9.73}$$

Substituting (9.73) into (9.69)-(9.71) and then substituting these results into (9.72) gives

$$\begin{bmatrix} I_1 \\ I_2 \end{bmatrix} = \begin{bmatrix} 0 \\ 0 \end{bmatrix} = \begin{bmatrix} y_{11} & y_{12} \\ y_{21} & y_{22} \end{bmatrix} \begin{bmatrix} V_1 \\ V_2 \end{bmatrix} + \begin{bmatrix} n_{K1} I_{y1} \\ n_{K'1} I_{y2} \end{bmatrix} \tag{9.74}$$

where

$$\begin{aligned} y_{11} &= y_{11(0)} + y_{11(1)} + y_{11(2)} \\ y_{12} &= y_{12(0)} + (n_{K1}/n_{K'1})\, y_{12(1)} + (n_{K2}/n_{K'2})\, y_{12(2)} \\ y_{21} &= y_{21(0)} + (n_{K1}/n_{K'1})\, y_{21(1)} + (n_{K'2}/n_{K2})\, y_{21(2)} \\ y_{22} &= y_{22(0)} + y_{22(1)} + y_{22(2)} \end{aligned} \tag{9.75}$$

In matrix form we write (9.74) as

$$0 = \mathbf{Y}\,\mathbf{V} + \mathbf{I}_s \tag{9.76}$$

which we solve for V to obtain

$$\mathbf{V} = -\mathbf{Y}^{-1}\,\mathbf{I}_s \tag{9.77}$$

or

$$\begin{bmatrix} V_1 \\ V_2 \end{bmatrix} = \frac{-1}{\det \mathbf{Y}} \begin{bmatrix} y_{22} & -y_{12} \\ -y_{21} & y_{11} \end{bmatrix} \begin{bmatrix} n_{K1} I_{y1} \\ n_{K'1} I_{y2} \end{bmatrix} \tag{9.78}$$

In this case we solve for the external port voltages which are easily converted to port voltages inside the transformations. Then, if the sequence network voltages are known, the networks can be completely solved. We illustrate the solution of a parallel-parallel fault condition by means of an example.

Example 9.7

Consider the system of Figure 9.11 for which the sequence impedances and source voltages are given in Example 9.6. Find the phase currents and voltages at the breaker locations for the following fault conditions:

Fault F: line b open

Fault F': 2LG fault on b and c

Solution

As in Example 9.6 we have a series fault at F and a shunt fault at F'. Therefore, the sequence networks are the same as before and are given by Figure 9.12. Here the similarity in the two problems ends. The symmetry in the fault condition of this example is given by the following:

Fault F: line b open

$$n_{K0} = 1, \quad n_{K1} = a^2, \quad n_{K2} = a$$

Fault F': 2LG fault on b and c

$$n_{K'0} = n_{K'1} = n_{K'2} = 1$$

For the sequence networks themselves we need to write current equations similar to (9.66)-(9.68). We again use the results of Example 9.6 to good advantage. From Table 9.3 we note that the **Y** matrix is the inverse of the **Z** matrix, or

$$\mathbf{Y}_0 = \mathbf{Z}_0^{-1} = \begin{bmatrix} j1.35 & -j0.25 \\ -j0.25 & j0.25 \end{bmatrix}^{-1} = \begin{bmatrix} -j0.91 & -j0.91 \\ -j0.91 & -j4.90 \end{bmatrix}$$

$$\mathbf{Y}_1 = \mathbf{Z}_1^{-1} = \begin{bmatrix} j0.99 & -j0.27 \\ -j0.27 & j0.27 \end{bmatrix}^{-1} = \begin{bmatrix} -j1.39 & -j1.39 \\ -j1.39 & -j5.09 \end{bmatrix}$$

and $\mathbf{Y}_2 = \mathbf{Y}_1$. Also, from Table 9.5 we note that

$$\mathbf{I}_y = -\mathbf{Z}_1^{-1}\mathbf{V}_z = \mathbf{Y}_1\mathbf{V}_z - \begin{bmatrix} -j1.39 & -j1.39 \\ -j1.39 & -j5.09 \end{bmatrix} \begin{bmatrix} 0.047 - j0.55 \\ -1.0 \end{bmatrix}$$

$$= \begin{bmatrix} +0.764 - j1.32 \\ +0.764 - j5.03 \end{bmatrix} = \begin{bmatrix} 1.528\underline{/-60.00^\circ} \\ 5.085\underline{/-81.36^\circ} \end{bmatrix}$$

Then we write

$$\begin{bmatrix} I_{K0} \\ I_{K'0} \end{bmatrix} = \begin{bmatrix} -j0.91 & -j0.91 \\ -j0.91 & -j4.90 \end{bmatrix} \begin{bmatrix} V_{K0} \\ V_{K'0} \end{bmatrix}$$

$$\begin{bmatrix} I_{K1} \\ I_{K'1} \end{bmatrix} = \begin{bmatrix} -j1.39 & -j1.39 \\ -j1.39 & -j5.09 \end{bmatrix} \begin{bmatrix} V_{K1} \\ V_{K'1} \end{bmatrix} + \begin{bmatrix} 1.528\underline{/-60.00^\circ} \\ 5.085\underline{/-81.36^\circ} \end{bmatrix}$$

$$\begin{bmatrix} I_{K2} \\ I_{K'2} \end{bmatrix} = \begin{bmatrix} -j1.39 & -j1.39 \\ -j1.39 & -j5.09 \end{bmatrix} \begin{bmatrix} V_{K2} \\ V_{K'2} \end{bmatrix}$$

These current equations are reflected across the phase shifters by (9.69)-(9.71) and combined according to (9.74) with the result

$$\begin{bmatrix} 0 \\ 0 \end{bmatrix} = \begin{bmatrix} y_{11} & y_{12} \\ y_{21} & y_{22} \end{bmatrix} \begin{bmatrix} V_1 \\ V_2 \end{bmatrix} + \begin{bmatrix} n_{K1} I_{y1} \\ n_{K'1} I_{y2} \end{bmatrix}$$

where from (9.75)

$$y_{11} = y_{11(0)} + y_{11(1)} + y_{11(2)}$$
$$= -j0.91 - j1.39 - j1.39 = -j3.69$$

$$y_{12} = y_{12(0)} + (n_{K1}/n_{K'1})\, y_{12(1)} + (n_{K2}/n_{K'2})\, y_{12(2)}$$
$$= -j0.91 + a^2(-j1.39) + a(-j1.39)$$
$$= -j0.91 + j1.39 = +j0.48$$

$$y_{21} = y_{21(0)} + (n_{K'1}/n_{K1})\, y_{21(1)} + (n_{K'2}/n_{K2})\, y_{21(2)}$$
$$= -j0.91 + a(-j1.39) + a^2(-j1.39) = +j0.48$$

$$y_{22} = y_{22(0)} + y_{22(1)} + y_{22(2)}$$
$$= -j4.90 - j5.09 - j5.09 = -j15.08$$

$$\det \mathbf{Y} = y_{11}y_{22} - y_{12}y_{21} = -55.65 + 0.23 = -55.41$$

Then from (9.78)

$$\begin{bmatrix} V_1 \\ V_2 \end{bmatrix} = \frac{1}{55.41}\begin{bmatrix} -j15.08 & -j0.48 \\ -j0.48 & -j3.69 \end{bmatrix}\begin{bmatrix} 1.528\underline{/-60.00^\circ + 240.0^\circ} \\ 5.085\underline{/-81.36^\circ} \end{bmatrix}$$

$$= \begin{bmatrix} 0.4118\underline{/96.06^\circ} \\ 0.3365\underline{/-173.59^\circ} \end{bmatrix} = \begin{bmatrix} V_{10} \\ V_{20} \end{bmatrix} = \begin{bmatrix} V_{11} \\ V_{21} \end{bmatrix} = \begin{bmatrix} V_{12} \\ V_{22} \end{bmatrix}$$

From (9.51) we compute

$$\begin{bmatrix} V_{K0} \\ V_{K'0} \end{bmatrix} = \begin{bmatrix} V_{10} \\ V_{20} \end{bmatrix} = \begin{bmatrix} 0.4118\underline{/96.06^\circ} \\ 0.3365\underline{/-173.59^\circ} \end{bmatrix}$$

$$\begin{bmatrix} V_{K1} \\ V_{K'1} \end{bmatrix} = \begin{bmatrix} aV_{11} \\ V_{21} \end{bmatrix} = \begin{bmatrix} 0.4118\underline{/216.06^\circ} \\ 0.3365\underline{/-173.59^\circ} \end{bmatrix}$$

$$\begin{bmatrix} V_{K2} \\ V_{K'2} \end{bmatrix} = \begin{bmatrix} a^2 V_{12} \\ V_{22} \end{bmatrix} = \begin{bmatrix} 0.4118\underline{/-23.94^\circ} \\ 0.3365\underline{/-173.59^\circ} \end{bmatrix}$$

Finally then we compute the voltages at F to be (using h = 1)

$$\begin{bmatrix} V_a \\ V_b \\ V_c \end{bmatrix} - \mathrm{A}\begin{bmatrix} 0.4118\underline{/98.74^\circ} \\ 0.4118\underline{/214.74^\circ} \\ 0.4118\underline{/-25.26^\circ} \end{bmatrix} = \begin{bmatrix} 0 \\ 1.235\underline{/-83.93^\circ} \\ 0 \end{bmatrix}$$

and at F'

$$\begin{bmatrix} V_{a'} \\ V_{b'} \\ V_{c'} \end{bmatrix} - \mathrm{A}\begin{bmatrix} 0.3365\underline{/186.27^\circ} \\ 0.3365\underline{/186.27^\circ} \\ 0.3365\underline{/186.27^\circ} \end{bmatrix} = \begin{bmatrix} 1.009\underline{/6.41^\circ} \\ 0 \\ 0 \end{bmatrix}$$

9.6 Series-Parallel Connection (H-type Faults)

The third and last connection required to solve for common simultaneous fault conditions is the series-parallel connection shown in Figure 9.14. This connection may be used for all series-parallel faults and, by reversing the defined left-port and right-port definitions, for all parallel-series faults as well. From Table 9.6 the series-parallel H-type fault configurations include:

1. A SLG fault at F and a 2LG fault at $F'(\mathrm{H}_{NN})$.
2. A SLG fault at F and one line open at $F'(\mathrm{H}_{NM})$.
3. Two lines open at F and a 2LG fault at $F'(\mathrm{H}_{MN})$.
4. Two lines open at F and one line open at $F'(\mathrm{H}_{MM})$.

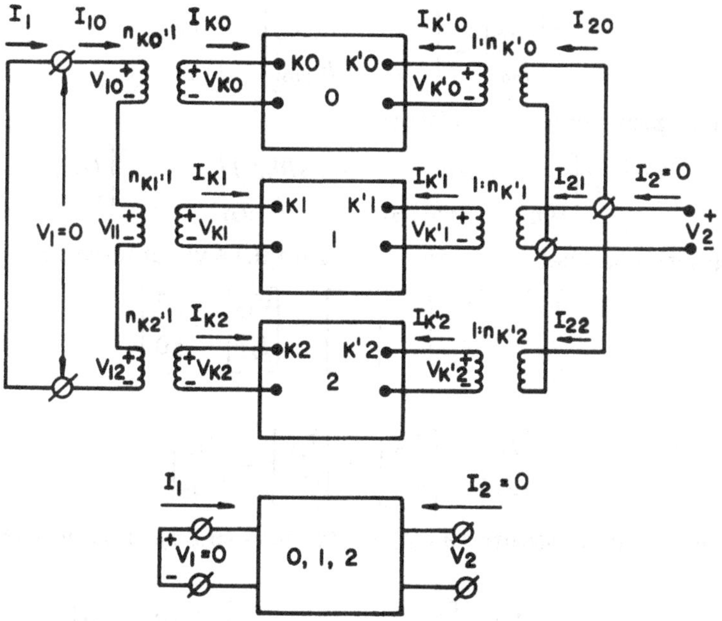

Fig. 9.14. Sequence network connection for simultaneous H-type faults.

For these conditions we write the two-port hybrid equations. For the active positive sequence network we have

$$\begin{bmatrix} V_{K1} \\ I_{K'1} \end{bmatrix} = \begin{bmatrix} h_{11(1)} & h_{12(1)} \\ h_{21(1)} & h_{22(1)} \end{bmatrix} \begin{bmatrix} I_{K1} \\ V_{K'1} \end{bmatrix} + \begin{bmatrix} V_{h1} \\ I_{h2} \end{bmatrix} \tag{9.79}$$

where the last term is the independent source term viewed from the $K1$ and $K'1$ ports. For the passive sequence networks we have

$$\begin{bmatrix} V_{K0} \\ I_{K'0} \end{bmatrix} = \begin{bmatrix} h_{11(0)} & h_{12(0)} \\ h_{21(0)} & h_{22(0)} \end{bmatrix} \begin{bmatrix} I_{K0} \\ V_{K'0} \end{bmatrix} \tag{9.80}$$

and

$$\begin{bmatrix} V_{K2} \\ I_{K'2} \end{bmatrix} = \begin{bmatrix} h_{11(2)} & h_{12(2)} \\ h_{21(2)} & h_{22(2)} \end{bmatrix} \begin{bmatrix} I_{K2} \\ V_{K'2} \end{bmatrix} \tag{9.81}$$

Proceeding as before, we now write these equations in terms of the quantities outside the transformers by premultiplying by

$$\begin{bmatrix} n_{Ki} & 0 \\ 0 & n_{K'i} \end{bmatrix}$$

for $i = 1, 0$, and 2 respectively. The result for the positive sequence equation is

$$\begin{bmatrix} V_{11} \\ I_{21} \end{bmatrix} = \begin{bmatrix} h_{11(1)} & (n_{K1}/n_{K'1})\, h_{12(1)} \\ (n_{K'1}/n_{K1})\, h_{21(1)} & h_{22(1)} \end{bmatrix} \begin{bmatrix} I_{11} \\ V_{21} \end{bmatrix} + \begin{bmatrix} n_{K1} V_{h1} \\ n_{K'1} I_{h2} \end{bmatrix} \tag{9.82}$$

For the zero sequence network with $n_{K0} = n_{K'0} = 1$,

$$\begin{bmatrix} V_{10} \\ I_{20} \end{bmatrix} = \begin{bmatrix} h_{11(0)} & h_{12(0)} \\ h_{21(0)} & h_{22(0)} \end{bmatrix} \begin{bmatrix} I_{10} \\ V_{20} \end{bmatrix} \tag{9.83}$$

And for the negative sequence network,

$$\begin{bmatrix} V_{12} \\ I_{22} \end{bmatrix} = \begin{bmatrix} h_{11(2)} & (n_{K2}/n_{K'2})\, h_{12(2)} \\ (n_{K'2}/n_{K2})\, h_{21(2)} & h_{22(2)} \end{bmatrix} \begin{bmatrix} I_{12} \\ V_{22} \end{bmatrix} \tag{9.84}$$

For the series-parallel connection of Figure 9.14 we observe that

$$\begin{bmatrix} V_1 \\ I_2 \end{bmatrix} = \begin{bmatrix} V_{10} \\ I_{20} \end{bmatrix} + \begin{bmatrix} V_{11} \\ I_{21} \end{bmatrix} + \begin{bmatrix} V_{12} \\ I_{22} \end{bmatrix} = \begin{bmatrix} 0 \\ 0 \end{bmatrix} \tag{9.85}$$

and

$$\begin{bmatrix} I_1 \\ V_2 \end{bmatrix} = \begin{bmatrix} I_{10} \\ V_{20} \end{bmatrix} = \begin{bmatrix} I_{11} \\ V_{21} \end{bmatrix} = \begin{bmatrix} I_{12} \\ V_{22} \end{bmatrix} \tag{9.86}$$

Therefore we may substitute (9.82)-(9.84) into (9.85), making use of (9.86) to write

$$\begin{bmatrix} V_1 \\ I_2 \end{bmatrix} = \begin{bmatrix} 0 \\ 0 \end{bmatrix} = \begin{bmatrix} h_{11} & h_{12} \\ h_{21} & h_{22} \end{bmatrix} \begin{bmatrix} I_1 \\ V_2 \end{bmatrix} + \begin{bmatrix} n_{K1} V_{h1} \\ n_{K'1} I_{h2} \end{bmatrix} \tag{9.87}$$

where

$$\begin{aligned} h_{11} &= h_{11(0)} + h_{11(1)} + h_{11(2)} \\ h_{12} &= h_{12(0)} + (n_{K1}/n_{K'1})\, h_{12(1)} + (n_{K2}/n_{K'2}) h_{12(2)} \\ h_{21} &= h_{21(0)} + (n_{K'1}/n_{K1})\, h_{21(1)} + (n_{K'2}/n_{K2})\, h_{21(2)} \\ h_{22} &= h_{22(0)} + h_{22(1)} + h_{22(2)} \end{aligned} \tag{9.88}$$

In matrix form (9.87) may be written as

$$\mathbf{0} = \mathbf{HN} + \mathbf{M}_s \tag{9.89}$$

where N is defined by (9.14), $\mathbf{M}_s$ by (9.87), and the matrix H by (9.88). Solving (9.89), we have

$$\mathbf{N} = -\,\mathbf{H}^{-1}\mathbf{M}_s \tag{9.90}$$

or

$$\begin{bmatrix} I_1 \\ V_2 \end{bmatrix} = \frac{-1}{\det \mathbf{H}} \begin{bmatrix} h_{22} & -h_{12} \\ -h_{21} & h_{11} \end{bmatrix} \begin{bmatrix} n_{K1} V_{h1} \\ n_{K'1} I_{h2} \end{bmatrix} \tag{9.91}$$

Knowing I_1 and V_2, we may solve completely for all port quantities and then for internal network quantities.

II. SIMULTANEOUS FAULTS BY MATRIX TRANSFORMATIONS

We now develop a second method for computing simultaneous faults which is based on the matrix transformation technique developed for a simple fault in sections 8.9-8.11. This method is also described briefly by Lewis and Pryce [13], which is recommended for further reading on the subject.

9.7 Constraint Matrix for Z-type Faults

Z-type faults include any combination of SLG shunt faults and series faults with two lines open and are characterized by a series-series connection of the sequence networks. This connection is conveniently described by the two-port Z parameters according to the equations

$$\begin{bmatrix} V_{Ki} \\ V_{K'i} \end{bmatrix} = \begin{bmatrix} z_{11(i)} & z_{12(i)} \\ z_{21(i)} & z_{22(i)} \end{bmatrix} \begin{bmatrix} I_{Ki} \\ I_{K'i} \end{bmatrix} + \delta_{1i} \begin{bmatrix} V_{z1} \\ V_{z2} \end{bmatrix}, \quad i = 0, 1, 2 \tag{9.92}$$

where

$$\begin{aligned} \delta_{1i} &= 0, \quad i \neq 1 \\ &= 1, \quad i = 1 \end{aligned}$$

But, by definition

$$\begin{bmatrix} I_{Ki} \\ I_{K'i} \end{bmatrix} = \begin{bmatrix} I_{ai} \\ I_{a'i} \end{bmatrix}, \quad i = 0, 1, 2 \tag{9.93}$$

and the port equations can be written in terms of the sequence currents. From section 8.8 we write

$$\mathbf{I}_{\text{old}} = \mathbf{K}\mathbf{I}_{\text{new}}$$

and, if we can find a matrix $\mathbf{K}$ which constrains $\mathbf{I}_{\text{old}}$, the sequence currents, to a series connection then we can use this same constraint matrix to find the new voltage vector from $\mathbf{V}_{\text{new}} = \mathbf{K}^{*t}\mathbf{V}_{\text{old}}$. For the problem at hand we define

$$\mathbf{I}_{\text{old}} = \mathbf{I} = \begin{bmatrix} I_{a0} \\ I_{a1} \\ I_{a2} \\ \hline I_{a'0} \\ I_{a'1} \\ I_{a'2} \end{bmatrix} \tag{9.94}$$

and write (9.92) in 0-1-2 order as

$$\begin{bmatrix} V_{K0} \\ V_{K1} \\ V_{K2} \\ \hline V_{K'0} \\ V_{K'1} \\ V_{K'2} \end{bmatrix} = \left[\begin{array}{ccc|ccc} z_{11(0)} & 0 & 0 & z_{12(0)} & 0 & 0 \\ 0 & z_{11(1)} & 0 & 0 & z_{12(1)} & 0 \\ 0 & 0 & z_{11(2)} & 0 & 0 & z_{12(2)} \\ \hline z_{21(0)} & 0 & 0 & z_{22(0)} & 0 & 0 \\ 0 & z_{21(1)} & 0 & 0 & z_{22(1)} & 0 \\ 0 & 0 & z_{21(2)} & 0 & 0 & z_{22(2)} \end{array}\right] \begin{bmatrix} I_{a0} \\ I_{a1} \\ I_{a2} \\ \hline I_{a'0} \\ I_{a'1} \\ I_{a'2} \end{bmatrix} + \begin{bmatrix} 0 \\ V_{z1} \\ 0 \\ \hline 0 \\ V_{z2} \\ 0 \end{bmatrix} \tag{9.95}$$

If we define, according to Figure 9.5,

$$\mathbf{I}_{\text{new}} \triangleq \hat{\mathbf{I}} = \begin{bmatrix} \hat{I}_{a0} \\ \hat{I}_{a'0} \end{bmatrix} = \begin{bmatrix} I_1 \\ I_2 \end{bmatrix} \tag{9.96}$$

where we use the circumflex to identify $\mathbf{I}_{\text{new}}$ since the prime has a different meaning here. Then $\mathbf{I} = \mathbf{K}\hat{\mathbf{I}}$, or for a series-series connection

$$\mathbf{I} = \begin{bmatrix} I_{a0} \\ I_{a1} \\ I_{a2} \\ \hline I_{a'0} \\ I_{a'1} \\ I_{a'2} \end{bmatrix} = \begin{matrix} \\ 0 \\ 1 \\ 2 \\ 0' \\ 1' \\ 2' \end{matrix} \begin{matrix} \hat{0} & \hat{0}' \\ \left[\begin{matrix} 1 & 0 \\ n_2 & 0 \\ n_1 & 0 \\ \hline 0 & 1 \\ 0 & n_2' \\ 0 & n_1' \end{matrix}\right] \end{matrix} \begin{bmatrix} \hat{I}_{a0} \\ \hat{I}_{a'0} \end{bmatrix} \tag{9.97}$$

Therefore, obviously,

$$\mathbf{K} = \begin{bmatrix} 1 & 0 \\ n_2 & 0 \\ n_1 & 0 \\ \hline 0 & 1 \\ 0 & n_2' \\ 0 & n_1' \end{bmatrix} \tag{9.98}$$

We easily show that $n_2^* = n_1$ and $n_1^* = n_2$ so that

$$\mathbf{K}^{*t} = \left[\begin{array}{ccc|ccc} 1 & n_1 & n_2 & 0 & 0 & 0 \\ 0 & 0 & 0 & 1 & n_1' & n_2' \end{array}\right] \tag{9.99}$$

Then

$$\hat{\mathbf{V}} = \mathbf{K}^{*t}\mathbf{V} \tag{9.100}$$

or

$$\hat{\mathbf{V}} = \begin{bmatrix} V_1 \\ V_2 \end{bmatrix} = \left[\begin{array}{ccc|ccc} 1 & n_1 & n_2 & 0 & 0 & 0 \\ 0 & 0 & 0 & 1 & n_1' & n_2' \end{array}\right] \begin{bmatrix} V_{K0} \\ V_{K1} \\ V_{K2} \\ \hline V_{K'0} \\ V_{K'1} \\ V_{K'2} \end{bmatrix} = \begin{bmatrix} V_{K0} + n_1 V_{K1} + n_2 V_{K2} \\ V_{K'0} + n_1' V_{K'1} + n_2' V_{K'2} \end{bmatrix} \tag{9.101}$$

Now (9.95) may be written in matrix form, using **I** and **V** as defined in (9.94) and (9.100) respectively, as

$$\mathbf{V} = \mathbf{Z}\,\mathbf{I} + \mathbf{V}_z \tag{9.102}$$

Premultiplying by $\mathbf{K}^{*t}$ gives

$$\hat{\mathbf{V}} = \mathbf{K}^{*t}\mathbf{Z}\,\mathbf{I} + \mathbf{K}^{*t}\mathbf{V}_z = (\mathbf{K}^{*t}\mathbf{Z}\,\mathbf{K})\,\hat{\mathbf{I}} + \mathbf{K}^{*t}\mathbf{V}_z = \hat{\mathbf{Z}}\,\hat{\mathbf{I}} + \hat{\mathbf{V}}_z \tag{9.103}$$

But, for a Z-type fault, $\hat{\mathbf{V}} = \mathbf{0}$ so that we compute, assuming $\hat{\mathbf{Z}}^{-1}$ exists,

$$\hat{\mathbf{I}} = -\,\hat{\mathbf{Z}}^{-1}\,\hat{\mathbf{V}}_z \tag{9.104}$$

We easily show that $\hat{\mathbf{Z}}^{-1}$ does exist since

$$\hat{\mathbf{Z}} = \mathbf{K}^{*t}\mathbf{Z}\,\mathbf{K} = \begin{bmatrix} z_{11} & z_{12} \\ z_{21} & z_{22} \end{bmatrix} \tag{9.105}$$

where z_{11}, z_{12}, z_{21}, and z_{22} are defined by equation (9.62) in section 9.4. Also

$$\hat{\mathbf{V}}_z = \mathbf{K}^{*t} \begin{bmatrix} 0 \\ V_{z1} \\ 0 \\ \hline 0 \\ V_{z2} \\ 0 \end{bmatrix} = \begin{bmatrix} n_1 V_{z1} \\ n_1' V_{z2} \end{bmatrix} \tag{9.106}$$

The sequence currents at the fault are then easily found by premultiplying (9.104) by K, i.e.,

$$\mathbf{I} = -\,\mathbf{K}\,\hat{\mathbf{Z}}^{-1}\,\hat{\mathbf{V}}_z \tag{9.107}$$

9.8 Constraint Matrix for Y-type and H-type Faults

Proceeding in exactly the same manner as above, we may easily establish the constraint matrix K for Y-type faults. For this type of fault it is convenient to write

$$\begin{bmatrix} I_{ai} \\ I_{a'i} \end{bmatrix} = \begin{bmatrix} I_{Ki} \\ I_{K'i} \end{bmatrix} = \begin{bmatrix} y_{11(i)} & y_{12(i)} \\ y_{21(i)} & y_{22(i)} \end{bmatrix} \begin{bmatrix} V_{Ki} \\ V_{K'i} \end{bmatrix} + \delta_{1i} \begin{bmatrix} I_{y1} \\ I_{y2} \end{bmatrix}, \qquad i = 0, 1, 2 \tag{9.108}$$

We then develop the constraint

$$\mathbf{V} = \mathbf{K}\,\hat{\mathbf{V}} \tag{9.109}$$

or

$$\begin{bmatrix} V_{K0} \\ V_{K1} \\ V_{K2} \\ \hline V_{K'0} \\ V_{K'1} \\ V_{K'2} \end{bmatrix} = \begin{matrix} \\ 0 \\ 1 \\ 2 \\ \\ 0' \\ 1' \\ 2' \end{matrix} \overset{\begin{matrix} \hat{0} & \hat{0}' \end{matrix}}{\begin{bmatrix} 1 & 0 \\ n_2 & 0 \\ n_1 & 0 \\ \hline 0 & 1 \\ 0 & n_2' \\ 0 & n_1' \end{bmatrix}} \begin{bmatrix} \hat{V}_{K0} \\ \hline \hat{V}_{K'0} \end{bmatrix} \tag{9.110}$$

where we define according to Figure 9.13,

$$\begin{bmatrix} \hat{V}_{K0} \\ \hat{V}_{K'0} \end{bmatrix} = \begin{bmatrix} V_1 \\ V_2 \end{bmatrix} \tag{9.111}$$

Using V as defined in (9.110), we write (9.108) as

$$\begin{bmatrix} I_{a0} \\ I_{a1} \\ I_{a2} \\ \hline I'_{a2} \\ I'_{a1} \\ I'_{a2} \end{bmatrix} = \left[\begin{array}{ccc|ccc} y_{11(0)} & 0 & 0 & y_{12(0)} & 0 & 0 \\ 0 & y_{11(1)} & 0 & 0 & y_{12(1)} & 0 \\ 0 & 0 & y_{11(2)} & 0 & 0 & y_{12(2)} \\ \hline y_{21(0)} & 0 & 0 & y_{22(0)} & 0 & 0 \\ 0 & y_{21(1)} & 0 & 0 & y_{22(1)} & 0 \\ 0 & 0 & y_{21(2)} & 0 & 0 & y_{22(2)} \end{array}\right] \begin{bmatrix} V_{K0} \\ V_{K1} \\ V_{K2} \\ \hline V_{K'0} \\ V_{K'1} \\ V_{K'2} \end{bmatrix} + \begin{bmatrix} 0 \\ I_{y1} \\ 0 \\ \hline 0 \\ I_{y2} \\ 0 \end{bmatrix} \tag{9.112}$$

From (9.110) we note that the constraint matrix **K** is exactly the same as that of (9.98) for the Z-type fault. For power invariance we compute

$$\hat{\mathbf{I}} = \mathbf{K}^{*t}\mathbf{I} \tag{9.113}$$

and write

$$\hat{\mathbf{I}} = \hat{\mathbf{Y}}\,\hat{\mathbf{V}} + \hat{\mathbf{I}}_y = 0 \tag{9.114}$$

Then

$$\hat{\mathbf{V}} = -\,\hat{\mathbf{Y}}^{-1}\,\hat{\mathbf{I}}_y \tag{9.115}$$

and from (9.109)

$$\mathbf{V} = -\,\mathbf{K}\,\hat{\mathbf{Y}}^{-1}\,\hat{\mathbf{I}}_y \tag{9.116}$$

Thus, as long as we use the **Y** parameter description, the **Y**-type fault is described by equations quite similar to those for the Z-type fault condition.

For the H-type fault it is easy to show that the constraint matrix **K** is again the same as in (9.98). In this case we write

$$\begin{bmatrix} V_{Ki} \\ I_{K'i} \end{bmatrix} = \begin{bmatrix} h_{11(i)} & h_{12(i)} \\ h_{21(i)} & h_{22(i)} \end{bmatrix} \begin{bmatrix} I_{Ki} \\ V_{K'i} \end{bmatrix} + \delta_{1i} \begin{bmatrix} V_{h1} \\ I_{h2} \end{bmatrix}, \quad i = 0, 1, 2 \tag{9.117}$$

or

$$\begin{bmatrix} V_{K0} \\ V_{K1} \\ V_{K2} \\ \hline I'_{a0} \\ I'_{a1} \\ I'_{a2} \end{bmatrix} = \left[\begin{array}{ccc|ccc} h_{11(0)} & 0 & 0 & h_{12(0)} & 0 & 0 \\ 0 & h_{11(1)} & 0 & 0 & h_{12(1)} & 0 \\ 0 & 0 & h_{11(2)} & 0 & 0 & h_{12(2)} \\ \hline h_{21(0)} & 0 & 0 & h_{22(0)} & 0 & 0 \\ 0 & h_{21(1)} & 0 & 0 & h_{22(1)} & 0 \\ 0 & 0 & h_{21(2)} & 0 & 0 & h_{22(2)} \end{array}\right] \begin{bmatrix} I_{a0} \\ I_{a1} \\ I_{a2} \\ \hline V_{K'0} \\ V_{K'1} \\ V_{K'2} \end{bmatrix} + \begin{bmatrix} 0 \\ V_{h1} \\ 0 \\ \hline 0 \\ I_{h2} \\ 0 \end{bmatrix} \tag{9.118}$$

Then we compute

$$\begin{bmatrix} I_{a0} \\ I_{a1} \\ I_{a2} \\ \hline V_{K'0} \\ V_{K'1} \\ V_{K'2} \end{bmatrix} = \mathrm{K}\begin{bmatrix} \hat{I}_{a0} \\ \hat{V}_{K'0} \end{bmatrix} = \mathrm{K}\begin{bmatrix} I_1 \\ V_2 \end{bmatrix} \tag{9.119}$$

and K is exactly the same as before. Using the notation of (9.11), we write

$$\mathrm{M} = \mathrm{H}\mathrm{N} + \mathrm{M}_h \tag{9.120}$$

But

$$\mathrm{N} = \mathrm{K}\hat{\mathrm{N}} \tag{9.121}$$

and for power invariance

$$\hat{\mathrm{M}} = \mathrm{K}^{*t}\mathrm{M} \tag{9.122}$$

Premultiplying (9.120) by K^{*t}, we compute

$$\hat{\mathrm{M}} = \hat{\mathrm{H}}\hat{\mathrm{N}} + \hat{\mathrm{M}}_h \tag{9.123}$$

where

$$\hat{\mathrm{H}} = \mathrm{K}^{*t}\mathrm{H}\mathrm{K}, \qquad \hat{\mathrm{M}}_h = \mathrm{K}^{*t}\hat{\mathrm{M}}_h \tag{9.124}$$

But $\hat{\mathrm{M}} = 0$ for this type fault, so that

$$\hat{\mathrm{N}} = -\hat{\mathrm{H}}^{-1}\hat{\mathrm{M}}_h \tag{9.125}$$

and $\mathrm{N} = -\mathrm{K}\hat{\mathrm{H}}^{-1}\hat{\mathrm{M}}_h$ or

$$\begin{bmatrix} I_1 \\ V_2 \end{bmatrix} = -\mathrm{K}\hat{\mathrm{H}}^{-1}\begin{bmatrix} n_1 V_{h1} \\ n_1' I_{h2} \end{bmatrix} \tag{9.126}$$

Problems

9.1. Compute the Y parameters for the two-port network of Figure 9.2. Check against the Z parameters of Example 9.1.

9.2. Compute the H parameters for the two-port network of Figure 9.2. Check against the Z parameters of Example 9.1.

9.3. Compute the A parameters for the two-port network of Figure 9.2. Check against the Z parameters of Example 9.1.

9.4. Compute the Z, Y, H, and A parameters for the two-port networks of Figure P9.4.

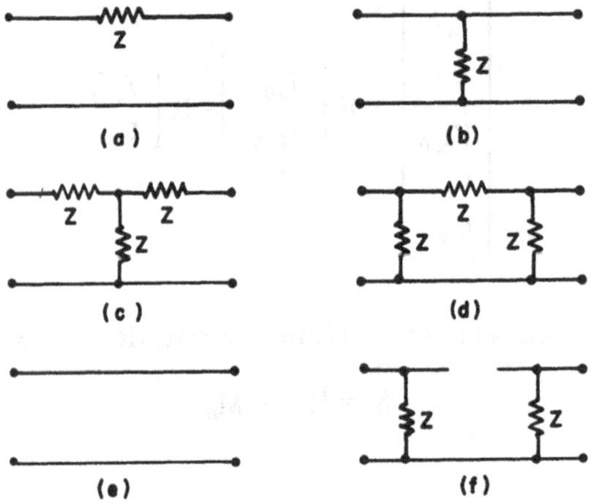

Fig. P9.4.

9.5. Compute the A parameters of the two-port networks of Figure P9.5.

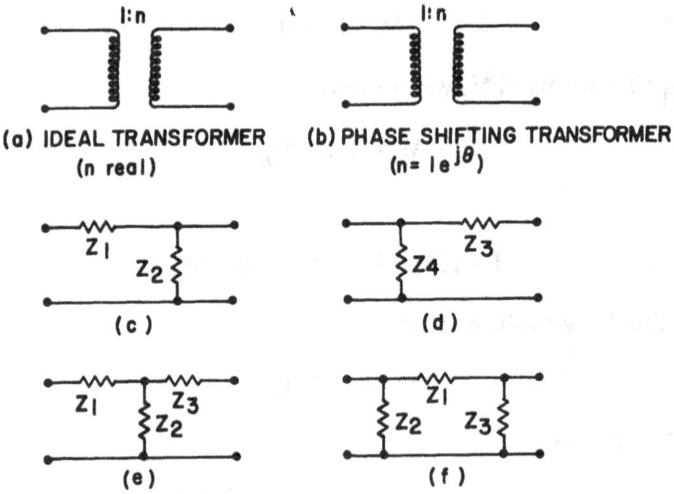

Fig. P9.5.

9.6. Compute the Z and Y parameters of the following two-port networks.

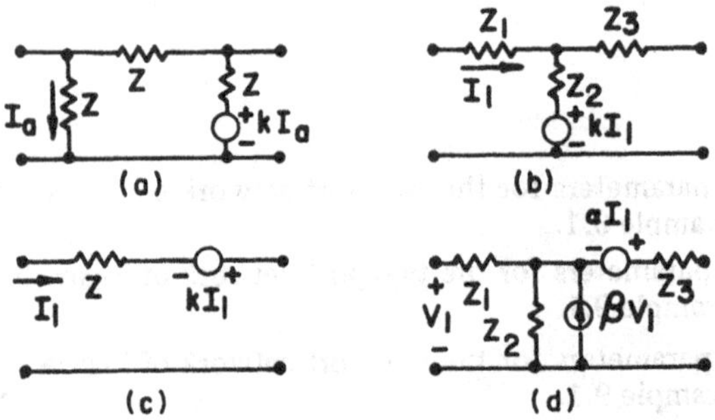

Fig. P9.6.

9.7. Compute the G and H parameters of the two-port networks of Figure P9.6.

9.8. Compute the Z and Y two-port equations for the networks of Figure P9.6 if the sources shown in each network are regarded as independent sources of magnitude V_s or I_s.

9.9. Repeat problem 9.8, solving for the G and H equations.

9.10. Repeat Example 9.3 if the controlled source has a value $I = kI_2$.

9.11. Verify the first column of Table 9.5.

9.12. Verify the second column of Table 9.5.

9.13. Verify the results of Example 9.5 by direct computation of the H equation.

9.14. Sketch the series connection of two-port networks of Figure P9.4 and determine the Z matrix of the interconnected network for the following combinations:

1. (a) in series with (b).
2. (a) in series with (c).
3. (b) in series with (c).
4. (c) in series with (d).
5. (a), (c), and (e) in series.
6. (d) in series with (f).

9.15. Investigate the parallel connection of the combinations named in problem 9.14.

9.16. Investigate the hybrid (series-parallel) connection of the combinations named in problem 9.14.

9.17. Compute the transmission parameters for a cascade connection (left to right) of the following two-port networks from Figure P9.4.

1. (a) and (b) with impedances labeled Z_1 and Z_2 respectively.
2. (b) and (a) with similar labeling.
3. (a), (b), and (a).
4. (b), (a), and (b).
5. (a) and a two-port network with matrix A.
6. A two-port network with matrix A and (a).

9.18. Consider the positive sequence network only and identify the currents and voltages defined in Figure 9.9 in terms of sequence currents and voltages of phase a.

(a) For a shunt unbalance at F, shunt unbalance at F'.
(b) For a series unbalance at F, shunt unbalance at F'.

9.19. Verify equation (9.54).

9.20. Beginning with the currents entering the series-series connection given by (9.65), develop an expression for the currents entering each of the ports of each sequence network (inside the transformers), and simplify as much as possible.

9.21. Repeat Example 9.6 if the fault condition is given as:

Fault F: lines b and c open
Fault F': SLG fault on a

9.22. Repeat Example 9.6 if the fault condition is given as:

Fault F: SLG fault on phase b
Fault F': SLG fault on phase c

9.23. Repeat Example 9.6 if the fault condition is given as:

Fault F: lines a and b open
Fault F': lines b and c open

9.24. Repeat Example 9.6 if the fault conditions F and F' are reversed.

9.25. Verify (9.69).

9.26. Beginning with (9.78) compute the voltages at each port of the sequence networks (inside the transformers) and show how these voltages can be used to completely solve the sequence networks.

9.27. Repeat Example 9.7 for the parallel-parallel fault condition:

Fault F: 2LG fault on phases a and b
Fault F': line b open

9.28. Repeat Example 9.7 for the parallel-parallel fault condition:

Fault F: 2LG fault on phases a and b
Fault F': 2LG fault on phases b and c

9.29. Repeat Example 9.7 for the parallel-parallel fault condition:

Fault F: line a open
Fault F': line c open

9.30. Beginning with (9.91) compute the sequence voltages and/or currents inside the transformers and show how these quantities can be used to completely solve the sequence networks.

9.31. Repeat Example 9.6 for the series-parallel fault condition:

Fault F: SLG fault on phase c
Fault F': 2LG fault on phases b and c

9.32. Repeat Example 9.6 for the series-parallel fault condition:

Fault F: SLG fault on phase b
Fault F': line b open

9.33. Repeat Example 9.6 for the series-parallel fault condition:

Fault F: lines a and b open
Fault F': line b open

9.34. Repeat Example 9.6 for the series-parallel fault condition:

Fault F: lines a and b open
Fault F': line c open

9.35. Compute the constraint matrix **K** for the fault condition of Example 9.6 using definition (9.98) for **K**. Then solve for the symmetrical components of the fault currents by using (9.107).

9.36. Compute the constraint matrix **K** for the fault condition of Example 9.7. Then solve for the symmetrical components of the voltages at the faults by using (9.116).

9.37. Compute the constraint matrix **K** for the fault condition of problem 9.31. Then solve for the symmetrical component quantities at the fault by using (9.126).

9.38. Compute the inverse of the **Z** matrix of (9.95). Is this the same as the **Y** matrix of (9.112)?

9.39. Suppose that for the Y-type fault we construct a constraint matrix based on the current equations of Table 8.2. Investigate using this constraint matrix to solve for the sequence quantities in a Y-type fault.

chapter 10

Analytical Simplifications

The preceding material on symmetrical components is based upon what is often called the three-component method, i.e., there are three sequence networks and three sets of sequence voltages and currents. This is emphasized in the computation of simple faults in Chapter 3 and is continued for more complex situations in Chapter 9. There is one concern, however, in using the three-component method, and that is the large number of network components required to describe the system. Recall that we begin with a three-phase system and simplify it by means of a single-phase (or per phase) representation. But in order to completely describe unbalanced situations, we need *three* such single-phase representations, namely the positive, negative, and zero sequence systems. What then has been gained? Perhaps we would have been just as well off to have retained the three phases a, b, and c.

This problem of the number of components needed to represent a system is crucial, whether the problem is to be solved by digital computer, by network analyzer, or by hand. In computer solutions, whether analog or digital, the network representation occupies the machine hardware or memory, and this is always limited by the size of the machine itself. Thus an arbitrary upper limit is always placed on the *size* of system which can be solved, and this limit is related to the computing device to be used.

10.1 Two-Component Method

The two-component method for solving unbalanced system conditions is based on the idea that the positive and negative sequence networks have nearly the same impedance. Recall that the positive sequence voltage is defined to have the phase sequence of the system generators. We arbitrarily label this sequence a-b-c. The negative sequence then has phase sequence a-c-b, and this is the only difference between the two. Both are balanced sets of voltages (or currents).

It is intuitively obvious that any passive network components, such as transmission lines or transformers, present the same impedance to either positive or negative sequence currents. Thus the impedances Z_1 and Z_2 for these components are equal. Only the rotating machines have different positive and negative sequence impedance. In synchronous machines, which contribute most of the fault current in a power system, these sequence impedances are nearly the same if we consider only the subtransient condition. Indeed, from Table 6.1 we see that for both round rotor and salient pole machines the subtransient and negative sequence reactances are exactly equal, i.e.,

$$x_d'' = x_2 \tag{10.1}$$

This is not exactly true for synchronous condensers or synchronous motors. However, these devices are usually small compared to the generator ratings, and some error in their fault computation can be tolerated.

Because of the equality or near-equality of all positive and negative sequence impedances we make the assumption that

$$Z_2 = Z_1 \tag{10.2}$$

for all circuit components; i.e., we will replace all negative sequence impedances by the corresponding positive sequence values.[1] Referring to Figure 2.5 with $Z_2 = Z_1$, we may rewrite the primitive sequence network equation (2.58) as

$$\begin{bmatrix} V_{a0} \\ V_{a1} \\ V_{a2} \end{bmatrix} = \begin{bmatrix} 0 \\ \mathrm{h}V_F \\ 0 \end{bmatrix} - \begin{bmatrix} Z_0 & 0 & 0 \\ 0 & Z_1 & 0 \\ 0 & 0 & Z_1 \end{bmatrix} \begin{bmatrix} I_{a0} \\ I_{a1} \\ I_{a2} \end{bmatrix} \tag{10.3}$$

If we add and subtract the second and third rows of (10.3) and replace these rows by the resulting sum and difference quantities, we have

$$\begin{bmatrix} V_{a0} \\ V_{a1} + V_{a2} \\ V_{a1} - V_{a2} \end{bmatrix} = \begin{bmatrix} 0 \\ \mathrm{h}V_F \\ \mathrm{h}V_F \end{bmatrix} - \begin{bmatrix} Z_0 & 0 & 0 \\ 0 & Z_1 & 0 \\ 0 & 0 & Z_1 \end{bmatrix} \begin{bmatrix} I_{a0} \\ I_{a1} + I_{a2} \\ I_{a1} - I_{a2} \end{bmatrix} \tag{10.4}$$

These equations are interesting because the last two rows are both positive sequence equations. This is confirmed by the fact that each contains V_F, a positive sequence source, and Z_1, a positive sequence impedance. These equations, therefore, describe the positive sequence Thevenin equivalent of Figure 2.5. If this is true, the sum and difference quantities $V_{a1} + V_{a2}$, $V_{a1} - V_{a2}$, and the corresponding currents are also positive sequence quantities. Thus under the assumption (10.2) we have created a new mathematical description consisting of one zero sequence equation and two positive sequence equations as in (10.4).

Because of the interest in positive-plus-negative and positive-minus-negative quantities it is also convenient to similarly rearrange the analysis and synthesis equations (2.21) and (2.23). For convenience we define the sum and difference quantities

$$\begin{bmatrix} V_{a0} \\ V_{a\Sigma} \\ V_{a\Delta} \end{bmatrix} \triangleq \begin{bmatrix} V_{a0} \\ V_{a1} + V_{a2} \\ V_{a1} - V_{a2} \end{bmatrix} \tag{10.5}$$

with a similar definition for currents. Then from

$$\begin{bmatrix} V_{a0} \\ V_{a1} \\ V_{a2} \end{bmatrix} = \frac{\mathrm{h}}{3} \begin{bmatrix} 1 & 1 & 1 \\ 1 & \mathrm{a} & \mathrm{a}^2 \\ 1 & \mathrm{a}^2 & \mathrm{a} \end{bmatrix} \begin{bmatrix} V_a \\ V_b \\ V_c \end{bmatrix} \tag{10.6}$$

we add and subtract the second and third rows to derive the new equation

[1] Some engineers use the average of Z_1 and Z_2 for machines.

$$\begin{bmatrix} V_{a0} \\ V_{a\Sigma} \\ V_{a\Delta} \end{bmatrix} = \frac{h}{3} \begin{bmatrix} 1 & 1 & 1 \\ 2 & -1 & -1 \\ 0 & j\sqrt{3} & -j\sqrt{3} \end{bmatrix} \begin{bmatrix} V_a \\ V_b \\ V_c \end{bmatrix} \tag{10.7}$$

or in matrix form

$$\mathbf{V}_{0\Sigma\Delta} = \mathbf{B}^{-1}\mathbf{V}_{abc} \tag{10.8}$$

where $\mathbf{V}_{0\Sigma\Delta}$ is defined as noted and

$$\mathbf{B}^{-1} \triangleq \frac{h}{3} \begin{bmatrix} 1 & 1 & 1 \\ 2 & -1 & -1 \\ 0 & j\sqrt{3} & -j\sqrt{3} \end{bmatrix} \tag{10.9}$$

From (10.8) we compute

$$\mathbf{V}_{abc} = \mathbf{B}\,\mathbf{V}_{0\Sigma\Delta} \tag{10.10}$$

where

$$\mathbf{B} \triangleq \frac{1}{h} \begin{bmatrix} 1 & 1 & 0 \\ 1 & -1/2 & -j\sqrt{3}/2 \\ 1 & -1/2 & +j\sqrt{3}/2 \end{bmatrix} \tag{10.11}$$

Using these relations, we may solve for any of the shunt or series unbalances as before since we may still write the same number of equations. The number of *networks*, however, is reduced from three to two since both the sigma and delta quantities are defined as voltages or currents associated with the positive sequence network. The two primitive sequence networks are shown in Figure 10.1.

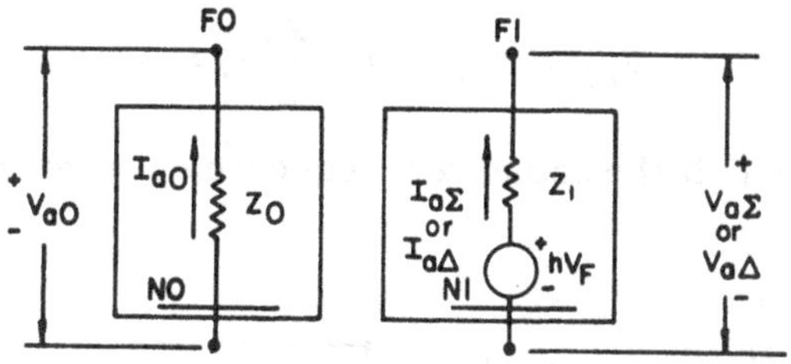

Fig. 10.1. Two sequence networks for the two-component method.

I. SHUNT FAULTS

We begin with an analysis of shunt faults which will parallel the three-component development of Chapter 3. Frequent comparison with the results of Chapter 3 is recommended.

10.2 Single-Line-to-Ground Fault

From Figure 3.2 we note that phase a is the symmetrical phase and that the boundary conditions are

$$I_b = I_c = 0, \qquad V_a = Z_f I_a \tag{10.12}$$

From (10.10) and (10.12) we compute

$$I_b = 0 = \frac{1}{h} I_{a0} - \frac{1}{2h} I_{a\Sigma} - j\frac{\sqrt{3}}{2h} I_{a\Delta}$$

$$I_c = 0 = \frac{1}{h} I_{a0} - \frac{1}{2h} I_{a\Sigma} + j\frac{\sqrt{3}}{2h} I_{a\Delta} \qquad (10.13)$$

Adding these equations, we compute for any h

$$2I_{a0} = I_{a\Sigma} \qquad (10.14)$$

and subtracting, we have

$$I_{a\Delta} = 0 \qquad (10.15)$$

From the second equation of (10.12) we write for any h

$$V_{a0} + (V_{a1} + V_{a2}) = Z_f(I_{a0} + I_{a1} + I_{a2})$$

and substituting I_{a0} from (10.14),

$$V_{a0} + V_{a\Sigma} = (3/2)\, Z_f I_{a\Sigma} \qquad (10.16)$$

Equation (10.15) requires that the positive sequence network be open at the fault point. This network is shown in Figure 10.2a and is recognized to be the normal (unfaulted) positive sequence network. All voltages measured in this normal network are to be labeled delta quantities as noted.

Equations (10.14) and (10.16) dictate that the positive and zero sequence networks be connected in series to compute the zero sequence and sigma quantities. This connection is shown in Figure 10.2b. We also note that since the current through the zero sequence network is $2I_{a0}$, we satisfy the first equation of (10.4) as follows:

$$V_{a0} = -Z_0 I_{a0} = -\frac{Z_0}{2}(2I_{a0}) \qquad (10.17)$$

Thus we use only one-half the actual zero sequence impedance values.

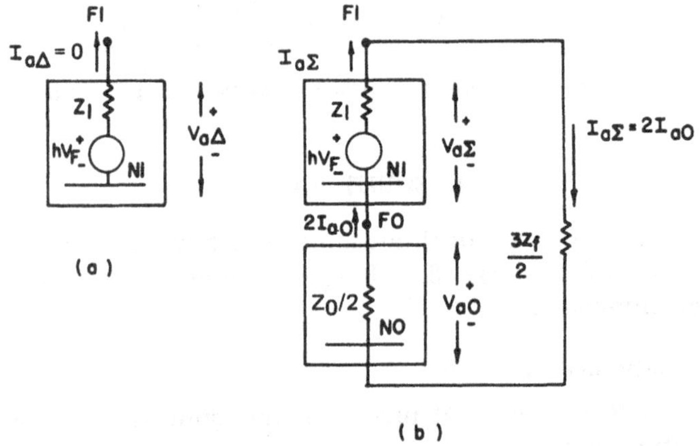

Fig. 10.2. Sequence network connections for a SLG fault on phase *a* with fault impedance Z_f: (a) delta, (b) sigma and zero.

From the networks of Figure 10.2 we easily compute the sequence voltages and currents at any point in the network. Then we apply equation (10.10) to synthesize phase quantities. At the fault the total current is $I_a = 3I_{a0}$. By inspection of Figure 10.2 we compute

$$2I_{a0} = \frac{hV_F}{(Z_0/2) + Z_1 + (3Z_f/2)}$$

or

$$I_{a0} = \frac{hV_F}{Z_0 + 2Z_1 + 3Z_f} \tag{10.18}$$

This is exactly the same as (3.5), the current computed by the three-component method, with Z_2 set equal to Z_1.

10.3 Line-to-Line Fault

From Figure 3.9 we note that phase a is the symmetrical phase. Also, from (3.6)–(3.8) and Figure 3.9 we write the boundary conditions

$$I_a = 0, \quad I_b = -I_c, \quad V_b - V_c = V_{bc} = Z_f I_b \tag{10.19}$$

Substituting (10.19) into the current version of (10.7), we compute

$$\begin{bmatrix} I_{a0} \\ I_{a\Sigma} \\ I_{a\Delta} \end{bmatrix} = \frac{h}{3}\begin{bmatrix} 1 & 1 & 1 \\ 2 & -1 & -1 \\ 0 & j\sqrt{3} & -j\sqrt{3} \end{bmatrix}\begin{bmatrix} 0 \\ I_b \\ -I_b \end{bmatrix} = \begin{bmatrix} 0 \\ 0 \\ j\dfrac{2\sqrt{3}}{3}hI_b \end{bmatrix} \tag{10.20}$$

The known voltage is V_{bc} which we compute from (10.10) as follows:

$$\begin{bmatrix} V_a \\ V_b \\ V_c \end{bmatrix} = \frac{1}{h}\begin{bmatrix} 1 & 1 & 0 \\ 1 & -1/2 & -j\sqrt{3}/2 \\ 1 & -1/2 & +j\sqrt{3}/2 \end{bmatrix}\begin{bmatrix} V_{a0} \\ V_{a\Sigma} \\ V_{a\Delta} \end{bmatrix}$$

Then

$$V_{bc} = V_b - V_c = -\frac{j\sqrt{3}}{h}V_{a\Delta} \tag{10.21}$$

From (10.20) we learn that both I_{a0} and $I_{a\Sigma}$ are zero. Thus the sigma currents and voltages are found from the normal network as shown in Figure 10.3a.

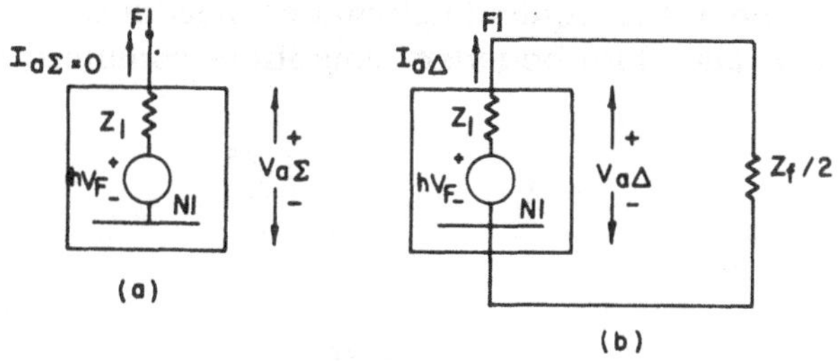

Fig. 10.3. Sequence network connections for a LL fault on phases b-c with fault impedance Z_f: (a) sigma, (b) delta.

From the third boundary condition (10.19) and equations (10.20) and (10.21) we compute for any h,

$$V_{a\Delta} = (Z_f/2)\, I_{a\Delta} \tag{10.22}$$

This equation is satisfied by the network of Figure 10.3b. From the two networks of Figure 10.3 the sequence voltages and currents may be completely determined.

The total sequence current at the fault may also be computed from Figure 10.3. From Figure 10.3a

$$I_{a\Sigma} = 0 \tag{10.23}$$

and from Figure 10.3b

$$I_{a\Delta} = \frac{\mathrm{h} V_F}{Z_1 + Z_f/2} \tag{10.24}$$

Adding (10.23) and (10.24), we easily compute

$$I_{a1} = \frac{\mathrm{h} V_F}{2Z_1 + Z_f} \tag{10.25}$$

This is exactly the same result found by the three-component method if we replace $Z_2 = Z_1$ in (3.11).

10.4 Double Line-to-Ground Fault

From Figure 3.13 we again note that phase a is the symmetrical phase. From (3.12)–(3.14) and Figure 3.13 the boundary conditions are

$$I_a = 0, \qquad V_b = Z_f I_b + Z_g(I_b + I_c), \qquad V_c = Z_f I_c + Z_g(I_b + I_c) \tag{10.26}$$

Since $I_a = 0$, we compute

$$\begin{bmatrix} I_{a0} \\ I_{a\Sigma} \\ I_{a\Delta} \end{bmatrix} = \frac{\mathrm{h}}{3} \begin{bmatrix} 1 & 1 & 1 \\ 2 & -1 & -1 \\ 0 & \mathrm{j}\sqrt{3} & -\mathrm{j}\sqrt{3} \end{bmatrix} \begin{bmatrix} 0 \\ I_b \\ I_c \end{bmatrix} = \frac{\mathrm{h}}{3} \begin{bmatrix} I_b + I_c \\ -(I_b + I_c) \\ \mathrm{j}\sqrt{3}(I_b - I_c) \end{bmatrix} \tag{10.27}$$

from which we observe that all sequence quantities will usually be nonzero since the right-hand side of (10.27) will be nonzero. This is because I_b and I_c are generally neither equal nor opposite for this type of fault. We also observe from (10.27) that for any h,

$$I_{a0} = -I_{a\Sigma} \tag{10.28}$$

and this establishes one of the required network connections.

From the second and third boundary conditions combined with (10.27) we write

$$V_b - V_c = Z_f(I_b - I_c) = -\frac{\mathrm{j}\sqrt{3}}{\mathrm{h}} Z_f I_{a\Delta}$$

But from (10.10)

$$V_b - V_c = -\frac{\mathrm{j}\sqrt{3}}{\mathrm{h}} V_{a\Delta}$$

Combining, we have for any h,

$$V_{a\Delta} = Z_f I_{a\Delta} \tag{10.29}$$

This completely determines the delta network connection shown in Figure 10.4a.

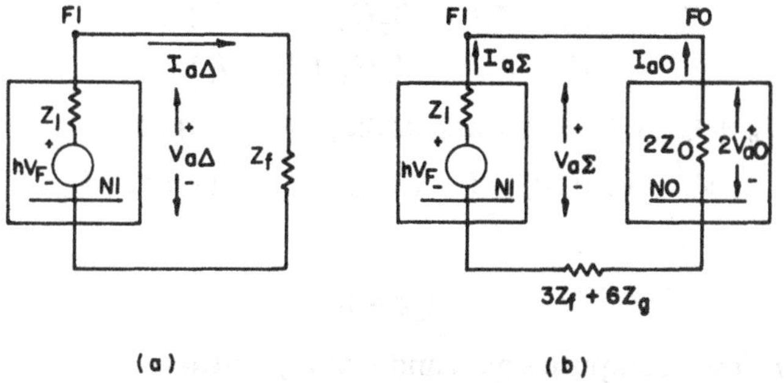

Fig. 10.4. Sequence network connections for a 2LG fault on phases *b-c* with phase impedance Z_f and ground impedance Z_g (see Figure 3.13): (a) delta, (b) sigma and zero.

Again from (10.26) and (10.27) we compute

$$V_b + V_c = Z_f(I_b + I_c) + 2Z_g(I_b + I_c)$$
$$= (Z_f + 2Z_g)(I_b + I_c) = (Z_f + 2Z_g)\left(\frac{3}{\text{h}} I_{a0}\right)$$

But from (10.10)

$$V_b + V_c = \frac{2}{\text{h}} V_{a0} - \frac{1}{\text{h}} V_{a\Sigma}$$

Combining, we have for any h

$$2V_{a0} - V_{a\Sigma} = (3Z_f + 6Z_g)I_{a0} \tag{10.30}$$

This equation is satisfied by the network connection of Figure 10.4b where twice the normal zero sequence impedance is set in order to force the voltage relationship of (10.30).

From Figure 10.4a and Figure 10.4b we compute

$$I_{a\Delta} = \frac{\text{h}V_F}{Z_1 + Z_f}, \qquad I_{a\Sigma} = \frac{\text{h}V_F}{Z_1 + 2Z_0 + 3Z_f + 6Z_g}$$

Adding, we compute

$$2I_{a1} = \text{h}V_F\left(\frac{1}{Z_1 + Z_f} + \frac{1}{Z_1 + 2Z_0 + 3Z_f + 6Z_g}\right)$$

which can be reduced to

$$I_{a1} = \frac{\text{h}V_F}{Z_1 + Z_f + \dfrac{(Z_1 + Z_f)(Z_0 + Z_f + 3Z_g)}{Z_0 + Z_1 + 2Z_f + 3Z_g}} \tag{10.31}$$

This equation is exactly the same as (3.24) which was derived from the three-component method if Z_2 in (3.24) is replaced by Z_1.

10.5 Three-Phase Fault

The general 3ϕ fault condition is shown in Figure 3.18 where we note the boundary conditions

$$\begin{aligned} V_a &= Z_f I_a + Z_g(I_a + I_b + I_c) \\ V_b &= Z_f I_b + Z_g(I_a + I_b + I_c) \\ V_c &= Z_f I_c + Z_g(I_a + I_b + I_c) \end{aligned} \tag{10.32}$$

Replacing $I_a + I_b + I_c$ by $(3/\mathrm{h}) I_{a0}$ and adding, we compute

$$V_a + V_b + V_c = 0 = (3/\mathrm{h}) Z_f I_{a0} + (9/\mathrm{h}) Z_g I_{a0}$$

or

$$I_{a0} = 0 \tag{10.33}$$

Also, from (10.7) we compute sigma and delta quantities

$$\begin{bmatrix} I_{a0} \\ I_{a\Sigma} \\ I_{a\Delta} \end{bmatrix} = \frac{\mathrm{h}}{3} \begin{bmatrix} 1 & 1 & 1 \\ 2 & -1 & -1 \\ 0 & \mathrm{j}\sqrt{3} & -\mathrm{j}\sqrt{3} \end{bmatrix} \begin{bmatrix} I_a \\ I_b \\ I_c \end{bmatrix} = \frac{\mathrm{h}}{3} \begin{bmatrix} 0 \\ 2I_a - (I_b + I_c) \\ \mathrm{j}\sqrt{3}(I_b - I_c) \end{bmatrix}$$

But, for this connection since $I_a + I_b + I_c = 0$, then $I_a = -(I_b + I_c)$ and

$$\begin{bmatrix} I_{a0} \\ I_{a\Sigma} \\ I_{a\Delta} \end{bmatrix} = \frac{\mathrm{h}}{3} \begin{bmatrix} 0 \\ 3I_a \\ \mathrm{j}\sqrt{3}(I_b - I_c) \end{bmatrix} \tag{10.34}$$

From the first of equations (10.32) we have, with $I_a + I_b + I_c = 0$,

$$V_a = Z_f I_a = (Z_f/\mathrm{h}) I_{a\Sigma}$$

but $V_a = (1/\mathrm{h})\, V_{a0} + (1/\mathrm{h})\, V_{a\Sigma} = (1/\mathrm{h})\, V_{a\Sigma}$ since $I_{a0} = 0$, or

$$V_{a\Sigma} = Z_f I_{a\Sigma} \tag{10.35}$$

Subtracting the second two of equations (10.32), we compute, with

$$I_a + I_b + I_c = 0$$

and incorporating (10.34),

$$V_{bc} = V_b - V_c = Z_f(I_b - I_c) = -\frac{\mathrm{j}\sqrt{3}}{\mathrm{h}} Z_f I_{a\Delta}$$

But from (10.7)

$$V_{a\Delta} = -\frac{\mathrm{h}(V_b - V_c)}{\mathrm{j}\sqrt{3}}$$

or for any h,

$$V_{a\Delta} = Z_f I_{a\Delta} \tag{10.36}$$

Thus the equations for the sigma and delta networks are exactly the same and are as shown in Figure 10.5.

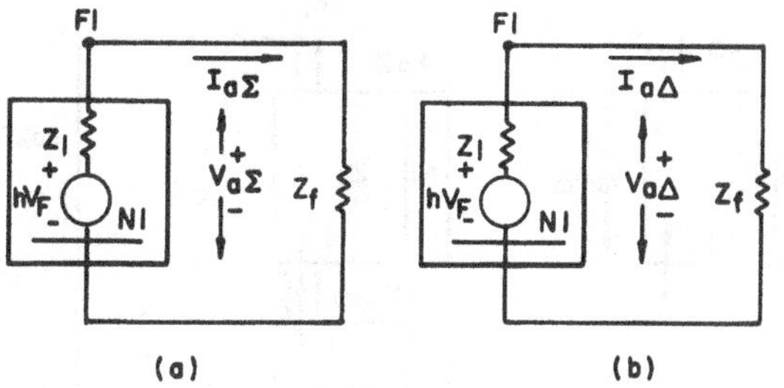

Fig. 10.5. Sequence network connections for a 3ϕ fault with impedance Z_f in each phase: (a) sigma, (b) delta.

By inspection of Figure 10.5 we compute

$$I_{a\Sigma} = \mathrm{h}V_F/(Z_1 + Z_f), \qquad I_{a\Delta} = \mathrm{h}V_F/(Z_1 + Z_f) \tag{10.37}$$

and adding we have $I_{a1} = \mathrm{h}V_F/(Z_1 + Z_f)$.

II. SERIES FAULTS

We next consider the application of the two-component method to series or longitudinal faults. Again, the network currents and voltages at the fault point are defined as indicated in Figure 3.20 with sequence networks as defined in Figure 3.21.

10.6 Two Lines Open (2LO)

The series unbalance for 2LO is shown schematically in Figure 3.32 for which the boundary conditions are seen to be

$$I_b = I_c = 0, \qquad V_{aa'} = ZI_a \tag{10.38}$$

where Z is the impedance in line a. Then from the current version of (10.7)

$$\begin{bmatrix} I_{a0} \\ I_{a\Sigma} \\ I_{a\Delta} \end{bmatrix} = \frac{\mathrm{h}}{3}\begin{bmatrix} 1 & 1 & 1 \\ 2 & -1 & -1 \\ 0 & \mathrm{j}\sqrt{3} & -\mathrm{j}\sqrt{3} \end{bmatrix}\begin{bmatrix} I_a \\ 0 \\ 0 \end{bmatrix} = \frac{\mathrm{h}}{3}\begin{bmatrix} I_a \\ 2I_a \\ 0 \end{bmatrix} \tag{10.39}$$

Thus the delta network must be open for this type of fault. Also, we observe that

$$2I_{a0} = I_{a\Sigma} \tag{10.40}$$

the same as for the SLG fault.

From the second boundary condition (10.38) incorporating (10.40) we write for any h,

$$V_{aa'0} + V_{aa'\Sigma} = Z(I_{a0} + I_{a\Sigma}) = (3/2)ZI_{a\Sigma} \tag{10.41}$$

Equations (10.40) and (10.41) fix the connection of sigma and zero sequence networks to be that of Figure 10.6. This connection has a striking resemblance to Figure 10.2 for a SLG shunt fault.

If we interpret $\mathrm{h}V_F$ to be the Thevenin equivalent open circuit voltage generated in the sigma network, we compute

$$2I_{a0} = I_{a\Sigma} = \frac{\mathrm{h}V_F}{Z_1 + Z_0/2 + (3/2)Z}$$

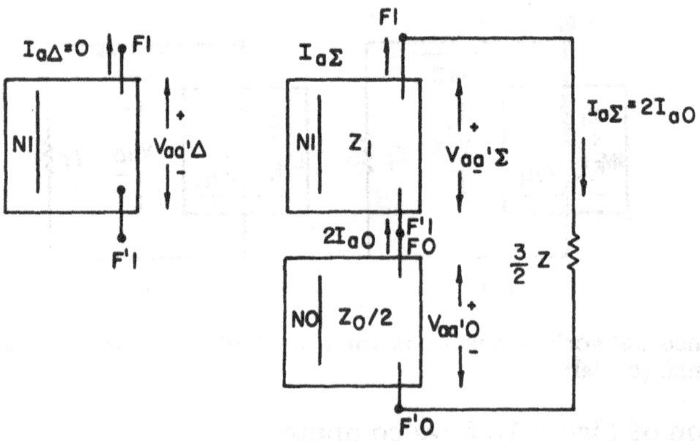

Fig. 10.6. Sequence network connections for phases b and c open and with impedance Z in phase a.

But we also know that $I_{a\Delta} = 0$. Adding these two equations, we have

$$2I_{a0} = 2I_{a1} = \frac{hV_F}{Z_1 + Z_0/2 + (3/2)Z}$$

or

$$I_{a0} = I_{a1} = \frac{hV_F}{2Z_1 + Z_0 + 3Z} \tag{10.42}$$

which agrees exactly with (3.86), derived by the three-component method, if we set $Z_1 = Z_2$ in that equation.

10.7 One Line Open (1LO)

If only one line is open and we assume that is line a, we have the boundary conditions

$$I_a = 0, \qquad V_{bb'} = ZI_b, \qquad V_{cc'} = ZI_c \tag{10.43}$$

where we assume an impedance Z in the sound lines.

Since $I_a = 0$, (10.27) applies and we immediately establish the similarity between this fault and the 2LG shunt fault where for any h,

$$I_{a0} = -I_{a\Sigma} \tag{10.44}$$

From the second two equations of (10.43) we compute the sigma and delta quantities

$$V_{bb'} + V_{cc'} = Z(I_b + I_c), \qquad V_{bb'} - V_{cc'} = Z(I_b - I_c) \tag{10.45}$$

But from (10.10) we compute

$$V_{bb'} + V_{cc'} = \frac{2}{h} V_{aa'0} - \frac{1}{h} V_{aa'\Sigma}$$

$$V_{bb'} - V_{cc'} = -\frac{j\sqrt{3}}{h} V_{aa'\Delta} \tag{10.46}$$

and from (10.27)

$$I_b + I_c = -\frac{3}{h} I_{a\Sigma}, \qquad I_b - I_c = -\frac{j\sqrt{3}}{h} I_{a\Delta} \tag{10.47}$$

Substituting (10.46) and (10.47) into (10.45), we compute for any h,

$$2V_{aa'0} - V_{aa'\Sigma} = -3ZI_{a\Sigma} \tag{10.48}$$

$$V_{aa'\Delta} = ZI_{a\Delta} \tag{10.49}$$

These two equations completely specify the network connections of Figure 10.7.

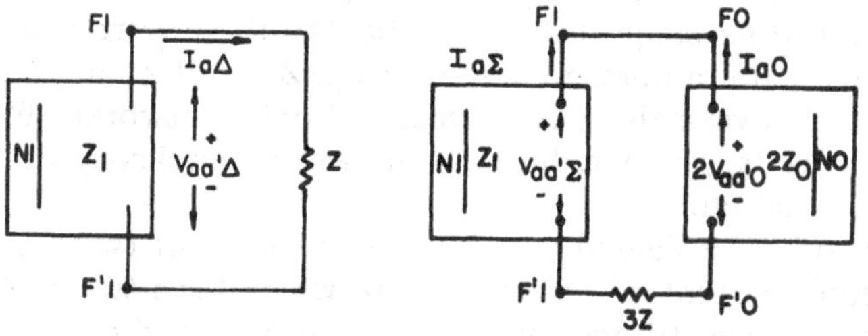

Fig. 10.7. Sequence network connections for line *a* open and an impedance *Z* in lines *b* and *c*.

This network connection is a parallel connection similar in construction to that derived for the 2LG shunt fault. By inspection we compute

$$I_{a\Delta} = \frac{hV_F}{Z_1 + Z}, \qquad I_{a\Sigma} = \frac{hV_F}{Z_1 + 2Z_0 + 3Z}$$

and adding we find, after algebraic reduction,

$$I_{a1} = \frac{hV_F}{Z_1 + Z + \dfrac{(Z_1 + Z)(Z_0 + Z)}{Z_0 + Z_1 + 2Z}} \tag{10.50}$$

This is exactly the same result computed by the three-component method and given by (3.78)-(3.80) if we set $Z_2 = Z_1$.

III. CHANGES IN SYMMETRY WITH TWO-COMPONENT CALCULATIONS

An obvious similarity exists between certain shunt and series fault connections computed by the two-component method. If we take advantage of these similarities, we should be able to extend the two-component computations to situations in which any phase is the symmetrical one.

It is again convenient to limit our analysis to common fault connections, namely, the shunt SLG and 2LG and the series 1LO and 2LO. These situations are shown in general form by Figures 8.1 and 8.5 for which we write the general equations, similar to (8.6)-(8.8) and (8.18), as

$$\begin{bmatrix} V_{a0} - 3Z_g I_{a0} \\ V_{a\Sigma} \\ V_{a\Delta} \end{bmatrix} = \frac{h}{3} \begin{bmatrix} Z_a & Z_b & Z_c \\ 2Z_a & -Z_b & -Z_c \\ 0 & +j\sqrt{3}\,Z_b & -j\sqrt{3}\,Z_c \end{bmatrix} \begin{bmatrix} I_a \\ I_b \\ I_c \end{bmatrix}$$

$$= \mathbf{B}^{-1} \begin{bmatrix} Z_a & 0 & 0 \\ 0 & Z_b & 0 \\ 0 & 0 & Z_c \end{bmatrix} \begin{bmatrix} I_a \\ I_b \\ I_c \end{bmatrix} \tag{10.51}$$

where $Z_g = 0$ for the series fault case. Then from (10.10) we may write

$$\begin{bmatrix} V_{aZ} \\ V_{a\Sigma} \\ V_{a\Delta} \end{bmatrix} = \mathbf{B}^{-1} \begin{bmatrix} Z_a & 0 & 0 \\ 0 & Z_b & 0 \\ 0 & 0 & Z_c \end{bmatrix} \mathbf{B} \begin{bmatrix} I_{a0} \\ I_{a\Sigma} \\ I_{a\Delta} \end{bmatrix} \tag{10.52}$$

where we define $V_{aZ} = V_{a0} - 3Z_g I_{a0}$.

The values of Z_a, Z_b, Z_c, and Z_g depend upon the type of fault, but in all cases (10.52) can be further simplified. In particular, if we restrict our consideration to the case where finite nonzero values of Z_a, Z_b, and Z_c are equal, there is no mutual coupling between the zero, sigma, and delta networks. We will make this simplification since it greatly reduces the network complexity while still retaining a realistic fault situation.

The faults to be considered may be recognized as of two types, those which have a single fault current of zero and those which have two zero currents. This information alone, substituted into (10.7), provides the interesting relations shown in Table 10.1.

Table 10.1. Sequence Current Relationships Due to Zero Phase Currents

Current	*Constraint*	*Current Relationships*
Simple zero	$I_a = 0$	$I_{a0} + I_{a\Sigma} = 0$ $-\mathrm{j}\sqrt{3}\, I_{a\Delta} = \mathrm{h}(I_b - I_c)$
	$I_b = 0$	$I_{a0} + I_{a\Sigma} = \mathrm{h}I_a$ $\mathrm{j}\sqrt{3}\, I_{a\Delta} = \mathrm{h}I_c$
	$I_c = 0$	$I_{a0} + I_{a\Sigma} = \mathrm{h}I_a$ $-\mathrm{j}\sqrt{3}\, I_{a\Delta} = \mathrm{h}I_b$
Double zero	$I_b = I_c = 0$	$2I_{a0} = I_{a\Sigma}$ $I_{a\Delta} = 0$
	$I_a = I_c = 0$	$I_{a0} + I_{a\Sigma} = 0$ $I_{a\Delta} = \mathrm{j}\sqrt{3}\, I_{a0}$
	$I_a = I_b = 0$	$I_{a0} + I_{a\Sigma} = 0$ $I_{a\Delta} = -\mathrm{j}\sqrt{3}\, I_{a0}$

10.8 Phase Shifting Transformer Relations

It is necessary to examine once again the use of a phase shifting transformer. In the cases in Chapters 8 and 9 the phase shifters were special in the sense that they always had a turns ratio of unity. Thus any voltage V_1 on the primary was changed in phase only but not in magnitude to a secondary value of $V_2 = e^{j\theta} V_1$ where θ is the phase shift in radians or degrees. Currents were shifted in phase by exactly the same amount.

In the case of zero, sigma, and delta networks it is necessary to use phase shifting transformers which change both the magnitude and angle of the primary voltage and current. Referring to Figure 10.8, let the turns ratio m be a complex

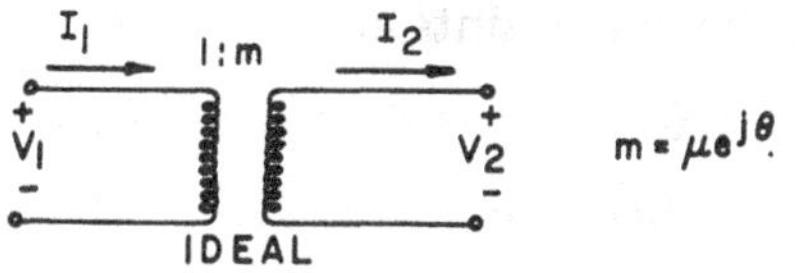

Fig. 10.8. An ideal transformer with complex turns ratio m.

number. Then by inspection we write

$$V_2 = mV_1 = \mu e^{j\theta} V_1 \tag{10.53}$$

and if $V_1 = |V_1|e^{j\delta_1}$ then (10.53) may be written as

$$V_2 = \mu |V_1| e^{j(\delta_1+\theta)} \tag{10.54}$$

The output V_2 is changed in magnitude by a factor μ and shifted in phase an amount $+\theta$ radians.

Since the transformer is ideal, we may also equate the input apparent power to the output apparent power, or $S_1 = S_2$, which we write as

$$V_1 I_1^* = V_2 I_2^* \tag{10.55}$$

Then, combining (10.53) and (10.55) $V_2/V_1 = I_1^*/I_2^* = m$ or

$$I_2 = (1/m^*) I_1 \tag{10.56}$$

If we let $I_1 = |I_1|e^{j\alpha_1}$, we compute $I_2 = (1/\mu e^{-j\theta})|I_1|e^{j\alpha_1}$ or

$$I_2 = \frac{1}{\mu} |I_1| e^{j(\alpha_1+\theta)} \tag{10.57}$$

Thus the current phasor I_2 is changed in magnitude an amount $1/\mu$ as in an ordinary transformer. It is also shifted in phase by $+\theta$ radians, which is exactly the same phase shift as the voltage phasor of (10.54).

10.9 SLG Faults with Arbitrary Symmetry

The SLG fault on phase a was investigated in section 10.2 with the solution shown in Figure 10.2. If the fault is on phase a, we may write the constraints from Figure 8.1

$$Z_a = 0, \qquad Z_b = Z_c = \infty, \qquad Z_g \neq 0 \tag{10.58}$$

and write the boundary conditions

$$V_a = Z_g I_a, \qquad I_b = I_c = 0 \tag{10.59}$$

Then from Table 10.1

$$2I_{a0} = I_{a\Sigma}, \qquad I_{a\Delta} = 0 \tag{10.60}$$

Also from (10.59)

$$V_a = Z_g I_a$$

$$(1/\mathrm{h})(V_{a0} + V_{a\Sigma}) = (Z_g/\mathrm{h})(I_{a0} + I_{a\Sigma})$$

or incorporating (10.60), we have for any h,

$$V_{aZ} + V_{a\Sigma} = 0 \tag{10.61}$$

If the fault is on phase b, the constraints are

$$Z_b = 0, \qquad Z_a = Z_c = \infty, \qquad Z_g \neq 0 \tag{10.62}$$

and the boundary conditions become

$$V_b = Z_g I_b, \qquad I_a = I_c = 0 \tag{10.63}$$

Then from Table 10.1

$$2I_{a0} = -2I_{a\Sigma} = -j\frac{2}{\sqrt{3}} I_{a\Delta} \tag{10.64}$$

Also from (10.63)

$$V_b = Z_g I_b$$

$$V_{a0} - \frac{1}{2} V_{a\Sigma} - j\frac{\sqrt{3}}{2} V_{a\Delta} = Z_g\left(I_{a0} - \frac{1}{2} I_{a\Sigma} - j\frac{\sqrt{3}}{2} I_{a\Delta}\right)$$

Combining with (10.64), we compute

$$V_{aZ} - \frac{1}{2} V_{a\Sigma} - j\frac{\sqrt{3}}{2} V_{a\Delta} = 0 \tag{10.65}$$

Results similar to (10.64) and (10.65) may be obtained for a SLG fault on phase c.

Upon closer examination of (10.60), (10.61), (10.64), and (10.65) a pattern is observed which leads us to the network connections of Figure 10.9. Note that for phase a symmetry $m_2 = 0$, which shorts the right side of that transformer and leaves the delta network open as in Figure 10.2.

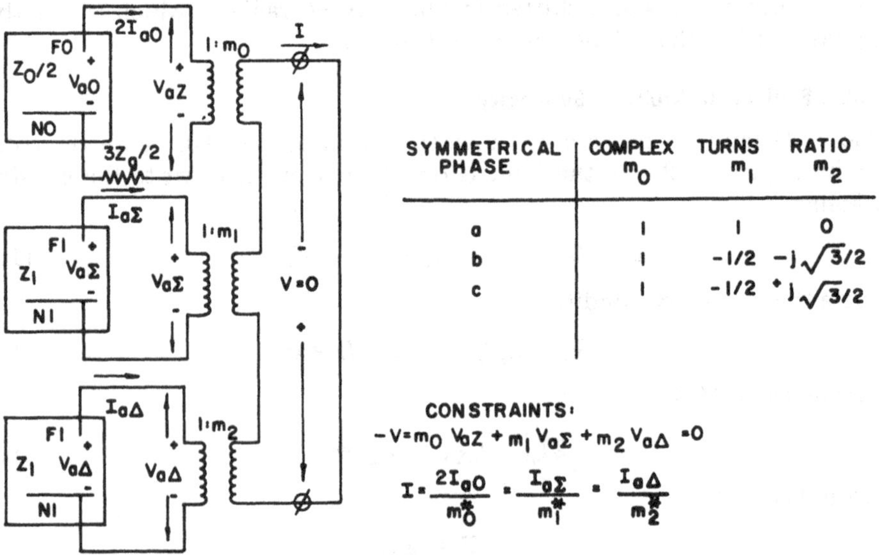

Fig. 10.9. Generalized fault diagram for a SLG fault on a zero-sigma-delta basis.

10.10 2LG Faults with Arbitrary Symmetry

Double line-to-ground faults for any symmetry may also be analyzed on a zero-sigma-delta basis. For the case of a 2LG fault on phase b and c we have the following constraints from Figure 8.1

$$Z_a = \infty, \quad Z_b = Z_c = Z, \quad Z_g \neq 0 \tag{10.66}$$

and may write the boundary conditions

$$I_a = 0, \quad V_b = ZI_b + \frac{3}{h} Z_g I_{a0}, \quad V_c = ZI_c + \frac{3}{h} Z_g I_{a0} \tag{10.67}$$

Then, from Table 10.1 for $I_a = 0$ we have

$$I_{a0} + I_{a\Sigma} = 0 \tag{10.68}$$

$$-j\sqrt{3}\, I_{a\Delta} = h(I_b - I_c) \tag{10.69}$$

Subtracting the voltage relations of (10.67), we compute

$$V_b - V_c = Z(I_b - I_c)$$

Then from (10.10) and (10.69) $-j\sqrt{3}\, V_{a\Delta} = -j\sqrt{3}\, ZI_{a\Delta}$ or

$$V_{a\Delta} - ZI_{a\Delta} = V_{Z\Delta} = 0 \tag{10.70}$$

where we define the quantity (10.70) to be $V_{Z\Delta}$.

Adding the voltage equations of (10.67), we compute

$$V_b + V_c = Z(I_b + I_c) + 6Z_g I_{a0}$$

which is rearranged to form

$$2[V_{a0} - (Z + 3Z_g)I_{a0}] = V_{a\Sigma} - ZI_{a\Sigma} \tag{10.71}$$

or

$$2V_{Z0} = V_{Z\Sigma} \tag{10.72}$$

where the quantities V_{Z0} and $V_{Z\Sigma}$ are defined by (10.71).

If the 2LG fault is on phases a and c, we write the constraints for Figure 8.1 as

$$Z_b = \infty, \quad Z_a = Z_c = Z, \quad Z_g \neq 0 \tag{10.73}$$

from which we have the boundary conditions

$$I_b = 0, \quad V_a = ZI_a + \frac{3}{h} Z_g I_{a0}, \quad V_c = ZI_c + \frac{3}{h} Z_g I_{a0} \tag{10.74}$$

From $I_b = 0$ we have from Table 10.1

$$hI_a = I_{a0} + I_{a\Sigma}, \quad hI_c = j\sqrt{3}\, I_{a\Delta} \tag{10.75}$$

But

$$I_{\text{fault}} = \frac{3}{h} I_{a0} = I_a + I_c = \frac{1}{h}(I_{a0} + I_{a\Sigma}) + \frac{j\sqrt{3}}{h} I_{a\Delta}$$

or for any h,

$$I_{a0} - \frac{1}{2} I_{a\Sigma} - j\frac{\sqrt{3}}{2} I_{a\Delta} = 0 \tag{10.76}$$

From the first voltage equation of (10.74) we write for any h,

$$V_{a0} + V_{a\Sigma} = Z(I_{a0} + I_{a\Sigma}) + 3Z_g I_{a0}$$

which is easily rearranged as

$$[V_{a0} - (Z + 3Z_g)I_{a0}] = -(V_{a\Sigma} - ZI_{a\Sigma}) \tag{10.77}$$

or

$$V_{Z0} = -V_{Z\Sigma} \tag{10.78}$$

From the remaining boundary condition on V_c we write

$$V_{a0} - \frac{1}{2}V_{a\Sigma} + j\frac{\sqrt{3}}{2}V_{a\Delta} = Z\left(I_{a0} - \frac{1}{2}I_{a\Sigma} + j\frac{\sqrt{3}}{2}I_{a\Delta}\right) + 3Z_gI_{a0}$$

from which we compute

$$[V_{a0} - (Z + 3Z_g)I_{a0}] = -j\frac{1}{\sqrt{3}}(V_{a\Delta} - ZI_{a\Delta}) \tag{10.79}$$

or

$$V_{Z0} = -j\frac{1}{\sqrt{3}}V_{Z\Delta} \tag{10.80}$$

If the fault is on phases a and b, a similar computation can be made. These results may be combined as shown in Figure 10.10. Note that when $m_2 = 0$ for phase a symmetry, this shorts the left side of that transformer leaving the delta network shorted on itself as in Figure 10.3. Also note the slight difference in defining V_{Z0}, $V_{Z\Sigma}$, and $V_{Z\Delta}$ compared to V_{aZ}, $V_{a\Sigma}$, and $V_{a\Delta}$ which is necessary due to the presence of the impedance Z in Figure 10.10.

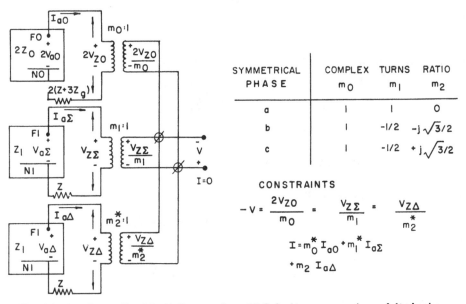

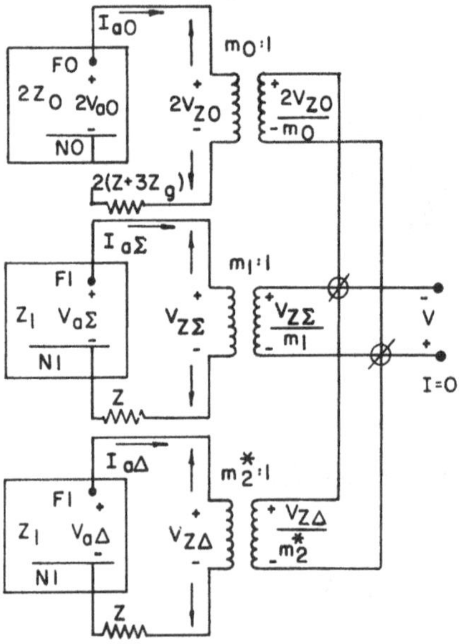

SYMMETRICAL PHASE	COMPLEX TURNS RATIO m_0	m_1	m_2
a	1	1	0
b	1	-1/2	$-j\sqrt{3}/2$
c	1	-1/2	$+j\sqrt{3}/2$

CONSTRAINTS

$$-V = \frac{2V_{Z0}}{m_0} = \frac{V_{Z\Sigma}}{m_1} = \frac{V_{Z\Delta}}{m_2^*}$$

$$I = m_0^* I_{a0} + m_1^* I_{a\Sigma} + m_2 I_{a\Delta}$$

Fig. 10.10. Generalized fault diagram for a 2LG fault on a zero-sigma-delta basis.

10.11 Series Faults with Arbitrary Symmetry

Series faults, both 1LO and 2LO, can be analyzed for any symmetry in exactly the same way as shunt faults. Details of this development are left to the interested reader. The results of this effort reveal the expected similarity to the shunt fault case.

For 2LO the network connections are similar to the SLG shunt fault situation except that the external impedances are different. In the 2LO case there is no impedance Z_g, so we set this to zero in Figure 10.9. There is an impedance Z, however, which is included as $Z/2$ in the zero network but as simply Z in the sigma and delta networks. Points labeled $N0$ and $N1$ are relabeled $M0$ and $M1$ where m is defined as the current input terminal (see Figure 8.5).

For the 1LO case the network connections are similar to Figure 10.10 except that the impedance Z_g is zero. Again, the points $N0$ and $N1$ must be relabeled $M0$ and $M1$ to indicate the point at which sequence currents enter the network.

The generalized fault diagrams for both shunt and series fault conditions are shown in Figure 10.11. Here, as in the three-component method, all common shunt and series faults may be described by two diagrams. For the series network connection we may write

$$-V = m_0 V_{Z0} + m_1 V_{Z\Sigma} + m_2 V_{Z\Delta} = 0$$

$$I = \frac{2I_{a0}}{m_0^*} = \frac{I_{a\Sigma}}{m_1^*} = \frac{I_{a\Delta}}{m_2^*} \tag{10.81}$$

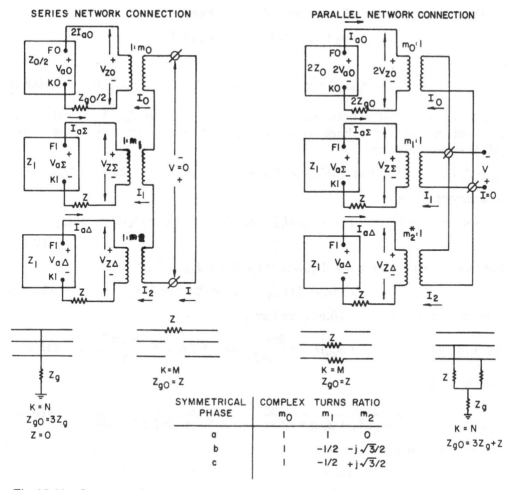

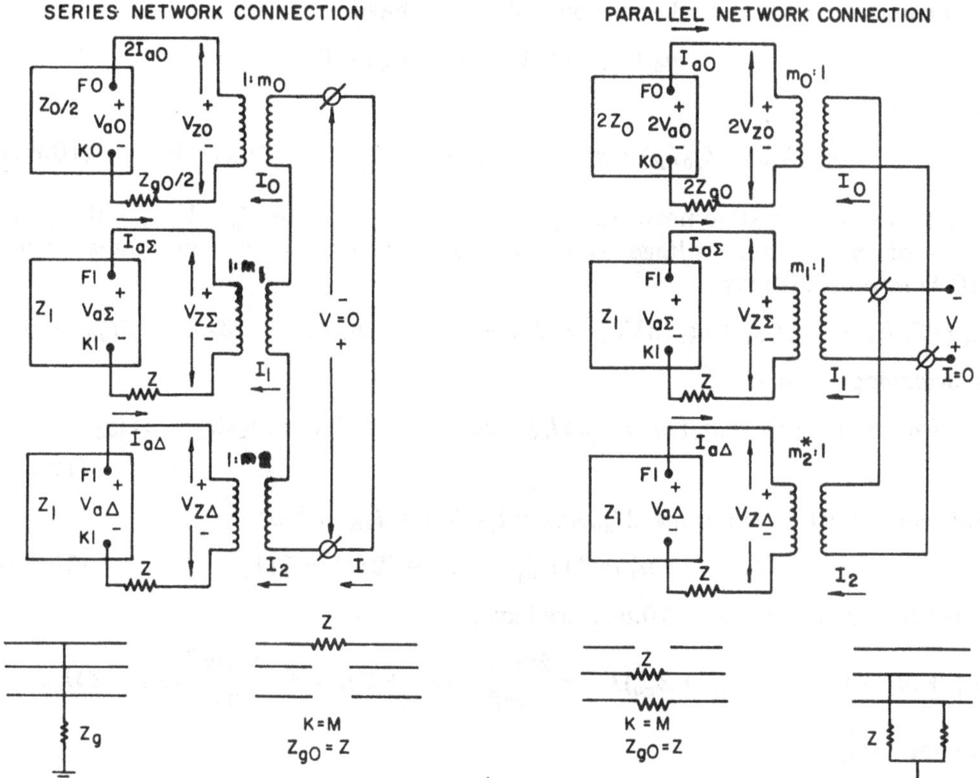

SYMMETRICAL PHASE	COMPLEX TURNS RATIO m_0	m_1	m_2
a	1	1	0
b	1	$-1/2$	$-j\sqrt{3}/2$
c	1	$-1/2$	$+j\sqrt{3}/2$

Fig. 10.11. Summary of zero-sigma-delta network connections for common fault conditions.

where m_0, m_1, and m_2 are defined in Figure 10.11. For parallel network connections we have

$$-V = 2V_{Z0}/m_0 = V_{Z\Sigma}/m_1 = V_{Z\Delta}/m_2^*$$
$$I = m_0^* I_{a0} + m_1^* I_{a\Sigma} + m_2 I_{a\Delta} = 0 \tag{10.82}$$

Equations (10.81) and (10.82) are easily verified in the cases examined in sections 10.9 and 10.10. Other cases are easily verified too, but this is left as an exercise for the reader.

Figure 10.11 is the result which permits us to analyze all common fault conditions by first making the assumption that $Z_1 = Z_2$. Then if we define the primitive zero, sigma, and delta networks as in Figure 10.11, we may solve these networks as an interconnection of one-port networks.

IV. SOLUTION OF THE GENERALIZED FAULT DIAGRAMS

Having developed the generalized fault diagrams of Figure 10.11, we may solve these networks for the sequence voltages and currents for any symmetry.

10.12 Series Network Connection—SLG and 2LO Faults

From Figure 10.11 and equation (10.81) we have

$$m_0 V_{Z0} + m_1 V_{Z\Sigma} + m_2 V_{Z\Delta} = 0$$

or

$$m_0(V_{a0} - Z_{g0}I_{a0}) + m_1(V_{a\Sigma} - ZI_{a\Sigma}) + m_2(V_{a\Delta} - ZI_{a\Delta}) = 0 \tag{10.83}$$

But, from the primitive network equation (10.4) we write V_{a0}, $V_{a\Sigma}$, and $V_{a\Delta}$ in terms of the prefault voltage $\text{h}V_F$ and the Thevenin impedances. Substituting (10.4) into (10.83) gives

$$m_0(-Z_0 I_{a0} - Z_{g0} I_{a0}) + m_1(\text{h}V_F - Z_1 I_{a\Sigma} - ZI_{a\Sigma}) + m_2(\text{h}V_F - Z_1 I_{a\Delta} - ZI_{a\Delta}) = 0$$

Rearranging, we have

$$(m_1 + m_2)\text{h}V_F = m_0(Z_0 + Z_{g0})I_{a0} + m_1(Z_1 + Z)I_{a\Sigma} + m_2(Z_1 + Z)I_{a\Delta} \tag{10.84}$$

But from 10.81 we compute $2I_{a0}/m_0^* = I_{a\Sigma}/m_1^* = I_{a\Delta}/m_2^*$ or

$$I_{a\Sigma} = (2m_1^*/m_0^*)I_{a0}, \qquad I_{a\Delta} = (2m_2^*/m_0^*)I_{a0} \tag{10.85}$$

Substituting (10.85) into (10.84), we have,

$$(m_1 + m_2)\text{h}V_F = m_0(Z_0 + Z_{g0})I_{a0} + \frac{2m_1 m_1^*}{m_0^*}(Z_1 + Z)I_{a0} + \frac{2m_2 m_2^*}{m_0^*}(Z_1 + Z)I_{a0}$$

Solving for I_{a0},

$$I_{a0} = \frac{\text{h}(m_1 + m_2)V_F}{m_0(Z_0 + Z_{g0}) + (2/m_0^*)(m_1 m_1^* + m_2 m_2^*)(Z_1 + Z)} \tag{10.86}$$

We note that for any symmetry

$$m_0 = m_0^* = 1, \qquad m_1 m_1^* = |m_1|^2$$
$$m_2 m_2^* = |m_2|^2, \qquad |m_1|^2 + |m_2|^2 = 1$$

so that

$$I_{a0} = \frac{h(m_1 + m_2)\,V_F}{(Z_0 + Z_{g0}) + 2(Z_1 + Z)} \tag{10.87}$$

which agrees with previous results.

From (10.85) we also compute

$$I_{a1} = \frac{m_1^* + m_2^*}{m_0^*} I_{a0}, \qquad I_{a2} = \frac{m_1^* - m_2^*}{m_0^*} I_{a0} \tag{10.88}$$

and the sequence currents are easily determined. Equation (10.87) may be combined with either (10.85) or (10.88) to synthesize the phase currents. Voltages may then be found from (10.4) and (10.9).

10.13 Parallel Network Connection—2LG and 1LO Faults

The parallel connection of Figure 10.11 may be analyzed from the basic equation (10.82)

$$m_0^* I_{a0} + m_1^* I_{a\Sigma} + m_2 I_{a\Delta} = 0 \tag{10.89}$$

We simplify this equation to solve for I_{a0} by combining the primitive equation (10.4) with the first of equations (10.82), viz.,

$$V_{Z\Sigma} = (2m_1/m_0)V_{Z0}, \qquad V_{Z\Delta} = (2m_2^*/m_0)V_{Z0}$$

to compute the relations

$$\begin{aligned} I_{a\Sigma} &= \frac{hV_F}{Z_1 + Z} + \frac{2m_1}{m_0}\left(\frac{Z_0 + Z_{g0}}{Z_1 + Z}\right) I_{a0} \\ I_{a\Delta} &= \frac{hV_F}{Z_1 + Z} + \frac{2m_2^*}{m_0}\left(\frac{Z_0 + Z_{g0}}{Z_1 + Z}\right) I_{a0} \end{aligned} \tag{10.90}$$

Then substituting (10.90) into (10.89), we have

$$I_{a0} = -\frac{h(m_1^* + m_2)\,V_F}{m_0^*(Z_1 + Z) + \dfrac{2(m_1 m_1^* + m_2 m_2^*)}{m_0}(Z_0 + Z_{g0})} \tag{10.91}$$

Substituting back into (10.90), we have the complete set of sequence quantities from which phase currents can be found.

Problems

10.1. Verify (10.8) and (10.10), particularly the matrices B and B^{-1}, by starting with (10.5).

10.2. Rework Example 3.1 for a SLG fault using the two-component method of section 10.2. Show that the results are equivalent.

10.3. Rework Example 3.2 for a LL fault using the two-component method of section 10.3.

10.4. Rework Example 3.3 for a 2LG fault using the two-component method of section 10.4.

10.5. Rework Example 3.4 for a 3ϕ fault using the two-component method of section 10.5.

10.6. Using the network of Figure 3.25 with series unbalance between F and F', use the two-component method to solve this system when
(a) Lines b and c are open with 1 ohm in line a (see section 10.6).
(b) Line a is open with 1 ohm in lines b and c (see section 10.7).

10.7. Repeat problem 10.2 using the two-component method if the SLG fault is on (a) phase *b* and (b) phase *c*.

10.8. Repeat problem 10.4 using the two-component method if the 2LG fault is on (a) phases *a* and *c* and (b) phases *a* and *b*.

10.9. Repeat problem 10.6(a) if the open lines are (a) lines *a* and *c* and (b) lines *a* and *b*.

10.10. Repeat problem 10.6(b) if the open line is (a) line *b* and (b) line *c*.

10.11. Verify Figure 10.9 for a SLG fault on phase *c*.

10.12. Verify Figure 10.10 for a 2LG fault on phases *a* and *b*.

10.13. Verify the series connections of Figure 10.11 for 2LO in any symmetry.

10.14. Verify the parallel connections of Figure 10.11 for 1LO in any symmetry.

10.15. Show that the results in (10.88) are identical with the currents computed by the three-component method for any symmetry.

10.16. Show that the results of (10.90) and (10.91) verify (3.24) for phase *a* symmetry.

10.17. Show that the results of (10.90) and (10.91) are identical with the currents computed by the three-component method for any symmetry.

chapter 11

Computer Solution Methods Using the Admittance Matrix

The electrical networks which power engineers must solve are usually large. This means that any solution "by hand" will often be out of the question and a computer will be used. Usually this will be a digital computer, and our emphasis in this chapter is on techniques which would be useful in performing a digital solution. However, the same network may also be solved by constructing a laboratory model of the physical network. A network analyzer is such a model and analyzers were used for many years in solving small to moderately sized networks.

There are several important differences in solving a network by digital computer as opposed to analyzer techniques. In the digital computer we solve *equations*. We are not concerned about the physical realizability of these equations. In the network analyzer, however, physical realizability is absolutely essential. The phase shifting transformers used in Chapters 9 and 10, for example, may be very difficult and expensive to build. In digital solutions this is of no concern whatever since we solve the equations which describe the system model with little regard to the concept of physically modeling the equations.

The digital computer has another distinct advantage in that it provides a means of solving large networks. Even the largest network analyzers are limited to networks of a hundred nodes or so. There is also a limit to the size of system which can be solved by a given digital computer. However, computers are commonly available which will solve networks of several hundred nodes. Furthermore, by clever usage of auxiliary memory in the form of magnetic discs, tapes, and drums the size of problem to be solved can be further expanded. The digital computer is therefore a flexible tool for the solution of large systems.

The size of the network should not be dismissed as unimportant, however. Large problems are costly to solve, and the engineer should know how to compute network equivalents or be able to use judgment to reduce any problem to its bare essentials.

Finally, the digital computer provides a means of solving large networks accurately. The hand solution of large networks, or even those of three or four nodes, involves considerable rounding error and often includes computational mistakes. These problems are not eliminated entirely by the digital computer, but they may be attacked in a systematic way to reduce errors to a minimum. Power system problems do not require 10 or 15 place accuracy; but being able to compute with such accuracy makes it possible to solve large systems with tolerable error, and this is important. Moreover, once a computer program is thoroughly "debugged" there is little likelihood that the solution obtained will include mechanical mistakes.

Because of the many advantages of digital computer solution of networks, we will consider in this chapter the way in which the system data and equations may be arranged for computer solution. The treatment here is not exhaustive by any means, and the interested reader is urged to consult the many fine references on the subject (see [64-76], for example).

11.1 Primitive Matrix

A power transmission or distribution system is a linear, passive, bilateral network of impedances (or admittances) which are interconnected in some specified way at various points called nodes (or buses). The impedances represent the per phase impedance of transmission lines and transformers. Usually we represent only one phase in three-phase systems since the impedance to the flow of positive, negative, or zero sequence currents is the same in each phase except at points of system unbalance, and these unbalances require special treatment. The rest of the network is made up of balanced impedances such that a per phase representation is possible. Indeed, if this were not the case, there would be no advantage in using symmetrical components.

The generators and loads are also connected to the nodes or buses exactly as these terminations occur on the physical system. In some studies the loads may be represented as constant impedances, in which case a per phase impedance (or admittance) may also be specified for the loads. Often, as in short circuit studies, the load currents are negligible when compared to the fault currents, and the load impedances are taken to be infinite. As pointed out in Chapter 6, the generators have a different impedance to the flow of positive, negative, and zero sequence currents. This is not the case with transformer and load impedances which are usually the same in each sequence network. Line impedances are the same in the positive and negative sequence networks but, as noted in Chapter 4 and 5, are different in the zero sequence network. Also, the topology of the sequence networks is often different in the positive and zero sequence cases.

In order to solve the network, it is necessary to organize the system impedance data and the network topology data in such a way that this information may be conveniently introduced and stored in the computer memory. The primitive network matrix provides a simple way of accomplishing the desired organization. The primitive matrix also defines an orientation or positive direction for each matrix element, and this too is useful.

In section 8.7 we defined Kron's primitive matrix for a network in which the branches were not mutually coupled. In power systems the branches of the network are often mutually coupled, especially in the zero sequence network. In such cases the primitive matrix contains off-diagonal terms for those mutually coupled elements. We write the primitive equation from (8.48) as

$$\mathbf{V} = \mathbf{Z}\mathbf{I} - \mathbf{E} \tag{11.1}$$

where

E = the column vector of branch source voltages, defined as a voltage rise in the direction of I

I = the column vector of branch currents with the direction chosen arbitrarily but thereafter used to define the branch orientation

$\mathcal{Z}$ = the primitive impedance matrix of self (diagonal) and mutual (off-diagonal) branch impedances

Note that we have used a script $\mathcal{Z}$ to indicate the primitive impedance. This will help us to distinguish clearly between primitive impedances and other important impedance quantities to be defined later.

If we number the primitive impedances in an arbitrary way, $\mathcal{Z}$ will appear as follows for a network with b branches.

$$\mathcal{Z} = \begin{array}{c} \\ 1 \\ 2 \\ \\ b \end{array} \begin{array}{c} \begin{array}{cccc} 1 & 2 & & b \end{array} \\ \begin{bmatrix} z_{11} & z_{12} & \cdots & z_{1b} \\ z_{21} & z_{22} & \cdots & z_{2b} \\ \cdots & \cdots & \cdots & \cdots \\ z_{b1} & z_{b2} & \cdots & z_{bb} \end{bmatrix} \end{array} \tag{11.2}$$

The elements of (11.2) are defined as follows:

z_{kk} = self impedance of branch k; $k = 1, \ldots, b$

$z_{k\ell}$ = mutual impedance between branches k and ℓ; $k, \ell = 1, \ldots, b$

In many networks the mutual impedance would be simply a mutual inductance. For mutually coupled transmission lines, however, every $z_{k\ell}$ will have both real imaginary parts as noted in Chapter 4.

Notice that the diagonal elements of (11.2) will always exist and will be finite and nonzero. Furthermore, if (11.1) represents the primitive equation for the positive, negative, or zero sequence network elements of a power system, the mutual impedance terms will always be smaller than the diagonal elements (Why?).

Since the diagonal elements of $\mathcal{Z}$ are all nonzero, the inverse of $\mathcal{Z}$ exists. Multiplying (11.1) by $\mathcal{Z}^{-1}$, we have

$$I = \mathcal{Y}V - J \tag{11.3}$$

where

$$\mathcal{Y} = \mathcal{Z}^{-1}, \qquad J = -\mathcal{Y}E \tag{11.4}$$

Equation (11.3) is the Norton equivalent of (11.1) with the current source term defined with positive direction as shown in Figure 11.1 (see [64]). Once the constraint matrix K for any network connection is specified, it is easy to show that

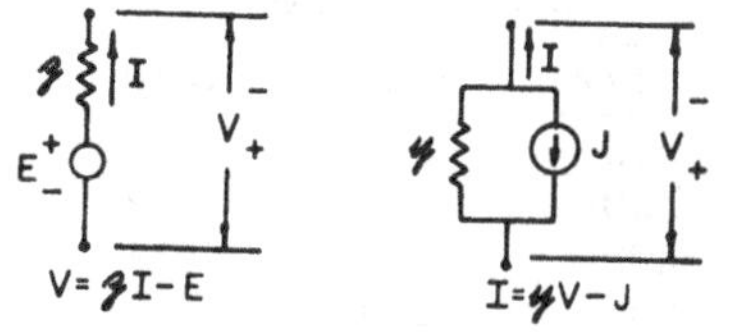

Fig. 11.1. Primitive network elements defined.

$$K^{*t}I = 0 \tag{11.5}$$

which asserts mathematically that the sum of the currents entering any node is zero. Therefore, when a transformation of the form

$$V = KV', \qquad I' = K^{*t}I \tag{11.6}$$

is used, the matrix expression

$$I' = Y'V' \tag{11.7}$$

will be found from the primitive equation (11.4) and the constraint (11.5). Because of (11.5) we may write the primitive equation as

$$J = \mathcal{Y}V \tag{11.8}$$

This is the admittance equivalent of the problem explored in section 8.6 where a voltage equation was derived which was the inverse of (11.7).

Usually for power systems the primitive matrix is sparse, i.e., it contains many zero elements (but never a zero-diagonal element). This is because the lines in a power system are mutually coupled only to those lines which are physically parallel and located in close proximity. This condition usually exists where several circuits share a common right-of-way.

Example 11.1

Compute the primitive $\mathcal{Y}$ and $\mathcal{Z}$ matrices for the small power system shown in Figure 11.2. The nodes are numbered in an arbitrary way and these numbers are

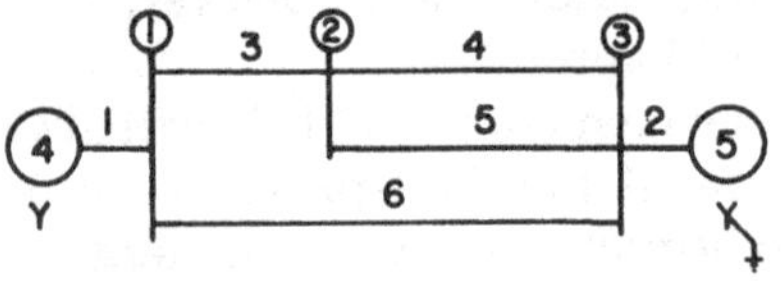

Fig. 11.2. A small power system with 5 nodes.

circled on the diagram. The circles numbered 4 and 5 are generators. The impedances, including generator impedances, are also numbered in an arbitrary way, beginning in this case with the generators and proceeding through the lines.

The impedances are taken to be pure inductive reactances for this example. These are all specified in Table 11.1, and all connecting nodes are also specified. This tabulated information completely describes the network for all sequences.

Table 11.1. Reactances of the System of Figure 11.2

Impedance Number	*Connecting Nodes*	*Self Impedance*			*Zero Sequence Mutual Impedance*	*Mutual Element*
		Positive	*Negative*	*Zero*		
1	4, 1	0.25	0.15	0.03	...	...
2	5, 3	0.20	0.12	0.02	...	...
3	1, 2	0.08	0.08	0.14	...	...
4	2, 3	0.06	0.06	0.10	0.05	5
5	2, 3	0.06	0.06	0.12	0.05	4
6	1, 3	0.13	0.13	0.17	...	...

We must somehow convey all this information to the computer in order to solve a problem.

Solution

We may write the primitive impedance matrices for all sequences by inspection. For the zero sequence we have

$$\mathbf{Z}_0 = \begin{array}{c} \\ 1 \\ 2 \\ 3 \\ 4 \\ 5 \\ 6 \end{array} \begin{array}{c} \begin{array}{cccccc} 1 & 2 & 3 & 4 & 5 & 6 \end{array} \\ \begin{bmatrix} j0.03 & 0 & 0 & 0 & 0 & 0 \\ 0 & j0.02 & 0 & 0 & 0 & 0 \\ 0 & 0 & j0.14 & 0 & 0 & 0 \\ 0 & 0 & 0 & j0.10 & j0.05 & 0 \\ 0 & 0 & 0 & j0.05 & j0.12 & 0 \\ 0 & 0 & 0 & 0 & 0 & j0.17 \end{bmatrix} \end{array}$$

and we note that the matrix is very sparse, having only two off-diagonal entries. Except for the 2 × 2 matrix in position 4–5, the inverse may be found by taking the reciprocal of each diagonal entry, i.e.,

$$y_{kk} = 1/z_{kk}$$

where element kk is not mutually coupled. For position 4–5 we readily compute

$$\begin{bmatrix} 0.10 & 0.05 \\ 0.05 & 0.12 \end{bmatrix}^{-1} = \begin{bmatrix} 12.632 & -5.263 \\ -5.263 & 10.526 \end{bmatrix}$$

to write the primitive admittance matrix for the zero sequence as

$$\mathcal{Y}_0 = \begin{array}{c} \\ 1 \\ 2 \\ 3 \\ 4 \\ 5 \\ 6 \end{array} \begin{array}{c} \begin{array}{cccccc} 1 & 2 & 3 & 4 & 5 & 6 \end{array} \\ \begin{bmatrix} -j33.333 & 0 & 0 & 0 & 0 & 0 \\ 0 & -j50.000 & 0 & 0 & 0 & 0 \\ 0 & 0 & -j7.143 & 0 & 0 & 0 \\ 0 & 0 & 0 & -j12.632 & j5.263 & 0 \\ 0 & 0 & 0 & j5.263 & -j10.526 & 0 \\ 0 & 0 & 0 & 0 & 0 & -j5.882 \end{bmatrix} \end{array}$$

The primitive matrices for the positive and negative sequences are diagonal. These are left as an exercise (see problem 11.10).

Since the primitive matrices are so sparse, we usually store only the nonzero terms in the computer memory. The tabulated form of the data in Table 11.1 is more efficient in terms of storage requirements and is the preferred storage format. It is implied that any element not specified is taken to be zero.

11.2 Node Incidence Matrix

We now develop a procedure whereby the primitive impedance or admittance matrix may be transformed into a form more useful for solution of the network. The primitive matrix itself provides no information concerning the way the various branches are connected to form a network. The connection information is known, however, and is easily tabulated as in column 2 of Table 11.1. We now arrange this tabulated information in matrix form to indicate the nodes to which the various branches are connected. We shall call this connection matrix the "augmented node incidence matrix" and denote this array by the symbol $\hat{A}$ where we define the $b \times m$ matrix

$$\hat{A} = \begin{array}{c} \\ 1 \\ 2 \\ \\ b \end{array} \begin{array}{c} \begin{array}{cccc} 1 & 2 & & m \end{array} \\ \begin{bmatrix} a_{11} & a_{12} & \cdots & a_{1m} \\ a_{21} & a_{22} & \cdots & a_{2m} \\ \cdots & \cdots & \cdots & \cdots \\ a_{b1} & a_{b2} & \cdots & a_{bm} \end{bmatrix} \end{array} \tag{11.9}$$

The rules for locating the nonzero elements of $\hat{A}$ are simple. We begin with a $b \times m$ array of zeros. Then we make nonzero entries for each row in columns corresponding to the two nodes to which that branch number (row number) is connected. For example, in Example 11.1 branch number 4 is connected to nodes two and three. Then in row 4 (branch 4) of $\hat{A}$ the only nonzero elements will be those in columns two and three, i.e., a_{42} and a_{43}.

It is convenient at this point to record an *orientation* for each branch as well as its connection. This means that we record the arbitrarily chosen positive current direction from Figure 11.1 for each branch. These current directions may be noted on the diagram or may be chosen according to some arbitrary convention such as from the smaller to the larger node number. In any event these directions or branch orientations may be recorded in the matrix $\hat{A}$ in the following way. Let

$$\begin{aligned} a_{pq} &= +1 \text{ if current in branch } p \text{ is leaving node } q \\ &= -1 \text{ if current in branch } p \text{ is entering node } q \\ &= 0 \text{ if branch } p \text{ is not connected to node } q \end{aligned}$$

$$p = 1, b, \qquad q = 1, m \tag{11.10}$$

The matrix $\hat{A}$ is recognized to be similar to the connection matrix defined by (8.43) except that it conveys nodal rather than mesh information.

Example 11.2

Form the augmented incidence matrix for the system of Example 11.1. Take the positive direction of branch current always to be from the smaller to the larger node number.

Solution

By inspection of Figure 11.2 or Table 11.1 we write

$$\hat{A} = \begin{array}{c|ccccc} b \backslash m & 1 & 2 & 3 & 4 & 5 \\ \hline 1 & 1 & 0 & 0 & -1 & 0 \\ 2 & 0 & 0 & 1 & 0 & -1 \\ 3 & 1 & -1 & 0 & 0 & 0 \\ 4 & 0 & 1 & -1 & 0 & 0 \\ 5 & 0 & 1 & -1 & 0 & 0 \\ 6 & 1 & 0 & -1 & 0 & 0 \end{array}$$

We now rewrite the primitive equation (11.3) with the subscript b to indicate that these are branch currents and voltages.

$$\mathbf{I}_b = \mathbf{\mathcal{Y}}\mathbf{V}_b - \mathbf{J}_b \tag{11.11}$$

If we premultiply by $\hat{\mathbf{A}}^t$, we have the interesting result

$$\hat{\mathbf{A}}^t\,\mathbf{I}_b = \mathbf{0} = \hat{\mathbf{A}}^t\,\mathbf{\mathcal{Y}}\,\mathbf{V}_b - \hat{\mathbf{A}}^t\,\mathbf{J}_b \tag{11.12}$$

This result is the zero (null) vector since

$$\hat{\mathbf{A}}^t\,\mathbf{I}_b = \text{sum of currents leaving each node} = 0$$

Again we illustrate by means of an example.

Example 11.3

Compute the product $\hat{\mathbf{A}}^t\,\mathbf{I}_b$ for the network of Example 11.1.

Solution

$$\hat{\mathbf{A}}^t\mathbf{I}_b = \begin{matrix} \\ 1 \\ 2 \\ 3 \\ 4 \\ 5 \end{matrix} \begin{matrix} \begin{matrix} 1 & 2 & 3 & 4 & 5 & 6 \end{matrix} \\ \begin{bmatrix} 1 & 0 & 1 & 0 & 0 & 1 \\ 0 & 0 & -1 & 1 & 1 & 0 \\ 0 & 1 & 0 & -1 & -1 & -1 \\ -1 & 0 & 0 & 0 & 0 & 0 \\ 0 & -1 & 0 & 0 & 0 & 0 \end{bmatrix} \end{matrix} \begin{bmatrix} I_{b1} \\ I_{b2} \\ I_{b3} \\ I_{b4} \\ I_{b5} \\ I_{b6} \end{bmatrix} = \begin{matrix} 1 \\ 2 \\ 3 \\ 4 \\ 5 \end{matrix} \begin{bmatrix} I_{b1} + I_{b3} + I_{b6} \\ -I_{b3} + I_{b4} + I_{b5} \\ I_{b2} - I_{b4} - I_{b5} - I_{b6} \\ -I_{b1} \\ -I_{b2} \end{bmatrix}$$

The first three rows of the result are obviously zero. Note that if nodes 4 and 5 are not connected to anything, I_{b1} and I_{b2} are also zero.

The product $\hat{\mathbf{A}}^t\mathbf{J}_b$ is the vector of all source currents entering each node. If we think of the network as being passive, these currents are usually thought of as external currents injected at each node, as indicated in Figure 11.3. These cur-

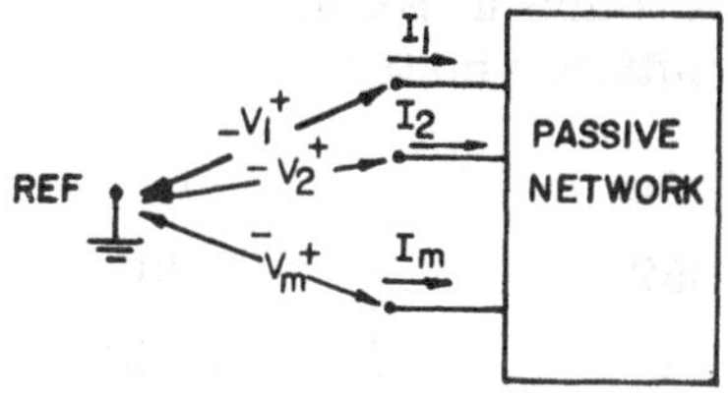

Fig. 11.3. Injected currents $\hat{\mathbf{A}}^t\,\mathbf{J}_b = \hat{\mathbf{I}}_m$.

rents are usually called simply the "node" currents and are conveniently designated by $\hat{\mathbf{I}}_m$. Thus

$$\hat{\mathbf{I}}_m = \hat{\mathbf{A}}^t\mathbf{J}_b = \hat{\mathbf{A}}^t\,\mathbf{\mathcal{Y}}\,\mathbf{V}_b \tag{11.13}$$

where we use the subscript m to remind us that there are m of these currents. Obviously the sum of these m nodal currents is equal to zero, so they are not independent. We will explore this property more fully later.

We may compute the apparent power delivered to the network in two ways.

From Tellegren's theorem [77] we may compute the power entering the terminals of Figure 11.3 or we may find the source power of each branch. These relations are given by

$$\hat{S}_m = \hat{\mathbf{V}}_m^t \hat{\mathbf{I}}_m^*$$

$$S_b = \mathbf{V}_b^t \mathbf{J}_b^* \qquad (11.14)$$

Since the network power is invariant then $\hat{S}_m = S_b$ or

$$\hat{\mathbf{V}}_m^t \hat{\mathbf{I}}_m^* = \mathbf{V}_b^t \mathbf{J}_b^* \qquad (11.15)$$

But from (11.13), since $\hat{\mathbf{A}}$ is real, $\hat{\mathbf{I}}_m^* = \hat{\mathbf{A}}^t \mathbf{J}_b^*$ and, substituting into (11.15), we compute $\hat{\mathbf{V}}_m^t \hat{\mathbf{I}}_m^* = \hat{\mathbf{V}}_m^t \hat{\mathbf{A}}^t \mathbf{J}_b^* = \mathbf{V}_b^t \mathbf{J}_b^*$ or $\hat{\mathbf{V}}_m^t \hat{\mathbf{A}}^t = \mathbf{V}_b^t$. Then the branch and node voltages are related by .

$$\mathbf{V}_b = \hat{\mathbf{A}} \hat{\mathbf{V}}_m \qquad (11.16)$$

and from (11.13)

$$\hat{\mathbf{I}}_m = \hat{\mathbf{A}}^t \mathfrak{Y} \hat{\mathbf{A}} \hat{\mathbf{V}}_m = \hat{\mathbf{Y}} \hat{\mathbf{V}}_m \qquad (11.17)$$

Equation (11.17) is called the *indefinite admittance matrix* description of the system [5]. It is indefinite because the voltage reference is not physically connected to the network. This description has some other interesting properties which will be investigated later [5].

In equation (11.17) we define the indefinite admittance matrix as the $m \times m$ matrix

$$\hat{\mathbf{Y}} = \hat{\mathbf{A}}^t \mathfrak{Y} \hat{\mathbf{A}} \qquad (11.18)$$

This computation is illustrated by an example.

Example 11.4

Compute the indefinite admittance matrix for the zero sequence admittances of the network of Example 11.1.

Solution

The primitive admittance matrix was computed in Example 11.1 and the node incidence matrix was found in Example 11.2. Using these results, we easily compute the indefinite admittance matrix $\hat{\mathbf{Y}}$.

$$\hat{\mathbf{Y}}_0 = -j \begin{array}{c} \\ 1 \\ 2 \\ 3 \\ 4 \\ 5 \end{array} \begin{array}{c} \begin{array}{ccccc} 1 & 2 & 3 & 4 & 5 \end{array} \\ \begin{bmatrix} 46.359 & -7.143 & -5.882 & -33.333 & 0 \\ -7.143 & 19.774 & -12.632 & 0 & 0 \\ -5.882 & -12.632 & 68.514 & 0 & -50.000 \\ -33.333 & 0 & 0 & 33.333 & 0 \\ 0 & 0 & -50.000 & 0 & 50.000 \end{bmatrix} \end{array}$$

In forming the indefinite admittance matrix, we have found a network description which involves only the external network connections. In doing this, we have completely suppressed the details of what exists within these external connections. This is a desirable feature since it permits us to describe the constraints among the external (or nodal) voltages and currents in a most efficient way. Since power networks are large and computer memory is limited, this efficiency is an important consideration.

We also recognize that the indefinite admittance matrix description is not exactly that required for a power system. Referring to Figure 11.3, we observe that the voltage reference is external to the network and the currents are not independent. Usually we would prefer to name some node in the network as the reference and measure all voltages with respect to that point. Having named a reference node, it is convenient to leave the current entering this node unspecified. Then the remaining $m-1$ nodal currents are independent.

11.3 Node Admittance and Impedance Matrices

Consider a network consisting of $n+1$ nodes, including the reference node as shown in Figure 11.4 where node 0 is chosen as the reference node. This net-

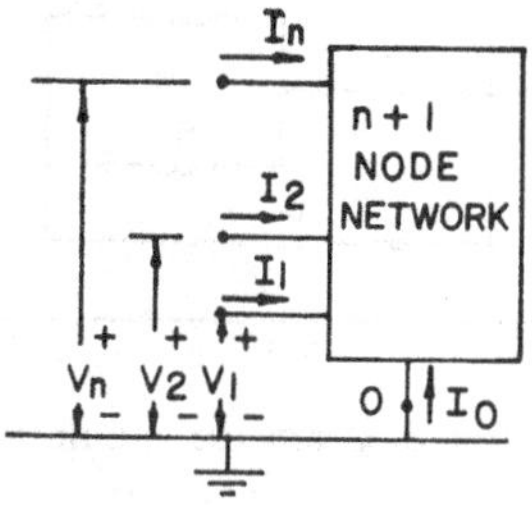

Fig. 11.4. An $n + 1$ node network with node 0 as reference.

work has the same primitive (impedance or admittance) matrix whether a reference is chosen or not. It has a different node incidence matrix, however, which we shall designate as A (without the hat). This network differs from the unreferenced case in that the n voltages and currents subscripted $1, 2, \ldots, n$ are independent sets. The reference node current is

$$I_0 = -(I_1 + I_2 + \cdots + I_n) \tag{11.19}$$

and this current is usually not required explicitly.

The node incidence matrix for the network of Figure 11.4 is formed using exactly the same rules as before except that no entries are made for the reference node at all, i.e.,

$$\begin{array}{cc} & \begin{array}{cccc} 1 & 2 & & n \end{array} \\ \mathrm{A} = \begin{array}{c} 1 \\ 2 \\ \\ b \end{array} & \begin{bmatrix} a_{11} & a_{12} & \cdots & a_{1n} \\ a_{21} & a_{22} & \cdots & a_{2n} \\ \cdots & \cdots & \cdots & \cdots \\ a_{b1} & a_{b2} & \cdots & a_{bn} \end{bmatrix} \end{array} \tag{11.20}$$

where

$$\begin{aligned} a_{pq} &= +1 \text{ if current in branch } p \text{ is leaving node } q \\ &= -1 \text{ if current in branch } p \text{ is entering node } q \\ &= 0 \text{ if branch } p \text{ is not connected to node } q \end{aligned}$$

$$p = 1, b; \qquad q = 1, n \tag{11.21}$$

This is the same as the augmented node incidence matrix $\hat{A}$ with all rows and columns deleted corresponding to nodes connected to the reference. The formation of A will be illustrated by an example.

Example 11.5

Find the node incidence matrix for the zero sequence network of the system specified in Example 11.1.

Solution

First we sketch the zero sequence network for the system of Figure 11.2 where we shall consider the *N*0 bus as the reference. Note that generator 4 (on the left), represented by impedance element 1, is Y-ungrounded and is therefore not connected to *N*0. Generator 5 is Y-grounded so its impedance, element 2, is connected to *N*0. See Figure 11.5.

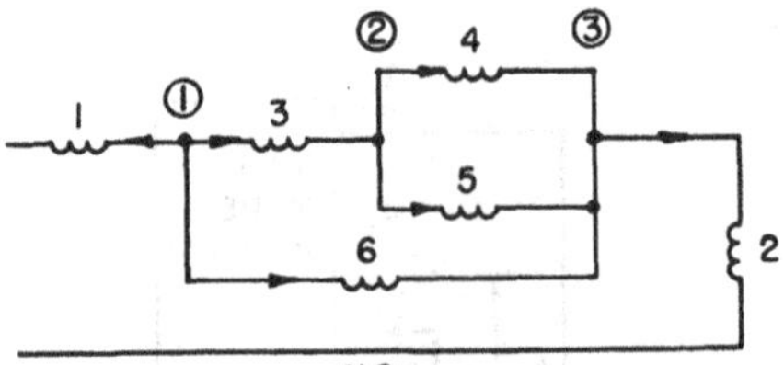

Fig. 11.5. Zero sequence diagram for the network of Example 11.1.

Proceeding according to the rules (11.21), we write

$$A = \begin{array}{c} \\ 2 \\ 3 \\ 4 \\ 5 \\ 6 \end{array} \begin{array}{c} \begin{array}{ccc} 1 & 2 & 3 \end{array} \\ \begin{bmatrix} 0 & 0 & 1 \\ 1 & -1 & 0 \\ 0 & 1 & -1 \\ 0 & 1 & -1 \\ 1 & 0 & -1 \end{bmatrix} \end{array}$$

where we have ignored branch 1 entirely since no zero sequence current will flow in this branch (actually, branch 1 could be omitted from the diagram as it does not enter into any computation). Since branch 1 is eliminated, the branches could be renumbered, but this is not essential. Note that this result is the same as $\hat{A}$ with row 1 and columns 4 and 5 deleted.

To compute the node admittance matrix, we proceed exactly as prescribed by (11.18) where we define the $n \times n$ matrix

$$Y = A^t \mathcal{Y} A \tag{11.22}$$

where now the primitive matrix must be that which actually includes only those elements used in the network. In Example 11.5 we note that branch 1 is not required, so this branch (row, column) must be eliminated from the primitive matrix and the incidence matrix (or set $y_{14} = y_1 = 0$ in all computations).

Example 11.6

Compute the node admittance matrix for the network of Example 11.5.

Solution

Equation (11.22) may be applied directly to compute

$$\mathbf{Y}_0 = -j \begin{array}{c} \\ 1 \\ 2 \\ 3 \end{array} \begin{array}{c} \begin{array}{ccc} 1 & 2 & 3 \end{array} \\ \begin{bmatrix} 13.025 & -7.143 & -5.882 \\ -7.143 & 19.774 & -12.632 \\ -5.882 & -12.632 & 68.514 \end{bmatrix} \end{array}$$

It is interesting to compare this result with $\hat{\mathbf{Y}}_0$ computed in Example 11.4. This result is observed to be the same as the 1-2-3 partition of $\hat{\mathbf{Y}}_0$ if $y_1 = y_{14} = -33.333$ is subtracted from position (1, 1). This is necessary since y_1 is not required in the zero sequence network and could have been omitted from the beginning. Rows (4, 5) and columns (4, 5) of $\hat{\mathbf{Y}}_0$ are eliminated since nodes 4 and 5 have become the reference node.

It is possible to develop (11.22) in a similar way to that by which (11.17) was derived. This is left as an exercise. (See problem 11.19.) The result is important however and may be stated as

$$\mathbf{I} = \mathbf{Y}\mathbf{V} \tag{11.23}$$

where Y is the node admittance matrix defined by (11.22) and V and I are column vectors of node voltages and currents defined in Figure 11.4. Since these voltages and currents are independent sets of variables, the inverse of Y exists, or we may write

$$\mathbf{V} = \mathbf{Z}\mathbf{I} \tag{11.24}$$

where

$$\mathbf{Z} = \mathbf{Y}^{-1} \tag{11.25}$$

Note that this last operation was not possible in the case of the indefinite admittance matrix since $\hat{\mathbf{Y}}$ is singular.

Many authors call Z the matrix of "open circuit driving point and transfer impedances" and Y the matrix of "short circuit driving point and transfer admittances." They are also called *Z*-BUS and *Y*-BUS by some authors [64]. Before examining these important matrices in detail, however, it is instructive to study the indefinite admittance matrix to learn the important properties of this matrix network description.

11.4 Indefinite Admittance Matrix

Consider the $n + 1$ terminal network of Figure 11.6 where the network behavior is to be described in terms of the currents entering at nodes $0, 1, \ldots, n$ and the voltages measured as voltage drops to an external reference point labeled REF. As noted previously, only n of the voltages and n of the currents are independent.

From (11.17) we write the indefinite admittance equation

$$\hat{\mathbf{I}} = \hat{\mathbf{Y}}\hat{\mathbf{V}} \tag{11.26}$$

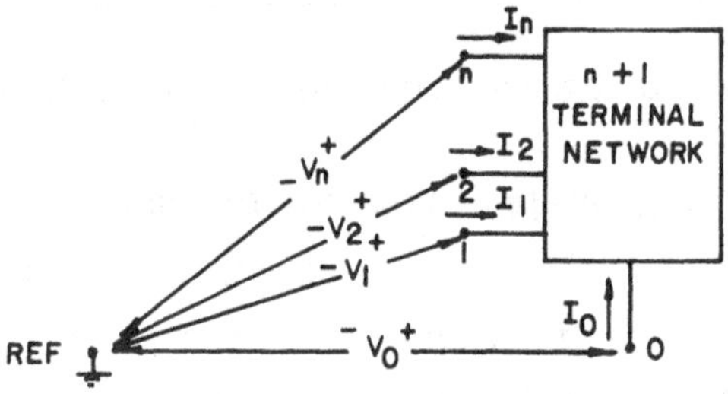

Fig. 11.6. An $n + 1$ terminal network with external voltage reference.

which provides information concerning the terminal behavior of the network. Expanding (11.26) we have

$$\begin{bmatrix} I_0 \\ I_1 \\ \cdots \\ I_n \end{bmatrix} = \begin{bmatrix} Y_{00} & Y_{01} & \cdots & Y_{0n} \\ Y_{10} & Y_{11} & \cdots & Y_{1n} \\ \cdots & \cdots & \cdots & \cdots \\ Y_{n0} & Y_{n1} & \cdots & Y_{nn} \end{bmatrix} \begin{bmatrix} V_0 \\ V_1 \\ \cdots \\ V_n \end{bmatrix} \tag{11.27}$$

This equation shows clearly the way in which the elements of $\hat{\mathbf{Y}}$ are defined. Each element is a "short circuit admittance." For example, the kth column is defined by

$$\begin{bmatrix} I_0 \\ I_1 \\ \cdots \\ I_n \end{bmatrix} = \begin{bmatrix} Y_{0k} \\ Y_{1k} \\ \cdots \\ Y_{nk} \end{bmatrix} V_k \tag{11.28}$$

with all terminals except k shorted.

For a given network we can let $V_k = 1.0$, for example, and short all other terminals. Then the kth column of admittances is equal to the currents entering the $n + 1$ nodes as specified by (11.28). This is a relatively simple task for most networks and can often be done by inspection.

In general we may state the rules for finding the elements of $\hat{\mathbf{Y}}$ as follows:

$$Y_{ik} = \frac{I_i}{V_k}, \; V_\ell = 0, \; \ell \neq k \tag{11.29}$$

This technique is more direct than forming the augmented node incidence matrix. The procedure will be illustrated by an example.

Example 11.7

Find the indefinite admittance matrix for the network of Figure 11.7 by the application of (11.29).

Solution

Figure 11.8a shows the network arranged with the condition

$$V_1 = 1.0, \qquad V_2 = V_3 = V_4 = 0$$

According to (11.29), the four terminal currents should be numerically equal to column 1 of $\hat{\mathbf{Y}}$. By inspection

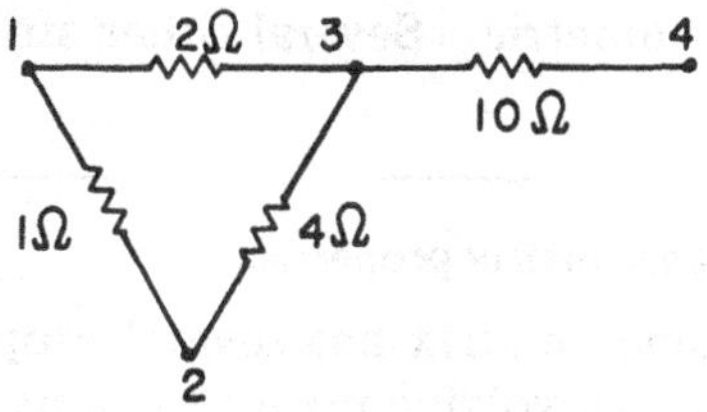

Fig. 11.7. Network for Example 11.7.

$$\begin{bmatrix} I_1 \\ I_2 \\ I_3 \\ I_4 \end{bmatrix} = \begin{bmatrix} Y_{11} \\ Y_{21} \\ Y_{31} \\ Y_{41} \end{bmatrix} = \begin{bmatrix} 1.5 \\ -1.0 \\ -0.5 \\ 0 \end{bmatrix}$$

The second column of $\hat{\mathbf{Y}}$ is found by setting

$$V_2 = 1.0, \qquad V_1 = V_3 = V_4 = 0$$

as shown in Figure 11.8b. Then we easily determine

$$\begin{bmatrix} I_1 \\ I_2 \\ I_3 \\ I_4 \end{bmatrix} = \begin{bmatrix} Y_{12} \\ Y_{22} \\ Y_{32} \\ Y_{42} \end{bmatrix} = \begin{bmatrix} -1.0 \\ 1.25 \\ -0.25 \\ 0 \end{bmatrix}$$

Columns 3 and 4 are found in exactly the same way with the result

$$\hat{\mathbf{Y}} = \begin{array}{c} \\ 1 \\ 2 \\ 3 \\ 4 \end{array} \begin{array}{c} \begin{array}{cccc} 1 & 2 & 3 & 4 \end{array} \\ \begin{bmatrix} 1.50 & -1.00 & -0.50 & 0 \\ -1.00 & 1.25 & -0.25 & 0 \\ -0.50 & -0.25 & 0.85 & -0.10 \\ 0 & 0 & -0.10 & 0.10 \end{bmatrix} \end{array}$$

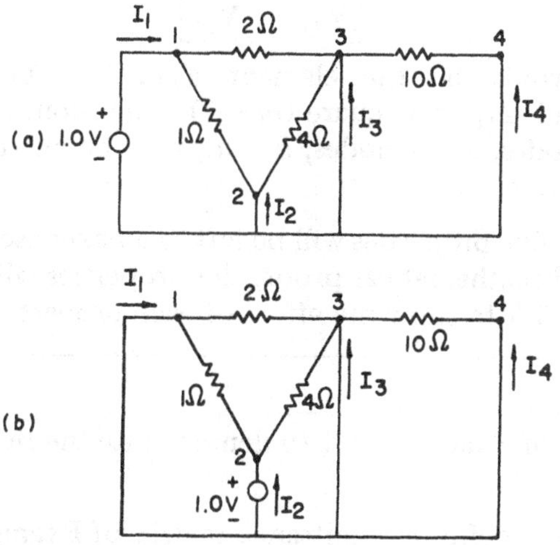

Fig. 11.8. Two steps in the solution of Figure 11.7.

Note that the matrix is symmetric. Several other striking features of this matrix will now be discussed.

11.4.1 Indefinite admittance matrix properties

The indefinite admittance matrix has several properties which are of considerable interest in applications involving power systems. There are five major properties, all of which will be stated without proof [5].

1. The summation of elements in any row or column is equal to zero.
2. If the jth terminal of an m terminal network is grounded (connected to the reference), the resulting admittance network description is that of an $m - 1$ terminal network and is found by deleting the jth row and jth column of the indefinite admittance matrix of the m terminal network. This resulting matrix description is simply the admittance matrix (or definite admittance matrix) of the j-referenced network.
3. Connecting any terminals j and k together in a given network yields a new network whose indefinite admittance matrix may be formed by adding rows and columns j and k of the original indefinite admittance matrix.
4. The effect of connecting an admittance y between any two terminals j and k of a network is to add y to the major diagonal elements jj and kk, and to subtract y from the off-diagonal elements jk and kj.
5. Any terminal k may be suppressed or eliminated from consideration (open circuited) by performing a matrix reduction on the indefinite admittance matrix, pivoting on element kk, with the result being an $m - 1$ dimensioned indefinite admittance matrix. This property can be easily visualized by rearranging the original matrix until the kth equation is the last (bottom) one. Then

$$\hat{\mathbf{Y}}_{\text{new}} = \mathbf{Y}_{11} - \mathbf{Y}_{12}\mathbf{Y}_{22}^{-1}\mathbf{Y}_{21} \tag{11.30}$$

where the original matrix is partitioned as

$$\hat{\mathbf{Y}} = \left[\begin{array}{c|c} \mathbf{Y}_{11} & \mathbf{Y}_{12} \\ \hline \mathbf{Y}_{21} & \mathbf{Y}_{22} \end{array}\right] \tag{11.31}$$

Then $\mathbf{Y}_{22}$ would be the element Y_{kk}, (1×1), $\mathbf{Y}_{12}$ is $m - 1 \times 1$, $\mathbf{Y}_{21} = \mathbf{Y}_{12}^t$, and $\mathbf{Y}_{11}$ is a square $(m - 1)$ dimensioned array. Actually any number of nodes, say k nodes, $k < m$, may be eliminated simultaneously by this technique.

The proof of the five properties will be left as an exercise (see problem 11.23). Instead of a detailed mathematical proof, the properties will be demonstrated by extending Example 11.7 to show the effect of each property.

Example 11.8

Use the network of Example 11.7 to demonstrate the five matrix properties.

Solution

Property 1. The indefinite admittance matrix of Example 11.7 was found to be

$$\hat{\mathbf{Y}} = \begin{array}{c} \\ 1 \\ 2 \\ 3 \\ 4 \end{array} \begin{array}{c} \begin{array}{cccc} 1 & 2 & 3 & 4 \end{array} \\ \begin{bmatrix} 1.50 & -1.00 & -0.50 & 0 \\ -1.00 & 1.25 & -0.25 & 0 \\ -0.50 & -0.25 & 0.85 & -0.10 \\ 0 & 0 & -0.10 & 0.10 \end{bmatrix} \end{array}$$

A careful inspection of this matrix verifies that the sum of elements in each row and each column is exactly zero. This is due to the fact that the four currents are not independent since they are constrained by Kirchhoff's law such that $-I_4 = I_1 + I_2 + I_3$. Thus the determinant of $\hat{\mathbf{Y}}$ is zero.

Property 2. Suppose we ground node 2 of Example 11.7 as shown in Figure 11.9. Then the (definite) admittance matrix may be found by deleting row 2 and

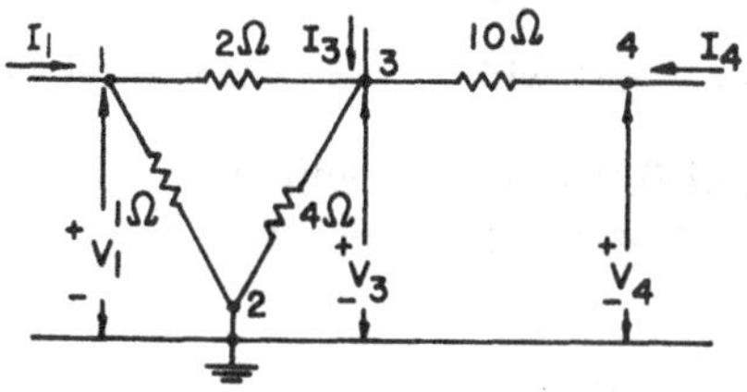

Fig. 11.9. Example network with node 2 grounded.

column 2 of the indefinite admittance matrix with the result

$$\mathbf{Y}_{(2grd)} = \begin{array}{c} \\ 1 \\ 3 \\ 4 \end{array} \begin{array}{c} \begin{array}{ccc} 1 & 3 & 4 \end{array} \\ \begin{bmatrix} 1.50 & -0.50 & 0 \\ -0.50 & 0.85 & -0.10 \\ 0 & -0.10 & 0.10 \end{bmatrix} \end{array}$$

This result may be readily checked by inspection of Figure 11.9 and by applying the equation $\mathbf{I} = \mathbf{Y}\,\mathbf{V}$ to that network.

Property 3. To illustrate property 3, we examine the network which results if nodes 1 and 3 are shorted together as shown in Figure 11.10. The third prop-

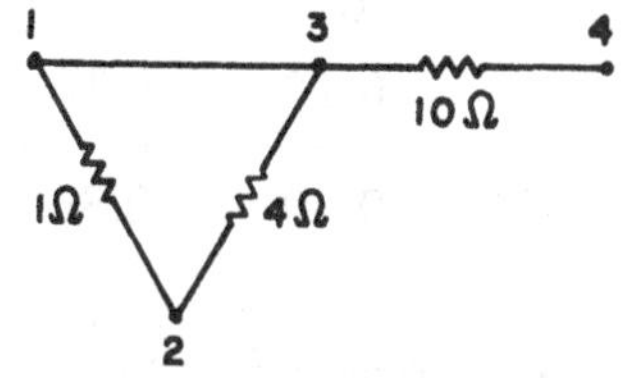

Fig. 11.10. Example network with nodes 1 and 3 connected.

erty indicates that $\hat{\mathbf{Y}}$ for this network may be obtained by replacing rows 1, 3 and columns 1, 3 by a new row and column formed by adding elements of these two rows and columns. The result is

$$\hat{Y}_{new} = \begin{array}{c} \\ 1,3 \\ 2 \\ 4 \end{array} \begin{array}{c} \begin{array}{ccc} 1,3 & 2 & 4 \end{array} \\ \begin{bmatrix} 1.35 & -1.25 & -0.10 \\ -1.25 & 1.25 & 0 \\ -0.10 & 0 & 0.10 \end{bmatrix} \end{array}$$

Note that the result is still indefinite and all rows and columns sum to zero.

Property 4. To demonstrate this property we consider the addition of a new admittance between nodes 2 and 4 as shown in Figure 11.11. According to

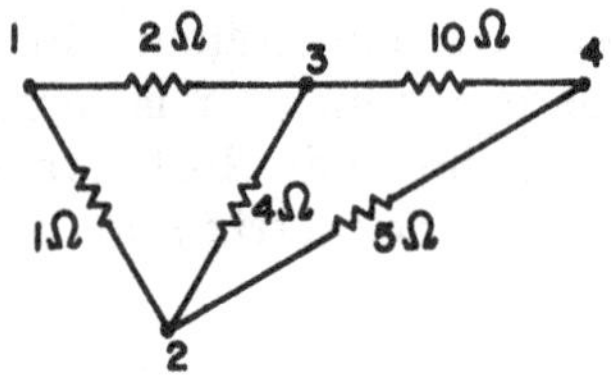

Fig. 11.11. Example network with 5 ohm resistor added between nodes 2 and 4.

property 4 we simply add 0.2 mho to elements 2, 2 and 4, 4 and then subtract 0.2 mho from elements 2, 4 and 4, 2. The result is

$$\hat{Y}_{new} = \begin{array}{c} \\ 1 \\ 2 \\ 3 \\ 4 \end{array} \begin{array}{c} \begin{array}{cccc} 1 & 2 & 3 & 4 \end{array} \\ \begin{bmatrix} 1.50 & -1.00 & -0.50 & 0 \\ -1.00 & 1.45 & -0.25 & -0.20 \\ -0.50 & -0.25 & 0.85 & -0.10 \\ 0 & -0.20 & -0.10 & 0.30 \end{bmatrix} \end{array}$$

This result may be readily verified by inspection of Figure 11.11.

Property 5. This property gives the rule for suppressing a node of the original matrix. Suppose that the external current entering node 3 of the original matrix is zero. Then we may reduce the indefinite admittance matrix description by a matrix reduction, pivoting on element 3, 3. The new circuit will be that defined by terminals 1, 2, 4 as shown in Figure 11.12. We may write the original matrix with ordering 1, 2, 4, 3 or

$$\hat{Y} = \begin{array}{c} \\ 1 \\ 2 \\ 4 \\ 3 \end{array} \begin{array}{c} \begin{array}{cccc} 1 & 2 & 4 & 3 \end{array} \\ \left[\begin{array}{ccc|c} 1.50 & -1.00 & 0 & -0.50 \\ -1.00 & 1.25 & 0 & -0.25 \\ 0 & 0 & 0.10 & -0.10 \\ \hline -0.50 & -0.25 & -0.10 & 0.85 \end{array}\right] \end{array} = \left[\begin{array}{c|c} Y_{11} & Y_{12} \\ \hline Y_{21} & Y_{22} \end{array}\right]$$

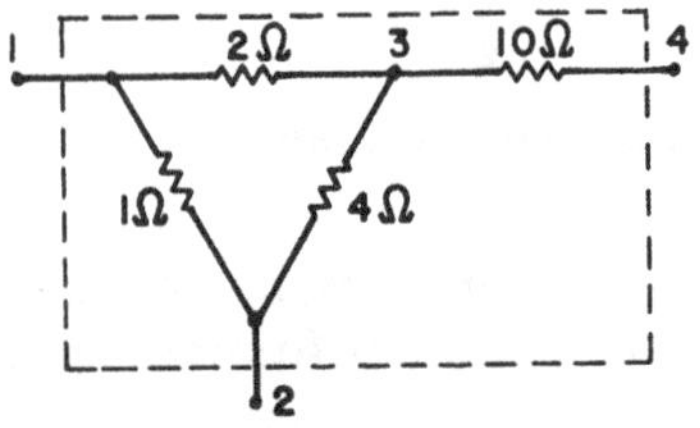

Fig. 11.12. Example network with node 3 suppressed.

where submatrices $\mathbf{Y}_{11}$, $\mathbf{Y}_{12}$, $\mathbf{Y}_{21}$, and $\mathbf{Y}_{22}$ defined in (11.31) have been identified. Then we compute

$$\hat{\mathbf{Y}}_{new} = \mathbf{Y}_{11} - \mathbf{Y}_{12}\mathbf{Y}_{22}^{-1}\mathbf{Y}_{21}$$

$$= \begin{matrix} & 1 & 2 & 4 \\ 1 \\ 2 \\ 4 \end{matrix}\begin{bmatrix} 1.50 & -1.00 & 0 \\ -1.00 & 1.25 & 0 \\ 0 & 0 & 0.10 \end{bmatrix} - \begin{bmatrix} -0.50 \\ -0.25 \\ -0.10 \end{bmatrix}\left[\frac{1}{0.85}\right][-0.50 \quad -0.25 \quad -0.10]$$

$$= \begin{bmatrix} 1.50 & -1.00 & 0 \\ -1.00 & 1.25 & 0 \\ 0 & 0 & 0.10 \end{bmatrix} - \begin{bmatrix} .2941 & .1470 & .0588 \\ .1470 & .0735 & .0294 \\ .0588 & .0294 & .0117 \end{bmatrix}$$

$$= \begin{bmatrix} 1.2059 & -1.1470 & -0.0588 \\ -1.1470 & 1.1765 & -0.0294 \\ -0.0588 & -0.0294 & 0.0883 \end{bmatrix}$$

The rows and columns still add to approximately zero but some round-off error is evident.

11.4.2 The indefinite admittance description of a power system

In the preceding section we developed a method for forming the indefinite admittance matrix of a network by constructing the primitive admittance matrix $\mathcal{Y}$ and the augmented node incidence matrix $\hat{\mathbf{A}}$ with the result

$$\hat{\mathbf{Y}} = \hat{\mathbf{A}}^t \mathcal{Y} \hat{\mathbf{A}} \tag{11.32}$$

This method proved to be wasteful of valuable computer memory since both $\hat{\mathbf{A}}$ and $\mathcal{Y}$ are extremely sparse matrices. Furthermore, the matrix multiplication indicated by (11.32) consists of a great many multiplications by zero.

A much more economical way of forming the indefinite admittance matrix is possible by the direct application of property 4. This property asserts that the effect on the matrix of adding any element to a network is to increase the diagonal elements and reduce the off-diagonal elements by exactly the branch admittance added. This suggests a direct method for building the indefinite admittance matrix, one element at a time, by starting with a matrix of zeros and simply adding each branch admittance as required by property 4 (see [65]).

Consider the network of Figure 11.13 which may be thought of as one of the sequence networks of the system at any stage of its development in computer memory, and where the element Y is to be added between nodes r and s. For the original $n+1$ terminal network prior to adding Y we write

$$\begin{aligned} I_0 &= Y_{00}V_0 + \cdots + Y_{0r}V_r + Y_{0s}V_s + \cdots + Y_{0n}V_n \\ \cdots &\quad \cdots \quad \cdots \quad \cdots \quad \cdots \quad \cdots \quad \cdots \\ I_r &= Y_{r0}V_0 + \cdots + Y_{rr}V_r + Y_{rs}V_s + \cdots + Y_{rn}V_n \\ I_s &= Y_{s0}V_0 + \cdots + Y_{sr}V_r + Y_{ss}V_s + \cdots + Y_{sn}V_n \\ \cdots &\quad \cdots \quad \cdots \quad \cdots \quad \cdots \quad \cdots \quad \cdots \\ I_n &= Y_{n0}V_0 + \cdots + Y_{nr}V_r + Y_{ns}V_s + \cdots + Y_{nn}V_n \end{aligned} \tag{11.33}$$

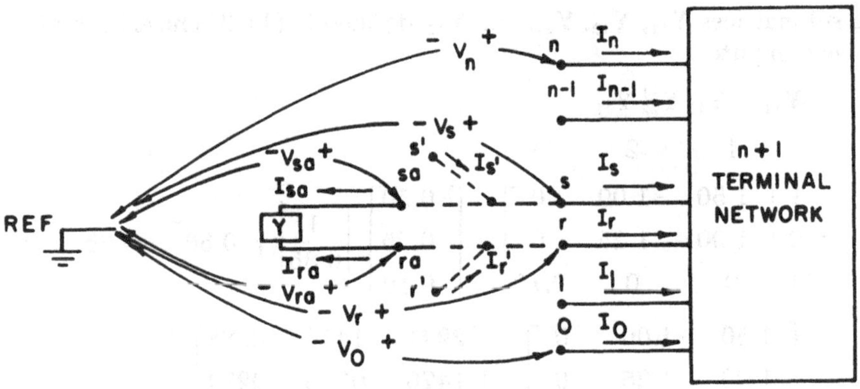

Fig. 11.13. Adding a new admittance Y from r to s.

But we may also write the indefinite equations for the element Y which is to be added as

$$I_{ra} = YV_{ra} - YV_{sa}$$

$$I_{sa} = -YV_{ra} + YV_{sa} \tag{11.34}$$

Now if the two networks are connected along the dashed lines of Figure 11.13, the node voltages and currents are constrained as

$$V_{ra} = V_r, \quad V_{sa} = V_s \tag{11.35}$$

$$I_r' = I_r + I_{ra}, \quad I_s' = I_s + I_{sa} \tag{11.36}$$

where we now define new terminals r' and s' with node currents I_r' and I_s'. Adding (11.33) and (11.34) according to (11.36) and making the voltage substitution (11.35) into (11.34), we have

$$\begin{aligned}
I_0 &= Y_{00}V_0 + \cdots + Y_{0r}V_r \qquad + Y_{0s}V_s \qquad + \cdots + Y_{0n}V_n \\
&\cdots \\
I_r' &= Y_{r0}V_0 + \cdots + (Y_{rr} + Y)V_r + (Y_{rs} - Y)V_s + \cdots + Y_{rn}V_n \\
I_s' &= Y_{s0}V_0 + \cdots + (Y_{sr} - Y)V_r + (Y_{ss} + Y)V_s + \cdots + Y_{sn}V_n \\
&\cdots \\
I_n &= Y_{n0}V_0 + \cdots + Y_{nr}V_r \qquad + Y_{ns}V_s \qquad + \cdots + Y_{nn}V_n
\end{aligned} \tag{11.37}$$

Equation (11.37) verifies property 4. Note that Y was added to the network from r to s. This changed $\hat{\mathbf{Y}}$ by adding Y to locations r, r and s, s and subtracting Y from locations r, s and s, r. This procedure is easily programmed for computer formation of $\hat{\mathbf{Y}}$, given only the tabulated line (or branch) data similar to that of Table 11.1. Furthermore, the line data can be in any arbitrary order which avoids costly sorting. Once $\hat{\mathbf{Y}}$ is formed, the row and column corresponding to the reference node ($N0$, $N1$, or $N2$) may be deleted to find the $\mathbf{Y}$ matrix, or these rows and columns could be ignored in the matrix formation.

11.4.3 Correcting for mutually coupled lines

Occasionally transmission lines are mutually coupled, especially in the zero sequence network. This situation presents no particular problem in the formation of the indefinite admittance matrix. In the development which follows only two mutually coupled lines are considered. The method is easily extended to any number of coupled lines, however, with no increase in the complexity of the matrix formation algorithm.

Consider the two mutually coupled transmission lines shown in Figure 11.14.

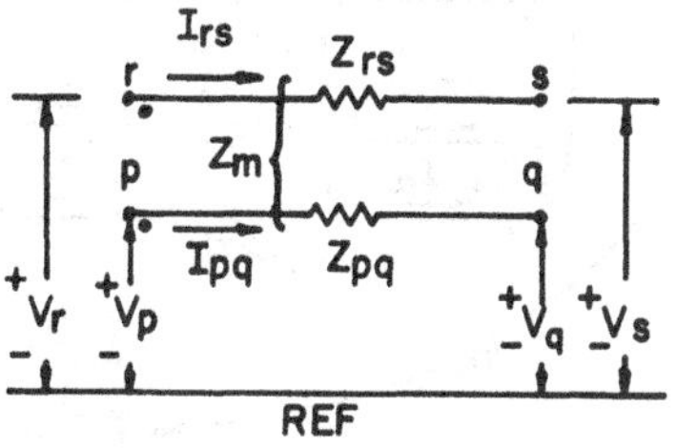

Fig. 11.14. Two mutually coupled lines.

From Chapter 4 we write the voltage drop equation

$$\begin{bmatrix} V_r - V_s \\ V_p - V_q \end{bmatrix} = \begin{bmatrix} V_{rs} \\ V_{pq} \end{bmatrix} = \begin{bmatrix} Z_{rs} & Z_m \\ Z_m & Z_{pq} \end{bmatrix} \begin{bmatrix} I_{rs} \\ I_{pq} \end{bmatrix} \tag{11.38}$$

Inverting and solving for the currents, we have

$$\begin{bmatrix} I_{rs} \\ I_{pq} \end{bmatrix} = \begin{bmatrix} Y'_{rs} & Y_m \\ Y_m & Y'_{pq} \end{bmatrix} \begin{bmatrix} V_r - V_s \\ V_p - V_q \end{bmatrix} \tag{11.39}$$

where the admittances are uniquely defined by the matrix inversion. Expanding, we have

$$\begin{aligned} I_{rs} &= (Y'_{rs} V_r - Y'_{rs} V_s) + (Y_m V_p - Y_m V_q) \\ I_{pq} &= (Y'_{pq} V_p - Y'_{pq} V_q) + (Y_m V_r - Y_m V_s) \end{aligned} \tag{11.40}$$

or in general for a coupled line, I_{rs} = (self admittance term) + (mutual admittance term), where the self admittance term is exactly the same as (11.34) for the uncoupled line. The mutual term specifies that Y_m be added in the rth row to the column having the same polarity as r, namely p, and subtracted from the column having opposite polarity, namely q.

Equation (11.40) may be expanded to the full indefinite admittance form by adding two more equations which are the negative of (11.40). The result is

$$\begin{aligned} I_{rs} &= (Y'_{rs} V_r - Y'_{rs} V_s) + (Y_m V_p - Y_m V_q) \\ I_{sr} &= (-Y'_{rs} V_r + Y'_{rs} V_s) + (-Y_m V_p + Y_m V_q) \\ I_{pq} &= (Y'_{pq} V_p - Y'_{pq} V_q) + (Y_m V_r - Y_m V_s) \\ I_{qp} &= (-Y'_{pq} V_p + Y'_{pq} V_q) + (-Y_m V_r + Y_m V_s) \end{aligned} \tag{11.41}$$

The network with elements to be added is shown in Figure 11.15.

When the coupled elements are added to the network, the following constraints are observed.

$$\begin{aligned} V_{ra} &= V_r, \quad & V_{sa} &= V_s \\ V_{pa} &= V_p, \quad & V_{qa} &= V_q \end{aligned} \tag{11.42}$$

and

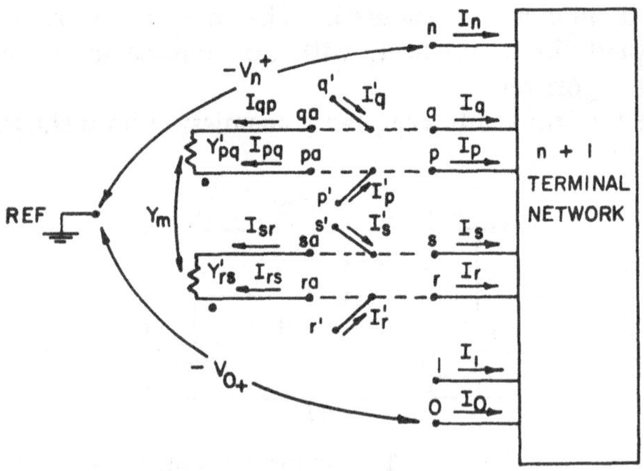

Fig. 11.15. Adding two mutually coupled elements.

$$
\begin{aligned}
I'_r &= I_r + I_{rs}, \qquad & I'_s &= I_s + I_{sr} \\
I'_p &= I_p + I_{pq}, \qquad & I'_q &= I_q + I_{qp}
\end{aligned}
\tag{11.43}
$$

Combining these constraints with the original indefinite admittance equations, we have

$$
\begin{array}{lllllll}
I_0 = Y_{00}V_0 + \cdots + Y_{0r}V_r & + Y_{0s}V_s & + \cdots + Y_{0p}V_p & + Y_{0q}V_q & + \cdots + Y_{0n}V_n \\
\cdots \quad \cdots \quad \cdots \quad \cdots & \cdots & \cdots \quad \cdots & \cdots & \cdots \quad \cdots \\
I'_r = Y_{r0}V_0 + \cdots + (Y_{rr} + Y'_{rs})V_r & + (Y_{rs} - Y'_{rs})V_s & + \cdots + (Y_{rp} + Y_m)V_p & + (Y_{rq} - Y_m)V_q & + \cdots + Y_{rn}V_n \\
I'_s = Y_{s0}V_0 + \cdots + (Y_{sr} - Y'_{rs})V_r & + (Y_{ss} + Y'_{rs})V_s & + \cdots + (Y_{sp} - Y_m)Y_p & + (Y_{sq} + Y_m)V_q & + \cdots + Y_{sn}V_n \\
\cdots \quad \cdots \quad \cdots \quad \cdots & \cdots & \cdots \quad \cdots & \cdots & \cdots \quad \cdots \\
I'_p = Y_{p0}V_0 + \cdots + (Y_{pr} + Y_m)V_r & + (Y_{ps} - Y_m)V_s & + \cdots + (Y_{pp} + Y'_{pq})V_p & + (Y_{pq} - Y'_{pq})V_q & + \cdots + Y_{pn}V_n \\
I'_q = Y_{q0}V_0 + \cdots + (Y_{qr} - Y_m)V_r & + (Y_{qs} + Y_m)V_s & + \cdots + (Y_{qp} - Y'_{pq})V_p & + (Y_{qq} + Y'_{pq})V_q & + \cdots + Y_{qn}V_n \\
\cdots \quad \cdots \quad \cdots \quad \cdots & \cdots & \cdots \quad \cdots & \cdots & \cdots \quad \cdots \\
I_n = Y_{n0}V_0 + \cdots + Y_{nr}V_r & + Y_{ns}V_s & + \cdots + Y_{np}V_p & + Y_{nq}V_q & + \cdots + Y_{nn}V_n
\end{array}
\tag{11.44}
$$

From Figure 11.15 we observe that terminals r and p are the dotted or polarized terminals. Therefore Y_m is *added* to all *like* (i.e., dotted or not dotted) locations namely rp and pr and also to locations sq and qs. By the same reasoning we *subtract* Y_m from all *unlike* locations, namely rq, qr, sp, and ps (see [65]).

Extension of the above to any number of mutually coupled elements follows exactly the same rules. Note that the mutual admittance always occurs between pairs of lines, and the rule for adding Y_m to the matrix is the same for every pair of mutually coupled branches.

Example 11.9

Compute the zero sequence indefinite admittance matrix of the system of Example 11.1 by the direct method discussed above.

Solution

From the data of Table 11.1 and the network topology of Figure 11.5 (with $y_{14} = y_1 = 0$) we record the self admittances by inspection as follows:

$$\mathbf{Y}_{self} = \begin{array}{c} \\ NO \\ 1 \\ 2 \\ 3 \end{array}\begin{array}{c} \begin{array}{cccc} NO & 1 & 2 & 3 \end{array} \\ \begin{bmatrix} y_2 & 0 & 0 & -y_2 \\ 0 & y_3 + y_6 & -y_3 & -y_6 \\ 0 & -y_3 & y_3 + y_4 + y_5 & -y_4 - y_5 \\ -y_2 & -y_6 & -y_4 - y_5 & y_2 + y_4 + y_5 + y_6 \end{bmatrix} \end{array}$$

where from Example 11.1

$$y_2 = -j50.000, \quad y_3 = -j7.143, \quad y_4 = -j12.632$$
$$y_5 = -j10.526, \quad y_6 = -j5.882$$

Then

$$\mathbf{Y}_{self} = \begin{array}{c} \\ NO \\ 1 \\ 2 \\ 3 \end{array}\begin{array}{c} \begin{array}{cccc} NO & 1 & 2 & 3 \end{array} \\ \begin{bmatrix} -j50.000 & 0 & 0 & j50.000 \\ 0 & -j13.025 & j7.143 & j5.882 \\ 0 & j7.143 & -j30.301 & j23.158 \\ j50.000 & j5.882 & j23.158 & -j79.040 \end{bmatrix} \end{array}$$

The mutual admittance between branches 4 and 5 is $y_m = +j5.263$. This admittance should be added and subtracted according to (11.44), i.e.,

Add y_m to $(r, p), (p, r), (s, q), (q, s)$

Subtract y_m from $(r, q), (q, r), (s, p), (p, s)$

In this example $r = p = 2$ and $s = q = 3$. Therefore we must

Add $2y_m$ to (2, 2), (3, 3)

Subtract $2y_m$ from (2, 3), (3, 2)

$$\hat{\mathbf{Y}} = \begin{array}{c} \\ NO \\ 1 \\ 2 \\ 3 \end{array}\begin{array}{c} \begin{array}{cccc} NO & 1 & 2 & 3 \end{array} \\ \begin{bmatrix} -j50.000 & 0 & 0 & j50.000 \\ 0 & -j13.025 & j7.143 & j5.882 \\ 0 & j7.143 & -j30.301 + 2y_m & j23.158 - 2y_m \\ j50.000 & j5.882 & j23.158 - 2y_m & -j79.040 + 2y_m \end{bmatrix} \end{array}$$

$$= -j \begin{array}{c} \\ NO \\ 1 \\ 2 \\ 3 \end{array}\begin{array}{c} \begin{array}{cccc} NO & 1 & 2 & 3 \end{array} \\ \begin{bmatrix} 50.000 & 0 & 0 & -50.000 \\ 0 & 13.025 & -7.143 & -5.882 \\ 0 & -7.143 & 19.775 & -12.632 \\ -50.000 & -5.882 & -12.632 & 68.514 \end{bmatrix} \end{array}$$

This is the same as the result in Example 11.4 if branch 1 is neglected in that example.

Property 4 provides a straightforward algorithm for computing the indefinite (or definite) admittance matrix for any network, including the effect of mutually coupled lines. There are other methods of accomplishing this same result such as the incidence matrix transformations (11.18) and (11.22).

11.5 Definite Admittance Matrix

The indefinite admittance matrix defines the relationship between node voltages and currents of a network in such a way that the details of the primitive admittances are suppressed. This is desirable since we seek a network solution where the total memory requirement for matrix storage is minimized. The indefinite admittance matrix is an efficient way of relating node voltages and currents and it may be formed with a minimum of effort. These are important considerations.

Another important feature of the indefinite admittance matrix is its sparsity. This depends entirely on the network represented, but power systems are usually quite sparse, having between 1.0 and 1.5 branches per node. This means that the indefinite admittance matrix itself has many zero elements and that considerable savings in memory requirements may be gained by storing only the nonzero elements in some simple tabular form.

In power system analysis we are interested in the *definite* admittance matrix or simply the node admittance matrix **Y** defined by the equation

$$\mathbf{I} = \mathbf{Y}\,\mathbf{V} \tag{11.45}$$

The admittance matrix is formed by crossing out the rows and columns of all nodes connected to the reference node. In a power system this will usually be the zero potential nodes $N0$, $N1$, or $N2$. Moreover the n currents and voltages of (11.45) are independent sets of complex quantities. Given the n independent voltages **V**, one can solve for the currents **I** by direct application of (11.45). Also, since **Y** is nonsingular, one can find **V** when **I** is given by computing

$$\mathbf{V} = \mathbf{Y}^{-1}\mathbf{I} = \mathbf{Z}\,\mathbf{I} \tag{11.46}$$

In applying (11.45) and (11.46) to the solution of faulted networks, we usually make the simplifying assumptions given in Table 11.2.

Table 11.2. Simplifying Assumptions Used in Fault Calculations

Assumption	*Comment*
1. All load currents negligible.	$\mathbf{I}_{\text{load}} = \begin{bmatrix} I_{\ell 1} \\ I_{\ell 2} \\ \cdots \\ I_{\ell\ell} \end{bmatrix} = 0$
2. All generated voltages are equal in phase and magnitude to the positive sequence prefault voltage, hV_f.	$\mathbf{V}_{\text{gen}} = \begin{bmatrix} V_{g1} \\ V_{g2} \\ \cdots \\ V_{gg} \end{bmatrix} = \begin{bmatrix} 1 \\ 1 \\ \cdots \\ 1 \end{bmatrix} hV_f$
3. The positive and negative sequence networks are identical.	$\mathbf{Z}_1 = \mathbf{Z}_2$
4. The networks are balanced except at fault points.	Z_{abc} are symmetric in all references for all lines.
5. All shunt admittance (line charging susceptance, etc.) negligible.	The only network reference connection is through generator impedances.

Since all generated voltages are equal in phase and magnitude, all generator nodes are connected together and a voltage hV_f is applied at that point as indicated by Figure 11.16 where only the positive sequence network is shown. Also,

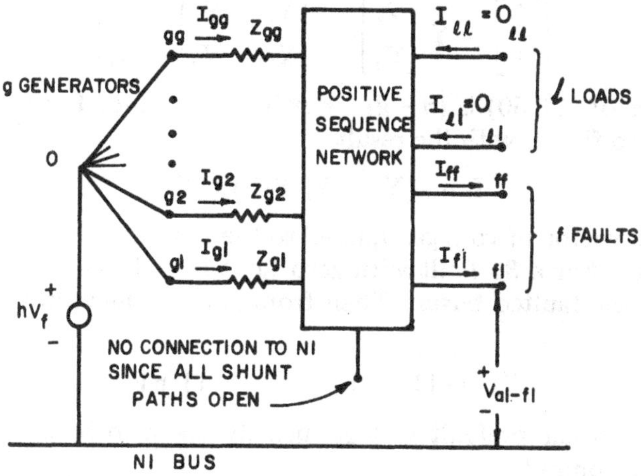

Fig. 11.16. Positive sequence network which results from the assumptions of Table 11.2.

since loads and shunt admittances are all neglected, the generator connections are the *only* connections to $N1$, the reference node. The negative and zero sequence networks are similar except that the voltage hV_f is zero and the generator nodes are connected directly to the reference.

Since there are f fault nodes, ℓ load nodes and g generator nodes, we may write the *indefinite* admittance equation for Figure 11.16 in terms of the positive sequence voltages as

$$\begin{bmatrix} -\mathbf{I}_f \\ \mathbf{I}_\ell \\ \mathbf{I}_g \end{bmatrix} = \hat{\mathbf{Y}} \begin{bmatrix} \mathbf{V}_{a1\text{-}f} \\ \mathbf{V}_{a1\text{-}\ell} \\ \mathbf{V}_{a1\text{-}g} \end{bmatrix} \tag{11.47}$$

We now consider a change in reference. Since the only reference connections are made through the generators, suppose we select node 0 as the reference (this is already the case in the negative or zero sequence networks). In terms of this reference all generator node voltages are zero and all load and fault voltages are the positive sequence voltage less hV_f. If we identify the node voltages with respect to node 0 by a single subscript, the node voltage at any node k may be written as

$$V_k = V_{a1\text{-}k} - hV_f, \quad k = 1, 2, \ldots, n$$

or in vector form

$$\mathbf{V} = \mathbf{V}_{a1} - [1 \quad 1 \quad \cdots \quad 1]^t hV_f = \mathbf{V}_{a1} + \mathbf{V}_F \tag{11.48}$$

Then adding $\hat{\mathbf{Y}}\,\mathbf{V}_F = \mathbf{0}$ from each side of (11.47), we compute

$$\begin{bmatrix} -\mathbf{I}_f \\ \mathbf{I}_\ell \\ \mathbf{I}_g \end{bmatrix} = \hat{\mathbf{Y}} \begin{bmatrix} \mathbf{V}_f \\ \mathbf{V}_\ell \\ \mathbf{0} \end{bmatrix} = \begin{bmatrix} \hat{\mathbf{Y}}_{ff} & \hat{\mathbf{Y}}_{f\ell} & \hat{\mathbf{Y}}_{fg} \\ \hat{\mathbf{Y}}_{\ell f} & \hat{\mathbf{Y}}_{\ell\ell} & \hat{\mathbf{Y}}_{\ell g} \\ \hat{\mathbf{Y}}_{gf} & \hat{\mathbf{Y}}_{g\ell} & \hat{\mathbf{Y}}_{gg} \end{bmatrix} \begin{bmatrix} \mathbf{V}_f \\ \mathbf{V}_\ell \\ \mathbf{0} \end{bmatrix} \tag{11.49}$$

Since nodes $g1, g2, \ldots, gg$ are connected to the reference, the last g rows and columns of $\hat{\mathbf{Y}}$ may be eliminated (property 2) to write the definite admittance equation

$$\begin{bmatrix} -\mathbf{I}_f \\ \mathbf{I}_\ell \end{bmatrix} = \mathbf{Y} \begin{bmatrix} \mathbf{V}_f \\ \mathbf{V}_\ell \end{bmatrix} = \begin{bmatrix} \mathbf{Y}_{ff} & \mathbf{Y}_{f\ell} \\ \mathbf{Y}_{\ell f} & \mathbf{Y}_{\ell\ell} \end{bmatrix} \begin{bmatrix} \mathbf{V}_f \\ \mathbf{V}_\ell \end{bmatrix} \tag{11.50}$$

where the order of (11.50) is $(n - g)$. Furthermore, since $\mathbf{I}_\ell = \mathbf{0}$, we may reduce this expression to find $\mathbf{I}_f$ with the result

$$\mathbf{I}_f = -(\mathbf{Y}_{ff} - \mathbf{Y}_{f\ell}\mathbf{Y}_{\ell\ell}^{-1}\mathbf{Y}_{\ell f})\mathbf{V}_f \tag{11.51}$$

where $\mathbf{V}_f$ is the vector of voltages applied to the faulted nodes, measured with respect to node 0. For a 3ϕ fault with zero fault impedance the sequence voltage $V_{a1\text{-}f}$ is zero at the faulted buses. Then from (11.48) the vector of voltages at all faulted buses is

$$\mathbf{V}_F = -[1 \quad 1 \quad \cdots \quad 1]^t \mathrm{h} V_f$$

where V_f is the scalar prefault voltage, usually taken to be unity. Substituting into (11.51), we compute

$$\mathbf{I}_f = +(\mathbf{Y}_{ff} - \mathbf{Y}_{f\ell}\mathbf{Y}_{\ell\ell}^{-1}\mathbf{Y}_{\ell f})\,[1 \quad 1 \quad \cdots \quad 1]^t \mathrm{h} V_f \tag{11.52}$$

The magnitude of $\mathbf{I}_f$ is simply scaled up or down in proportion to the magnitude of $\mathrm{h}V_f$.

Equation (11.52) provides a compact form of the solution for all fault currents $\mathbf{I}_f$ under 3ϕ fault conditions. If there is only one fault, this equation is scalar. Note that the procedure of finding $\mathbf{I}_f$ for each of the n nodes requires n different complex matrix inversions each of order $(n - 1)$. This is a serious disadvantage for this method and makes its use very expensive, even for small networks. For this reason an alternate method is sought. One alternative is to develop an iterative technique which will converge on the solution $\mathbf{I}_f$ in a stepwise procedure. A more direct alternative is to invert the admittance matrix and find the fault currents by working with the impedance matrix. Often the matrix inversion is avoided by forming the impedance matrix without first developing the admittance matrix [64]. This procedure has its own special problems as noted in Chapter 12.

Example 11.10

Compute the positive sequence admittance matrix for the network described in Example 11.1 and Table 11.1. Then compute the total fault current

1. For a 3ϕ fault at node 1.
2. For a 3ϕ fault at node 2.
3. For a 3ϕ fault at node 3.
4. For simultaneous 3ϕ faults at nodes 1 and 2.

Assume zero fault impedance in each case. Let $\mathrm{h} = 1$ and $\mathbf{V}_f = 1.0$.

Solution

From Example 11.6 we have for the positive sequence network

$$Y = \begin{array}{c} \\ 1 \\ 2 \\ 3 \end{array} \begin{array}{c} \begin{array}{ccc} 1 & 2 & 3 \end{array} \\ \begin{bmatrix} -j24.19 & j12.50 & j7.69 \\ j12.50 & -j45.84 & j33.34 \\ j7.69 & +j33.34 & -j46.03 \end{bmatrix} \end{array}$$

Since the faults are at different locations, (11.52) must be solved separately for each case. For example, for case 1 we compute

$$I_{f1} = (Y_{11} - \mathbf{Y}_{1\ell}\, \mathbf{Y}_{\ell\ell}^{-1}\, \mathbf{Y}_{\ell 1})\mathrm{h}V_f$$

or

$$I_{f1} = -j24.19 - [j12.50 \quad j7.69] \begin{bmatrix} -j45.84 & j33.34 \\ j33.34 & -j46.03 \end{bmatrix}^{-1} \begin{bmatrix} j12.50 \\ j7.69 \end{bmatrix} = -j7.852$$

Similarly for case 2 we have

$$I_{f2} = -j45.84 - [12.50 \quad j33.34] \begin{bmatrix} -j24.19 & j7.69 \\ j7.69 & -j46.03 \end{bmatrix}^{-1} \begin{bmatrix} j12.50 \\ j33.34 \end{bmatrix} = -j7.436$$

and for case 3

$$I_{f3} = -j46.03 - [j7.69 \quad j33.34] \begin{bmatrix} -j24.19 & j12.50 \\ j12.50 & -j45.84 \end{bmatrix}^{-1} \begin{bmatrix} j7.69 \\ j33.34 \end{bmatrix} = -j8.230$$

Case 4 is a simultaneous fault on nodes 1 and 2, so the solution is

$$\begin{bmatrix} I_{f1} \\ I_{f2} \end{bmatrix} = \left\{ \begin{bmatrix} -j24.19 & j12.50 \\ j12.50 & -j45.84 \end{bmatrix} - \begin{bmatrix} j7.69 \\ j33.34 \end{bmatrix} [-j46.03]^{-1} [j7.69 \quad j33.34] \right\} \begin{bmatrix} 1.0 \\ 1.0 \end{bmatrix}$$

$$= \begin{bmatrix} -j4.835 \\ -j3.622 \end{bmatrix}$$

The simultaneous faults are less severe than the isolated faults on buses 1 or 2 because one contribution from the adjacent bus is zero.

Example 11.10 illustrates the major difficulty of using the admittance matrix to solve for fault currents, namely, that every different fault location requires the inversion of a different matrix. To avoid this numerical problem, in Chapter 12 we shall examine the impedance matrix as a means of finding fault currents and develop a method of finding the impedance matrix which avoids a matrix inversion.

Problems

11.1. Construct the positive sequence primitive impedance matrix for the network of Figure P11.1, using symbols z_{13}, etc., for the branch impedances.

11.2. Construct the positive sequence primitive impedance matrix for the network of Figure P11.2, using symbols z_{13}, etc., for the branch impedances.

11.3. Construct the positive sequence primitive impedance matrix for the network of Figure P11.3, using symbols z_{13}, etc., for the branch impedances. Repeat for the zero sequence and assume that z_{24} and z_{25} are mutually coupled by an amount z_m.

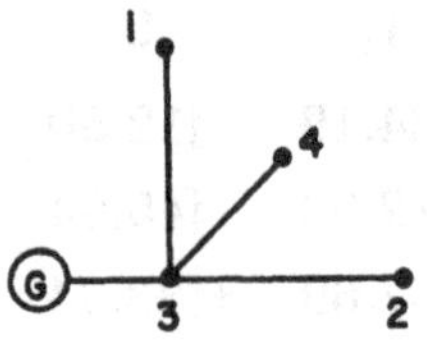

Fig. P11.1.

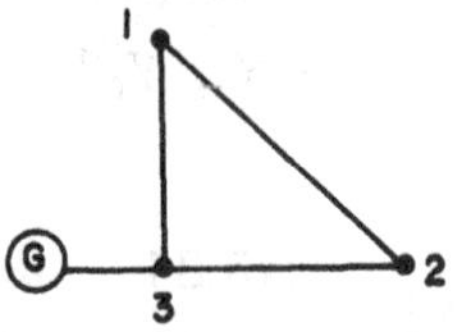

Fig. P11.2.

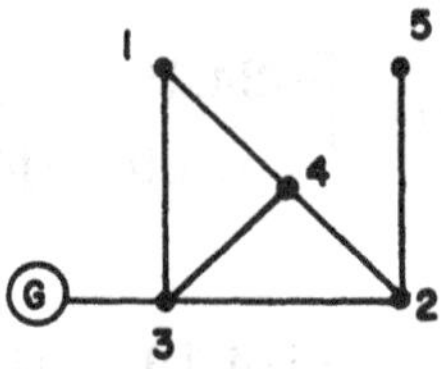

Fig. P11.3.

11.4. Construct the positive and zero sequence primitive impedance matrices for the network of Figure P11.4, using the line data tabulated below and ignoring all resistances.

Branch Number	Branch Terminals	*Positive Sequence*		*Zero Sequence*		*Mutual*	
		R_1	X_1	R_0	X_0	X_m	to
1	G-3	0	j0.20	0	j0.05	...	...
2	1-3	0.10	j0.30	0.30	j0.90	...	...
3	3-4	0.11	j0.25	0.40	j0.80	...	...
4	2-3	0.08	j0.20	0.30	j0.75	...	...
5	1-4	0.10	j0.15	0.35	j0.35	...	...
6	2-4	0.20	j0.50	1.00	j1.20	j0.20	8
7	1-5	0.15	j0.30	∞	∞	...	...
8	2-5	0.25	j0.40	0.70	j1.10	j0.20	6
9	5-6	0.35	j0.90	1.00	j2.50	...	...

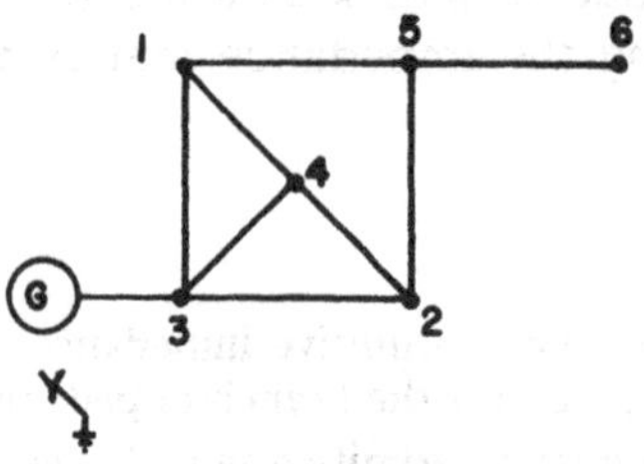

Fig. P11.4.

11.5. Repeat problem 11.4, including the effect of resistances.

11.6. Invert the positive and zero sequence primitive reactance matrices of problem 11.4 to find the primitive admittance matrices $\mathcal{Y}_1$, and $\mathcal{Y}_0$ respectively.

11.7. Invert the positive and zero sequence primitive impedance matrices of problem 11.5 to find the primitive admittance matrices y_1 and y_0 respectively.

11.8. Find the zero sequence primitive impedance matrix for the network of Figure P11.4 if the lines 1-3, 3-4, and 3-2 are all mutually coupled as follows:

$$z_{m(2,3)} = j0.15, \quad z_{m(2,4)} = j0.10, \quad z_{m(3,4)} = j0.20$$

11.9. Invert the zero sequence impedance matrix found in problem 11.8 to find the new $\mathcal{Y}_0$.

11.10. Compute the positive and negative sequence primitive $\mathcal{Z}$ and $\mathcal{Y}$ matrices for the system of Example 11.1.

11.11. Develop the augmented node incidence matrices $\hat{A}$ for the following networks with positive branch orientation from smaller to larger node numbers.
(a) The circuit of Figure P11.1.
(b) The circuit of Figure P11.2.
(c) The circuit of Figure P11.3.
(d) The positive sequence circuit of Figure P11.4.
(e) The zero sequence circuit of Figure P11.4.
Use data from problem 11.4 for all networks.

11.12. Develop a zero sequence equivalent network for the system of Figure 11.14 where the mutually coupled line pair is replaced by an equivalent circuit of self impedances.

11.13. Compute the indefinite admittance matrix $\hat{Y}$ for the following networks using the augmented node incidence matrices found in problem 11.11 and applying (11.18).
(a) The circuit of Figure P11.1.
(b) The circuit of Figure P11.2.
(c) The circuit of Figure P11.3.
(d) The positive sequence circuit of Figure P11.4.
(e) The zero sequence circuit of Figure P11.4.

11.14. Check the result of Example 11.4 by applying (11.18).

11.15. Verify (11.17) beginning with (11.13) and (11.14).

11.16. Compute the node incidence matrices A for the zero sequence networks given by (a) Figure P11.1, (b) Figure P11.2, (c) Figure P11.3, and (d) Figure P11.4.

11.17. Using the results of problem 11.16, compute the corresponding zero sequence definite admittance matrices Y, using (11.22).

11.18. Verify the computed result of Example 11.6.

11.19. Verify (11.22).

11.20. Using a digital computer, invert the admittance matrices of each circuit of Figures P11.1-P11.4 to find the zero sequence impedance matrices for these circuits.

11.21. Compute the positive sequence indefinite admittance matrices of the following networks by applying (11.29). Neglect all resistances.
(a) The circuit of Figure P11.1.
(b) The circuit of Figure P11.2.
(c) The circuit of Figure P11.3.
(d) The circuit of Figure P11.4.

11.22. Repeat problem 11.21 including the effect of resistances.

11.23. Prove the five properties of the indefinite admittance matrix stated in section 11.4.1 (see [5]).

11.24. Beginning with the indefinite admittance matrices computed in problem 11.13, find the $\hat{Y}$ matrix description after the following circuit changes are made:
(a) Grounding node 2 of Figure P11.2.
(b) Grounding node 5 of Figure P11.4.
(c) Connecting nodes 1 and 5 of Figure P11.3.
(d) Connecting nodes 2 and 6 of Figure P11.4.
(e) Connecting an impedance j0.1 between nodes 1 and 5 of Figure P11.3.
(f) Suppressing terminal 4 of Figure 11.3.
(g) Suppressing terminals 4 and 5 of Figure 11.4.

11.25. Compute the positive sequence indefinite admittance matrices for the following networks, using "property 4" and (11.37) to build the matrix one element at a time. Compare the result with that of problem 11.13.

(a) The circuit of Figure P11.1.
(b) The circuit of Figure P11.2.
(c) The circuit of Figure P11.3.
(d) The circuit of Figure P11.4.

11.26. Compute the zero sequence indefinite admittance matrices for the network of Figure P11.4, using the result (11.44). Build the matrix one element at a time in the order given in problem 11.4.

11.27. Compute the 3ϕ fault current using the admittance matrix technique (11.52) for the following systems (neglect resistance).

(a) Figure P11.1 with a fault at node 4.
(b) Figure P11.1 with faults at nodes 1 and 4.
(c) Figure P11.2 with a fault at node 2.
(d) Figure P11.3 with a fault at node 4.
(e) Figure P11.3 with a fault at node 5.
(f) Figure P11.4 with a fault at node 4.
(g) Figure P11.4 with a fault at node 5.
(h) Figure P11.4 with faults at node 4 and 6.
(i) Figure P11.4 with another generator at node 6 with impedance 0 + j0.1 and a fault at node 4.
(j) Same as (i) with faults at nodes 4 and 5.
(k) Same as (i) with a fault at node 4 with fault resistance of 1.0 pu.

chapter 12

Computer Solution Methods Using the Impedance Matrix

In Chapter 11 an admittance matrix method was devised which could be used to compute fault currents in a power system. This method, although simple to implement, was observed to have certain disadvantages in its application to large networks. It was also observed that the impedance matrix, although more difficult to derive, has certain advantages for fault computations. This is primarily due to the impedance matrix being an "open circuit" network description, and this coincides with the open circuit approximation usually used in fault studies. This chapter will be devoted to a study of impedance matrix methods for use in fault studies. We shall also develop an algorithm for finding the impedance matrix which is more direct and cheaper to implement than performing an inversion of the admittance matrix.

12.1 Impedance Matrix in Shunt Fault Computations

Consider the network shown in Figure 12.1 where the network could be the positive, negative, or zero sequence network but where the reference is always node 0 of Figure 11.16 or the common generator node. We arbitrarily define all currents to be entering the network at nodes 1, 2, . . . , n and all voltages to be the voltage drops from each node to the reference. Then we define the impedance matrix by the equation

$$\mathbf{V} = \mathbf{Z}\,\mathbf{I} \tag{12.1}$$

In expanded form (12.1) is written as

$$\begin{bmatrix} V_1 \\ V_2 \\ \cdots \\ V_n \end{bmatrix} = \begin{bmatrix} Z_{11} & Z_{12} & \cdots & Z_{1n} \\ Z_{21} & Z_{22} & \cdots & Z_{2n} \\ \cdots & \cdots & \cdots & \cdots \\ Z_{n1} & Z_{n2} & \cdots & Z_{nn} \end{bmatrix} \begin{bmatrix} I_1 \\ I_2 \\ \cdots \\ I_n \end{bmatrix} \tag{12.2}$$

If every current except that entering the kth node is zero, we have from (12.2)

$$\begin{bmatrix} V_1 \\ V_2 \\ \cdots \\ V_n \end{bmatrix} = \begin{bmatrix} Z_{1k} \\ Z_{2k} \\ \cdots \\ Z_{nk} \end{bmatrix} I_k, \qquad I_j = 0, j \neq k \tag{12.3}$$

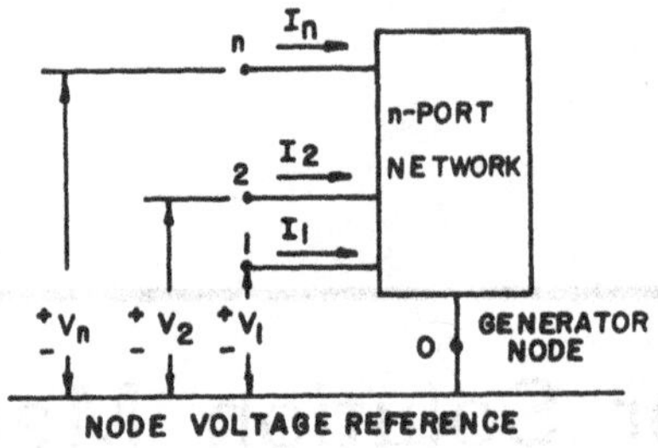

Fig. 12.1. An n-port sequence network.

This *defines* the kth column of the impedance matrix since we may solve (12.3) to compute

$$Z_{ik} = (V_i/I_k)_{I_j=0,\, j\neq k}, \qquad i = 1, 2, \ldots, n \tag{12.4}$$

Since all impedance elements are defined with all nodes open except one, the impedance elements are called the "open circuit driving point and transfer impedances" or

$$Z_{ik} = \text{open circuit driving point impedance, } i = k$$
$$= \text{open circuit transfer impedance, } i \neq k$$

For small networks these impedances may conveniently be found by injecting 1.0 ampere at node k, i.e., I_k = 1.0 ampere (or 1.0 pu), and solving (12.4) for the voltages.

The impedance matrix can be used directly for fault computation if we apply the generator voltage to the faulted node in the positive sequence network as noted in Figure 11.16.

12.1.1 Three-phase fault computations

Three-phase faults are computed by connecting the faulted node to the $N1$ bus through a suitable fault impedance Z_f as noted in Chapter 3 and in Figure 3.19. When the generator node 0 is used as a reference, the positive sequence network is as shown in Figure 12.2. Note that the fault current I_{fk} is the negative of

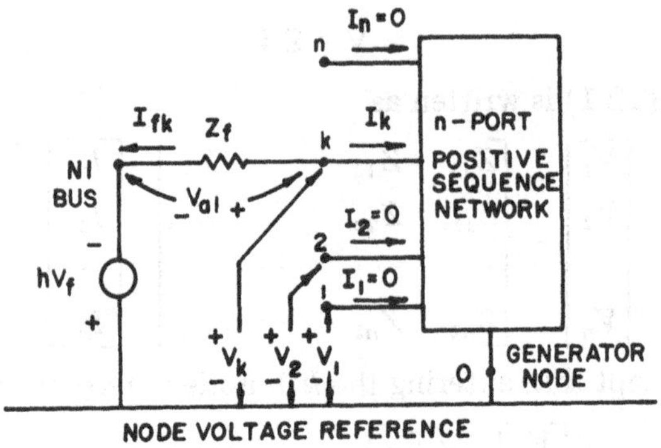

Fig. 12.2. Network connections for a 3ϕ fault on node k.

I_k, the current entering node k. Since all currents except I_k are zero, (12.4) describes the network exactly. Solving this equation for I_k, we have

$$I_k = V_k / Z_{kk} = -I_{fk} \tag{12.5}$$

From the way in which node k is terminated we also may write

$$V_k = Z_f I_{fk} - \mathrm{h} V_f \tag{12.6}$$

which we may substitute into (12.5) to eliminate V_k and solve for I_{fk} or

$$I_{fk} = \mathrm{h} V_f / Z_T \tag{12.7}$$

where $Z_T = Z_{kk} + Z_f$. Thus the 3ϕ fault current at node k depends upon the open circuit driving point impedance at node k. This is the diagonal impedance element Z_{kk} of the impedance matrix and may be thought of as the impedance seen looking into the network at node k with all nodes except the kth node open.

Once the total fault current I_{fk} is known, we may readily find all the node voltages from the impedance matrix equation or, more simply, from (12.3). Thus for any node i we may write

$$V_i = Z_{ik} I_k = -Z_{ik} \mathrm{h} V_f / Z_T, \qquad i = 1, 2, \ldots, n \tag{12.8}$$

These are the node voltages measured with respect to the reference node 0. The positive sequence voltages are measured with respect to node $N1$ and are found by adding $\mathrm{h} V_f$ to (12.8), or at node i we have

$$V_{a1-i} = \mathrm{h} V_f \left(1 - \frac{Z_{ik}}{Z_T}\right) \tag{12.9}$$

Knowing the voltage at each node, we may compute the current flowing in all lines in the network. This requires knowledge of the primitive impedances. For any pair of nodes i and j we may compute the current flowing in line $\mathfrak{z}_{ij}$ as

$$I_{ij} = \frac{V_i - V_j}{\mathfrak{z}_{ij}} = \frac{\mathrm{h} V_f (Z_{jk} - Z_{ik})}{\mathfrak{z}_{ij} Z_T} \tag{12.10}$$

Note carefully that the capital letter Z with two subscripts refers to matrix elements whereas the script $\mathfrak{z}_{ij}$ is the primitive impedance element between nodes i and j.

In summary we note that the matrix diagonal element is used in finding the total fault current, and the other elements of the kth column are used to find node voltages and branch currents.

Example 12.1

Find the positive sequence impedance matrix for the system of Example 11.1. Then compute the total fault currents for faults on each bus. The fault impedance is zero and V_f is unity. Let h = 1.

Solution

The positive sequence network is shown in Figure 12.3 where the two generator impedances have been connected to node 0. According to (12.3) we can find the first column of the Z matrix by injecting 1.0 pu current into node 1. Then the three resulting voltages V_1, V_2, and V_3 will correspond to the first column impedances Z_{11}, Z_{21}, and Z_{31} respectively. This situation is shown in Figure 12.4 with the result

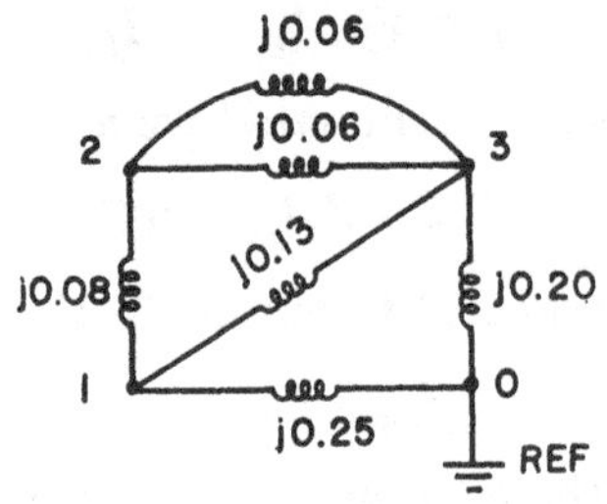

Fig. 12.3. Positive sequence network.

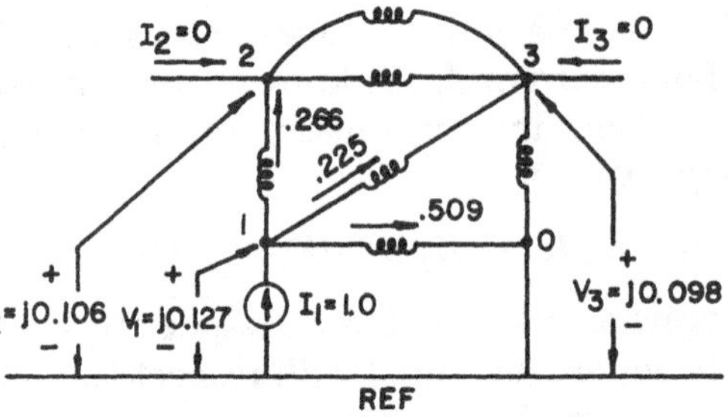

Fig. 12.4. Network solution when $I_1 = 1.0$.

$$\begin{bmatrix} V_1 \\ V_2 \\ V_3 \end{bmatrix} = \begin{bmatrix} Z_{11} \\ Z_{21} \\ Z_{31} \end{bmatrix} = \begin{bmatrix} j0.1274 \\ j0.1061 \\ j0.0981 \end{bmatrix}$$

The second and third columns may be found in the same way with the result

$$\mathbf{Z} = \begin{array}{c} \\ 1 \\ 2 \\ 3 \end{array} \begin{array}{c} \begin{array}{ccc} 1 & 2 & 3 \end{array} \\ \begin{bmatrix} j0.1274 & j0.1061 & j0.0981 \\ j0.1061 & j0.1345 & j0.1151 \\ j0.0981 & j0.1151 & j0.1215 \end{bmatrix} \end{array}$$

The 3ϕ fault currents are computed from (12.7). With $V_f = 1.0$, h = 1, and $Z_f = 0$ we compute

$$I_{f1} = 1.0/Z_{11} = 1.0/j0.1274 = -j7.852 \quad \text{pu}$$
$$I_{f2} = 1.0/Z_{22} = 1.0/j0.1345 = -j7.436 \quad \text{pu}$$
$$I_{f3} = 1.0/Z_{33} = 1.0/j0.1215 = -j8.230 \quad \text{pu}$$

These results check exactly with the results of Example 11.10. Equations (12.9) and (12.10) could now be used to compute the system positive sequence voltages and line contributions for each fault location. This is left as an exercise.

12.1.2 Single-line-to-ground fault computation

We may extend the impedance matrix solution to situations which involve the negative and zero sequence networks as well as the positive sequence network. In all sequence networks we assume that the common generator connection or node 0 of Figure 11.16 is taken as the reference. Then we write **V** = **Z I** for all networks where **I** is the vector of currents entering each node in the sequence networks and the voltages are the voltage drops from each node to the nodal reference bus 0 as shown in Figure 12.5. The notation used to distinguish between sequences is also given in Figure 12.5. In general the nodal subscripts read (sequence, node) in all cases.

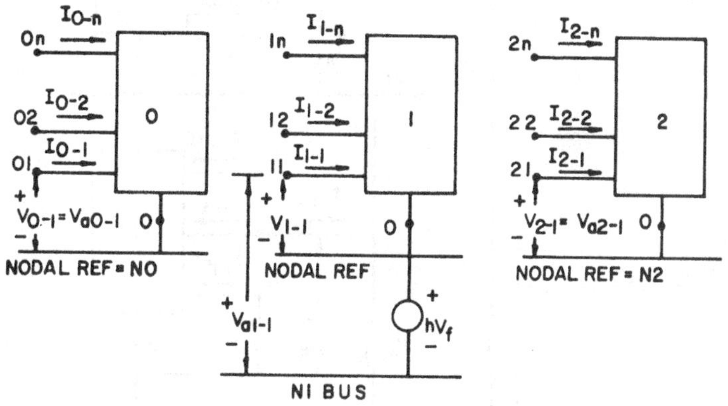

Fig. 12.5. Sequence quantities defined for nodal equations.

We further identify the *sequence* voltages by the subscript a where necessary, especially in the positive sequence network. Using this notation, we may write the nodal equations in matrix form as

$$\mathbf{V}_0 = \mathbf{Z}_0 \mathbf{I}_0, \qquad \mathbf{V}_1 = \mathbf{Z}_1 \mathbf{I}_1, \qquad \mathbf{V}_2 = \mathbf{Z}_2 \mathbf{I}_2 \tag{12.11}$$

where the meaning of these equations is clear from Figure 12.5. Specifically, the positive sequence equation is

$$\begin{bmatrix} V_{1\text{-}1} \\ V_{1\text{-}2} \\ \cdots \\ V_{1\text{-}n} \end{bmatrix} = \begin{bmatrix} Z_{1\text{-}11} & Z_{1\text{-}12} & \cdots & Z_{1\text{-}1n} \\ Z_{1\text{-}21} & Z_{1\text{-}22} & \cdots & Z_{1\text{-}2n} \\ \cdots & \cdots & \cdots & \cdots \\ Z_{1\text{-}n1} & Z_{1\text{-}n2} & \cdots & Z_{1\text{-}nn} \end{bmatrix} \begin{bmatrix} I_{1\text{-}1} \\ I_{1\text{-}2} \\ \cdots \\ I_{1\text{-}n} \end{bmatrix} \tag{12.12}$$

We note one additional problem concerning notation. Here we define the sequence currents to be those *entering* the network. At the faulted node, however, we identify the sequence currents I_{a0}, I_{a1}, and I_{a2} to be those *leaving* the network. Where necessary to identify these fault currents we shall add the subscript a in the usual way.

For a SLG fault on node k the sequence networks are arranged as in Figure 12.6. Since the driving point impedance at k is known, we write

$$I_{a0\text{-}k} = I_{a1\text{-}k} = I_{a2\text{-}k} = \frac{\mathrm{h}V_f}{Z_{0\text{-}kk} + Z_{1\text{-}kk} + Z_{2\text{-}kk} + 3Z_f} \triangleq \frac{\mathrm{h}V_f}{Z_T} \tag{12.13}$$

where we define

$$Z_T = Z_{0\text{-}kk} + Z_{1\text{-}kk} + Z_{2\text{-}kk} + 3Z_f \tag{12.14}$$

Then the voltages at all nodes may be computed from the currents of (12.13) and the Z matrix. From (12.8) for any sequence s we may write

$$V_{s\text{-}i} = -Z_{s\text{-}ik}\,\mathrm{h}V_f/Z_T \tag{12.15}$$

This equation gives the sequence voltage in the zero and negative sequence net-

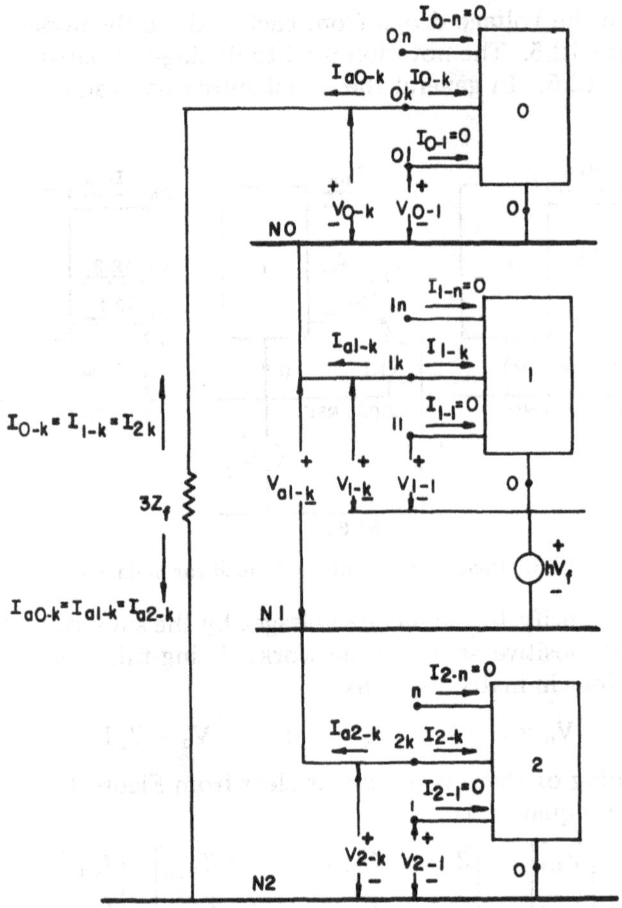

Fig. 12.6. Sequence network connections for a SLG fault on bus *k*.

works. In the positive sequence network the voltage hV_f must be added as in (12.9). The resulting sequence voltages are

$$\begin{bmatrix} V_{a0\text{-}1} \\ V_{a0\text{-}2} \\ \cdots \\ V_{a0\text{-}n} \end{bmatrix} = \begin{bmatrix} V_{0\text{-}1} \\ V_{0\text{-}2} \\ \cdots \\ V_{0\text{-}n} \end{bmatrix} = -\frac{hV_f}{Z_T} \begin{bmatrix} Z_{0\text{-}1k} \\ Z_{0\text{-}2k} \\ \cdots \\ Z_{0\text{-}nk} \end{bmatrix} \tag{12.16}$$

$$\begin{bmatrix} V_{a1\text{-}1} \\ V_{a1\text{-}2} \\ \cdots \\ V_{a1\text{-}n} \end{bmatrix} = \begin{bmatrix} V_{1\text{-}1} \\ V_{1\text{-}2} \\ \cdots \\ V_{1\text{-}n} \end{bmatrix} + h \begin{bmatrix} V_f \\ V_f \\ \cdots \\ V_f \end{bmatrix} = h \begin{bmatrix} V_f \\ V_f \\ \cdots \\ V_f \end{bmatrix} - \frac{hV_f}{Z_T} \begin{bmatrix} Z_{1\text{-}1k} \\ Z_{1\text{-}2k} \\ \cdots \\ Z_{1\text{-}nk} \end{bmatrix} \tag{12.17}$$

$$\begin{bmatrix} V_{a2\text{-}1} \\ V_{a2\text{-}2} \\ \cdots \\ V_{a2\text{-}n} \end{bmatrix} = \begin{bmatrix} V_{2\text{-}1} \\ V_{2\text{-}2} \\ \cdots \\ V_{2\text{-}n} \end{bmatrix} = -\frac{hV_f}{Z_T} \begin{bmatrix} Z_{2\text{-}1k} \\ Z_{2\text{-}2k} \\ \cdots \\ Z_{2\text{-}nk} \end{bmatrix} \tag{12.18}$$

The currents flowing in the branches of the positive and negative sequence networks may be computed from equation (12.10). To compute the current in branch *i-j* in the zero sequence network which is not mutually coupled, again use (12.10). If this element is mutually coupled to another element or elements, (12.10) must be solved simultaneously for all coupled currents. In full double-subscript notation the solution becomes very messy to write. For example if elements *i-j*, *m-n*, and *p-q* are mutually coupled in the primitive matrix, the solution for all coupled currents involves the inverse of that partition of the primitive impedance matrix with the result

$$\begin{bmatrix} I_{ij} \\ I_{mn} \\ I_{pq} \end{bmatrix} = \begin{bmatrix} \mathfrak{z}_{ij\text{-}ij} & \mathfrak{z}_{ij\text{-}mn} & \mathfrak{z}_{ij\text{-}pq} \\ \mathfrak{z}_{mn\text{-}ij} & \mathfrak{z}_{mn\text{-}mn} & \mathfrak{z}_{mn\text{-}pq} \\ \mathfrak{z}_{pq\text{-}ij} & \mathfrak{z}_{pq\text{-}mn} & \mathfrak{z}_{pq\text{-}pq} \end{bmatrix}^{-1} \begin{bmatrix} Z_{jk} - Z_{ik} \\ Z_{nk} - Z_{mk} \\ Z_{qk} - Z_{pk} \end{bmatrix} \frac{hV_f}{Z_T} \tag{12.19}$$

where all impedances denoted $\mathfrak{z}$ are primitive zero sequence impedances. As in the 3ϕ fault we note that the diagonal element of the Z matrix helps us find the total fault current, and the remaining matrix elements from the kth column help us find the node voltages and hence the branch currents.

Usually we save matrix storage by assuming that the positive and negative sequence networks are identical. This changes Z_T to

$$Z_T = Z_{0\text{-}kk} + 2Z_{1\text{-}kk} + 3Z_f \tag{12.20}$$

and changes the Z subscripts in (12.18) to indicate positive sequence impedances.

Example 12.2

Compute the zero sequence impedance matrix for the system of Example 11.1. Then, using the positive sequence Z matrix from Example 12.1, compute the total fault current for a SLG fault on bus 2 and find all voltages and branch line currents. Let h = 1.

Solution

The zero sequence network is shown in Figure 12.7. We find the impedance matrix this time by inverting the admittance matrix of Example 11.6 with the result

$$Z_0 = \begin{array}{c} \\ 1 \\ 2 \\ 3 \end{array} \begin{array}{c} \begin{array}{ccc} 1 & 2 & 3 \end{array} \\ \begin{bmatrix} j0.1157 & j0.0546 & j0.0200 \\ j0.0546 & j0.0831 & j0.0200 \\ j0.0200 & j0.0200 & j0.0200 \end{bmatrix} \end{array}$$

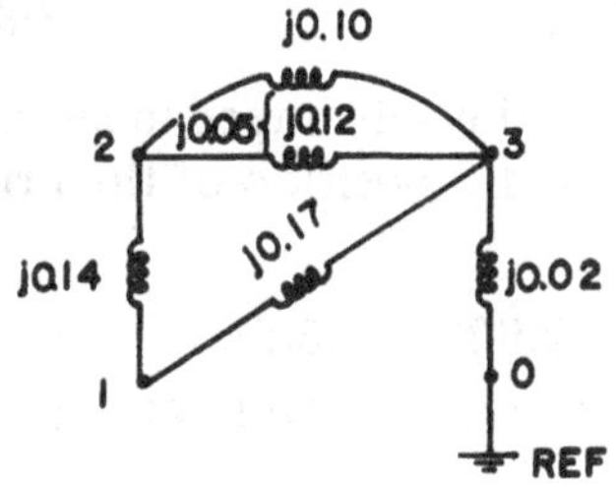

Fig. 12.7. Zero sequence network.

Notice that the technique of injecting 1.0 pu current into node 1 to find column 1 poses a more difficult problem than in Example 12.1 because of the mutually coupled elements. The solution to this problem will be considered in sections 12.7 and 12.8.

In the following computations we assume that the fault impedance is zero and the negative sequence impedances are equal to the corresponding positive sequence impedances. Then from (12.14) with $k = 2$

$$Z_T = Z_{0\text{-}22} + 2Z_{1\text{-}22} = j0.0831 + j0.2690 = j0.3521$$

Then from (12.13) the sequence fault currents are

$$I_{a0\text{-}2} = I_{a1\text{-}2} = I_{a2\text{-}2} = 1.0/j0.3521 = -j2.841 \quad \text{pu}$$

The sequence voltages are computed from (12.16)-(12.18) as

$$\begin{bmatrix} V_{a0\text{-}1} \\ V_{a0\text{-}2} \\ V_{a0\text{-}3} \end{bmatrix} = j2.841 \begin{bmatrix} j0.0546 \\ j0.0831 \\ j0.0200 \end{bmatrix} = \begin{bmatrix} -0.1551 \\ -0.2360 \\ -0.0568 \end{bmatrix} \quad \text{pu}$$

$$\begin{bmatrix} V_{a1\text{-}1} \\ V_{a1\text{-}2} \\ V_{a1\text{-}3} \end{bmatrix} = \begin{bmatrix} 1.0 \\ 1.0 \\ 1.0 \end{bmatrix} + j2.841 \begin{bmatrix} j0.1061 \\ j0.1345 \\ j0.1151 \end{bmatrix} = \begin{bmatrix} 0.6986 \\ 0.6180 \\ 0.6730 \end{bmatrix} \quad \text{pu}$$

$$\begin{bmatrix} V_{a2\text{-}1} \\ V_{a2\text{-}2} \\ V_{a2\text{-}3} \end{bmatrix} = j2.841 \begin{bmatrix} j0.1061 \\ j0.1345 \\ j0.1151 \end{bmatrix} = \begin{bmatrix} -0.3014 \\ -0.3820 \\ -0.3270 \end{bmatrix} \quad \text{pu}$$

There are three lines contributing to the total fault current at bus 2, one line from bus 1, and two parallel lines from bus 3 (see Figures 12.3 and 12.7). The contribution from bus 1 is computed from (12.10).

$$I_{1\text{-}12} = I_{2\text{-}12} = \frac{hV_f(Z_{1\text{-}22} - Z_{1\text{-}12})}{Z_T \mathfrak{z}_{1\text{-}12}} = -\frac{j2.841\ (j0.0284)}{j0.08} = -j1.008 \quad \text{pu}$$

$$I_{0\text{-}12} = \frac{hV_f(Z_{0\text{-}22} - Z_{0\text{-}12})}{Z_T \mathfrak{z}_{0\text{-}12}} = -\frac{j2.841\ (j0.0285)}{j0.14} = -j0.578 \quad \text{pu}$$

For the parallel lines in the positive sequence network we again apply (12.10). Since the positive sequence line impedances for $\mathfrak{z}_{23}$ are equal, the current divides equally.

$$I_{1\text{-}32} = I_{2\text{-}32} = \frac{hV_f(Z_{1\text{-}22} - Z_{1\text{-}32})}{Z_T \mathfrak{z}_{32}} = -\frac{j2.841\ (j0.0194)}{j0.06} = -j0.916 \quad \text{pu}$$

In the zero sequence network the self impedances are unequal and the lines are mutually coupled. We use the 4-5 position of the primitive admittance matrix to compute from (12.19)

$$\begin{bmatrix} I_{0\text{-}32(4)} \\ I_{0\text{-}32(5)} \end{bmatrix} = -j2.841 \begin{bmatrix} -j12.632 & j5.263 \\ j5.263 & -j10.526 \end{bmatrix} \begin{bmatrix} j0.0631 \\ j0.0631 \end{bmatrix} = \begin{bmatrix} -j1.320 \\ -j0.943 \end{bmatrix} \quad \text{pu}$$

12.2 An Impedance Matrix Algorithm

The impedance matrix is an excellent network description to use in the solution of faulted networks. If load currents are neglected, the fault information needed, both for total fault current and branch currents, is easily computed from the impedance matrix elements.

Although it is easy to use for fault computations, the impedance matrix is not as easy to form as the admittance matrix. Furthermore, whereas the admittance matrix is extremely sparse, the impedance matrix is completely filled. It is symmetric, however, so that only $n(n + 1)/2$ impedances need to be saved in computer memory. This requires $n(n + 1)$ memory locations if the impedances saved are complex. However, many systems have high x/r ratios so that sufficient accuracy is obtained by using only the reactances.

One difficulty associated with forming the impedance matrix is the ordering of the primitive elements. Ordering is no problem in forming the admittance matrix. This is because the indefinite admittance matrix exists and all matrix elements are defined when the voltage reference is arbitrarily selected. This is not true for an impedance matrix description. Even the simplest possible network consisting of one element has no finite impedance matrix description *unless* the element is *connected to the voltage reference.* This is because of the way the elements of the matrix are defined as given by the equation

$$Z_{ik} = (V_i/I_k)_{I_j=0,\, j\neq k}, \qquad i = 1, 2, \ldots, n \tag{12.21}$$

If an impedance element is not connected in any way to the reference, V_i is clearly infinite when I_k is given any value (such as 1.0). This characteristic of the impedance matrix suggests that the primitive matrix impedances should be *sorted* beginning with an element which *is connected* to the voltage reference (bus 0) and with each succeeding element connected directly to one of the previously entered elements. Thus an algorithm for building the impedance matrix in a step-by-step manner is required. Much has been written on this subject, as noted in [67-72] which describe early attempts at implementing the impedance matrix in fault studies. These works are recommended reading on the subject, particularly the work of Brown et al. [70, 71] and El-Abiad [72]. The analysis that follows should be regarded as tutorial and not necessarily the best or the only way to construct an impedance matrix.

In the remainder of this chapter we will consider the various ways a network may be changed by adding a new radial branch or closing a loop thereby changing the "previously defined" Z matrix made up of elements added prior to the one under consideration.

12.3 Adding a Radial Impedance to the Reference Node

Consider a network in which p nodes have already been defined and the $(p \times p)$ Z matrix has been computed. We now introduce a new branch of impedance $\mathfrak{z}$ from the reference to a previously undefined node q as shown in Figure 12.8. Applying (12.21) with $I_k = I_q$ we can determine the qth column of the Z matrix. Since the impedance $\mathfrak{z}$ is connected only to the reference and all currents except I_q are zero, all voltages are zero except V_q which is given by (12.21); or the new impedance matrix is

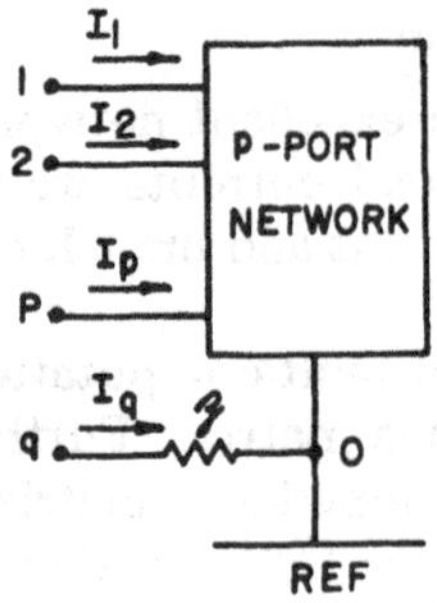

Fig. 12.8. Adding a new branch radial from the reference.

$$\mathbf{Z} = \begin{array}{c} \\ 1 \\ 2 \\ \\ p \\ q \end{array} \begin{array}{c} \begin{array}{cccccc} 1 & 2 & & p & q \end{array} \\ \begin{bmatrix} Z_{11} & Z_{12} & \cdots & Z_{1p} & 0 \\ Z_{21} & Z_{22} & \cdots & Z_{2p} & 0 \\ \cdots & \cdots & \cdots & \cdots & \cdots \\ Z_{p1} & Z_{p2} & \cdots & Z_{pp} & 0 \\ 0 & 0 & \cdots & 0 & \mathfrak{z} \end{bmatrix} \end{array} \tag{12.22}$$

We summarize as follows:

Conditions:
1. Branch added from reference to q.
2. q previously undefined.
3. New branch not mutually coupled to other elements.

Rule:
1. Set $Z_{qq} = \mathfrak{z}$.

12.4 Adding a Radial Branch to a New Node

We now consider the addition of a radial branch with self impedance $\mathfrak{z}$ from a previously defined node k to form a new node q as shown in Figure 12.9. Again we use the defining equation (12.21) to find the qth column. If we let $I_q = 1.0$, the voltages $V_1, V_2, \ldots, V_p, V_q$ will be numerically equal to the qth column impedances. But letting $I_q = 1.0$ gives the same voltage distribution for the new network as letting $I_k = 1.0$ in the old network (without the added branch). These voltages are precisely the kth column impedances. Element Z_{qq} is the driving

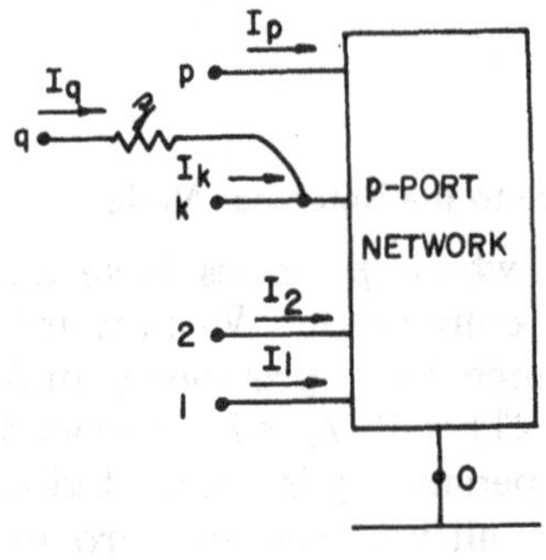

Fig. 12.9. Adding a new branch radial from node k.

point impedance at q which by inspection is $(Z_{kk} + \mathfrak{z})$. Therefore the new matrix is given by

$$
\mathbf{Z} = \begin{array}{c} \\ 1 \\ 2 \\ \\ k \\ \\ p \\ q \end{array}
\begin{array}{c}
\begin{array}{ccccccccc} 1 & 2 & & k & & p & q \end{array} \\
\left[\begin{array}{ccccccc}
Z_{11} & Z_{12} & \cdots & Z_{1k} & \cdots & Z_{1p} & Z_{1k} \\
Z_{21} & Z_{22} & \cdots & Z_{2k} & \cdots & Z_{2p} & Z_{2k} \\
\cdots & \cdots & \cdots & \cdots & \cdots & \cdots & \cdots \\
Z_{k1} & Z_{k2} & \cdots & Z_{kk} & \cdots & Z_{kp} & Z_{kk} \\
\cdots & \cdots & \cdots & \cdots & \cdots & \cdots & \cdots \\
Z_{p1} & Z_{p2} & \cdots & Z_{pk} & \cdots & Z_{pp} & Z_{pk} \\
Z_{k1} & Z_{k2} & \cdots & Z_{kk} & \cdots & Z_{kp} & (Z_{kk} + \mathfrak{z})
\end{array}\right]
\end{array}
\qquad (12.23)
$$

Conditions:

1. Branch added from k to q.
2. k previously defined.
3. q previously undefined.
4. k is not the reference node.
5. p nodes previously defined.
6. New branch not mutually coupled to other elements.

Rules:

1. Set $Z_{iq} = Z_{ik}$, $i = 1, 2, \ldots, p$.
2. Set $Z_{qi} = Z_{ki}$, $i = 1, 2, \ldots, p$.
3. Set $Z_{qq} = Z_{kk} + \mathfrak{z}$.

12.5 Closing a Loop to the Reference

Now consider a case in which an element is added from the reference to a node k which has been previously defined. In other words, the element being added is a link branch since its addition will close a loop in the network. The added element is not mutually coupled to other network elements.

We shall do this in two steps. First we add the branch radially from node k to a new node q. The result of this step is given by (12.23). Then as a second step we connect node q to the reference as indicated by the dashed line in Figure 12.10. If the impedance matrix for the original p-port network was $p \times p$, the

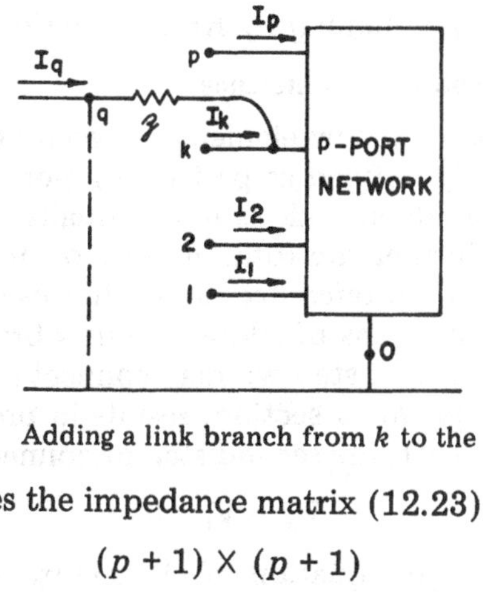

Fig. 12.10. Adding a link branch from k to the reference.

branch added radially gives the impedance matrix (12.23) which is

$$(p + 1) \times (p + 1)$$

If we write out the voltage equations in matrix form, we have following the first step

$$\begin{bmatrix} V_1 \\ V_2 \\ \cdots \\ V_k \\ \cdots \\ V_p \\ \hline V_q \end{bmatrix} = \left[\begin{array}{cccccccc|c} Z_{11} & Z_{12} & \cdots & Z_{1k} & \cdots & Z_{1p} & & Z_{1k} \\ Z_{21} & Z_{22} & \cdots & Z_{2k} & \cdots & Z_{2p} & & Z_{2k} \\ \cdots & \cdots & \cdots & \cdots & \cdots & \cdots & & \cdots \\ Z_{k1} & Z_{k2} & \cdots & Z_{kk} & \cdots & Z_{kp} & & Z_{kk} \\ \cdots & \cdots & \cdots & \cdots & \cdots & \cdots & & \cdots \\ Z_{p1} & Z_{p2} & \cdots & Z_{pk} & \cdots & Z_{pp} & & Z_{pk} \\ \hline Z_{k1} & Z_{k2} & \cdots & Z_{kk} & \cdots & Z_{kp} & & (Z_{kk} + z) \end{array}\right] \begin{bmatrix} I_1 \\ I_2 \\ \cdots \\ I_k \\ \cdots \\ I_p \\ \hline I_q \end{bmatrix} \tag{12.24}$$

Then, upon connecting node q to the reference, $V_q = 0$, a Kron reduction may be performed to eliminate I_q as a variable, reducing the matrix once more to $p \times p$. In matrix notation (12.24) may be written as

$$\begin{bmatrix} \mathbf{V} \\ 0 \end{bmatrix} = \begin{bmatrix} \mathbf{Z}_a & \mathbf{Z}_b \\ \mathbf{Z}_c & Z_d \end{bmatrix} \begin{bmatrix} \mathbf{I} \\ I_q \end{bmatrix}$$

from which we compute $\mathbf{V} = [\mathbf{Z}_a - \mathbf{Z}_b \; Z_d^{-1} \; \mathbf{Z}_c]\,\mathbf{I}$. Then the new impedance matrix may be written as

$$\mathbf{Z} = \begin{bmatrix} Z_{11} & \cdots & Z_{1p} \\ \cdots & \cdots & \cdots \\ Z_{p1} & \cdots & Z_{pp} \end{bmatrix} - \begin{bmatrix} Z_{1k} \\ \cdots \\ Z_{pk} \end{bmatrix} Z_d^{-1} \; [Z_{k1} \ldots Z_{kp}] \tag{12.25}$$

where $Z_d = Z_{kk} + z$.

Conditions:

1. New link branch added from k to reference.
2. k previously defined as one of p nodes.
3. New branch not mutually coupled to other network elements.

Rules:

1. Set $q = p + 1$.
2. Set $Z_{iq} = Z_{ik}$, $i = 1, \ldots, p$.
3. Set $Z_{qi} = Z_{ki}$, $i = 1, \ldots, p$.
4. Set $Z_{qq} = Z_{kk} + z$.
5. Eliminate row q, column q by Kron reduction.

12.6 Closing a Loop Not Involving the Reference

Consider the addition of a new element between nodes i and k where nodes i and k are both previously defined as part of a p-port network shown in Figure 12.11. Assume that the added node is not mutually coupled to any other network element. Assume further the total number of nodes defined to be $p \geqslant i, k$ and that neither i nor k is the reference node. In this case as in the previous one it is convenient to think in terms of adding the new branch from k to a new temporary node q. Then as a final step we may connect q to i. The first step, being the same as that of the previous section, results in precisely the same equation (12.24). In this case, however, the second step of connecting q to i gives

$$V_q - V_i = 0 \tag{12.26}$$

Therefore, we replace the qth equation of (12.24) by the difference between the

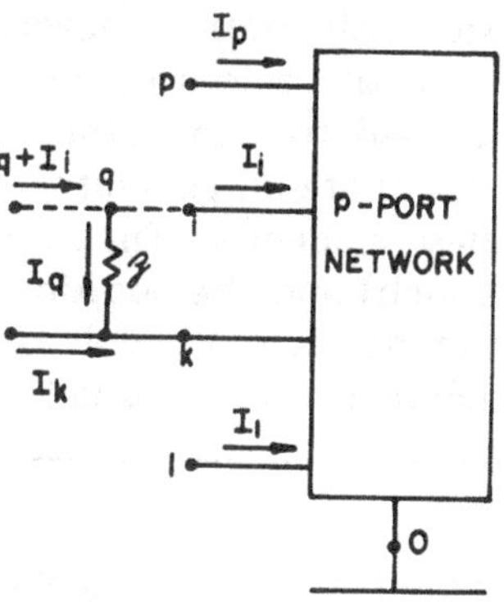

Fig. 12.11. Adding a link branch from i to k.

qth and ith equations. At the same time we recognize that the current I_i shown in Figure 12.11 is not the new nodal current at this node. Once q is connected to i the total current entering this node is $(I_q + I_i)$. Changing all current variables in the ith row from I_i to $(I_q + I_i)$ requires subtracting column i from column q. The result of both operations on (12.24) is

$$\begin{bmatrix} V_1 \\ \cdots \\ V_i \\ \cdots \\ V_k \\ \cdots \\ V_p \\ V_q - V_i \end{bmatrix} = \begin{array}{c} 1 \\ \\ i \\ \\ k \\ \\ p \\ q \end{array} \overset{\begin{array}{ccccccccc} 1 & & i & & k & & p & & q \end{array}}{\begin{bmatrix} Z_{11} & \cdots & Z_{1i} & \cdots & Z_{1k} & \cdots & Z_{1p} & (Z_{1k} - Z_{1i}) \\ \cdots & \cdots & \cdots & \cdots & \cdots & \cdots & \cdots & \cdots \\ Z_{i1} & \cdots & Z_{ii} & \cdots & Z_{ik} & \cdots & Z_{ip} & (Z_{ik} - Z_{ii}) \\ \cdots & \cdots & \cdots & \cdots & \cdots & \cdots & \cdots & \cdots \\ Z_{k1} & \cdots & Z_{ki} & \cdots & Z_{kk} & \cdots & Z_{kp} & (Z_{kk} - Z_{ki}) \\ \cdots & \cdots & \cdots & \cdots & \cdots & \cdots & \cdots & \cdots \\ Z_{p1} & \cdots & Z_{pi} & \cdots & Z_{pk} & \cdots & Z_{pp} & (Z_{pk} - Z_{pi}) \\ (Z_{k1} - Z_{i1}) & \cdots & (Z_{ki} - Z_{ii}) & \cdots & (Z_{kk} - Z_{ik}) & \cdots & (Z_{kp} - Z_{ip}) & Z_d \end{bmatrix}} \begin{bmatrix} I_1 \\ \cdots \\ I_q + I_i \\ \cdots \\ I_k \\ \cdots \\ I_p \\ I_q \end{bmatrix} \quad (12.27)$$

where $Z_d = Z_{ii} + Z_{kk} - Z_{ik} - Z_{ki} + \mathfrak{z} = Z_{ii} + Z_{qq} - Z_{iq} - Z_{qi}$ (of 12.23).

Since $V_q - V_i = 0$, the last row and column are eliminated by Kron reduction with the result

$$\mathbf{Z} = \begin{bmatrix} Z_{11} & \cdots & Z_{1p} \\ \cdots & \cdots & \cdots \\ Z_{p1} & \cdots & Z_{pp} \end{bmatrix} - \begin{bmatrix} Z_{1k} - Z_{1i} \\ \cdots \\ Z_{pk} - Z_{pi} \end{bmatrix} Z_d^{-1} \left[(Z_{k1} - Z_{i1}) \ldots (Z_{kp} - Z_{ip}) \right] \quad (12.28)$$

where $Z_d = Z_{ii} + Z_{kk} - Z_{ik} - Z_{ki} + \mathfrak{z}$.

Conditions:

1. Branch added from i to k.
2. i and k both previously defined.
3. p nodes defined in all.
4. Added branch not mutually coupled to other network elements.

Rules:

1. Set $q = p + 1$.
2. Set $Z_{jq} = Z_{jk} - Z_{ji}, j = 1, \ldots, p$.
3. Set $Z_{qj} = Z_{kj} - Z_{ij}, j = 1, \ldots, p$.
4. Set $Z_{qq} = Z_{ii} + Z_{kk} - Z_{ik} - Z_{ki} + \mathfrak{z} = Z_{ii} + Z_{qq} - Z_{iq} - Z_{qi}$.
5. Eliminate row q and column q by Kron reduction.

This completes the four basic configurations needed to add a new element to the positive sequence network. In the zero sequence network, however, the added branch may be mutually coupled to some other network element. Since connections to the reference node 0 are all generator impedances, we need not consider mutual coupling for these elements. The cases of interest then are the addition of a radial line (tree branch) and the addition of a line which closes a loop (link branch) where the reference is not involved. These cases are investigated following Example 12.3 which involves no mutual coupling.

Example 12.3

Consider the positive sequence network given in Table 11.1 and shown in Figure 12.3. Take the branches in the order given in Table 11.1 and construct the impedance matrix using the rules of sections 12.3–12.6.

Solution

From Table 11.1 we have the positive sequence data as follows:

Branch Number	*Connecting Nodes*	*Positive Sequence Impedance*
1	0-1	j0.25
2	0-3	j0.20
3	1-2	j0.08
4	2-3	j0.06
5	2-3	j0.06
6	1-3	j0.13

If we introduce the elements in the order 1, 2, . . . , 6, we will build up the network piecemeal as shown by the six steps of Figure 12.12. Sketching the network is not necessary, but it is helpful in describing the process. The actual

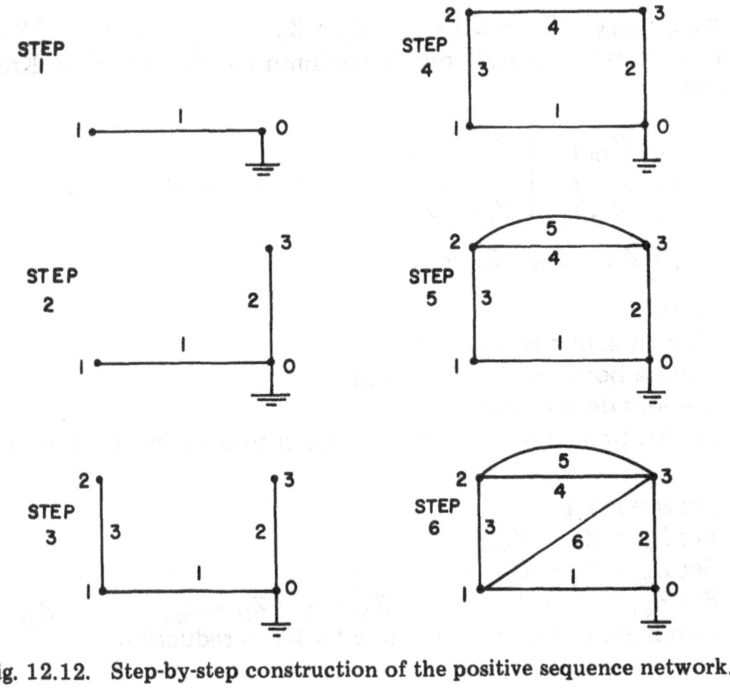

Fig. 12.12. Step-by-step construction of the positive sequence network.

decision-making process needed at each step can be done by computer programming techniques.

We begin with a computer impedance array which is initialized to zero. Since the resistance of all elements is negligible, only one such array is required and it should be at least 4 × 4.

Step 1. Introduce element 1.

$$\mathfrak{z}_{01} = \mathrm{j}0.25$$

Since this is the first element, it is a radial element from the reference to node 1 as shown in step 1 of Figure 12.12. If this first element were not connected to the reference, it would be necessary to sort the elements to find one which was connected to 0.

The rule for introducing an element radial from the reference is given in (12.22). Thus we complete step 1 by adding 0.25 in location q-q or 1-1 with the result.

$$\mathbf{Z} = \mathrm{j}\begin{array}{c} \\ 1 \\ 2 \\ 3 \end{array}\!\!\begin{array}{c} \begin{array}{ccc} 1 & 2 & 3 \end{array} \\ \begin{bmatrix} 0.25 & 0 & 0 \\ 0 & 0 & 0 \\ 0 & 0 & 0 \end{bmatrix} \end{array}$$

Step 2. The addition of $\mathfrak{z}_{03} = \mathrm{j}0.20$ is performed exactly the same way as the previous step except that the diagonal term affected is element 3-3.

$$\mathbf{Z} = \mathrm{j}\begin{array}{c} \\ 1 \\ 2 \\ 3 \end{array}\!\!\begin{array}{c} \begin{array}{ccc} 1 & 2 & 3 \end{array} \\ \begin{bmatrix} 0.25 & 0 & 0 \\ 0 & 0 & 0 \\ 0 & 0 & 0.20 \end{bmatrix} \end{array}$$

Step 3. We now add a radial element $\mathfrak{z}_{12} = \mathrm{j}0.08$ as shown in Figure 12.12. The rule for doing this is given by (12.23) where $k = 1$, $q = 2$. Thus the first row is added to the second row and the first column to the second column. Then Z_{22} is set equal to $Z_{11} + 0.08$.

$$\mathbf{Z} = \mathrm{j}\begin{array}{c} \\ 1 \\ 2 \\ 3 \end{array}\!\!\begin{array}{c} \begin{array}{ccc} 1 & 2 & 3 \end{array} \\ \begin{bmatrix} 0.25 & 0.25 & 0 \\ 0.25 & 0.33 & 0 \\ 0 & 0 & 0.20 \end{bmatrix} \end{array}$$

Step 4. This addition of the fourth element of $\mathfrak{z}_{23} = \mathrm{j}0.06$ closes a loop not involving the reference. The rule for this situation is given by (12.28) which is a Kron reduction of (12.27). We begin with the 4 × 4 array where $k = 2$, $i = 3$. Thus we have an augmented matrix

$$\hat{\mathbf{Z}} = \mathrm{j}\begin{array}{c} \\ 1 \\ 2 \\ 3 \\ q \end{array}\!\!\begin{array}{c} \begin{array}{cccc} 1 & 2 & 3 & q \end{array} \\ \left[\begin{array}{ccc|c} 0.25 & 0.25 & 0 & 0.25 \\ 0.25 & 0.33 & 0 & 0.33 \\ 0 & 0 & 0.20 & -0.20 \\ \hline 0.25 & 0.33 & -0.20 & 0.59 \end{array}\right] \end{array}$$

where (col 4) = (col 2) - (col 3) and similarly for row 4. Element 4-4 is

$$Z_d = Z_{22} + Z_{33} - Z_{23} - Z_{32} + j0.06 = j0.59$$

Performing the Kron reduction to eliminate row q and column q, we have

$$\mathbf{Z} = j \begin{array}{c} 1 \\ 2 \\ 3 \end{array} \overset{\begin{array}{ccc} 1 & 2 & 3 \end{array}}{\begin{bmatrix} 0.1441 & 0.1102 & 0.0847 \\ 0.1102 & 0.1454 & 0.1119 \\ 0.0847 & 0.1119 & 0.1322 \end{bmatrix}}$$

Step 5. The fifth element $\mathfrak{z}_{23}$ = j0.06 calls for exactly the same operation as the previous step. We begin with the 4 X 4 array. Again we have $k = 2, i = 3$.

$$\hat{\mathbf{Z}} = j \begin{array}{c} 1 \\ 2 \\ 3 \\ q \end{array} \overset{\begin{array}{cccc} 1 & 2 & 3 & q \end{array}}{\left[\begin{array}{ccc|c} 0.1441 & 0.1102 & 0.0847 & 0.0255 \\ 0.1102 & 0.1454 & 0.1119 & 0.0335 \\ 0.0847 & 0.1119 & 0.1322 & -0.0203 \\ \hline 0.0255 & 0.0335 & -0.0203 & 0.1138 \end{array}\right]}$$

This array is reduced to find the matrix at this step to be

$$\mathbf{Z} = j \begin{array}{c} 1 \\ 2 \\ 3 \end{array} \overset{\begin{array}{ccc} 1 & 2 & 3 \end{array}}{\begin{bmatrix} 0.1384 & 0.1027 & 0.0893 \\ 0.1027 & 0.1356 & 0.1178 \\ 0.0893 & 0.1178 & 0.1286 \end{bmatrix}}$$

Step 6. The last element to be introduced is $\mathfrak{z}_{13}$ = j0.13. This element also closes a loop, so we begin with the 4 X 4 array, with $k = 1, i = 3$.

$$\hat{\mathbf{Z}} = j \begin{array}{c} 1 \\ 2 \\ 3 \\ q \end{array} \overset{\begin{array}{cccc} 1 & 2 & 3 & q \end{array}}{\left[\begin{array}{ccc|c} 0.1384 & 0.1027 & 0.0893 & 0.0491 \\ 0.1027 & 0.1356 & 0.1178 & -0.0151 \\ 0.0893 & 0.1178 & 0.1286 & -0.0393 \\ \hline 0.0491 & -0.0151 & -0.0393 & 0.2184 \end{array}\right]}$$

After Kron reduction, pivoting on element q-q, we have the positive sequence impedance matrix:

$$\mathbf{Z}_1 = j \begin{bmatrix} 0.1273 & 0.1061 & 0.0981 \\ 0.1061 & 0.1345 & 0.1151 \\ 0.0981 & 0.1151 & 0.1215 \end{bmatrix}$$

12.7 Adding a Mutually Coupled Radial Element

We now consider the addition of a radial element from node p to a new node q where the added element is mutually coupled to the network element from m to n. We assume that nodes m, n, and p have been previously defined in a network with p total nodes. (Note there is no loss in generality by letting p be the highest

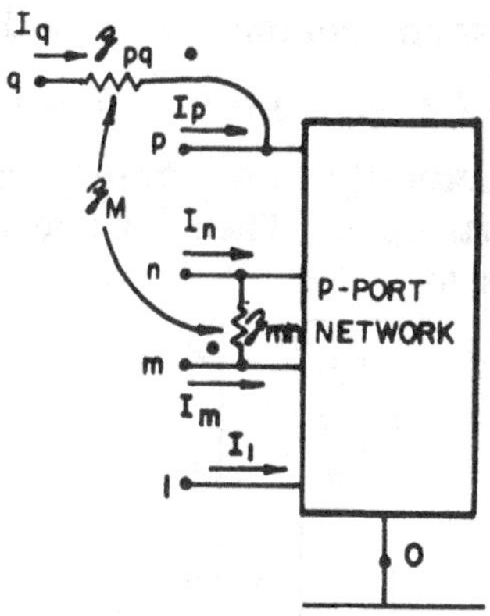

Fig. 12.13. Adding a mutually coupled radial element.

numbered node.) Assume that there is an impedance $\mathfrak{z}_{mn}$ between nodes m and n and the new impedance $\mathfrak{z}_{pq}$ is mutually coupled to $\mathfrak{z}_{mn}$ by a mutual impedance $\mathfrak{z}_M$. The network connection is shown in Figure 12.13.

The voltage drop for two mutually coupled lines is given by (4.14). For the two elements under consideration here we write

$$\begin{bmatrix} V_{mn} \\ V_{pq} \end{bmatrix} = \begin{bmatrix} V_m - V_n \\ V_p - V_q \end{bmatrix} = \begin{bmatrix} \mathfrak{z}_{mn} & \mathfrak{z}_M \\ \mathfrak{z}_M & \mathfrak{z}_{pq} \end{bmatrix} \begin{bmatrix} I_{mn} \\ I_{pq} = -I_q \end{bmatrix} \tag{12.29}$$

Solving for the currents, we write

$$\begin{bmatrix} I_{mn} \\ I_{pq} = -I_q \end{bmatrix} = \begin{bmatrix} \mathfrak{y}_{mn} & \mathfrak{y}_M \\ \mathfrak{y}_M & \mathfrak{y}_{pq} \end{bmatrix} \begin{bmatrix} V_{mn} \\ V_{pq} \end{bmatrix} \tag{12.30}$$

where the admittance matrix of (12.30) is the inverse of the impedance matrix of (12.29).

Suppose we write the equation for the new network of Figure 12.13 in symbolic form, i.e.,

$$\begin{bmatrix} V_1 \\ \cdots \\ V_k \\ \cdots \\ V_m \\ V_n \\ \cdots \\ V_p \\ V_q \end{bmatrix} = \begin{array}{c} \\ 1 \\ \\ k \\ \\ m \\ n \\ \\ p \\ q \end{array} \begin{array}{c} \begin{array}{cccccccccccc} 1 & & k & & m & n & & p & q \end{array} \\ \begin{bmatrix} Z_{11} & \cdots & Z_{1k} & \cdots & Z_{1m} & Z_{1n} & \cdots & Z_{1p} & \tilde{Z}_{1q} \\ \cdots & \cdots & \cdots & \cdots & \cdots & \cdots & \cdots & \cdots & \cdots \\ Z_{k1} & \cdots & Z_{kk} & \cdots & Z_{km} & Z_{kn} & \cdots & Z_{kp} & \tilde{Z}_{kq} \\ \cdots & \cdots & \cdots & \cdots & \cdots & \cdots & \cdots & \cdots & \cdots \\ Z_{m1} & \cdots & Z_{mk} & \cdots & Z_{mm} & Z_{mn} & \cdots & Z_{mp} & \tilde{Z}_{mq} \\ Z_{n1} & \cdots & Z_{nk} & \cdots & Z_{nm} & Z_{nn} & \cdots & Z_{np} & \tilde{Z}_{nq} \\ \cdots & \cdots & \cdots & \cdots & \cdots & \cdots & \cdots & \cdots & \cdots \\ Z_{p1} & \cdots & Z_{pk} & \cdots & Z_{pm} & Z_{pn} & \cdots & Z_{pp} & \tilde{Z}_{pq} \\ \tilde{Z}_{q1} & \cdots & \tilde{Z}_{qk} & \cdots & \tilde{Z}_{qm} & \tilde{Z}_{qn} & \cdots & \tilde{Z}_{qp} & \tilde{Z}_{qq} \end{bmatrix} \end{array} \begin{bmatrix} I_1 \\ \cdots \\ I_k \\ \cdots \\ I_m \\ I_n \\ \cdots \\ I_p \\ I_q \end{bmatrix} \tag{12.31}$$

where the elements of the qth row and column are completely unknown as yet. This fact is acknowledged by the tilde placed over these elements. The matrix elements of Z's without tilde's are the original Z's. Finally we note that caution must be observed to clearly distinguish between branch impedances $\mathfrak{z}_{mn}$ and $\mathfrak{z}_{pq}$ and matrix elements Z_{mn} and Z_{pq}.

To determine the qth row and column we make the following test. Let

$$I_k = 1, \quad I_j = 0, \quad j \neq k, \quad k = 1, 2, \ldots, p \tag{12.32}$$

This requires that we methodically move from port to port injecting $I_k = 1$ at each port with all other ports open. Then for injection at the kth node we compute the voltages at all nodes as

$$\begin{bmatrix} V_1 \\ \cdots \\ V_k \\ \cdots \\ V_m \\ V_n \\ \cdots \\ V_p \\ V_q \end{bmatrix} = \begin{bmatrix} Z_{1k} \\ \cdots \\ Z_{kk} \\ \cdots \\ Z_{mk} \\ Z_{nk} \\ \cdots \\ Z_{pk} \\ \tilde{Z}_{qk} \end{bmatrix} \tag{12.33}$$

Note that $I_q = 0$ for all tests. Therefore, the voltage drop from m to n is a function only of I_{mn} or

$$V_{mn} = \mathfrak{z}_{mn} I_{mn} + \mathfrak{z}_M (0) \tag{12.34}$$

The injection of $I_k = 1$ may cause a current I_{mn} to flow through $\mathfrak{z}_{mn}$. If so, this current will induce a voltage V_{pq} because of the mutual coupling. From (12.30) with $I_q = 0$ we compute

$$\begin{aligned} \begin{bmatrix} I_{mn} \\ 0 \end{bmatrix} &= \begin{bmatrix} \mathfrak{y}_{mn} & \mathfrak{y}_M \\ \mathfrak{y}_M & \mathfrak{y}_{pq} \end{bmatrix} \begin{bmatrix} V_{mn} \\ V_{pq} \end{bmatrix} = \begin{bmatrix} \mathfrak{y}_{mn} & \mathfrak{y}_M \\ \mathfrak{y}_M & \mathfrak{y}_{pq} \end{bmatrix} \begin{bmatrix} V_m - V_n \\ V_p - V_q \end{bmatrix} \\ &= \begin{bmatrix} \mathfrak{y}_{mn} & \mathfrak{y}_M \\ \mathfrak{y}_M & \mathfrak{y}_{pq} \end{bmatrix} \begin{bmatrix} Z_{mk} - Z_{nk} \\ Z_{pk} - \tilde{Z}_{qk} \end{bmatrix} \end{aligned} \tag{12.35}$$

The second equation of (12.35) may be written as

$$0 = \mathfrak{y}_M (Z_{mk} - Z_{nk}) + \mathfrak{y}_{pq} (Z_{pk} - \tilde{Z}_{qk})$$

from which we compute

$$\tilde{Z}_{qk} = Z_{pk} + (\mathfrak{y}_M / \mathfrak{y}_{pq})(Z_{mk} - Z_{nk}) \tag{12.36}$$

However, from (12.29) and (12.30) we have the relations

$$\mathfrak{y}_{mn} = \frac{\mathfrak{z}_{pq}}{\Delta}, \quad \mathfrak{y}_{pq} = \frac{\mathfrak{z}_{mn}}{\Delta}$$

$$\mathfrak{y}_M = -\frac{\mathfrak{z}_M}{\Delta}, \quad \Delta = \mathfrak{z}_{mn}\mathfrak{z}_{pq} - \mathfrak{z}_M^2 \tag{12.37}$$

from which other forms of the solution are possible. One convenient form is the following:

$$\tilde{Z}_{qk} = Z_{pk} - \frac{\mathfrak{z}_M}{\mathfrak{z}_{mn}} (Z_{mk} - Z_{nk}) \tag{12.38}$$

This equation may be used to find the first $q - 1$ elements of the qth row and, by symmetry, the qth column. The only unknown then is the element $\tilde{Z}_{qq}$.

To find $\tilde{Z}_{qq}$, we devise the following test. Let

$$I_q = 1, \quad I_1 = I_2 = \cdots = I_p = 0 \tag{12.39}$$

Then from (12.31) we have

$$\begin{bmatrix} V_1 \\ \cdots \\ V_m \\ V_n \\ \cdots \\ V_p \\ V_q \end{bmatrix} = \begin{bmatrix} Z_{1q} \\ \cdots \\ Z_{mq} \\ Z_{nq} \\ \cdots \\ Z_{pq} \\ \tilde{Z}_{qq} \end{bmatrix} \tag{12.40}$$

where the tilde's have been removed from the first p impedance elements as they are now known. Also, since $I_{pq} = -I_q = -1$, we may write from (12.30)

$$\begin{bmatrix} I_{mn} \\ -1 \end{bmatrix} = \begin{bmatrix} y_{mn} & y_M \\ y_M & y_{pq} \end{bmatrix} \begin{bmatrix} V_m - V_n \\ V_p - V_q \end{bmatrix}$$

or $-1 = y_M (V_m - V_n) + y_{pq}(V_p - V_q)$. Then

$$V_q = V_p + 1/y_{pq} + (y_M/y_{pq})(V_m - V_n) \tag{12.41}$$

This may be changed to the form $V_q = V_p + z_{pq} - (z_M/z_{mn})(z_M + V_m - V_n)$ or, combining with (12.40),

$$\tilde{Z}_{qq} = Z_{pq} + z_{pq} - (z_M/z_{mn})(z_M + Z_{mq} - Z_{nq}) \tag{12.42}$$

This completes the formulation of the new impedance matrix.

We should be able to test this result to see if it agrees with the case of adding an uncoupled radial line where $z_M = 0$. In this case, from (12.38) we have $\tilde{Z}_{qk} = Z_{pk}$ and from (12.42) $\tilde{Z}_{qq} = Z_{pq} + z_{pq}$. This is the same result as found earlier for the uncoupled radial line.

Conditions:

1. Branch added radially from p to q.
2. Branch mn previously defined.
3. Branch pq previously undefined.
4. p is not the reference node.
5. Branch pq mutually coupled to branch mn.

Rules:

1. Set $Z_{qk} = Z_{pk} - (z_M/z_{mn})(Z_{mk} - Z_{nk})$, $k \neq q$.
2. Set $Z_{kq} = Z_{qk}$, $k \neq q$.
3. Set $Z_{qq} = Z_{pq} + z_{pq} - (z_M/z_{mn})(z_M + Z_{mq} - Z_{nq})$.

Having considered the addition of a mutually coupled radial element, there is no need to study separately the addition of a mutually coupled link branch. This case can always be handled by first adding the mutually coupled branch radially to a fictitious node q and then closing the loop from q to the appropriate node by means of a Kron reduction to eliminate row q and column q.

The case just considered will solve a great many of the mutually coupled zero sequence network problems. Occasionally, however, a situation is encountered where several lines occupy the same right-of-way and a number of circuits are mutually coupled. This case is considered in section 12.8.

Example 12.4

Consider the zero sequence network given in Table 11.1 and shown in Figure 12.7. Consider the branches in the order listed in the table and construct the zero sequence impedance matrix.

Solution

From Table 11.1 and Example 11.1 we find the following zero sequence data.

Branch Number	*Connecting Nodes*	*Zero Sequence Impedance*	*Mutual Impedance*
1	open	. . .	. . .
2	0-3	j0.02	. . .
3	1-2	j0.14	. . .
4	2-3	j0.10	j0.05
5	2-3	j0.12	j0.05
6	1-3	j0.17	. . .

We proceed in much the same fashion as for the positive sequence network until we reach the mutually coupled element.

Step 1. Introduce element 2.

$$\mathfrak{z}_{03} = j0.02$$

$$Z = j \begin{array}{c} \\ 1 \\ 2 \\ 3 \end{array} \begin{array}{c} \begin{array}{ccc} 1 & 2 & 3 \end{array} \\ \begin{bmatrix} 0 & 0 & 0 \\ 0 & 0 & 0 \\ 0 & 0 & 0.02 \end{bmatrix} \end{array}$$

Step 2. Introduce element 3.

$$\mathfrak{z}_{12} = j0.14$$

Neither of the nodes 1 or 2 have been entered yet. Since a radial element must be connected to the network in some way, we must skip this element and pick it up later.

Step 3. Introduce element 4.

$$\mathfrak{z}_{23} = j0.10$$

This is a radial element which connects node 2 to a new node 3. Thus with $k = 3$, $q = 2$ we have

$$Z = j \begin{array}{c} \\ 1 \\ 2 \\ 3 \end{array} \begin{array}{c} \begin{array}{ccc} 1 & 2 & 3 \end{array} \\ \begin{bmatrix} 0 & 0 & 0 \\ 0 & 0.12 & 0.02 \\ 0 & 0.02 & 0.02 \end{bmatrix} \end{array}$$

Step 4. Introduce element 5.

$$\mathfrak{z}_{23} = j0.12 \text{ with } \mathfrak{z}_M = j0.05$$

This is a mutually coupled element which closes a loop. However we add it as a radial element from node 2 to node q and will later eliminate node q.

The qth row and column elements are computed from (12.38). Thus

$$\tilde{Z}_{qk} = Z_{pk} - (\mathfrak{z}_M/\mathfrak{z}_{mn})(Z_{mk} - Z_{nk})$$

with

$$m = 2, \quad n = 3, \quad p = 2, \quad k = 1, 2, 3$$

For $k = 1$,

$$\tilde{Z}_{q1} = Z_{21} - (\mathfrak{z}_M/\mathfrak{z}_{mn})(Z_{21} - Z_{31}) = 0$$

$k = 2$,

$$\tilde{Z}_{q2} = Z_{22} - (\mathfrak{z}_M/\mathfrak{z}_{mn})(Z_{22} - Z_{32}) = \mathrm{j}0.12 - \mathrm{j}\frac{0.05}{0.10}(0.12 - 0.02) = \mathrm{j}0.07$$

$k = 3$,

$$\tilde{Z}_{q3} = Z_{23} - (\mathfrak{z}_M/\mathfrak{z}_{mn})(Z_{23} - Z_{33}) = \mathrm{j}0.02 - \mathrm{j}\frac{0.05}{0.10}(0.02 - 0.02) = \mathrm{j}0.02$$

For the diagonal term we use (12.42) to compute

$$\begin{aligned}\tilde{Z}_{qq} &= Z_{2q} + \mathfrak{z}_{2q} - (\mathfrak{z}_M/\mathfrak{z}_{mn})(\mathfrak{z}_M + Z_{2q} - Z_{3q}) \\ &= \mathrm{j}0.07 + \mathrm{j}0.12 - \mathrm{j}\frac{0.05}{0.10}(0.05 + 0.07 - 0.02) = \mathrm{j}0.14\end{aligned}$$

We now have the matrix

$$\hat{\mathbf{Z}} = \mathrm{j}\begin{array}{c} \\ 1 \\ 2 \\ 3 \\ q \end{array}\begin{array}{c} \begin{array}{cccc} 1 & 2 & 3 & q \end{array} \\ \left[\begin{array}{ccc|c} 0 & 0 & 0 & 0 \\ 0 & 0.12 & 0.02 & 0.07 \\ 0 & 0.02 & 0.02 & 0.02 \\ \hline 0 & 0.07 & 0.02 & 0.14 \end{array}\right]\end{array}$$

where a new node q extends radially from node 3. We close q to 2 by the following steps: (1) Replace row q by (row 2) - (row q), (2) replace column q by (col 2) - (col q), and (3) replace element qq by $Z_{qq} + Z_{22} - Z_{2q} - Z_{q2}$ as noted in (12.27) for closing a loop. This gives us the new matrix

$$\hat{\mathbf{Z}} = \begin{array}{c} \\ 1 \\ 2 \\ 3 \\ q \end{array}\begin{array}{c} \begin{array}{cccc} 1 & 2 & 3 & q \end{array} \\ \left[\begin{array}{ccc|c} 0 & 0 & 0 & 0 \\ 0 & 0.12 & 0.02 & 0.05 \\ 0 & 0.02 & 0.02 & 0 \\ \hline 0 & 0.05 & 0 & 0.12 \end{array}\right]\end{array}$$

Now reduce $\hat{\mathbf{Z}}$, pivoting on Z_{qq} to compute

$$\mathbf{Z} = \mathrm{j}\begin{array}{c} \\ 1 \\ 2 \\ 3 \end{array}\begin{array}{c} \begin{array}{ccc} 1 & 2 & 3 \end{array} \\ \left[\begin{array}{ccc} 0 & 0 & 0 \\ 0 & 0.0992 & 0.02 \\ 0 & 0.02 & 0.02 \end{array}\right]\end{array}$$

Step 5. Introduce element 6.

$$z_{13} = j0.17$$

This element is radial from node 3 to the newly defined node 1. Therefore by inspection of (12.23) with $k = 3$, $q = 1$ we have

$$Z = j \begin{array}{c} \\ 1 \\ 2 \\ 3 \end{array} \begin{array}{c} \begin{array}{ccc} 1 & 2 & 3 \end{array} \\ \begin{bmatrix} 0.1900 & 0.0200 & 0.0200 \\ 0.0200 & 0.0992 & 0.0200 \\ 0.0200 & 0.0200 & 0.0200 \end{bmatrix} \end{array}$$

Step 6. Introduce element 3 which was passed earlier.

$$z_{12} = j0.14$$

This element closes a loop from node 1 to node 2. From (12.23) with $k = 1$, $i = 2$ we compute

$$\hat{Z} = \begin{array}{c} \\ 1 \\ 2 \\ 3 \\ q \end{array} \begin{array}{c} \begin{array}{cccc} 1 & 2 & 3 & q \end{array} \\ \left[\begin{array}{ccc|c} 0.1900 & 0.0200 & 0.0200 & 0.1900 \\ 0.0200 & 0.0992 & 0.0200 & 0.0200 \\ 0.0200 & 0.0200 & 0.0200 & 0.0200 \\ \hline 0.1900 & 0.0200 & 0.0200 & 0.3300 \end{array}\right] \end{array}$$

This adds the new element radially from node 1 to a new node q. Then from (12.27) we form the intermediate result

$$\hat{Z}_{new} = j \begin{array}{c} \\ 1 \\ 2 \\ 3 \\ q \end{array} \begin{array}{c} \begin{array}{cccc} 1 & 2 & 3 & q \end{array} \\ \left[\begin{array}{ccc|c} 0.1900 & 0.0200 & 0.0200 & 0.1700 \\ 0.0200 & 0.0992 & 0.0200 & -0.0792 \\ 0.0200 & 0.0200 & 0.0200 & 0 \\ \hline 0.1700 & -0.0792 & 0 & 0.3892 \end{array}\right] \end{array}$$

Finally we make a Kron reduction, pivoting on element qq to compute

$$Z_0 = j \begin{array}{c} \\ 1 \\ 2 \\ 3 \end{array} \begin{array}{c} \begin{array}{ccc} 1 & 2 & 3 \end{array} \\ \begin{bmatrix} 0.1157 & 0.0546 & 0.0200 \\ 0.0546 & 0.0831 & 0.0200 \\ 0.0200 & 0.0200 & 0.0200 \end{bmatrix} \end{array}$$

Example 12.5

Using the impedance matrices of Examples 12.3 and 12.4, compute the total fault current I_a for a SLG fault on phase a of each node with zero fault impedance. Let h = 1.

Solution

The total fault current is computed from (12.13) as

$$I_{ak} = 3I_{a0\text{-}k} = 3\,\mathrm{h}V_f/Z_T$$

Setting $hV_f = 1.0$, we easily select the appropriate diagonal impedances from the positive and zero sequence impedance matrices to compute the following values.

Faulted Node	I_a *(pu)*	V_{a0} *(pu)*	V_{a1} *(pu)*	V_{a2} *(pu)*
1	-j8.098	-0.3124	0.6562	-0.3438
2	-j8.522	-0.2360	0.6180	-0.3820
3	-j11.406	-0.0760	0.5380	-0.4620

It is interesting to note that these fault currents are greater in every case than the corresponding 3ϕ bus fault currents computed in Example 12.1. This is not unusual for systems with no neutral impedance in the generators.

12.8 Adding a Group of Mutually Coupled Lines

We expand the foregoing analysis to the case where a group of mutually coupled lines are to be added to a network. We will arbitrarily assume that the first self impedance $\mathfrak{z}_{ij}$ of the group has been introduced. The remainder of the mutually coupled group is to be added as radial connections to nodes $k \ldots n$. These radials may later be connected to other nodes by Kron reduction, thereby forming closed loops if this is necessary.

Consider the network shown in Figure 12.14 where the branch $\mathfrak{z}_{ij}$ has been

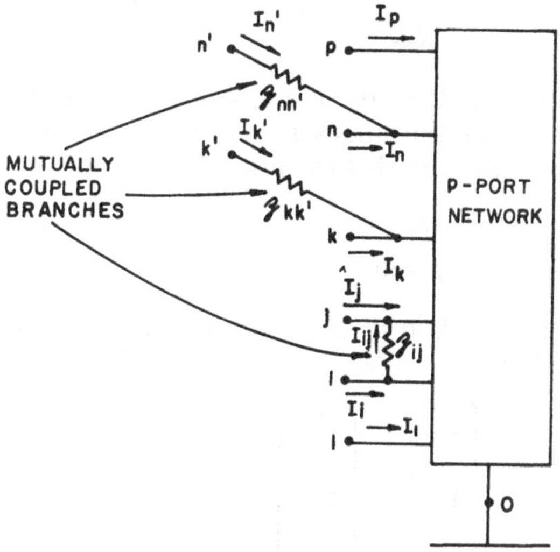

Fig. 12.14. Adding a group of mutually coupled elements.

incorporated into the p-port network and we now wish to add the mutually coupled branches $\mathfrak{z}_{kk'} \ldots \mathfrak{z}_{nn'}$ and where this entire group of elements is a mutually coupled group consisting of, say, m elements. Taking the group of m mutually coupled elements from the primitive matrix, we write this submatrix as

$$\begin{bmatrix} V_{ij} \\ V_{kk'} \\ \cdots \\ V_{nn'} \end{bmatrix} = \begin{matrix} \\ 1 \\ 2 \\ \\ m \end{matrix} \begin{matrix} 1 & 2 & & m \\ \left[\begin{matrix} \mathfrak{z}_{ij} & \mathfrak{z}_{ik'} & \cdots & \mathfrak{z}_{in'} \\ \mathfrak{z}_{kj} & \mathfrak{z}_{kk'} & \cdots & \mathfrak{z}_{kn'} \\ \cdots & \cdots & \cdots & \cdots \\ \mathfrak{z}_{nj} & \mathfrak{z}_{nk'} & \cdots & \mathfrak{z}_{nn'} \end{matrix}\right] \end{matrix} \begin{bmatrix} I_{ij} \\ I_{kk'} = -I_{k'} \\ \cdots \\ I_{nn'} = -I_{n'} \end{bmatrix} \tag{12.43}$$

where we recognize that branch currents in the radial elements can be related to the nodal currents entering. Since the inverse of this primitive matrix exists, we compute it, writing the result in partitioned form as

$$\begin{array}{c} 1 \\ 2 \\ \\ m \end{array}\left[\begin{array}{c} I_{ij} \\ \hline -I_{k'} \\ \cdots \\ -I_{n'} \end{array}\right] = \left[\begin{array}{c|c} \mathcal{Y}_{ij} & \mathcal{Y}_b \\ \hline & \\ \mathcal{Y}_c & \mathcal{Y}_d \\ & \end{array}\right] \left[\begin{array}{c} V_i - V_j \\ \hline V_k - V_{k'} \\ \cdots \\ V_n - V_{n'} \end{array}\right] \tag{12.44}$$

The impedance matrix which we wish to find may be written in symbolic form with a tilde over all unknown elements as follows.

$$\left[\begin{array}{c} V_1 \\ \cdots \\ V_i \\ V_j \\ \cdots \\ V_k \\ \cdots \\ V_n \\ \cdots \\ V_p \\ \hline V_{k'} \\ \cdots \\ V_{n'} \end{array}\right] = \left[\begin{array}{cccccccccccc|ccc} Z_{11} & \cdots & Z_{1i} & Z_{1j} & \cdots & Z_{1k} & \cdots & Z_{1n} & \cdots & Z_{1p} & \tilde{Z}_{1k'} & \cdots & \tilde{Z}_{1n'} \\ \cdots & \cdots & \cdots & \cdots & \cdots & \cdots & \cdots & \cdots & \cdots & \cdots & \cdots & \cdots & \cdots \\ Z_{i1} & \cdots & Z_{ii} & Z_{ij} & \cdots & Z_{ik} & \cdots & Z_{in} & \cdots & Z_{ip} & \tilde{Z}_{ik'} & \cdots & \tilde{Z}_{in'} \\ Z_{j1} & \cdots & Z_{ji} & Z_{jj} & \cdots & Z_{jk} & \cdots & Z_{jn} & \cdots & Z_{jp} & \tilde{Z}_{jk'} & \cdots & \tilde{Z}_{jn'} \\ \cdots & \cdots & \cdots & \cdots & \cdots & \cdots & \cdots & \cdots & \cdots & \cdots & \cdots & \cdots & \cdots \\ Z_{k1} & \cdots & Z_{ki} & Z_{kj} & \cdots & Z_{kk} & \cdots & Z_{kn} & \cdots & Z_{kp} & \tilde{Z}_{kk'} & \cdots & \tilde{Z}_{kn'} \\ \cdots & \cdots & \cdots & \cdots & \cdots & \cdots & \cdots & \cdots & \cdots & \cdots & \cdots & \cdots & \cdots \\ Z_{n1} & \cdots & Z_{ni} & Z_{nj} & \cdots & Z_{nk} & \cdots & Z_{nn} & \cdots & Z_{np} & \tilde{Z}_{nk'} & \cdots & \tilde{Z}_{nn'} \\ \cdots & \cdots & \cdots & \cdots & \cdots & \cdots & \cdots & \cdots & \cdots & \cdots & \cdots & \cdots & \cdots \\ Z_{p1} & \cdots & Z_{pi} & Z_{pj} & \cdots & Z_{pk} & \cdots & Z_{pn} & \cdots & Z_{pp} & \tilde{Z}_{pk'} & \cdots & \tilde{Z}_{pn'} \\ \hline \tilde{Z}_{k'1} & \cdots & \tilde{Z}_{k'i} & \tilde{Z}_{k'j} & \cdots & \tilde{Z}_{k'k} & \cdots & \tilde{Z}_{k'n} & \cdots & \tilde{Z}_{k'p} & \tilde{Z}_{k'k'} & \cdots & \tilde{Z}_{k'n'} \\ \cdots & \cdots & \cdots & \cdots & \cdots & \cdots & \cdots & \cdots & \cdots & \cdots & \cdots & \cdots & \cdots \\ \tilde{Z}_{n'1} & \cdots & \tilde{Z}_{n'i} & \tilde{Z}_{n'j} & \cdots & \tilde{Z}_{n'k} & \cdots & \tilde{Z}_{n'n} & \cdots & \tilde{Z}_{n'p} & \tilde{Z}_{n'k'} & \cdots & \tilde{Z}_{n'n'} \end{array}\right] \left[\begin{array}{c} I_1 \\ \cdots \\ I_i \\ I_j \\ \cdots \\ I_k \\ \cdots \\ I_n \\ \cdots \\ I_p \\ \hline I_{k'} \\ \cdots \\ I_{n'} \end{array}\right] \tag{12.45}$$

We can find the unknown elements by performing two tests. First, we inject 1.0 unit of current at each of the nodes $1 \ldots p$ while holding all other currents at zero, including the currents entering the primed nodes, i.e.,

$$\begin{aligned} &I_s = 1.0, \quad I_t = 0, \quad t \neq s, \quad s = 1, 2, \ldots, p \\ &I_{k'} = \cdots = I_{n'} = 0 \end{aligned} \tag{12.46}$$

By inspection of (12.45) this test equates the port voltages to the impedance elements of the sth column, i.e.,

$$\left[\begin{array}{c} V_1 \\ \cdots \\ V_i \\ V_j \\ \cdots \\ V_k \\ \cdots \\ V_n \\ \cdots \\ V_p \\ \hline V_{k'} \\ \cdots \\ V_{n'} \end{array}\right] = \left[\begin{array}{c} Z_{1s} \\ \cdots \\ Z_{is} \\ Z_{js} \\ \cdots \\ Z_{ks} \\ \cdots \\ Z_{ns} \\ \cdots \\ Z_{ps} \\ \hline \tilde{Z}_{k's} \\ \cdots \\ \tilde{Z}_{n's} \end{array}\right], \quad s \neq k', \ldots, n' \tag{12.47}$$

Now, returning to (12.44), we note that the currents entering the primed nodes are all zero and we may write the $(m - 1)$ equations

$$0 = \mathcal{Y}_c (V_i - V_j) + \mathcal{Y}_d \begin{bmatrix} V_k - V_{k'} \\ \cdots \\ V_n - V_{n'} \end{bmatrix}$$

Solving for the primed voltages, we have

$$\begin{bmatrix} V_{k'} \\ \cdots \\ V_{n'} \end{bmatrix} = \begin{bmatrix} V_k \\ \cdots \\ V_n \end{bmatrix} + \mathcal{Y}_d^{-1} \mathcal{Y}_c (V_i - V_j)$$

Now, from (12.47) we substitute the appropriate impedance values in place of the voltages to write

$$\begin{bmatrix} \tilde{Z}_{k's} \\ \cdots \\ \tilde{Z}_{n's} \end{bmatrix} = \begin{bmatrix} Z_{ks} \\ \\ Z_{ns} \end{bmatrix} + \mathcal{Y}_d^{-1} \mathcal{Y}_c (Z_{is} - Z_{js}), \quad s = 1, \ldots, p \tag{12.48}$$

This determines all unknown elements except the lower right partition of (12.45).

The second test consists of causing a current of 1.0 to enter each of the primed nodes in turn while holding all other port currents to zero. Let

$$\begin{aligned} &I_{s'} = 1, \quad I_{t'} = 0, \quad t' \neq s' \\ &s' = k', \ldots, n' \\ &I_1 = I_2 = \cdots = I_p = 0 \end{aligned} \tag{12.49}$$

Then (12.45) becomes

$$\begin{bmatrix} V_1 \\ \cdots \\ V_i \\ V_j \\ \cdots \\ V_k \\ \cdots \\ V_n \\ \cdots \\ V_p \\ \hline V_{k'} \\ \cdots \\ V_{n'} \end{bmatrix} = \begin{bmatrix} Z_{1s'} \\ \cdots \\ Z_{is'} \\ Z_{js'} \\ \cdots \\ Z_{ks'} \\ \cdots \\ Z_{ns'} \\ \cdots \\ Z_{ps'} \\ \hline \tilde{Z}_{k's'} \\ \cdots \\ \tilde{Z}_{n's'} \end{bmatrix}, \quad s' = k', \ldots, n' \tag{12.50}$$

where the upper p elements are shown without the tilde since these are now known. From the primitive equation (12.44) we have under test conditions (12.49)

$$\begin{bmatrix} I_{ij} \\ \hline 0 \\ \cdots \\ 0 \\ -I_{s'} = -1 \\ 0 \\ \cdots \\ 0 \end{bmatrix} \triangleq \begin{bmatrix} I_{ij} \\ \hline \\ \\ \\ -\mathbf{I}_{s'} \\ \\ \\ \\ \end{bmatrix} = \left[\begin{array}{c|c} \mathcal{Y}_{ij} & \mathcal{Y}_b \\ \hline \\ \mathcal{Y}_c & \mathcal{Y}_d \\ \\ \end{array}\right] \begin{bmatrix} V_i - V_j \\ \hline V_k - V_{k'} \\ \cdots \\ \cdots \\ V_s - V_{s'} \\ \cdots \\ \cdots \\ V_n - V_{n'} \end{bmatrix} \tag{12.51}$$

Solving for the primed voltages, we have

$$\begin{bmatrix} V_{k'} \\ \cdots \\ V_{n'} \end{bmatrix} = \begin{bmatrix} V_k \\ \cdots \\ V_n \end{bmatrix} + \mathcal{Y}_d^{-1}\,\mathbf{I}_{s'} + \mathcal{Y}_d^{-1}\mathcal{Y}_c(V_i - V_j)$$

Finally, substituting the impedances (12.50) for these voltages, we determine the s'th column of unknowns

$$\begin{bmatrix} \tilde{Z}_{k's'} \\ \cdots \\ \tilde{Z}_{n's'} \end{bmatrix} = \begin{bmatrix} Z_{ks'} \\ \cdots \\ Z_{ns'} \end{bmatrix} + \mathcal{Y}_d^{-1}\,\mathbf{I}_{s'} + \mathcal{Y}_d^{-1}\mathcal{Y}_c(Z_{is'} - Z_{js'}) \tag{12.52}$$

Now repeat with s' taking all values from k' to n' to complete the impedance matrix.

The process described for mutually coupled elements is certainly not the only one which may be devised. A number of excellent papers have been written on this subject and the interested reader is encouraged to study some of these. Recent "build and discard" techniques have also been devised which attempt to retain only certain required nodes of very large systems [73-75]. An interesting review of the literature has been made by Kruempel [76] who carefully traces the development of these techniques.

12.9 Comparison of Admittance and Impedance Matrix Techniques

From the discussion presented in Chapters 11 and 12 it should be apparent that power system faults may be solved by computer, using either an admittance or an impedance matrix technique. It is also apparent that each technique has its problems: the admittance method in requiring a new inversion for each new fault simulation and the impedance method in its difficult and time-consuming matrix formation method. But there are other important considerations which should be mentioned such as computer memory utilization and the methods of changing the network representation corresponding to network changes.

In our formulation of the admittance matrix and in its use for fault calculation we made no use of the matrix sparsity. Tinney and others [81, 82] have made a thorough analysis of sparsity in their optimally ordered triangular factorization technique. This technique permits the use of the admittance matrix, stored in sparse triangular factored form, to solve network problems. This formulation always requires much less computer storage than the impedance matrix, an advantage that increases as the square of the network size. Furthermore, their studies show that the sparse matrix technique exhibits a definite advantage in computer time required on all problems except faults on small networks where there is only one nonzero current in the I vector. In such cases the impedance matrix is faster. Since the trend is toward the solution of very large networks, the optimally ordered triangular factorization technique has definite advantages because it can effectively and economically solve networks at least 10 times as large as the impedance matrix technique [83].

Another important practical consideration in solving faulted networks by computer is the ability to change the network to represent added or removed lines without completely rebuilding the stored admittance or impedance matrix. As noted in Chapter 11, the addition or removal of a branch or group of branches in the Y matrix formulation, whether mutually coupled or not, is an almost trivial problem of adding or subtracting a few primitive admittances. A number of algorithms have been devised for making changes to the impedance matrix [84-86] which make it possible to consider limited changes in the network without altering the Z matrix. A method has also been devised [83] to permit similar changes in an ordered triangular factorization without changing that factorization.

A detailed consideration of the method of computer solution of faulted networks is beyond the scope of this book on symmetrical components, but the discussion of Chapters 11 and 12 should serve as an introduction.

Problems

12.1. Extend Example 12.1 to find the voltages at each node and all line contributions (a) for a 3ϕ fault on node 1, (b) for a 3ϕ fault on node 2, and (c) for a 3ϕ fault on node 3.

12.2. Repeat Example 12.1 if the fault impedance Z_f is 0.1 pu (pure resistance) and with a prefault voltage of 1.05 pu.

12.3. Use the method of "unit current injection" described in Example 12.1 to find the impedance matrices for the following positive sequence networks: (a) Figure P11.1, (b) Figure P11.2, (c) Figure P11.3, and (d) Figure P11.4. Use data given in problem 11.4. Neglect the resistance.

12.4. Repeat problem 12.3 to find the zero sequence impedance matrices.

12.5. Consider the network of Figure P11.4. Add a Δ generator at node 6 with $Z_1 = 0 + j0.1$ and find the impedance matrix by the unit current injection method. Neglect resistance.

12.6. Compute the total 3ϕ fault currents for the case of zero fault impedance, considering each node in turn as a faulted node, for the networks of problem 12.3.

12.7. Compute the total 3ϕ fault current at each node of the system of problem 12.5. Compare results with those of problem 12.6(d).

12.8. Compute the line current contributions and bus sequence voltages for the following faults. The fault is a 3ϕ bus fault in each case.
(a) Fault on node 2, Figure P11.1.
(b) Fault on node 1, Figure P11.2.
(c) Fault on node 4, Figure P11.3.
(d) Fault on node 5, Figure P11.3.
(e) Fault on node 5, Figure P11.4.
(f) Fault on node 2, Figure P11.4.
(g) Fault on node 2, system of problem 12.5.

12.9. Repeat Example 12.2 to find the complete network solution for a SLG fault
(a) On bus 1 with $Z_f = 0.1 + j0$, $V_f = 1.05 + j0$.
(b) On bus 2 with $Z_f = 0.1 + j0$, $V_f = 1.05 + j0$.
(c) On bus 3 with $Z_f = 0.05 + j0.05$, $V_f = 1.10 + j0$.
(d) On bus 3 with $Z_f = 0.1 + j0$, $V_f = 1.05 + j0$.

12.10. Compute the total SLG fault current for the fault conditions described in problem 12.8.

12.11. Compute the sequence voltages of all buses adjacent to the fault and all contributing line currents for the faults of problem 12.10.

12.12. Verify equation (12.27). In particular, show that changing the ith row current from I_i to $I_q + I_i$ requires taking impedance differences as shown in column q.

12.13. Repeat Example 12.3 except introduce the elements in the order
(a) 1, 3, 4, 5, 6, 2.
(b) 1, 6, 2, 4, 5, 3.
(c) 2, 4, 5, 6, 1, 3.
(d) 1, 4, 6, 5, 3, 2.

12.14. Use the Z matrix algorithm of sections 12.3-12.6 to find the positive sequence impedance matrices of the following networks: (a) Figure P11.1, (b) Figure P11.2, (c) Figure P11.3, (d) Figure P11.4, and (e) systems of problem 12.5.

12.15. Use the algorithms of sections 12.3-12.8 to find the zero sequence impedance matrices of the networks specified in problem 12.14.

12.16. Show that (12.38) may also be written as

$$\text{(a)}\ \tilde{Z}_{qk} = Z_{pk} - \frac{(1 - \mathfrak{z}_{mn}\mathfrak{y}_{mn})(Z_{mk} - Z_{nk})}{\mathfrak{z}_{mn}\mathfrak{y}_{M}}$$

$$\text{(b)}\ \tilde{Z}_{qk} = Z_{pk} - \frac{\mathfrak{z}_{M}\mathfrak{y}_{mn}(Z_{mk} - Z_{nk})}{1 - \mathfrak{z}_{M}\mathfrak{y}_{M}}$$

12.17. Verify the steps leading to (12.42).

12.18. Verify (12.48).

12.19. Verify (12.52).

12.20. Show that (12.48) and (12.52) reduce to (12.38) and (12.42) when there are only two mutually coupled lines.

12.21. Develop formulas similar to (12.48) and (12.52) for the special case of three mutually coupled lines.

appendix A

Matrix Algebra

The analysis of power systems under faulted conditions by analytical techniques is complex because of the large number of simultaneous equations that must be formed and solved. Matrix algebra, with its capability for the writing and manipulating of large numbers of equations in terms of a small number of symbols, is a desirable, almost indispensable mathematical tool.

The purpose of this appendix is to review the appropriate concepts of matrix algebra.[1] First, we consider simultaneous linear equations and their solution by means of determinants.

A.1 Simultaneous Linear Equations

We often work with equations of the form

$$\begin{aligned} x_1 &= a_{11}y_1 + a_{12}y_2 + \cdots + a_{1n}y_n \\ x_2 &= a_{21}y_1 + a_{22}y_2 + \cdots + a_{2n}y_n \\ \cdots & \quad \cdots \quad \cdots \quad \cdots \quad \cdots \\ x_n &= a_{n1}y_1 + a_{n2}y_2 + \cdots + a_{nn}y_n \end{aligned} \tag{A.1}$$

Here the x's and y's are variables and the a's are constant (either real or complex) coefficients. Note particularly the method of subscripting. In the problems of interest, the x's will often be voltages (or currents), the y's will be currents (or voltages), and the a's will be impedances (or admittances).

Equation (A.1) can be solved for the y's by means of determinants to get

$$\begin{aligned} y_1 &= f_{11}x_1 + f_{12}x_2 + \cdots + f_{1n}x_n \\ y_2 &= f_{21}x_1 + f_{22}x_2 + \cdots + f_{2n}x_n \\ \cdots & \quad \cdots \quad \cdots \quad \cdots \quad \cdots \\ y_n &= f_{n1}x_1 + f_{n2}x_2 + \cdots + f_{nn}x_n \end{aligned} \tag{A.2}$$

where

$$f_{ij} = C_{ji}/\Delta \tag{A.3}$$

[1] Much of the material for this appendix is taken from notes, "Review of Matrix Algebra," prepared by H. W. Hale and used for teaching short courses in Power Systems Engineering at Iowa State University, Ames. It is used with permission of the author.

where

$$\Delta = \det \begin{bmatrix} a_{11} & a_{12} & \cdots & a_{1n} \\ \cdots & \cdots & \cdots & \cdots \\ \cdots & \cdots & \cdots & \cdots \\ a_{n1} & a_{n2} & \cdots & a_{nn} \end{bmatrix} = \sum_{j=1}^{n} a_{ij} C_{ij} \tag{A.4}$$

and

C_{ij} = the cofactor ij = $(-1)^{i+j} M_{ij}$

M_{ij} = the minor determinant resulting from the deletion of row i and column j

(A.5)

The general form of (A.1) and of its solution (A.2) will be pertinent to some of the following material.

A.2 Definitions Pertaining to Matrices

Following are some of the basic definitions and terminology pertaining to matrices.

1. *Definition of a matrix.* A matrix is a rectangular (including square as a special case) array of elements (symbols or numbers) arranged in rows and columns. For example

$$\begin{bmatrix} a_{11} & a_{12} & a_{13} \\ a_{21} & a_{22} & a_{22} \end{bmatrix}$$

is a matrix of two rows and three columns. Note that in this general form the subscripts attached to the elements (a's) indicate the row (first subscript) and column (second subscript) in which it is found. Thus, the term "a_{ij}" or "element ij" specifies a definite location at which the element is found. For a specific matrix of numerical values such as

$$\begin{bmatrix} 1 & -2 \\ 3 & 2 \end{bmatrix}$$

subscripts are not attached to the elements; however, it is apparent that

$$a_{11} = 1, \quad a_{12} = -2, \quad a_{21} = 3, \quad a_{22} = 2$$

2. *Notation.* We will indicate arrays as matrices by enclosure in brackets as shown in definition 1. Other conventions are in use and may be encountered in the literature. We can also indicate a matrix symbolically by a single symbol such as

$$\mathbf{A} = \begin{bmatrix} a_{11} & a_{12} \\ a_{21} & a_{22} \end{bmatrix}$$

Although other conventions, such as underscoring or script type, are used, we will use boldface type for matrices. In most cases it is clear from the context which quantities are matrices and which are not. We will see later that we can write equations in terms of symbolic representation of matrices and manipulate the symbols as long as the appropriate rules are followed.

3. *Order of a matrix.* A matrix with m rows and n columns is referred to as an $m \times n$ or (m, n) matrix. In the special case that the matrix is square, $m = n$, it is referred to as an nth order matrix.

4. *Column and row matrices.* A matrix with only one column is referred to as a column matrix, while one with only one row is a row matrix. They are also often referred to as column and row vectors respectively.

5. *Transpose of a matrix.* The new matrix formed by interchanging the rows and columns of a given matrix is called the transpose of the original matrix. Thus

$$\begin{bmatrix} 1 & -1 & 2 \\ 1 & 3 & -2 \end{bmatrix} \text{ is the transpose of } \begin{bmatrix} 1 & 1 \\ -1 & 3 \\ 2 & -2 \end{bmatrix}$$

Obviously, an $m \times n$ matrix has a transpose that is an $n \times m$ matrix. We indicate the transpose of a matrix $\mathbf{A}$ by $\mathbf{A}'$ or $\mathbf{A}^T$ or $\mathbf{A}^t$.

6. *Conjugate of a matrix.* If all elements of a matrix are replaced by their complex conjugates, the new matrix is the conjugate of the original one. Thus

$$\begin{bmatrix} (1 - j1) & 1 \\ (3 - j3) & (2 + j2) \end{bmatrix} \text{ is the conjugate of } \begin{bmatrix} (1 + j1) & 1 \\ (3 + j3) & (2 - j2) \end{bmatrix}$$

The conjugate of a matrix $\mathbf{A}$ is indicated symbolically as $\mathbf{A}^*$.

7. *Symmetric matrix.* A square matrix that has $a_{ji} = a_{ij}$ for all i and j is symmetric. Thus

$$\begin{bmatrix} 1 & 2 \\ 2 & 3 \end{bmatrix} \text{ is symmetric, while } \begin{bmatrix} 1 & 3 \\ 2 & 3 \end{bmatrix} \text{ is not symmetric}$$

Another way of stating the definition is that a symmetric matrix is equal to its own transpose. Only square matrices can be symmetric.

8. *Principal diagonal.* The elements $a_{11}, a_{22}, a_{33}, \ldots, a_{nn}$ of a square matrix form its principal diagonal.

9. *Unit matrix.* A square matrix that has +1's on its principal diagonal and zeros elsewhere is a unit matrix. It is denoted symbolically as $\mathbf{I}$, $\mathbf{U}$, $\mathbf{E}$, or $\mathbf{1}$. Sometimes a subscript is attached to indicate the order. Thus,

$$\mathbf{U} = \mathbf{1} = \mathbf{E}_3 = \begin{bmatrix} 1 & 0 & 0 \\ 0 & 1 & 0 \\ 0 & 0 & 1 \end{bmatrix} \tag{A.6}$$

In electrical engineering we usually avoid using $\mathbf{I}$ for the unit matrix.

10. *Null matrix.* A matrix with all elements equal to zero is a null matrix.

11. *Scalars.* The elements of matrices are called scalars. A scalar is a 1×1 matrix.

12. *Submatrix.* If some rows and columns of a matrix are deleted, the result is a submatrix. For example, if

$$\mathbf{A} = \begin{bmatrix} 1 & 2 & 3 & 2 \\ 4 & 2 & 1 & 3 \end{bmatrix}$$

has columns 2 and 4 deleted, the result is a new matrix

$$\mathbf{B} = \begin{bmatrix} 1 & 3 \\ 4 & 1 \end{bmatrix}$$

that is a submatrix of A.

13. *Determinant of a matrix.* It is important to note that a matrix *is not* a determinant. However, the determinant of a square matrix has meaning and is taken to be the (scalar) quantity

$$\det \mathbf{A} = \det \begin{bmatrix} 2 & 5 \\ 4 & 6 \end{bmatrix} = 12 - 20 = -8$$

The determinant of a matrix that is not square is not defined.

14. *Nonsingular and singular matrices.* If a square matrix has a nonzero determinant it is *nonsingular.* All other matrices, including those that are not square, are *singular.*

15. *Rank of a matrix.* The rank of a matrix is the order (number of rows or columns) of the largest nonsingular square submatrix. Thus

$$\mathbf{A} = \begin{bmatrix} 1 & 2 & 4 \\ 3 & 1 & 3 \end{bmatrix}$$

has rank 2 because

$$\det \begin{bmatrix} 1 & 2 \\ 3 & 1 \end{bmatrix} = -5 \neq 0$$

while

$$\mathbf{B} = \begin{bmatrix} 1 & 1 & -1 \\ 1 & 1 & -1 \end{bmatrix}$$

has rank 1.

16. *Adjoint of the matrix* A. The adjoint of the matrix A, written adj A, is defined as

$$\text{adj } \mathbf{A} = \begin{bmatrix} C_{11} & C_{21} & \cdots & C_{n1} \\ C_{12} & C_{22} & \cdots & C_{n2} \\ \cdots & \cdots & \cdots & \cdots \\ C_{1n} & C_{2n} & \cdots & C_{nn} \end{bmatrix} \tag{A.7}$$

where

$$C_{ij} = \text{cofactor of } a_{ij}$$

17. *Inverse of the matrix* A. The inverse of a square matrix A of order n is defined by the equation

$$\mathbf{A}\,\mathbf{A}^{-1} = \mathbf{A}^{-1}\,\mathbf{A} = \mathbf{U} \tag{A.8}$$

In terms of elementary operations the inverse of A may be computed from the formula

$$\mathbf{A}^{-1} = \text{adj } \mathbf{A}/\det \mathbf{A} \tag{A.9}$$

Only nonsingular matrices ($\det \mathbf{A} \neq 0$) have inverses.

18. *Skew-symmetric matrix.* A skew-symmetric matrix A is defined by the relation

$$a_{ij} = -a_{ji} \tag{A.10}$$

The major diagonal elements of a skew-symmetric matrix must be zero since for $i = j$ the only way (A.10) may be satisfied is for $a_{ij} = 0$. An example of a skew-symmetric matrix is

$$\mathbf{A} = \begin{bmatrix} 0 + j0 & 3 - j & 2 + j \\ -3 + j & 0 + j0 & 4 + j0 \\ -2 - j & -4 + j0 & 0 + j0 \end{bmatrix}$$

19. *Hermitian matrix.* A hermitian matrix A is defined by the relation

$$a_{ij} = a_{ji}^* \tag{A.11}$$

The major diagonal elements of a hermitian matrix must be real, for with $i = j$ the only way (A.11) can be satisfied is for a_{ij} to be real. Only square matrices may be classified as hermitian. An example of a hermitian matrix is

$$\mathbf{A} = \begin{bmatrix} 2 + j0 & 2 - j & 0 + j0 \\ 2 + j & -5 + j0 & 1 + j3 \\ 0 + j0 & 1 - j3 & 0 + j0 \end{bmatrix}$$

20. *Skew-hermitian matrix.* A skew-hermitian matrix A is defined by the relation

$$a_{ij} = -a_{ji}^* \tag{A.12}$$

and this property applies only when A is square. The terms of the major diagonal $i = j$ may be zero or pure imaginary, but not real. An example of a skew-hermitian matrix is given by

$$\mathbf{A} = \begin{bmatrix} 0 + j2 & 5 + j0 & 3 + j2 \\ -5 + j0 & 0 + j0 & -1 - j7 \\ -3 + j2 & 1 - j7 & 0 - j6 \end{bmatrix}$$

21. *Dominant matrix.* A dominant matrix A is defined as a symmetric matrix which has the additional property that

$$a_{ii} \geqq \sum_{j=1}^{n} |a_{ij}|,\ j \neq i \tag{A.13}$$

for all values of i. An example of a dominant matrix is given by

$$\mathbf{A} = \begin{bmatrix} 4 & 2 & -1 \\ 2 & 7 & 3 \\ -1 & 3 & 6 \end{bmatrix}$$

where (A.13) is satisfied or the elements of the major diagonal are greater than the sum of the absolute value of all elements in that row (or column). For example, the Y matrix of a power system is always dominant.

A.3 Rules of Matrix Algebra

Matrix algebra has rules that must be followed explicitly. The basic rules are:

1. *Equality of matrices.* Two matrices are equal if and only if each of their corresponding elements are equal. Thus G = H only if $g_{ij} = h_{ij}$ for all i and j. Only matrices having the same number of rows and columns can be equal.

2. *Addition of matrices.* Two matrices are added by adding corresponding elements to obtain the corresponding element in the sum. Thus

$$\begin{bmatrix} 1 & 0 & 2 \\ 2 & -1 & 3 \end{bmatrix} + \begin{bmatrix} 2 & -1 & 0 \\ -1 & 3 & -1 \end{bmatrix} = \begin{bmatrix} 3 & -1 & 2 \\ 1 & 2 & 2 \end{bmatrix}$$

Addition is defined only when both matrices have the same number of rows and columns. Addition is commutative, i.e.,

$$\mathbf{A} + \mathbf{B} = \mathbf{B} + \mathbf{A}$$

and is also associative, i.e.,

$$\mathbf{A} + (\mathbf{B} + \mathbf{C}) = (\mathbf{A} + \mathbf{B}) + \mathbf{C}$$

3. *Subtraction.* A matrix is subtracted from a second matrix by subtracting the elements of the first from the elements of the second. Subtraction is defined only when both matrices have the same number of rows and columns.

4. *Multiplication of a matrix by a scalar.* A matrix is multiplied by a scalar by multiplying each element of the matrix by the scalar, e.g.,

$$k\begin{bmatrix} 2 & 1 \\ -1 & 3 \end{bmatrix} = \begin{bmatrix} 2k & k \\ -k & 3k \end{bmatrix}$$

5. *Multiplication of matrices.* Multiplication is defined by the following:

$$\mathbf{G} = \mathbf{A} \times \mathbf{B} \text{ (or } \mathbf{A}\,\mathbf{B})$$

implies that the elements of G are related to the elements of A and B by

$$g_{ij} = \sum_{k=1}^{n} a_{ik} b_{kj} \tag{A.14}$$

where n is the number of columns of A and the number of rows of B. Multiplication is defined if and only if A has the same number of columns as B has rows. Two such matrices are said to be *conformable*. Thus

$$\underset{2\times 2}{\begin{bmatrix} g_{11} & g_{12} \\ g_{21} & g_{22} \end{bmatrix}} = \underset{2\times 3}{\begin{bmatrix} a_{11} & a_{12} & a_{13} \\ a_{21} & a_{22} & a_{23} \end{bmatrix}} \underset{3\times 2}{\begin{bmatrix} b_{11} & b_{12} \\ b_{21} & b_{22} \\ b_{31} & b_{32} \end{bmatrix}}$$

where, for example,

$$g_{12} = \sum_{k=1}^{3} a_{1k} b_{k2} = a_{11} b_{12} + a_{12} b_{22} + a_{13} b_{32}$$

It follows that in the product

$$C = A \times B \tag{A.15}$$

where A is an $m \times n$ matrix and B is an $n \times p$ matrix, C is an $m \times p$ matrix. That is, the number of rows and columns of C will be equal to the number of rows of A and the number of columns of B respectively. It is sometimes convenient to check conformability by writing the order (row $\times$ column) above each matrix. Thus (A.15) would be written

$$\overset{m \times p}{C} = \overset{m \times n}{A} \times \overset{n \times p}{B} \tag{A.16}$$

The adjacent terms of the superscripted numbers must agree ($n = n$) to have conformability, and the outer terms determine the order of C, viz., $m \times p$. This will work for products involving any number of matrices.

It is also apparent that multiplication is *not* commutative, i.e.,

$$A \times B \neq B \times A \tag{A.17}$$

It is, however, associative and is distributive over addition, i.e.,

$$A(BC) = (AB)C \tag{A.18}$$

and

$$A \times (B + C) = AB + AC \tag{A.19}$$

Because multiplication is not commutative, it is vital that the order of multiplication be clear. Thus in the product $A \times B$ we say that B is premultiplied by A or that A is postmultiplied by B.

6. *The product* UA. From the multiplication rule it is apparent that the product of an identity matrix of order n and an $n \times p$ matrix A is the matrix A, i.e.,

$$UA = A \tag{A.20}$$

If A is square, i.e., of order n, then $U \times A = A = A \times U$. Thus in this very special case multiplication is commutative.

7. *The inverse matrix and the solution of equations.* Suppose that given the matrix A there exists a matrix A^{-1} such that $A^{-1} \times A = U$. Then A^{-1} is called the inverse of A. The use of and the properties of the inverse are pointed out in terms of (A.1)–(A.5). Consider (A.1) which can be written in matrix from as $x = Ay$ where x, A, and y are $n \times 1$, $n \times n$, and $n \times 1$ matrices respectively. Assuming that the inverse of A exists, we may premultiply both sides of the equation by A^{-1}, with the result that

$$A^{-1}x = A^{-1}Ay = Uy = y$$

Now (A.2) can be written as $y = Fx$. It follows that $F = A^{-1}$ and the rules for determining the inverse of a square matrix result from a comparison of F (as in equations A.3, A.4, and A.5) and A^{-1}. Thus given

$$A = \begin{bmatrix} a_{11} & a_{12} & a_{13} \\ a_{21} & a_{22} & a_{23} \\ a_{31} & a_{32} & a_{33} \end{bmatrix}$$

the inverse of A is

$$\mathrm{A}^{-1} = \frac{1}{\Delta}\begin{bmatrix} C_{11} & C_{21} & C_{31} \\ C_{12} & C_{22} & C_{32} \\ C_{13} & C_{23} & C_{33} \end{bmatrix} = \frac{\text{adj A}}{\det \mathrm{A}} \qquad \text{(A.21)}$$

where Δ is the determinant of A, and C_{ij} is the appropriate cofactor. The extension to higher order matrices is apparent. It is also apparent that only square matrices can have an inverse and that a square matrix has an inverse if and only if it has a nonzero determinant, i.e., if it is nonsingular.

A.4 Some Useful Theorems on Matrices

We state without proof several theorems which are useful, even indispensable, in dealing with matrices and matrix equations.

1. The inverse of the product of two square matrices is equal to the product of the inverse taken in reverse order, i.e.,

$$(\mathbf{AB})^{-1} = \mathbf{B}^{-1}\mathbf{A}^{-1} \qquad \text{(A.22)}$$

2. The transpose of the product of two matrices is equal to the product of the transposes taken in reverse order, i.e.,

$$(\mathbf{AB})^t = \mathbf{B}^t\mathbf{A}^t \qquad \text{(A.23)}$$

3. The transpose of the sum of two matrices is equal to the sum of the transposes, i.e.,

$$(\mathbf{A} + \mathbf{B})^t = \mathbf{A}^t + \mathbf{B}^t \qquad \text{(A.24)}$$

4. The inverse of the transpose of a square matrix is equal to the transpose of the inverse, i.e.,

$$(\mathbf{A}^t)^{-1} = (\mathbf{A}^{-1})^t \qquad \text{(A.25)}$$

5. If a square matrix is equal to its transpose, it is symmetric,

$$\mathbf{A} = \mathbf{A}^t$$
$$\longleftrightarrow \mathbf{A} \text{ is symmetric} \qquad \text{(A.26)}$$

6. If a square matrix is equal to the negative of its transpose, it is skew-symmetric, i.e.,

$$\mathbf{A} = -\mathbf{A}^t$$
$$\longleftrightarrow \mathbf{A} \text{ is skew-symmetric} \qquad \text{(A.27)}$$

7. Any square matrix **A** with all *real* elements may be expressed as the sum of a symmetric matrix and a skew-symmetric matrix (this is also true if the elements are functions of the Laplace transform variable s). Stated mathematically,

$$\mathbf{A} \text{ is square}$$
$$\longleftrightarrow \mathbf{A} = \mathbf{B} + \mathbf{C}$$
$$\text{where } \mathbf{B} \text{ is symmetric}$$
$$\mathbf{C} \text{ is skew-symmetric} \qquad \text{(A.28)}$$

From theorems 5 and 6 we compute

$$\mathbf{A}^t = \mathbf{B}^t + \mathbf{C}^t = \mathbf{B} - \mathbf{C} \tag{A.29}$$

Then we may add and subtract (A.28) and (A.29) to obtain, respectively, the symmetric matrix

$$\mathbf{B} = (1/2)\,(\mathbf{A} + \mathbf{A}^t) \tag{A.30}$$

and the skew-symmetric matrix

$$\mathbf{C} = (1/2)\,(\mathbf{A} - \mathbf{A}^t) \tag{A.31}$$

8. A hermitian matrix is a square matrix which is equal to the conjugate of its transpose and vice versa, i.e.,

$$\mathbf{A} = (\mathbf{A}^t)^*$$
$$\longleftrightarrow \mathbf{A} \text{ is hermitian} \tag{A.32}$$

9. A skew-hermitian matrix is equal to the negative of the conjugate of its transpose and vice versa, i.e.,

$$\mathbf{A} = -(\mathbf{A}^t)^*$$
$$\longleftrightarrow \mathbf{A} \text{ is skew-hermitian} \tag{A.33}$$

10. Any square matrix **A** with real and complex elements may be written as the sum of a hermitian matrix and a skew-hermitian matrix, i.e.,

$$\mathbf{A} \text{ is square and real or complex}$$
$$\longleftrightarrow \mathbf{A} = \mathbf{B} + \mathbf{C}$$
$$\text{where } \mathbf{B} \text{ is hermitian}$$
$$\mathbf{C} \text{ is skew-hermitian} \tag{A.34}$$

From theorems 3, 8, and 9 we easily show that we may write the hermitian matrix as

$$\mathbf{B} = (1/2)\,[\mathbf{A} + (\mathbf{A}^t)^*] \tag{A.35}$$

and the skew-hermitian matrix as

$$\mathbf{C} = (1/2)\,[\mathbf{A} - (\mathbf{A}^t)^*] \tag{A.36}$$

A.5 Matrix Partitioning

When equations involving large matrices are to be manipulated, it is frequently useful to partition these matrices into smaller submatrices. Partitioning may simplify the procedures and quite frequently makes the proof of general results much simpler. Consider the following example:

$$\mathbf{C} = \begin{bmatrix} c_1 \\ c_2 \\ c_3 \end{bmatrix} = \left[\begin{array}{cc|cc} a_{11} & a_{12} & a_{13} & a_{14} \\ a_{21} & a_{22} & a_{23} & a_{24} \\ a_{31} & a_{32} & a_{33} & a_{34} \end{array}\right] \left[\begin{array}{c} d_1 \\ d_2 \\ \hline d_3 \\ d_4 \end{array}\right] = \mathbf{A}\,\mathbf{D} \tag{A.37}$$

If we partition A and D by dashed lines as shown we can write

$$C = A D = [A_1 \, A_2] \begin{bmatrix} D_1 \\ D_2 \end{bmatrix} = [A_1 D_1 + A_2 D_2] \qquad (A.38)$$

where the elements of A and D are now themselves matrices and are submatrices of the original A and D as determined by the partitioning. It is important to note that the partitioning of A and D must be consistent; i.e., all products and sums of matrices must be defined. It would not be possible to partition A between the second and third columns and D between the first and second rows and have a meaningful result. In other words, the partitioning of A by columns determines the partitioning of D by rows, or vice versa.

We can carry the partitioning another step forward as shown below.

$$C = \begin{bmatrix} c_1 \\ c_2 \\ \hline c_3 \end{bmatrix} = \begin{bmatrix} a_{11} & a_{12} & a_{13} & a_{14} \\ a_{21} & a_{22} & a_{23} & a_{24} \\ \hline a_{31} & a_{32} & a_{33} & a_{34} \end{bmatrix} \begin{bmatrix} d_1 \\ d_2 \\ \hline d_3 \\ d_4 \end{bmatrix} \qquad (A.39)$$

and now write

$$C = A D = \begin{bmatrix} C_1 \\ C_2 \end{bmatrix} = \begin{bmatrix} A_{11} & A_{12} \\ A_{21} & A_{22} \end{bmatrix} \begin{bmatrix} D_1 \\ D_2 \end{bmatrix} \qquad (A.40)$$

or

$$C_1 = A_{11} D_1 + A_{12} D_2, \qquad C_2 = A_{21} D_1 + A_{22} D_2$$

Again the partitioning of the matrices must be consistent; i.e., all products and sums must be defined. As before, the partitioning of A by columns determines the partitioning of D by rows. In addition, the partitioning of A by rows determines the partitioning of C by rows or vice versa.

These results can be generalized into multiple partitionings as shown below.

$$
\begin{array}{c}
\begin{matrix} 1 & 2 & 3 & 4 & 5 \end{matrix} \\
\begin{matrix} 1 \\ 2 \\ 3 \\ 4 \\ 5 \\ 6 \\ 7 \\ 8 \end{matrix}
\begin{bmatrix} & & & \\ & C_{11} & & C_{12} \\ & & & \\ \hline & C_{21} & & C_{22} \\ & & & \end{bmatrix}
\end{array}
=
\begin{array}{c}
\begin{matrix} 1 & 2 & 3 & 4 & 5 & 6 & 7 & 8 & 9 & 10 \end{matrix} \\
\begin{matrix} 1 \\ 2 \\ 3 \\ 4 \\ 5 \\ 6 \\ 7 \\ 8 \end{matrix}
\begin{bmatrix} & & & \\ A_{11} & A_{12} & A_{13} \\ \hline & & \\ A_{21} & A_{22} & A_{23} \\ & & \end{bmatrix}
\end{array}
\begin{array}{c}
\begin{matrix} 1 & 2 & 3 & 4 & 5 \end{matrix} \\
\begin{matrix} 1 \\ 2 \\ 3 \\ 4 \\ 5 \\ 6 \\ 7 \\ 8 \\ 9 \\ 10 \end{matrix}
\begin{bmatrix} D_{11} & D_{12} \\ \hline D_{21} & D_{22} \\ \hline D_{31} & D_{32} \end{bmatrix}
\end{array}
$$

$$(A.41)$$

A.6 Matrix Reduction (Kron Reduction)

In physical problems the following type of equation, after rearranging and partitioning, is often obtained.

$$\begin{bmatrix} C_1 \\ 0 \end{bmatrix} = \begin{bmatrix} A_{11} & A_{12} \\ A_{21} & A_{22} \end{bmatrix} \begin{bmatrix} D_1 \\ D_2 \end{bmatrix} \tag{A.42}$$

Here, the nature of the problems are such that A_{22} is usually nonsingular and the variables in D_2 are unwanted and can be eliminated. If we expand the above equation, the result is

$$\begin{aligned} C_1 &= A_{11} D_1 + A_{12} D_2 \\ 0 &= A_{21} D_1 + A_{22} D_2 \end{aligned} \tag{A.43}$$

Remembering that the quantities in these equations are matrices and must be treated as such, we can solve the second equation for D_2 with the result that $D_2 = -A_{22}^{-1} A_{21} D_1$ and substitute into the first of equations (A.43) with the result

$$C_1 = (A_{11} - A_{12} A_{22}^{-1} A_{21}) D_1 \tag{A.44}$$

This last result is particularly useful and is a graphic illustration of the utility of matrix methods.

A.7 Transformation of Coordinates

It is often desirable in electrical engineering problems to transform a matrix equation from one coordinate system to another in which a solution may be more readily obtained. For example, we often find it convenient to transform from the *a-b-c* phase domain into the 0-1-2 symmetrical component domain or the *0-d-q* Park domain. The result is a *similarity transformation* of the impedances as will be shown.

Let T be any unique (i.e., nonsingular) transformation which relates currents (and voltages) between the *a-b-c* domain and the α-β-γ domain according to the equation

$$\begin{bmatrix} I_a \\ I_b \\ I_c \end{bmatrix} = T \begin{bmatrix} I_\alpha \\ I_\beta \\ I_\gamma \end{bmatrix} \tag{A.45}$$

and similarly for voltages. We write (A.45) in matrix notation as

$$I_{abc} = T\, I_{\alpha\beta\gamma} \tag{A.46}$$

Suppose we are given a matrix voltage equation, expressing the voltage drop in lines *a*, *b*, and *c* as

$$V_{abc} = Z_{abc} I_{abc} \tag{A.47}$$

Premultiplying both sides of (A.47) by T^{-1} we have (since T^{-1} exists)

$$T^{-1} V_{abc} = T^{-1} Z_{abc} I_{abc}$$

If we now insert the product $T\,T^{-1} = U$ between Z_{abc} and I_{abc}, we have

$$T^{-1} V_{abc} = T^{-1} Z_{abc}\, T\, T^{-1} I_{abc}$$

But by (A.46) this equation is

$$V_{\alpha\beta\gamma} = T^{-1} Z_{abc}\, T\, I_{\alpha\beta\gamma} \tag{A.48}$$

Thus under the (linear) transformation T, the impedance has been changed to

$$\mathbf{Z}_{\alpha\beta\gamma} = \mathbf{T}^{-1}\,\mathbf{Z}_{abc}\mathbf{T} \tag{A.49}$$

or $\mathbf{Z}_{abc}$ has undergone a *similarity transformation*. This is an important transformation of linear algebra, and we say that the two matrices $\mathbf{Z}_{abc}$ and $\mathbf{Z}_{\alpha\beta\gamma}$ are similar.

A.8 Computer Programs

In working with matrices of order greater than two it is extremely laborious to perform any computations by hand. Matrix inversions, for example, require something greater than n^3 multiplications for an $n \times n$ matrix. Very clever manipulations can save perhaps half the work, but nothing like an n-fold saving is possible. Similarly, a great deal of computation is required for matrix multiplications and even additions.

Our attitude here will be that a digital computer is usually necessary and is available for matrix problems. Since FORTRAN is nearly a universal computer language, solutions will be given in this language.

Two difficult matrix computations often required are the *matrix inversion* and the *matrix reduction*. Both operations can be performed by the same FORTRAN subroutine, and an example of a program to perform these operations follows.[2]

```
C     MAIN PROGRAM FOR SHIPLEY-COLEMAN-JOHNSON COMPLEX INVERTER-REDUCER.
C     N IS ONE DIMENSION OF THE SQUARE INPUT MATRIX - MXIN.
C     KGO IS THE ARRAY OF ROWS AND/OR COLUMN INDICES OF ROWS AND
C     COLUMNS TO BE ELIMINATED IN THE REDUCTION.  NOT USED FOR INVERSION.
C     INV = 1 FOR INVERSION,  INV = 0 FOR REDUCTION.
      COMPLEX * 8 MXIN(100), MXOUT(100), INVERT, REDUCE, OUT
      INTEGER * 4 N, KGO(10), INV
      DATA INVERT / 'INVERTED'/, REDUCE/ 'REDUCED '/
    1 OUT = INVERT
C
      READ(5,10)N, INV
   10 FORMAT(2I5)
      NN = N * N
      IF( INV .EQ. 1) GO TO 30
      OUT = REDUCE
      READ(5,20) KGO
   20 FORMAT(10I5)
   30 READ(5,40) (MXIN(I), I = 1,NN)
C
C     MXIN IN ROW ORDER (REAL,IMAG).
C
   40 FORMAT(8F10.0)
      WRITE(6,50) OUT
   50 FORMAT('1THE MATRIX BEING ', 2A4/)
      DO 80  J = 1,N
      JMIN = (J - 1) * N
      DO 70  I = 1, N
      IJ = I + JMIN
      WRITE(6,60) J, I, MXIN(IJ)
   60 FORMAT(' MXIN(', I2, ',', I2, ') = ', 1PE14.6, ' +J * ', 1PE14.6)
   70 CONTINUE
   80 CONTINUE
      CALL CSHP(MXIN, MXOUT, N, KGO, INV, &110, &90)
   90 WRITE(6,100)
  100 FORMAT('0OOPS')
```

[2] This program was written by G. N. Johnson and is used with his permission.

```
      GO TO 1
110   WRITE(6,120) OUT
120   FORMAT('0THE ', 2A4, ' MATRIX' /)
      DO 150  J = 1, N
      JM1N = (J - 1) * N
      DO 140  I = 1, N
      IJ = I + JM1N
      WRITE(6,130) J, I, MXOUT(IJ)
130   FORMAT(' MXOUT(', I2, ',', I2, ') = ', 1PE14.6, '+J * ', 1PE14.6)
140   CONTINUE
150   CONTINUE
      GO TO 1
                                     END


      SUBROUTINE CSHP(MXIN, MXOUT, N, KGO, INV, *, *)
C
C     INV = 1 FOR INVERSION, INV = 0 FOR KRON REDUCTION.
C     MXIN AND MXOUT ARE THE INPUT AND OUTPUT MATRICES RESPECTIVELY.
C     MXIN AND MXOUT ARE SINGLE SUBSCRIPTED AND IN ROW ORDER.
C     N IS ONE DIMENSION OF THE SQUARE MATRIX MXIN.
C
      COMPLEX * 8 MXIN(100), MXOUT(100), DIAGEL, CONJG
      INTEGER*4 L(10),K, KGO(10)
C
C     L IS A WORK ARRAY FOR KEEPING TRACK OF ROWS DONE DURING INVERSION.
C
      COMPLEX * 16 PIVINV, AA, BB, CC, DD, DOUBLE(100)
      REAL * 4 BIGSTA, TRY
      NN = N * N
      IF(INV .EQ. 0) GO TO 220
      KGOTOT = N
      GO TO 1
C
C     REDUCE A GIVEN MATRIX OF NTH ORDER BY ELIMINATING KGOTOT ROWS
C     AND COLUMNS SPECIFIED IN THE ARRAY KGO(I).
220   DO 230 KGOTOT = 1, N
      IF( KGO (KGOTOT) .EQ. 0) GO TO 250
230   CONTINUE
      WRITE(6,240)
240   FORMAT('0*** YOU WANT TO ELIMINATE ALL OF THE MATRIX? ***')
      RETURN2
250   KGOTOT = KGOTOT - 1
1     DO 5  I = 1, NN
C
C     MOVE INPUT MATRIX INTO DOUBLE PRECISION WORKING MATRIX.
C
5     DOUBLE(I) = MXIN(I)
9     DO 100  III = 1, KGOTOT
      IF(INV .EQ. 0) GO TO 35
C
C     PICK LARGEST UNUSED DIAGONAL ELEMENT FOR PIVOT FOR INVERSION.
C
      BIGSTA = 0.0
      DO 20 I = 1, N
      DO 10 J = 1, III
      IF(J .EQ. III) GO TO 10
      IF(I .EQ. L(J)) GO TO 20
10    CONTINUE
30    DIAGEL = DOUBLE((I-1) * N + I)
      TRY = REAL(DIAGEL * CONJG(DIAGEL))
      IF(BIGSTA .GT. TRY) GO TO 20
      K = I
      L(III) = I
      BIGSTA = TRY
```

```
 20    CONTINUE
       IF(BIGSTA .LT. 1.E-26) GO TO 120
C
C      LARGEST DIAGONAL ELEMENT CHOSEN.
C
       GO TO 37
C
C      GOIN G TO DO A KRON REDUCTION WHEN INV = 0...
C
 35    K = KGO(III)
 37    DD = DOUBLE((K-1) * N + K)
       PIVINV = 1.0D00 / DD
       KM1N = (K-1) * N
       DO 50  I = 1, N
       IF(I .EQ. K) GO TO 50
       BB = DOUBLE(KM1N + I)
       DO 40  J = 1, N
       IF(J .EQ. K) GO TO 40
C
C      MOST ACCURATE WAY TO CALC A(I,J) = A(I,J) -(A(I,K)* A(K,J))/A(K,K)
C      IF A WAS THE MATRIX BEING INVERTED OR REDUCED.
C
       IJ = (J-1) * N + I
       AA = DOUBLE(IJ)
       CC = DOUBLE(IJ - I + K)
       AA =(AA * DD - BB * CC) * PIVINV
       DOUBLE(IJ) = AA
 40    CONTINUE
 50    CONTINUE
       IF(INV .EQ. 0) GO TO 100
C
C      DIVIDE PIVOT ROW BY -A(K,K) ON INVERSION ONLY.
C
       DO 55  I = 1, N
       IF(I .EQ. K) GO TO 55
       KM1NI = KM1N + I
       AA = DOUBLE(KM1NI)
       AA = -AA * PIVINV
       DOUBLE(KM1NI) = AA
 55    CONTINUE
C
C      DIVIDE PIVOT COLUMN BY -A(K,K) ON INVERSION ONLY.
C
       DO 60  J = 1, N
       IF(J .EQ. K) GO TO 60
       KJ = (J-1) * N + K
       AA = DOUBLE(KJ)
       AA = -AA * PIVINV
       DOUBLE(KJ) = AA
 60    CONTINUE
C
C      REPLACE A(K,K) BY -1.0/A(K,K) ON INVERSION ONLY.
C
       DOUBLE(KM1N + K) = -PIVINV
 100   CONTINUE
       IF(INV .EQ. 0) GO TO 150
C
C      NEGATE ENTIRE RESULTANT MATRIX TO REALIZE INVERSE.
C
       DO 110  I = 1, NN
 110   MXOUT(I) = -DOUBLE(I)
       RETURN1
 120   WRITE (6,130)
 130   FORMAT ('0***THIS MATRIX IS SINGULAR***')
       RETURN2
```

```
C
C     COMPLEX SHIPLEY MATRIX INVERTOR MODIFIED & PGMD BY G.N.JOHNSON P.E.
C
 150   NM = N - KGOTOT
C
C      KRON REDUCTION - REGROUPING OUTPUT MATRIX.
C
        K = 0
        DO 170  I = 1, N
        DO 160  J = 1, KGOTOT
        IF(I .EQ. KGO(J)) GO TO 170
  160   CONTINUE
        K = K + 1
        L(K) = I
  170   CONTINUE
        DO 180  I = 1, NM
        IL = L(I)
        IM1MN = (I-1) * NM
        ILM1N = (IL-1) * N
        DO 180  J = 1, NM
        JL = L(J)
  180   MXOUT(IM1MN + J) = DOUBLE(ILM1N + JL)
        N = NM
        RETURN1
        END
```

appendix B

Line Impedance Tables

The following tables provide all the data normally required for determining the normalized impedance of overhead transmission or distribution lines.

1. Tables B.1–B.3 give base currents, base impedances, and base admittances for a wide range of voltages and for four convenient MVA bases.
2. Tables B.4–B.11 give the characteristics of overhead wires, presented in a convenient format. They are used with permission of the Westinghouse Corporation and are taken from [14] and [54].
3. Tables B.12–B.23 give characteristics often needed for bundled conductor EHV lines. These tables also use the convenient form of the previous tables. They are used with permission of the Edison Electric Institute and come from [78].
4. Tables B.24 and B.25 give the 60 Hz spacing factor for inductive reactance x_d and capacitive reactance x'_d respectively for a wide range of wire spacings.

Table B.1. Base Current in Amperes

BASE KILOVOLTS	BASE MEGAVOLT-AMPERES 50.00	100.00	200.00	250.00
2.30	12551.0928	25102.1856	50204.3712	62755.4640
2.40	12028.1306	24056.2612	48112.5224	60140.6530
4.00	7216.8784	14433.7567	28867.5135	36084.3918
4.16	6939.3061	13878.6122	27757.2245	34696.5306
4.40	6560.7985	13121.5970	26243.1941	32803.9926
4.80	6014.0653	12028.1306	24056.2612	30070.3265
6.60	4373.8657	8747.7314	17495.4627	21869.3284
6.90	4183.6976	8367.3952	16734.7904	20918.4880
7.20	4009.3769	8018.7537	16037.5075	20046.8843
11.00	2624.3194	5248.6388	10497.2776	13121.5970
11.45	2521.1802	5042.3604	10084.7209	12605.9011
12.00	2405.6261	4811.2522	9622.5045	12028.1306
12.47	2314.9570	4629.9139	9259.8279	11574.7849
13.20	2186.9328	4373.8657	8747.7314	10934.6642
13.80	2091.8488	4183.6976	8367.3952	10459.2440
14.40	2004.6884	4009.3769	8018.7537	10023.4422
22.00	1312.1597	2624.3194	5248.6388	6560.7985
24.94	1157.4785	2314.9570	4629.9139	5787.3924
33.00	874.7731	1749.5463	3499.0925	4373.8657
34.50	836.7395	1673.4790	3346.9581	4183.6976
44.00	656.0799	1312.1597	2624.3194	3280.3993
55.00	524.8639	1049.7278	2099.4555	2624.3194
60.00	481.1252	962.2504	1924.5009	2405.6261
66.00	437.3866	874.7731	1749.5463	2186.9328
69.00	418.3698	836.7395	1673.4790	2091.8488
88.00	328.0399	656.0799	1312.1597	1640.1996
100.00	288.6751	577.3503	1154.7005	1443.3757
110.00	262.4319	524.8639	1049.7278	1312.1597
115.00	251.0219	502.0437	1004.0874	1255.1093
132.00	218.6933	437.3866	874.7731	1093.4664
138.00	209.1849	418.3698	836.7395	1045.9244
154.00	187.4514	374.9028	749.8055	937.2569
161.00	179.3013	358.6027	717.2053	896.5066
220.00	131.2160	262.4319	524.8639	656.0799
230.00	125.5109	251.0219	502.0437	627.5546
275.00	104.9728	209.9456	419.8911	524.8639
330.00	87.4773	174.9546	349.9093	437.3866
345.00	83.6740	167.3479	334.6958	418.3698
360.00	80.1875	160.3751	320.7501	400.9377
362.00	79.7445	159.4890	318.9780	398.7226
420.00	68.7322	137.4643	274.9287	343.6609
500.00	57.7350	115.4701	230.9401	288.6751
525.00	54.9857	109.9715	219.9430	274.9287
550.00	52.4864	104.9728	209.9456	262.4319
700.00	41.2393	82.4786	164.9572	206.1965
735.00	39.2755	78.5511	157.1021	196.3776
750.00	38.4900	76.9800	153.9601	192.4501
765.00	37.7353	75.4706	150.9412	188.6766
1000.00	28.8675	57.7350	115.4701	144.3376
1100.00	26.2432	52.4864	104.9728	131.2160
1200.00	24.0563	48.1125	96.2250	120.2813
1300.00	22.2058	44.4116	88.8231	111.0289
1400.00	20.6197	41.2393	82.4786	103.0983
1500.00	19.2450	38.4900	76.9800	96.2250

Table B.2. Base Impedance in Ohms

BASE KILOVOLTS	BASE MEGAVOLT-AMPERES 50.00	100.00	200.00	250.00
2.30	0.1058	0.0529	0.0264	0.0212
2.40	0.1152	0.0576	0.0288	0.0230
4.00	0.3200	0.1600	0.0800	0.0640
4.16	0.3461	0.1731	0.0865	0.0692
4.40	0.3872	0.1936	0.0968	0.0774
4.80	0.4608	0.2304	0.1152	0.0922
6.60	0.8712	0.4356	0.2178	0.1742
6.90	0.9522	0.4761	0.2381	0.1904
7.20	1.0368	0.5184	0.2592	0.2074
11.00	2.4200	1.2100	0.6050	0.4840
11.45	2.6221	1.3110	0.6555	0.5244
12.00	2.8800	1.4400	0.7200	0.5760
12.47	3.1100	1.5550	0.7775	0.6220
13.20	3.4848	1.7424	0.8712	0.6970
13.80	3.8088	1.9044	0.9522	0.7618
14.40	4.1472	2.0736	1.0368	0.8294
22.00	9.6800	4.8400	2.4200	1.9360
24.94	12.4401	6.2200	3.1100	2.4880
33.00	21.7800	10.8900	5.4450	4.3560
34.50	23.8050	11.9025	5.9513	4.7610
44.00	38.7200	19.3600	9.6800	7.7440
55.00	60.5000	30.2500	15.1250	12.1000
60.00	72.0000	36.0000	18.0000	14.4000
66.00	87.1200	43.5600	21.7800	17.4240
69.00	95.2200	47.6100	23.8050	19.0440
88.00	154.8800	77.4400	38.7200	30.9760
100.00	200.0000	100.0000	50.0000	40.0000
110.00	242.0000	121.0000	60.5000	48.4000
115.00	264.5000	132.2500	66.1250	52.9000
132.00	348.4800	174.2400	87.1200	69.6960
138.00	380.8800	190.4400	95.2200	76.1760
154.00	474.3200	237.1600	118.5800	94.8640
161.00	518.4200	259.2100	129.6050	103.6840
220.00	968.0000	484.0000	242.0000	193.6000
230.00	1058.0000	529.0000	264.5000	211.6000
275.00	1512.5000	756.2500	378.1250	302.5000
330.00	2178.0000	1089.0000	544.5000	435.6000
345.00	2380.5000	1190.2500	595.1250	476.1000
360.00	2592.0000	1296.0000	648.0000	518.4000
362.00	2620.8800	1310.4400	655.2200	524.1760
420.00	3528.0000	1764.0000	882.0000	705.6000
500.00	5000.0000	2500.0000	1250.0000	1000.0000
525.00	5512.5000	2756.2500	1378.1250	1102.5000
550.00	6050.0000	3025.0000	1512.5000	1210.0000
700.00	9800.0000	4900.0000	2450.0000	1960.0000
735.00	10804.5000	5402.2500	2701.1250	2160.9000
750.00	11250.0000	5625.0000	2812.5000	2250.0000
765.00	11704.5000	5852.2500	2926.1250	2340.9000
1000.00	20000.0000	10000.0000	5000.0000	4000.0000
1100.00	24200.0000	12100.0000	6050.0000	4840.0000
1200.00	28800.0000	14400.0000	7200.0000	5760.0000
1300.00	33800.0000	16900.0000	8450.0000	6760.0000
1400.00	39200.0000	19600.0000	9800.0000	7840.0000
1500.00	45000.0000	22500.0000	11250.0000	9000.0000

Table B.3. Base Admittance in Micromhos

BASE KILOVOLTS	BASE MEGAVOLT-AMPERES 50.00	100.00	200.00	250.00
2.30	9451795.8412	18903591.6824	37807183.3648	47258979.2060
2.40	8680555.5556	17361111.1111	34722222.2222	43402777.7778
4.00	3125000.0000	6250000.0000	12500000.0000	15625000.0000
4.16	2889238.1657	5778476.3314	11556952.6627	14446190.8284
4.40	2582644.6281	5165289.2562	10330578.5124	12913223.1405
4.80	2170138.8889	4340277.7778	8680555.5556	10850694.4444
6.60	1147842.0569	2295684.1139	4591368.2277	5739210.2847
6.90	1050199.5379	2100399.0758	4200798.1516	5250997.6896
7.20	964506.1728	1929012.3457	3858024.6914	4822530.8642
11.00	413223.1405	826446.2810	1652892.5620	2066115.7025
11.45	381380.9805	762761.9611	1525523.9221	1906904.9027
12.00	347222.2222	694444.4444	1388888.8889	1736111.1111
12.47	321541.5473	643083.0947	1286166.1894	1607707.7367
13.20	286960.5142	573921.0285	1147842.0569	1434802.5712
13.80	262549.8845	525099.7690	1050199.5379	1312749.4224
14.40	241126.5432	482253.0864	964506.1728	1205632.7160
22.00	103305.7851	206611.5702	413223.1405	516528.9256
24.94	80385.3868	160770.7737	321541.5473	401926.9342
33.00	45913.6823	91827.3646	183654.7291	229568.4114
34.50	42007.9815	84015.9630	168031.9261	210039.9076
44.00	25826.4463	51652.8926	103305.7851	129132.2314
55.00	16528.9256	33057.8512	66115.7025	82644.6281
60.00	13888.8889	27777.7778	55555.5556	69444.4444
66.00	11478.4206	22956.8411	45913.6823	57392.1028
69.00	10501.9954	21003.9908	42007.9815	52509.9769
88.00	6456.6116	12913.2231	25826.4463	32283.0579
100.00	5000.0000	10000.0000	20000.0000	25000.0000
110.00	4132.2314	8264.4628	16528.9256	20661.1570
115.00	3780.7183	7561.4367	15122.8733	18903.5917
132.00	2869.6051	5739.2103	11478.4206	14348.0257
138.00	2625.4988	5250.9977	10501.9954	13127.4942
154.00	2108.2813	4216.5627	8433.1253	10541.4066
161.00	1928.9379	3857.8759	7715.7517	9644.6896
220.00	1033.0579	2066.1157	4132.2314	5165.2893
230.00	945.1796	1890.3592	3780.7183	4725.8979
275.00	661.1570	1322.3140	2644.6281	3305.7851
330.00	459.1368	918.2736	1836.5473	2295.6841
345.00	420.0798	840.1596	1680.3193	2100.3991
360.00	385.8025	771.6049	1543.2099	1929.0123
362.00	381.5512	763.1025	1526.2049	1907.7562
420.00	283.4467	566.8934	1133.7868	1417.2336
500.00	200.0000	400.0000	800.0000	1000.0000
525.00	181.4059	362.8118	725.6236	907.0295
550.00	165.2893	330.5785	661.1570	826.4463
700.00	102.0408	204.0816	408.1633	510.2041
735.00	92.5540	185.1081	370.2161	462.7701
750.00	88.8889	177.7778	355.5556	444.4444
765.00	85.4372	170.8744	341.7489	427.1861
1000.00	50.0000	100.0000	200.0000	250.0000
1100.00	41.3223	82.6446	165.2893	206.6116
1200.00	34.7222	69.4444	138.8889	173.6111
1300.00	29.5858	59.1716	118.3432	147.9290
1400.00	25.5102	51.0204	102.0408	127.5510
1500.00	22.2222	44.4444	88.8889	111.1111

Table B.4. Characteristics of Copper Conductors, Hard Drawn, 97.3 Percent Conductivity

Size of Conductor: Circular Mils	Size of Conductor: A.W.G. or B. & S.	Number of Strands	Diameter of Individual Strands Inches	Outside Diameter Inches	Breaking Strength Pounds	Weight Pounds per Mile	Approx. Current Carrying Capacity* Amps	Geometric Mean Radius at 60 Cycles Feet	r_a Resistance Ohms per Conductor per Mile, 25°C. (77°F.): d-c	25°C.: 25 cycles	25°C.: 50 cycles	25°C.: 60 cycles	50°C. (122°F.): d-c	50°C.: 25 cycles	50°C.: 50 cycles	50°C.: 60 cycles	x_a Inductive Reactance Ohms per Conductor Per Mile At 1 Ft. Spacing: 25 cycles	x_a: 50 cycles	x_a: 60 cycles	x_a' Shunt Capacitive Reactance Megohms per Conductor Per Mile At 1 Ft Spacing: 25 cycles	x_a': 50 cycles	x_a': 60 cycles
1 000 000	...	37	0.1644	1.151	43 830	16 300	1 300	0.0368	0.0585	0.0594	0.0620	0.0634	0.0640	0.0648	0.0672	0.0685	0.1666	0.333	0.400	0.216	0.1081	0.0901
900 000	...	37	0.1560	1.092	39 510	14 670	1 220	0.0349	0.0650	0.0658	0.0682	0.0695	0.0711	0.0718	0.0740	0.0752	0.1693	0.339	0.406	0.220	0.1100	0.0916
800 000	...	37	0.1470	1.029	35 120	13 040	1 130	0.0329	0.0731	0.0739	0.0760	0.0772	0.0800	0.0806	0.0826	0.0837	0.1722	0.344	0.413	0.224	0.1121	0.0934
750 000	...	37	0.1424	0.997	33 400	12 230	1 090	0.0319	0.0780	0.0787	0.0807	0.0818	0.0853	0.0859	0.0878	0.0888	0.1739	0.348	0.417	0.226	0.1132	0.0943
700 000	...	37	0.1375	0.963	31 170	11 410	1 040	0.0308	0.0836	0.0842	0.0861	0.0871	0.0914	0.0920	0.0937	0.0947	0.1759	0.352	0.422	0.229	0.1145	0.0954
600 000	...	37	0.1273	0.891	27 020	9 781	940	0.0285	0.0975	0.0981	0.0997	0.1006	0.1066	0.1071	0.1086	0.1095	0.1799	0.360	0.432	0.235	0.1173	0.0977
500 000	...	37	0.1162	0.814	22 510	8 151	840	0.0260	0.1170	0.1175	0.1188	0.1196	0.1280	0.1283	0.1296	0.1303	0.1845	0.369	0.443	0.241	0.1205	0.1004
500 000	...	19	0.1622	0.811	21 590	8 151	840	0.0256	0.1170	0.1175	0.1188	0.1196	0.1280	0.1283	0.1296	0.1303	0.1853	0.371	0.445	0.241	0.1206	0.1005
450 000	...	19	0.1539	0.770	19 750	7 336	780	0.0243	0.1300	0.1304	0.1316	0.1323	0.1422	0.1426	0.1437	0.1443	0.1879	0.376	0.451	0.245	0.1224	0.1020
400 000	...	19	0.1451	0.726	17 560	6 521	730	0.0229	0.1462	0.1466	0.1477	0.1484	0.1600	0.1603	0.1613	0.1619	0.1909	0.382	0.458	0.249	0.1245	0.1038
350 000	...	19	0.1357	0.679	15 590	5 706	670	0.0214	0.1671	0.1675	0.1684	0.1690	0.1828	0.1831	0.1840	0.1845	0.1943	0.389	0.466	0.254	0.1269	0.1058
350 000	...	12	0.1708	0.710	15 140	5 706	670	0.0225	0.1671	0.1675	0.1684	0.1690	0.1828	0.1831	0.1840	0.1845	0.1918	0.384	0.460	0.251	0.1253	0.1044
300 000	...	19	0.1257	0.629	13 510	4 891	610	0.01987	0.1950	0.1953	0.1961	0.1966	0.213	0.214	0.214	0.215	0.1982	0.396	0.476	0.259	0.1296	0.1080
300 000	...	12	0.1581	0.657	13 170	4 891	610	0.0208	0.1950	0.1953	0.1961	0.1966	0.213	0.214	0.214	0.215	0.1957	0.392	0.470	0.256	0.1281	0.1068
250 000	...	19	0.1147	0.574	11 360	4 076	540	0.01813	0.234	0.234	0.235	0.235	0.256	0.256	0.257	0.257	0.203	0.406	0.487	0.266	0.1329	0.1108
250 000	....	12	0.1443	0.600	11 130	4 076	540	0.01902	0.234	0.234	0.235	0.235	0.256	0.256	0.257	0.257	0.200	0.401	0.481	0.263	0.1313	0.1094
211 600	4/0	19	0.1055	0.528	9 617	3 450	480	0.01668	0.276	0.277	0.277	0.278	0.302	0.303	0.303	0.303	0.207	0.414	0.497	0.272	0.1359	0.1132
211 600	4/0	12	0.1328	0.552	9 483	3 450	490	0.01750	0.276	0.277	0.277	0.278	0.302	0.303	0.303	0.303	0.205	0.409	0.491	0.269	0.1343	0.1119
211 600	4/0	7	0.1739	0.522	9 154	3 450	480	0.01579	0.276	0.277	0.277	0.278	0.302	0.303	0.303	0.303	0.210	0.420	0.503	0.273	0.1363	0.1136
167 800	3/0	12	0.1183	0.492	7 556	2 736	420	0.01559	0.349	0.349	0.349	0.350	0.381	0.381	0.382	0.382	0.210	0.421	0.505	0.277	0.1384	0.1153
167 800	3/0	7	0.1548	0.464	7 366	2 736	420	0.01404	0.349	0.349	0.349	0.350	0.381	0.381	0.382	0.382	0.216	0.431	0.518	0.281	0.1405	0.1171
133 100	2/0	7	0.1379	0.414	5 926	2 170	360	0.01252	0.440	0.440	0.440	0.440	0.481	0.481	0.481	0.481	0.222	0.443	0.532	0.289	0.1445	0.1205
105 500	1/0	7	0.1228	0.368	4 752	1 720	310	0.01113	0.555	0.555	0.555	0.555	0.606	0.607	0.607	0.607	0.227	0.455	0.546	0.298	0.1488	0.1240
83 690	1	7	0.1093	0.328	3 804	1 364	270	0.00992	0.699	0.699	0.699	0.699	0.765				0.233	0.467	0.560	0.306	0.1528	0.1274
83 690	1	3	0.1670	0.360	3 620	1 351	270	0.01016	0.692	0.692	0.692	0.692	0.757				0.232	0.464	0.557	0.299	0.1495	0.1246
66 370	2	7	0.0974	0.292	3 045	1 082	230	0.00883	0.881	0.882	0.882	0.882	0.964				0.239	0.478	0.574	0.314	0.1570	0.1308
66 370	2	3	0.1487	0.320	2 913	1 071	240	0.00903	0.873				0.955				0.238	0.476	0.571	0.307	0.1537	0.1281
66 370	2	1	......	0.258	3 003	1 061	220	0.00836	0.864				0.945				0.242	0.484	0.581	0.323	0.1614	0.1345
52 630	3	7	0.0867	0.260	2 433	858	200	0.00787	1.112				1.216				0.245	0.490	0.588	0.322	0.1611	0.1343
52 630	3	3	0.1325	0.285	2 359	850	200	0.00805	1.101				1.204				0.244	0.488	0.585	0.316	0.1578	0.1315
52 630	3	1	......	0.229	2 439	841	190	0.00745	1.090		Same as d-c		1.192		Same as d-c		0.248	0.496	0.595	0.331	0.1656	0.1380
41 740	4	3	0.1180	0.254	1 879	674	180	0.00717	1.388				1.518				0.250	0.499	0.599	0.324	0.1619	0.1349
41 740	4	1	......	0.204	1 970	667	170	0.00663	1.374				1.503				0.254	0.507	0.609	0.339	0.1697	0.1415
33 100	5	3	0.1050	0.226	1 505	534	150	0.00638	1.750				1.914				0.256	0.511	0.613	0.332	0.1661	0.1384
33 100	5	1	......	0.1819	1 591	529	140	0.00590	1.733				1.895				0.260	0.519	0.623	0.348	0.1738	0.1449
26 250	6	3	0.0935	0.201	1 205	424	130	0.00568	2.21				2.41				0.262	0.523	0.628	0.341	0.1703	0.1419
26 250	6	1	......	0.1620	1 280	420	120	0.00526	2.18				2.39				0.265	0.531	0.637	0.356	0.1779	0.1483
20 820	7	1	......	0.1443	1 030	333	110	0.00468	2.75				3.01				0.271	0.542	0.651	0.364	0.1821	0.1517
16 510	8	1	......	0.1285	826	264	90	0.00417	3.47				3.80				0.277	0.554	0.665	0.372	0.1862	0.1552

* For conductor at 75°C., air at 25°C., wind 1.4 miles per hour (2 ft/sec), frequency = 60 cycles.

Table B.5. Characteristics of Anaconda Hollow Copper Conductors (Anaconda Wire & Cable Company)

Design Number	Size of Conductor Circular Mils or A.W.G.	Wires: Number	Wires: Diameter Inches	Outside Diameter Inches	Breaking Strain Pounds	Weight Pounds per Mile	Geometric Mean Radius at 60 Cycles Feet	Approx. Current Carrying Capacity Amps‡	r_a Resistance Ohms per Conductor per Mile, 25°C. (77°F.) d-c 25 cycles	25°C. (77°F.) 50 cycles 60 cycles	50°C. (122°F.) d-c 25 cycles	50°C. (122°F.) 50 cycles 60 cycles	x_a Inductive Reactance Ohms per Conductor per Mile at 1 Ft. Spacing, 25 cycles	50 cycles	60 cycles	x_a' Shunt Capacitive Reactance Megohms per Conductor per Mile at 1 Ft. Spacing, 25 cycles	50 cycles	60 cycles
966	890 500	28	0.1610	1.650	36 000	15 085	0.0612	1395	0.0671	0.0676	0.0734	0.0739	0.1412	0.282	0.339	0.1907	0.0953	0.0794
96R1	750 000	42	0.1296	1.155	34 200	12 345	0.0408	1160	0.0786	0.0791	0.0860	0.0865	0.1617	0.323	0.388	0.216	0.1080	0.0900
939	650 000	50	0.1097	1.126	29 500	10 761	0.0406	1060	0.0909	0.0915	0.0994	0.1001	0.1621	0.324	0.389	0.218	0.1089	0.0908
360R1	600 000	50	0.1053	1.007	27 500	9 905	0.0387	1020	0.0984	0.0991	0.1077	0.1084	0.1644	0.329	0.395	0.221	0.1105	0.0921
938	550 000	50	0.1009	1.036	25 200	9 103	0.0373	960	0.1076	0.1081	0.1177	0.1183	0.1663	0.333	0.399	0.224	0.1119	0.0932
4R5	510 000	50	0.0970	1.000	22 700	8 485	0.0360	910	0.1173	0.1178	0.1283	0.1289	0.1681	0.336	0.404	0.226	0.1131	0.0943
892R3	500 000	18	0.1558	1.080	21 400	8 263	0.0394	900	0.1178	0.1184	0.1289	0.1296	0.1630	0.326	0.391	0.221	0.1104	0.0920
933	450 000	21	0.1353	1.074	19 300	7 476	0.0398	850	0.1319	0.1324	0.1443	0.1448	0.1630	0.326	0.391	0.221	0.1106	0.0922
924	400 000	21	0.1227	1.014	17 200	6 642	0.0376	810	0.1485	0.1491	0.1624	0.1631	0.1658	0.332	0.398	0.225	0.1126	0.0939
925R1	380 500	22	0.1211	1.003	16 300	6 331	0.0373	780	0.1565	0.1572	0.1712	0.1719	0.1663	0.333	0.399	0.226	0.1130	0.0942
565R1	350 000	21	0.1196	0.950	15 100	5 813	0.0353	750	0.1695	0.1700	0.1854	0.1860	0.1691	0.338	0.406	0.230	0.1150	0.0958
936	350 000	15	0.1444	0.860	15 400	5 776	0.0311	740	0.1690	0.1695	0.1849	0.1854	0.1754	0.351	0.421	0.237	0.1185	0.0988
378R1	350 000	30	0.1059	0.736	16 100	5 739	0.0253	700	0.1685	0.1690	0.1843	0.1849	0.1860	0.372	0.446	0.248	0.1241	0.1034
954	321 000	22	0.1113	0.920	13 850	5 343	0.0340	700	0.1851	0.1856	0.202	0.203	0.1710	0.342	0.410	0.232	0.1161	0.0968
935	300 000	18	0.1205	0.839	13 100	4 984	0.0307	670	0.1980	0.1985	0.216	0.217	0.1761	0.352	0.423	0.239	0.1194	0.0995
903R1	300 000	15	0.1338	0.797	13 200	4 953	0.0289	660	0.1969	0.1975	0.215	0.216	0.1793	0.359	0.430	0.242	0.1212	0.1010
178R2	300 000	12	0.1507	0.750	13 050	4 937	0.0266	650	0.1964	0.1969	0.215	0.216	0.1833	0.367	0.440	0.247	0.1234	0.1028
926	250 000	18	0.1100	0.766	10 950	4 155	0.0279	600	0.238	0.239	0.260	0.261	0.1810	0.362	0.434	0.245	0.1226	0.1022
915R1	250 000	15	0.1214	0.725	11 000	4 148	0.0266	590	0.237	0.238	0.259	0.260	0.1834	0.367	0.440	0.249	0.1246	0.1038
24R1	250 000	12	0.1368	0.683	11 000	4 133	0.0245	580	0.237	0.238	0.259	0.260	0.1876	0.375	0.450	0.253	0.1267	0.1066
923	4/0	18	0.1005	0.700	9 300	3 521	0.0255	530	0.281	0.282	0.307	0.308	0.1855	0.371	0.445	0.252	0.1258	0.1049
922	4/0	15	0.1109	0.663	9 300	3 510	0.0238	520	0.281	0.282	0.307	0.308	0.1889	0.378	0.453	0.256	0.1278	0.1065
50R2	4/0	14	0.1152	0.650	9 300	3 510	0.0234	520	0.280	0.281	0.306	0.307	0.1898	0.380	0.455	0.257	0.1285	0.1071
158R1	3/0	16	0.0961	0.606	7 500	2 785	0.0221	460	0.354	0.355	0.387	0.388	0.1928	0.386	0.463	0.262	0.1310	0.1091
495R1	3/0	15	0.0996	0.595	7 600	2 785	0.0214	460	0.353	0.354	0.386	0.387	0.1943	0.389	0.466	0.263	0.1316	0.1097
570R2	3/0	12	0.1123	0.560	7 600	2 772	0.0201	450	0.352	0.353	0.385	0.386	0.1976	0.395	0.474	0.268	0.1338	0.1115
909R2	2/0	15	0.0880	0.530	5 950	2 213	0.0191	370	0.446	0.446	0.487	0.487	0.200	0.400	0.481	0.271	0.1357	0.1131
412R2	2/0	14	0.0913	0.515	6 000	2 207	0.0184	370	0.446	0.446	0.487	0.487	0.202	0.404	0.485	0.274	0.1368	0.1140
937	2/0	13	0.0950	0.505	6 000	2 203	0.0181	370	0.446	0.446	0.487	0.487	0.203	0.406	0.487	0.275	0.1375	0.1146
930	125 600	14	0.0885	0.500	5 650	2 083	0.0180	360	0.473	0.473	0.517	0.517	0.203	0.406	0.487	0.276	0.1378	0.1149
934	121 300	15	0.0836	0.500	5 400	2 015	0.0179	350	0.491	0.491	0.537	0.537	0.203	0.407	0.488	0.276	0.1378	0.1149
901	119 400	12	0.0936	0.470	5 300	1 979	0.0165	340	0.507	0.507	0.555	0.555	0.207	0.415	0.498	0.280	0.1400	0.1167

‡For conductor at 75°C., air at 25°C., wind 1.4 miles per hour (2 ft/sec), frequency = 60 cycles, average tarnished surface.

Table B.6. Characteristics of General Cable Type HH Hollow Copper Conductors (General Cable Corporation)

Conductor Size Circular Mils or A. W. G.	Outside[1] Diameter Inches	Wall Thickness Inches	Weight Pounds per Mile	Breaking Strength Pounds	Geometric Mean Radius Feet	Approx. Current Carrying Capacity[2] Amps	r_a Resistance Ohms per Conductor per Mile								x_a Inductive Reactance Ohms per Conductor per Mile at 1 Foot Spacing			x_a' Shunt Capacitive Reactance Megohms per Conductor per Mile at 1 Foot Spacing		
							25°C. (77°F.)				50°C. (122°F.)									
							d-c	25 cycles	50 cycles	60 cycles	d-c	25 cycles	50 cycles	60 cycles	25 cycles	50 cycles	60 cycles	25 cycles	50 cycles	60 cycles
1 000 000	2.103	0.150*	16 160	43 190	0.0833	1620	0.0576	0.0576	0.0577	0.0577	0.0630	0.0630	0.0631	0.0631	0.1257	0.251	0.302	0.1734	0.0867	0.0722
950 000	2.035	0.147*	15 350	41 030	0.0805	1565	0.0606	0.0606	0.0607	0.0607	0.0663	0.0664	0.0664	0.0664	0.1274	0.255	0.306	0.1757	0.0879	0.0732
900 000	1.966	0.144*	14 540	38 870	0.0778	1505	0.0640	0.0640	0.0641	0.0641	0.0700	0.0701	0.0701	0.0701	0.1291	0.258	0.310	0.1782	0.0891	0.0742
850 000	1.901	0.140*	13 730	36 710	0.0751	1450	0.0677	0.0678	0.0678	0.0678	0.0741	0.0742	0.0742	0.0742	0.1309	0.262	0.314	0.1805	0.0903	0.0752
800 000	1.820	0.137*	12 920	34 550	0.0722	1390	0.0720	0.0720	0.0720	0.0721	0.0788	0.0788	0.0788	0.0788	0.1329	0.266	0.319	0.1833	0.0917	0.0764
790 000	1.650	0.131†	12 760	34 120	0.0646	1335	0.0729	0.0729	0.0730	0.0730	0.0797	0.0798	0.0799	0.0799	0.1385	0.277	0.332	0.1906	0.0953	0.0794
750 000	1.750	0.133*	12 120	32 390	0.0691	1325	0.0768	0.0768	0.0768	0.0769	0.0840	0.0840	0.0841	0.0841	0.1351	0.270	0.324	0.1864	0.0932	0.0777
700 000	1.686	0.130*	11 310	30 230	0.0665	1265	0.0822	0.0823	0.0823	0.0823	0.0900	0.0900	0.0901	0.0901	0.1370	0.274	0.329	0.1891	0.0945	0.0788
650 000	1.610	0.126*	10 500	28 070	0.0635	1200	0.0886	0.0886	0.0886	0.0887	0.0969	0.0970	0.0970	0.0970	0.1394	0.279	0.335	0.1924	0.0962	0.0802
600 000	1.558	0.123*	9 692	25 910	0.0615	1140	0.0959	0.0960	0.0960	0.0960	0.1050	0.1051	0.1051	0.1051	0.1410	0.282	0.338	0.1947	0.0974	0.0811
550 000	1.478	0.119*	8 884	23 750	0.0583	1075	0.1047	0.1048	0.1048	0.1048	0.1146	0.1146	0.1147	0.1147	0.1437	0.287	0.345	0.1985	0.0992	0.0827
512 000	1.400	0.115*	8 270	22 110	0.0551	1020	0.1124	0.1125	0.1125	0.1125	0.1230	0.1230	0.1231	0.1231	0.1466	0.293	0.352	0.202	0.1012	0.0843
500 000	1.390	0.115*	8 076	21 590	0.0547	1005	0.1151	0.1151	0.1152	0.1152	0.1259	0.1260	0.1260	0.1260	0.1469	0.294	0.353	0.203	0.1014	0.0845
500 000	1.268	0.109†	8 074	21 590	0.0494	978	0.1151	0.1152	0.1152	0.1152	0.1259	0.1260	0.1260	0.1261	0.1521	0.304	0.365	0.209	0.1047	0.0872
500 000	1.100	0.130†	8 068	21 590	0.0420	937	0.1150	0.1151	0.1152	0.1153	0.1258	0.1259	0.1260	0.1260	0.1603	0.321	0.385	0.219	0.1098	0.0915
500 000	1.020	0.144†	8 063	21 590	0.0384	915	0.1150	0.1150	0.1152	0.1152	0.1258	0.1259	0.1260	0.1261	0.1648	0.330	0.396	0.225	0.1124	0.0937
450 000	1.317	0.111*	7 268	19 430	0.0518	939	0.1279	.01280	0.1280	0.1280	0.1400	0.1401	0.1401	0.1401	0.1496	0.299	0.359	0.207	0.1033	0.0861
450 000	1.188	0.105†	7 266	19 430	0.0462	910	0.1278	0.1279	0.1279	0.1280	0.1399	0.1400	0.1400	0.1401	0.1554	0.311	0.373	0.214	0.1070	0.0892
400 000	1.218	0.106*	6 460	17 270	0.0478	864	0.1439	0.1440	0.1440	0.1440	0.1575	0.1576	0.1576	0.1576	0.1537	0.307	0.369	0.212	0.1061	0.0884
400 000	1.103	0.100†	6 458	17 270	0.0428	838	0.1438	0.1439	0.1439	0.1440	0.1574	0.1575	0.1575	0.1576	0.1593	0.319	0.382	0.219	0.1097	0.0914
350 000	1.128	0.102*	5 653	15 110	0.0443	790	0.1644	0.1645	0.1645	0.1645	0.1799	0.1800	0.1800	0.1800	0.1576	0.315	0.378	0.218	0.1089	0.0907
350 000	1.014	0.096†	5 650	15 110	0.0393	764	0.1644	0.1645	0.1645	0.1646	0.1799	0.1800	0.1800	0.1801	0.1637	0.328	0.393	0.225	0.1127	0.0939
300 000	1.020	0.096*	4 845	12 950	0.0399	709	0.1918	0.1919	0.1919	0.1919	0.210	0.210	0.210	0.210	0.1628	0.326	0.391	0.225	0.1124	0.0937
300 000	0.919	0.091†	4 843	12 950	0.0355	687	0.1917	0.1918	0.1918	0.1919	0.210	0.210	0.210	0.210	0.1688	0.338	0.405	0.232	0.1162	0.0968
250 000	0.914	0.091*	4 037	10 790	0.0357	626	0.230	0.230	0.230	0.230	0.252	0.252	0.252	0.252	0.1685	0.337	0.404	0.233	0.1163	0.0970
250 000	0.818	0.086†	4 036	10 790	0.0315	606	0.230	0.230	0.230	0.230	0.252	0.252	0.252	0.252	0.1748	0.350	0.420	0.241	0.1203	0.1002
250 000	0.766	0.094†	4 034	10 790	0.0292	594	0.230	0.230	0.230	0.230	0.252	0.252	0.252	0.252	0.1787	0.357	0.429	0.245	0.1226	0.1022
214 500	0.650	0.098†	3 459	9 265	0.0243	524	0.268	0.268	0.268	0.268	0.293	0.293	0.293	0.294	0.1879	0.376	0.451	0.257	0.1285	0.1071
4/0	0.733	0.082†	3 415	9 140	0.0281	539	0.272	0.272	0.272	0.272	0.297	0.297	0.298	0.298	0.1806	0.361	0.433	0.248	0.1242	0.1035
3/0	0.608	0.080†	2 707	7 240	0.0230	454	0.343	0.343	0.343	0.343	0.375	0.375	0.375	0.375	0.1907	0.381	0.458	0.262	0.1309	0.1091
2/0	0.500	0.080†	2 146	5 750	0.0186	382	0.432	0.432	0.432	0.432	0.472	0.473	0.473	0.473	0.201	0.403	0.483	0.276	0.1378	0.1149

Notes: *Thickness at edges of interlocked segments. †Thickness uniform throughout.

(1) Conductors of smaller diameter for given cross-sectional area also available; in the naught sizes, some additional diameter expansion is possible.

(2) For conductor at 75°C., air at 25°C., wind 1.4 miles per hour (2 ft/sec), frequency = 60 cycles.

Table B.7. Characteristics of Aluminum Conductors, Hard Drawn, 61 Percent Conductivity (Aluminum Company of America)

Size of Conductor Circular Mils or A. W. G.	No. of Strands	Diameter of Individual Strands Inches	Outside Diameter Inches	Ultimate Strength Pounds	Weight Pounds Per Mile	Geometric Mean Radius at 60 Cycles Feet	Approx. Current Carrying Capacity* Amps	r_a Resistance, Ohms per Conductor per Mile, 25°C. (77°F.) d-c	25°C. 25 cycles	25°C. 50 cycles	25°C. 60 cycles	50°C. (122°F.) d-c	50°C. 25 cycles	50°C. 50 cycles	50°C. 60 cycles	x_a Inductive Reactance Ohms per Conductor per Mile at 1 Ft. Spacing 25 cycles	x_a 50 cycles	x_a 60 cycles	x_a' Shunt Capacitive Reactance Megohms per Conductor per Mile at 1 Ft. Spacing 25 cycles	x_a' 50 cycles	x_a' 60 cycles
6	7	0.0612	0.184	528	130	0.00556	100	3.56	3.56	3.56	3.56	3.91	3.91	3.91	3.91	0.2626	0.5251	0.6301	0.3468	0.1734	0.1445
4	7	0.0772	0.232	826	207	0.00700	134	2.24	2.24	2.24	2.24	2.46	2.46	2.46	2.46	0.2509	0.5017	0.6201	0.3302	0.1651	0.1376
3	7	0.0867	0.260	1022	261	0.00787	155	1.77	1.77	1.77	1.77	1.95	1.95	1.95	1.95	0.2450	0.4899	0.5879	0.3221	0.1610	0.1342
2	7	0.0974	0.292	1266	329	0.00883	180	1.41	1.41	1.41	1.41	1.55	1.55	1.55	1.55	0.2391	0.4782	0.5739	0.3139	0.1570	0.1308
1	7	0.1094	0.328	1537	414	0.00992	209	1.12	1.12	1.12	1.12	1.23	1.23	1.23	1.23	0.2333	0.4665	0.5598	0.3055	0.1528	0.1273
1/0	7	0.1228	0.368	1865	523	0.01113	242	0.885	0.8851	0.8853	0.885	0.973	0.9731	0.9732	0.973	0.2264	0.4528	0.5434	0.2976	0.1488	0.1240
1/0	19	0.0745	0.373	2090	523	0.01177	244	0.885	0.8851	0.8853	0.885	0.973	0.9731	0.9732	0.973	0.2246	0.4492	0.5391	0.2964	0.1482	0.1235
2/0	7	0.1379	0.414	2350	659	0.01251	282	0.702	0.7021	0.7024	0.702	0.771	0.7711	0.7713	0.771	0.2216	0.4431	0.5317	0.2890	0.1445	0.1204
2/0	19	0.0837	0.419	2586	659	0.01321	283	0.702	0.7021	0.7024	0.702	0.771	0.7711	0.7713	0.771	0.2188	0.4376	0.5251	0.2882	0.1441	0.1201
3/0	7	0.1548	0.464	2845	832	0.01404	327	0.557	0.5571	0.5574	0.558	0.612	0.6121	0.6124	0.613	0.2157	0.4314	0.5177	0.2810	0.1405	0.1171
3/0	19	0.0940	0.470	3200	832	0.01483	328	0.557	0.5571	0.5574	0.558	0.612	0.6121	0.6124	0.613	0.2129	0.4258	0.5110	0.2801	0.1400	0.1167
4/0	7	0.1739	0.522	3590	1049	0.01577	380	0.441	0.4411	0.4415	0.442	0.485	0.4851	0.4855	0.486	0.2099	0.4196	0.5036	0.2726	0.1363	0.1136
4/0	19	0.1055	0.528	3890	1049	0.01666	381	0.441	0.4411	0.4415	0.442	0.485	0.4851	0.4855	0.486	0.2071	0.4141	0.4969	0.2717	0.1358	0.1132
250 000	37	0.0822	0.575	4860	1239	0.01841	425	0.374	0.3741	0.3746	0.375	0.411	0.4111	0.4115	0.412	0.2020	0.4040	0.4848	0.2657	0.1328	0.1107
266 800	7	0.1953	0.586	4525	1322	0.01771	441	0.350	0.3502	0.3506	0.351	0.385	0.3852	0.3855	0.386	0.2040	0.4079	0.4895	0.2642	0.1321	0.1101
266 800	37	0.0849	0.594	5180	1322	0.01902	443	0.350	0.3502	0.3506	0.351	0.385	0.3852	0.3855	0.386	0.2004	0.4007	0.4809	0.2633	0.1316	0.1097
300 000	19	0.1257	0.629	5300	1487	0.01983	478	0.311	0.3112	0.3117	0.312	0.342	0.3422	0.3426	0.343	0.1983	0.3965	0.4758	0.2592	0.1296	0.1080
300 000	37	0.0900	0.630	5830	1487	0.02017	478	0.311	0.3112	0.3117	0.312	0.342	0.3422	0.3426	0.343	0.1974	0.3947	0.4737	0.2592	0.1296	0.1080
336 400	19	0.1331	0.666	5940	1667	0.02100	514	0.278	0.2782	0.2788	0.279	0.306	0.3062	0.3067	0.307	0.1953	0.3907	0.4688	0.2551	0.1276	0.1063
336 400	37	0.0954	0.668	6400	1667	0.02135	514	0.278	0.2782	0.2788	0.279	0.306	0.3062	0.3067	0.307	0.1945	0.3890	0.4668	0.2549	0.1274	0.1062
350 000	37	0.0973	0.681	6680	1735	0.02178	528	0.267	0.2672	0.2678	0.268	0.294	0.2942	0.2947	0.295	0.1935	0.3870	0.4644	0.2537	0.1268	0.1057
397 500	19	0.1447	0.724	6880	1967	0.02283	575	0.235	0.2352	0.2359	0.236	0.258	0.2582	0.2589	0.259	0.1911	0.3822	0.4587	0.2491	0.1246	0.1038
477 000	19	0.1585	0.793	8090	2364	0.02501	646	0.196	0.1963	0.1971	0.198	0.215	0.2153	0.2160	0.216	0.1865	0.3730	0.4476	0.2429	0.1214	0.1012
500 000	19	0.1623	0.812	8475	2478	0.02560	664	0.187	0.1873	0.1882	0.189	0.206	0.2062	0.2070	0.208	0.1853	0.3707	0.4448	0.2412	0.1206	0.1005
500 000	37	0.1162	0.813	9010	2478	0.02603	664	0.187	0.1873	0.1882	0.189	0.206	0.2062	0.2070	0.208	0.1845	0.3689	0.4427	0.2410	0.1205	0.1004
556 500	19	0.1711	0.856	9440	2758	0.02701	710	0.168	0.1683	0.1693	0.170	0.185	0.1853	0.1862	0.187	0.1826	0.3652	0.4383	0.2374	0.1187	0.0989
636 000	37	0.1311	0.918	11240	3152	0.02936	776	0.147	0.1474	0.1484	0.149	0.162	0.1623	0.1633	0.164	0.1785	0.3569	0.4283	0.2323	0.1162	0.0968
715 500	37	0.1391	0.974	12640	3546	0.03114	817	0.137	0.1314	0.1326	0.133	0.144	0.1444	0.1455	0.146	0.1754	0.3508	0.4210	0.2282	0.1141	0.0951
750 000	37	0.1424	0.997	12980	3717	0.03188	864	0.125	0.1254	0.1267	0.127	0.137	0.1374	0.1385	0.139	0.1743	0.3485	0.4182	0.2266	0.1133	0.0944
750 000	61	0.1109	0.998	13510	3717	0.03211	864	0.125	0.1254	0.1267	0.127	0.137	0.1374	0.1385	0.139	0.1739	0.3477	0.4173	0.2263	0.1132	0.0943
795 000	37	0.1466	1.026	13770	3940	0.03283	897	0.117	0.1175	0.1188	0.120	0.129	0.1294	0.1306	0.131	0.1728	0.3455	0.4146	0.2244	0.1122	0.0935
874 500	37	0.1538	1.077	14830	4334	0.03443	949	0.107	0.1075	0.1089	0.110	0.118	0.1185	0.1198	0.121	0.1703	0.3407	0.4088	0.2210	0.1105	0.0921
954 000	37	0.1606	1.024	16180	4728	0.03596	1000	0.0979	0.0985	0.1002	0.100	0.108	0.1085	0.1100	0.111	0.1682	0.3363	0.4036	0.2179	0.1090	0.0908
1 000 000	61	0.1280	1.152	17670	4956	0.03707	1030	0.0934	0.0940	0.0956	0.0966	0.103	0.1035	0.1050	0.106	0.1666	0.3332	0.3998	0.2162	0.1081	0.0901
1 000 000	91	0.1048	1.153	18380	4956	0.03720	1030	0.0934	0.0940	0.0956	0.0966	0.103	0.1035	0.1050	0.106	0.1664	0.3328	0.3994	0.2160	0.1080	0.0900
1 033 500	37	0.1672	1.170	18260	5122	0.03743	1050	0.0904	0.0910	0.0927	0.0936	0.0994	0.0999	0.1015	0.102	0.1661	0.3322	0.3987	0.2150	0.1075	0.0896
1 113 000	61	0.1351	1.216	19660	5517	0.03910	1110	0.0839	0.0845	0.0864	0.0874	0.0922	0.0928	0.0945	0.0954	0.1639	0.3278	0.3934	0.2124	0.1062	0.0885
1 192 500	61	0.1398	1.258	21000	5908	0.04048	1160	0.0783	0.0790	0.0810	0.0821	0.0860	0.0866	0.0884	0.0895	0.1622	0.3243	0.3892	0.2100	0.1050	0.0875
1 192 500	91	0.1145	1.259	21400	5908	0.04062	1160	0.0783	0.0790	0.0810	0.0821	0.0860	0.0866	0.0884	0.0895	0.1620	0.3240	0.3888	0.2098	0.1049	0.0874
1 272 000	61	0.1444	1.300	22000	6299	0.04180	1210	0.0734	0.0741	0.0762	0.0774	0.0806	0.0813	0.0832	0.0843	0.1606	0.3211	0.3853	0.2076	0.1038	0.0865
1 351 500	61	0.1489	1.340	23400	6700	0.04309	1250	0.0691	0.0699	0.0721	0.0733	0.0760	0.0767	0.0787	0.0798	0.1590	0.3180	0.3816	0.2054	0.1027	0.0856
1 431 000	61	0.1532	1.379	24300	7091	0.04434	1300	0.0653	0.0661	0.0685	0.0697	0.0718	0.0725	0.0747	0.0759	0.1576	0.3152	0.3782	0.2033	0.1016	0.0847
1 510 500	61	0.1574	1.417	25600	7487	0.04556	1320	0.0618	0.0627	0.0651	0.0665	0.0679	0.0687	0.0710	0.0722	0.1562	0.3123	0.3748	0.2014	0.1007	0.0839
1 590 000	61	0.1615	1.454	27000	7883	0.04674	1380	0.0587	0.0596	0.0622	0.0636	0.0645	0.0653	0.0677	0.0690	0.1549	0.3098	0.3718	0.1997	0.0998	0.0832
1 590 000	91	0.1322	1.454	28100	7883	0.04691	1380	0.0587	0.0596	0.0622	0.0636	0.0645	0.0653	0.0677	0.0690	0.1547	0.3094	0.3713	0.1997	0.0998	0.0832

*For conductor at 75°C, wind 1.4 miles per hour (2 ft./sec), frequency = 60 cycles.

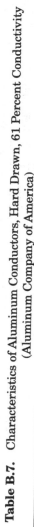

Table B.8. Characteristics of Aluminum Cable, Steel Reinforced (Aluminum Company of America)

Circular Mils or A.W.G. Aluminum	Aluminum			Steel		Outside Diameter Inches	Copper Equivalent* Circular Mils or A. W. G.	Ultimate Strength Pounds	Weight Pounds per Mile	Geometric Mean Radius at 60 Cycles Feet	Approx. Current Carrying Capacity† Amps	r_a Resistance Ohms per Conductor per Mile: 25°C. (77°F.) Small Currents				r_a Resistance: 50°C. (122°F.) Current Approx. 75% Capacity‡				x_a Inductive Reactance Ohms per Conductor per Mile at 1 Ft. Spacing All Currents			x_a' Shunt Capacitive Reactance Megohms per Conductor per Mile at 1 Ft. Spacing		
	Strands	Layers	Strand Dia. Inches	Strands	Strand Dia. Inches							d-c	25 cycles	50 cycles	60 cycles	d-c	25 cycles	50 cycles	60 cycles	25 cycles	50 cycles	60 cycles	25 cycles	50 cycles	60 cycles
1 590 000	54	3	0.1716	19	0.1030	1.545	1 000 000	56 000	10 777	0.0520	1 380	0.0587	0.0588	0.0590	0.0591	0.0646	0.0656	0.0675	0.0684	0.1495	0.299	0.359	0.1953	0.0977	0.0814
1 510 500	54	3	0.1673	19	0.1004	1.506	950 000	53 200	10 237	0.0507	1 340	0.0618	0.0619	0.0621	0.0622	0.0680	0.0690	0.0710	0.0720	0.1508	0.302	0.362	0.1971	0.0986	0.0821
1 431 000	54	3	0.1628	19	0.0977	1.465	900 000	50 400	9 699	0.0493	1 300	0.0652	0.0653	0.0655	0.0656	0.0718	0.0729	0.0749	0.0760	0.1522	0.304	0.365	0.1991	0.0996	0.0830
1 351 000	54	3	0.1582	19	0.0949	1.424	850 000	47 600	9 160	0.0479	1 250	0.0691	0.0692	0.0694	0.0695	0.0761	0.0771	0.0792	0.0803	0.1536	0.307	0.369	0.201	0.1006	0.0838
1 272 000	54	3	0.1535	19	0.0921	1.382	800 000	44 800	8 621	0.0465	1 200	0.0734	0.0735	0.0737	0.0738	0.0808	0.0819	0.0840	0.0851	0.1551	0.310	0.372	0.203	0.1016	0.0847
1 192 500	54	3	0.1486	19	0.0892	1.338	750 000	43 100	8 082	0.0450	1 160	0.0783	0.0784	0.0786	0.0788	0.0862	0.0872	0.0894	0.0906	0.1568	0.314	0.376	0.206	0.1028	0.0857
1 113 000	54	3	0.1436	19	0.0862	1.293	700 000	40 200	7 544	0.0435	1 110	0.0839	0.0840	0.0842	0.0844	0.0924	0.0935	0.0957	0.0969	0.1585	0.317	0.380	0.208	0.1040	0.0867
1 033 500	54	3	0.1384	7	0.1384	1.246	650 000	37 100	7 019	0.0420	1 060	0.0903	0.0905	0.0907	0.0908	0.0994	0.1005	0.1025	0.1035	0.1603	0.321	0.385	0.211	0.1053	0.0878
954 000	54	3	0.1329	7	0.1329	1.196	600 000	34 200	6 479	0.0403	1 010	0.0979	0.0980	0.0981	0.0982	0.1078	0.1088	0.1118	0.1128	0.1624	0.325	0.390	0.214	0.1068	0.0890
900 000	54	3	0.1291	7	0.1291	1.162	566 000	32 300	6 112	0.0391	970	0.104	0.104	0.104	0.104	0.1145	0.1155	0.1175	0.1185	0.1639	0.328	0.393	0.216	0.1078	0.0898
874 500	54	3	0.1273	7	0.1273	1.146	550 000	31 400	5 940	0.0386	950	0.107	0.107	0.107	0.108	0.1178	0.1188	0.1218	0.1228	0.1646	0.329	0.395	0.217	0.1083	0.0903
795 000	54	3	0.1214	7	0.1214	1.093	500 000	28 500	5 399	0.0368	900	0.117	0.118	0.118	0.119	0.1288	0.1308	0.1358	0.1378	0.1670	0.334	0.401	0.220	0.1100	0.0917
795 000	26	2	0.1749	7	0.1360	1.108	500 000	31 200	5 770	0.0375	900	0.117	0.117	0.117	0.117	0.1288	0.1288	0.1288	0.1288	01660	0.332	0.399	0.219	0.1095	0.0912
795 000	30	2	0.1628	19	0.0977	1.140	500 000	38 400	6 517	0.0393	910	0.117	0.117	0.117	0.117	0.1288	0.1288	0.1288	0.1288	0.1637	0.327	0.393	0.217	0.1085	0.0904
715 500	54	3	0.1151	7	0.1151	1.036	450 000	26 300	4 859	0.0349	830	0.131	0.131	0.131	0.132	0.1442	0.1452	0.1472	0.1482	0.1697	0.339	0.407	0.224	0.1119	0.0932
715 500	26	2	0.1659	7	0.1290	1.051	450 000	28 100	5 193	0.0355	840	0.131	0.131	0.131	0.131	0.1442	0.1442	0.1442	0.1442	0.1687	0.337	0.405	0.223	0.1114	0.0928
715 500	30	2	0.1544	19	0.0926	1.081	450 000	34 600	5 865	0.0372	840	0.131	0.131	0.131	0.131	0.1442	0.1442	0.1442	0.1442	0.1664	0.333	0.399	0.221	0.1104	0.0920
666 600	54	3	0.1111	7	0.1111	1.000	419 000	24 500	4 527	0.0337	800	0.140	0.140	0.141	0.141	0.1541	0.1571	0.1591	0.1601	0.1715	0.343	0.412	0.226	0.1132	0.0943
636 000	54	3	0.1085	7	0.1085	0.977	400 000	23 600	4 319	0.0329	770	0.147	0.147	0.148	0.148	0.1618	0.1638	0.1678	0.1688	0.1726	0.345	0.414	0.228	0.1140	0.0950
636 000	26	2	0.1564	7	0.1216	0.990	400 000	25 000	4 616	0.0335	780	0.147	0.147	0.147	0.147	0.1618	0.1618	0.1618	0.1618	0.1718	0.344	0.412	0.227	0.1135	0.0946
636 000	30	2	0.1456	19	0.0874	1.019	400 000	31 500	5 213	0.0351	780	0.147	0.147	0.147	0.147	0.1618	0.1618	0.1618	0.1618	0.1693	0.339	0.406	0.225	0.1125	0.0937
605 000	54	3	0.1059	7	0.1059	0.953	380 500	22 500	4 109	0.0321	750	0.154	0.155	0.155	0.155	0.1695	0.1715	0.1755	0.1775	0.1739	0.348	0.417	0.230	0.1149	0.0957
605 000	26	2	0.1525	7	0.1186	0.966	380 500	24 100	4 391	0.0327	760	0.154	0.154	0.154	0.154	0.1700	0.1720	0.1720	0.1720	0.1730	0.346	0.415	0.229	0.1144	0.0953
556 500	26	2	0.1463	7	0.1138	0.927	350 000	22 400	4 039	0.0313	730	0.168	0.168	0.168	0.168	0.1849	0.1859	0.1859	0.1859	0.1751	0.350	0.420	0.232	0.1159	0.0965
556 500	30	2	0.1362	7	0.1362	0.953	350 000	27 200	4 588	0.0328	730	0.168	0.168	0.168	0.168	0.1849	0.1859	0.1859	0.1859	0.1728	0.345	0.415	0.230	0.1149	0.0957
500 000	30	2	0.1291	7	0.1291	0.904	314 500	24 400	4 122	0.0311	690	0.187	0.187	0.187	0.187	0.206				0.1754	0.351	0.421	0.234	0.1167	0.0973
477 000	26	2	0.1355	7	0.1054	0.858	300 000	19 430	3 462	0.0290	670	0.196	0.196	0.196	0.196	0.216				0.1790	0.353	0.430	0.237	0.1186	0.0988
477 000	30	2	0.1261	7	0.1261	0.883	300 000	23 300	3 933	0.0304	670	0.196	0.196	0.196	0.196	0.216				0.1766	0.353	0.424	0.235	0.1176	0.0980
397 500	26	2	0.1236	7	0.0961	0.783	250 000	16 190	2 885	0.0265	590	0.235				0.259				0.1836	0.367	0.441	0.244	0.1219	0.1015
397 500	30	2	0.1151	7	0.1151	0.806	250 000	19 980	3 277	0.0278	600	0.235		Same as d-c		0.259		Same as d-c		0.1812	0.362	0.435	0.242	0.1208	0.1006
336 400	26	2	0.1138	7	0.0885	0.721	4/0	14 050	2 442	0.0244	530	0.278				0.306				0.1872	0.376	0.451	0.250	0.1248	0.1039
336 400	30	2	0.1059	7	0.1059	0.741	4/0	17 040	2 774	0.0255	530	0.278				0.306				0.1855	0.371	0.445	0.248	0.1238	0.1032
300 000	26	2	0.1074	7	0.0835	0.680	188 700	12 650	2 178	0.0230	490	0.311				0.342				0.1908	0.382	0.458	0.254	0.1269	0.1057
300 000	30	2	0.1000	7	0.1000	0.700	188 700	15 430	2 473	0.0241	500	0.311				0.342				0.1883	0.377	0.452	0.252	0.1258	0.1049
266 800	26	2	0.1013	7	0.0788	0.642	3/0	11 250	1 936	0.0217	460	0.350				0.385				0.1936	0.387	0.465	0.258	0.1289	0.1074

Circular Mils or A.W.G. Aluminum	Al Strands	Al Layers	Al Strand Dia. Inches	Steel Strands	Steel Strand Dia. Inches	Outside Diameter Inches	Copper Equivalent*	Ultimate Strength Pounds	Weight Pounds per Mile	Geometric Mean Radius For Current Approx. 75% Capacity ‡	Approx. Current Carrying Capacity† Amps	25°C d-c	25°C 25 cycles	25°C 50 cycles	25°C 60 cycles	50°C d-c	50°C 25 cycles	50°C 50 cycles	50°C 60 cycles	x_a Single Layer Conductors, Small Currents 25 cycles	Small Currents 50 cycles	Small Currents 60 cycles	Current Approx. 75% Capacity‡ 25 cycles	Current Approx. 75% Capacity‡ 50 cycles	Current Approx. 75% Capacity‡ 60 cycles	x_a' 25 cycles	x_a' 50 cycles	x_a' 60 cycles
266 800	6	1	0.2109	7	0.0703	0.633	3/0	9 645	1 802	0.00684	460	0.351	0.351	0.351	0.352	0.386	0.430	0.510	0.552	0.194	0.388	0.466	0.252	0.504	0.605	0.259	0.1294	0.1079
4/0	6	1	0.1878	1	0.1878	0.563	2/0	8 420	1 542	0.00814	340	0.441	0.442	0.444	0.445	0.485	0.514	0.567	0.592	0.218	0.437	0.524	0.242	0.484	0.581	0.267	0.1336	0.1113
3/0	6	1	0.1672	1	0.1672	0.502	1/0	6 675	1 223	0.00600	300	0.556	0.557	0.559	0.560	0.612	0.642	0.697	0.723	0.225	0.450	0.540	0.259	0.517	0.621	0.275	0.1377	0.1147
2/0	6	1	0.1490	1	0.1490	0.447	1	5 345	970	0.00510	270	0.702	0.702	0.704	0.706	0.773	0.806	0.866	0.895	0.231	0.462	0.554	0.267	0.534	0.641	0.284	0.1418	0.1182
1/0	6	1	0.1327	1	0.1327	0.398	2	4 280	769	0.00446	230	0.885	0.885	0.887	0.888	0.974	1.01	1.08	1.12	0.237	0.473	0.568	0.273	0.547	0.656	0.292	0.1460	0.1216
1	6	1	0.1182	1	0.1182	0.355	3	3 480	610	0.00418	200	1.12	1.12	1.12	1.12	1.23	1.27	1.34	1.38	0.242	0.483	0.580	0.277	0.554	0.665	0.300	0.1500	0.1250
2	6	1	0.1052	1	0.1052	0.316	4	2 790	484	0.00418	180	1.41	1.41	1.41	1.41	1.55	1.59	1.66	1.69	0.247	0.493	0.592	0.277	0.554	0.665	0.308	0.1542	0.1285
2	7	1	0.0974	1	0.1299	0.325	4	3 525	566	0.00504	180	1.41	1.41	1.41	1.41	1.55	1.59	1.62	1.65	0.247	0.493	0.592	0.267	0.535	0.642	0.306	0.1532	0.1276
3	6	1	0.0937	1	0.0937	0.281	5	2 250	384	0.00430	160	1.78	1.78	1.78	1.78	1.95	1.95	2.04	2.07	0.252	0.503	0.604	0.275	0.551	0.661	0.317	0.1583	0.1320
4	6	1	0.0834	1	0.0834	0.250	6	1 830	304	0.00437	140	2.24	2.24	2.24	2.24	2.47	2.50	2.54	2.57	0.257	0.514	0.611	0.274	0.549	0.659	0.325	0.1627	0.1355
4	7	1	0.0772	1	0.1029	0.257	6	2 288	356	0.00452	140	2.24	2.24	2.24	2.24	2.47	2.50	2.53	2.55	0.257	0.515	0.618	0.273	0.545	0.655	0.323	0.1615	0.1346
5	6	1	0.0743	1	0.0743	0.223	7	1 460	241	0.00416	120	2.82	2.82	2.82	2.82	3.10	3.12	3.16	3.18	0.262	0.525	0.630	0.279	0.557	0.665	0.333	0.1666	0.1388
6	6	1	0.0661	1	0.0661	0.198	8	1 170	191	0.00394	100	3.56	3.56	3.56	3.56	3.92	3.94	3.97	3.98	0.268	0.536	0.643	0.281	0.561	0.673	0.342	0.1708	0.1423

* Based on copper 97 percent, aluminum 61 percent conductivity. † For conductor at 75°C., air at 25°C., wind 1.4 miles per hour (2 ft/sec), frequency = 60 cycles
‡ "Current Approx. 75% Capacity" is 75% of the "Approx. Current Carrying Capacity in Amps." and is approximately the current which will produce 50°C. conductor temp. (25°C. rise) with 25°C. air temp., wind 1.4 miles per hour.

Table B.9. Characteristics of "Expanded" Aluminum Cable, Steel Reinforced (Aluminum Company of America)

Circular Mils or A.W.G. Aluminum	Aluminum			Steel		Filler Section Aluminum		Filler Section Paper		Outside Diameter Inches	Copper Equivalent Circular Mils or A. W. G.	Ultimate Strength Pounds	Weight Pounds per Mile	Geometric Mean Radius at 60 Cycles Feet	Approx. Current Carrying Capacity Amps	r_a Resistance Ohms per Conductor per Mile: 25°C. (77°F.) Small Currents				50°C. (122°F.) Current Approx. 75% Capacity‡				x_a Inductive Reactance Ohms per Conductor per Mile at 1 Ft. Spacing All Currents			x_a' Shunt Capacitive Reactance Megohms per Conductor per Mile at 1 Ft. Spacing		
	Strands	Layers	Strand Dia. Inches	Strands	Strand Dia. Inches	Strands	Strand Dia. Inches	Strands	Layers							d-c	25 cycles	50 cycles	60 cycles	d-c	25 cycles	50 cycles	60 cycles	25 cycles	50 cycles	60 cycles	25 cycles	50 cycles	60 cycles
850 000	54	2	0.1255	19	0.0834	4	0.1182	23	2	1.38	534 000	35 371	7 200	(1)	(1)	(1)				(1)				(1)	(1)	(1)	(1)	(1)	(1)
1 150 000	54	2	0.1409	19	0.0921	4	0.1353	24	2	1.55	724 000	41 900	9 070																
1 338 000	66	2	0.1350	19	0.100	4	0.184	18	2	1.75	840 000	49 278	11 340																

(1) Electrical Characteristics not available until laboratory measurements are completed.

Table B.10. Characteristics of Copperweld-Copper Conductors (Copperweld Steel Company)

Size of Conductor: Nominal Designation	Size of Conductor: Number and Diameter of Wires: Copperweld	Size of Conductor: Number and Diameter of Wires: Copper	Outside Diameter Inches	Copper Equivalent Circular Mils or A.W.G.	Rated Breaking Load Lbs.	Weight Lbs. per Mile	Geometric Mean Radius at 60 Cycles Feet	Approx. Current Carrying Capacity at 60 Cycles Amps*	r_a Resistance Ohms per Conductor per Mile at 25°C. (77°F.) Small Currents: d-c	25 cycles	50 cycles	60 cycles	r_a Resistance Ohms per Conductor per Mile at 50°C. (122°F.) Current Approx. 75% of Capacity**: d-c	25 cycles	50 cycles	60 cycles	x_a Inductive Reactance Ohms per Conductor per Mile One ft. Spacing Average Currents: 25 cycles	50 cycles	60 cycles	x_a' Capacitive Reactance Megohms per Conductor per Mile One ft. Spacing: 25 cycles	50 cycles	60 cycles
350 E	7x.1576″	12x.1576″	0.788	350 000	32 420	7 409	0.0220	660	0.1658	0.1728	0.1789	0.1812	0.1812	0.1915	0.201	0.204	0.1929	0.386	0.463	0.243	0.1216	0.1014
350 EK	4x.1470″	15x.1470″	0.735	350 000	23 850	6 536	0.0245	680	0.1658	0.1682	0.1700	0.1705	0.1812	0.1845	0.1873	0.1882	0.1875	0.375	0.450	0.248	0.1241	0.1034
350 V	3x.1751″	9x.1893″	0.754	350 000	23 480	6 578	0.0226	650	0.1655	0.1725	0.1800	0.1828	0.1809	0.1910	0.202	0.206	0.1915	0.383	0.460	0.246	0.1232	0.1027
300 E	7x.1459″	12x.1459″	0.729	300 000	27 770	6 351	0.0204	600	0.1934	0.200	0.207	0.209	0.211	0.222	0.232	0.235	0.1969	0.394	0.473	0.249	0.1244	0.1037
300 EK	4x.1361″	15x.1361″	0.680	300 000	20 960	5 602	0.0227	610	0.1934	0.1958	0.1976	0.198	0.211	0.215	0.218	0.219	0.1914	0.383	0.460	0.254	0.1269	0.1057
300 V	3x.1621″	9x.1752″	0.698	300 000	20 730	5 639	0.0209	590	0.1930	0.200	0.208	0.210	0.211	0.222	0.233	0.237	0.1954	0.391	0.469	0.252	0.1259	0.1050
250 E	7x.1332″	12x.1332″	0.666	250 000	23 920	5 292	0.01859	540	0.232	0.239	0.245	0.248	0.254	0.265	0.275	0.279	0.202	0.403	0.484	0.255	0.1276	0.1604
250 EK	4x.1242″	15x.1242″	0.621	250 000	17 840	4 669	0.0207	540	0.232	0.235	0.236	0.237	0.254	0.258	0.261	0.261	0.1960	0.392	0.471	0.260	0.1301	0.1084
250 V	3x.1480″	9x.1600″	0.637	250 000	17 420	4 699	0.01911	530	0.232	0.239	0.246	0.249	0.253	0.264	0.276	0.281	0.200	0.400	0.480	0.258	0.1292	0.1077
4/0 E	7x.1225″	12x.1225″	0.613	4/0	20 730	4 479	0.01711	480	0.274	0.281	0.287	0.290	0.300	0.312	0.323	0.326	0.206	0.411	0.493	0.261	0.1306	0.1088
4/0 G	2x.1944″	5x.1944″	0.583	4/0	15 640	4 168	0.01409	460	0.273	0.284	0.294	0.298	0.299	0.318	0.336	0.342	0.215	0.431	0.517	0.265	0.1324	0.1103
4/0 EK	4x.1143″	15x.1143″	0.571	4/0	15 370	3 951	0.01903	490	0.274	0.277	0.278	0.279	0.300	0.304	0.307	0.308	0.200	0.401	0.481	0.266	0.1331	0.1109
4/0 V	3x.1361″	9x.1472″	0.586	4/0	15 000	3 977	0.01758	470	0.274	0.281	0.288	0.291	0.299	0.311	0.323	0.328	0.204	0.409	0.490	0.264	0.1322	0.1101
4/0 F	1x.1833″	6x.1833″	0.550	4/0	12 290	3 750	0.01558	470	0.273	0.280	0.285	0.287	0.299	0.309	0.318	0.322	0.210	0.421	0.505	0.269	0.1344	0.1220
3/0 E	7x.1091″	12x.1091″	0.545	3/0	16 800	3 552	0.01521	420	0.346	0.353	0.359	0.361	0.378	0.391	0.402	0.407	0.212	0.423	0.508	0.270	0.1348	0.1123
3/0 J	3x.1851″	4x.1851″	0.555	3/0	16 170	3 732	0.01156	410	0.344	0.356	0.367	0.372	0.377	0.398	0.419	0.428	0.225	0.451	0.541	0.268	0.1341	0.1118
3/0 G	2x.1731″	2x.1731″	0.519	3/0	12 860	3 305	0.01254	400	0.344	0.355	0.365	0.369	0.377	0.397	0.416	0.423	0.221	0.443	0.531	0.273	0.1365	0.1137
3/0 EK	4x.1018″	4x.1018″	0.509	3/0	12 370	3 134	0.01697	420	0.346	0.348	0.350	0.351	0.378	0.382	0.386	0.386	0.206	0.412	0.495	0.274	0.1372	0.1143
3/0 V	3x.1311″	9x.1311″	0.522	3/0	12 220	3 154	0.01566	410	0.345	0.352	0.360	0.362	0.377	0.390	0.403	0.406	0.210	0.420	0.504	0.273	0.1363	0.1136
3/0 F	1x.1632″	6x.1632″	0.490	3/0	9 980	2 974	0.01388	410	0.344	0.351	0.356	0.358	0.377	0.388	0.397	0.401	0.216	0.432	0.519	0.277	0.1385	0.1155
2/0 K	4x.1780″	3x.1780″	0.534	2/0	17 600	3 411	0.00912	360	0.434	0.447	0.459	0.466	0.475	0.499	0.524	0.535	0.237	0.475	0.570	0.271	0.1355	0.1129
2/0 J	3x.1648″	4x.1648″	0.494	2/0	13 430	2 960	0.01029	350	0.434	0.446	0.457	0.462	0.475	0.498	0.520	0.530	0.231	0.463	0.555	0.277	0.1383	0.1152
2/0 G	2x.1542″	5x.1542″	0.463	2/0	10 510	2 622	0.01119	350	0.434	0.445	0.456	0.459	0.475	0.497	0.518	0.525	0.227	0.454	0.545	0.281	0.1406	0.1171
2/0 V	3x.1080″	9x.1167″	0.465	2/0	9 846	2 502	0.01395	360	0.435	0.442	0.450	0.452	0.476	0.489	0.504	0.509	0.216	0.432	0.518	0.281	0.1404	0.1170
2/0 F	1x.1454″	6x.1454″	0.436	2/0	8 094	2 359	0.01235	350	0.434	0.441	0.446	0.448	0.475	0.487	0.497	0.501	0.222	0.444	0.533	0.285	0.1427	0.1189
1/0 K	4x.1585″	3x.1585″	0.475	1/0	14 490	2 703	0.00812	310	0.548	0.560	0.573	0.579	0.599	0.625	0.652	0.664	0.243	0.487	0.584	0.279	0.1397	0.1164
1/0 J	3x.1467″	4x.1467″	0.440	1/0	10 970	2 346	0.00917	310	0.548	0.559	0.570	0.576	0.599	0.624	0.648	0.659	0.237	0.474	0.569	0.285	0.1423	0.1186
1/0 G	2x.1373″	5x.1373″	0.412	1/0	8 563	2 078	0.00996	310	0.548	0.559	0.568	0.573	0.599	0.623	0.645	0.654	0.233	0.466	0.559	0.289	0.1447	0.1206
1/0 F	1x.1294″	6x.1294″	0.388	1/0	6 536	1 870	0.01099	310	0.548	0.554	0.559	0.562	0.599	0.612	0.622	0.627	0.228	0.456	0.547	0.294	0.1469	0.1224
1 N	5x.1546″	2x.1546″	0.464	1	15 410	2 541	0.00638	280	0.691	0.705	0.719	0.726	0.755	0.787	0.818	0.832	0.256	0.512	0.614	0.281	0.1405	0.1171
1 K	4x.1412″	3x.1412″	0.423	1	11 900	2 144	0.00723	270	0.691	0.704	0.716	0.722	0.755	0.784	0.813	0.825	0.249	0.498	0.598	0.288	0.1438	0.1198
1 J	3x.1307″	4x.1307″	0.392	1	9 000	1 861	0.00817	270	0.691	0.703	0.714	0.719	0.755	0.783	0.806	0.820	0.243	0.486	0.583	0.293	0.1465	0.1221
1 G	2x.1222″	5x.1222″	0.367	1	6 956	1 649	0.00887	260	0.691	0.702	0.712	0.716	0.755	0.781	0.805	0.815	0.239	0.478	0.573	0.298	0.1488	0.1240
1 F	1x.1153″	6x.1153″	0.346	1	5 266	1 483	0.00980	270	0.691	0.698	0.704	0.705	0.755	0.769	0.781	0.786	0.234	0.468	0.561	0.302	0.1509	0.1258
2 P	6x.1540″	1x.1540″	0.462	2	16 870	2 487	0.00501	250	0.871	0.886	0.901	0.909	0.952	0.988	1.024	1.040	0.268	0.536	0.643	0.281	0.1406	0.1172
2 N	5x.1377″	2x.1377″	0.413	2	12 680	2 015	0.00568	240	0.871	0.885	0.899	0.906	0.952	0.986	1.020	1.035	0.261	0.523	0.627	0.289	0.1446	0.1205
2 K	4x.1257″	3x.1257″	0.377	2	9 730	1 701	0.00644	240	0.871	0.884	0.896	0.902	0.952	0.983	1.014	1.028	0.255	0.510	0.612	0.296	0.1479	0.1232
2 J	3x.1164″	4x.1164″	0.349	2	7 322	1 476	0.00727	230	0.871	0.883	0.894	0.899	0.952	0.982	1.010	1.022	0.249	0.498	0.598	0.301	0.1506	0.1255
2 A	1x.1699″	2x.1699″	0.366	2	5 876	1 356	0.00763	240	0.869	0.875	0.880	0.882	0.950	0.962	0.973	0.979	0.247	0.493	0.592	0.298	0.1489	0.1241
2 G	2x.1089″	5x.1089″	0.327	2	5 626	1 307	0.00790	230	0.871	0.882	0.892	0.896	0.952	0.980	1.006	1.016	0.245	0.489	0.587	0.306	0.1529	0.1275
2 F	1x.1026″	6x.1026″	0.308	2	4 233	1 176	0.00873	230	0.871	0.878	0.884	0.885	0.952	0.967	0.979	0.985	0.230	0.479	0.575	0.310	0.1551	0.1292
P	6x.1371″	1x.1371″	0.411	3	13 910	1 973	0.00445	220	1.098	1.113	1.127	1.136	1.200	1.239	1.273	1.296	0.274	0.547	0.657	0.290	0.1448	0.1207
N	5x.1226″	2x.1226″	0.368	3	10 390	1 598	0.00506	210	1.098	1.112	1.126	1.133	1.200	1.237	1.273	1.289	0.267	0.534	0.641	0.298	0.1487	0.1239
K	4x.1120″	3x.1120″	0.336	3	7 910	1 349	0.00574	210	1.098	1.111	1.123	1.129	1.200	1.233	1.267	1.281	0.261	0.522	0.626	0.304	0.1520	0.1266
J	3x.1036″	4x.1036″	0.311	3	5 955	1 171	0.00648	200	1.098	1.110	1.121	1.126	1.200	1.232	1.262	1.275	0.255	0.509	0.611	0.309	0.1547	0.1289
A	1x.1513″	2x.1513″	0.326	3	4 810	1 075	0.00679	210	1.096	1.102	1.107	1.109	1.198	1.211	1.225	1.229	0.252	0.505	0.606	0.306	0.1531	0.1275
P	6x.1221″	1x.1221″	0.366	4	11 420	1 564	0.00397	190	1.385	1.400	1.414	1.423	1.514	1.555	1.598	1.616	0.280	0.559	0.671	0.298	0.1489	0.1241
N	5x.1092″	2x.1092″	0.328	4	8 460	1 267	0.00451	180	1.385	1.399	1.413	1.420	1.514	1.554	1.593	1.610	0.273	0.546	0.655	0.306	0.1528	0.1274
D	2x.1615″	1x.1615″	0.348	4	7 340	1 191	0.00566	190	1.382	1.389	1.396	1.399	1.511	1.529	1.544	1.542	0.262	0.523	0.628	0.301	0.1507	0.1256
A	1x.1347″	2x.1347″	0.290	4	3 938	853	0.00604	180	1.382	1.388	1.393	1.395	1.511	1.525	1.540	1.545	0.258	0.517	0.620	0.314	0.1572	0.1310
P	6x.1087″	1x.1087″	0.326	5	9 311	1 240	0.00353	160	1.747	1.762	1.776	1.785	1.909	1.954	2.00	2.02	0.285	0.571	0.685	0.306	0.1531	0.1275
D	2x.1438″	1x.1438″	0.310	5	6 035	944	0.00504	160	1.742	1.749	1.756	1.759	1.905	1.924	1.941	1.939	0.268	0.535	0.642	0.310	0.1548	0.1290
A	1x.1200″	2x.1200″	0.258	5	3 193	676	0.00538	160	1.742	1.748	1.753	1.755	1.905	1.920	1.936	1.941	0.264	0.528	0.634	0.323	0.1614	0.1345
D	2x.1281″	1x.1281″	0.276	6	4 942	749	0.00449	140	2.20	2.21	2.21	2.22	2.40	2.42	2.44	2.44	0.273	0.547	0.656	0.318	0.1590	0.1325
A	1x.1068″	2x.1068″	0.230	6	2 585	536	0.00479	140	2.20	2.20	2.21	2.21	2.40	2.42	2.44	2.44	0.270	0.540	0.648	0.331	0.1655	0.1379
C	1x.1046″	2x.1046″	0.225	6	2 143	514	0.00469	130	2.20	2.20	2.21	2.21	2.40	2.42	2.44	2.44	0.271	0.542	0.651	0.333	0.1663	0.1386
D	2x.1141″	1x.1141″	0.246	7	4 022	594	0.00400	120	2.77	2.78	2.79	2.79	3.03	3.05	3.07	3.07	0.279	0.558	0.670	0.326	0.1631	0.1359
A	1x.1266″	2x.0895″	0.223	7	2 754	495	0.00441	120	2.77	2.78	2.78	2.78	3.03	3.05	3.07	3.07	0.274	0.548	0.658	0.333	0.1666	0.1388
D	2x.1016″	1x.1016″	0.219	8	3 256	471	0.00356	110	3.49	3.50	3.51	3.51	3.82	3.84	3.86	3.86	0.285	0.570	0.684	0.334	0.1672	0.1393
A	1x.1127″	2x.0797″	0.199	8	2 233	392	0.00394	100	3.49	3.50	3.51	3.51	3.82	3.84	3.86	3.87	0.280	0.560	0.672	0.341	0.1706	0.1422
C	1x.0808″	2x.0834″	0.179	8	1 362	320	0.00373	100	3.49	3.50	3.51	3.51	3.82	3.84	3.86	3.86	0.283	0.565	0.679	0.349	0.1744	0.1453
½ D	2x.0808″	1x.0808″	0.174	9½	1 743	298	0.00283	85	4.91	4.92	4.92	4.93	5.37	5.39	5.42	5.42	0.297	0.598	0.712	0.351	0.1754	0.1462

*Based on a conductor temperature of 75°C. and an ambient of 25°C., wind 1.4 miles per hour (2 ft/sec.), frequency = 60 cycles, average tarnished surface.

**Resistances at 50°C. total temperature, based on an ambient of 25°C. plus 25°C, rise due to heating effect of current. The approximate magnitude of current necessary to produce the 25°C. rise is 75% of the "Approximate Current Carrying Capacity at 60 cycles."

Table B.11. Characteristics of Copperweld Conductors (Copperweld Steel Company)

Nominal Conductor Size	Number and Size of Wires	Outside Diameter Inches	Area of Conductor Circular Mils	Rated Breaking Load Pounds: High Strength	Rated Breaking Load Pounds: Extra High Strength	Weight Pounds per Mile	Geometric Mean Radius at 60 Cycles and Average Currents Feet	Approx. Current Carrying Capacity* Amps at 60 Cycles	r_a Resistance Ohms per Conductor per Mile at 25°C. (77°F.) Small Currents: d-c	25 cycles	50 cycles	60 cycles	r_a Resistance Ohms per Conductor per Mile at 75°C. (167°F.) Current Approx. 75% of Capacity**: d-c	25 cycles	50 cycles	60 cycles	x_a Inductive Reactance Ohms per Conductor per Mile One Ft. Spacing Average Currents: 25 cycles	50 cycles	60 cycles	x_a' Capacitive Reactance Megohms per Conductor per Mile One Ft. Spacing: 25 cycles	50 cycles	60 cycles
30% Conductivity																						
7/8"	19 No. 5	0.910	628 900	55 570	66 910	9 344	0.00758	620	0.306	0.316	0.326	0.331	0.363	0.419	0.476	0.499	0.261	0.493	0.592	0.233	0.1165	0.0971
13/16"	19 No. 6	0.810	498 800	45 830	55 530	7 410	0.00675	540	0.386	0.396	0.406	0.411	0.458	0.518	0.580	0.605	0.267	0.505	0.606	0.241	0.1206	0.1005
23/32"	19 No. 7	0.721	395 500	37 740	45 850	5 877	0.00601	470	0.486	0.496	0.506	0.511	0.577	0.643	0.710	0.737	0.273	0.517	0.621	0.250	0.1248	0.1040
21/32"	19 No. 8	0.642	313 700	31 040	37 690	4 660	0.00535	410	0.613	0.623	0.633	0.638	0.728	0.799	0.872	0.902	0.279	0.529	0.635	0.258	0.1289	0.1074
9/16"	19 No. 9	0.572	248 800	25 500	30 610	3 696	0.00477	360	0.773	0.783	0.793	0.798	0.917	0.995	1.075	1.106	0.285	0.541	0.649	0.266	0.1330	0.1109
5/8"	7 No. 4	0.613	292 200	24 780	29 430	4 324	0.00511	410	0.656	0.664	0.672	0.676	0.778	0.824	0.870	0.887	0.281	0.533	0.640	0.261	0.1306	0.1088
9/16"	7 No. 5	0.546	231 700	20 470	24 650	3 429	0.00455	360	0.827	0.835	0.843	0.847	0.981	1.030	1.080	1.099	0.287	0.545	0.654	0.269	0.1347	0.1122
1/2"	7 No. 6	0.486	183 800	16 890	20 460	2 719	0.00405	310	1.042	1.050	1.058	1.062	1.237	1.290	1.343	1.364	0.293	0.557	0.668	0.278	0.1388	0.1157
7/16"	7 No. 7	0.433	145 700	13 910	16 890	2 157	0.00361	270	1.315	1.323	1.331	1.335	1.560	1.617	1.675	1.697	0.299	0.569	0.683	0.286	0.1429	0.1191
3/8"	7 No. 8	0.385	115 600	11 440	13 890	1 710	0.00321	230	1.658	1.666	1.674	1.678	1.967	2.03	2.09	2.12	0.305	0.581	0.697	0.294	0.1471	0.1226
11/32"	7 No. 9	0.343	91 650	9 393	11 280	1 356	0.00286	200	2.09	2.10	2.11	2.11	2.48	2.55	2.61	2.64	0.311	0.592	0.711	0.303	0.1512	0.1260
5/16"	7 No. 10	0.306	72 680	7 758	9 196	1 076	0.00255	170	2.64	2.64	2.65	2.66	3.13	3.20	3.27	3.30	0.316	0.604	0.725	0.311	0.1553	0.1294
3 No. 5	3 No. 5	0.392	99 310	9 262	11 860	1 467	0.00457	220	1.926	1.931	1.936	1.938	2.29	2.31	2.34	2.35	0.289	0.545	0.654	0.293	0.1465	0.1221
3 No. 6	3 No. 6	0.349	78 750	7 639	9 754	1 163	0.00407	190	2.43	2.43	2.44	2.44	2.88	2.91	2.94	2.95	0.295	0.556	0.668	0.301	0.1506	0.1255
3 No. 7	3 No. 7	0.311	62 450	6 291	7 922	922.4	0.00363	160	3.06	3.07	3.07	3.07	3.63	3.66	3.70	3.71	0.301	0.568	0.682	0.310	0.1547	0.1289
3 No. 8	3 No. 8	0.277	49 530	5 174	6 282	731.5	0.00323	140	3.86	3.87	3.87	3.87	4.58	4.61	4.65	4.66	0.307	0.580	0.696	0.318	0.1589	0.1324
3 No. 9	3 No. 9	0.247	39 280	4 250	5 129	580.1	0.00288	120	4.87	4.87	4.88	4.88	5.78	5.81	5.85	5.86	0.313	0.591	0.710	0.326	0.1629	0.1358
3 No. 10	3 No. 10	0.220	31 150	3 509	4 160	460.0	0.00257	110	6.14	6.14	6.15	6.15	7.28	7.32	7.36	7.38	0.319	0.603	0.724	0.334	0.1671	0.1392
40% Conductivity																						
7/8"	19 No. 5	0.910	628 900	50 240	......	9 344	0.01175	690	0.229	0.239	0.249	0.254	0.272	0.321	0.371	0.391	0.236	0.449	0.539	0.233	0.1165	0.0971
13/16"	19 No. 6	0.810	498 800	41 600	......	7 410	0.01046	610	0.289	0.299	0.309	0.314	0.343	0.396	0.450	0.472	0.241	0.461	0.553	0.241	0.1206	0.1005
23/32"	19 No. 7	0.721	395 500	34 390	......	5 877	0.00931	530	0.365	0.375	0.385	0.390	0.433	0.490	0.549	0.573	0.247	0.473	0.567	0.250	0.1248	0.1040
21/32"	19 No. 8	0.642	313 700	28 380	......	4 660	0.00829	470	0.460	0.470	0.480	0.485	0.546	0.608	0.672	0.698	0.253	0.485	0.582	0.258	0.1289	0.1074
9/16"	19 No. 9	0.572	248 800	23 390	......	3 696	0.00739	410	0.580	0.590	0.600	0.605	0.688	0.756	0.826	0.753	0.259	0.496	0.595	0.266	0.1330	0.1109
5/8"	7 No. 4	0.613	292 200	22 310	......	4 324	0.00792	470	0.492	0.500	0.508	0.512	0.584	0.624	0.664	0.680	0.255	0.489	0.587	0.261	0.1306	0.1088
9/16"	7 No. 5	0.546	231 700	18 510	......	3 429	0.00705	410	0.620	0.628	0.636	0.640	0.736	0.780	0.843	0.840	0.261	0.501	0.601	0.269	0.1347	0.1122
1/2"	7 No. 6	0.486	183 800	15 330	......	2 719	0.00628	350	0.782	0.790	0.798	0.802	0.928	0.975	1.021	1.040	0.267	0.513	0.615	0.278	0.1388	0.1157
7/16"	7 No. 7	0.433	145 700	12 670	......	2 157	0.00559	310	0.986	0.994	1.002	1.006	1.170	1.220	1.271	1.291	0.273	0.524	0.629	0.286	0.1429	0.1191
3/8"	7 No. 8	0.385	115 600	10 460	......	1 710	0.00497	270	1.244	1.252	1.260	1.264	1.476	1.530	1.584	1.606	0.279	0.536	0.644	0.294	0.1471	0.1226
11/32"	7 No. 9	0.343	91 650	8 616	......	1 356	0.00443	230	1.568	1.576	1.584	1.588	1.861	1.919	1.978	2.00	0.285	0.548	0.658	0.303	0.1512	0.1260
5/16"	7 No. 10	0.306	72 680	7 121	......	1 076	0.00395	200	1.978	1.986	1.994	1.998	2.35	2.41	2.47	2.50	0.291	0.559	0.671	0.311	0.1553	0.1294
3 No. 5	3 No. 5	0.392	99 310	8 373	......	1 467	0.00621	250	1.445	1.450	1.455	1.457	1.714	1.738	1.762	1.772	0.269	0.514	0.617	0.293	0.1465	0.1221
3 No. 6	3 No. 6	0.349	78 750	6 934	......	1 163	0.00553	220	1.821	1.826	1.831	1.833	2.16	2.19	2.21	2.22	0.275	0.526	0.631	0.301	0.1506	0.1255
3 No. 7	3 No. 7	0.311	62 450	5 732	......	922.4	0.00492	190	2.30	2.30	2.31	2.31	2.73	2.75	2.78	2.79	0.281	0.537	0.645	0.310	0.1547	0.1289
3 No. 8	3 No. 8	0.277	49 530	4 730	......	731.5	0.00439	160	2.90	2.90	2.91	2.91	3.44	3.47	3.50	3.51	0.286	0.549	0.659	0.318	0.1589	0.1324
3 No. 9	3 No. 9	0.247	39 280	3 898	......	580.1	0.00391	140	3.65	3.66	3.66	3.66	4.33	4.37	4.40	4.41	0.292	0.561	0.673	0.326	0.1629	0.1358
3 No. 10	3 No. 10	1.220	31 150	3 221	......	460.0	0.00348	120	4.61	4.61	4.62	4.62	5.46	5.50	5.53	5.55	0.297	0.572	0.687	0.334	0.1671	0.1392
3 No. 12	3 No. 12	0.174	19 590	2 236	......	289.3	0.00276	90	7.32	7.33	7.33	7.34	8.69	8.73	8.77	8.78	0.310	0.596	0.715	0.351	0.1754	0.1462

*Based on conductor temperature of 125°C. and an ambient of 25°C.

**Resistance at 75°C. total temperature, based on an ambient of 25°C. plus 50°C. rise due to heating effect of current. The approximate magnitude of current necessary to produce the 50°C. rise is 75% of the "Approximate Current Carrying Capacity at 60 Cycles."

Table B.12. Inductive Reactance of ACSR Bundled Conductors at 60 Hz with 1, 2, and 3 Conductors per Bundle

CODE	AREA CMIL	STRANDS AL	STRANDS ST	DIA. IN.	GMR FT.	60 HZ INDUCTIVE REACTANCE [1] x_a IN OHMS/MILE FOR 1 FOOT RADIUS: SINGLE COND.	2 - CONDUCTOR SPACING (IN.) 6	2 - CONDUCTOR SPACING (IN.) 9	2 - CONDUCTOR SPACING (IN.) 12	2 - CONDUCTOR SPACING (IN.) 15	2 - CONDUCTOR SPACING (IN.) 18	3 - CONDUCTOR SPACING (IN.) 6	3 - CONDUCTOR SPACING (IN.) 9	3 - CONDUCTOR SPACING (IN.) 12	3 - CONDUCTOR SPACING (IN.) 15	3 - CONDUCTOR SPACING (IN.) 18
EXPANDED	3108000	62/8	19	2.500	0.0900	0.2922	0.1881	0.1635	0.1461	0.1326	0.1215	0.1535	0.1207	0.0974	0.0793	0.0646
EXPANDED	2294000	66/6	19	2.320	0.0858	0.2980	0.1910	0.1664	0.1490	0.1355	0.1244	0.1554	0.1226	0.0993	0.0813	0.0665
EXPANDED	1414000	58/4	19	1.750	0.0640	0.3336	0.2088	0.1842	0.1668	0.1532	0.1422	0.1673	0.1345	0.1112	0.0931	0.0784
EXPANDED	1275000	50/4	19	1.600	0.0578	0.3459	0.2150	0.1904	0.1730	0.1594	0.1484	0.1714	0.1386	0.1153	0.0973	0.0825
KIWI	2167000	72	7	1.737	0.0571	0.3474	0.2158	0.1912	0.1737	0.1602	0.1491	0.1719	0.1391	0.1158	0.0977	0.0830
BLUEBIRD	2156000	84	19	1.762	0.0588	0.3438	0.2140	0.1894	0.1719	0.1584	0.1473	0.1707	0.1379	0.1146	0.0966	0.0818
CHUKAR	1780000	84	19	1.602	0.0536	0.3551	0.2196	0.1950	0.1775	0.1640	0.1529	0.1744	0.1416	0.1184	0.1003	0.0856
FALCON	1590000	54	19	1.545	0.0523	0.3580	0.2211	0.1965	0.1790	0.1655	0.1544	0.1754	0.1426	0.1193	0.1013	0.0866
LAPWING	1590000	45	7	1.502	0.0498	0.3640	0.2241	0.1995	0.1820	0.1685	0.1574	0.1774	0.1446	0.1213	0.1033	0.0885
PARROT	1510500	54	19	1.506	0.0506	0.3621	0.2231	0.1985	0.1810	0.1675	0.1564	0.1768	0.1440	0.1207	0.1026	0.0879
NUTHATCH	1510500	45	7	1.466	0.0486	0.3670	0.2255	0.2009	0.1835	0.1699	0.1589	0.1784	0.1456	0.1223	0.1043	0.0895
PLOVER	1431000	54	19	1.465	0.0494	0.3650	0.2245	0.1999	0.1825	0.1689	0.1579	0.1777	0.1449	0.1217	0.1036	0.0889
BOBOLINK	1431000	45	7	1.427	0.0470	0.3710	0.2276	0.2030	0.1855	0.1720	0.1609	0.1797	0.1469	0.1237	0.1056	0.0909
MARTIN	1351500	54	19	1.424	0.0482	0.3680	0.2260	0.2014	0.1840	0.1704	0.1594	0.1787	0.1459	0.1227	0.1046	0.0899
DIPPER	1351500	45	7	1.385	0.0459	0.3739	0.2290	0.2044	0.1869	0.1734	0.1623	0.1807	0.1479	0.1246	0.1066	0.0918
PHEASANT	1272000	54	19	1.382	0.0466	0.3721	0.2281	0.2035	0.1860	0.1725	0.1614	0.1801	0.1473	0.1240	0.1060	0.0912
BITTERN	1272000	45	7	1.345	0.0444	0.3779	0.2310	0.2064	0.1890	0.1754	0.1644	0.1820	0.1492	0.1260	0.1079	0.0932
GRACKLE	1192500	54	19	1.333	0.0451	0.3760	0.2301	0.2055	0.1880	0.1745	0.1634	0.1814	0.1486	0.1253	0.1073	0.0925
BUNTING	1192500	45	7	1.302	0.0429	0.3821	0.2331	0.2085	0.1910	0.1775	0.1664	0.1834	0.1506	0.1274	0.1093	0.0946
FINCH	1113000	54	19	1.293	0.0436	0.3801	0.2321	0.2075	0.1901	0.1765	0.1655	0.1828	0.1500	0.1267	0.1087	0.0939
BLUEJAY	1113000	45	7	1.259	0.0415	0.3861	0.2351	0.2105	0.1931	0.1795	0.1685	0.1848	0.1520	0.1287	0.1107	0.0959
CURLEW	1033500	54	7	1.246	0.0420	0.3847	0.2344	0.2098	0.1923	0.1788	0.1677	0.1843	0.1515	0.1282	0.1102	0.0954
ORTOLAN	1033500	45	7	1.213	0.0402	0.3900	0.2370	0.2124	0.1950	0.1815	0.1704	0.1861	0.1533	0.1300	0.1119	0.0972
TANAGER	1033500	36	1	1.186	0.0384	0.3955	0.2398	0.2152	0.1978	0.1842	0.1732	0.1879	0.1551	0.1318	0.1138	0.0990
CARDINAL	954000	54	7	1.196	0.0402	0.3900	0.2370	0.2124	0.1950	0.1815	0.1704	0.1861	0.1533	0.1300	0.1119	0.0972
RAIL	954000	45	7	1.165	0.0386	0.3949	0.2395	0.2149	0.1975	0.1839	0.1729	0.1877	0.1549	0.1318	0.1136	0.0988
CATBIRD	954000	36	1	1.140	0.0370	0.4000	0.2421	0.2175	0.2000	0.1865	0.1754	0.1894	0.1566	0.1333	0.1153	0.1005
CANARY	900000	54	7	1.162	0.0392	0.3930	0.2386	0.2140	0.1965	0.1830	0.1719	0.1871	0.1543	0.1310	0.1130	0.0982
RUDDY	900000	45	7	1.131	0.0374	0.3987	0.2414	0.2168	0.1994	0.1858	0.1748	0.1890	0.1562	0.1329	0.1149	0.1001
MALLARD	795000	30	19	1.140	0.0392	0.3930	0.2386	0.2140	0.1965	0.1830	0.1719	0.1871	0.1543	0.1310	0.1130	0.0982
DRAKE	795000	26	7	1.108	0.0373	0.3991	0.2416	0.2170	0.1995	0.1860	0.1749	0.1891	0.1563	0.1330	0.1150	0.1002
CONDOR	795000	54	7	1.093	0.0370	0.4000	0.2421	0.2175	0.2000	0.1865	0.1754	0.1894	0.1566	0.1333	0.1153	0.1005
CUCKOO	795000	24	7	1.092	0.0366	0.4014	0.2427	0.2181	0.2007	0.1871	0.1761	0.1899	0.1571	0.1338	0.1157	0.1010
TERN	795000	45	7	1.063	0.0352	0.4061	0.2451	0.2205	0.2030	0.1895	0.1784	0.1914	0.1586	0.1354	0.1173	0.1026
COOT	795000	36	1	1.040	0.0377	0.3978	0.2409	0.2163	0.1989	0.1853	0.1743	0.1887	0.1559	0.1326	0.1145	0.0998
REDWING	715500	30	19	1.081	0.0373	0.3991	0.2416	0.2170	0.1995	0.1860	0.1749	0.1891	0.1563	0.1330	0.1150	0.1002
STARLING	715500	26	7	1.051	0.0355	0.4051	0.2446	0.2200	0.2025	0.1890	0.1779	0.1911	0.1583	0.1350	0.1170	0.1022
STILT	715500	24	7	1.036	0.0347	0.4078	0.2460	0.2214	0.2039	0.1904	0.1793	0.1920	0.1592	0.1359	0.1179	0.1031
GANNET	666600	26	7	1.014	0.0343	0.4092	0.2467	0.2221	0.2046	0.1911	0.1800	0.1925	0.1597	0.1364	0.1184	0.1036
FLAMINGO	666600	24	7	1.000	0.0355	0.4121	0.2481	0.2235	0.2061	0.1925	0.1815	0.1934	0.1606	0.1374	0.1193	0.1046
-------	653900	18	3	0.953	0.0308	0.4223	0.2532	0.2286	0.2111	0.1976	0.1865	0.1968	0.1640	0.1408	0.1227	0.1080
EGRET	636000	30	19	1.019	0.0352	0.4061	0.2451	0.2205	0.2030	0.1895	0.1784	0.1914	0.1586	0.1354	0.1173	0.1026
GROSBEAK	636000	26	7	0.990	0.0335	0.4121	0.2481	0.2235	0.2061	0.1925	0.1815	0.1934	0.1606	0.1374	0.1193	0.1046
ROOK	636000	24	7	0.977	0.0327	0.4150	0.2496	0.2250	0.2075	0.1940	0.1829	0.1944	0.1616	0.1383	0.1203	0.1055
KINGBIRD	636000	18	1	0.940	0.0304	0.4239	0.2540	0.2294	0.2119	0.1984	0.1873	0.1974	0.1646	0.1413	0.1232	0.1085
SWIFT	636000	36	1	0.930	0.0301	0.4251	0.2546	0.2300	0.2125	0.1990	0.1879	0.1978	0.1650	0.1417	0.1236	0.1089
TEAL	605000	30	19	0.994	0.0341	0.4099	0.2470	0.2224	0.2050	0.1914	0.1804	0.1927	0.1599	0.1366	0.1186	0.1038
SQUAB	605000	26	7	0.966	0.0327	0.4150	0.2496	0.2250	0.2075	0.1940	0.1829	0.1944	0.1616	0.1383	0.1203	0.1055
PEACOCK	605000	24	7	0.953	0.0319	0.4180	0.2511	0.2265	0.2090	0.1955	0.1844	0.1954	0.1626	0.1393	0.1213	0.1065
EAGLE	556500	30	7	0.953	0.0327	0.4150	0.2496	0.2250	0.2075	0.1940	0.1829	0.1944	0.1616	0.1383	0.1203	0.1055
DOVE	556500	26	7	0.927	0.0314	0.4200	0.2520	0.2274	0.2100	0.1964	0.1854	0.1961	0.1633	0.1400	0.1219	0.1072
PARAKEET	556500	24	7	0.914	0.0306	0.4231	0.2536	0.2290	0.2115	0.1980	0.1869	0.1971	0.1643	0.1410	0.1230	0.1082
OSPREY	556500	18	1	0.879	0.0284	0.4321	0.2581	0.2335	0.2161	0.2025	0.1915	0.2001	0.1673	0.1440	0.1260	0.1112
HEN	477000	30	7	0.883	0.0304	0.4239	0.2540	0.2294	0.2119	0.1984	0.1873	0.1974	0.1646	0.1413	0.1232	0.1085
HAWK	477000	26	7	0.858	0.0289	0.4300	0.2571	0.2325	0.2150	0.2015	0.1904	0.1994	0.1666	0.1433	0.1253	0.1105
FLICKER	477000	24	7	0.846	0.0284	0.4321	0.2581	0.2335	0.2161	0.2025	0.1915	0.2001	0.1673	0.1440	0.1260	0.1112
PELICAN	477000	18	1	0.814	0.0264	0.4410	0.2626	0.2380	0.2205	0.2070	0.1959	0.2031	0.1703	0.1470	0.1289	0.1142
LARK	397500	30	7	0.806	0.0277	0.4352	0.2596	0.2350	0.2176	0.2040	0.1930	0.2011	0.1683	0.1451	0.1270	0.1123
IBIS	397500	26	7	0.783	0.0264	0.4410	0.2626	0.2380	0.2205	0.2070	0.1959	0.2031	0.1703	0.1470	0.1289	0.1142
BRANT	397500	24	7	0.772	0.0258	0.4438	0.2639	0.2394	0.2219	0.2084	0.1973	0.2040	0.1712	0.1479	0.1299	0.1151
CHICKADEE	397500	18	1	0.743	0.0241	0.4521	0.2681	0.2435	0.2260	0.2125	0.2014	0.2068	0.1740	0.1507	0.1326	0.1179
ORIOLE	336400	30	7	0.741	0.0255	0.4452	0.2647	0.2401	0.2226	0.2091	0.1980	0.2045	0.1717	0.1484	0.1304	0.1156
LINNET	336400	26	7	0.721	0.0243	0.4511	0.2676	0.2430	0.2255	0.2120	0.2009	0.2064	0.1736	0.1504	0.1323	0.1176
MERLIN	336400	18	1	0.684	0.0222	0.4620	0.2731	0.2485	0.2310	0.2175	0.2064	0.2101	0.1773	0.1540	0.1360	0.1212
OSTRICH	300000	26	7	0.680	0.0229	0.4583	0.2712	0.2466	0.2291	0.2156	0.2045	0.2068	0.1760	0.1528	0.1347	0.1203

Table B.13. Inductive Reactance of ACSR Bundled Conductors at 60 Hz with 4 and 6 Conductors per Bundle

CODE	AREA CMIL	STRANDS AL	STRANDS ST	DIA. IN.	GMR FT.	60 Hz INDUCTIVE REACTANCE [1] X_A IN OHMS/MILE FOR 1 FOOT RADIUS — 4 - CONDUCTOR SPACING (IN.) 6	9	12	15	18	6 - CONDUCTOR SPACING (IN.) 6	9	12	15	18
EXPANDED	3108000	62/8	19	2.500	0.0900	0.1256	0.0887	0.0625	0.0422	0.0256	0.0826	0.0416	0.0125	-0.0101	-0.0285
EXPANDED	2294000	66/6	19	2.320	0.0858	0.1271	0.0902	0.0640	0.0437	0.0271	0.0835	0.0425	0.0134	-0.0091	-0.0276
EXPANDED	1414000	58/4	19	1.750	0.0640	0.1360	0.0991	0.0729	0.0526	0.0360	0.0894	0.0484	0.0194	-0.0032	-0.0216
EXPANDED	1275000	50/4	19	1.600	0.0578	0.1390	0.1021	0.0760	0.0557	0.0391	0.0915	0.0505	0.0214	-0.0011	-0.0196
KIWI	2167000	72	7	1.737	0.0571	0.1394	0.1025	0.0763	0.0560	0.0394	0.0918	0.0508	0.0217	-0.0009	-0.0193
BLUEBIRD	2156000	84	19	1.762	0.0588	0.1385	0.1016	0.0754	0.0551	0.0385	0.0912	0.0502	0.0211	-0.0015	-0.0199
CHUKAR	1780000	84	19	1.602	0.0536	0.1413	0.1044	0.0783	0.0579	0.0414	0.0930	0.0520	0.0229	0.0004	-0.0181
FALCON	1590000	54	19	1.545	0.0523	0.1421	0.1052	0.0790	0.0587	0.0421	0.0935	0.0525	0.0234	0.0009	-0.0176
LAPWING	1590000	45	7	1.502	0.0498	0.1436	0.1067	0.0805	0.0602	0.0436	0.0945	0.0535	0.0244	0.0019	-0.0166
PARROT	1510500	54	19	1.506	0.0506	0.1431	0.1062	0.0800	0.0597	0.0431	0.0942	0.0532	0.0241	0.0015	-0.0169
NUTHATCH	1510500	45	7	1.466	0.0486	0.1443	0.1074	0.0812	0.0609	0.0443	0.0950	0.0540	0.0249	0.0024	-0.0161
PLOVER	1431000	54	19	1.465	0.0494	0.1438	0.1069	0.0807	0.0604	0.0438	0.0947	0.0537	0.0246	0.0020	-0.0164
BOBOLINK	1431000	45	7	1.427	0.0470	0.1453	0.1084	0.0822	0.0619	0.0453	0.0957	0.0547	0.0256	0.0030	-0.0154
MARTIN	1351500	54	19	1.424	0.0482	0.1446	0.1077	0.0815	0.0612	0.0446	0.0952	0.0542	0.0251	0.0025	-0.0159
DIPPER	1351500	45	7	1.385	0.0459	0.1460	0.1091	0.0830	0.0627	0.0461	0.0962	0.0552	0.0261	0.0035	-0.0149
PHEASANT	1272000	54	19	1.382	0.0466	0.1456	0.1087	0.0825	0.0622	0.0456	0.0959	0.0549	0.0258	0.0032	-0.0152
BITTERN	1272000	45	7	1.345	0.0444	0.1470	0.1101	0.0840	0.0637	0.0471	0.0968	0.0558	0.0268	0.0042	-0.0142
GRACKLE	1192500	54	19	1.333	0.0451	0.1466	0.1097	0.0835	0.0632	0.0466	0.0965	0.0555	0.0264	0.0039	-0.0146
BUNTING	1192500	45	7	1.302	0.0429	0.1481	0.1112	0.0850	0.0647	0.0481	0.0975	0.0565	0.0274	0.0049	-0.0136
FINCH	1113000	54	19	1.293	0.0436	0.1476	0.1107	0.0845	0.0642	0.0476	0.0972	0.0562	0.0271	0.0046	-0.0139
BLUEJAY	1113000	45	7	1.259	0.0415	0.1491	0.1122	0.0860	0.0657	0.0491	0.0982	0.0572	0.0281	0.0056	-0.0129
CURLEW	1033500	54	7	1.246	0.0420	0.1487	0.1118	0.0857	0.0653	0.0488	0.0980	0.0570	0.0279	0.0053	-0.0131
ORTOLAN	1033500	45	7	1.213	0.0402	0.1501	0.1132	0.0870	0.0667	0.0501	0.0989	0.0579	0.0288	0.0062	-0.0122
TANAGER	1033500	36	1	1.186	0.0384	0.1515	0.1146	0.0884	0.0681	0.0515	0.0998	0.0588	0.0297	0.0071	-0.0113
CARDINAL	954000	54	7	1.196	0.0402	0.1501	0.1132	0.0870	0.0667	0.0501	0.0989	0.0579	0.0288	0.0062	-0.0122
RAIL	954000	45	7	1.165	0.0386	0.1513	0.1144	0.0882	0.0679	0.0513	0.0997	0.0587	0.0296	0.0070	-0.0114
CATBIRD	954000	36	1	1.140	0.0370	0.1526	0.1157	0.0895	0.0692	0.0526	0.1005	0.0595	0.0304	0.0079	-0.0106
CANARY	900000	54	7	1.162	0.0392	0.1508	0.1139	0.0877	0.0674	0.0508	0.0994	0.0584	0.0293	0.0067	-0.0117
RUDDY	900000	45	7	1.131	0.0374	0.1523	0.1154	0.0892	0.0689	0.0523	0.1003	0.0593	0.0302	0.0077	-0.0108
MALLARD	795000	30	19	1.140	0.0392	0.1508	0.1139	0.0877	0.0674	0.0508	0.0994	0.0584	0.0293	0.0067	-0.0117
DRAKE	795000	26	7	1.108	0.0373	0.1523	0.1154	0.0893	0.0689	0.0524	0.1004	0.0594	0.0303	0.0077	-0.0107
CONDOR	795000	54	7	1.093	0.0370	0.1526	0.1157	0.0895	0.0692	0.0526	0.1005	0.0595	0.0304	0.0079	-0.0106
CUCKOO	795000	24	7	1.092	0.0366	0.1529	0.1160	0.0898	0.0695	0.0529	0.1007	0.0597	0.0307	0.0081	-0.0103
TERN	795000	45	7	1.063	0.0352	0.1541	0.1172	0.0910	0.0707	0.0541	0.1015	0.0605	0.0314	0.0089	-0.0096
COOT	795000	36	1	1.040	0.0377	0.1520	0.1151	0.0889	0.0686	0.0520	0.1001	0.0591	0.0301	0.0075	-0.0109
REDWING	715500	30	19	1.081	0.0373	0.1523	0.1154	0.0893	0.0689	0.0524	0.1004	0.0594	0.0303	0.0077	-0.0107
STARLING	715500	26	7	1.051	0.0355	0.1538	0.1169	0.0908	0.0704	0.0539	0.1014	0.0604	0.0313	0.0087	-0.0097
STILT	715500	24	7	1.036	0.0347	0.1545	0.1176	0.0914	0.0711	0.0545	0.1018	0.0608	0.0317	0.0092	-0.0093
GANNET	666600	26	7	1.014	0.0343	0.1549	0.1180	0.0918	0.0715	0.0549	0.1021	0.0611	0.0320	0.0094	-0.0090
FLAMINGO	666600	24	7	1.000	0.0355	0.1556	0.1187	0.0925	0.0722	0.0556	0.1025	0.0615	0.0324	0.0099	-0.0086
--------	653900	18	3	0.953	0.0308	0.1581	0.1212	0.0951	0.0748	0.0582	0.1042	0.0632	0.0341	0.0116	-0.0069
EGRET	636000	30	19	1.019	0.0352	0.1541	0.1172	0.0910	0.0707	0.0541	0.1015	0.0605	0.0314	0.0089	-0.0096
GROSBEAK	636000	26	7	0.990	0.0335	0.1556	0.1187	0.0925	0.0722	0.0556	0.1025	0.0615	0.0324	0.0099	-0.0086
ROOK	636000	24	7	0.977	0.0327	0.1563	0.1194	0.0932	0.0729	0.0563	0.1030	0.0620	0.0329	0.0104	-0.0081
KINGBIRD	636000	18	1	0.940	0.0304	0.1585	0.1216	0.0955	0.0752	0.0586	0.1045	0.0635	0.0344	0.0118	-0.0066
SWIFT	636000	36	1	0.930	0.0301	0.1588	0.1219	0.0958	0.0755	0.0589	0.1047	0.0637	0.0346	0.0120	-0.0064
TEAL	605000	30	19	0.994	0.0341	0.1551	0.1182	0.0920	0.0717	0.0551	0.1022	0.0612	0.0321	0.0095	-0.0089
SQUAB	605000	26	7	0.966	0.0327	0.1563	0.1194	0.0932	0.0729	0.0563	0.1030	0.0620	0.0329	0.0104	-0.0081
PEACOCK	605000	24	7	0.953	0.0319	0.1571	0.1202	0.0940	0.0737	0.0571	0.1035	0.0625	0.0334	0.0109	-0.0076
EAGLE	556500	30	7	0.953	0.0327	0.1563	0.1194	0.0932	0.0729	0.0563	0.1030	0.0620	0.0329	0.0104	-0.0081
DOVE	556500	26	7	0.927	0.0314	0.1576	0.1207	0.0945	0.0742	0.0576	0.1038	0.0628	0.0338	0.0112	-0.0072
PARAKEET	556500	24	7	0.914	0.0306	0.1583	0.1214	0.0953	0.0750	0.0584	0.1044	0.0634	0.0343	0.0117	-0.0067
OSPREY	556500	18	1	0.879	0.0284	0.1606	0.1237	0.0975	0.0772	0.0606	0.1059	0.0649	0.0358	0.0132	-0.0052
HEN	477000	30	7	0.883	0.0304	0.1585	0.1216	0.0955	0.0752	0.0586	0.1045	0.0635	0.0344	0.0118	-0.0066
HAWK	477000	26	7	0.858	0.0289	0.1601	0.1232	0.0970	0.0767	0.0601	0.1055	0.0645	0.0354	0.0129	-0.0056
FLICKER	477000	24	7	0.846	0.0284	0.1606	0.1237	0.0975	0.0772	0.0606	0.1059	0.0649	0.0358	0.0132	-0.0052
PELICAN	477000	18	1	0.814	0.0264	0.1628	0.1259	0.0997	0.0794	0.0628	0.1074	0.0664	0.0373	0.0147	-0.0037
LARK	397500	30	7	0.806	0.0277	0.1614	0.1245	0.0983	0.0780	0.0614	0.1064	0.0654	0.0363	0.0137	-0.004
IBIS	397500	26	7	0.783	0.0264	0.1628	0.1259	0.0997	0.0794	0.0628	0.1074	0.0664	0.0373	0.0147	-0.003
BRANT	397500	24	7	0.772	0.0258	0.1635	0.1266	0.1004	0.0801	0.0635	0.1078	0.0668	0.0377	0.0152	-0.003
CHICKADEE	397500	18	1	0.743	0.0241	0.1656	0.1287	0.1025	0.0822	0.0656	0.1092	0.0682	0.0391	0.0165	-0.001
ORIOLE	336400	30	7	0.741	0.0255	0.1639	0.1270	0.1008	0.0805	0.0639	0.1081	0.0671	0.0380	0.0154	-0.003
LINNET	336400	26	7	0.721	0.0243	0.1653	0.1284	0.1023	0.0819	0.0654	0.1090	0.0680	0.0389	0.0164	-0.002
MERLIN	336400	18	1	0.684	0.0222	0.1681	0.1312	0.1050	0.0847	0.0681	0.1109	0.0699	0.0408	0.0182	-0.000
OSTRICH	300000	26	7	0.680	0.0229	0.1671	0.1302	0.1041	0.0837	0.0672	0.1102	0.0692	0.0401	0.0176	-0.000

Table B.14. Inductive Reactance of ACAR Bundled Conductors at 60 Hz with 1, 2, and 3 Conductors per Bundle

AREA 62% EQ. EC - AL CMIL	STRANDS EC/6201		DIA. IN.	GMR FT.	60 Hz INDUCTIVE REACTANCE [1] X_A IN OHMS/MILE FOR 1 FOOT RADIUS										
					SINGLE COND.	2 - CONDUCTOR SPACING (IN.)					3 - CONDUCTOR SPACING (IN.)				
						6	9	12	15	18	6	9	12	15	18
2413000	72	19	1.821	0.0596	0.3422	0.2132	0.1886	0.1711	0.1576	0.1465	0.1701	0.1373	0.1141	0.0960	0.0813
2375000	63	28	1.821	0.0596	0.3422	0.2132	0.1886	0.1711	0.1576	0.1465	0.1701	0.1373	0.1141	0.0960	0.0813
2338000	54	37	1.821	0.0599	0.3416	0.2128	0.1882	0.1708	0.1573	0.1462	0.1699	0.1371	0.1139	0.0958	0.0811
2297000	54	7	1.762	0.0571	0.3474	0.2158	0.1912	0.1737	0.1602	0.1491	0.1719	0.1391	0.1158	0.0977	0.0830
2262000	48	13	1.762	0.0578	0.3459	0.2150	0.1904	0.1730	0.1594	0.1484	0.1714	0.1386	0.1153	0.0973	0.0825
2226000	42	19	1.762	0.0578	0.3459	0.2150	0.1904	0.1730	0.1594	0.1484	0.1714	0.1386	0.1153	0.0973	0.0825
2227000	54	7	1.735	0.0561	0.3495	0.2168	0.1922	0.1748	0.1612	0.1502	0.1726	0.1398	0.1165	0.0985	0.0837
2193000	48	13	1.735	0.0530	0.3564	0.2203	0.1957	0.1782	0.1647	0.1536	0.1749	0.1421	0.1188	0.1008	0.0860
2159000	42	19	1.735	0.0568	0.3480	0.2161	0.1915	0.1740	0.1605	0.1494	0.1721	0.1393	0.1160	0.0980	0.0832
1899000	54	7	1.602	0.0519	0.3590	0.2215	0.1969	0.1795	0.1660	0.1549	0.1757	0.1429	0.1197	0.1016	0.0869
1870000	48	13	1.602	0.0490	0.3660	0.2250	0.2004	0.1830	0.1694	0.1584	0.1781	0.1453	0.1220	0.1039	0.0892
1841000	42	19	1.602	0.0526	0.3574	0.2207	0.1961	0.1787	0.1651	0.1541	0.1752	0.1424	0.1191	0.1011	0.0863
1673000	54	7	1.504	0.0486	0.3670	0.2255	0.2009	0.1835	0.1699	0.1589	0.1784	0.1456	0.1223	0.1043	0.0895
1647000	48	13	1.504	0.0461	0.3734	0.2287	0.2041	0.1867	0.1731	0.1621	0.1805	0.1477	0.1245	0.1064	0.0917
1622000	42	19	1.504	0.0495	0.3647	0.2244	0.1998	0.1824	0.1688	0.1578	0.1776	0.1448	0.1216	0.1035	0.0888
1337000	54	7	1.345	0.0436	0.3801	0.2321	0.2075	0.1901	0.1765	0.1655	0.1828	0.1500	0.1267	0.1087	0.0939
1296000	42	19	1.345	0.0440	0.3790	0.2316	0.2070	0.1895	0.1760	0.1649	0.1824	0.1496	0.1263	0.1083	0.0935
243000	30	7	1.302	0.0421	0.3844	0.2342	0.2096	0.1922	0.1786	0.1676	0.1842	0.1514	0.1281	0.1101	0.0953
211000	24	13	1.302	0.0417	0.3855	0.2348	0.2102	0.1928	0.1792	0.1682	0.1846	0.1518	0.1285	0.1105	0.0957
179000	18	19	1.302	0.0426	0.3829	0.2335	0.2089	0.1915	0.1779	0.1669	0.1837	0.1509	0.1276	0.1096	0.0948
163000	30	7	1.259	0.0407	0.3885	0.2363	0.2117	0.1942	0.1807	0.1696	0.1856	0.1528	0.1295	0.1114	0.0967
133000	24	13	1.259	0.0408	0.3882	0.2361	0.2115	0.1941	0.1806	0.1695	0.1855	0.1527	0.1294	0.1113	0.0966
104000	18	19	1.259	0.0412	0.3870	0.2356	0.2110	0.1935	0.1800	0.1689	0.1851	0.1523	0.1290	0.1109	0.0962
153000	33	4	1.246	0.0401	0.3903	0.2372	0.2126	0.1951	0.1816	0.1705	0.1862	0.1534	0.1301	0.1120	0.0973
138000	30	7	1.246	0.0403	0.3897	0.2369	0.2123	0.1948	0.1813	0.1702	0.1860	0.1532	0.1299	0.1118	0.0971
109000	24	13	1.246	0.0405	0.3891	0.2366	0.2120	0.1945	0.1810	0.1699	0.1858	0.1530	0.1297	0.1116	0.0969
080000	18	19	1.246	0.0407	0.3885	0.2363	0.2117	0.1942	0.1807	0.1696	0.1856	0.1528	0.1295	0.1114	0.0967
077000	30	7	1.212	0.0393	0.3927	0.2384	0.2138	0.1964	0.1828	0.1718	0.1870	0.1542	0.1309	0.1129	0.0981
049000	24	13	1.212	0.0389	0.3940	0.2390	0.2144	0.1970	0.1834	0.1724	0.1874	0.1546	0.1313	0.1133	0.0985
022000	18	19	1.212	0.0396	0.3918	0.2380	0.2134	0.1959	0.1824	0.1713	0.1867	0.1539	0.1306	0.1125	0.0978
050000	30	7	1.196	0.0388	0.3943	0.2392	0.2146	0.1971	0.1836	0.1725	0.1875	0.1547	0.1314	0.1134	0.0986
023000	24	13	1.196	0.0384	0.3955	0.2398	0.2152	0.1978	0.1842	0.1732	0.1879	0.1551	0.1318	0.1138	0.0990
996000	18	19	1.196	0.0391	0.3933	0.2387	0.2141	0.1967	0.1831	0.1721	0.1872	0.1544	0.1311	0.1131	0.0983
994800	30	7	1.165	0.0376	0.3981	0.2411	0.2165	0.1990	0.1855	0.1744	0.1888	0.1560	0.1327	0.1146	0.0999
954600	30	7	1.141	0.0369	0.4004	0.2422	0.2176	0.2002	0.1866	0.1756	0.1895	0.1567	0.1335	0.1154	0.1007
969300	24	13	1.165	0.0374	0.3987	0.2414	0.2168	0.1994	0.1858	0.1748	0.1890	0.1562	0.1329	0.1149	0.1001
958000	24	13	1.158	0.0371	0.3997	0.2419	0.2173	0.1999	0.1863	0.1753	0.1893	0.1565	0.1332	C.1152	0.1004
943900	18	19	1.165	0.0381	0.3965	0.2403	0.2157	0.1982	0.1847	0.1736	0.1882	0.1554	0.1322	0.1141	0.0994
900300	30	7	1.108	0.0358	0.4040	0.2441	0.2195	0.2020	0.1885	0.1774	0.1908	0.1580	0.1347	0.1166	0.1019
795000	30	7	1.042	0.0334	0.4125	0.2483	0.2237	0.2062	0.1927	0.1816	0.1936	0.1608	0.1375	0.1194	0.1047
877300	24	13	1.108	0.0355	0.4051	0.2446	0.2200	0.2025	0.1890	0.1779	0.1911	0.1583	0.1350	0.1170	0.1022
795000	24	13	1.055	0.0339	0.4107	0.2474	0.2228	0.2053	0.1918	0.1807	0.1930	0.1602	0.1369	0.1188	0.1041
854200	18	19	1.108	0.0361	0.4030	0.2436	0.2190	0.2015	0.1880	0.1769	0.1904	0.1576	0.1343	0.1163	0.1015
795000	18	19	1.069	0.0349	0.4071	0.2456	0.2210	0.2036	0.1900	0.1790	0.1918	0.1590	0.1357	0.1177	0.1029
829000	30	7	1.063	0.0343	0.4092	0.2467	0.2221	0.2046	0.1911	0.1800	0.1925	0.1597	0.1364	0.1184	0.1036
807700	24	13	1.063	0.0342	0.4096	0.2468	0.2222	0.2048	0.1913	0.1802	0.1926	0.1598	0.1365	0.1185	0.1037
786500	18	19	1.063	0.0348	0.4075	0.2458	0.2212	0.2037	0.1902	0.1791	0.1919	0.1591	0.1358	0.1178	0.1030
727500	33	4	0.990	0.0319	0.4180	0.2511	0.2265	0.2090	0.1955	0.1844	0.1954	0.1626	0.1393	0.1213	0.1065
718300	30	7	0.990	0.0320	0.4177	0.2509	0.2263	0.2088	0.1953	0.1842	0.1953	0.1625	0.1392	0.1212	0.1064
700000	24	13	0.990	0.0317	0.4188	0.2515	0.2269	0.2094	0.1959	0.1848	0.1957	0.1629	0.1396	0.1215	0.1068
681600	18	19	0.990	0.0324	0.4162	0.2501	0.2255	0.2081	0.1945	0.1835	0.1948	0.1620	0.1387	0.1207	0.1059
632000	15	4	0.927	0.0296	0.4271	0.2556	0.2310	0.2136	0.2000	0.1890	0.1984	0.1656	0.1424	0.1243	0.1096
616200	12	7	0.927	0.0291	0.4292	0.2566	0.2320	0.2146	0.2011	0.1900	0.1991	0.1663	0.1431	0.1250	0.1103
587400	15	4	0.814	0.0260	0.4429	0.2635	0.2389	0.2214	0.2079	0.1968	0.2037	0.1709	0.1476	0.1296	0.1148
575200	12	7	0.814	0.0261	0.4424	0.2632	0.2386	0.2212	0.2077	0.1966	0.2035	0.1707	0.1475	0.1294	0.1147
543600	15	4	0.684	0.0221	0.4626	0.2733	0.2487	0.2313	0.2177	0.2067	0.2103	0.1775	0.1542	0.1361	0.1214
535000	12	7	0.684	0.0219	0.4637	0.2739	0.2493	0.2318	0.2183	0.2072	0.2106	0.1778	0.1546	0.1365	0.1218

Table B.15. Inductive Reactance of ACAR Bundled Conductors at 60 Hz with 4 and 6 Conductors per Bundle

AREA 62% EQ. EC - AL CMIL	STRANDS EC/6201		DIA. IN.	GMR FT.	60 Hz INDUCTIVE REACTANCE (1) X_A IN OHMS/MILE FOR 1 FOOT RADIUS — 4 - CONDUCTOR SPACING (IN.)					6 - CONDUCTOR SPACING (IN.)				
					6	9	12	15	18	6	9	12	15	18
2413000	72	19	1.821	0.0596	0.1381	0.1012	0.0750	0.0547	0.0381	0.0909	0.0499	0.0208	-0.0018	-0.0202
2375000	63	28	1.821	0.0596	0.1381	0.1012	0.0750	0.0547	0.0381	0.0909	0.0499	0.0208	-0.0018	-0.0202
2338000	54	37	1.821	0.0599	0.1380	0.1011	0.0749	0.0546	0.0380	0.0908	0.0498	0.0207	-0.0019	-0.0203
2297000	54	7	1.762	0.0571	0.1394	0.1025	0.0763	0.0560	0.0394	0.0918	0.0508	0.0217	-0.0009	-0.0193
2262000	48	13	1.762	0.0578	0.1390	0.1021	0.0760	0.0557	0.0391	0.0915	0.0505	0.0214	-0.0011	-0.0196
2226000	42	19	1.762	0.0578	0.1390	0.1021	0.0760	0.0557	0.0391	0.0915	0.0505	0.0214	-0.0011	-0.0196
2227000	54	7	1.735	0.0561	0.1400	0.1031	0.0769	0.0566	0.0400	0.0921	0.0511	0.0220	-0.0005	-0.0190
2193000	48	13	1.735	0.0530	0.1417	0.1048	0.0786	0.0583	0.0417	0.0933	0.0523	0.0232	0.0006	-0.0178
2159000	42	19	1.735	0.0568	0.1396	0.1027	0.0765	0.0562	0.0396	0.0919	0.0509	0.0218	-0.0008	-0.0192
1899000	54	7	1.602	0.0519	0.1423	0.1054	0.0792	0.0589	0.0423	0.0937	0.0527	0.0236	0.0010	-0.0174
1870000	48	13	1.602	0.0490	0.1441	0.1072	0.0810	0.0607	0.0441	0.0948	0.0538	0.0248	0.0022	-0.0162
1841000	42	19	1.602	0.0526	0.1419	0.1050	0.0788	0.0585	0.0419	0.0934	0.0524	0.0233	0.0008	-0.0177
1673000	54	7	1.504	0.0486	0.1443	0.1074	0.0812	0.0609	0.0443	0.0950	0.0540	0.0249	0.0024	-0.0161
1647000	48	13	1.504	0.0461	0.1459	0.1090	0.0828	0.0625	0.0459	0.0961	0.0551	0.0260	0.0034	-0.0150
1622000	42	19	1.504	0.0495	0.1438	0.1068	0.0807	0.0604	0.0438	0.0946	0.0536	0.0246	0.0020	-0.0164
1337000	54	7	1.345	0.0436	0.1476	0.1107	0.0845	0.0642	0.0476	0.0972	0.0562	0.0271	0.0046	-0.0139
1296000	42	19	1.345	0.0440	0.1473	0.1104	0.0842	0.0639	0.0473	0.0970	0.0560	0.0269	0.0044	-0.0141
1243000	30	7	1.302	0.0421	0.1487	0.1118	0.0856	0.0653	0.0487	0.0979	0.0569	0.0278	0.0053	-0.0132
1211000	24	13	1.302	0.0417	0.1490	0.1121	0.0859	0.0656	0.0490	0.0981	0.0571	0.0280	0.0055	-0.0130
1179000	18	19	1.302	0.0426	0.1483	0.1114	0.0852	0.0649	0.0483	0.0977	0.0567	0.0276	0.0050	-0.0134
1163000	30	7	1.259	0.0407	0.1497	0.1128	0.0866	0.0663	0.0497	0.0986	0.0576	0.0285	0.0059	-0.0125
1133000	24	13	1.259	0.0408	0.1496	0.1127	0.0865	0.0662	0.0496	0.0986	0.0576	0.0285	0.0059	-0.0125
1104000	18	19	1.259	0.0412	0.1493	0.1124	0.0862	0.0659	0.0493	0.0984	0.0574	0.0283	0.0057	-0.0127
1153000	33	4	1.246	0.0401	0.1501	0.1132	0.0871	0.0667	0.0502	0.0989	0.0579	0.0288	0.0062	-0.0122
1138000	30	7	1.246	0.0403	0.1500	0.1131	0.0869	0.0666	0.0500	0.0988	0.0578	0.0287	0.0061	-0.0123
1109000	24	13	1.246	0.0405	0.1498	0.1129	0.0868	0.0664	0.0499	0.0987	0.0577	0.0286	0.0060	-0.0124
1080000	18	19	1.246	0.0407	0.1497	0.1128	0.0866	0.0663	0.0497	0.0986	0.0576	0.0285	0.0059	-0.0125
1077000	30	7	1.212	0.0393	0.1507	0.1138	0.0877	0.0674	0.0508	0.0993	0.0583	0.0292	0.0067	-0.0118
1049000	24	13	1.212	0.0389	0.1511	0.1142	0.0880	0.0677	0.0511	0.0995	0.0585	0.0294	0.0069	-0.0116
1022000	18	19	1.212	0.0396	0.1505	0.1136	0.0874	0.0671	0.0505	0.0992	0.0582	0.0291	0.0065	-0.0119
1050000	30	7	1.196	0.0388	0.1511	0.1142	0.0881	0.0677	0.0512	0.0996	0.0586	0.0295	0.0069	-0.0115
1023000	24	13	1.196	0.0384	0.1515	0.1146	0.0884	0.0681	0.0515	0.0998	0.0588	0.0297	0.0071	-0.0113
996000	18	19	1.196	0.0391	0.1509	0.1140	0.0878	0.0675	0.0509	0.0994	0.0584	0.0293	0.0068	-0.0117
994800	30	7	1.165	0.0376	0.1521	0.1152	0.0890	0.0687	0.0521	0.1002	0.0592	0.0301	0.0075	-0.0109
954600	30	7	1.141	0.0369	0.1527	0.1158	0.0896	0.0693	0.0527	0.1006	0.0596	0.0305	0.0079	-0.0105
969300	24	13	1.165	0.0374	0.1523	0.1154	0.0892	0.0689	0.0523	0.1003	0.0593	0.0302	0.0077	-0.0108
958000	24	13	1.158	0.0371	0.1525	0.1156	0.0894	0.0691	0.0525	0.1005	0.0595	0.0304	0.0078	-0.0106
943900	18	19	1.165	0.0381	0.1517	0.1148	0.0886	0.0683	0.0517	0.0999	0.0589	0.0298	0.0073	-0.0112
900300	30	7	1.108	0.0358	0.1536	0.1167	0.0905	0.0702	0.0536	0.1012	0.0602	0.0311	0.0085	-0.0099
795000	30	7	1.042	0.0334	0.1557	0.1188	0.0926	0.0723	0.0557	0.1026	0.0616	0.0325	0.0099	-0.0085
877300	24	13	1.108	0.0355	0.1538	0.1169	0.0908	0.0704	0.0539	0.1014	0.0604	0.0313	0.0087	-0.0097
795000	24	13	1.055	0.0339	0.1552	0.1183	0.0922	0.0718	0.0553	0.1023	0.0613	0.0322	0.0096	-0.0388
854200	18	19	1.108	0.0361	0.1533	0.1164	0.0902	0.0699	0.0533	0.1010	0.0600	0.0309	0.0084	-0.010
795000	18	19	1.069	0.0349	0.1544	0.1175	0.0913	0.0710	0.0544	0.1017	0.0607	0.0316	0.0091	-0.0094
829000	30	7	1.063	0.0343	0.1549	0.1180	0.0918	0.0715	0.0549	0.1021	0.0611	0.0320	0.0094	-0.0090
807700	24	13	1.063	0.0342	0.1550	0.1181	0.0919	0.0716	0.0550	0.1021	0.0611	0.0320	0.0095	-0.0090
786500	18	19	1.063	0.0348	0.1544	0.1175	0.0914	0.0710	0.0545	0.1018	0.0608	0.0317	0.0091	-0.0093
727500	33	4	0.990	0.0319	0.1571	0.1202	0.0940	0.0737	0.0571	0.1035	0.0625	0.0334	0.0109	-0.007
718300	30	7	0.990	0.0320	0.1570	0.1201	0.0939	0.0736	0.0570	0.1035	0.0625	0.0334	0.0108	-0.007
700000	24	13	0.990	0.0317	0.1573	0.1204	0.0942	0.0739	0.0573	0.1037	0.0627	0.0336	0.0110	-0.007
681600	18	19	0.990	0.0324	0.1566	0.1197	0.0935	0.0732	0.0566	0.1032	0.0622	0.0331	0.0106	-0.007
632000	15	4	0.927	0.0296	0.1593	0.1224	0.0963	0.0760	0.0594	0.1050	0.0640	0.0350	0.0124	-0.006
616200	12	7	0.927	0.0291	0.1599	0.1230	0.0968	0.0765	0.0599	0.1054	0.0644	0.0353	0.0127	-0.005
487400	15	4	0.814	0.0260	0.1633	0.1264	0.1002	0.0799	0.0633	0.1077	0.0667	0.0376	0.0150	-0.003
475200	12	7	0.814	0.0261	0.1632	0.1263	0.1001	0.0798	0.0632	0.1076	0.0666	0.0375	0.0149	-0.003
343600	15	4	0.684	0.0221	0.1682	0.1313	0.1051	0.0848	0.0682	0.1109	0.0699	0.0409	0.0183	-0.000
335000	12	7	0.684	0.0219	0.1685	0.1316	0.1054	0.0851	0.0685	0.1111	0.0701	0.0410	0.0185	0.000

Table B.16 Capacitive Reactance of ACSR Bundled Conductors at 60 Hz with 1, 2, and 3 Conductors per Bundle

CODE	AREA CMIL	STRANDS AL	STRANDS ST	DIA. IN.	60 Hz CAPACITIVE REACTANCE (1) x_A' IN MEGOHM-MILES FOR 1 FOOT RADIUS: SINGLE COND.	2 - CONDUCTOR SPACING (IN.) 6	9	12	16	18	3 - CONDUCTOR SPACING (IN.) 6	9	12	15	18
EXPANDED	3108000	62/8	19	2.500	0.0671	0.0438	0.0378	0.0336	0.0302	0.0275	0.0361	0.0281	0.0224	0.0180	0.0143
EXPANDED	2294000	66/6	19	2.320	0.0693	0.0449	0.0389	0.0347	0.0313	0.0285	0.0368	0.0288	0.0231	0.0187	0.0151
EXPANDED	1414000	58/4	19	1.750	0.0777	0.0491	0.0431	0.0388	0.0356	0.0328	0.0396	0.0316	0.0259	0.0215	0.0179
EXPANDED	1275000	50/4	19	1.600	0.0803	0.0505	0.0444	0.0402	0.0369	0.0342	0.0405	0.0325	0.0268	0.0224	0.0188
KIWI	2167000	72	7	1.737	0.0779	0.0492	0.0432	0.0390	0.0356	0.0329	0.0397	0.0317	0.0260	0.0216	0.0179
BLUEBIRD	2156000	84	19	1.762	0.0775	0.0490	0.0430	0.0387	0.0354	0.0327	0.0395	0.0315	0.0258	0.0214	0.0178
CHUKAR	1780000	84	19	1.602	0.0803	0.0504	0.0444	0.0402	0.0368	0.0341	0.0405	0.0325	0.0268	0.0224	0.0187
FALCON	1590000	54	19	1.545	0.0814	0.0510	0.0450	0.0407	0.0374	0.0347	0.0408	0.0328	0.0271	0.0227	0.0191
LAPWING	1590000	45	7	1.502	0.0822	0.0514	0.0454	0.0411	0.0378	0.0351	0.0411	0.0331	0.0274	0.0230	0.0194
PARROT	1510500	54	19	1.506	0.0821	0.0514	0.0453	0.0411	0.0378	0.0351	0.0411	0.0331	0.0274	0.0230	0.0194
NUTHATCH	1510500	45	7	1.466	0.0829	0.0518	0.0457	0.0415	0.0382	0.0355	0.0414	0.0333	0.0276	0.0232	0.0196
PLOVER	1431000	54	19	1.465	0.0830	0.0518	0.0457	0.0415	0.0382	0.0355	0.0414	0.0333	0.0277	0.0232	0.0196
BOBOLINK	1431000	45	7	1.427	0.0837	0.0522	0.0461	0.0419	0.0386	0.0359	0.0416	0.0336	0.0279	0.0235	0.0199
MARTIN	1351500	54	19	1.424	0.0838	0.0522	0.0462	0.0419	0.0386	0.0359	0.0416	0.0336	0.0279	0.0235	0.0199
DIPPER	1351500	45	7	1.385	0.0846	0.0526	0.0466	0.0423	0.0390	0.0363	0.0419	0.0339	0.0282	0.0238	0.0202
PHEASANT	1272000	54	19	1.382	0.0847	0.0526	0.0466	0.0423	0.0390	0.0363	0.0419	0.0339	0.0282	0.0238	0.0202
BITTERN	1272000	45	7	1.345	0.0855	0.0530	0.0470	0.0427	0.0394	0.0367	0.0422	0.0342	0.0285	0.0241	0.0205
GRACKLE	1192500	54	19	1.333	0.0858	0.0532	0.0471	0.0429	0.0396	0.0369	0.0423	0.0343	0.0286	0.0242	0.0206
BUNTING	1192500	45	7	1.302	0.0865	0.0535	0.0475	0.0432	0.0399	0.0372	0.0425	0.0345	0.0288	0.0244	0.0208
FINCH	1113000	54	19	1.293	0.0867	0.0536	0.0476	0.0433	0.0400	0.0373	0.0426	0.0346	0.0289	0.0245	0.0209
BLUEJAY	1113000	45	7	1.259	0.0875	0.0540	0.0480	0.0437	0.0404	0.0377	0.0429	0.0348	0.0292	0.0247	0.0211
CURLEW	1033500	54	7	1.246	0.0878	0.0542	0.0481	0.0439	0.0406	0.0379	0.0430	0.0349	0.0293	0.0248	0.0712
ORTOLAN	1033500	45	7	1.213	0.0886	0.0546	0.0485	0.0443	0.0410	0.0383	0.0432	0.0352	0.0295	0.0251	0.0215
TANAGER	1033500	36	1	1.186	0.0892	0.0549	0.0489	0.0446	0.0413	0.0386	0.0435	0.0354	0.0297	0.0253	0.0217
CARDINAL	954000	54	7	1.196	0.0890	0.0548	0.0488	0.0445	0.0412	0.0385	0.0434	0.0353	0.0297	0.0252	0.0216
RAIL	954000	45	7	1.165	0.0898	0.0552	0.0491	0.0449	0.0416	0.0389	0.0436	0.0356	0.0299	0.0255	0.0219
CATBIRD	954000	36	1	1.140	0.0904	0.0555	0.0495	0.0452	0.0419	0.0392	0.0438	0.0358	0.0301	0.0257	0.0221
CANARY	900000	54	7	1.162	0.0898	0.0552	0.0492	0.0449	0.0416	0.0389	0.0437	0.0356	0.0299	0.0255	0.0219
RUDDY	900000	45	7	1.131	0.0906	0.0556	0.0496	0.0453	0.0420	0.0393	0.0439	0.0359	0.0302	0.0258	0.0222
MALLARD	795000	30	19	1.140	0.0904	0.0555	0.0495	0.0452	0.0419	0.0392	0.0438	0.0358	0.0301	0.0257	0.0221
DRAKE	795000	26	7	1.108	0.0912	0.0559	0.0499	0.0456	0.0423	0.0396	0.0441	0.0361	0.0304	0.0260	0.0224
CONDOR	795000	54	7	1.093	0.0916	0.0561	0.0501	0.0458	0.0425	0.0398	0.0443	0.0362	0.0305	0.0261	0.0225
CUCKOO	795000	24	7	1.092	0.0917	0.0561	0.0501	0.0458	0.0425	0.0398	0.0443	0.0362	0.0306	0.0261	0.0225
TERN	795000	45	7	1.063	0.0925	0.0565	0.0505	0.0462	0.0429	0.0402	0.0445	0.0365	0.0308	0.0264	0.0228
COOT	795000	36	1	1.040	0.0931	0.0568	0.0508	0.0466	0.0433	0.0405	0.0448	0.0367	0.0310	0.0266	0.0230
REDWING	715500	30	19	1.081	0.0920	0.0563	0.0503	0.0460	0.0427	0.0400	0.0444	0.0363	0.0307	0.0262	0.0226
STARLING	715500	26	7	1.051	0.0928	0.0567	0.0507	0.0464	0.0431	0.0404	0.0446	0.0366	0.0309	0.0265	0.0229
STILT	715500	24	7	1.036	0.0932	0.0569	0.0509	0.0466	0.0433	0.0406	0.0448	0.0368	0.0311	0.0267	0.0231
GANNET	666600	26	7	1.014	0.0939	0.0572	0.0512	0.0469	0.0436	0.0409	0.0450	0.0370	0.0313	0.0269	0.0233
FLAMINGO	666600	24	7	1.000	0.0943	0.0574	0.0514	0.0471	0.0438	0.0411	0.0451	0.0371	0.0314	0.0270	0.0234
--------	653900	18	3	0.953	0.0957	0.0581	0.0521	0.0479	0.0445	0.0418	0.0456	0.0376	0.0319	0.0275	0.0239
EGRET	636000	30	19	1.019	0.0937	0.0571	0.0511	0.0469	0.0436	0.0408	0.0450	0.0369	0.0312	0.0268	0.0232
GROSBEAK	636000	26	7	0.990	0.0946	0.0576	0.0516	0.0473	0.0440	0.0413	0.0452	0.0372	0.0315	0.0271	0.0235
ROOK	636000	24	7	0.977	0.0950	0.0578	0.0518	0.0475	0.0442	0.0415	0.0454	0.0373	0.0317	0.0272	0.0236
KINGBIRD	636000	18	1	0.940	0.0961	0.0583	0.0523	0.0481	0.0448	0.0420	0.0458	0.0377	0.0320	0.0276	0.0240
SWIFT	636000	36	1	0.930	0.0964	0.0585	0.0525	0.0482	0.0449	0.0422	0.0459	0.0378	0.0321	0.0277	0.0241
TEAL	605000	30	19	0.994	0.0945	0.0575	0.0515	0.0472	0.0439	0.0412	0.0452	0.0372	0.0315	0.0271	0.0235
SQUAB	605000	26	7	0.966	0.0953	0.0579	0.0519	0.0477	0.0443	0.0416	0.0455	0.0375	0.0318	0.0274	0.0238
PEACOCK	605000	24	7	0.953	0.0957	0.0581	0.0521	0.0479	0.0445	0.0418	0.0456	0.0376	0.0319	0.0275	0.0239
EAGLE	556500	30	7	0.953	0.0957	0.0581	0.0521	0.0479	0.0445	0.0418	0.0456	0.0376	0.0319	0.0275	0.0239
DOVE	556500	26	7	0.927	0.0965	0.0586	0.0525	0.0483	0.0450	0.0423	0.0459	0.0379	0.0322	0.0278	0.0242
PARAKEET	556500	24	7	0.914	0.0970	0.0588	0.0527	0.0485	0.0452	0.0425	0.0460	0.0380	0.0323	0.0279	0.0243
OSPREY	556500	18	1	0.879	0.0981	0.0593	0.0533	0.0491	0.0457	0.0430	0.0464	0.0384	0.0327	0.0283	0.0247
HEN	477000	30	7	0.883	0.0980	0.0593	0.0533	0.0490	0.0457	0.0430	0.0464	0.0383	0.0327	0.0282	0.0246
HAWK	477000	26	7	0.858	0.0988	0.0597	0.0537	0.0494	0.0461	0.0434	0.0467	0.0386	0.0329	0.0285	0.0249
LICKER	477000	24	7	0.846	0.0992	0.0599	0.0539	0.0496	0.0463	0.0436	0.0468	0.0388	0.0331	0.0287	0.0251
ELICAN	477000	18	1	0.814	0.1004	0.0605	0.0545	0.0502	0.0469	0.0442	0.0472	0.0392	0.0335	0.0291	0.0254
ARK	397500	30	7	0.806	0.1007	0.0606	0.0546	0.0503	0.0470	0.0443	0.0473	0.0393	0.0336	0.0291	0.0255
BIS	397500	26	7	0.783	0.1015	0.0611	0.0550	0.0508	0.0475	0.0448	0.0476	0.0395	0.0339	0.0294	0.0258
RANT	397500	24	7	0.772	0.1020	0.0613	0.0552	0.0510	0.0477	0.0450	0.0477	0.0397	0.0340	0.0296	0.0260
HICKADEE	397500	18	1	0.743	0.1031	0.0618	0.0558	0.0516	0.0482	0.0455	0.0481	0.0401	0.0344	0.0300	0.0263
ORIOLE	336400	30	7	0.741	0.1032	0.0619	0.0559	0.0516	0.0483	0.0456	0.0481	0.0401	0.0344	0.0300	0.0264
INNET	336400	26	7	0.721	0.1040	0.0623	0.0563	0.0520	0.0487	0.0460	0.0484	0.0404	0.0347	0.0303	0.0266
ERLIN	336400	18	1	0.684	0.1056	0.0631	0.0570	0.0528	0.0495	0.0468	0.0489	0.0409	0.0352	0.0308	0.0272
OSTRICH	300000	26	7	0.680	0.1057	0.0631	0.0571	0.0529	0.0496	0.0468	0.0490	0.0409	0.0352	0.0308	0.0272

Table B.17. Capacitive Reactance of ACSR Bundled Conductors at 60 Hz with 4 and 6 Conductors per Bundle

CODE	AREA CMIL	STRANDS AL	STRANDS ST	DIA. IN.	60 Hz CAPACITIVE REACTANCE (1) X_A IN MEGOHM-MILES FOR 1 FOOT RADIUS — 4 - CONDUCTOR SPACING (IN.) 6	9	12	15	18	6 - CONDUCTOR SPACING (IN.) 6	9	12	15	18
EXPANDED	3108000	62/8	19	2.500	0.0296	0.0206	0.0142	0.0092	0.0052	0.0195	0.0094	0.0023	-0.0032	-0.0077
EXPANDED	2294000	66/6	19	2.320	0.0302	0.0212	0.0148	0.0098	0.0057	0.0198	0.0098	0.0027	-0.0028	-0.0073
EXPANDED	1414000	58/4	19	1.750	0.0323	0.0233	0.0169	0.0119	0.0078	0.0212	0.0112	0.0041	-0.0014	-0.0059
EXPANDED	1275000	50/4	19	1.600	0.0329	0.0239	0.0175	0.0126	0.0085	0.0217	0.0116	0.0045	-0.0010	-0.0055
KIWI	2167000	72	7	1.737	0.0323	0.0233	0.0169	0.0119	0.0079	0.0213	0.0112	0.0041	-0.0014	-0.0059
BLUEBIRD	2156000	84	19	1.762	0.0322	0.0232	0.0168	0.0118	0.0078	0.0212	0.0112	0.0041	-0.0015	-0.0060
CHUKAR	1780000	84	19	1.602	0.0329	0.0239	0.0175	0.0125	0.0085	0.0217	0.0116	0.0045	-0.0010	-0.0055
FALCON	1590000	54	19	1.545	0.0332	0.0242	0.0178	0.0128	0.0088	0.0218	0.0118	0.0047	-0.0008	-0.0053
LAPWING	1590000	45	7	1.502	0.0334	0.0244	0.0180	0.0130	0.0090	0.0220	0.0120	0.0048	-0.0007	-0.0052
PARROT	1510500	54	19	1.506	0.0334	0.0244	0.0180	0.0130	0.0089	0.0220	0.0119	0.0048	-0.0007	-0.0052
NUTHATCH	1510500	45	7	1.466	0.0336	0.0246	0.0182	0.0132	0.0091	0.0221	0.0121	0.0050	-0.0006	-0.0051
PLOVER	1431000	54	19	1.465	0.0336	0.0246	0.0182	0.0132	0.0091	0.0221	0.0121	0.0050	-0.0006	-0.0051
BOBOLINK	1431000	45	7	1.427	0.0338	0.0248	0.0184	0.0134	0.0093	0.0222	0.0122	0.0051	-0.0004	-0.0049
MARTIN	1351500	54	19	1.424	0.0338	0.0248	0.0184	0.0134	0.0094	0.0222	0.0122	0.0051	-0.0004	-0.0049
DIPPER	1351500	45	7	1.385	0.0340	0.0250	0.0186	0.0136	0.0096	0.0224	0.0124	0.0052	-0.0003	-0.0048
PHEASANT	1272000	54	19	1.382	0.0340	0.0250	0.0186	0.0136	0.0096	0.0224	0.0124	0.0053	-0.0003	-0.0048
BITTERN	1272000	45	7	1.345	0.0342	0.0252	0.0188	0.0138	0.0098	0.0225	0.0125	0.0054	-0.0001	-0.0046
GRACKLE	1192500	54	19	1.333	0.0343	0.0253	0.0189	0.0139	0.0098	0.0226	0.0125	0.0054	-0.0001	-0.0046
BUNTING	1192500	45	7	1.307	0.0345	0.0254	0.0190	0.0141	0.0100	0.0227	0.0127	0.0055	0.0000	-0.0045
FINCH	1113000	54	19	1.293	0.0345	0.0255	0.0191	0.0141	0.0101	0.0227	0.0127	0.0056	0.0001	-0.0044
BLUEJAY	1113000	45	7	1.259	0.0347	0.0257	0.0193	0.0143	0.0103	0.0229	0.0128	0.0057	0.0002	-0.0043
CURLEW	1033500	54	7	1.246	0.0348	0.0258	0.0194	0.0144	0.0103	0.0229	0.0129	0.0058	0.0003	-0.0043
ORTOLAN	1033500	45	7	1.213	0.0350	0.0260	0.0196	0.0146	0.0105	0.0230	0.0130	0.0059	0.0004	-0.0041
TANAGER	1033500	36	1	1.186	0.0352	0.0261	0.0197	0.0148	0.0107	0.0231	0.0131	0.0060	0.0005	-0.0040
CARDINAL	954000	54	7	1.196	0.0351	0.0261	0.0197	0.0147	0.0107	0.0231	0.0131	0.0060	0.0005	-0.0041
RAIL	954000	45	7	1.165	0.0353	0.0263	0.0199	0.0149	0.0108	0.0232	0.0132	0.0061	0.0006	-0.0039
CATBIRD	954000	36	1	1.140	0.0355	0.0264	0.0200	0.0151	0.0110	0.0233	0.0133	0.0062	0.0007	-0.0038
CANARY	900000	54	7	1.162	0.0353	0.0263	0.0199	0.0149	0.0109	0.0232	0.0132	0.0061	0.0006	-0.0039
RUDDY	900000	45	7	1.131	0.0355	0.0265	0.0201	0.0151	0.0111	0.0234	0.0134	0.0062	0.0007	-0.0038
MALLARD	795000	30	19	1.140	0.0355	0.0264	0.0200	0.0151	0.0110	0.0233	0.0133	0.0062	0.0007	-0.0038
DRAKE	795000	26	7	1.108	0.0357	0.0266	0.0202	0.0153	0.0112	0.0235	0.0135	0.0063	0.0008	-0.0037
CONDOR	795000	54	7	1.093	0.0358	0.0267	0.0203	0.0154	0.0113	0.0236	0.0135	0.0064	0.0009	-0.0036
CUCKOO	795000	24	7	1.092	0.0358	0.0267	0.0203	0.0154	0.0113	0.0236	0.0135	0.0064	0.0009	-0.0036
TERN	795000	45	7	1.063	0.0360	0.0269	0.0205	0.0156	0.0115	0.0237	0.0137	0.0066	0.0010	-0.0035
COOT	795000	36	1	1.040	0.0361	0.0271	0.0207	0.0157	0.0117	0.0238	0.0138	0.0067	0.0011	-0.0034
REDWING	715500	30	19	1.081	0.0358	0.0268	0.0204	0.0155	0.0114	0.0235	0.0136	0.0065	0.0010	-0.0036
STARLING	715500	26	7	1.051	0.0361	0.0270	0.0206	0.0157	0.0116	0.0237	0.0137	0.0066	0.0011	-0.0034
STILT	715500	24	7	1.036	0.0362	0.0271	0.0207	0.0158	0.0117	0.0238	0.0138	0.0067	0.0012	-0.0033
GANNET	666600	26	7	1.014	0.0363	0.0273	0.0209	0.0159	0.0119	0.0239	0.0139	0.0068	0.0013	-0.0032
FLAMINGO	666600	24	7	1.000	0.0364	0.0274	0.0210	0.0160	0.0120	0.0240	0.0140	0.0069	0.0013	-0.0032
-------	653900	18	3	0.953	0.0368	0.0278	0.0214	0.0164	0.0123	0.0242	0.0142	0.0071	0.0016	-0.0029
EGRET	636000	30	19	1.019	0.0363	0.0273	0.0209	0.0159	0.0118	0.0239	0.0139	0.0068	0.0012	-0.0033
GROSBEAK	636000	26	7	0.990	0.0365	0.0275	0.0211	0.0161	0.0121	0.0240	0.0140	0.0069	0.0014	-0.0031
ROOK	636000	24	7	0.977	0.0366	0.0276	0.0212	0.0162	0.0122	0.0241	0.0141	0.0070	0.0015	-0.0031
KINGBIRD	636000	18	1	0.940	0.0369	0.0279	0.0215	0.0165	0.0124	0.0243	0.0143	0.0072	0.0016	-0.0029
SWIFT	636000	36	1	0.930	0.0370	0.0279	0.0215	0.0166	0.0125	0.0244	0.0143	0.0072	0.0017	-0.0028
TEAL	605000	30	19	0.994	0.0365	0.0274	0.0210	0.0161	0.0120	0.0240	0.0140	0.0069	0.0014	-0.0031
SQUAB	605000	26	7	0.966	0.0367	0.0277	0.0213	0.0163	0.0122	0.0242	0.0141	0.0070	0.0015	-0.0030
PEACOCK	605000	24	7	0.953	0.0368	0.0278	0.0214	0.0164	0.0123	0.0242	0.0142	0.0071	0.0016	-0.0029
EAGLE	556500	30	7	0.953	0.0368	0.0278	0.0214	0.0164	0.0123	0.0242	0.0142	0.0071	0.0016	-0.0029
DOVE	556500	26	7	0.927	0.0370	0.0280	0.0216	0.0166	0.0125	0.0244	0.0143	0.0072	0.0017	-0.0028
PARAKEET	556500	24	7	0.914	0.0371	0.0281	0.0217	0.0167	0.0126	0.0244	0.0144	0.0073	0.0018	-0.0027
OSPREY	556500	18	1	0.879	0.0374	0.0284	0.0220	0.0170	0.0129	0.0246	0.0146	0.0075	0.0020	-0.0025
HEN	477000	30	7	0.883	0.0373	0.0283	0.0219	0.0170	0.0129	0.0246	0.0146	0.0075	0.0020	-0.0026
HAWK	477000	26	7	0.858	0.0376	0.0285	0.0221	0.0172	0.0131	0.0247	0.0147	0.0076	0.0021	-0.0024
FLICKER	477000	24	7	0.846	0.0377	0.0286	0.0222	0.0173	0.0132	0.0248	0.0148	0.0077	0.0022	-0.0023
PELICAN	477000	18	1	0.814	0.0380	0.0289	0.0225	0.0176	0.0135	0.0250	0.0150	0.0079	0.0024	-0.0022
LARK	397500	30	7	0.806	0.0380	0.0290	0.0226	0.0176	0.0136	0.0251	0.0150	0.0079	0.0024	-0.0021
IBIS	397500	26	7	0.783	0.0382	0.0292	0.0228	0.0179	0.0138	0.0252	0.0152	0.0081	0.0025	-0.0020
BRANT	397500	24	7	0.772	0.0383	0.0293	0.0229	0.0180	0.0139	0.0253	0.0152	0.0081	0.0026	-0.0019
CHICKADEE	397500	18	1	0.743	0.0386	0.0296	0.0232	0.0182	0.0142	0.0255	0.0154	0.0083	0.0028	-0.0017
ORIOLE	336400	30	7	0.741	0.0386	0.0296	0.0232	0.0183	0.0142	0.0255	0.0154	0.0083	0.0028	-0.0017
LINNET	336400	26	7	0.721	0.0389	0.0298	0.0234	0.0185	0.0144	0.0256	0.0156	0.0085	0.0030	-0.0016
MERLIN	336400	18	1	0.684	0.0392	0.0302	0.0238	0.0189	0.0148	0.0259	0.0158	0.0087	0.0032	-0.0013
OSTRICH	300000	26	7	0.680	0.0393	0.0303	0.0239	0.0139	0.0148	0.0259	0.0159	0.0088	0.0032	-0.0013

Table B.18. Capacitive Reactance of ACAR Bundled Conductors at 60 Hz with 1, 2, and 3 Conductors per Bundle

AREA 62% EQ. EC - AL CMIL	STRANDS EC/6201		DIA. IN.	60 Hz CAPACITIVE REACTANCE (1) X_A' IN MEGOHM-MILES FOR 1 FOOT RADIUS										
				SINGLE COND.	2 - CONDUCTOR SPACING (IN.)					3 - CONDUCTOR SPACING (IN.)				
					6	9	12	15	18	6	9	12	15	18
2413000	72	19	1.821	0.0765	0.0485	0.0425	0.0383	0.0349	0.0322	0.0392	0.0312	0.0255	0.0211	0.0175
2375000	63	28	1.821											
2338000	54	37	1.821											
2297000	54	7	1.762	0.0775	0.0490	0.0430	0.0387	0.0354	0.0327	0.0395	0.0315	0.0258	0.0214	0.0178
2262000	48	13	1.762											
2226000	42	19	1.762											
2227000	54	7	1.735	0.0779	0.0493	0.0432	0.0390	0.0357	0.0330	0.0397	0.0317	0.0260	0.0216	0.0180
2193000	48	13	1.735											
2159000	42	19	1.735											
1899000	54	7	1.602	0.0803	0.0504	0.0444	0.0402	0.0368	0.0341	0.0405	0.0325	0.0268	0.0224	0.0187
1870000	48	13	1.602											
1841000	42	19	1.602											
1673000	54	7	1.504	0.0822	0.0514	0.0454	0.0411	0.0378	0.0351	0.0411	0.0331	0.0274	0.0230	0.0194
1647000	48	13	1.504											
1622000	42	19	1.504											
1337000	54	7	1.345	0.0855	0.0530	0.0470	0.0427	0.0394	0.0367	0.0422	0.0342	0.0285	0.0241	0.0205
1296000	42	19	1.345											
1243000	30	7	1.302	0.0865	0.0535	0.0475	0.0432	0.0399	0.0372	0.0425	0.0345	0.0288	0.0244	0.0208
1211000	24	13	1.302											
1179000	18	19	1.302											
1163000	30	7	1.259	0.0875	0.0540	0.0480	0.0437	0.0404	0.0377	0.0429	0.0348	0.0292	0.0247	0.0211
1133000	24	13	1.259											
1104000	18	19	1.259											
1153000	33	4	1.246	0.0878	0.0542	0.0481	0.0439	0.0406	0.0379	0.0430	0.0349	0.0293	0.0248	0.0212
1138000	30	7	1.246											
1109000	24	13	1.246											
1080000	18	19	1.246											
1077000	30	7	1.212	0.0886	0.0546	0.0486	0.0443	0.0410	0.0383	0.0432	0.0352	0.0295	0.0251	0.0215
1049000	24	13	1.212											
1022000	18	19	1.212											
050000	30	7	1.196	0.0890	0.0548	0.0488	0.0445	0.0412	0.0385	0.0434	0.0353	0.0297	0.0252	0.0216
023000	24	13	1.196											
996000	18	19	1.196											
994800	30	7	1.165	0.0898	0.0552	0.0491	0.0449	0.0416	0.0389	0.0436	0.0356	0.0299	0.0255	0.0219
954600	30	7	1.141	0.0904	0.0555	0.0495	0.0452	0.0419	0.0392	0.0438	0.0358	0.0301	0.0257	0.0221
969300	24	13	1.165	0.0898	0.0552	0.0491	0.0449	0.0416	0.0389	0.0436	0.0356	0.0299	0.0255	0.0219
958000	24	13	1.158	0.0899	0.0552	0.0492	0.0450	0.0417	0.0390	0.0437	0.0357	0.0300	0.0256	0.0220
943900	18	19	1.165	0.0898	0.0552	0.0491	0.0449	0.0416	0.0389	0.0436	0.0356	0.0299	0.0255	0.0219
900300	30	7	1.108	0.0912	0.0559	0.0499	0.0456	0.0423	0.0396	0.0441	0.0361	0.0304	0.0260	0.0224
795000	30	7	1.042	0.0931	0.0568	0.0508	0.0465	0.0432	0.0405	0.0447	0.0367	0.0310	0.0266	0.0230
877300	24	13	1.108	0.0912	0.0559	0.0499	0.0456	0.0423	0.0396	0.0441	0.0361	0.0304	0.0260	0.0224
795000	24	13	1.055	0.0927	0.0566	0.0506	0.0463	0.0430	0.0403	0.0446	0.0366	0.0309	0.0265	0.0229
854200	18	19	1.108	0.0912	0.0559	0.0499	0.0456	0.0423	0.0396	0.0441	0.0361	0.0304	0.0260	0.0224
795000	18	19	1.069	0.0923	0.0564	0.0504	0.0462	0.0428	0.0401	0.0445	0.0365	0.0308	0.0264	0.0227
829000	30	7	1.063	0.0925	0.0565	0.0505	0.0462	0.0429	0.0402	0.0445	0.0365	0.0308	0.0264	0.0228
807700	24	13	1.063											
786500	18	19	1.063											
727500	33	4	0.990	0.0946	0.0576	0.0516	0.0473	0.0440	0.0413	0.0452	0.0372	0.0315	0.0271	0.0235
718300	30	7	0.990											
00000	24	13	0.990											
81600	18	19	0.990											
32000	15	4	0.927	0.0965	0.0586	0.0525	0.0483	0.0450	0.0423	0.0459	0.0379	0.0322	0.0278	0.0242
16200	12	7	0.927											
87400	15	4	0.814	0.1004	0.0605	0.0545	0.0502	0.0469	0.0442	0.0472	0.0392	0.0335	0.0291	0.0254
75200	12	7	0.814											
43600	15	4	0.684	0.1056	0.0631	0.0570	0.0528	0.0495	0.0468	0.0489	0.0409	0.0352	0.0308	0.0272
35000	12	7	0.684											

Table B.19. Capacitive Reactance of ACAR Bundled Conductors at 60 Hz with 4 and 6 Conductors per Bundle

AREA 62% EQ. EC - AL CMIL	STRANDS EC/6201		DIA. IN.	60 Hz CAPACITIVE REACTANCE (1) X_A' IN MEGOHM-MILES FOR 1 FOOT RADIUS — 4 - CONDUCTOR SPACING (IN.)					6 - CONDUCTOR SPACING (IN.)				
				6	9	12	15	18	6	9	12	15	18
2413000	72	19	1.821	0.0320	0.0230	0.0166	0.0116	0.0075	0.0210	0.0110	0.0039	-0.0016	-0.0061
2375000	63	28	1.821										
2338000	54	37	1.821										
2297000	54	7	1.762	0.0322	0.0232	0.0168	0.0118	0.0078	0.0212	0.0112	0.0041	-0.0015	-0.0060
2262000	48	13	1.762										
2226000	42	19	1.762										
2227000	54	7	1.735	0.0323	0.0233	0.0169	0.0119	0.0079	0.0213	0.0112	0.0041	-0.0014	-0.0059
2193000	48	13	1.735										
2159000	42	19	1.735										
1899000	54	7	1.602	0.0329	0.0239	0.0175	0.0125	0.0085	0.0217	0.0116	0.0045	-0.0010	-0.0055
1870000	48	13	1.602										
1841000	42	19	1.602										
1673000	54	7	1.504	0.0334	0.0244	0.0180	0.0130	0.0090	0.0220	0.0119	0.0048	-0.0007	-0.0052
1647000	48	13	1.504										
1622000	42	19	1.504										
1337000	54	7	1.345	0.0342	0.0252	0.0188	0.0138	0.0098	0.0225	0.0125	0.0054	-0.0001	-0.0046
1296000	42	19	1.345										
1243000	30	7	1.302	0.0345	0.0254	0.0190	0.0141	0.0100	0.0227	0.0127	0.0055	0.0000	-0.0045
1211000	24	13	1.302										
1179000	18	19	1.302										
1163000	30	7	1.259	0.0347	0.0257	0.0193	0.0143	0.0103	0.0229	0.0128	0.0057	0.0002	-0.0043
1133000	24	13	1.259										
1104000	18	19	1.259										
1153000	33	4	1.246	0.0348	0.0258	0.0194	0.0144	0.0103	0.0229	0.0129	0.0058	0.0003	-0.0043
1138000	30	7	1.246										
1109000	24	13	1.246										
1080000	18	19	1.246										
1077000	30	7	1.212	0.0350	0.0260	0.0196	0.0146	0.0106	0.0230	0.0130	0.0059	0.0004	-0.0041
1049000	24	13	1.212										
1022000	18	19	1.212										
1050000	30	7	1.196	0.0351	0.0261	0.0197	0.0147	0.0107	0.0231	0.0131	0.0060	0.0005	-0.0041
1023000	24	13	1.196										
996000	18	19	1.196										
994800	30	7	1.165	0.0353	0.0263	0.0199	0.0149	0.0108	0.0232	0.0132	0.0061	0.0006	-0.0039
954600	30	7	1.141	0.0354	0.0264	0.0200	0.0151	0.0110	0.0233	0.0133	0.0062	0.0007	-0.0038
969300	24	13	1.165	0.0353	0.0263	0.0199	0.0149	0.0108	0.0232	0.0132	0.0061	0.0006	-0.0039
958000	24	13	1.158	0.0353	0.0263	0.0199	0.0149	0.0109	0.0233	0.0132	0.0061	0.0006	-0.0039
943900	18	19	1.165	0.0353	0.0263	0.0199	0.0149	0.0108	0.0232	0.0132	0.0061	0.0006	-0.0039
900300	30	7	1.108	0.0357	0.0266	0.0202	0.0153	0.0112	0.0235	0.0135	0.0063	0.0008	-0.0037
795000	30	7	1.042	0.0361	0.0271	0.0207	0.0157	0.0117	0.0238	0.0138	0.0067	0.0011	-0.0034
877300	24	13	1.108	0.0357	0.0266	0.0202	0.0153	0.0112	0.0235	0.0135	0.0063	0.0008	-0.0037
795000	24	13	1.055	0.0360	0.0270	0.0206	0.0156	0.0116	0.0237	0.0137	0.0066	0.0011	-0.0034
854200	18	19	1.108	0.0357	0.0266	0.0202	0.0153	0.0112	0.0235	0.0135	0.0063	0.0008	-0.0037
795000	18	19	1.069	0.0359	0.0269	0.0205	0.0155	0.0115	0.0237	0.0136	0.0065	0.0010	-0.0035
829000	30	7	1.063	0.0360	0.0269	0.0205	0.0156	0.0115	0.0237	0.0137	0.0066	0.0010	-0.0035
807700	24	13	1.063										
786500	18	19	1.063										
727500	33	4	0.990	0.0365	0.0275	0.0211	0.0161	0.0121	0.0240	0.0140	0.0069	0.0014	-0.0031
718300	30	7	0.990										
700000	24	13	0.990										
681600	18	19	0.990										
632000	15	4	0.927	0.0370	0.0280	0.0216	0.0166	0.0125	0.0244	0.0143	0.0072	0.0017	-0.0028
616200	12	7	0.927										
487400	15	4	0.814	0.0380	0.0289	0.0225	0.0176	0.0135	0.0250	0.0150	0.0079	0.0024	-0.0022
475200	12	7	0.814										
343600	15	4	0.684	0.0392	0.0302	0.0238	0.0189	0.0148	0.0259	0.0158	0.0087	0.0032	-0.0013
335000	12	7	0.684										

Table B.20. Resistance of ACSR Conductors (ohms per mile) (Courtesy Aluminum Company of America)

CODE	AREA CMIL	STRANDS AL	STRANDS ST	DIA. IN.	DC 25°C	AC - 60 Hz 25°C	AC - 60 Hz 50°C	AC - 60 Hz 75°C	AC - 60 Hz 100°C
EXPANDED	3108000	62/8	19	2.500	0.0294	0.0333	0.0362	0.0389	0.0418
EXPANDED	2294000	66/6	19	2.320	0.0399	0.0412	0.0453	0.0493	0.0533
EXPANDED	1414000	58/4	19	1.750	0.0644	0.0663	0.0728	0.0793	0.0859
EXPANDED	1275000	50/4	19	1.600	0.0716	0.0736	0.0808	0.0881	0.0953
KIWI	2167000	72	7	1.737	0.0421	0.0473	0.0515	0.0552	0.0593
BLUEBIRD	2156000	84	19	1.762	0.0420	0.0464	0.0507	0.0545	0.0586
CHUKAR	1780000	84	19	1.602	0.0510	0.0548	0.0599	0.0647	0.0696
FALCON	1590000	54	19	1.545	0.0567	0.0594	0.0653	0.0707	0.0763
LAPWING	1590000	45	7	1.502	0.0571	0.0608	0.0664	0.0719	0.0774
PARROT	1510500	54	19	1.506	0.0597	0.0625	0.0686	0.0744	0.0802
NUTHATCH	1510500	45	7	1.466	0.0602	0.0636	0.0697	0.0755	0.0813
PLOVER	1431000	54	19	1.465	0.0630	0.0657	0.0721	0.0782	0.0843
BOBOLINK	1431000	45	7	1.427	0.0636	0.0668	0.0733	0.0794	0.0856
MARTIN	1351500	54	19	1.424	0.0667	0.0692	0.0760	0.0825	0.0890
DIPPER	1351500	45	7	1.385	0.0672	0.0705	0.0771	0.0836	0.0901
PHEASANT	1272000	54	19	1.382	0.0709	0.0732	0.0805	0.0874	0.0944
BITTERN	1272000	45	7	1.345	0.0715	0.0746	0.0817	0.0886	0.0956
GRACKLE	1192500	54	19	1.333	0.0756	0.0778	0.0855	0.0929	0.1000
BUNTING	1192500	45	7	1.302	0.0762	0.0792	0.0867	0.0942	0.1002
FINCH	1113000	54	19	1.293	0.0810	0.0832	0.0914	0.0993	0.1080
BLUEJAY	1113000	45	7	1.259	0.0818	0.0844	0.0926	0.1010	0.1090
CURLEW	1033500	54	7	1.246	0.0871	0.0893	0.0979	0.1070	0.1150
ORTOLAN	1033500	45	7	1.213	0.0881	0.0905	0.0994	0.1080	0.1170
TANAGER	1033500	36	1	1.186	0.0885	0.0915	0.1010	0.1090	0.1160
CARDINAL	954000	54	7	1.196	0.0944	0.0963	0.1060	0.1150	0.1250
RAIL	954000	45	7	1.165	0.0954	0.0978	0.1080	0.1170	0.1260
CATBIRD	954000	36	1	1.140	0.0959	0.0987	0.1090	0.1180	0.1270
CANARY	900000	54	7	1.162	0.1000	0.1020	0.1120	0.1220	0.1320
RUDDY	900000	45	7	1.131	0.1010	0.1030	0.1130	0.1230	0.1340
MALLARD	795000	30	19	1.140	0.111	0.114	0.125	0.137	0.147
DRAKE	795000	26	7	1.108	0.112	0.114	0.125	0.137	0.147
CONDOR	795000	54	7	1.093	0.113	0.115	0.127	0.138	0.149
CUCKOO	795000	24	7	1.092	0.113	0.114	0.127	0.137	0.148
TERN	795000	45	7	1.063	0.114	0.116	0.128	0.139	0.150
COOT	795000	36	1	1.040	0.115	0.117	0.129	0.141	0.152
REDWING	715500	30	19	1.081	0.124	0.126	0.139	0.151	0.164
STARLING	715500	26	7	1.051	0.125	0.126	0.139	0.151	0.164
STILT	715500	24	7	1.036	0.126	0.127	0.141	0.153	0.165
GANNET	666600	26	7	1.014	0.134	0.135	0.149	0.162	0.176
FLAMINGO	666600	24	7	1.000	0.135	0.137	0.151	0.164	0.177
--------	653900	18	3	0.953	0.140	0.142	0.156	0.171	0.184
EGRET	636000	30	19	1.019	0.139	0.143	0.157	0.172	0.186
GROSBEAK	636000	26	7	0.990	0.140	0.142	0.156	0.170	0.184
ROOK	636000	24	7	0.977	0.142	0.143	0.157	0.172	0.186
KINGBIRD	636000	18	1	0.940	0.143	0.145	0.160	0.174	0.188
SWIFT	636000	36	1	0.930	0.144	0.146	0.161	0.175	0.189
TEAL	605000	30	19	0.994	0.146	0.150	0.165	0.180	0.195
SQUAB	605000	26	7	0.966	0.147	0.149	0.164	0.179	0.193
PEACOCK	605000	24	7	0.953	0.149	0.150	0.165	0.180	0.195
EAGLE	556500	30	7	0.953	0.158	0.163	0.179	0.196	0.212
DOVE	556500	26	7	0.927	0.160	0.162	0.178	0.194	0.211
PARAKEET	556500	24	7	0.914	0.162	0.163	0.179	0.196	0.212
OSPREY	556500	18	1	0.879	0.163	0.166	0.183	0.199	0.215
HEN	477000	30	7	0.883	0.185	0.190	0.209	0.228	0.247
HAWK	477000	26	7	0.858	0.187	0.188	0.207	0.226	0.245
FLICKER	477000	24	7	0.846	0.189	0.190	0.209	0.228	0.247
PELICAN	477000	18	1	0.814	0.191	0.193	0.212	0.232	0.250
LARK	397500	30	7	0.806	0.222	0.227	0.250	0.273	0.295
IBIS	397500	26	7	0.783	0.224	0.226	0.249	0.271	0.294
BRANT	397500	24	7	0.772	0.226	0.227	0.250	0.273	0.295
CHICKADEE	397500	18	1	0.743	0.229	0.231	0.254	0.277	0.300
ORIOLE	336400	30	7	0.741	0.262	0.268	0.295	0.322	0.349
LINNET	336400	26	7	0.721	0.265	0.267	0.294	0.321	0.347
MERLIN	336400	18	1	0.684	0.270	0.273	0.300	0.328	0.355
OSTRICH	300000	26	7	0.680	0.297	0.299	0.329	0.359	0.389

Table B.21. Resistance of ACAR Conductors (ohms per mile)
(Courtesy Kaiser Aluminum and Chemical Sales Inc.)

AREA 62% EQ. EC - AL CMIL	STRANDS EC/6201		DIA. IN.	DC 20°C	AC - 60 Hz 30°C			
					30°C	50°C	75°C	100°C
2413000	72	19	1.821	0.0373	0.0456	0.0483	0.0516	0.0565
2375000	63	28	1.821	0.0379	0.0462	0.0488	0.0521	0.0555
2338000	54	37	1.821	0.0385	0.0467	0.0493	0.0527	0.0561
2297000	54	7	1.762	0.0392	0.0474	0.0502	0.0538	0.0573
2262000	48	13	1.762	0.0399	0.0479	0.0507	0.0543	0.0580
2226000	42	19	1.762	0.0405	0.0485	0.0513	0.0549	0.0585
2227000	54	7	1.735	0.0405	0.0485	0.0514	0.0551	0.0589
2193000	48	13	1.735	0.0411	0.0491	0.0520	0.0557	0.0595
2159000	42	19	1.735	0.0417	0.0497	0.0526	0.0562	0.0601
1899000	54	7	1.602	0.0474	0.0550	0.0585	0.0629	0.0674
1870000	48	13	1.602	0.0482	0.0557	0.0592	0.0636	0.0682
1841000	42	19	1.602	0.0489	0.0564	0.0599	0.0644	0.0689
1673000	54	7	1.504	0.0539	0.0611	0.0651	0.0702	0.0754
1647000	48	13	1.504	0.0546	0.0619	0.0659	0.0711	0.0762
1622000	42	19	1.504	0.0555	0.0627	0.0667	0.0719	0.0771
1337000	54	7	1.345	0.0674	0.0742	0.0794	0.0860	0.0925
1296000	42	19	1.345	0.0695	0.0763	0.0815	0.0881	0.0947
1243000	30	7	1.302	0.0725	0.0793	0.0849	0.0919	0.0989
1211000	24	13	1.302	0.0744	0.0812	0.0868	0.0937	0.1008
1179000	18	19	1.302	0.0764	0.0831	0.0887	0.0957	0.1028
1163000	30	7	1.259	0.0775	0.0842	0.0902	0.0977	0.1052
1133000	24	13	1.259	0.0795	0.0862	0.0922	0.0997	0.1073
1104000	18	19	1.259	0.0816	0.0882	0.0942	0.1018	0.1095
1153000	33	4	1.246	0.0781	0.0850	0.0910	0.0987	0.1064
1138000	30	7	1.246	0.0791	0.0859	0.0920	0.0997	0.1074
1109000	24	13	1.246	0.0812	0.0880	0.0941	0.1017	0.1095
1080000	18	19	1.246	0.0834	0.0900	0.0962	0.1039	0.1117
1077000	30	7	1.212	0.0836	0.0904	0.0969	0.1050	0.1132
1049000	24	13	1.212	0.0859	0.0926	0.0991	0.1072	0.1154
1022000	18	19	1.212	0.0882	0.0948	0.1013	0.1096	0.1177
1050000	30	7	1.196	0.0859	0.0926	0.0993	0.1076	0.1160
1023000	24	13	1.196	0.0881	0.0948	0.1015	0.1099	0.1183
996000	18	19	1.196	0.0904	0.0971	0.1038	0.1123	0.1207
994800	30	7	1.165	0.0906	0.0974	0.1044	0.1133	0.1221
954600	30	7	1.141	0.0944	0.1012	0.1086	0.1178	0.1271
969300	24	13	1.165	0.0929	0.0997	0.1068	0.1156	0.1246
958000	24	13	1.158	0.0941	0.1008	0.1080	0.1170	0.1260
943900	18	19	1.165	0.0955	0.1021	0.1092	0.1182	0.1271
900300	30	7	1.108	0.1001	0.1070	0.1148	0.1246	0.1345
795000	30	7	1.042	0.1133	0.1204	0.1293	0.1404	0.1516
877300	24	13	1.108	0.1027	0.1096	0.1174	0.1272	0.1372
795000	24	13	1.055	0.1133	0.1204	0.1290	0.1400	0.1509
854200	18	19	1.108	0.1054	0.1123	0.1201	0.1300	0.1400
795000	18	19	1.069	0.1134	0.1203	0.1288	0.1394	0.1501
829000	30	7	1.063	0.1087	0.1157	0.1243	0.1350	0.1456
807700	24	13	1.063	0.1115	0.1185	0.1271	0.1378	0.1486
786500	18	19	1.063	0.1145	0.1216	0.1302	0.1410	0.1518
727500	33	4	0.990	0.1238	0.1312	0.1411	0.1534	0.1658
718300	30	7	0.990	0.1254	0.1327	0.1427	0.1550	0.1675
700000	24	13	0.990	0.1287	0.1360	0.1459	0.1584	0.1709
681600	18	19	0.990	0.1322	0.1394	0.1494	0.1619	0.1743
632000	15	4	0.927	0.1425	0.1503	0.1615	0.1756	0.1897
616200	12	7	0.927	0.1462	0.1539	0.1652	0.1794	0.1935
487400	15	4	0.814	0.1849	0.1938	0.2085	0.2268	0.2453
475200	12	7	0.814	0.1896	0.1986	0.2133	0.2317	0.2501
343600	15	4	0.684	0.2623	0.2739	0.2948	0.3209	0.3470
335000	12	7	0.684	0.2690	0.2806	0.3015	0.3278	0.3540

Table B.22. Electrical Characteristics of Overhead Ground Wires

PART A: ALUMOWELD[1] STRAND

STRAND (AWG)	RESISTANCE (OHMS/MILE) SMALL CURRENTS 25°C DC	SMALL CURRENTS 25°C 60 Hz	75% OF CAP. 75°C DC	75% OF CAP. 75°C 60 Hz	60 Hz REACTANCE FOR 1 FOOT RADIUS INDUCTIVE OHMS/MILE	CAPACITIVE MEGOHM-MILES	60 Hz GEOMETRIC MEAN RADIUS FEET
7 NO. 5	1.217	1.240	1.432	1.669	0.707	0.1122	0.002958
7 NO. 6	1.507	1.536	1.773	2.010	0.721	0.1157	0.002633
7 NO. 7	1.900	1.937	2.240	2.470	0.735	0.1191	0.002345
7 NO. 8	2.400	2.440	2.820	3.060	0.749	0.1226	0.002085
7 NO. 9	3.020	3.080	3.560	3.800	0.763	0.1260	0.001858
7 NO.10	3.810	3.880	4.480	4.730	0.777	0.1294	0.001658
3 NO. 5	2.780	2.780	3.270	3.560	0.707	0.1221	0.002940
3 NO. 6	3.510	3.510	4.130	4.410	0.721	0.1255	0.002618
3 NO. 7	4.420	4.420	5.210	5.470	0.735	0.1289	0.002333
3 NO. 8	5.580	5.580	6.570	6.820	0.749	0.1324	0.002078
3 NO. 9	7.040	7.040	8.280	8.520	0.763	0.1358	0.001853
3 NO.10	8.870	8.870	10.440	10.670	0.777	0.1392	0.001650

PART B: SINGLE LAYER ACSR[2]

CODE	RESISTANCE (OHMS/MILE) DC 25°C	60 Hz & 75°C I=0[3]	60 Hz & 75°C I=100	60 Hz & 75°C I=200	60 Hz REACTANCE FOR 1 FOOT RADIUS INDUCTIVE OHMS/MILE AT 75°C I=0	INDUCTIVE I=100	INDUCTIVE I=200	CAPACITIVE MEGOHM-MILES
BRAHMA	0.394	0.470	0.510	0.565	0.500	0.520	0.545	0.1043
COCHIN	0.400	0.480	0.520	0.590	0.505	0.515	0.550	0.1065
DORKING	0.443	0.535	0.575	0.650	0.515	0.530	0.565	0.1079
DOTTEREL	0.479	0.565	0.620	0.705	0.515	0.530	0.575	0.1091
GUINEA	0.531	0.630	0.685	0.780	0.520	0.545	0.590	0.1106
LEGHORN	0.630	0.760	0.810	0.930	0.530	0.550	0.605	0.1131
MINORCA	0.765	0.915	0.980	1.130	0.540	0.570	0.640	0.1160
PETREL	0.830	1.000	1.065	1.220	0.550	0.580	0.655	0.1172
GROUSE	1.080	1.295	1.420	1.520	0.570	0.640	0.675	0.1240

PART C: STEEL CONDUCTORS[4]

GRADE (7-STRAND)	DIA. IN.	RESISTANCE (OHMS/MILE) 60 Hz I=0[3]	60 Hz I=30	60 Hz I=60	60 Hz REACTANCE FOR 1 FOOT RADIUS INDUCTIVE OHMS/MILE I=0	INDUCTIVE I=30	INDUCTIVE I=60	CAPACITIVE MEGOHM-MILES
ORDINARY	1/4	9.5	11.4	11.3	1.3970	3.7431	3.4379	0.1354
ORDINARY	9/32	7.1	9.2	9.0	1.2027	3.0734	2.5146	0.1319
ORDINARY	5/16	5.4	7.5	7.8	0.8382	2.5146	2.0409	0.1288
ORDINARY	3/8	4.3	6.5	6.6	0.8382	2.2352	1.9687	0.1234
ORDINARY	1/2	2.3	4.3	5.0	0.7049	1.6893	1.4236	0.1148
E.B.	1/4	8.0	12.0	10.1	1.2027	4.4704	3.1565	0.1354
E.B.	9/32	6.0	10.0	8.7	1.1305	3.7783	2.6255	0.1319
E.B.	5/16	4.9	8.0	7.0	0.9843	2.9401	2.5146	0.1288
E.B.	3/8	3.7	7.0	6.3	0.8382	2.5997	2.4303	0.1234
E.B.	1/2	2.1	4.9	5.0	0.7049	1.8715	1.7615	0.1148
E.B.B.	1/4	7.0	12.8	10.9	1.6764	5.1401	3.9482	0.1354
E.B.B.	9/32	5.4	10.9	8.7	1.1305	4.4833	3.7783	0.1319
E.B.B.	5/16	4.0	9.0	6.8	0.9843	3.6322	3.0734	0.1288
E.B.B.	3/8	3.5	7.9	6.0	0.8382	3.1168	2.7940	0.1234
E.B.B.	1/2	2.0	5.7	4.7	0.7049	2.3461	2.2352	0.1148

1. DATA COMPILED FROM E.D. 3015 - COPPERWELD STEEL COMPANY.
2. DATA COMPILED FROM "RESISTANCE AND REACTANCE OF ALUMINUM CONDUCTORS" - ALCOA.
3. CONDUCTOR CURRENT IN AMPERES.
4. DATA COMPILED FROM "SYMMETRICAL COMPONENTS" WAGNER & EVANS (BOOK) McGRAW-HILL.

Table B.23. Typical EHV Line Characteristics

Line to Line Voltage (kV)	Cond. per Phase at 18 (Inches)	Conductor Code and Diameter (Inches)	Phase Spacing (Feet)	GMD (Feet)	60 Hz Inductive Reactance in Ohms per Mile			60 Hz Capacitive Reactance in Megohm-Miles			Z_0 (Ohms)	SIL (MVA)
					X_A	X_D	$X_A + X_D$	X_A'	X_D'	$X_A' + X_D'$		
345	1	Expanded — 1.750	28	35.3	0.3336	0.4325	0.7661	0.0777	0.1057	0.1834	374.8	318
345	2	*Curlew — 1.246	28	35.3	0.1677	0.4325	0.6002	0.0379	0.1057	0.1436	293.6	405
500	1	Expanded — 2.500	38	47.9	0.2922	0.4694	0.7616	0.0671	0.1147	0.1818	372.1	672
500	2	*Chukar — 1.602	38	47.9	0.1529	0.4694	0.6223	0.0341	0.1147	0.1488	304.3	822
500	3	Rail — 1.165	38	47.9	0.0988	0.4694	0.5682	0.0219	0.1147	0.1366	278.6	897
500	4	*Parakeet — 0.914	38	47.9	0.0584	0.4694	0.5278	0.0126	0.1147	0.1273	259.2	965
735	3	Expanded — 1.750	56	70.6	0.0784	0.5166	0.5950	0.0179	0.1263	0.1442	292.9	1 844
735	4	Pheasant — 1.382	56	70.6	0.0456	0.5166	0.5622	0.0096	0.1263	0.1359	276.4	1 955

* Nearest to base-case diameter.

Table B.24. Inductive Reactance Spacing Factor x_d (ohms/conductor/mile) at 60 Hz

	0.0	0.1	0.2	0.3	0.4	0.5	0.6	0.7	0.8	0.9
0	------	-0.2794	-0.1953	-0.1461	-0.1112	-0.0841	-0.0620	-0.0433	-0.0271	-0.0128
1	0.0	0.0116	0.0221	0.0318	0.0408	0.0492	0.0570	0.0644	0.0713	0.0779
2	0.0841	0.0900	0.0957	0.1011	0.1062	0.1112	0.1159	0.1205	0.1249	0.1292
3	0.1333	0.1373	0.1411	0.1449	0.1485	0.1520	0.1554	0.1588	0.1620	0.1651
4	0.1682	0.1712	0.1741	0.1770	0.1798	0.1825	0.1852	0.1878	0.1903	0.1928
5	0.1953	0.1977	0.2001	0.2024	0.2046	0.2069	0.2090	0.2112	0.2133	0.2154
6	0.2174	0.2194	0.2214	0.2233	0.2252	0.2271	0.2290	0.2308	0.2326	0.2344
7	0.2361	0.2378	0.2395	0.2412	0.2429	0.2445	0.2461	0.2477	0.2493	0.2508
8	0.2523	0.2538	0.2553	0.2568	0.2582	0.2597	0.2611	0.2625	0.2639	0.2653
9	0.2666	0.2680	0.2693	0.2706	0.2719	0.2732	0.2744	0.2757	0.2769	0.2782
10	0.2794	0.2806	0.2818	0.2830	0.2842	0.2853	0.2865	0.2876	0.2887	0.2899
11	0.2910	0.2921	0.2932	0.2942	0.2953	0.2964	0.2974	0.2985	0.2995	0.3005
12	0.3015	0.3025	0.3035	0.3045	0.3055	0.3065	0.3074	0.3084	0.3094	0.3103
13	0.3112	0.3122	0.3131	0.3140	0.3149	0.3158	0.3167	0.3176	0.3185	0.3194
14	0.3202	0.3211	0.3219	0.3228	0.3236	0.3245	0.3253	0.3261	0.3270	0.3278
15	0.3286	0.3294	0.3302	0.3310	0.3318	0.3326	0.3334	0.3341	0.3349	0.3357
16	0.3364	0.3372	0.3379	0.3387	0.3394	0.3402	0.3409	0.3416	0.3424	0.3431
17	0.3438	0.3445	0.3452	0.3459	0.3466	0.3473	0.3480	0.3487	0.3494	0.3500
18	0.3507	0.3514	0.3521	0.3527	0.3534	0.3540	0.3547	0.3554	0.3560	0.3566
19	0.3573	0.3579	0.3586	0.3592	0.3598	0.3604	0.3611	0.3617	0.3623	0.3629
20	0.3635	0.3641	0.3647	0.3653	0.3659	0.3665	0.3671	0.3677	0.3683	0.3688
21	0.3694	0.3700	0.3706	0.3711	0.3717	0.3723	0.3728	0.3734	0.3740	0.3745
22	0.3751	0.3756	0.3762	0.3767	0.3773	0.3778	0.3783	0.3789	0.3794	0.3799
23	0.3805	0.3810	0.3815	0.3820	0.3826	0.3831	0.3836	0.3841	0.3846	0.3851
24	0.3856	0.3861	0.3866	0.3871	0.3876	0.3881	0.3886	0.3891	0.3896	0.3901
25	0.3906	0.3911	0.3916	0.3920	0.3925	0.3930	0.3935	0.3939	0.3944	0.3949
26	0.3953	0.3958	0.3963	0.3967	0.3972	0.3977	0.3981	0.3986	0.3990	0.3995
27	0.3999	0.4004	0.4008	0.4013	0.4017	0.4021	0.4026	0.4030	0.4035	0.4039
28	0.4043	0.4048	0.4052	0.4056	0.4061	0.4065	0.4069	0.4073	0.4078	0.4082
29	0.4086	0.4090	0.4094	0.4098	0.4103	0.4107	0.4111	0.4115	0.4119	0.4123
30	0.4127	0.4131	0.4135	0.4139	0.4143	0.4147	0.4151	0.4155	0.4159	0.4163
31	0.4167	0.4171	0.4175	0.4179	0.4182	0.4186	0.4190	0.4194	0.4198	0.4202

Table B.24 *(continued)*

	0.0	0.1	0.2	0.3	0.4	0.5	0.6	0.7	0.8	0.9
32	0.4205	0.4209	0.4213	0.4217	0.4220	0.4224	0.4228	0.4232	0.4235	0.4239
33	0.4243	0.4246	0.4250	0.4254	0.4257	0.4261	0.4265	0.4268	0.4272	0.4275
34	0.4279	0.4283	0.4286	0.4290	0.4293	0.4297	0.4300	0.4304	0.4307	0.4311
35	0.4314	0.4318	0.4321	0.4324	0.4328	0.4331	0.4335	0.4338	0.4342	0.4345
36	0.4348	0.4352	0.4355	0.4358	0.4362	0.4365	0.4368	0.4372	0.4375	0.4378
37	0.4382	0.4385	0.4388	0.4391	0.4395	0.4398	0.4401	0.4404	0.4408	0.4411
38	0.4414	0.4417	0.4420	0.4423	0.4427	0.4430	0.4433	0.4436	0.4439	0.4442
39	0.4445	0.4449	0.4452	0.4455	0.4458	0.4461	0.4464	0.4467	0.4470	0.4473
40	0.4476	0.4479	0.4482	0.4485	0.4488	0.4491	0.4494	0.4497	0.4500	0.4503
41	0.4506	0.4509	0.4512	0.4515	0.4518	0.4521	0.4524	0.4527	0.4530	0.4532
42	0.4535	0.4538	0.4541	0.4544	0.4547	0.4550	0.4553	0.4555	0.4558	0.4561
43	0.4564	0.4567	0.4570	0.4572	0.4575	0.4578	0.4581	0.4584	0.4586	0.4589
44	0.4592	0.4595	0.4597	0.4600	0.4603	0.4606	0.4608	0.4611	0.4614	0.4616
45	0.4619	0.4622	0.4624	0.4627	0.4630	0.4632	0.4635	0.4638	0.4640	0.4643
46	0.4646	0.4648	0.4651	0.4654	0.4656	0.4659	0.4661	0.4664	0.4667	0.4669
47	0.4672	0.4674	0.4677	0.4680	0.4682	0.4685	0.4687	0.4690	0.4692	0.4695
48	0.4697	0.4700	0.4702	0.4705	0.4707	0.4710	0.4712	0.4715	0.4717	0.4720
49	0.4722	0.4725	0.4727	0.4730	0.4732	0.4735	0.4737	0.4740	0.4742	0.4744
50	0.4747	0.4749	0.4752	0.4754	0.4757	0.4759	0.4761	0.4764	0.4766	0.4769
51	0.4771	0.4773	0.4776	0.4778	0.4780	0.4783	0.4785	0.4787	0.4790	0.4792
52	0.4795	0.4797	0.4799	0.4801	0.4804	0.4806	0.4808	0.4811	0.4813	0.4815
53	0.4818	0.4820	0.4822	0.4824	0.4827	0.4829	0.4831	0.4834	0.4836	0.4838
54	0.4840	0.4843	0.4845	0.4847	0.4849	0.4851	0.4854	0.4856	0.4858	0.4860
55	0.4863	0.4865	0.4867	0.4869	0.4871	0.4874	0.4876	0.4878	0.4880	0.4882
56	0.4884	0.4887	0.4889	0.4891	0.4893	0.4895	0.4897	0.4900	0.4902	0.4904
57	0.4906	0.4908	0.4910	0.4912	0.4914	0.4917	0.4919	0.4921	0.4923	0.4925
58	0.4927	0.4929	0.4931	0.4933	0.4935	0.4937	0.4940	0.4942	0.4944	0.4946
59	0.4948	0.4950	0.4952	0.4954	0.4956	0.4958	0.4960	0.4962	0.4964	0.4966
60	0.4968	0.4970	0.4972	0.4974	0.4976	0.4978	0.4980	0.4982	0.4984	0.4986
61	0.4988	0.4990	0.4992	0.4994	0.4996	0.4998	0.5000	0.5002	0.5004	0.5006
62	0.5008	0.5010	0.5012	0.5014	0.5016	0.5018	0.5020	0.5022	0.5023	0.5025
63	0.5027	0.5029	0.5031	0.5033	0.5035	0.5037	0.5039	0.5041	0.5043	0.5045
64	0.5046	0.5048	0.5050	0.5052	0.5054	0.5056	0.5058	0.5060	0.5062	0.5063

65	0.5065	0.5067	0.5069	0.5071	0.5073	0.5075	0.5076	0.5078	0.5080	0.5082
66	0.5084	0.5086	0.5087	0.5089	0.5091	0.5093	0.5095	0.5097	0.5098	0.5100
67	0.5102	0.5104	0.5106	0.5107	0.5109	0.5111	0.5113	0.5115	0.5116	0.5118
68	0.5120	0.5122	0.5124	0.5125	0.5127	0.5129	0.5131	0.5132	0.5134	0.5136
69	0.5138	0.5139	0.5141	0.5143	0.5145	0.5147	0.5148	0.5150	0.5152	0.5153
70	0.5155	0.5157	0.5159	0.5160	0.5162	0.5164	0.5166	0.5167	0.5169	0.5171
71	0.5172	0.5174	0.5176	0.5178	0.5179	0.5181	0.5183	0.5184	0.5186	0.5188
72	0.5189	0.5191	0.5193	0.5194	0.5196	0.5198	0.5199	0.5201	0.5203	0.5204
73	0.5206	0.5208	0.5209	0.5211	0.5213	0.5214	0.5216	0.5218	0.5219	0.5221
74	0.5223	0.5224	0.5226	0.5228	0.5229	0.5231	0.5232	0.5234	0.5236	0.5237
75	0.5239	0.5241	0.5242	0.5244	0.5245	0.5247	0.5249	0.5250	0.5252	0.5253
76	0.5255	0.5257	0.5258	0.5260	0.5261	0.5263	0.5265	0.5266	0.5268	0.5269
77	0.5271	0.5272	0.5274	0.5276	0.5277	0.5279	0.5280	0.5282	0.5283	0.5285
78	0.5287	0.5288	0.5290	0.5291	0.5293	0.5294	0.5296	0.5297	0.5299	0.5300
79	0.5302	0.5304	0.5305	0.5307	0.5308	0.5310	0.5311	0.5313	0.5314	0.5316
80	0.5317	0.5319	0.5320	0.5322	0.5323	0.5325	0.5326	0.5328	0.5329	0.5331
81	0.5332	0.5334	0.5335	0.5337	0.5338	0.5340	0.5341	0.5343	0.5344	0.5346
82	0.5347	0.5349	0.5350	0.5352	0.5353	0.5355	0.5356	0.5358	0.5359	0.5360
83	0.5362	0.5363	0.5365	0.5366	0.5368	0.5369	0.5371	0.5372	0.5374	0.5375
84	0.5376	0.5378	0.5379	0.5381	0.5382	0.5384	0.5385	0.5387	0.5388	0.5389
85	0.5391	0.5392	0.5394	0.5395	0.5396	0.5398	0.5399	0.5401	0.5402	0.5404
86	0.5405	0.5406	0.5408	0.5409	0.5411	0.5412	0.5413	0.5415	0.5416	0.5418
87	0.5419	0.5420	0.5422	0.5423	0.5425	0.5426	0.5427	0.5429	0.5430	0.5432
88	0.5433	0.5434	0.5436	0.5437	0.5438	0.5440	0.5441	0.5442	0.5444	0.5445
89	0.5447	0.5448	0.5449	0.5451	0.5452	0.5453	0.5455	0.5456	0.5457	0.5459
90	0.5460	0.5461	0.5463	0.5464	0.5466	0.5467	0.5468	0.5470	0.5471	0.5472
91	0.5474	0.5475	0.5476	0.5478	0.5479	0.5480	0.5482	0.5483	0.5484	0.5486
92	0.5487	0.5488	0.5489	0.5491	0.5492	0.5493	0.5495	0.5496	0.5497	0.5499
93	0.5500	0.5501	0.5503	0.5504	0.5505	0.5506	0.5508	0.5509	0.5510	0.5512
94	0.5513	0.5514	0.5515	0.5517	0.5518	0.5519	0.5521	0.5522	0.5523	0.5524
95	0.5526	0.5527	0.5528	0.5530	0.5531	0.5532	0.5533	0.5535	0.5536	0.5537
96	0.5538	0.5540	0.5541	0.5542	0.5544	0.5545	0.5546	0.5547	0.5549	0.5550
97	0.5551	0.5552	0.5554	0.5555	0.5556	0.5557	0.5559	0.5560	0.5561	0.5562
98	0.5563	0.5565	0.5566	0.5567	0.5568	0.5570	0.5571	0.5572	0.5573	0.5575
99	0.5576	0.5577	0.5578	0.5579	0.5581	0.5582	0.5583	0.5584	0.5586	0.5587
100	0.5588	0.5589	0.5590	0.5592	0.5593	0.5594	0.5595	0.5596	0.5598	0.5599

See note p. 462.

Table B.24 *(continued)*

Zero-Sequence Resistance & Inductive Factors (r_e, x_e)*
(ohms/conductor/mile)

	ρ (ohm-meter)	r_e, x_e (f = 60 Hz)
r_e	All	0.2860
x_e	1	2.050
	5	2.343
	10	2.469
	50	2.762
	100†	2.888†
	500	3.181
	1000	3.307
	5000	3.600
	10,000	3.726

*From formulas:

$$r_e = 0.004764f$$

$$x_e = 0.006985f \log_{10} 4{,}665{,}600 \frac{\rho}{f}$$

where f = frequency
ρ = resistivity (ohm-meter)

†This is an average value which may be used in the absence of definite information.

Fundamental Equations:

$$z_1 = z_2 = r_a + j(x_a + x_d)$$

$$z_0 = r_a + r_e + j(x_a + x_e - 2x_d)$$

$$x_d = \omega k \ln d$$

d = separation, ft

Table B.25. Shunt Capacitive Reactance Spacing Factor x'_d (megohms/conductor/mile) at 60 Hz

	0.0	0.1	0.2	0.3	0.4	0.5	0.6	0.7	0.8	0.9
0	------	-0.0683	-0.0477	-0.0357	-0.0272	-0.0206	-0.0152	-0.0106	-0.0066	-0.0031
1	0.0000	0.0028	0.0054	0.0078	0.0100	0.0120	0.0139	0.0157	0.0174	0.0190
2	0.0206	0.0220	0.0234	0.0247	0.0260	0.0272	0.0283	0.0295	0.0305	0.0316
3	0.0326	0.0336	0.0345	0.0354	0.0363	0.0372	0.0380	0.0388	0.0396	0.0404
4	0.0411	0.0419	0.0426	0.0433	0.0440	0.0446	0.0453	0.0459	0.0465	0.0471
5	0.0477	0.0483	0.0489	0.0495	0.0500	0.0506	0.0511	0.0516	0.0521	0.0527
6	0.0532	0.0536	0.0541	0.0546	0.0551	0.0555	0.0560	0.0564	0.0569	0.0573
7	0.0577	0.0581	0.0586	0.0590	0.0594	0.0598	0.0602	0.0606	0.0609	0.0613
8	0.0617	0.0621	0.0624	0.0628	0.0631	0.0635	0.0638	0.0642	0.0645	0.0649
9	0.0652	0.0655	0.0658	0.0662	0.0665	0.0668	0.0671	0.0674	0.0677	0.0680
10	0.0683	0.0686	0.0689	0.0692	0.0695	0.0698	0.0700	0.0703	0.0706	0.0709
11	0.0711	0.0714	0.0717	0.0719	0.0722	0.0725	0.0727	0.0730	0.0732	0.0735
12	0.0737	0.0740	0.0742	0.0745	0.0747	0.0749	0.0752	0.0754	0.0756	0.0759
13	0.0761	0.0763	0.0765	0.0768	0.0770	0.0772	0.0774	0.0776	0.0779	0.0781
14	0.0783	0.0785	0.0787	0.0789	0.0791	0.0793	0.0795	0.0797	0.0799	0.0801
15	0.0803	0.0805	0.0807	0.0809	0.0811	0.0813	0.0815	0.0817	0.0819	0.0821
16	0.0823	0.0824	0.0826	0.0828	0.0830	0.0832	0.0833	0.0835	0.0837	0.0839
17	0.0841	0.0842	0.0844	0.0846	0.0847	0.0849	0.0851	0.0852	0.0854	0.0856
18	0.0857	0.0859	0.0861	0.0862	0.0864	0.0866	0.0867	0.0869	0.0870	0.0872
19	0.0874	0.0875	0.0877	0.0878	0.0880	0.0881	0.0883	0.0884	0.0886	0.0887
20	0.0889	0.0890	0.0892	0.0893	0.0895	0.0896	0.0898	0.0899	0.0900	0.0902
21	0.0903	0.0905	0.0906	0.0907	0.0909	0.0910	0.0912	0.0913	0.0914	0.0916
22	0.0917	0.0918	0.0920	0.0921	0.0922	0.0924	0.0925	0.0926	0.0928	0.0929
23	0.0930	0.0931	0.0933	0.0934	0.0935	0.0937	0.0938	0.0939	0.0940	0.0942
24	0.0943	0.0944	0.0945	0.0947	0.0948	0.0949	0.0950	0.0951	0.0953	0.0954
25	0.0955	0.0956	0.0957	0.0958	0.0960	0.0961	0.0962	0.0963	0.0964	0.0965
26	0.0967	0.0968	0.0969	0.0970	0.0971	0.0972	0.0973	0.0974	0.0976	0.0977
27	0.0978	0.0979	0.0980	0.0981	0.0982	0.0983	0.0984	0.0985	0.0986	0.0987
28	0.0989	0.0990	0.0991	0.0992	0.0993	0.0994	0.0995	0.0996	0.0997	0.0998
29	0.0999	0.1000	0.1001	0.1002	0.1003	0.1004	0.1005	0.1006	0.1007	0.1008
30	0.1009	0.1010	0.1011	0.1012	0.1013	0.1014	0.1015	0.1016	0.1017	0.1018
31	0.1019	0.1020	0.1021	0.1022	0.1023	0.1023	0.1024	0.1025	0.1026	0.1027

Table B.25 *(continued)*

	0.0	0.1	0.2	0.3	0.4	0.5	0.6	0.7	0.8	0.9
32	0.1028	0.1029	0.1030	0.1031	0.1032	0.1033	0.1034	0.1035	0.1035	0.1036
33	0.1037	0.1038	0.1039	0.1040	0.1041	0.1042	0.1043	0.1044	0.1044	0.1045
34	0.1046	0.1047	0.1048	0.1049	0.1050	0.1050	0.1051	0.1052	0.1053	0.1054
35	0.1055	0.1056	0.1056	0.1057	0.1058	0.1059	0.1060	0.1061	0.1061	0.1062
36	0.1063	0.1064	0.1065	0.1066	0.1066	0.1067	0.1068	0.1069	0.1070	0.1070
37	0.1071	0.1072	0.1073	0.1074	0.1074	0.1075	0.1076	0.1077	0.1078	0.1078
38	0.1079	0.1080	0.1081	0.1081	0.1082	0.1083	0.1084	0.1085	0.1085	0.1086
39	0.1087	0.1088	0.1088	0.1089	0.1090	0.1091	0.1091	0.1092	0.1093	0.1094
40	0.1094	0.1095	0.1096	0.1097	0.1097	0.1098	0.1099	0.1100	0.1100	0.1101
41	0.1102	0.1102	0.1103	0.1104	0.1105	0.1105	0.1106	0.1107	0.1107	0.1108
42	0.1109	0.1110	0.1110	0.1111	0.1112	0.1112	0.1113	0.1114	0.1114	0.1115
43	0.1116	0.1117	0.1117	0.1118	0.1119	0.1119	0.1120	0.1121	0.1121	0.1122
44	0.1123	0.1123	0.1124	0.1125	0.1125	0.1126	0.1127	0.1127	0.1128	0.1129
45	0.1129	0.1130	0.1131	0.1131	0.1132	0.1133	0.1133	0.1134	0.1135	0.1135
46	0.1136	0.1136	0.1137	0.1138	0.1138	0.1139	0.1140	0.1140	0.1141	0.1142
47	0.1142	0.1143	0.1143	0.1144	0.1145	0.1145	0.1146	0.1147	0.1147	0.1148
48	0.1148	0.1149	0.1150	0.1150	0.1151	0.1152	0.1152	0.1153	0.1153	0.1154
49	0.1155	0.1155	0.1156	0.1156	0.1157	0.1158	0.1158	0.1159	0.1159	0.1160
50	0.1161	0.1161	0.1162	0.1162	0.1163	0.1164	0.1164	0.1165	0.1165	0.1166
51	0.1166	0.1167	0.1168	0.1168	0.1169	0.1169	0.1170	0.1170	0.1171	0.1172
52	0.1172	0.1173	0.1173	0.1174	0.1174	0.1175	0.1176	0.1176	0.1177	0.1177
53	0.1178	0.1178	0.1179	0.1180	0.1180	0.1181	0.1181	0.1182	0.1182	0.1183
54	0.1183	0.1184	0.1184	0.1185	0.1186	0.1186	0.1187	0.1187	0.1188	0.1188
55	0.1189	0.1189	0.1190	0.1190	0.1191	0.1192	0.1192	0.1193	0.1193	0.1194
56	0.1194	0.1195	0.1195	0.1196	0.1196	0.1197	0.1197	0.1198	0.1198	0.1199
57	0.1199	0.1200	0.1200	0.1201	0.1202	0.1202	0.1203	0.1203	0.1204	0.1204
58	0.1205	0.1205	0.1206	0.1206	0.1207	0.1207	0.1208	0.1208	0.1209	0.1209
59	0.1210	0.1210	0.1211	0.1211	0.1212	0.1212	0.1213	0.1213	0.1214	0.1214
60	0.1215	0.1215	0.1216	0.1216	0.1217	0.1217	0.1218	0.1218	0.1219	0.1219
61	0.1220	0.1220	0.1221	0.1221	0.1221	0.1222	0.1222	0.1223	0.1223	0.1224
62	0.1224	0.1225	0.1225	0.1226	0.1226	0.1227	0.1227	0.1228	0.1228	0.1229
63	0.1229	0.1230	0.1230	0.1231	0.1231	0.1231	0.1232	0.1232	0.1233	0.1233

64	0.1234	0.1234	0.1235	0.1235	0.1236	0.1236	0.1237	0.1237	0.1237	0.1238
65	0.1238	0.1239	0.1239	0.1240	0.1240	0.1241	0.1241	0.1242	0.1242	0.1242
66	0.1243	0.1243	0.1244	0.1244	0.1245	0.1245	0.1246	0.1246	0.1247	0.1247
67	0.1247	0.1248	0.1248	0.1249	0.1249	0.1250	0.1250	0.1250	0.1251	0.1251
68	0.1252	0.1252	0.1253	0.1253	0.1254	0.1254	0.1254	0.1255	0.1255	0.1256
69	0.1256	0.1257	0.1257	0.1257	0.1258	0.1258	0.1259	0.1259	0.1260	0.1260
70	0.1260	0.1261	0.1261	0.1262	0.1262	0.1262	0.1263	0.1263	0.1264	0.1264
71	0.1265	0.1265	0.1265	0.1266	0.1266	0.1267	0.1267	0.1268	0.1268	0.1268
72	0.1269	0.1269	0.1270	0.1270	0.1270	0.1271	0.1271	0.1272	0.1272	0.1272
73	0.1273	0.1273	0.1274	0.1274	0.1274	0.1275	0.1275	0.1276	0.1276	0.1276
74	0.1277	0.1277	0.1278	0.1278	0.1278	0.1279	0.1279	0.1280	0.1280	0.1280
75	0.1281	0.1281	0.1282	0.1282	0.1282	0.1283	0.1283	0.1284	0.1284	0.1284
76	0.1285	0.1285	0.1286	0.1286	0.1286	0.1287	0.1287	0.1288	0.1288	0.1288
77	0.1289	0.1289	0.1289	0.1290	0.1290	0.1291	0.1291	0.1291	0.1292	0.1292
78	0.1292	0.1293	0.1293	0.1294	0.1294	0.1294	0.1295	0.1295	0.1296	0.1296
79	0.1296	0.1297	0.1297	0.1297	0.1298	0.1298	0.1299	0.1299	0.1299	0.1300
80	0.1300	0.1300	0.1301	0.1301	0.1301	0.1302	0.1302	0.1303	0.1303	0.1303
81	0.1304	0.1304	0.1304	0.1305	0.1305	0.1306	0.1306	0.1306	0.1307	0.1307
82	0.1307	0.1308	0.1308	0.1308	0.1309	0.1309	0.1309	0.1310	0.1310	0.1311
83	0.1311	0.1311	0.1312	0.1312	0.1312	0.1313	0.1313	0.1313	0.1314	0.1314
84	0.1314	0.1315	0.1315	0.1316	0.1316	0.1316	0.1317	0.1317	0.1317	0.1318
85	0.1318	0.1318	0.1319	0.1319	0.1319	0.1320	0.1320	0.1320	0.1321	0.1321
86	0.1321	0.1322	0.1322	0.1322	0.1323	0.1323	0.1324	0.1324	0.1324	0.1325
87	0.1325	0.1325	0.1326	0.1326	0.1326	0.1327	0.1327	0.1327	0.1328	0.1328
88	0.1328	0.1329	0.1329	0.1329	0.1330	0.1330	0.1330	0.1331	0.1331	0.1331
89	0.1332	0.1332	0.1332	0.1333	0.1333	0.1333	0.1334	0.1334	0.1334	0.1335
90	0.1335	0.1335	0.1336	0.1336	0.1336	0.1337	0.1337	0.1337	0.1338	0.1338
91	0.1338	0.1339	0.1339	0.1339	0.1340	0.1340	0.1340	0.1340	0.1341	0.1341
92	0.1341	0.1342	0.1342	0.1342	0.1343	0.1343	0.1343	0.1344	0.1344	0.1344
93	0.1345	0.1345	0.1345	0.1346	0.1346	0.1346	0.1347	0.1347	0.1347	0.1348
94	0.1348	0.1348	0.1348	0.1349	0.1349	0.1349	0.1350	0.1350	0.1350	0.1351
95	0.1351	0.1351	0.1352	0.1352	0.1352	0.1353	0.1353	0.1353	0.1353	0.1354
96	0.1354	0.1354	0.1355	0.1355	0.1355	0.1356	0.1356	0.1356	0.1357	0.1357
97	0.1357	0.1357	0.1358	0.1358	0.1358	0.1359	0.1359	0.1359	0.1360	0.1360
98	0.1360	0.1361	0.1361	0.1361	0.1361	0.1362	0.1362	0.1362	0.1363	0.1363
99	0.1363	0.1364	0.1364	0.1364	0.1364	0.1365	0.1365	0.1365	0.1366	0.1366
100	0.1366	0.1366	0.1367	0.1367	0.1367	0.1368	0.1368	0.1368	0.1369	0.1369

See note p. 466.

Table B.25 (*continued*)

Zero-Sequence Shunt Capacitive Reactance Factor x'_0 (megohms/conductor/mi)

Conductor Height above Ground (ft)	x'_0 (f = 60 Hz)
10	0.267
15	0.303
20	0.328
25	0.318
30	0.364
40	0.390
50	0.410
60	0.426
70	0.440
80	0.452
90	0.462
100	0.472

$$x'_0 = \frac{12.30}{f} \log_{10} 2h$$

where h = height above ground
f = frequency

Fundamental Equations:

$$x'_1 = x'_2 = x'_a = x'_d$$

$$x'_0 = x'_a + x'_c - 2x'_d$$

$$x'_d = (1/\omega k') \ln d$$

d = separation, ft

appendix C

Trigonometric Identities for Three-Phase Systems

In solving problems involving three-phase systems, the engineer encounters a large number of trigonometric functions involving the angles ±120°. Some of these are listed here to save others the time and effort of computing these same quantities over and over. Although the degree symbol (°) has been omitted from angles ±120°, it is always implied.

$$\sin(\theta \pm 120) = -\frac{1}{2}\sin\theta \pm \frac{\sqrt{3}}{2}\cos\theta \tag{C.1}$$

$$\cos(\theta \pm 120) = -\frac{1}{2}\cos\theta \mp \frac{\sqrt{3}}{2}\sin\theta \tag{C.2}$$

$$\begin{aligned}\sin^2(\theta \pm 120) &= \frac{1}{4}\sin^2\theta + \frac{3}{4}\cos^2\theta \mp \frac{\sqrt{3}}{2}\sin\theta\cos\theta \\ &= \frac{1}{2} + \frac{1}{4}\cos 2\theta \mp \frac{\sqrt{3}}{4}\sin 2\theta \end{aligned} \tag{C.3}$$

$$\begin{aligned}\cos^2(\theta \pm 120) &= \frac{1}{4}\cos^2\theta + \frac{3}{4}\sin^2\theta \pm \frac{\sqrt{3}}{2}\sin\theta\cos\theta \\ &= \frac{1}{2} - \frac{1}{4}\cos 2\theta \pm \frac{\sqrt{3}}{4}\sin 2\theta \end{aligned} \tag{C.4}$$

$$\begin{aligned}\sin\theta\sin(\theta \pm 120) &= -\frac{1}{2}\sin^2\theta \pm \frac{\sqrt{3}}{2}\sin\theta\cos\theta \\ &= -\frac{1}{4} + \frac{1}{4}\cos 2\theta \pm \frac{\sqrt{3}}{4}\sin 2\theta \end{aligned} \tag{C.5}$$

$$\begin{aligned}\cos\theta\cos(\theta \pm 120) &= -\frac{1}{2}\cos^2\theta \mp \frac{\sqrt{3}}{2}\sin\theta\cos\theta \\ &= -\frac{1}{4} - \frac{1}{4}\cos 2\theta \mp \frac{\sqrt{3}}{4}\sin 2\theta \end{aligned} \tag{C.6}$$

$$\begin{aligned}\sin\theta\cos(\theta \pm 120) &= -\frac{1}{2}\sin\theta\cos\theta \mp \frac{\sqrt{3}}{2}\sin^2\theta \\ &= -\frac{1}{4}\sin 2\theta \pm \frac{\sqrt{3}}{4}\cos 2\theta \mp \frac{\sqrt{3}}{4} \end{aligned} \tag{C.7}$$

$$\cos\theta\ \sin(\theta \pm 120) = -\frac{1}{2}\sin\theta\ \cos\theta \pm \frac{\sqrt{3}}{2}\cos^2\theta$$

$$= -\frac{1}{4}\sin 2\theta \pm \frac{\sqrt{3}}{4}\cos 2\theta \pm \frac{\sqrt{3}}{4} \qquad (C.8)$$

$$\sin(\theta + 120)\cos(\theta + 120) = -\frac{1}{2}\sin\theta\ \cos\theta - \frac{\sqrt{3}}{4}\cos^2\theta + \frac{\sqrt{3}}{4}\sin^2\theta$$

$$= -\frac{1}{4}\sin 2\theta - \frac{\sqrt{3}}{4}\cos 2\theta \qquad (C.9)$$

$$\sin(\theta + 120)\cos(\theta - 120) = \sin\theta\ \cos\theta - \frac{\sqrt{3}}{4} = \frac{1}{2}\sin 2\theta - \frac{\sqrt{3}}{4} \qquad (C.10)$$

$$\sin(\theta - 120)\cos(\theta + 120) = \sin\theta\ \cos\theta + \frac{\sqrt{3}}{4} = \frac{1}{2}\sin 2\theta + \frac{\sqrt{3}}{4} \qquad (C.11)$$

$$\sin(\theta - 120)\cos(\theta - 120) = -\frac{1}{2}\sin\theta\ \cos\theta + \frac{\sqrt{3}}{4}\cos^2\theta - \frac{\sqrt{3}}{4}\sin^2\theta$$

$$= -\frac{1}{4}\sin 2\theta + \frac{\sqrt{3}}{4}\cos 2\theta \qquad (C.12)$$

$$\sin(\theta + 120)\sin(\theta - 120) = \frac{1}{4}\sin^2\theta - \frac{3}{4}\cos^2\theta = -\frac{1}{4} - \frac{1}{2}\cos 2\theta \qquad (C.13)$$

$$\cos(\theta + 120)\cos(\theta - 120) = \frac{1}{4}\cos^2\theta - \frac{3}{4}\sin^2\theta = -\frac{1}{4} + \frac{1}{2}\cos 2\theta \qquad (C.14)$$

$$\sin(2\theta \pm 120) = -\frac{1}{2}\sin 2\theta \pm \frac{\sqrt{3}}{2}\cos 2\theta \qquad (C.15)$$

$$\cos(2\theta \pm 120) = -\frac{1}{2}\cos 2\theta \mp \frac{\sqrt{3}}{2}\sin 2\theta \qquad (C.16)$$

$$\sin\theta + \sin(\theta - 120) + \sin(\theta + 120) = 0 \qquad (C.17)$$

$$\cos\theta + \cos(\theta - 120) + \cos(\theta + 120) = 0 \qquad (C.18)$$

$$\sin^2\theta + \sin^2(\theta - 120) + \sin^2(\theta + 120) = \frac{3}{2} \qquad (C.19)$$

$$\cos^2\theta + \cos^2(\theta - 120) + \cos^2(\theta + 120) = \frac{3}{2} \qquad (C.20)$$

$$\sin\theta\ \cos\theta + \sin(\theta - 120)\cos(\theta - 120) + \sin(\theta + 120)\cos(\theta + 120) = 0 \qquad (C.21)$$

In addition to the above the following commonly used identities are often required.

$$\sin^2\theta + \cos^2\theta = 1$$

$$\sin\theta\ \cos\theta = \frac{1}{2}\sin 2\theta$$

$$\cos^2\theta - \sin^2\theta = \cos 2\theta$$

$$\cos^2\theta = \frac{1 + \cos 2\theta}{2}$$

$$\sin^2\theta = \frac{1 - \cos 2\theta}{2}$$

appendix D

Self Inductance of a Straight Finite Cylindrical Wire

In Chapter 4 much use is made of the primitive inductance of a straight finite line of cylindrical cross section which appears in equation (4.20) as[1]

$$L = \frac{\mu_w s}{8\pi} + \frac{\mu_m s}{2\pi}\left(\ln\frac{2s}{r} - 1\right) \quad \text{H} \tag{D.1}$$

μ_w = permeability of the wire, H/m
μ_m = permeability of the surrounding medium, H/m

This expression is due to Rosa and Grover [25] who made an intensive study of the inductance of common physical circuit arrangements. Although their work is recommended reading, the modern engineer will be troubled by its use of the CGS system of units. A more modern treatment of the subject leading to the same result is that of Attwood [23]. Here the MKS rationalized system of units is used, and the subject is thoroughly explored.

In studying material like that of Chapter 4, the engineer often rebels at having unfamiliar facts given without proof. Since (D.1) is not familiar to all, it will be developed here, following a derivation similar to that of Attwood [23].

The self inductance of a wire is divided into two parts—that due to flux inside the wire, called L_i, and that due to flux external to the wire, called L_e. Consider a long, straight, cylindrical wire of radius r and length s, both in meters, made of a material having permeability μ_w henry/meter. To compute the internal inductance, we assume that the current is uniformly distributed such that by the Biot-Savart law we may write the field intensity at a point u meters from the center as

$$H_u = \frac{i \text{ enclosed}}{2\pi u} = \frac{iu}{2\pi r^2} \quad \text{A/m} \tag{D.2}$$

Then

$$B_u = \mu_w H_u = \frac{iu\mu_w}{2\pi r^2} \quad \text{Wb/m}^2 \tag{D.3}$$

where

$\mu_w = 4\pi \times 10^{-7} \mu_{wr}$ H/m
μ_{wr} = relative permeability of the wire material

[1] Both μ_w and μ_m may be written as the product of $\mu_o = 4\pi \times 10^{-7}$ H/m, the permeability of free space and a relative permeability μ_{wr} or μ_{mr}, a numeric.

Since the wire is s meters long, it contains a finite amount of energy which can be related to the inductance by the familiar expression

$$W = (1/2)L_i i^2 \quad \text{J} \tag{D.4}$$

If we can compute W, we can find L_i. Attwood [23] points out that the energy density is proportional to B^2, which is known, and is often given by

$$\text{energy density} = B^2/2\mu_w \quad \text{J/m}^3 \tag{D.5}$$

Suppose we consider a thin cylindrical shell of the conductor with radius u, thickness du, and length s. Then

$$dW = (B_u^2/2\mu_w)\,(2\pi us\,du) \tag{D.6}$$

or, substituting for B_u from (D.3)

$$dW = \frac{i^2 \mu_w s}{4\pi r^4}\, u^3\, du$$

and

$$W = \frac{i^2 \mu_w s}{4\pi r^4} \int_0^r u^3\, du = \frac{i^2 \mu_w s}{16\pi} \quad \text{J} \tag{D.7}$$

But from (D.4) the inductance due to internal flux is

$$L_i = 2W/i^2 = \mu_w s/8\pi \quad \text{H} \tag{D.8}$$

The external partial inductance is easier to compute by the familiar expression

$$L_e = \lambda_e/i = N\phi_e/i \tag{D.9}$$

where λ_e is the external flux linkage, N is the number of turns, and ϕ_e is the external flux. Since we are considering the inductance of only one long straight wire, the concept of a flux linkage is troublesome to many engineers. This may be explained by the circuit configuration shown in Figure D.1 where the length s of the wire under consideration is part of a loop of wire. However, it is apparent that if we can compute the self inductance of only that length s we have found the inductance of an "isolated" wire. The inductance of such an isolated wire is of no value whatever since the current i in (D.9) must begin and end at the same point (the generator). However, as shown in Chapter 4, the concept of an isolated wire is extremely valuable in determining the impedance matrix of a group of parallel wires.

To find the external inductance, we must first find the flux density B_p at the external point P, u meters perpendicular from the wire and a meters from one end,

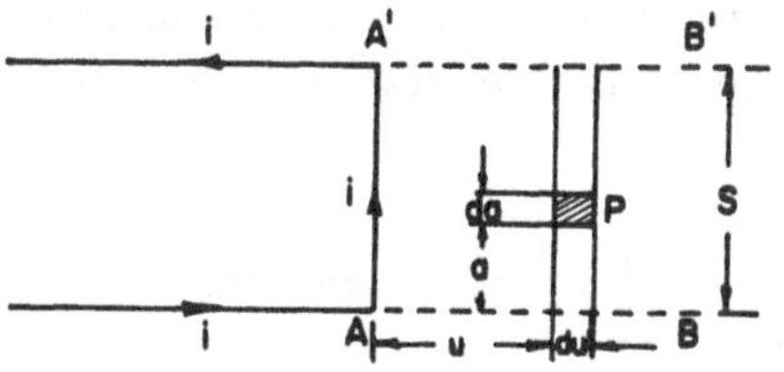

Fig. D.1. Inductance of wire section AA' of length s. (*From Electric and Magnetic Fields*, 3rd ed., Stephen S. Attwood, John Wiley and Sons, Inc., New York, 1949. Used with permission.)

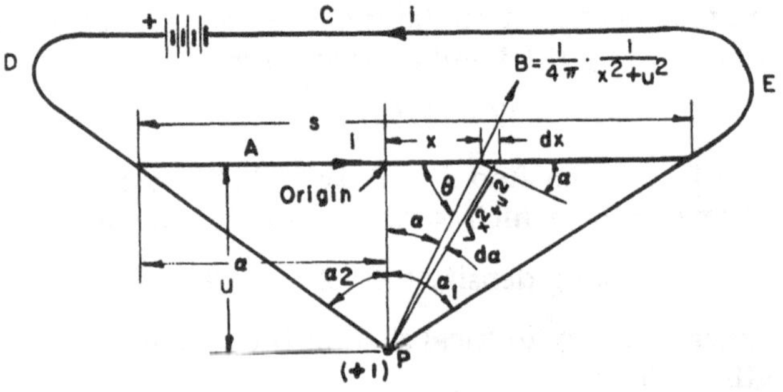

Fig. D.2. Intensity of a straight current section. (From *Electric and Magnetic Fields*, 3rd ed., Stephen S. Attwood, John Wiley and Sons, Inc., New York, 1949. Used with permission.)

as shown in Figure D.1. This may be determined in turn by finding the intensity H_P at P due to i which is equal numerically to the force on the wire due to a unit magnetic pole at P [23].

From Figure D.2 we write the force dF due to i in element dx as

$$dF = (Bi \sin\theta)\, dx \tag{D.10}$$

But if there is a unit pole at P, the flux density at line segment dx is

$$B = \frac{1}{4\pi} \times \frac{1}{x^2 + u^2} = \frac{\cos^2\alpha}{4\pi u^2}$$

Then, since $\sin\theta = \cos\alpha$, we write (D.10) term by term as

$$dF = (\cos^2\alpha/4\pi u^2) \times (i) \times (\cos\alpha) \times (u d\alpha/\cos^2\alpha) = i \cos\alpha\, d\alpha/4\pi u$$

and

$$F = \frac{i}{4\pi u} \int_{\alpha_2}^{\alpha_1} \cos\alpha\, d\alpha = \frac{i}{4\pi u} (\sin\alpha_1 \pm \sin\alpha_2) = H_P \tag{D.11}$$

Choosing the plus sign, (D.11) may be written as

$$B_P = \frac{i\mu_m}{4\pi u} \left(\frac{s - a}{\sqrt{(s-a)^2 + u^2}} + \frac{a}{\sqrt{a^2 + u^2}} \right) \text{ Wb/m}^2 \tag{D.12}$$

where $\mu_m = 4\pi \times 10^{-7} \mu_{mr}$ is the permeability of the medium surrounding the wire. Here B_P is a vector pointing downward in Figures D.1 or D.2 and normal to the plane containing the wire and P. The total flux in the area $AB\,B'\,A'$ of Figure D.1 may be found by integrating B_P over this area, noting carefully that B_P is everywhere normal to this plane. Also note that we are finding only the flux in a strip s meters long and extending radially from the wire. This is precisely the flux ϕ_e needed in (D.9) to compute the external inductance. Integrating (D.12) we have

$$\phi_e = \int_{u=r}^{u=\infty} \int_{a=0}^{a=s} B_P\, du\, da \text{ Wb}$$

$$= \frac{i\mu_m}{2\pi} \left(\sqrt{s^2 + u^2} - u - s \ln \frac{s + \sqrt{s^2 + u^2}}{u} \right)_{u=r}^{u=\infty} \tag{D.13}$$

Attwood shows that this expression may be written as

$$\phi_e = \frac{\mu_m i}{2\pi}\left(\frac{s^2}{\sqrt{s^2+u^2}+u} - s \ln \frac{-1}{(s/u)-\sqrt{(s^2/u^2)+1}}\right)_{u=r}^{u=\infty}$$

and the limits then inserted to find

$$\phi_e = \frac{\mu_m i}{2\pi}\left(s \ln \frac{s+\sqrt{s^2+r^2}}{r} - \sqrt{s^2+r^2} + r\right) \text{ Wb} \tag{D.14}$$

Then

$$L_e = \frac{\mu_m}{2\pi}\left(s \ln \frac{s+\sqrt{s^2+r^2}}{r} - \sqrt{s^2+r^2} + r\right) \text{ H}$$

and since $s >> r$ in every case of interest, we write this equation as

$$L_e = (\mu_m s/2\pi)\left(\ln \frac{2s}{r} - 1\right) \text{ H} \tag{D.15}$$

or per unit of length,

$$\ell_e = (\mu_m/2\pi)\left(\ln \frac{2s}{r} - 1\right) \text{ H/m} \tag{D.16}$$

It seems strange that the inductance per unit of length is a function of the length s. However, as shown in Chapter 4, the argument of the logarithm may have *any* numerator since it always disappears when circuits containing a return path for the current are used. From (D.8) and (D.16) we have the desired result, viz.,

$$\begin{aligned} \ell &= \ell_i + \ell_e \\ &= (\mu_w/8\pi) + (\mu_m/2\pi)\left(\ln \frac{2s}{r} - 1\right) \text{ H/m} \end{aligned} \tag{D.17}$$

The mutual inductance is computed in a similar way except in this case the limits of integration are from D to ∞, where D is the distance to the linked current i_2. Then

$$M = \lambda_e/i_2 = (\mu_m s/2\pi)\left(\ln \frac{2s}{D} - 1\right) \text{ H} \tag{D.18}$$

or

$$m = (\mu_m/2\pi)\left(\ln \frac{2s}{D} - 1\right) \text{ H/m} \tag{D.19}$$

These results for the self and mutual inductance for a cylindrical wire of length s are both strange and interesting. They are strange because portions of the solution such as the $2s$ in the logarithm and the (-1) in the parentheses never appear in any practical problem involving a "return" wire. The results are of interest, however, because these primitive building blocks provide powerful tools for analyzing complex problems involving many wires.

appendix E

Solved Examples

It is often useful for the student of symmetrical components to apply the techniques being studied to a typical system. The purpose of this appendix is to specify small sample systems along with their complete solutions for both normal and faulted conditions.

Following are solutions for 3-node, 6-node, and 14-node networks. The 3-node network used here is the same system shown in Figure 11.2 and used in the examples of Chapters 11 and 12. The 6-node network is a well-known circuit introduced by Ward and Hale [79] to which a table of zero sequence data has been added. The 14-node network was adapted from the IBM assembler load flow described in [80]. These three networks will provide useful data for checking hand computations and computer programs.

E.1 A 3-Node Network

Consider the 3-node network shown in Figure E.1. This is the same network as that of Figure 11.2 and is redrawn here for convenience. A normal load flow for the 3-node network is given in Table E.1 where an arbitrary load and generation pattern is assumed.

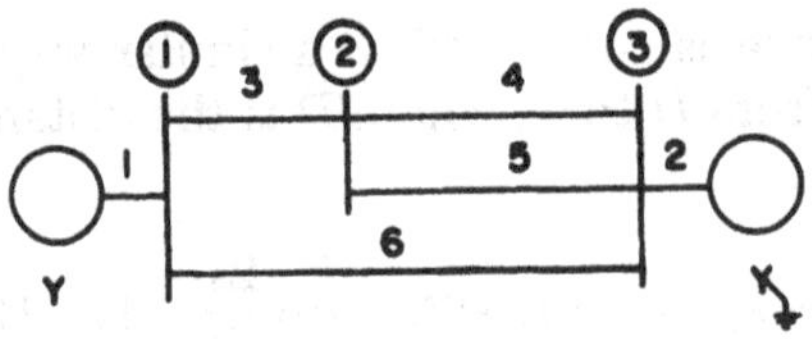

Fig. E.1. A 3-node network.

The primitive impedance data is given in Table 11.1, where it is noted that the positive and negative sequence data are not equal for the two generators. The usual practice, especially in digital computer solutions, is to assume that the positive and negative sequence impedances are equal and to provide only one set of positive sequence data. When these impedances are not equal, as in the system of Figure E.1, one can either ignore the negative sequence data and set $Z_2 = Z_1$ or average the two impedances. The first alternative will give correct results for 3ϕ faults but will result in slight errors on all computations which involve the negative sequence network. Averaging Z_1 and Z_2 will make SLG faults correct and 3ϕ faults incorrect. Here we shall *average* the positive and negative sequence data and will call this result the positive sequence data.

Table E.1. A 3-Node Network Load Flow Study

ICWA STATE UNIVERSITY VERSION OF 360 LOADFLOW PROGRAM

TITLE-3 BUS EXAMPLE

TRANSMISSION LINE AND TRANSFORMER DATA ASSEMBLY

LINE P	LINE Q	NO.	ACTUAL R(PCT)	ACTUAL X(PCT)	CONVERTED NO TAP EFFECT KVAC	BC/2(PU)	G(PU)	B(PU)	MVA RATING	TAP RATIO	TAP LIMITS TMIN	TAP LIMITS TMAX	PHASE SHIFT	MAP DATA FLOW PG	FLOW LOC Q	TAP LOC	REV FLOW PG	REV FLOW LOC Q
1	2	0	0.0	8.00	0.0	0.0	0.0	-12.5000	0.				0.0	0	0		0	0
2	3	0	0.0	6.00	0.0	0.0	0.0	-16.6667	0.				0.0	0	0		0	C
2	3	1	0.0	6.00	0.0	0.0	0.0	-16.6667	0.				C.0	0	0		0	0
1	3	0	0.0	13.00	0.0	0.0	0.0	-7.6923	0.				0.0	0	0		0	0

BASE CASE BUS DATA ENTERED

BUS NO.	BUS NAME	AREA	REG	VOLTAGE MAG(PU)	VOLTAGE ANG(DEG)	LOAD MW	LOAD MVAR	GENERATION MW	GENERATION MVAR	QMIN MVAR	QMAX MVAR	REACTOR KVAR	MAP DATA PAGE	VOLT LOC A	LOAD LOC Q	GEN LOC Q	REACTOR LOC QS
1	BUS 1	64	2	1.030	0.0	0.0	0.0	0.0	0.0	0.0	0.0	0.	0	0	0	0	0
2	BUS 2	64	0	1.000	0.0	50.0	20.0	0.0	0.0	0.0	0.0	0.	0	0	0	0	0
3	BUS 3	64	1	1.020	0.0	80.0	40.0	100.0	0.0	0.0	70.0	0.	0	0	0	0	0

SUMMARY

LINE AND BUS TOTALS	ACTUAL	MAX
TRANSMISSION LINES	4	750
TRANSFORMERS - FIXED	0	250
- LTC	0	250
PHASE SHIFTERS	0	0
TOTAL LINES	4	750
BUSES - NON REG (INCLUDING SWING)	2	500
- GENERATOR	1	190
TOTAL BUSES	3	500
CAPACITORS OR REACTORS	0	70

	MW	MVAR
TOTAL LOAD	130.000	60.000
TOTAL LOSSES	0.000	0.929
LINE CHARGING		0.0
FIXED CAP/REAC		0.0
SYSTEM MISMATCH	-0.000	0.006
TOTAL GENERATION	130.000	60.934
ITERATIONS BETWEEN BUS-ORDER SORTS		50

MISCELLANEOUS CONSTANTS		
ACTUAL ITERATIONS	4	
MAXIMUM ITERATIONS	10	
TOLERANCE - REAL	1.000E-03	
- IMAG	1.000E-03	
ACC FACT. - REAL	1.7	1.1
- IMAG	1.8	1.4
LTC START	10	
SKIP	4	
END	400	
ACC FACT	1.2	

THERE ARE NO VOLTAGES UNDER 0.950

THERE ARE NO VOLTAGES OVER 1.050

Table E.1. (continued)

IOWA STATE UNIVERSITY VERSION OF 360 LOADFLOW PROGRAM

TITLE-3 BUS EXAMPLE

REPORT OF LOADFLOW CALCULATIONS   TOTAL ITERATIONS = 4   SWING BUS = 1 BUS 1 , AREA 64.

X------------------------- B U S - D A T A -------------------------X   X----------- L I N E - D A T A -----------X
                                                  X------- T O -------X

AREA NO.	BUS	NAME	MISMATCH MW	MVAR	VOLTS	ANGLE	LOAD MW	MVAR	GENERATION MW	MVAR	CAP/REAC MVAR	BUS NAME	AREA NO.	PAR NO.	LINE FLOW MW	MVAR	PCT CAP	TAP
64	1	BUS 1	0.0	0.0	1.030	0.0	0.0	0.0	30.0	23.1								
												2-BUS 2	64	0	22.53	15.16		
												3-BUS 3	64	0	7.47	7.96		
64	2	BUS 2	-0.000	0.004	1.018	-1.0	50.0	20.0	0.0	0.0								
												1-BUS 1	64	0	-22.53	-14.60		
												3-BUS 3	64	0	-13.74	-2.70		
												3-BUS 3	64	1	-13.74	-2.70		
64	3	BUS 3	0.000	0.002	1.020	-0.5	80.0	40.0	100.0	37.8R								
												1-BUS 1	64	0	-7.47	-7.81		
												2-BUS 2	64	0	13.74	2.81		
												2-BUS 2	64	1	13.74	2.81		

END OF REPORT FOR THIS CASE

Table E.2. A 3-Node Network Fault Study

3 BUS EXAMPLE
TRANSMISSION LINE AND TRANSFORMER DATA ASSEMBLY

DATA FOR POSITIVE SEQUENCE

SELF								
ACTUAL					CONVERTED			
P	Q	NO.	R(PCT)	X(PCT)	R (PU)	X (PU)	G (PU)	B (PU)
0	1	0	0.0	20.00	0.0	0.2000	0.0	-5.0000
0	3	0	0.0	16.00	0.0	0.1600	0.0	-6.2500
1	2	0	0.0	8.00	0.0	0.0800	0.0	-12.5000
2	3	0	0.0	6.00	0.0	0.0600	0.0	-16.6667
2	3	1	0.0	6.00	0.0	0.0600	0.0	-16.6667
1	3	0	0.0	13.00	0.0	0.1300	0.0	-7.6923

EXISTING Z MATRIX FOR THE POSITIVE SEQUENCE NETWORK

BUS	BUS	R	X
1	1	0.0	0.10467
1	2	0.0	0.08401
1	3	0.0	0.07627
2	2	0.0	0.11221
2	3	0.0	0.09279
3	3	0.0	0.09899

DATA FOR ZERO SEQUENCE

SELF									MUTUAL						
ACTUAL					CONVERTED				ACTUAL					CONVERTED	
P	Q	NO.	R(PCT)	X(PCT)	R (PU)	X (PU)	G (PU)	B (PU)	P	Q	NO.	RM(PCT)	XM(PCT)	RM (PU)	XM (PU)
0	3	0	0.0	2.00	0.0	0.0200	0.0	-50.0000							
1	2	0	0.0	14.00	0.0	0.1400	0.0	-7.1429							
2	3	0	0.0	10.00	0.0	0.1000			2	3	1	0.0	5.00	0.0	0.0500
2	3	1	0.0	12.00	0.0	0.1200			2	3	0	0.0	5.00	0.0	0.0500
1	3	0	0.0	17.00	0.0	0.1700	0.0	-5.8824							

EXISTING MUTUAL ADMITTANCE MATRIX

SELF			COUPLED LINE			SELF		MUTUAL	
BUS	BUS	NO.	BUS	BUS	NO.	G	B	GM	BM
2	3	0				0.0	-12.63158		
2	3	0	2	3	1			0.0	5.26316
2	3	1				0.0	-10.52633		

Table E.2. *(continued)*

EXISTING Z MATRIX FOR THE ZERO SEQUENCE NETWORK

BUS	BUS	R	X
1	1	0.0	0.11574
1	2	0.0	0.05458
1	3	0.0	0.02000
2	2	0.0	0.08306
2	3	0.0	0.02000
3	3	0.0	0.02000

3 BUS EXAMPLE
REPORT OF FAULT CALCULATIONS

B U S - D A T A

BUS	NAME	3-PHASE AMPS	DEGREES	S-L-G VOLTS	DEGREES	AMPS	DEGREES
1	BUS 1	9.554	-90.0			9.229	-90.0T
				0.6780	0.0	3.076	-90.0P
				0.3220	180.0	3.076	-90.0N
				0.3560	180.0	3.076	-90.0Z

L I N E - D A T A

BUS NO.	3-PHASE VOLTS	DEGREES	AMPS	DEGREES	S-L-G VOLTS	DEGREES	AMPS	DEGREES
0- 0	1.0000	0.0	5.000	-90.0			3.220	-90.0T
					1.0000	0.0	1.610	-90.0P
					0.0	0.0	1.610	-90.0N
					0.0	0.0	0.0	0.0Z
2- 0	0.1973	0.0	2.467	-90.0			2.932	-90.0T
					0.7416	0.0	0.794	-90.0P
					0.2584	180.0	0.794	-90.0N
					0.1679	180.0	1.344	-90.0Z
3- 0	0.2713	0.0	2.087	-90.0			3.077	-90.0T
					0.7654	0.0	0.672	-90.0P
					0.2346	180.0	0.672	-90.0N
					0.0615	180.0	1.732	-90.0Z

BUS	NAME	3-PHASE AMPS	DEGREES	S-L-G VOLTS	DEGREES	AMPS	DEGREES
2	BUS 2	8.912	-90.0			9.756	-90.0T
				0.6351	0.0	3.252	-90.0P
				0.3649	180.0	3.252	-90.0N
				0.2701	180.0	3.252	-90.0Z

BUS NO.	3-PHASE VOLTS	DEGREES	AMPS	DEGREES	S-L-G VOLTS	DEGREES	AMPS	DEGREES
1- 0	0.2513	0.0	3.142	-90.0			2.955	-90.0T
					0.7268	0.0	1.146	-90.0P
					0.2732	180.0	1.146	-90.0N
					0.1775	180.0	0.662	-90.0Z
3- 0	0.1731	0.0	2.885	-90.0			3.617	-90.0T
					0.6982	0.0	1.053	-90.0P
					0.3018	180.0	1.053	-90.0N
					0.0650	180.0	1.511	-90.0Z
3- 1	0.1731	0.0	2.885	-90.0			3.185	-90.0T
					0.6982	0.0	1.053	-90.0P
					0.3018	180.0	1.053	-90.0N
					0.0650	180.0	1.079	-90.0Z

```
3 BUS EXAMPLE
REPORT OF FAULT CALCULATIONS

X------------------- B U S - D A T A --------------------X   X----------------------- L I N E - D A T A -----------------------X
             X-- 3-PHASE ---X---------- S-L-G -----------X           X--------- 3-PHASE -----------X----------- S-L-G -----------X
 BUS   NAME     AMPS DEGREES  VOLTS DEGREES   AMPS DEGREES         BUS NO. VOLTS DEGREES   AMPS DEGREES  VOLTS DEGREES  AMPS DEGREES

   3 BUS 3     10.102  -90.0                 13.763  -90.0T-------------------------------------------------------------------------
                              0.5459    0.0   4.588  -90.0P
                              0.4541  180.0   4.588  -90.0N
                              0.0918  180.0   4.588  -90.0Z
                                                                   0- 0  1.0000    0.0  6.250  -90.0                  10.264  -90.0T
                                                                                                     1.0000     0.0   2.838  -90.0P
                                                                                                     0.0        0.0   2.838  -90.0N
                                                                                                     0.0        0.0   4.588  -90.0Z
                                                                                                     ----------------------------
                                                                   1- 0  0.2295    0.0  1.766  -90.0                   1.604  -90.0T
                                                                                                     0.6501     0.0   0.802  -90.0P
                                                                                                     0.3499   180.0   0.802  -90.0N
                                                                                                     0.0918   180.0   0.000    0.0Z
                                                                                                     ----------------------------
                                                                   2- 0  0.0626    0.0  1.043  -90.0                   0.948  -90.0T
                                                                                                     0.5743     0.0   0.474  -90.0P
                                                                                                     0.4257   180.0   0.474  -90.0N
                                                                                                     0.0918   180.0   0.000    0.0Z
                                                                                                     ----------------------------
                                                                   2- 1  0.0626    0.0  1.043  -90.0                   0.948  -90.0T
                                                                                                     0.5743     0.0   0.474  -90.0P
                                                                                                     0.4257   180.0   0.474  -90.0N
                                                                                                     0.0918   180.0   0.000    0.0Z
                                                                                                     ----------------------------
```

A complete 3ϕ and SLG fault study for the 3-node network is given in Table E.2. This table includes all primitive impedance data, the impedance matrices for both sequences, and the detailed fault data.

E.2 A 6-Node Network

Consider the 6-node network shown in Figure E.2. This network is more completely specified than the 3-node network. The system data includes load data and line susceptances in addition to the line, transformer, and generator impedances. The line impedance data is given on a 50 MVA base in Table E.3. The computer program used to solve this network will change the base to 100 MVA before solving.

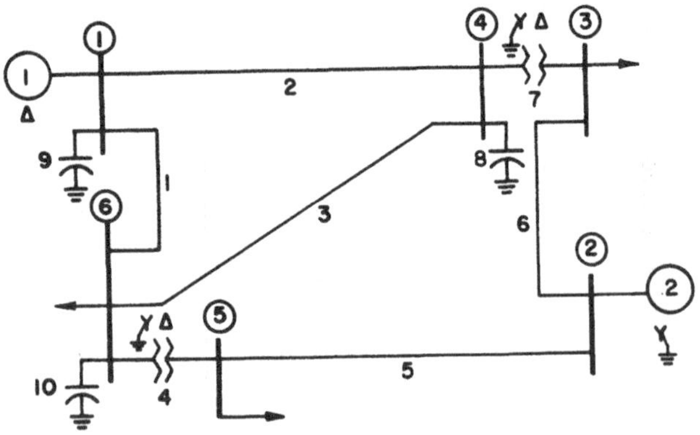

Fig. E.2. A 6-node network [79].

Table E.3. 6-Node Network with Impedances in pu on a 50 MVA Base

Impedance Number	*Connecting Nodes*	*Self Impedance* $Z_1 = Z_2$	*Self Impedance* Z_0	*Mutual Impedance* Z_M	*Mutual Impedance* Branch
1	1-6	0.123 + j0.518	0.492 + j1.042	...	...
2	1-4	0.080 + j0.370	0.400 + j0.925	0.250 + j0.475	3
3	4-6	0.097 + j0.407	0.450 + j1.030	0.250 + j0.475	2
4	5-6	0.000 + j0.300	0.000 + j0.300	...	...
5	2-5	0.282 + j0.640	1.410 + j1.920	...	...
6	2-3	0.723 + j1.050	1.890 + j2.630	...	...
7	3-4	0.000 + j0.133	0.000 + j0.133	...	...
8	0-4	0.000 − j34.100	...	...	...
9	0-1	0.000 − j29.500	...	...	...
10	0-6	0.000 − j28.500	...	...	...
11	Gen 1	0.010 + j0.120	...	...	...
12	Gen 2	0.015 + j0.240	0.000 + j0.016	...	...

A complete load flow result is given in Table E.4 where all line and generator flow data is given in megawatts and megavars. All voltages are given in pu and angles are in degrees.

A complete bus fault study is given in Table E.5.

Table E.4. A 6-Node Network Load Flow Study

IOWA STATE UNIVERSITY VERSION OF 360 LOADFLOW PROGRAM

TITLE-SIX BUS SYSTEM

TRANSMISSION LINE AND TRANSFORMER DATA ASSEMBLY

LINE P	LINE Q	NO.	ACTUAL R(PCT)	ACTUAL X(PCT)	ACTUAL KVAC	CONVERTED NO TAP EFFECT BC/2(PU)	G(PU)	B(PU)	MVA RATING	TAP RATIO	TAP LIMITS TMIN	TAP LIMITS TMAX	PHASE SHIFT	MAP DATA FLOW PG	FLOW LOC Q	TAP LOC	REV FLOW PG	REV FLOW LOC Q
1	4	0	16.00	74.00	1406.80	0.0070	0.2791	-1.2910	0.				0.0	0	0		0	C
1	6	0	24.60	103.60	1983.00	0.0099	0.2173	-0.9137	0.				0.0	0	0		0	0
2	3	0	144.60	210.00	0.0	0.0	0.2224	-0.3230	0.				0.0	0	0		0	0
2	5	0	56.40	128.00	0.0	0.0	0.2383	-0.6542	0.				0.0	0	0		0	0
4	3	0	0.0	26.60	0.0	0.0	0.0	-3.7594	0.	0.909			0.0	0	0	0	0	0
4	6	0	19.40	81.40	1525.80	0.0076	0.2771	-1.1625	0.				0.0	0	0		0	0
6	5	0	0.0	60.00	0.0	0.0	0.0	-1.6667	0.	0.976			0.0	0	0	0	0	0

BASE CASE BUS DATA ENTERED

BUS NO.	BUS NAME	AREA	REG	VOLTAGE MAG(PU)	VOLTAGE ANG(DEG)	LOAD MW	LOAD MVAR	GENERATION MW	GENERATION MVAR	QMIN MVAR	QMAX MVAR	REACTOR KVAR	MAP DATA PAGE	VOLT LOC A	LOAD LOC Q	GEN LOC Q	REACTOR LOC QS
1	ONE	64	2	1.050	0.0	0.0	0.0	0.0	0.0	0.0	0.0	0.	0	0	0	0	0
2	TWO	64	1	1.100	0.0	0.0	0.0	25.0	0.0	-12.5	12.5	0.	0	0	0	0	0
3	THREE	64	0	1.000	0.0	27.5	6.5	0.0	0.0	0.0	0.0	0.	0	0	0	0	0
4	FOUR	64	0	1.000	0.0	0.0	0.0	0.0	0.0	0.0	0.0	0.	0	0	0	0	0
5	FIVE	64	0	1.000	0.0	15.0	9.0	0.0	0.0	0.0	0.0	0.	0	0	0	0	0
6	SIX	64	0	1.000	0.0	25.0	2.5	0.0	0.0	0.0	0.0	0.	0	0	0	0	0

SUMMARY

LINE AND BUS TOTALS	ACTUAL	MAX
TRANSMISSION LINES	5	750
TRANSFORMERS - FIXED	2	250
- LTC	0	250
PHASE SHIFTERS	0	0
TOTAL LINES	7	750
BUSES - NON REG (INCLUDING SWING)	5	500
- GENERATOR	1	190
TOTAL BUSES	6	500
CAPACITORS OR REACTORS	0	70

	MW	MVAR
TOTAL LOAD	67.500	18.000
TOTAL LOSSES	5.112	17.666
LINE CHARGING		-4.618
FIXED CAP/REAC		0.0
SYSTEM MISMATCH	-0.002	-0.001
TOTAL GENERATION	72.610	31.047

MISCELLANEOUS CONSTANTS	
ACTUAL ITERATIONS	3
MAXIMUM ITERATIONS	10
TOLERANCE - REAL	1.000E-03
- IMAG	1.000E-03

Table E.4. (continued)

IOWA STATE UNIVERSITY VERSION OF 360 LOADFLOW PROGRAM

TITLE-SIX BUS SYSTEM

REPORT OF LOADFLOW CALCULATIONS   TOTAL ITERATIONS = 3 SWING BUS = 1 ONE ,AREA 64.

BUS - DATA												LINE - DATA (TO)						
AREA NO.	BUS	NAME	MISMATCH MW	MISMATCH MVAR	VOLTS	ANGLE	LOAD MW	LOAD MVAR	GENERATION MW	GENERATION MVAR	CAP/REAC MVAR	BUS NAME	AREA NO.	PAR NO.	LINE FLOW MW	LINE FLOW MVAR	PCT CAP	TAP
64	1	ONE	0.0	0.0	1.050	0.0	0.0	0.0	47.6	21.8		4-FOUR	64	0	25.40	12.74		
												6-SIX	64	0	22.15	9.02		
64	2	TWO	-0.000	0.000	1.100	-3.4	0.0	0.0	25.0	9.3R		3-THREE	64	0	3.58	-0.00		
												5-FIVE	64	0	16.42	9.28		
64	3	THREE	-0.002	0.000	1.001	-12.8	27.5	6.5	0.0	0.0		2-TWO	64	0	-7.70	1.28		
												4-FOUR	64	0	-19.79	-7.78		
64	4	FOUR	0.001	-0.001	0.930	-9.8	0.0	0.0	0.0	0.0		1-ONE	64	0	-24.25	-8.55		
												3-THREE	64	0	19.79	8.98		0.909
												6-SIX	64	0	4.46	-0.43		
64	5	FIVE	-0.001	-0.000	0.919	-12.3	15.0	9.0	0.0	0.0		2-TWO	64	0	-14.76	-5.52		
												6-SIX	64	0	-0.24	-3.48		
64	6	SIX	-0.001	-0.000	0.919	-12.2	25.0	2.5	0.0	0.0		1-ONE	64	0	-20.83	-5.39		
												4-FOUR	64	0	-4.41	-0.68		
												5-FIVE	64	0	0.24	3.56		0.976

END OF REPORT FOR THIS CASE

LOW VOLTAGE SUMMARY -- BUS VOLTAGES BELOW 0.950

BUS NO.	BUS NAME	AREA NO.	VOLTS	ANGLE	BUS NO.	BUS NAME	AREA NO.	VOLTS	ANGLE	BUS NO.	BUS NAME	AREA NO.	VOLTS	ANGLE	BUS NO.	BUS NAME	AREA NO.	VOLTS	ANGLE
4	FOUR	64	0.930	-9.8	5	FIVE	64	0.919	-12.3	6	SIX	64	0.919	-12.2					

HIGH VOLTAGE SUMMARY -- BUS VOLTAGES ABOVE 1.050

BUS NO.	BUS NAME	AREA NO.	VOLTS	ANGLE	BUS NO.	BUS NAME	AREA NO.	VOLTS	ANGLE	BUS NO.	BUS NAME	AREA NO.	VOLTS	ANGLE	BUS NO.	BUS NAME	AREA NO.	VOLTS	ANGLE
2	TWO	64	1.100	-3.4															

NO LINES OVERLOADED

Table E.5. A 6-Node Network Fault Study

SIX BUS SYSTEM
TRANSMISSION LINE AND TRANSFORMER DATA ASSEMBLY

GROUP 1 — INPUT BASE: 0.0 KV, 0. MVA — CONVERTED BASE: 0.0 KV, 100. MVA

LINE

DATA FOR POSITIVE SEQUENCE

SELF									
ACTUAL					CONVERTED				
P	Q	NO.	R(PCT)	X(PCT)	R (PU)	X (PU)	G (PU)	B (PU)	
0	1	0	2.00	24.00	0.0200	0.2400	0.3448	-4.1379	
0	2	0	3.00	48.00	0.0300	0.4800	0.1297	-2.0752	
1	4	0	16.00	74.00	0.1600	0.7400	0.2791	-1.2910	
1	6	0	24.60	103.60	0.2460	1.0360	0.2170	-0.9137	
2	3	0	144.60	210.00	1.4460	2.1000	0.2224	-0.3230	
2	5	0	56.40	128.00	0.5640	1.2800	0.2883	-0.6542	
3	4	0	0.0	26.60	0.0	0.2660	0.0	-3.7594	
6	4	0	19.40	81.40	0.1940	0.8140	0.2771	-1.1625	
5	6	0	0.0	60.00	0.0	0.6000	0.0	-1.6667	

SIX BUS SYSTEM
EXISTING Z MATRIX FOR THE POSITIVE SEQUENCE NETWORK

BUS	BUS	R	X
1	1	0.02253	0.21503
1	2	-0.00609	0.04974
1	3	0.02635	0.16117
1	4	0.02254	0.17266
1	5	0.02118	0.12825
1	6	0.01831	0.16379
2	2	0.04422	0.38094
2	3	-0.01594	0.15713
2	4	-0.00786	0.13434
2	5	-0.00698	0.22306
2	6	0.00023	0.15223
3	3	0.16244	0.73912
3	4	0.14333	0.53368
3	5	0.06192	0.27007
3	6	0.07295	0.32786
4	4	0.13269	0.57694
4	5	0.06506	0.27818
4	6	0.06881	0.34726
5	5	0.16569	0.80649
5	6	0.13256	0.46538
6	6	0.13034	0.61119

Table E.5. *(continued)*

SIX BUS SYSTEM
TRANSMISSION LINE AND TRANSFORMER DATA ASSEMBLY

GROUP 1   INPUT BASE 0.0 KV 0. MVA   CONVERTED BASE 0.0 KV 100. MVA

LINE

DATA FOR ZERO SEQUENCE

SELF										MUTUAL						
ACTUAL					CONVERTED				ACTUAL					CONVERTED		
P	Q	NO.	R(PCT)	X(PCT)	R (PU)	X (PU)	G (PU)	B (PU)	P	Q	NO.	RM(PCT)	XM(PCT)	RM (PU)	XM (PU)
0	2	0	0.0	3.20	0.0	0.0320	0.0	-31.2500							
0	4	0	0.0	26.60	0.0	0.2660	0.0	-3.7594							
0	6	0	0.0	60.00	0.0	0.6000	0.0	-1.6667							
1	4	0	80.00	185.00	0.8000	1.8500			6	4	0	50.00	95.00	0.5000	0.9500
1	6	0	98.40	208.40	0.9840	2.0840	0.1853	-0.3924							
2	3	0	378.00	526.00	3.7800	5.2600	0.0901	-0.1254							
2	5	0	282.00	384.00	2.8200	3.8400	0.1242	-0.1692							
6	4	0	90.00	206.00	0.9000	2.0600			1	4	0	50.00	95.00	0.5000	0.9500

SIX BUS SYSTEM
EXISTING MUTUAL ADMITTANCE MATRIX

SELF			COUPLED LINE			SELF		MUTUAL	
BUS	BUS	NO.	BUS	BUS	NO.	G	B	GM	BM
1	4	0				0.23184	-0.62012		
1	4	0	6	4	0			-0.08895	0.30339
6	4	0				0.20995	-0.55521		

SIX BUS SYSTEM
EXISTING Z MATRIX FOR THE ZERO SEQUENCE NETWORK

BUS	BUS	R	X
1	1	0.36392	1.11336
1	2	0.0	0.0
1	3	0.0	0.0
1	4	-0.01120	0.11474
1	5	0.0	0.0
1	6	0.02526	0.34120
2	2	-0.00000	0.03200
2	3	-0.00000	0.03200
2	4	0.0	0.0
2	5	-0.00000	0.03200
2	6	0.0	0.0
3	3	3.78000	5.29200
3	4	0.0	0.0
3	5	-0.00000	0.03200
3	6	0.0	0.0
4	4	0.00756	0.24138
4	5	0.0	0.0
4	6	-0.01706	0.05554
5	5	2.82000	3.87200
5	6	0.0	0.0

SIX BUS SYSTEM
REPORT OF FAULT CALCULATIONS

BUS DATA								LINE DATA								
		3-PHASE		S-L-G					3-PHASE				S-L-G			
BUS	NAME	AMPS	DEGREES	VOLTS	DEGREES	AMPS	DEGREES	BUS NO.	VOLTS	DEGREES	AMPS	DEGREES	VOLTS	DEGREES	AMPS	DEGREES
1	ONE	4.625	-84.0			1.879	-75.2T									
				0.8665	-1.4	0.626	-75.2P									
				0.1354	-171.1	0.626	-75.2N									
				0.7336	176.7	0.626	-75.2Z									
								0- 0	1.0000	0.0	4.152	-85.2			1.125	-76.4T
													1.0000	0.0	0.562	-76.4P
													0.0	0.0	0.562	-76.4N
													0.0	0.0	0.0	0.0Z
								4- 0	0.1960	6.0	0.259	-71.8			0.466	-74.1T
													0.8920	-0.9	0.035	-62.9P
													0.1091	-172.6	0.035	-62.9N
													0.0722	-159.6	0.397	-76.1Z
								6- 0	0.2378	1.3	0.223	-75.4			0.289	-72.1T
													0.8980	-1.0	0.030	-66.5P
													0.1032	-171.5	0.030	-66.5N
													0.2143	-169.4	0.229	-73.6Z
2	TWO	2.608	-83.4			3.756	-83.6T									
				0.5199	0.2	1.252	-83.6P									
				0.4801	179.7	1.252	-83.6N									
				0.0401	-173.6	1.252	-83.6Z									
								0- 0	1.0000	0.0	2.079	-86.4			3.247	-85.5T
													1.0000	0.0	0.998	-86.7P
													0.0	0.0	0.998	-86.7N
													0.0	0.0	1.252	-83.6Z
								3- 0	0.6043	-8.4	0.237	-63.9			0.228	-64.1T
													0.8078	-3.0	0.114	-64.1P
													0.1977	-167.9	0.114	-64.1N
													0.0401	-173.6	0.000	0.0Z
								5- 0	0.4328	-11.3	0.309	-77.6			0.297	-77.8T
													0.7245	-3.1	0.149	-77.8P
													0.2794	-171.9	0.149	-77.8N
													0.0401	-173.6	0.000	0.0Z
3	THREE	1.321	-77.6			0.379	-58.8T									
				0.9101	-1.9	0.126	-58.8P									
				0.0956	-161.2	0.126	-58.8N									
				0.8214	175.7	0.126	-58.8Z									
								2- 0	0.8044	-4.6	0.315	-60.1			0.185	-53.1T
													0.9841	-0.7	0.030	-41.3P
													0.0199	-143.0	0.030	-41.3N
													0.0040	-148.8	0.126	-58.8Z
								4- 0	0.2726	7.1	1.025	-82.9			0.196	-64.1T
													0.9332	-1.2	0.098	-64.1P
													0.0698	-163.8	0.098	-64.1N
													0.0	0.0	0.0	0.0Z

Table E.5. *(continued)*

4 FOUR	1.689	-77.0			2.110	-78.9T									
			0.5840	1.3	0.703	-78.9P									
			0.4164	178.1	0.703	-78.9N									
			0.1699	-170.7	0.703	-78.9Z									
							0- 0	0.0	0.0	0.0	0.0			0.639	-80.7T
												0.0	0.0	0.0	0.0P
												0.0	0.0	0.0	0.0N
												0.0	0.0	0.639	-80.7Z
							1- 0	0.7078	-2.3	0.935	-80.1			0.795	-81.5T
												0.8778	-0.5	0.389	-82.0P
												0.1225	-176.4	0.389	-82.0N
												0.0811	-163.4	0.018	-59.3Z
							3- 0	0.0752	26.8	0.283	-63.2			0.236	-65.1T
												0.6128	2.5	0.118	-65.1P
												0.3887	176.0	0.118	-65.1N
												0.0	0.0	0.0	0.0Z
							6- 0	0.4027	-2.6	0.481	-79.2			0.449	-79.1T
												0.7510	0.0	0.200	-81.1P
												0.2490	179.9	0.200	-81.1N
												0.0409	-151.9	0.050	-62.8Z
5 FIVE	1.215	-78.4			0.474	-60.1T									
			0.8774	-2.7	0.158	-60.1P									
			0.1302	-161.7	0.158	-60.1N									
			0.7572	173.8	0.158	-60.1Z									
							2- 0	0.7390	-4.9	0.528	-71.1			0.295	-56.7T
												0.9702	-1.1	0.069	-52.8P
												0.0353	-148.3	0.069	-52.8N
												0.0051	-150.1	0.158	-60.1Z
							6- 0	0.4163	6.1	0.694	-83.9			0.181	-65.7T
												0.9260	-1.1	0.090	-65.7P
												0.0765	-166.0	0.090	-65.7N
												0.0	0.0	0.0	0.0Z
6 SIX	1.600	-78.0			1.741	-80.0T									
			0.6377	1.2	0.580	-80.0P									
			0.3626	178.0	0.580	-80.0N									
			0.2764	-174.6	0.580	-80.0Z									
							0- 0	0.0	0.0	0.0	0.0			0.461	-84.6T
												0.0	0.0	0.0	0.0P
												0.0	0.0	0.0	0.0N
												0.0	0.0	0.461	-84.6Z
							1- 0	0.7380	-2.0	0.693	-78.7			0.535	-79.4T
												0.9046	-0.4	0.251	-80.7P
												0.0956	-176.4	0.251	-80.7N
												0.1985	-174.2	0.034	-60.4Z
							4- 0	0.4337	-1.1	0.518	-77.7			0.466	-76.6T
												0.7946	0.3	0.188	-79.7P
												0.2054	178.8	0.188	-79.7N
												0.0337	-152.9	0.093	-63.8Z
							5- 0	0.2334	12.9	0.389	-77.1			0.282	-79.1T
												0.7213	2.3	0.141	-79.1P
												0.2808	174.1	0.141	-79.1N
												0.0	0.0	0.0	0.0Z

E.3 A 14-Node Network

The final example network is the 14-node network given in Figure E.3. A load flow study for this network, including the given line, load, and generation data, is given in Table E.6. A fault study is given in Table E.7.

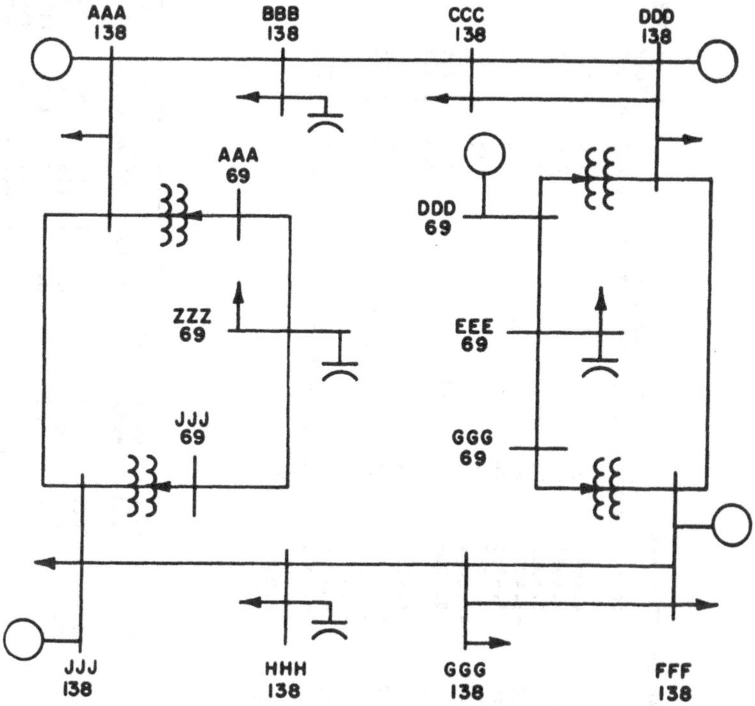

Fig. E.3. A 14-node network.

Table E.6. A 14-Node Network Load Flow Study

IOWA STATE UNIVERSITY VERSION OF 360 LOADFLOW PROGRAM

TITLE-PROBLEM FROM IBM ASSEMBLER LDF

TRANSMISSION LINE AND TRANSFORMER DATA ASSEMBLY

LINE P	LINE Q	NO.	ACTUAL R(PCT)	ACTUAL X(PCT)	ACTUAL BC(PCT)	CONVERTED NO TAP EFFECT BC/2(PU)	G(PU)	B(PU)	MVA RATING	TAP RATIO	TAP LIMITS TMIN	TAP LIMITS TMAX	PHASE SHIFT	MAP DATA FLOW PG	FLOW LOC Q	TAP LOC	REV FLOW PG	REV FLOW LOC Q
1	7	0	6.70	20.00	4.20	0.0210	1.5060	-4.4955	0.				0.0	1	322		1	706
1	9	0	6.70	20.00	4.20	0.0210	1.5060	-4.4955	0.				0.0	1	84		1	86
2	1	0	0.0	12.00	0.0	0.0	0.0	-8.3333	0.	0.990			0.0	1	214	[illegible]	1	212
2	11	0	35.00	42.00	0.70	0.0035	1.1710	-1.4052	0.				0.0	1	296		1	456
3	5	0	6.70	20.00	4.20	0.0210	1.5060	-4.4955	0.				0.0	1	272		1	752
3	10	0	6.70	20.00	4.20	0.0210	1.5060	-4.4955	0.				0.0	1	46		1	44
3	10	1	6.70	20.00	4.20	0.0210	1.5060	-4.4955	0.				0.0	1	94		1	92
4	3	0	0.0	12.00	0.0	0.0	0.0	-8.3333	0.	0.990			0.0	1	220	173	1	222
4	12	0	35.00	42.00	0.70	0.0035	1.1710	-1.4052	0.				0.0	1	331		1	395
5	14	0	6.70	20.00	4.20	0.0210	1.5060	-4.4955	0.				0.0	1	830		1	828
5	14	1	6.70	20.00	4.20	0.0210	1.5060	-4.4955	0.				0.0	1	878		1	876
6	5	0	0.0	12.00	0.0	0.0	0.0	-8.3333	0.	1.000	0.900	1.100	0.0	1	764	717	1	766
6	12	0	35.00	42.00	0.70	0.0035	1.1710	-1.4052	0.				0.0	1	651		1	491
7	13	0	6.70	20.00	4.20	0.0210	1.5060	-4.4955	0.				0.0	1	868		1	870
8	7	0	0.0	12.00	0.0	0.0	0.0	-8.3333	0.	1.000	0.900	1.100	0.0	1	806	757	1	804
8	11	0	35.00	42.00	0.70	0.0035	1.1710	-1.4052	0.				0.0	1	663		1	535
9	10	0	3.40	10.00	2.10	0.0105	3.0477	-8.9638	0.				0.0	1	88		1	90
13	14	0	3.40	10.00	2.10	0.0105	3.0477	-8.9638	0.				0.0	1	872		1	874

BASE CASE BUS DATA ENTERED

BUS NO.	NAME	AREA	REG	VOLTAGE MAG(PU)	ANG(DEG)	LOAD MW	LOAD MVAR	GENERATION MW	GENERATION MVAR	QMIN MVAR	QMAX MVAR	REACTOR MVAR	MAP DATA PAGE	VOLT LOC A	LOAD LOC Q	GEN LOC Q	REACTOR LOC QS
1	AAA138	212	1	1.020	0.0	100.0	50.0	200.0	0.0	0.0	100.0	0.	1	18	146	82	0
2	AAA69	212	0	1.000	0.0	0.0	0.0	0.0	0.0	0.0	0.0	0.	1	216	0	0	0
3	DDD138	213	0	1.000	0.0	100.0	50.0	200.0	56.2	0.0	0.0	0.	1	32	176	112	0
4	DDD69	213	1	1.000	0.0	0.0	0.0	0.0	0.0	0.0	100.0	0.	1	282	0	234	0
5	FFF138	213	1	1.020	0.0	100.0	50.0	200.0	0.0	0.0	100.0	0.	1	942	944	880	0
6	FFF69	213	0	1.000	0.0	0.0	0.0	0.0	0.0	0.0	0.0	0.	1	714	0	0	0
7	JJJ138	212	2	1.040	0.0	100.0	50.0	0.0	0.0	0.0	0.0	0.	1	930	818	882	0
8	JJJ69	212	0	1.000	0.0	0.0	0.0	0.0	0.0	0.0	0.0	0.	1	712	0	0	0
9	BBB	212	0	1.000	0.0	50.0	25.0	0.0	0.0	0.0	0.0	20.	1	22	150	0	152
10	CCC	213	0	1.000	0.0	50.0	25.0	0.0	0.0	0.0	0.0	0.	1	26	154	0	0
11	ZZZ	212	0	1.000	0.0	25.0	20.0	0.0	0.0	0.0	0.0	10.	1	501	454	0	537
12	EEE	213	0	1.000	0.0	25.0	20.0	0.0	0.0	0.0	0.0	10.	1	442	397	0	477
13	HHH	212	0	1.000	0.0	50.0	25.0	0.0	0.0	0.0	0.0	20.	1	808	934	0	936
14	GGG	213	0	1.000	0.0	50.0	25.0	0.0	0.0	0.0	0.0	0.	1	938	940	0	0

IOWA STATE UNIVERSITY VERSION OF 360 LOADFLOW PROGRAM

TITLE-PROBLEM FROM IBM ASSEMBLER LDF

REPORT OF LOADFLOW CALCULATIONS

TOTAL ITERATIONS = 3 SWING BUS = 7 JJJ138 ,AREA 212.

BUS - DATA												LINE - DATA						
AREA NO.	BUS	NAME	MISMATCH MW	MISMATCH MVAR	VOLTS	ANGLE	LOAD MW	LOAD MVAR	GENERATION MW	GENERATION MVAR	CAP/REAC MVAR	TO BUS NAME	AREA NO.	PAR NO.	LINE FLOW MW	LINE FLOW MVAR	PCT CAP	TAP
212	1	AAA138	-0.000	-0.000	1.020	6.6	100.0	50.0	200.0	38.0R		2-AAA69	212	0	20.66	2.18		
												7-JJJ138	212	0	52.62	-26.52		
												9-BBB	212	0	26.73	12.38		
212	2	AAA69	0.000	0.001	1.008	5.2	0.0	0.0	0.0	0.0		1-AAA138	212	0	-20.66	-1.68		0.990
												11-ZZZ	212	0	20.66	1.68		
213	3	DDD138	0.003	-0.013	1.013	9.8	100.0	50.0	200.0	56.2		4-DDD69	213	0	14.56	2.44		
												5-FFF138	213	0	9.35	-8.78		
												10-CCC	213	0	38.04	6.28		
												10-CCC	213	1	38.04	6.28		
213	4	DDD69	0.001	0.000	1.000	8.8	0.0	0.0	0.0	2.8R		3-DDD138	213	0	-14.56	-2.19		0.990
												12-EEE	213	0	14.56	4.96		
213	5	FFF138	-0.002	-0.000	1.020	8.5	100.0	50.0	200.0	67.8R		3-DDD138	213	0	-9.27	4.70		
												6-FFF69	213	0	11.97	7.19		
												14-GGG	213	0	48.65	2.95		
												14-GGG	213	1	48.65	2.95		
213	6	FFF69	-0.000	0.001	1.000	7.7	0.0	0.0	0.0	0.0		5-FFF138	213	0	-11.97	-6.96		0.988R
												12-EEE	213	0	11.97	6.96		
212	7	JJJ138	0.0	0.0	1.040	0.0	100.0	50.0	63.0	112.2		1-AAA138	212	0	-50.45	28.52		
												8-JJJ69	212	0	6.41	11.01		
												13-HHH	212	0	7.02	22.65		
212	8	JJJ69	0.000	0.001	1.000	-0.4	0.0	0.0	0.0	0.0		7-JJJ138	212	0	-6.41	-10.83		0.973R
												11-ZZZ	212	0	6.41	10.83		
212	9	BBB	-0.001	0.000	0.975	4.1	50.0	25.0	0.0	0.0	-19.01	1-AAA138	212	0	-26.13	-14.78		
												10-CCC	213	0	-23.87	8.78		
213	10	CCC	-0.001	-0.002	0.974	5.7	50.0	25.0	0.0	0.0		3-DDD138	213	0	-37.05	-7.46		
												3-DDD138	213	1	-37.05	-7.46		
												9-BBB	212	0	24.11	-10.08		
212	11	ZZZ	0.000	0.000	0.931	0.3	25.0	20.0	0.0	0.0	-8.66	2-AAA69	212	0	-19.17	-0.55		
												8-JJJ69	212	0	-5.83	-10.78		
213	12	EEE	-0.002	-0.002	0.928	6.2	25.0	20.0	0.0	0.0	-8.61	4-DDD69	213	0	-13.72	-4.61		
												6-FFF69	213	0	-11.28	-6.79		
212	13	HHH	0.000	0.000	0.988	0.1	50.0	25.0	0.0	0.0	-19.51	7-JJJ138	212	0	-6.60	-25.73		
												14-GGG	213	0	-43.40	20.24		
213	14	GGG	0.000	-0.001	0.982	3.1	50.0	25.0	0.0	0.0		5-FFF138	213	0	-47.11	-2.56		

Table E.6. *(continued)*

```
IOWA STATE UNIVERSITY VERSION OF 360 LOADFLOW PROGRAM

TITLE-PROBLEM FROM IBM ASSEMBLER LDF.

REPORT OF LOADFLOW CALCULATIONS                                       TOTAL ITERATIONS =   3 SWING BUS =    7 JJJ138 ,AREA 212.
X------------------------------- B U S - D A T A ------------------------------X------------ L I N E - D A T A ------------X
                                                                              X--------------- T O -----------------X
AREA                  MISMATCH                       LOAD        GENERATION    CAP/REAC                AREA PAR   LINE FLOW     PCT
NO.  BUS  NAME        MW    MVAR   VOLTS  ANGLE    MW     MVAR    MW     MVAR     MVAR  BUS  NAME     NO.  NO.    MW    MVAR    CAP    TAP

                                                                                          5-FFF138  213   1   -47.11   -2.56
                                                                                         13-HHH     212   0    44.21  -19.88

                                              END OF REPORT FOR THIS CASE




                                              LOW VOLTAGE SUMMARY -- BUS VOLTAGES BELOW 0.950

   BUS       AREA                          BUS       AREA                           BUS       AREA                          BUS       AREA
NO.  NAME    NO.  VOLTS  ANGLE          NO.  NAME    NO.  VOLTS  ANGLE           NO.  NAME    NO.  VOLTS  ANGLE          NO.  NAME    NO.  VOLTS  ANGLE

 11 ZZZ      212  0.931    0.3           12 EEE      213  0.928    6.2



THERE ARE NO VOLTAGES OVER 1.050



NO LINES OVERLOADED
```

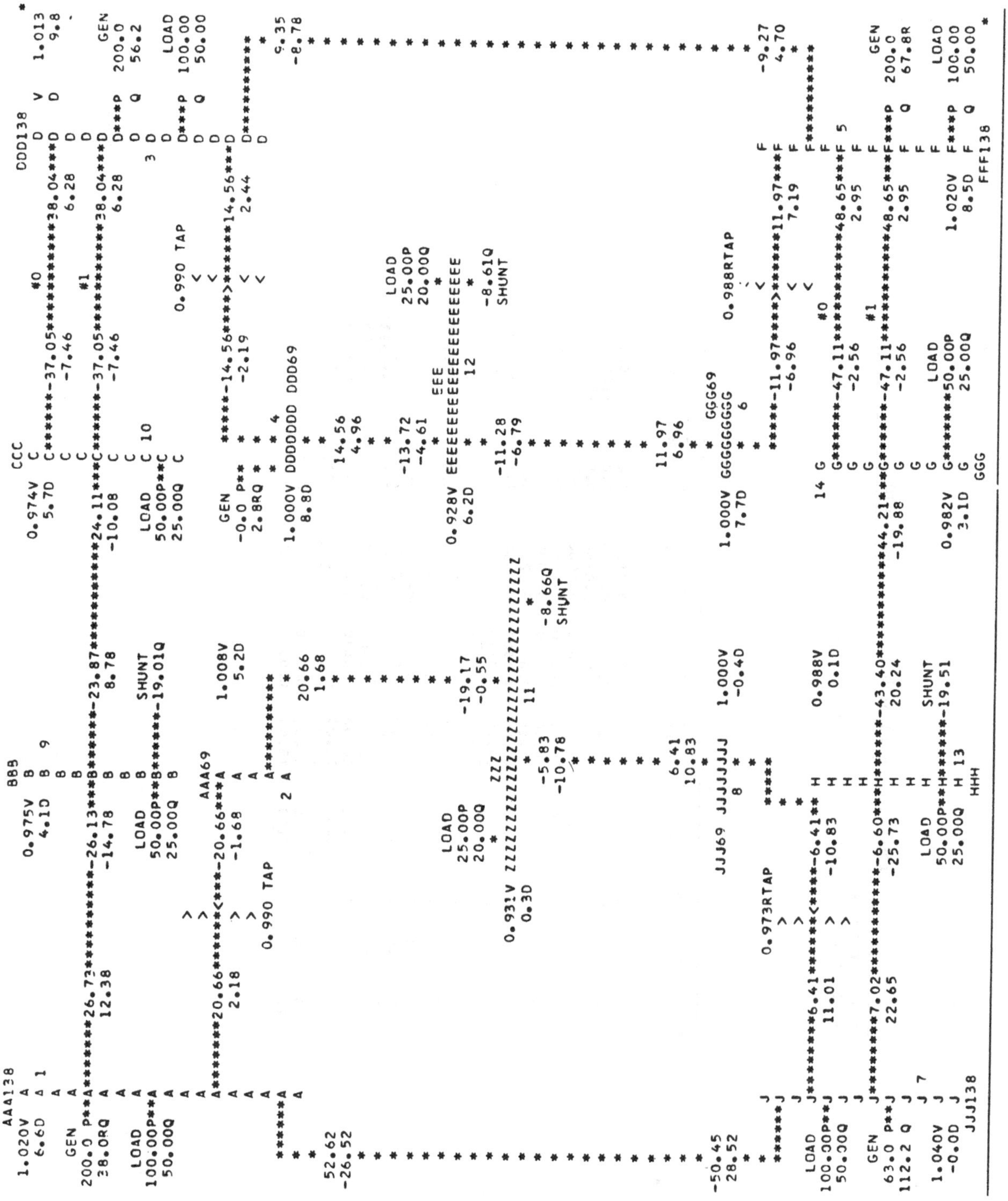

Table E.7. A 14-Node Network Fault Study

PROBLEM FROM IBM ASSEMBLER LDF
TRANSMISSION LINE AND TRANSFORMER DATA ASSEMBLY

DATA FOR POSITIVE SEQUENCE

SELF									
ACTUAL					CONVERTED				
P	Q	NO.	R(PCT)	X(PCT)	R (PU)	X (PU)	G (PU)	B (PU)
0	1	0	0.0	1.00	0.0	0.0100	0.0	-100.0000
0	3	0	0.0	1.00	0.0	0.0100	0.0	-100.0000
0	5	0	0.0	1.00	0.0	0.0100	0.0	-100.0000
0	7	0	0.20	10.00	0.0020	0.1000	0.1999	-9.9960
1	7	0	6.70	20.00	0.0670	0.2000	1.5060	-4.4955
1	9	0	6.70	20.00	0.0670	0.2000	1.5060	-4.4955
2	1	0	0.0	12.00	0.0	0.1200	0.0	-8.3333
2	11	0	35.00	42.00	0.3500	0.4200	1.1710	-1.4052
3	5	0	6.70	20.00	0.0670	0.2000	1.5060	-4.4955
3	10	0	6.70	20.00	0.0670	0.2000	1.5060	-4.4955
3	10	1	6.70	20.00	0.0670	0.2000	1.5060	-4.4955
4	3	0	0.0	12.00	0.0	0.1200	0.0	-8.3333
4	12	0	35.00	42.00	0.3500	0.4200	1.1710	-1.4052
5	14	0	6.70	20.00	0.0670	0.2000	1.5060	-4.4955
5	14	1	6.70	20.00	0.0670	0.2000	1.5060	-4.4955
6	5	0	0.0	12.00	0.0	0.1200	0.0	-8.3333
6	12	0	35.00	42.00	0.3500	0.4200	1.1710	-1.4052
7	13	0	6.70	20.00	0.0670	0.2000	1.5060	-4.4955
8	7	0	0.0	12.00	0.0	0.1200	0.0	-8.3333
8	11	0	35.00	42.00	0.3500	0.4200	1.1710	-1.4052
9	10	0	3.40	10.00	0.0340	0.1000	3.0477	-8.9638
13	14	0	3.40	10.00	0.0340	0.1000	3.0477	-8.9638

PROBLEM FROM IBM ASSEMBLER LDF
EXISTING Z MATRIX FOR THE POSITIVE SEQUENCE NETWORK

BUS	BUS	R	X
1	1	0.00016	0.00944
1	2	0.00043	0.00890
1	3	-0.00006	0.00020
1	4	-0.00005	0.00019
1	5	-0.00004	0.00007
1	6	-0.00005	0.00008
1	7	-0.00053	0.00291
1	8	-0.00080	0.00345
1	9	0.00005	0.00482
1	10	-0.00001	0.00251
1	11	-0.00019	0.00617
1	12	-0.00005	0.00013
1	13	-0.00028	0.00149
1	14	-0.00016	0.00078
2	2	0.00635	0.11926
2	3	-0.00006	0.00020
2	4	-0.00006	0.00019
2	5	-0.00011	0.00015
2	6	-0.00011	0.00016
2	7	-0.00253	0.00762
2	8	-0.00845	0.01726
2	9	0.00019	0.00455
2	10	0.00006	0.00237
2	11	-0.00105	0.06826
2	12	-0.00008	0.00017
2	13	-0.00132	0.00389
2	14	-0.00072	0.00202
3	3	0.00022	0.00933
3	4	0.00064	0.00862
3	5	-0.00015	0.00046
3	6	-0.00057	0.00117
3	7	-0.00006	0.00012
3	8	-0.00006	0.00012
3	9	0.00007	0.00476
3	10	0.00015	0.00705
3	11	-0.00006	0.00016
3	12	0.00003	0.00489
3	13	-0.00010	0.00029
3	14	-0.00013	0.00037
4	4	0.00701	0.11858
4	5	-0.00057	0.00117
4	6	-0.00694	0.01120
4	7	-0.00013	0.00020
4	8	-0.00012	0.00020
4	9	0.00029	0.00440
4	10	0.00047	0.00651
4	11	-0.00009	0.00020
4	12	0.00003	0.06489
4	13	-0.00035	0.00068
4	14	-0.00046	0.00093
5	5	0.00021	0.00935
5	6	0.00063	0.00864
5	7	-0.00019	0.00126
5	8	-0.00012	0.00118
5	9	-0.00009	0.00026
5	10	-0.00012	0.00036
5	11	-0.00011	0.00067
5	12	0.00003	0.00490
5	13	0.00000	0.00531
5	14	0.00011	0.00733
6	6	0.00700	0.11860
6	7	-0.00012	0.00118
6	8	-0.00006	0.00110
6	9	-0.00031	0.00062
6	10	-0.00044	0.00090
6	11	-0.00008	0.00063
6	12	0.00003	0.06490
6	13	0.00025	0.00491
6	14	0.00044	0.00677
7	7	0.00889	0.05699
7	8	0.01090	0.05228
7	9	-0.00029	0.00151
7	10	-0.00018	0.00081
7	11	0.00418	0.02995
7	12	-0.00012	0.00069
7	13	0.00439	0.02913
7	14	0.00207	0.01519
8	8	0.01854	0.15848
8	9	-0.00043	0.00179
8	10	-0.00025	0.00095
8	11	0.00505	0.08787
8	12	-0.00009	0.00065
8	13	0.00542	0.02673
8	14	0.00262	0.01395
9	9	0.03369	0.10479
9	10	0.01675	0.05478
9	11	-0.00012	0.00317
9	12	-0.00001	0.00251
9	13	-0.00019	0.00089
9	14	-0.00014	0.00057
10	10	0.02527	0.08091
10	11	-0.00009	0.00166
10	12	0.00001	0.00370
10	13	-0.00015	0.00059
10	14	-0.00013	0.00047
11	11	0.17700	0.28806
11	12	-0.00009	0.00041
11	13	0.00205	0.01531
11	14	0.00095	0.00798
12	12	0.17503	0.27490
12	13	-0.00005	0.00230
12	14	-0.00001	0.00385
13	13	0.03583	0.11722
13	14	0.01778	0.06126
14	14	0.02575	0.08430

Table E.7. *(continued)*

PROBLEM FROM IBM ASSEMBLER LDF
TRANSMISSION LINE AND TRANSFORMER DATA ASSEMBLY

DATA FOR ZERO SEQUENCE

SELF									MUTUAL						
ACTUAL					CONVERTED				ACTUAL					CONVERTED	
P	Q	NO.	R(PCT)	X(PCT)	R (PU)	X (PU)	G (PU)	B (PU)	P	Q	NO.	RM(PCT)	XM(PCT)	RM (PU)	XM (PU)
0	1	0	0.10	6.00	0.0010	0.0600	0.2777	-16.6620							
0	3	0	0.10	7.00	0.0010	0.0700	0.2040	-14.2828							
0	5	0	0.10	5.00	0.0010	0.0500	0.3998	-19.9920							
1	7	0	13.50	65.00	0.1350	0.6500	0.3063	-1.4748							
1	9	0	13.40	60.00	0.1340	0.6000	0.3545	-1.5875							
2	0	0	0.0	12.00	0.0	0.1200	0.0	-8.3333							
2	11	0	13.40	60.00	0.1340	0.6000	0.3545	-1.5875							
3	5	0	17.00	60.00	0.1700	0.6000	0.4371	-1.5428							
3	10	0	15.00	70.00	0.1500	0.7000			3	10	1	0.0	20.00	0.0	0.2000
3	10	1	15.00	70.00	0.1500	0.7000			3	10	0	0.0	20.00	0.0	0.2000
4	0	0	1.00	30.00	0.0100	0.3000	0.1110	-3.3296							
4	12	0	70.00	120.00	0.7000	1.2000	0.3627	-0.6218							
5	14	0	15.00	65.00	0.1500	0.6500			5	14	1	0.0	22.00	0.0	0.2200
5	14	1	17.50	70.00	0.1750	0.7000			5	14	0	0.0	22.00	0.0	0.2200
6	0	0	1.50	35.00	0.0150	0.3500	0.1222	-2.8519							
6	12	0	75.00	120.00	0.7500	1.2000	0.3745	-0.5993							
7	13	0	15.00	65.50	0.1500	0.6550	0.3322	-1.4506							
8	0	0	0.0	12.00	0.0	0.1200	0.0	-8.3333							
8	11	0	70.00	125.00	0.7000	1.2500	0.3410	-0.6090							
9	10	0	8.50	35.00	0.0850	0.3500	0.6552	-2.6980							
13	14	0	7.50	30.00	0.0750	0.3000	0.7843	-3.1373							

PROBLEM FROM IBM ASSEMBLER LDF
EXISTING MUTUAL ADMITTANCE MATRIX

SELF			COUPLED LINE			SELF		MUTUAL	
BUS	BUS	NO.	BUS	BUS	NO.	G	B	GM	BM
3	10	0				0.36532	-1.45797		
3	10	0	3	10	1			-0.18514	0.37689
3	10	1				0.36532	-1.45797		
5	14	0				0.44964	-1.58181		
5	14	0	5	14	1			-0.24998	0.43465
5	14	1				0.44222	-1.45462		

PROBLEM FROM IBM ASSEMBLER LDF
EXISTING Z MATRIX FOR THE ZERO SEQUENCE NETWORK

BUS	BUS	R	X
1	1	0.00160	0.05622
1	2	0.0	0.0
1	3	-0.00036	0.00248
1	4	0.0	0.0
1	5	-0.00024	0.00137
1	6	0.0	0.0
1	7	0.00114	0.03887
1	8	0.0	0.0
1	9	0.00046	0.03314
1	10	-0.00046	0.01963
1	11	0.0	0.0
1	12	0.0	0.0
1	13	0.00031	0.02132
1	14	-0.00023	0.01325
2	2	0.00237	0.11406
2	3	0.0	0.0
2	4	0.0	0.0
2	5	0.0	0.0
2	6	0.0	0.0
2	7	0.0	0.0
2	8	-0.00237	0.00594
2	9	0.0	0.0
2	10	0.0	0.0
2	11	0.00759	0.08169
2	12	0.0	0.0
2	13	0.0	0.0
2	14	0.0	0.0
3	3	0.00266	0.06093
3	4	0.0	0.0
3	5	-0.00085	0.00441
3	6	0.0	0.0
3	7	-0.00052	0.00309
3	8	0.0	0.0
3	9	0.00126	0.02758
3	10	0.00248	0.04227
3	11	0.0	0.0
3	12	0.0	0.0
3	13	-0.00067	0.00371
3	14	-0.00074	0.00400
4	4	0.01996	0.27534
4	5	0.0	0.0
4	6	-0.01135	0.02889
4	7	0.0	0.0
4	8	0.0	0.0
4	9	0.0	0.0
4	10	0.0	0.0
4	11	0.0	0.0
4	12	0.00633	0.15301
4	13	0.0	0.0
4	14	0.0	0.0
5	5	0.00179	0.04570
5	6	0.0	0.0
5	7	0.00031	0.01540
5	8	0.0	0.0
5	9	-0.00049	0.00268
5	10	-0.00062	0.00345
5	11	0.0	0.0
5	12	0.0	0.0
5	13	0.00115	0.02958
5	14	0.00166	0.03610
6	6	0.02792	0.31616
6	7	0.0	0.0
6	8	0.0	0.0
6	9	0.0	0.0
6	10	0.0	0.0
6	11	0.0	0.0
6	12	0.00591	0.17150
6	13	0.0	0.0
6	14	0.0	0.0
7	7	0.09463	0.47554
7	8	0.0	0.0
7	9	0.00023	0.02350
7	10	-0.00047	0.01451
7	11	0.0	0.0
7	12	0.0	0.0
7	13	0.04823	0.26051
7	14	0.02491	0.16200
8	8	0.00237	0.11406
8	9	0.0	0.0
8	10	0.0	0.0
8	11	-0.00759	0.03831
8	12	0.0	0.0
8	13	0.0	0.0
8	14	0.0	0.0
9	9	0.07391	0.37365
9	10	0.03562	0.22231
9	11	0.0	0.0
9	12	0.0	0.0
9	13	-0.00021	0.01376
9	14	-0.00050	0.00928
10	10	0.05895	0.34063
10	11	0.0	0.0
10	12	0.0	0.0
10	13	-0.00059	0.00933
10	14	-0.00070	0.00695
11	11	0.13678	0.48166
11	12	0.0	0.0
11	13	0.0	0.0
11	14	0.0	0.0
12	12	0.36837	0.76236
12	13	0.0	0.0
12	14	0.0	0.0
13	13	0.10164	0.50158
13	14	0.05342	0.31201
14	14	0.06791	0.38077

Table E.7. *(continued)*

PROBLEM FROM IBM ASSEMBLER LDF
REPORT OF FAULT CALCULATIONS

BUS	NAME	BUS-DATA 3-PHASE AMPS	DEGREES	BUS-DATA S-L-G VOLTS	DEGREES	AMPS	DEGREES	LINE-DATA BUS NO.	3-PHASE VOLTS	DEGREES	AMPS	DEGREES	S-L-G VOLTS	DEGREES	AMPS	DEGREES
1	AAA138	105.903	-89.1			39.931	-88.5T									
				0.8743	-0.1	13.310	-88.5P									
				0.1257	-179.5	13.310	-88.5N									
				0.7486	179.8	13.310	-88.5Z									
								0- 0	1.0000	0.0100.000		-90.0			37.612	-89.4T
													1.0000	0.0	12.568	-89.5P
													0.0	0.0	12.568	-89.5N
													0.0	0.0	12.476	-89.2Z
								2- 0	0.0649	27.9	0.541	-62.1			0.136	-61.6T
													0.8815	0.2	0.068	-61.6P
													0.1185	178.7	0.068	-61.6N
													0.0	0.0	0.0	0.0Z
								7- 0	0.6958	-5.0	3.299	-76.5			1.177	-76.7T
													0.9615	-0.5	0.415	-76.0P
													0.0393	-168.3	0.415	-76.0N
													0.5176	179.8	0.348	-78.3Z
								9- 0	0.4892	-0.3	2.319	-71.8			1.081	-74.8T
													0.9358	-0.1	0.291	-71.3P
													0.0642	-179.2	0.291	-71.3N
													0.4412	-179.3	0.500	-78.8Z
2	AAA69	8.373	-87.0			8.501	-87.6T									
				0.6616	0.3	2.834	-87.6P									
				0.3384	179.4	2.834	-87.6N									
				0.3233	-178.7	2.834	-87.6Z									
								0- 0	0.0	0.0	0.0	0.0			2.694	-88.7T
													0.0	0.0	0.0	0.0P
													0.0	0.0	0.0	0.0N
													0.0	0.0	2.694	-88.7Z
								1- 0	0.9254	-0.0	7.712	-90.0			5.220	-90.6T
													0.9748	0.0	2.610	-90.6P
													0.0252	179.7	2.610	-90.6N
													0.0	0.0	0.0	0.0Z
								11- 0	0.4315	-5.2	0.789	-55.4			0.684	-58.2T
													0.8070	-0.8	0.267	-56.0P
													0.1934	-176.7	0.267	-56.0N
													0.2325	177.1	0.151	-65.8Z
3	DDD138	107.162	-88.7			37.664	-87.8T									
				0.8829	-0.1	12.555	-87.8P									
				0.1172	-179.1	12.555	-87.8N									
				0.7657	179.7	12.555	-87.8Z									
								0- 0	1.0000	0.0100.000		-90.0			34.369	-89.2T
													1.0000	0.0	11.716	-89.1P
													0.0	0.0	11.716	-89.1N
													0.0	0.0	10.938	-89.5Z

PROBLEM FROM IBM ASSEMBLER LDF
REPORT OF FAULT CALCULATIONS

BUS	NAME	3-PHASE AMPS	3-PHASE DEGREES	S-L-G VOLTS	S-L-G DEGREES	S-L-G AMPS	S-L-G DEGREES
4	DDD69	8.418	-86.6			5.841	-86.2T
				0.7687	-0.1	1.947	-86.2P
				0.2313	-179.6	1.947	-86.2N
				0.5375	179.6	1.947	-86.2Z
5	FFF138	106.918	-88.7			46.557	-88.0T
				0.8549	-0.1	15.519	-88.0P
				0.1452	-179.3	15.519	-88.0N
				0.7097	179.7	15.519	-88.0Z

LINE - DATA

BUS NO.	3-PHASE VOLTS	3-PHASE DEGREES	3-PHASE AMPS	3-PHASE DEGREES	S-L-G VOLTS	S-L-G DEGREES	S-L-G AMPS	S-L-G DEGREES
4- 0	0.0887	31.9	0.739	-58.1			0.173	-57.2T
					0.8916	0.2	0.087	-57.2P
					0.1085	178.0	0.087	-57.2N
					0.0	0.0	0.0	0.0Z
5- 0	0.9516	-1.0	4.512	-72.5			2.196	-73.6T
					0.9943	-0.1	0.529	-71.6P
					0.0060	-159.8	0.529	-71.6N
					0.0564	-166.8	1.140	-75.5Z
10- 0	0.2445	-0.3	1.159	-71.8			0.527	-74.8T
					0.9115	-0.1	0.136	-70.9P
					0.0885	-179.0	0.136	-70.9N
					0.5316	178.9	0.257	-78.9Z
10- 1	0.2445	-0.3	1.159	-71.8			0.527	-74.8T
					0.9115	-0.1	0.136	-70.9P
					0.0885	-179.0	0.136	-70.9N
					0.5316	178.9	0.257	-78.9Z
0- 0	0.0	0.0	0.0	0.0			1.791	-88.4T
					0.0	0.0	0.0	0.0P
					0.0	0.0	0.0	0.0N
					0.0	0.0	1.791	-88.4Z
3- 0	0.9273	0.1	7.727	-89.9			3.574	-89.5T
					0.9832	0.0	1.787	-89.5P
					0.0168	179.6	1.787	-89.5N
					0.0	0.0	0.0	0.0Z
12- 0	0.4558	-4.0	0.834	-54.2			0.557	-56.4T
					0.8740	-0.5	0.193	-53.8P
					0.1263	-176.2	0.193	-53.8N
					0.2982	-178.6	0.172	-62.3Z
0- 0	1.0000	0.0	100.000	-90.0			43.222	-89.2T
					1.0000	0.0	14.515	-89.3P
					0.0	0.0	14.515	-89.3N
					0.0	0.0	14.192	-89.1Z

Table E.7. *(continued)*

PROBLEM FROM IBM ASSEMBLER LDF
REPORT OF FAULT CALCULATIONS

BUS DATA								LINE DATA								
		3-PHASE		S-L-G					3-PHASE				S-L-G			
BUS	NAME	AMPS	DEGREES	VOLTS	DEGREES	AMPS	DEGREES	BUS NO.	VOLTS	DEGREES	AMPS	DEGREES	VOLTS	DEGREES	AMPS	DEGREES
								3- 0	0.9517	-1.0	4.512	-72.5			2.338	-73.6T
													0.9930	-0.1	0.655	-71.8P
													0.0075	-160.1	0.655	-71.8N
													0.0698	-167.1	1.029	-75.9Z
								6- 0	0.0887	31.9	0.739	-58.1			0.215	-57.4T
													0.8657	0.3	0.107	-57.4P
													0.1344	177.8	0.107	-57.4N
													0.0	0.0	0.0	0.0Z
								14- 0	0.2163	-1.5	1.026	-72.9			0.470	-74.7T
													0.8863	-0.1	0.149	-72.2P
													0.1138	-178.9	0.149	-72.2N
													0.5608	179.3	0.173	-78.9Z
								14- 1	0.2163	-1.5	1.026	-72.9			0.452	-74.3T
													0.8863	-0.1	0.149	-72.2P
													0.1138	-178.9	0.149	-72.2N
													0.5608	179.3	0.155	-78.1Z
6	FFF69	8.417	-86.6			5.406	-85.7T									
				0.7859	-0.3	1.802	-85.7P									
				0.2141	-179.0	1.802	-85.7N									
				0.5719	179.3	1.802	-85.7Z									
								0- 0	0.0	0.0	0.0	0.0			1.633	-88.3T
													0.0	0.0	0.0	0.0P
													0.0	0.0	0.0	0.0N
													0.0	0.0	1.633	-88.3Z
								5- 0	0.9271	0.1	7.726	-89.9			3.308	-89.0T
													0.9844	-0.0	1.654	-89.0P
													0.0156	-179.8	1.654	-89.0N
													0.0	0.0	0.0	0.0Z
								12- 0	0.4558	-4.0	0.834	-54.2			0.542	-56.4T
													0.8834	-0.6	0.178	-53.3P
													0.1170	-175.7	0.178	-53.3N
													0.3092	-177.6	0.186	-62.3Z
7	JJJ138	17.338	-81.1			4.999	-79.2T									
				0.9040	-0.2	1.666	-79.2P									
				0.0961	-178.1	1.666	-79.2N									
				0.8079	179.5	1.666	-79.2Z									
								0- 0	1.0000	0.0	9.998	-88.9			1.922	-86.9T
													1.0000	0.0	0.961	-86.9P
													0.0	0.0	0.961	-86.9N
													0.0	0.0	0.0	0.0Z

PROBLEM FROM IBM ASSEMBLER LDF
REPORT OF FAULT CALCULATIONS

BUS - DATA

BUS	NAME	3-PHASE AMPS	3-PHASE DEGREES	S-L-G VOLTS	S-L-G DEGREES	S-L-G AMPS	S-L-G DEGREES
8	JJJ69	6.267	-83.3			6.931	-84.8T
				0.6315	0.8	2.310	-84.8P
				0.3687	178.6	2.310	-84.8N
				0.2636	-176.0	2.310	-84.8Z

LINE - DATA

BUS NO.	3-PHASE VOLTS	3-PHASE DEGREES	3-PHASE AMPS	3-PHASE DEGREES	S-L-G VOLTS	S-L-G DEGREES	S-L-G AMPS	S-L-G DEGREES
1- 0	0.9518	-1.0	4.512	-72.5			1.982	-75.6T
					0.9954	-0.1	0.434	-70.6P
					0.0049	-158.9	0.434	-70.6N
					0.0648	-170.9	1.121	-79.6Z
8- 0	0.0887	31.9	0.740	-58.1			0.142	-56.1T
					0.9110	0.1	0.071	-56.1P
					0.0890	179.0	0.071	-56.1N
					0.0	0.0	0.0	0.0Z
13- 0	0.4893	-0.3	2.320	-71.8			0.989	-74.6T
					0.9510	-0.1	0.223	-69.9P
					0.0491	-177.8	0.223	-69.9N
					0.4415	-179.7	0.545	-78.5Z

BUS - DATA

BUS	NAME	3-PHASE AMPS	3-PHASE DEGREES	S-L-G VOLTS	S-L-G DEGREES	S-L-G AMPS	S-L-G DEGREES
9	BBB	9.085	-72.2			4.999	-76.4T
				0.8172	0.9	1.666	-76.4P
				0.1834	175.8	1.666	-76.4N
				0.6347	-177.6	1.666	-76.4Z

LINE - DATA

BUS NO.	3-PHASE VOLTS	3-PHASE DEGREES	3-PHASE AMPS	3-PHASE DEGREES	S-L-G VOLTS	S-L-G DEGREES	S-L-G AMPS	S-L-G DEGREES
0- 0	0.0	0.0	0.0	0.0			2.197	-86.0T
					0.0	0.0	0.0	0.0P
					0.0	0.0	0.0	0.0N
					0.0	0.0	2.197	-86.0Z
7- 0	0.6673	2.6	5.561	-87.4			4.100	-88.9T
					0.8775	0.9	2.050	-88.9P
					0.1234	173.5	2.050	-88.9N
					0.0	0.0	0.0	0.0Z
11- 0	0.4506	-4.1	0.824	-54.3			0.730	-57.0T
					0.7968	-0.5	0.304	-55.8P
					0.2033	-178.1	0.304	-55.8N
					0.0902	-163.6	0.123	-63.0Z
1- 0	0.9582	-0.8	4.543	-72.3			2.611	-76.3T
					0.9922	-0.1	0.833	-76.5P
					0.0080	-167.0	0.833	-76.5N
					0.0552	-167.2	0.944	-76.0Z
10- 0	0.4797	-0.9	4.542	-72.1			2.388	-76.5T
					0.9047	0.4	0.833	-76.3P
					0.0955	176.6	0.833	-76.3N
					0.3752	-175.5	0.722	-76.9Z

Table E.7. *(continued)*

PROBLEM FROM IBM ASSEMBLER LDF
REPORT OF FAULT CALCULATIONS

BUS - DATA		3-PHASE		S-L-G				LINE - DATA	3-PHASE				S-L-G			
BUS	NAME	AMPS	DEGREES	VOLTS	DEGREES	AMPS	DEGREES	BUS NO.	VOLTS	DEGREES	AMPS	DEGREES	VOLTS	DEGREES	AMPS	DEGREES
10	CCC	11.797	-72.7			5.834	-77.7T									
				0.8359	1.0	1.945	-77.7P									
				0.1648	175.0	1.945	-77.7N									
				0.6722	-177.5	1.945	-77.7Z									
								3- 0	0.9204	-1.4	4.364	-72.9			2.086	-78.3T
													0.9866	-0.2	0.719	-78.0P
													0.0137	-168.9	0.719	-78.0N
													0.0823	-171.1	0.647	-79.0Z
								3- 1	0.9204	-1.4	4.364	-72.9			2.086	-78.3T
													0.9866	-0.2	0.719	-78.0P
													0.0137	-168.9	0.719	-78.0N
													0.0823	-171.1	0.647	-79.0Z
								9- 0	0.3243	-0.7	3.070	-71.9			1.663	-76.3T
													0.8890	0.6	0.506	-77.0P
													0.1114	175.3	0.506	-77.0N
													0.4378	-176.8	0.651	-75.2Z
11	ZZZ	2.958	-58.4			2.573	-65.1T									
				0.7128	2.7	0.858	-65.1P									
				0.2899	173.3	0.858	-65.1N									
				0.4294	-171.0	0.858	-65.1Z									
								2- 0	0.8367	-7.4	1.530	-57.6			1.472	-66.7T
													0.9476	-1.5	0.444	-64.3P
													0.0585	-154.2	0.444	-64.3N
													0.0704	-160.4	0.586	-70.4Z
								8- 0	0.7806	-9.1	1.428	-59.3			1.102	-62.9T
													0.9302	-1.7	0.414	-66.0P
													0.0755	-158.4	0.414	-66.0N
													0.0335	-143.9	0.279	-53.9Z
12	EEE	3.069	-57.5			2.005	-61.3T									
				0.7828	1.1	0.668	-61.3P									
				0.2178	176.2	0.668	-61.3N									
				0.5660	-177.1	0.668	-61.3Z									
								4- 0	0.8388	-7.3	1.534	-57.5			1.010	-61.5T
													0.9622	-1.2	0.334	-61.3P
													0.0434	-151.3	0.334	-61.3N
													0.1024	-153.7	0.341	-61.8Z
								6- 0	0.8388	-7.3	1.534	-57.5			0.996	-61.1T
													0.9622	-1.2	0.334	-61.3P
													0.0434	-151.3	0.334	-61.3N
													0.1147	-153.3	0.327	-60.8Z
13	HHH	8.158	-73.0			3.967	-76.7T									
				0.8383	0.7	1.322	-76.7P									
				0.1621	176.3	1.322	-76.7N									
				0.6768	-178.2	1.322	-76.7Z									

PROBLEM FROM IBM ASSEMBLER LDF
REPORT OF FAULT CALCULATIONS

B U S - D A T A

BUS	NAME	3-PHASE AMPS	3-PHASE DEGREES	S-L-G VOLTS	S-L-G DEGREES	S-L-G AMPS	S-L-G DEGREES
14	GGG	11.345	-73.0			5.336	-77.7T
				0.8438	0.9	1.779	-77.7P
				0.1568	175.3	1.779	-77.7N
				0.6880	-177.8	1.779	-77.7Z

L I N E - D A T A

BUS NO.	3-PHASE VOLTS	3-PHASE DEGREES	3-PHASE AMPS	3-PHASE DEGREES	S-L-G VOLTS	S-L-G DEGREES	S-L-G AMPS	S-L-G DEGREES
7- 0	0.7631	-2.6	3.618	-74.1			1.659	-77.4T
					0.9612	-0.2	0.586	-77.9P
					0.0390	-175.3	0.586	-77.9N
					0.3504	-177.2	0.486	-76.3Z
14- 0	0.4797	-0.9	4.542	-72.1			2.309	-76.3T
					0.9158	0.3	0.736	-75.9P
					0.0844	177.1	0.736	-75.9N
					0.4186	-176.5	0.837	-77.0Z
5- 0	0.9204	-1.4	4.364	-72.9			2.095	-78.0T
					0.9872	-0.1	0.684	-77.6P
					0.0130	-168.6	0.684	-77.6N
					0.0643	-170.4	0.727	-78.7Z
5- 1	0.9204	-1.4	4.364	-72.9			2.016	-77.7T
					0.9872	-0.1	0.684	-77.6P
					0.0130	-168.6	0.684	-77.6N
					0.0643	-170.4	0.648	-77.9Z
13- 0	0.2765	-2.1	2.618	-73.3			1.225	-77.2T
					0.8868	0.5	0.410	-78.1P
					0.1135	176.1	0.410	-78.1N
					0.5631	-177.5	0.404	-75.6Z

appendix **F**

Δ-Y Transformations

Conversion of impedances from Δ to Y and Y to Δ are computed so often that the formulas for these operations are included here for convenience.

F.1 Y to Δ Transformation

Given three impedances Z_a, Z_b, and Z_c connected in Y, the equivalent Δ impedances Z_{ab}, Z_{bc}, and Z_{ca} are computed as follows:

$$\begin{bmatrix} Z_{ab} \\ Z_{bc} \\ Z_{ca} \end{bmatrix} = Y_\Sigma \begin{bmatrix} Z_b & 0 & 0 \\ 0 & Z_c & 0 \\ 0 & 0 & Z_a \end{bmatrix} \begin{bmatrix} Z_a \\ Z_b \\ Z_c \end{bmatrix} \tag{F.1}$$

where

$$Y_\Sigma = 1/Z_a + 1/Z_b + 1/Z_c$$

In the special case where $Z_a = Z_b = Z_c = Z$, then $Z_{ab} = Z_{bc} = Z_{ca} = 3Z$.

F.2 Δ to Y Transformation

Given three impedances Z_{ab}, Z_{bc}, and Z_{ca} connected in Δ, the equivalent Y impedances Z_a, Z_b, and Z_c are computed as follows:

$$\begin{bmatrix} Z_a \\ Z_b \\ Z_c \end{bmatrix} = \frac{1}{Z_\Sigma} \begin{bmatrix} Z_{ca} & 0 & 0 \\ 0 & Z_{ab} & 0 \\ 0 & 0 & Z_{bc} \end{bmatrix} \begin{bmatrix} Z_{ab} \\ Z_{bc} \\ Z_{ca} \end{bmatrix} \tag{F.2}$$

where

$$Z_\Sigma = Z_{ab} + Z_{bc} + Z_{ca}$$

In the special case where $Z_{ab} = Z_{bc} = Z_{ca} = Z$, then $Z_a = Z_b = Z_c = Z/3$.

Bibliography

1. Fortescue, C. L. Method of Symmetrical Coordinates Applied to the Solution of Polyphase Networks. *Trans. AIEE* 37: 1027–1140, 1918.
2. Electrical Engineering Staff, Iowa State University. Symmetrical Components. Unpubl. power system short course lecture notes, Ames, 1952–1968.
3. Nilsson, James W. *Introduction to Circuits, Instruments, and Electronics.* Harcourt, Brace, & World, New York, 1968.
4. Korn, Granino A., and Korn, Theresa M. *Mathematical Handbook for Scientists and Engineers*, 2nd ed. McGraw-Hill, New York, 1968.
5. Huelsman, L. P. *Circuits, Matrices and Linear Vector Spaces.* McGraw-Hill, New York, 1963.
6. Marcus, M., and Minc, H. *Introduction to Linear Algebra.* Macmillan, New York, 1965.
7. Hohn, Franz E. *Elementary Matrix Algebra.* Macmillan, New York, 1958.
8. Ogata, Katsuhiko. *State Space Analysis of Control Systems.* Prentice-Hall, Englewood Cliffs, N.J., 1965.
9. Stevenson, W. D., Jr. *Elements of Power System Analysis*, 2nd ed. McGraw-Hill, New York, 1962.
10. Wagner, C. F., and Evans, R. D. *Symmetrical Components.* McGraw-Hill, New York, 1933.
11. Clarke, Edith. *Circuit Analysis of A-C Power Systems*, 2 vols. General Electric Co., Schenectady, N.Y., 1950.
12. Boast, W. B. *Vector Fields.* Harper and Row, New York, 1964.
13. Lewis, W. E., and Pryce, D. G. *The Application of Matrix Theory to Electrical Engineering.* E. and F. N. Spon, London, 1965.
14. Westinghouse Electric Corporation. *Electrical Transmission and Distribution Reference Book*, 4th ed. East Pittsburgh, Pa., 1950.
15. Ward, J. B. Equivalent Circuits in Power System Studies. Res. Ser. 109, Eng. Exp. Sta., Purdue University, Lafayette, Ind., 1949.
16. Ferguson, W. H. Symmetrical Component Network Connections for the Solution of Phase-Interchange Faults. *Trans. AIEE* 78 (pt. 3): 948–50, 1959.
17. Kimbark, E. W. *Power System Stability*, vol. 1. Elements of Stability Calculations. Wiley, New York, 1948.
18. Kimbark, E. W. *Power System Stability*, vol. 2. Power Circuit Breakers and Protective Relays. Wiley, New York, 1950.
19. Kimbark, E. W. *Power System Stability*, vol. 3. Synchronous Machines. Wiley, New York, 1956.

20. General Electric Company. *GE Network Analyzers* (manual). Schenectady, N.Y., 1950.
21. Stevenson, W. D., Jr. *Elements of Power System Analysis*, 1st ed. McGraw-Hill, New York, 1955.
22. Woodruff, L. F. *Principles of Electric Power Transmission*, Wiley, New York, 1948.
23. Attwood, Stephen S. *Electric and Magnetic Fields*, 3rd ed. Wiley, New York, 1928.
24. Calabrese, G. O. *Symmetrical Components Applied to Electric Power Networks*. Ronald Press, New York, 1959.
25. Rosa, E. B., and Grover, F. W. Formulas and Tables for the Calculation of Mutual and Self-Inductance. Scientific paper 169, vol. 8, U.S. Bureau of Standards bull., Washington, D.C., 1912.
26. Lewis, W. W. *Transmission Line Engineering*. McGraw-Hill, New York, 1928.
27. Carson, John R. Wave Propagation in Overhead Wires with Ground Return. *Bell System Tech. J.* 5: 539–54, 1926.
28. Gross, E. T. B., and Hesse, M. H. Electromagnetic Unbalance of Untransposed Lines. *Trans. AIEE* 72: 1323–36, 1953.
29. Gross, E. T. B., and Nelson, S. W. Electromagnetic Unbalance of Untransposed Transmission Lines. II. Single Lines with Horizontal Conductor Arrangement. *Trans. AIEE* 74: 887–93, 1954.
30. Gross, E. T. B., Drinnan, J. H., and Jochum, E. Electromagnetic Unbalance of Untransposed Transmission Lines. III. Double Circuit Lines. *Trans. AIEE* 79: 1362–71, 1959.
31. Dwight, H. B. Resistance and Reactance of Commercial Steel Conductors. *Elec. J.*, p. 25, January 1919.
32. Hesse, M. H. Circulating Currents in Parallel Untransposed Multicircuit Lines. I. Numerical Evaluations. *Trans. IEEE* PAS-85: 802–11, 1966.
33. Hesse, M. H. Circulating Currents in Parallel Untransposed Multicircuit Lines. II. Methods for Estimating Current Unbalance. *Trans. IEEE* PAS-85: 812–20, 1966.
34. Gross, E. T. B., and Weston, A. H. Transposition of High-Voltage Overhead Lines and Elimination of Electrostatic Unbalance to Ground. *Trans. AIEE* 70: 1837–44, 1951.
35. Gross, E. T. B. Unbalances of Untransposed Overhead Lines. *J. Franklin Inst.*, pp. 487–97, December 1952.
36. Gross, E. T. B., and McNutt, W. J. Electrostatic Unbalance to Ground of Twin Conductor Lines. *Trans. AIEE* 72: 1288–97, 1953.
37. Starr, F. M. Equivalent Circuits. I. *Trans. AIEE*, vol. 51, June 1932.
38. Hesse, M. H. Electromagnetic and Electrostatic Transmission Line Parameters by Digital Computer. *Trans. IEEE* PAS-82: 282–91, 1963.
39. Lyon, W. V. *Applications of the Method of Symmetrical Components*. McGraw-Hill, New York, 1937.
40. Finkbeiner, Daniel T., II. *Introduction to Matrices and Linear Transformations*, 2nd ed. W. H. Freeman, San Francisco, 1966.
41. Crary, S. B., and Saline, L. E. Location of Series Capacitors in High-Voltage Transmission Systems. *Trans. AIEE* 72: 1140–51, 1953.
42. Holley, H., Colemen, D., and Shipley, R. B. Untransposed EHV Line Computations. *Trans. IEEE* PAS-83: 291–96, 1964.

43. Hesse, M. H. Electromagnetic and Electrostatic Transmission Line Parameters by Digital Computer. *Trans. IEEE* PAS-82: 282–91, 1963.
44. Neumann, R. *Symmetrical Component Analysis of Unsymmetrical Polyphase Systems.* Sir Isacc Pitman & Sons, London, 1939.
45. Concordia, Charles. *Synchronous Machines—Theory and Performance.* Wiley, New York, 1951.
46. Prentice, B. R. Fundamental Concepts of Synchronous Machine Reactances. *Trans. AIEE* 56(suppl.):1–21, 1937.
47. Park, R. H. Two-Reaction Theory of Synchronous Machines. I. Generalized Method of Analysis. *Trans. AIEE* 48: 716–30, 1929.
48. Lewis, W. A. A Basic Analysis of Synchronous Machines. I. *Trans. AIEE* PAS-77: 436–55, 1958.
49. Krause, P. C., and Thomas, C. H. Simulation of Symmetrical Induction Machinery. *Trans. IEEE* PAS-84: 1038–53, 1965.
50. Floyd, G. D., and Sills, H. R. Generator Design to Meet Long Distance Transmission Requirements in Canada. CIGRE rept. 131, 1948.
51. AIEE Switchgear Committee. Simplified Calculation of Fault Currents. *Trans. AIEE* 67 (pt. 2): 1433–35, 1948.
52. C57.12.70–1978 American National Standard Terminal Markings and Connections for Distribution and Power Transformers, IEEE Standards Association, P.O. Box 1331, Piscataway, NJ 08855-1331, USA.
53. Federal Power Commission. National Power Survey. I and II. USGPO, Washington, D.C., 1964.
54. Westinghouse Electric Corporation. *Electric Utility Engineering Reference Book*, vol. 3. Distribution Systems. East Pittsburgh, Pa., 1959.
55. Garin, A. N. Zero-Phase-Sequence Characteristics of Transformers. I and II. *General Electric Rev.* 43: (March, April): 131–36, 174–79, 1940.
56. Atabekov, G. I. *The Relay Protection of High Voltage Networks.* Pergamon Press, New York, 1960.
57. Kron, G. *Tensor Analysis of Networks.* Wiley, New York, 1939.
58. LeCorbeiller, P. *Matrix Analysis of Electric Networks.* Harvard University Press, Cambridge, Mass., 1950.
59. Hancock, N. N. *Matrix Analysis of Electrical Machinery.* Pergamon Press, London, 1964.
60. Stigant, S. Austin. *Modern Electrical Engineering Mathematics.* Hutchinson, London, 1946.
61. Synge, J. L. The Fundamental Theorem of Electrical Networks. *Quart. Appl. Math.*, vol. 9, no. 2, July, 1951.
62. Guillemin, Ernst A. Communication Networks, vol. 2. Wiley, New York, 1935.
63. Anderson, P. M. Analysis of Simultaneous Faults by Two-Port Network Theory. *Trans. IEEE*, PAS-90 (Sept./Oct.): 2199–2205, 1971.
64. Stagg, Glenn W. and El-Abiad, Ahmed H. *Computer Methods in Power System Analysis.* McGraw-Hill, New York, 1968.
65. Anderson, P. M., Bowen, D. W., and Shah, A. P. An Indefinite Admittance Network Description for Fault Computation. *Trans. IEEE*, PAS-89 (July/August): 1215–19, 1970.
66. Shildneck, L. P., Synchronous Machine Reactances, a Fundamental and Physical Viewpoint. *General Electric Rev.*, 35 (November): 560–65, 1932.
67. Glimn, A. F., Habermann, R., Jr., Henderson. J. M., and Kirchmayer, L. K.

Digital Calculation of Network Impedances. *Trans. AIEE* 74 (pt. 3): 1285–97, 1955.

68. Hale, H. W., and Ward, J. B. Digital Computation of Driving Point and Transfer Impedances. *Trans. AIEE* 76 (pt. 3): 476–81, 1957.
69. Toalston, A. L. Digital Solution of Short-Circuit Currents for Networks Including Mutual Impedances. *Trans. AIEE* 78 (pt. 3B):1720–23, 1959.
70. Brown, H. E., and Person, C. E. Digital Calculations of Single-Phase to Ground Faults. *Trans. AIEE* 79 (pt. 3): 657–60, 1960.
71. Brown, H. E., Person, C. E., Kirchmayer, L. K., and Stagg, G. W. Digital Calculation of Three-Phase Short Circuits by Matrix Method. *Trans. AIEE* 79 (pt. 3): 1277–82, 1960.
72. El-Abiad, A. H. Digital Calculation of Line-to-Ground Short Circuits by Matrix Method. *Trans. AIEE* 79 (pt. 3): 323–32, 1960.
73. Baumann, R. Some Remarks on Building Algorithms. Proc. Power Systems Computation Conference, London, 1963.
74. Baumann, R. Some New Aspects of Load Flow Calculations. I. Impedance Matrix Generation Controlled by Network Typology. *Trans. IEEE* PAS-85: 1164–76, 1966.
75. Brown, H. E., and Person, C. E. Short Circuit Studies of Large Systems. Proc. Second Power Systems Computation Conference, pt. 2, report 4.11, Stockholm, 1966.
76. Kruempel, K. C. A Study of the Building Algorithm for the Bus Impedance Matrix. Unpubl. Ph.D. thesis, University of Wisconsin. University Microfilms, Ann Arbor, Mich., January 1970.
77. Penfield, Paul, Spence, Robert, and Duinker, S. *Tellegren's Theorem and Electrical Networks*. MIT Press, Cambridge, 1970.
78. General Electric Company. Project EHV. *EHV Transmission Line Reference Book*. Edison Electric Institute, New York, 1968.
79. Ward, J. B., and Hale, H. W. Digital Solution of Power-Flow Problems. *Trans. AIEE* 75 (pt. 3): 398–404, 1956.
80. Allen, R. F., Zakos, R. J., and Lowd, R. IBM System/360 Electric Power System Load Flow Program. Operating System Version, Order File 360D-16.4.005 and 360D-16.4.006. IBM Corp., Chicago, Ill.
81. Tinney, W. F., and Walker, J. W. Direct Solution of Sparse Network Equations by Optimally Ordered Triangular Factorization. *Proc. IEEE* 55 (November): 1801–09, 1967.
82. Sato, N., and Tinney, W. F. Techniques for Exploiting the Sparsity of the Network Admittance Matrix. *Trans. IEEE* PAS-82 (December): 944-50,1963.
83. Tinney, William F. Compensation Methods for Network Solutions by Optimally Ordered Triangular Factorization. *Trans. IEEE* PAS-91 (January/February): 123–27, 1972.
84. DyLiacco, T. E., and Ramarao, K. A. Short-Circuit Calculations for Multi-Link Switching and End Faults. *Trans. IEEE* PAS-89 (July/August): 1226-37,1970.
85. Reitan, D. K., and Kreumpel, K. C. Modification of the Bus Impedance Matrix for System Changes Involving Mutual Couplings. *Proc. IEEE*, 57 (August): 1432–33, 1969.
86. Storry, J. O., and Brown, H. E. An improved method of incorporating mutual couplings in single-phase short-circuit calculations. *Trans. IEEE* PAS-89 (January): 71–77, 1970.

Index

ACSR
  effect of magnetic material, 133
  inductance, 76
Admittance matrix
  definite
    assumptions in fault calculations, 386-87
    fault calculations, 388
    power system description, 386-89
  indefinite, 372, 375-76
    building algorithm, 381-82
    correction for mutual coupling, 382-85
    example illustrating properties, 378-81
    finding elements, 376
    power system description, 381-82
    properties, 378
  node, 374-75
    making changes, 419
    sparsity, 418
  primitive, 367
Algebraic equations
  machine calculations, 183, 184
  power system computation, 3, 4
Alumoweld conductor
  effect of magnetic material, 133
A matrix
  n-phase system, 21-23
  power invariance, 26, 27
  three-phase system, 25
American National Standards Institute (ANSI). *See* Standards, U.S.A.
Analysis equation for three-phase systems, 24
a-operator, 16-17, 20
  n-phase system, 20
  table, 17
Atebekov, G. I., 273, 275
Attwood, Stephen S., 77

Base admittance, 6, 7, 8
Base current, 6, 7, 8
Base impedance, 6, 7, 8
Base quantity
  pu calculations, 5, 6
  selection, 5
Base value, 5-7
  change, 7
  table, 8
  three-phase systems, 7-10
Base voltage, 6, 7
Base voltampere, 6
Bessel functions in inductance calculations, 78
B matrix, two-component method, 347
Boast, W. B., 5
Boundary conditions for shunt faults, 36
Brown, H. E., 401
Bundled conductors, impedance of lines, 106-12

Calabrese, G. O., 77, 158
Capacitance
  coefficients, 160, 161
  effect of conductor height, 155
  to ground, transposed lines
    positive and negative sequence, 152-55
    using conductor height, 155
    zero sequence with ground wires, 157-58
    zero sequence without ground wires, 156
  method of images, 158-59
  multiplying constant, k′, 152-53
  mutual, 158-63
  self and mutual, 158-77
    double circuit lines, 172-77
    three-phase line with ground wires, 168-70
    three-phase line without ground wires, 163
  sequence
    similarity transformation, 166-67, 170
    three-phase line with ground wires, 170, 172
    three-phase line without ground wires, 166-68
  three-phase lines
    double circuit lines, 172-77
    phase, with ground wires, 168-72
    phase, without ground wires, 163-66
    sequence, with ground wires, 170, 172
    sequence, without ground wires, 166-68
Carson, John R., 78
Carson's line, 78-80
Charging current of transmission lines, 162, 174, 177, 178
Circuit diagram for shunt faults, 36-37
Circulant matrix in machine calculations, 31
Clarke, Edith, 71, 158, 225, 226, 228, 243, 245, 246, 254, 257, 258, 259

C matrix, symmetrical component transformation. *See also* Capacitance
  n-phase systems, 21
  three-phase systems, 24
Coefficients of electrostatic induction, 160
Cofactor of a matrix element, 161
Completely transposed line. *See* Impedance, transposed lines
Computer solutions, 3, 365–66
  using the admittance matrix, 365–89
  using the impedance matrix, 393–419
Constraint matrix, K, 286–88
Copperweld conductor, effect of magnetic material, 133
Current, momentary, synchronous machine, 221–22
Current envelope, synchronous machine, 219–21
Current unbalance, incomplete transpositions, 102–6. *See also* Unbalance

D_m. *See* GMD
D_s. *See* GMD
dc offset, synchronous machines, 183, 185, 188, 221–22
Delta
  network, 347
  quantities, 346–47
Diagonal matrix, impedances, 30
Differential equations
  machine calculations, 183
  power system calculations, 3, 4
Digital computer applications, 4, 5, 365–419
Dot convention, mutually coupled circuits, 74
d-q transformation. *See* Park's transformation
Dwight, H. B., 135
Dynamic problems, 3, 4

Earth resistivity, effect on inductance, 79–80
Eigenvalues, circulant matrix, 31
Eigenvectors, circulant matrix, 31
El-Abiad, A. H., 401
Electromagnetic unbalance. *See* Unbalance
Electrostatic unbalance. *See* Unbalance
Equivalent circuit
  induction motor, negative sequence, 224
  induction motor, positive sequence, 223
  synchronous machine, *d-q*, 205–7
  synchronous machine, *d-q*, steady state, 213, 214
  synchronous machine, *d-q*, subtransient, 219
  synchronous machine, T, 207
Equivalent spacing, 73
  computing sequence line impedance, 92
Evans, R. D., 71, 79, 80, 129, 134, 135, 158, 228, 241, 242

Fault, 31, 36
  analysis procedure, 36–37
  assumptions, 386
  diagram, series and parallel connections, 283
  longitudinal. *See* Fault, series
  point
    series faults, 53–54
    shunt faults, 31, 37
  series, 36, 53–66
    arbitrary symmetry, two-component method, 360–62
    circuit diagram, 61
    generalized fault diagram, 278–82
    1LO, generalized fault diagram, 282
    1LO, three-component method, 63–64
    1LO, two-component method, 354–55
    sequence network connections, 53–66
    sequence two-port networks, 55–60
    three-component method, 53–66
    two-component method, 353–55
    2LO, generalized fault diagram, 281
    2LO, three-component method, 64–66
    2LO, two-component method, 353–54
    two-port Thevenin equivalent, 55–60
    unequal Z, three-component method, 61–63
  severity, 49, 52, 184
  shunt, 36–53
    circuit diagram, 36–37
    frequency of occurrence, 52
    generalized fault diagram, 273–78
    LL, three-component method, 42–44
    LL, two-component method, 349–50
    relative severity, 49, 52
    sequence network connections, 37–53
    SLG, arbitrary symmetry, two-component method, 357–58
    SLG, generalized fault diagram, 275–77
    SLG, three-component method, 37–41
    SLG, two-component method, 347–49
    3ϕ, generalized fault diagram, 273–75
    3ϕ, three-component method, 49–52
    3ϕ, two-component method, 352–53
    2LG, arbitrary symmetry, two-component method, 358–60
    2LG, generalized fault diagram, 277–78
    2LG, three-component method, 44–49
    2LG, two-component method, 350–51
  simultaneous, 308–44
    connection of sequence networks, 323–36
    constraint matrix, H-type faults, 340–41
    constraint matrix, Y-type faults, 339–40
    constraint matrix, Z-type faults, 337–39
    four cases of interest, 308, 324
    matrix transformation, 336–41
    parallel-parallel connection, 330–34
    series-parallel connection, 334–36
    series-series connection, 325–30
    two-port combinations, 324
    two-port network theory applied, 308–36
  transformations
    series, 304
    shunt, summary, 300–301
    shunt, with fault impedance, 298–304
    shunt, without fault impedance, 294–98
  transverse. *See* Fault, shunt
Faulted network, 3, 31, 36
Federal Power Commission (U.S.A.), 243

Flux linkages
constant, concept, 183, 184, 186, 187, 188, 214, 222
cylindrical wires
external, 75-76
internal, 75
synchronous machines, 185-89
Fortescue, C. L., 4, 19, 27, 71
Frame of reference, for synchronous machine phasors
arbitrary, 208-9
d-q, 209

General Electric Co., 71, 255, 256
Generalized fault diagram
solution, two-component method, 362-63
three-component method, 283
two-component method, 361
Geometric mean distance. *See* GMD
GMD
composite conductors, 99
self and mutual, 72
GMD method for transposed lines
capacitance
positive and negative sequence, 152-55
zero sequence, with ground wires, 157-58
zero sequence, without ground wires, 156-57
impedance
positive and negative sequence, 71-73, 98
zero sequence, with ground wires, 129-33
zero sequence, without ground wires, 98-100
GMR. *See* GMD
Gross, E. T. B., 104, 144, 178
Ground wires. *See also* Impedance, sequence
current division with earth, 118-19
effect on sequence impedance, 114-18
Grover, F. W., 77
Guillemin, E. A., 308

Hancock, N. N., 284
Hesse, M. H., 104, 138
Hueslman, L. P., 209, 308, 309, 319

Impedance
effect of steel ground wires, 133-37
induction motor, table, 225
primitive, untransposed lines, 82-83
self, untransposed lines, 82-83
sequence
completely transposed lines, 98
lines with bundled conductors, 106-12
lines with *n* ground wires, 128-29
lines with one ground wire, 112-18
lines with two ground wires, 123-28
transposed line with ground wires, 119-23
transposed line with two ground wires, 126-28
untransposed lines with one ground wire, 114-18
untransposed lines with two ground wires, 125
transposed lines
general equation, 98
positive and negative sequence, 71-73, 98-100
untransposed lines
primitive equations, 81-83
sequence currents, 104
unbalance due to incomplete transpositions, 102-6
zero sequence
current basis, 133
effect of steel ground wires, example, 135-37
GMD method for transposed lines, 99-100, 129-33
Impedance matrix
building, 401
adding group of mutually coupled lines, 415-18
adding mutually coupled radial branch, 408-11
adding radial branch, 402-3
adding radial branch to reference, 401-2
closing loop not involving reference, 404-8
closing loop to reference, 403-4
example, positive sequence, 406-8
example, zero sequence, 412-14
sorting of primitive impedances, 401
node, 375, 393-94
advantages for fault computation, 393
building algorithm, 401-18
branch currents for three-phase faults, 395
comparison with admittance method, 418-19
currents in mutually coupled branches, 399-400
node voltages for three-phase faults, 395
making changes in, 419
open circuit, 393-94
sequence quantities for SLG faults, 397-98
shunt fault calculations, 393-400
SLG fault calculations, 396-400
three-phase fault calculations, 394-96
primitive Z, 367
sequence, unequal series Z, 27-30
table of special cases, 30
Incidence matrix. *See* Node incidence matrix
Indefinite admittance matrix. *See* Admittance matrix, indefinite
Inductance
leakage, synchronous machine, 207
mutual, parallel cylindrical wires, 76
self, parallel cylindrical wires, 75-76
synchronous machine
direct axis subtransient, 197-98
direct axis synchronous, 195
direct axis transient, 198
negative sequence, 200-201
phase domain, 186-87
quadrature axis subtransient, 199

Inductance (*continued*)
quadrature axis synchronous, 195–97
quadrature axis transient, 199
table, 203
zero sequence, 201, 202
Inductance multiplying constant, k, 76–77
Induction motor, 222–28
equivalent circuit, 222–24
fault contribution, 226–27
general considerations, 222
operation with one phase open, 227–28
torque-speed characteristic, 224–25
Initial-plus-change notation, machine subtransient condition, 215
Internal sources in two-port networks, 57
Interrupting duty for circuit breakers, 221

Kimbark, E. W., 71, 183, 185, 188, 189, 195, 200, 203, 204, 224
Kirchhoff's law, 53, 79
Krause, P. C., 208
Kron, G., 228, 273, 284, 286, 288, 294, 304, 308, 366
Kron reduction. *See* Matrix reduction
Kron's constraint matrix, connection matrix, 286–88
Kron's primitive network, 288–89
Kron's transformation matrix, 286–88
Kruempel, K. C., 418

Lagerstrom, J. E., 5
LeCorbeiller, P., 284
Lenz's law
mutually coupled wires, 74
synchronous machines, 214
transformer windings, 235
Lewis, W. A., 100, 206
Lewis, W. E., 30, 284
Lewis, W. W., 77
Line impedance, sequence, 71–151
Load representation, 12, 13
Lyon, W. V., 158

Machine
dynamics, 184
excitation, 222
impedances, symmetrical components, 30, 31
circulant matrix, 31
induction, 222–28
symmetrical components, 30, 31
synchronous, 183–222
Machines
induction, 222–28
synchronous, 183–222
Matrix
augmented node incidence, A, 370
constraint, Kron's, K, 286
constraint, table for shunt faults, 300–301
Kron's connection. *See* Matrix, constraint
Kron's transformation. *See* Matrix, constraint
primitive admittance, $\mathfrak{Y}$, 367
primitive impedance, $\mathfrak{Z}$, 288–90, 367
Matrix reduction, 108, 109, 114, 124, 170, 378, 381, 388, 411, 414, 416, 418

Maxwell's coefficients, 160, 181
capacitance calculations, 160, 163, 166, 169
Minor of a matrix element, 161, 163
Momentary duty for circuit breakers, 222
Motors, induction, 222–28
Mutual coupling
admittance matrix building, 382–85
equations, 74
nonreciprocal, between sequences, 30, 114
parallel wires, 73–75
between primitive matrix elements, 366
between sequences, 29, 98, 119
Mutual impedances
passive series networks, 27–30
symmetrical component calculations, 28

Negative sequence, 24
Nelson, S. W., 104
Network
primitive
Kron's, 288–89
sequence, 294, 299
two-port. *See* Two-port network
Nilsson, J. W., 5, 16
Node admittance matrix. *See* Admittance matrix, node
Node impedance matrix. *See* Impedance matrix, node
Node incidence matrix, 373–74
augmented, 369–73
Nonreciprocal coupling between sequences, 29–30
Nonsymmetric matrix of impedances, 30
Normalization of system quantities, 5–10. *See also* pu calculations
n-phase system, 19–23

Open circuit *Z* parameters, two-port networks, 55–56
Optimally ordered triangular factorization, 418–19
Orthogonal transformation, 28, 189

Parallel lines
effect of transposition, 137–43
optimizing wire arrangement, 143–45
Park, R. H., 189, 190, 200, 201, 228, 229
Park's transformation, 189
inverse defined, 190
synchronous machines, 189–94
Per phase representation, 4, 19
Per unit. *See* pu
Peterson coil, 270–72
Phasor, 15–16
diagram
frames of reference in machines, 208–10
synchronous machine, steady state, 211–13
synchronous machine, subtransient, 214–18
notation, 15
quantity, 15
transformation, $\mathcal{P}$, 16
Port, 309
Positive-minus-negative quantity. *See* Delta

Positive-plus-negative quantity. *See* Sigma
Positive sequence, 24, 32
Potential coefficients, 159-60, 163, 172
matrix for untransposed line, 163
Power
coupling between sequences, 26
invariance
matrix calculations, 284-86
symmetrical components, 26
transformation for, 26, 27, 285
from symmetrical components, 25-27
Prentice, B. R., 189, 196
Prime mover, 184
Primitive matrix, Kron's, 366-68
Primitive network. *See* Network
Proximity effect, in inductance calculations, 77-78
Pryce, D. G., 30, 284, 336
pu
admittance
on 100 MVA base for 1.0 mi, 10-11
three-phase system, 9
calculations, 5-10
examples, 12-15
conversion to system values, 10
impedance, 6, 7
on 100 MBA base for 1.0 mi, 10-11
three-phase system, 9
quantity, 5
value, 5

Reactance, synchronous machine
direct axis, 195
negative sequence, 199-201
quadrature axis, 195-97
subtransient, 197-99
transient, 197-99
zero sequence, 201-42
Reciprocity in sequence impedances, 29-30
Residual current, 177
Rosa, E. B., 77
Rotation
cycle, of transmission line, 87-90
linear transformation, 86
line conductors, 85-94
matrix
R_ϕ, clockwise rotation, 85
R_ϕ^{-1}, counterclockwise rotation, 86
R_{012}, sequence domain, 93

Self impedance of passive networks, 28
Sequence
current distribution in sequence networks, 53
impedance. *See* Impedance, sequence
machines, 30-31
passive series networks, 28
to flow of sequence currents, 32
network, 31, 32
primitive, with fault impedance, 299
primitive, without fault impedance, 294
Series fault. *See* Fault
Series impedance
unbalanced, 27
sequence components, 28
Short circuit. *See* Fault
Short circuit *Y* parameters, two-port networks, 56
Shunt fault. *See* Fault
Sigma
network, 347
quantities, 346-47
Similarity transformation
capacitance calculations, 166, 167, 170
circulant matrix, 31
impedances, 28, 31
Simultaneous faults. *See* Fault
Skin effect, in inductance calculations, 77, 78
Sources, internal, in two-port networks, 57
Spacing factor, 73
Sparsity
methods of exploiting, 418
node admittance matrix, 386
primitive matrix, 368-69
Speed voltage, in synchronous machines, 191, 192
Standards
I.E.E.E., circuit breakers, 221-22
I.E.E.E., synchronous machine parameters, 186, 188
U.S.A., machine letter symbols, 186
U.S.A., transformer terminal markings, 234-36
U.S.A., three-phase transformer terminal markings, 247-48
Steady state equations, 3, 4
Steel conductors, 133-37
Stevenson, W. D., Jr., 72, 77, 78, 100, 129, 152, 155, 156, 203, 249, 250, 251
Stigant, S., 284
Subtransient
component of machine currents, 220
solution of synchronous machine equations, 214-19
Susceptance, capacitive, transmission lines, 152, 153, 155, 157, 178
Symmetrical components
analysis equation, 24
currents, 25
LL voltages, 25
n-phase system, 19-23
three-phase LN voltages, 23-25
Symmetry
changes in, 273-307
two-component calculations, 355-62
creating by labeling, 273
generalized
1LO, 282
series faults, 278-84
shunt faults, 273-78
SLG faults, 277
summary of network connections, 283
2LG faults, 278
2LO faults, 281
transformations. *See* Fault
Synchronous
component of machine currents, 219
machine analysis, 183-222
Synthesis
equations, 24, 25
sequence quantities, 25

Tellegren's theorem, 372
Terminal pair. *See* Port
Thevenin
 equivalent
  delta network, 347
  sequence networks, 32, 324
  sigma and delta networks, 346-47
  sigma network, 347
  synchronous machine, 184-85, 214, 219
  two-port, 55-60
 impedance
  to compute system unbalance, 105
  to flow of sequence currents, 32
 theorem, 32, 55, 219
 voltage
  at fault point, 53
  positive sequence network, 32
  prefault at F, 32
  system adjacent to unbalance, 105
Thomas, C. H., 208
Three-component method, 36-66
Time constant, synchronous machine
 armature, 204
 field, 204
 T_d', 204
 T_{d0}', 202-4
 T_{d0}'', 205
 T_d'', 205
 T_{q0}', 205
 T_q', 205
 T_{q0}'', 205
 T_q'', 205
 table, 203
Tinney, W. F., 418
Torque equation, induction motor, 224
Torque-speed characteristics, 225
Transformation
 sequence
  Kron's technique, 292
  series faults, 304
  shunt faults, 294-304
  shunt faults with impedance, 298-99
  SLG, by Kron's technique, 295-96
  2LG, by Kron's technique, 296-98
 similarity, 28
Transformer, 231-65
 auto, 239-43
  equivalent circuit, 240-41
  impedances, table, 243
  three-winding, 241-42
 banks, of single-phase units, 243-47
  three-phase Δ-Y connection, 244-45
  three-phase, single-phase units, 243-47
  three-phase, three-winding autotransformers, 246
  three-phase, zero sequence equivalents, 244-45
  three-winding autotransformers, 246
 core losses, 231-32
 excitation, 231-32
 grounding, 255-57
  Y-Δ, 255-56
  zigzag, 257
 interconnected star-delta, 257
 load tap changing, 262-63
 off-nominal turns ratio, 260-65
 phase shifting
  complex turns ratio, 275
  parameters derived, 356-57
 pi equivalent, 262-65
 single-phase. *See* Transformer, two-winding
 three-phase, 247-60
  core-type, 252-53
  grounding, 255-57
  shell-type, 253-54
  terminal markings, 247
  Y-Δ phase shift, 247-51
  Y-Y and Δ-Δ phase shift, 247
  zero sequence equivalents, 255-56
  zero sequence impedance, 251-55
  zigzag-Δ, 257-60
 three-winding, 236-39
  equivalent circuit, 236-37
  impedances, 238
  normalization, 238-39
 turns ratio, 231-32
  off-nominal, 260-65
 two-winding, 231-36
  equivalent circuit, 232
  impedance, table, 234, 235
  normalization, 233
  polarity, 234-36
  single-phase, 231-32
  terminal markings, 234-36
 zigzag-Δ, 257-60
Transient component, machine currents, 219-20
Transmission line impedance. *See* Impedance
Transposition, 84-98
 rotation, 84-94
 twisting, 95-98
Triangular factorization, 418-19
Twisting of line conductors, 85-95
Twist matrix
 phase domain, 95
 sequence domain, 97-98
 T_ϕ, 97
 T_{012}, 97-98
 three definitions, 95
Two-component method
 fault calculations, 345-47
 sequence current constraints, 356
 sequence networks, 347
 series faults, arbitrary symmetry, 360-61
 SLG fault on any phase, 357-58
 solution of generalized diagram, 362-63
 2LG fault on any phase, 358-60
Two-port network, 308-23
 cascade connection, 323
 hybrid connection, 321-22
 independent sources, 57, 316
 internal sources, 57, 313-18
 open circuit Z parameters, 55-56
 parallel connection, 320-21
 parameters, table, 309
 relationship among parameters, 312
 relationship among source parameters, 317
 series connection, 319-20

short circuit Y parameters, 56
simultaneous faults, 323-25
termination impedance, 310
test for finding parameters, 315-16
uncoupled, in positive sequence, 59-60
Two-port parameters, 55-56
Two-port Thevenin equivalent. *See* Thevenin, equivalent

Unbalance
electromagnetic
double circuit, circulating, 137-43
double circuit, through, 137-43
examples, 106
factor, 104
optimizing, in parallel circuits, 143-45
electrostatic
factor, 179
ground displacement, 178-79
summary of factors affecting, 180
transmission lines, 177-81
ungrounded neutral, 179
factors, unequal series impedances, 104-5
series impedance, 27-30
Untransposed lines. *See* Impedance, untransposed line
U.S.A. Standards. *See* Standards, U.S.A.

Vandermonde matrix, 22
Voltage
drop
line with rotation cycle, 89
mutually coupled wires, 74
unequal phase impedances, 28
equation, primitive, 81
positive sequence, 32

Wagner, C. F., 71, 79, 80, 129, 134, 135, 158, 228, 241, 242
Ward, J. B., 55
Westinghouse Electric Corp., 52, 71, 100, 201, 234, 235, 236, 237, 249, 250, 252, 253, 257
Woodruff, L. F., 72, 100

Y-bus matrix. *See* Admittance matrix, node
Y matrix. *See* Admittance matrix, node

Z-bus matrix. *See* Impedance matrix, node
0-d-q frame of reference, 191
Zero potential bus, 32
Zero sequence, or zero-phase sequence, 19, 24
Z Matrix. *See* Impedance matrix, node

PER UNIT CALCULATIONS

$$Z = \frac{S_{B\text{-}1\phi}}{V^2_{B\text{-}LN}}(Z \text{ ohm}) = \frac{S_{B\text{-}3\phi}}{V^2_{B\text{-}LL}}(Z \text{ ohm}) \text{ pu} = \frac{\text{Base MVA}_{3\phi}}{(\text{Base kV}_{LL})^2}(Z \text{ ohm}) \text{ pu}$$

$$Y = \frac{V^2_{B\text{-}LN}}{S_{B\text{-}1\phi}}(Y \text{ mho}) = \frac{V^2_{B\text{-}LL}}{S_{B\text{-}3\phi}}(Y \text{ mho}) \text{ pu} = \frac{(\text{Base kV}_{LL})^2\ (Y\ \mu\text{mho})}{(\text{Base MVA}_{3\phi})\ (10^6)} = \frac{(\text{Base kV}_{LL})^2\ (10^{-6})}{(\text{Base MVA}_{3\phi})\ (Z\ \text{M}\Omega)} \text{ pu}$$

$$Z_n = Z_o \left(\frac{V_{Bo}}{V_{Bn}}\right)^2 \left(\frac{S_{Bn}}{S_{Bo}}\right) \text{ pu}$$

PHASOR TRANSFORMATION

$$\mathcal{P}[a(t)] = \mathcal{P}[\sqrt{2}|A|\cos(\omega t + \alpha)] = |A|e^{j\alpha} = A$$

$$\mathcal{P}^{-1}[A] = \mathcal{R}e(\sqrt{2}A\, e^{j\omega t}) = a(t)$$

SYNTHESIS EQUATION

$$\mathbf{V}_{abc} = \mathbf{A}\,\mathbf{V}_{012}$$

$$\begin{bmatrix} V_a \\ V_b \\ V_c \end{bmatrix} = \frac{1}{h}\begin{bmatrix} 1 & 1 & 1 \\ 1 & a^2 & a \\ 1 & a & a^2 \end{bmatrix}\begin{bmatrix} V_{a0} \\ V_{a1} \\ V_{a2} \end{bmatrix}$$

ANALYSIS EQUATION

$$\mathbf{V}_{012} = \mathbf{A}^{-1}\,\mathbf{V}_{abc}$$

$$\begin{bmatrix} V_{a0} \\ V_{a1} \\ V_{a2} \end{bmatrix} = \frac{h}{3}\begin{bmatrix} 1 & 1 & 1 \\ 1 & a & a^2 \\ 1 & a^2 & a \end{bmatrix}\begin{bmatrix} V_a \\ V_b \\ V_c \end{bmatrix}$$

Printed in the USA/Agawam, MA
December 7, 2015
627417.004